T0184366

# CISM COURSES AND LECTURES

*Series Editors:*

*The Rectors of CISM*
Sandor Kaliszky - Budapest
Mahir Sayir - Zurich
Wilhelm Schneider - Wien

*The Secretary General of CISM*
Giovanni Bianchi - Milan

*Executive Editor*
Carlo Tasso - Udine

The series presents lecture notes, monographs, edited works and proceedings in the field of Mechanics, Engineering, Computer Science and Applied Mathematics.
Purpose of the series is to make known in the international scientific and technical community results obtained in some of the activities organized by CISM, the International Centre for Mechanical Sciences.

# INTERNATIONAL CENTRE FOR MECHANICAL SCIENCES

COURSES AND LECTURES - No. 372

# ADVANCED MANUFACTURING SYSTEMS AND TECHNOLOGY

EDITED BY

E. KULJANIC
UNIVERSITY OF UDINE

Springer-Verlag Wien GmbH

Le spese di stampa di questo volume sono in parte coperte da
contributi del Consiglio Nazionale delle Ricerche.

This volume contains 488 illustrations

This work is subject to copyright.
All rights are reserved,
whether the whole or part of the material is concerned
specifically those of translation, reprinting, re-use of illustrations,
broadcasting, reproduction by photocopying machine
or similar means, and storage in data banks.
© 1996 by Springer-Verlag Wien
Originally published by Springer-Verlag Wien New York in 1996

In order to make this volume available as economically and as
rapidly as possible the authors' typescripts have been
reproduced in their original forms. This method unfortunately
has its typographical limitations but it is hoped that they in no
way distract the reader.

ISBN 978-3-211-82808-3          ISBN 978-3-7091-2678-3 (eBook)
DOI 10.1007/978-3-7091-2678-3

# MATERIALS SCIENCE AND THE SCIENCE
# OF MANUFACTURING - INCREASING PRODUCTIVITY
# MAKING PRODUCTS MORE RELIABLE AND LESS EXPENSIVE

**ORGANIZERS**

University of Udine - Faculty of Engineering - Department of
Electrical, Managerial and Mechanical Engineering - Italy
Centre International des Sciences Mécaniques, CISM - Udine - Italy
University of Rijeka - Technical Faculty - Croatia

**CONFERENCE VENUE**

CISM - PALAZZO DEL TORSO
Piazza Garibaldi, 18 - UDINE

# PREFACE

*The International Conference on Advanced Manufacturing Systems and Technology - AMST is held every third year. The First International Conference - AMST'87 was held in Opatija (Croatia) in October 1987. The Second International Conference - AMST'90 was held in Trento (Italy) in June 1990 and the Third International Conference - AMST'93 was held in Udine (Italy) in April 1993.*

*The Fourth International Conference on Advanced Manufacturing Systems and Technology - AMST'96 aims at presenting trend and an up-to-date information on the latest developments - research results and industrial experience in the field of machining of conventional and advanced materials CIM non-conventional machining processes forming and quality assurance thus providing an international forum for a beneficial exchange of ideas, and furthering a favorable cooperation between research and industry.*

*E. Kuljanic*

## HONOUR COMMITTEE

S. CECOTTI, President of Giunta Regione Autonoma Friuli-Venezia Giulia
M. STRASSOLDO DI GRAFFEMBERGO, Rector of the University of Udine G.
BIANCHI, General Secretary of CISM
S. DEL GIUDICE, Dean of the Faculty of Engineering, University of Udine
J. BRNIC, Dean of the Technical Faculty, University of Rijeka
C. MELZI, President of the Associazione Industriali della Provincia di Udine

## SCIENTIFIC COMMITTEE

E. KULJANIC (Chairman), University of Udine, Italy
N. ALBERTI, University of Palermo, Italy
A. ALTO, Polytechnic of Bari, Italy
P. BARIANI, University of Padova, Italy
G. BIANCHI, CISM, Udine, Italy
A. BUGINI, University of Brescia, Italy
R. CEBALO, University of Zagreb, Croatia
G. CHRISSOLOURIS, University of Patras, Greece
M.F. DE VRIES, University of Wisconsin Madison, U.S.A.
R. IPPOLITO, Polytechnic of Torino, Italy
F. JOVANE, Polytechnic of Milano, Italy
I. KATAVIC, University of Rijeka, Croatia
H.J.J. KALS, University of Twente, The Netherlands
F. KLOCKE, T.H. Aachen, Germany
W. KONIG, T.H. Aachen, Germany
F. LE MAITRE, Ecole Nationale Superieure de Mechanique, France
E. LENZ, Technion, Israel
R. LEVI, Polytechnic of Torino, Italy
B. LINDSTROM, Royal Institute of Technology, Sweden
V. MATKOVIC, Croatian Academy of Science and Arts, Croatia
J.A. Mc GEOUGH, University of Edimburg, UK
M.E. MERCHANT, IAMS, Ohio, U.S.A.
G.F. MICHELETTI, Polytechic of Torino, Italy
B. MILCIC, INAS, Zagreb, Croatia
S. NOTO LA DIEGA, University of Palermo, Italy
J. PEKLENIK,University of Ljubljana, Slovenia
H. SCHULZ, T.H. Darmstadt, Germany
N.P. SUH, MIT, Mass., U.S.A.
H.K. TONSHOFF, University of Hannover, Germany
B.F. von TURKOVICH, University of Vermont, U.S.A.
K. UEHARA, University of Toyo, Japan
A. VILLA, Polytechnic of Torino, Italy

## ORGANIZING COMMITEE

E. KULJANIC (Chairman)
M. NICOLICH (Secretary)
C. BANDERA, F. COSMI, F. DE BONA, M. GIOVAGNONI, F. MIANI,
M. PEZZETTA, P. PASCOLO, M. REINI, A. STROZZI, G. CUKOR


## SPONSORSHIP ORGANIZATIONS

Presidente della Giunta Regione Autonoma Friuli-Venezia Giulia
Croatian Academy of Science and Arts, Zagreb
C.U.M. Community of Mediterranean Universities


## SUPPORTING ORGANIZATIONS

Comitato per la promozione degli studi tecnico-scientifici
University of Udine
Pietro Rosa T.B.M. s.r.l., Maniago

# CONTENTS

## Part III - Forming

**Part IV - Flexible Machining Systems**

# WORLD TRENDS IN THE ENGINEERING OF THE TECHNOLOGICAL AND HUMAN RESOURCES OF MANUFACTURING

M. E. Merchant

Institute of Advanced Manufacturing Sciences, Cincinnati, OH, U.S.A.

## KEYNOTE PAPER

KEY WORDS: Trends, CIM, Manufacturing Technology, Human-Resource Factors, Manufacturing Engineering Education, New Approaches

ABSTRACT: In the period from the beginning of organized manufacturing in the 1700s to the 1900s, increasing disparate departmentalization in the developing manufacturing companies resulted in a long-term evolutionary trend toward an increasingly splintered, "bits-and-pieces" type operational approach to manufacturing. Then in the 1950s, the "watershed" event of the advent of digital computer technology and its application to manufacturing offered tremendous promise and potential to enable the integration of those bits-and-pieces, and thus, through computer integrated manufacturing (CIM) to operate manufacturing as a system. This initiated a long-term technological trend toward realization of that promise and potential. However, as the technology to do that developed, it was discovered in the late 1980s that that technology would only live up to its full potential if the engineering of it was integrated with effective engineering of the human-resource factors associated with the utilization of the technology in the operation of the overall system of manufacturing in manufacturing companies (enterprises). This new socio-technological approach to the engineering and operation of manufacturing has resulted in a powerful new long-term trend -- one toward realistic and substantial accomplishment of total integration of both *technological* and *human-resource* factors in the engineering and operation of the overall system of manufacturing in manufacturing enterprises. A second consequence of this new approach to the engineering of manufacturing is a strong imperative for change in the programs of education of manufacturing engineers. As a result, the higher education of manufacturing professionals is now beginning to respond to that imperative.

Published in: E. Kuljanic (Ed.) *Advanced Manufacturing Systems and Technology*, CISM Courses and Lectures No. 372, Springer Verlag, Wien New York, 1996.

## 1. INTRODUCTION

In my keynote paper presented at AMST'93 I discussed broad socio-technical and specific manufacturing long-term trends which had evolved over the years and were at work to shape manufacturing in the 21st century. Since that time those trends have not only evolved further, but are playing an even more active and better understood role in shaping manufacturing. In addition, they are shaping not only manufacturing itself, but also the education of tomorrow's manufacturing engineers. Therefore, in this paper, which is somewhat of a sequel to my 1993 paper, we will explore the nature and implications of that further evolution and understanding.

What has happened in these areas is strongly conditioned by all of the long-term trends in manufacturing that have gone before, since the very beginnings of manufacturing as an organized industrial activity. Therefore, we will begin with a brief review of those trends, duplicating somewhat the review of these which was presented in the 1993 paper.

## 2. THE EARLY TREND

Manufacturing as an organized industrial activity was spawned by the Industrial Revolution at the close of the 18th century. Manufacturing technology played a key role in this, since it was Wilkinson's invention of a "precision" boring machine which made it possible to bore a large cylinder to an accuracy less than "the thickness of a worn shilling". That precision was sufficient to produce a cylinder for an invention which James Watt had conceived, but had been unable to embody in workable form, namely the steam engine. Because of Wilkinson's invention, production of such engines then became a reality, providing power for factories.

As factories grew in size, managing the various functions needed to carry on the operation of a manufacturing company grew more and more difficult, leading to establishment of functional departments within a company. However, the unfortunate result of this was that, because communication between these specialized disparate departments was not only poor but difficult, these departments gradually became more and more isolated from one another. This situation finally lead to a "bits-and-pieces" approach to the creation of products, throughout the manufacturing industry.

## 3. A WATERSHED EVENT

Then, in the 1950s, there occurred a technological event having major potential to change that situation, namely the invention of the digital computer. This was indeed a watershed event for manufacturing though not recognized as such at the time. However, by the 1960s, as digital computer technology gradually began to be applied to manufacturing in various ways (as for example in the form of numerical control of machine tools) the potential of the digital computer for manufacturing slowly began to be understood. It gradually began to be recognized as an extremely  powerful tool -- a *systems* tool -- capable of integrating manufacturing's former "bits-and-pieces" to operate it as a *system*. This recognition spawned a new understanding of the nature of manufacturing, namely that manufacturing is fundamentally a *system*. Thus, with the aid of the digital computer, it should be able to be operated as a system.

Out of this recognition grew a wholly new concept, namely that of the Computer Integrated Manufacturing (CIM) System -- a system having capability not only to flexibly *automate* and on-line *optimize* manufacturing, but also to *integrate* it and thus operate it as a system. By the end of the 1960s this concept had led to initial understanding of the basic components of the CIM system and their inter-relationship, as illustrated, for example, in Figure 1.

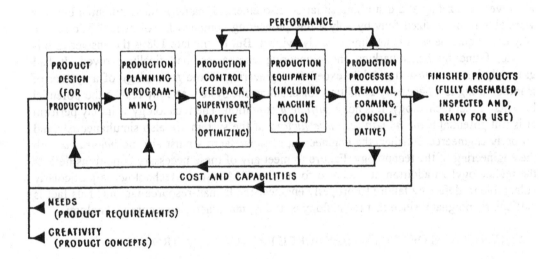

Figure 1. Initial Concept of the Computer Integrated Manufacturing System, 1969

## 4. NEW INSIGHT EMERGES

What followed during the 1970s and early 1980s was a long, frustrating struggle to develop and implement CIM system technology in order to reduce it to practice in industry and thus reap its inherent potential benefits. It is important to note, however, that the focus and thrust of this struggle was almost totally on the technology of the system. As the struggle progressed, and the technology finally began to be implemented more and more widely in the manufacturing industry, observation of the most successful cases of its reduction to practice began to make clear and substantiate the very substantial benefits which CIM technology has the potential to bring to manufacturing. The most significant of these were found to be the following.

Greatly:
- increased product quality
- decreased lead times
- increased worker satisfaction
- increased customer satisfaction
- decreased costs
- increased productivity
- increased flexibility (agility)
- increased product producibility

However, a puzzling and disturbing situation also emerged, namely, these potential benefits were able to be realized fully by only a few pioneering companies, worldwide! The reason why this should be so was not immediately evident. But by the late 1980s the answer to this puzzle, found by benchmarking the pioneering companies, had finally evolved. It had gradually become clear that while excellent engineering of the *technology* of a system of manufacturing is a *necessary* condition for enabling the system to fully realize the potential benefits of that technology, it is not a *sufficient* condition. The technology will only perform at is full potential if the *human-resource factors* of the system are also simultaneously and properly engineered. Further, the engineering of those factors must also be *integrated* with the engineering of the technology. Failure to meet any of these necessary conditions defeats the technology! In addition, it was also found the CIM systems technology is particularly vulnerable to defeat by failure to properly engineer the human-resource factors. This fact is particularly poignant, since that technology is, today, manufacturing's core technology.

## 5. ENGINEERING OF HUMAN-RESOURCE FACTORS IS INTRODUCED

Efforts to develop methodology for proper engineering of human-resource factors in modern systems of manufacturing gradually began to be discovered and developed. Although this process is still continuing, some of the more effective methodologies which have already emerged and been put into practice include:

- **empower** individuals with the full authority and knowledge necessary to the carrying out of their responsibilities
- use empowered **multi-disciplinary teams** (both managerial and operational) to carry out the functions required to realize products
- empower a company's collective human resources to fully **communicate** and **cooperate with** each other.

Further, an important principle underlying the *joint* engineering of the technology *and* the human-resource factors of modern systems of manufacturing has recently become apparent. This can be stated as follows:

> So develop and apply the technology that it will support the *user*, rather than, that the user will have to support the *technology*.

## 6. A NEW APPROACH TO THE ENGINEERING OF MANUFACTURING EMERGES

Emergence of such new understanding as that described in the two preceding sections is resulting in substantial re-thinking of earlier concepts, not only of the CIM system, but also of the manufacturing enterprise in general. In particular, this had lead to the recognition that these concepts should be broadened to include both the *technological* and the *human-resource-oriented* operations of a manufacturing enterprise. Thus the emerging focus of that concept is no longer purely technological.

This new integrated socio-technological approach to the engineering and operation of the system of manufacturing is resulting in emergence of a powerful long-term overall trend in world industry. That trend can be characterized as one toward realistic and substantial accomplishment of total integration of both *technological* and *human-resource* factors in the engineering and operation of an overall manufacturing enterprise.

The trend thus comprises two parallel sets of mutually integrated activities. The first of these is devoted to development and implementation of new, integrated *technological* approaches to the engineering and operation of manufacturing enterprises. The second is devoted to the development and implementation of new, integrated highly *human-resource-oriented* approaches to the engineering and operation of such enterprises. To ensure maximum success in the ongoing results of this overall endeavor, both sets of activities must be integrated with each other and jointly pursued, hand-in-hand.

## 7. IMPLICATIONS FOR EDUCATION OF MANUFACTURING ENGINEERS

Because the new approach and long-term trend described above are having a revolutionary impact on the engineering of manufacturing, these also have very considerable implications for the education of future manufacturing engineers. Quite evidently, these professionals must not only be educated in how to engineer today's and tomorrow's manufacturing technology, as presently. They must now also be educated in how to engineer the human-resource factors involved in development, application and use of that technology in practice. Further, they must also be educated in how to effectively engineer the interactions and the integration of the two.

The imperative that they be so educated stems from the fact that, if they are not, the technology which they engineer will fail to perform at is full potential, or may even fail completely. Thus, if we do not so educate them, we send these engineers out into industry lacking the knowledge required to be *successful* manufacturing engineers.

The higher education of manufacturing professionals is now beginning to respond to this new basic imperative. For example, consider the tone and content of the SME International Conference on Preparing World Class Manufacturing Professionals held in San Diego, California in March of this year. (It was attended by 265 persons from 27 different countries.) The titles of some of the conference sessions are indicative of the conference's tone and content; for instance:

- New Concepts for Manufacturing Education
- The Holistic Manufacturing Professional
- Customer-Drive Curricular Development
- Teaching the Manufacturing Infrastructure
- Future View of Manufacturing Education.


## 8. CONCLUSION

A radical metamorphosis is now underway in the engineering and operation of manufacturing throughout the world; much of that is still in its infancy. The main engine driving that metamorphosis is the growing understanding that, for the engineering of manufacturing's technologies to be successful, it must intimately include the engineering of manufacturing's human-resource factors as well. Understanding and methodology for accomplishing such engineering are still in early stages of development. Programs of education of manufacturing engineers to equip them to be successful in practicing this new approach to the engineering and operation of manufacturing are even more rudimentary at this stage today.

However, this new approach is already beginning to show strong promise of being able to make manufacturing enterprises far more productive and "human-friendly" than they have ever been before.

That poses to all of us, as manufacturing professionals, an exciting *challenge*!

# ADVANCED MACHINING OF TITANIUM- AND NICKEL-BASED ALLOYS

F. Klocke, W. König and K. Gerschwiler

Lab. for Machine Tools and Production Engineering (WZL)
RWTH, Aachen, Germany

## KEYNOTE PAPER

KEY WORDS: Titanium alloys, nickel-based alloys, turning, ceramics, PCD, PCBN

ABSTRACT: At present, the majority of tools used for turning titanium- and nickel-based alloys are made of carbide. An exceptionally interesting alternative is the use of PCD, whisker-reinforced cutting ceramic or PCBN tools. Turning nickel-based alloys with whisker-reinforced cutting ceramics is of great interest, mainly for commercial reasons. A change from carbides to PCD for turning operations on titanium-based alloys and to PCBN for nickel-based alloys should invariably be considered if the advantages of using these cutting materials, e.g. higher cutting speeds, shorter process times, longer tool lives or better surface quality outweigh the higher tool costs.

## 1. INTRODUCTION

Titanium- and nickel-based alloys are the materials most frequently used for components exposed to a combination of high dynamic stresses and high operating temperatures. They are the preferred materials for blades, wheels and housing components in the hot sections of fixed gas turbines and aircraft engines (Fig. 1). Current application limits are roughly 600 °C for titanium-based alloys, 650 °C for nickel-based forging alloys and 1050 °C for nickel-based casting alloys [1].

Because of their physical and mechanical properties, titanium- and nickel-based alloys are among the most difficult materials to machine. Cutting operations are carried out mainly with HSS or carbide tools. Owing to the high thermal and mechanical stresses involved, these cutting materials must be used at relatively low cutting speeds.

Published in: E. Kuljanic (Ed.) *Advanced Manufacturing Systems and Technology*, CISM Courses and Lectures No. 372, Springer Verlag, Wien New York, 1996.

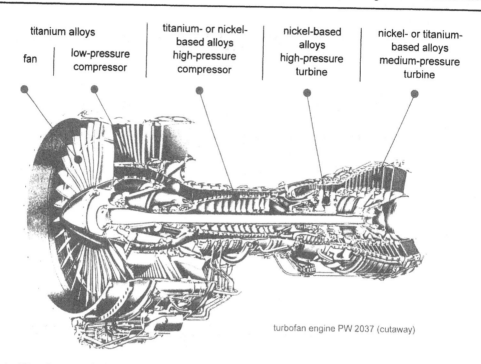

titanium alloys

fan | low-pressure compressor

titanium- or nickel-based alloys high-pressure compressor

nickel-based alloys high-pressure turbine

nickel- or titanium-based alloys medium-pressure turbine

turbofan engine PW 2037 (cutaway)

Fig. 1: Titanium- and nickel-based alloys in turbine construction [MTU]

Polycrystalline cubic diamond (PCD) represents an alternative for turning titanium-based alloys, cutting ceramics and polycrystalline cubic boron nitride (PCBN) for nickel-based alloys. Cutting materials from these three groups are characterized by great hardness and wear resistance. They can be used at higher cutting speeds than carbides, significantly reducing overall process times while achieving equal or superior machining quality.

Reliable, cost-effective use of these cutting materials is, however, dependent on a very careful matching of the cutting material category and the cutting parameters to the task in hand. This in turn demands the most precise possible knowledge of the machining properties of the relevant work material and the wear and performance behaviour to be expected from the cutting materials under the given constraints.

## 2. TURNING TiAl6V4 TITANIUM ALLOY WITH PCD

The TiAl6V4 titanium alloy most frequently used for turbine construction is a heterogeneous two-phase material. The hexagonal $\alpha$-phase is relatively hard, brittle, difficult to form and strongly susceptible to strain hardening. This phase acts on the contacting tool cutting edge like the wear-promoting cementite lamellae in the pearlite cores of carbon steels. The machining behaviour of the cubic body-centred ß-phase closely resembles that of ferrite, which also crystallizes in the cubic body-centred lattice; it is easily formed, relatively ductile and has a strong tendency to adhere [3].

The great technical significance of the titanium alloys is based not only on their great strength but, above all, on the yield-point/density ratio, which no other metallic material has

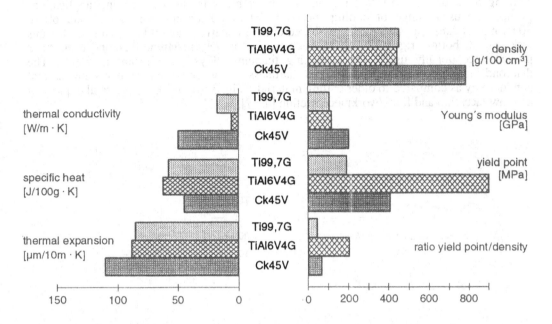

Fig. 2: Physical and mechanical properties of pure titanium, TiAl6V4 and Ck45 tempering steel [3]

yet come close to attaining (Fig. 2). Even high-strength steels with yield point values ofapproximately 1,000 MPa only achieve about half the ratio reached by TiAl6V4 titanium alloy [3].

One important physical property governing the machinability of titanium alloys is their low thermal conductivity, amounting to only about 10 - 20 % that of steel (Fig. 2). In consequence, only a small proportion of the generated heat is removed via the chips. As compared to operations on Ck45 steel, some 20 - 30 % more heat must be dissipated via the tool when working TiAl6V4 titanium alloy, depending on the thermal conductivity of the cutting material (Fig. 3, top left). This results in exceptionally high thermal stresses on the cutting tools, significantly exceeding those encountered when machining steel (Fig. 3, top right). In terms of cutting operations on titanium alloys, this means that the cutting tools are subjected not only to substantial mechanical stresses but also to severe thermal stress [3, 4].

Another characteristic feature of cutting operations on titanium alloys under conventional cutting parameters is the formation of lamellar chips. These are caused by a constant alternation between upsetting and slipping phenomena in the shearing zone (Fig. 3, bottom left). Owing to this disconinuous chip formation, tools are exposed to cyclic mechanical and thermal stresses whose frequency and amplitude depend directly on the cutting parameters. The dynamic components of cutting force may amount to some 20 - 35 % of the static components. The mechanical and thermal swelling stress may promote tool fatigue or failure through crack initiation, shell-shaped spalling, chipping out of cutting material particles or cutting edge chipping [3, 4].

Turning processes on TiAl6V4 rely mainly on carbides in the K20 cutting applications group. The usual range of cutting speeds is 50 to 60 m/min for roughing and 60 to 80 m/min for finishing. Oxide-ceramic-based cutting materials cannot be considered for this task (Fig. 3, bottom right) [2 - 4]. Monocrystalline or polycrystalline diamond tools have proved exceptionally useful for machining titanium alloys (Fig. 3, bottom right). The diamond tools are characterized by great hardness and wear resistance, excellent thermal conductivity as compared to other cutting materials (Fig. 3, top left), low thermal expansion and low face/chip and flank/workpiece friction [4 - 7].

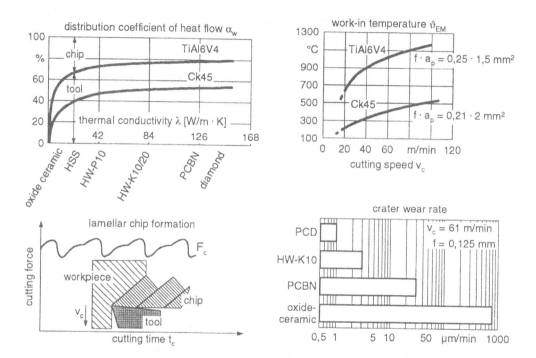

Fig. 3: Wear-relevant properties of TiAl6V4 and comparison of cutting materials [3, 4]

The range of cutting speeds for turning operations on TiAl6V4 with PCD tools extends from $v_c = 100$ to 200 m/min. Crater wear increases with rising cutting speed. Because crater wear is greatly reduced by the use of a cooling lubricant, PCD turning processes on TiAl6V4 should be wet operations.

The wear-determining interactions between the work material and the cutting material during PCD machining of titanium alloys are extraordinarily complex. They are characterized by diffusion and graphitization phenomena, thermally-induced crack initiation, surface destruction due to lamellar chip formation and possible formation of a wear-inhibiting titanium carbide reaction film on the diamond grains [5, 6].

Owing to these varied interactions between the cutting and work materials, the performance potential of PCD cutting materials in titanium machining processes is heavily dependent on the composition of the cutting material. Of particular interest are the composition of the binder phase, its volumetric proportion and the size of the diamond grains [6].

Crater wear is a main criterion for assessing the performance potential of a PCD cutting material in a titanium machining operation. Flank wear is of subordinate importance, especially at high cutting speeds.

The lowest crater wear in plain turning tests on TiAl6V4 titanium alloy was measured for a PCD grade with SiC as the binder. Crater wear was heavily influenced by binder content and grain size in the case of PCD grades with cobalt-containing binders (Fig. 4). The greatest crater wear was observed for the PCD grade with the highest binder content and the smallest grain size [6, 7].

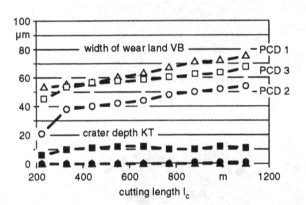

| composition | PCD 1 △▲ | PCD 2 ○● | PCD 3 □■ |
|---|---|---|---|
| diamond | 92 % | 92 % | 80 % |
| binder | 8 % | 8 % | 20 % |
| grit size | 6-10 µm | 2-6 µm | 0,5-1 µm |

binder phase: WC-Co
percentage volumes

| process: | plain turning | cutting speed: | $v_c$ = 110 m/min |
|---|---|---|---|
| material: | TiAl6V4 | depth of cut: | $a_p$ = 2,0 mm |
| tool: | ○△ SCGW 120408 | feed: | f = 0,1 mm |
| | □ SPGN 120308 | cooling lubricant: | emulsion |

Fig. 4: Crater and flank wear in turning operations on TiAl6V4 as a function of the PCD grade [6, 7]

Because of its catalyzing effect, cobalt also encourages graphitization of the diamond. The result is low resistance of the cutting material to abrasive wear. These is demonstrated very clearly by scratch marks in PCD cutting materials annealed at different temperatures. Unlike low-binder, large-grain types, high-binder, small-grain types leave a clear scratch diamond track on a specimen annealed at 800 °C (Fig. 5). Cobalt and diamond also have different coefficients of thermal expansion, favouring the development of thermal expansion cracks. This is particularly observable with fine-grained types. In combination with dynamic stressing of the cutting material through lamellar chip formation, these cracks make it easier for single PCD grains or even complete grain clusters to detach from the binder [5 - 7].

The low crater wear on the SiC-containing or large-grain, cobalt-containing PCD grains may be due to the formation of a wear-inhibiting titanium carbide film on the diamond grains. It is suspected that a diffusion-led reaction occurs between titanium from the work material and carbon from the tool in the crater zone of the face at the beginning of the machining process. The resulting titanium carbide reaction film adheres firmly to the face of the diamond tool and remains there throughout the remainder of the machining operation. Since the diffusion rate of carbon in titanium carbide is lower by several powers of ten than that of carbon in titanium, tool wear is slowed down substantially [3, 8].

To achieve the lowest possible crater wear, PCD grades with large diamond grains, low cobalt content or a ß-SiC binder phase are therefore preferable for turning operations on titanium alloys.

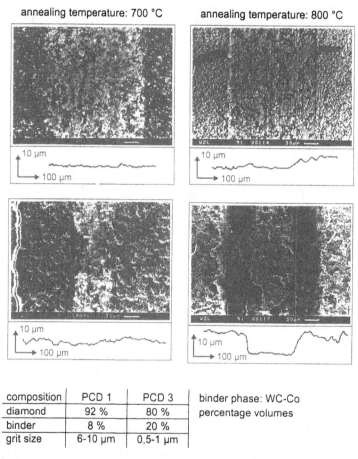

| composition | PCD 1 | PCD 3 |
|---|---|---|
| diamond | 92 % | 80 % |
| binder | 8 % | 20 % |
| grit size | 6-10 μm | 0,5-1 μm |

binder phase: WC-Co
percentage volumes

Fig. 5: Overall views and cross-section profiles of the scratch track on the surface of PCD cutting materials as a function of annealing temperature [6, 7]

## 2. MACHINING NICKEL-BASED ALLOYS WITH CERAMICS AND PCD

Inconel 718 and Waspaloy are among the most important and frequently-used nickel-based alloys. Both materials are vacuum-melted and precipitation-hardenable. They are characterized by their great high-temperature resistance, distinctly above that of steels and titanium alloys (Fig. 6).

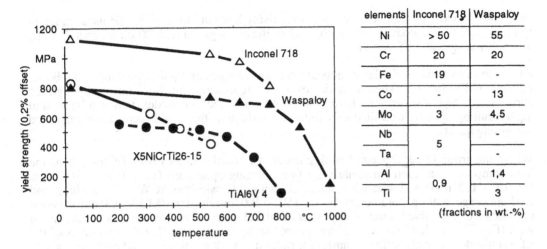

| elements | Inconel 718 | Waspaloy |
|---|---|---|
| Ni | > 50 | 55 |
| Cr | 20 | 20 |
| Fe | 19 | - |
| Co | - | 13 |
| Mo | 3 | 4,5 |
| Nb | 5 | - |
| Ta | | - |
| Al | 0,9 | 1,4 |
| Ti | | 3 |

(fractions in wt.-%)

Fig. 6: Chemical composition of Inconel 718 and Waspaloy and comparison of their high temperature strength with that of TiAl6V4 and high-alloyed steel

In general, the nickel-based alloys belong to the group of hard-to-machine materials. Their low specific heat and thermal conductivity as compared to steels, their pronounced tendency to built-up edge formation and strain hardening and the abrasive effect of carbides and intermetallic phases result in exceptionally high mechanical and thermal stresses on the cutting edge during machining. Owing to the high cutting temperatures which occur, high-speed

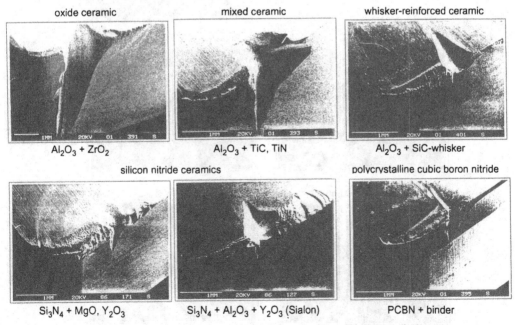

Fig. 7: Alternatives to carbide as cutting material for turning nickel-based alloys

steel and carbide tools can be used only at relatively low cutting speeds. The usual range of turning speeds for uncoated carbides of ISO applications group K10/20 on Inconel 718 and Waspaloy is $v_c$ = 20 - 35 m/min.

Alternatives to carbides for lathe tools are cutting ceramics and polycrystalline cubic boron nitride (PCBN) [9 - 12]. These two classes of cutting material are characterized by high red hardness and high resistance to thermal wear. As compared to carbides, they can be used at higher cutting speeds, with distinctly reduced production times and identical or improved machining quality.

Within the group of $Al_2O_3$-based cutting materials, mixed ceramics with TiC or TiN as the hard component are used particularly for finish turning operations ($v_c$ = 150 - 400 m/min, f = 0.1 - 0.2 mm) and ceramics ductilized with SiC whiskers (CW) for finishing and medium-range cutting parameters ($v_c$ = 150 - 300 m/min, f = 0.12 - 0.3 mm). Oxide ceramics are unsuitable for machining work on nickel-based alloys, owing to intensive notch wear (Fig. 7). The Sialon materials have proved to be the most usable representatives of the silicon nitride group of cutting ceramics for roughing work on nickel-based alloys ($v_c$ = 100

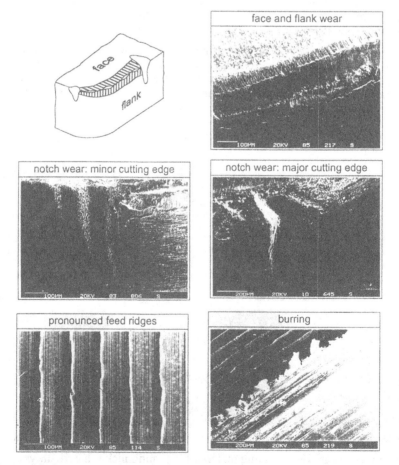

Fig. 8: Characteristic wear modes during turning of nickel-based alloys with ceramics

to 200 m/min, f = 0.2 to 0.4 mm). PCBN cutting materials are used mainly for finishing work on nickel-based alloys.

Characteristic for the turning of nickel-based alloys with cutting ceramics or PCBN is the occurrence of notch wear on the major and minor cutting edges of the tools (Fig. 8). In many applications, notch wear is decisive for tool life. Notching on the minor cutting edge leads to a poorer surface surface finish, notching on the major cutting edge to burring on the edge of the workpiece. Apart from the cutting material and cutting parameters, one of the main influences on notching of the major cutting edge is the tool cutting edge angle. This should be as small as possible. A tool cutting edge angle of $\kappa_r = 45^0$ has proved favourable for turning operations with cutting ceramics and PCBN.

Current state-of-the-art technology for turning Inconel 718 and Waspaloy generally relies on whisker-reinforced cutting ceramics. They have almost completely replaced $Al_2O_3$-based ceramics with TiC/TiN or Sialon for both finishing and roughing operations. This trend is due to the superior toughness and wear behaviour of the whisker-reinforced cutting ceramics and the higher cutting speeds which can be used. These advantages result in longer reproducible tool lives, greater process reliability and product quality and a drastic reduction in machining times as compared to carbides (Fig. 9). The arcuate tool-life curve is typical for turning operations on nickel-based alloys with cutting ceramic or PCBN. It results from various mechanisms which dominate wear, depending on the cutting speed. Notch wear on the major cutting edge tends to determine tool life in the lower range of cutting speeds, chip and flank wear in the upper range. The arcuate tool-life curves indicate that there is an optimum range of cutting speeds. The closer together the ascending and

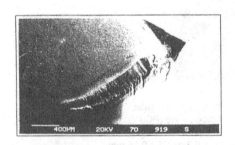

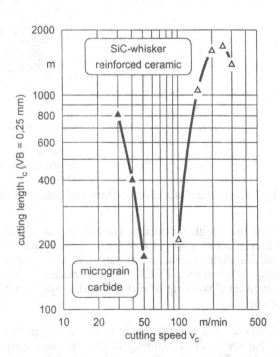

work material: Inconel 718 (solution annealed)
cutting parameters:    $a_p$ = 3 mm, f = 0,25 mm
coolant:            emulsion

Fig. 9: Comparative tool lives: turning Inconel 718 with carbide and whisker-reinforced ceramic

descending arms of the tool-life curves, the more important will it be to work in the narrowest possible range near the tool-life optimum.

Excellent machining results are obtained with PCBN-based tools in finish turning work on nickel-based alloys. Because of their great hardness and wear resistance, PCBN cutting materials can be used at higher cutting speeds. These range from 300 to 600 m/min for finish turning on Inconel 718 and Waspaloy (Fig. 10). As shown by the SEM scans in Fig. 11, the wear behaviour of PCBN tools at these high cutting speeds is no longer determined by notch wear, but chiefly by progressive chip and flank wear. The high performance of the tools is assisted by the specialized tool geometry. Of interest here are the large corner radius, which together with the low depth of cut ensures a small effective tool cutting edge angle of $\kappa_{eff} = 30^{\circ}$, the cutting edge geometry, which is not bevelled but has an edge rounding in the order of $r_n = 25 - 50$ μm and the tool orthogonal rake of $\gamma_o = 0^{\circ}$.

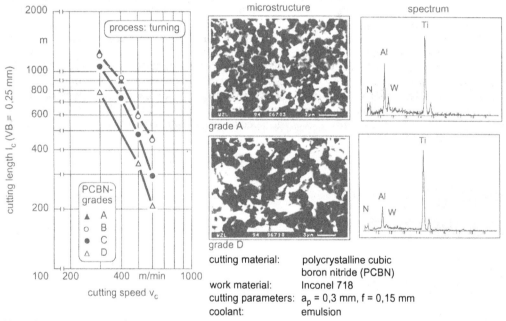

Fig. 10: Influence of microstructure and composition on the performance of PCBN grades

Apart from higher available cutting speeds and excellent wear behaviour, PCBN cutting materials achieve longer tool lives, allowing parts to be finished in a single cut and reliably attaining high accuracies-to-shape-and-size over a long machining time. Because of their high performance, PCBN cutting materials represent a cost-effective alternative to conventional working of nickel-based alloys with carbides or cutting ceramics, despite high tool prices.

The choice of a material grade suited to the specific machining task is of special importance for the successful machining of nickel-based alloys with PCBN cutting materials (Fig. 10 and 11). There are often substantial differences between the PCBN cutting materials available on the market in terms of the modification and the fraction of boron nitride, the grain size and the structure of the binding phase. The resulting chemical, physical and mechanical properties have a decisive influence on the wear and performance behaviour of PCBN tools. Fine-grained PCBN grades with a TiC- or TiN-based binder and a binder fraction of 30 to 50 vol.-% have proved suitable for finishing operations on Inconel 718 and Waspaloy.

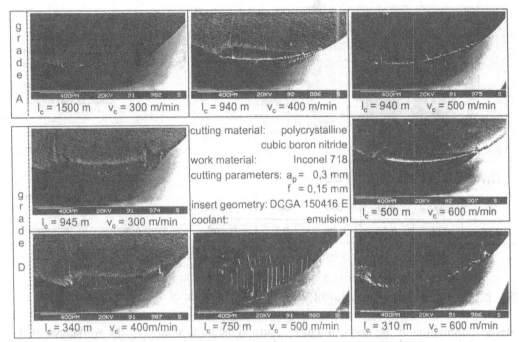

Fig. 11: Wear profiles of PCBN cutting edges

Apart from longitudinal and face turning, the manufacture of turbine discs requires a large number of grooving operations. Depending on groove width and depth, these are characterized by the use of slender tools, by unfavourable contact parameters and by difficult chip forming and chip removal conditions. Because tools are subject to high stresses, carbide tools are generally used for such operations.

Studies of grooving operations on Waspaloy turbine discs have shown that whisker-reinforced cutting ceramic or PCBN are also excellently suited for this machining task. Both types of cutting material can be used at much higher cutting speeds ($v_c$ = 200 - 300 m/min), with drastically reduced production times as compared to carbides (Fig. 12). In the present case, production time was reduced from 5.6 min with carbide to 0.56 min with ceramic or PCBN. The PCBN tools were characterized primarily by high process reliability and relatively low wear, enabling several grooves to be cut with each cutting edge. Special attention must be paid to intensive cooling for the PCBN materials, to prevent softening of the solder used to join the PCBN blank to the carbide substrate.

Components produced with PCBN or whisker-reinforced ceramic generally achieve an excellent surface finish. As shown by results for grooving operations on Waspaloy, grooves machined with PCBN or SiC-whisker reinforced cutting ceramic attain distinctly better surface roughness values than carbide-machined equivalents (Fig. 13).

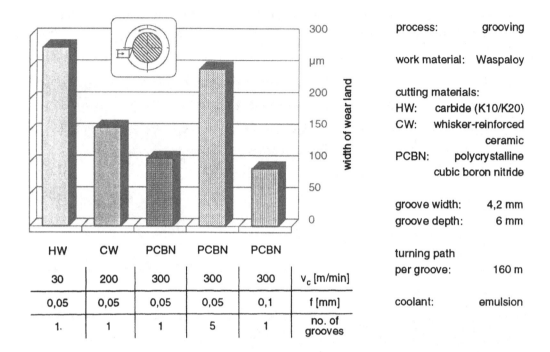

The following table accompanies the figure:

| | HW | CW | PCBN | PCBN | PCBN | |
|---|---|---|---|---|---|---|
| $v_c$ [m/min] | 30 | 200 | 300 | 300 | 300 | |
| f [mm] | 0,05 | 0,05 | 0,05 | 0,05 | 0,1 | |
| no. of grooves | 1. | 1 | 1 | 5 | 1 | |

process: grooving

work material: Waspaloy

cutting materials:
HW: carbide (K10/K20)
CW: whisker-reinforced ceramic
PCBN: polycrystalline cubic boron nitride

groove width: 4,2 mm
groove depth: 6 mm

turning path per groove: 160 m

coolant: emulsion

Fig. 12: Tool wear during grooving of Waspaloy as a function of the cutting material

Plastic deformation of the microstructure occurs in the surface zone of the workpiece as a result of the machining operation, causing hardening and increased final hardness of the work material. The extent of deformation and the size of the hardness increase are dependent on the cutting parameters, tool geometry and tool wear.

Plastic deformation of the surface zone is associated with a change in grain shape. Characteristic for the turning of nickel-based alloys is a pronounced arcuate deformation of the grain boundaries against the workpiece rotation (Fig. 14). Numerous similarly arcuate slip lines occur within the individual grains due to slipping of atomic layers along specific crystallographic planes. The plastic deformation of the work material microstructure visible under the optical microscope generally extends to a depth of 10 or 20 µm from the surface. It is usually confined to grains lying directly at the surface, but may extend over several grain layers at a greater workpiece depth, depending on the severity of thermal effects in the surface zone.

The surface zone hardness increase caused by plastic deformation can be determined by means of microhardness measurements. Inclusions, grain boundaries and other micro-inhomogeneities result in scatter of the individual measurements. Especially where there are successive machining operations, hardening of the work material may lead to increased stress on the tool and to greater wear.

Microhardness measurements reveal a significant increase in hardness in the surface zone (Fig. 14). This amounts to roughly 100 - 200 Vickers units as compared to the hardness of the uninfluenced base material, depending on the cutting parameters.

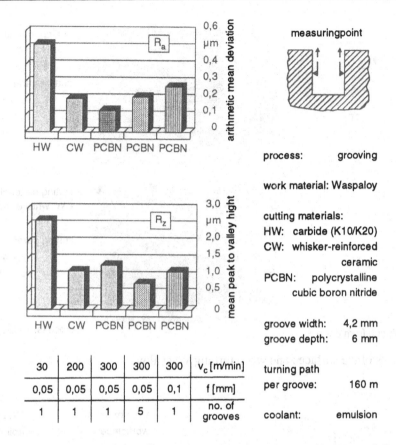

Fig. 13: Surface finish of grooved Waspaloy as a function of the cutting material

Manufacturing-induced changes in the surface zone can have a substantial effect on component properties. This applies particularly to the fatigue strength of dynamically-stressed parts. In view of the high standards of safety and reliability demanded for jet engine parts, the extent to which any change in cutting material and cutting parameters affects part properties is of decisive importance.

Comparative studies of components produced with PCBN and cutting ceramic tools at high cutting speeds showed no significant negative effects of these cutting parameters on the surface zone structure as compared to machining with carbide tools. This conclusion applies not only to influencing of the the surface zone but also to dynamic stressing of the components. This is evident from a comparison of the mean values and standard deviations in the number of cycles to failure in pulsating stress tensile strength tests on fatigue test specimens (Fig. 15). Under the test conditions, the use of PCBN tools leads to a demonstrable but slight increase in the number of cycles to failure. The smaller number of cycles to failure for the specimens machined with whisker-reinforced oxide ceramics is due principally to the increase in the feed rate by a factor of three.

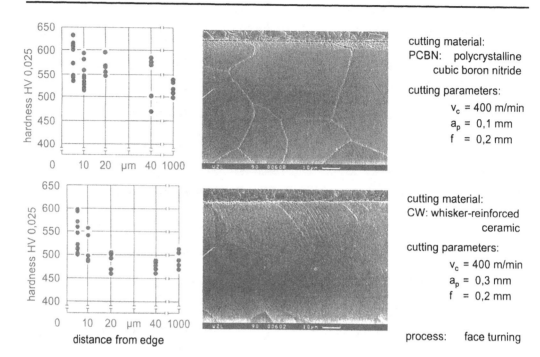

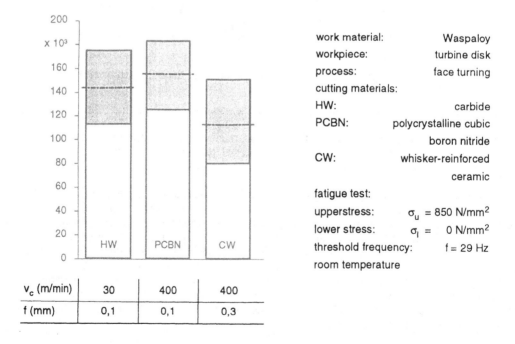

Fig. 14: Effects of the surface zone when turning Waspaloy

Fig. 15: Fatigue strength under pulsating tensile stress of turned Waspaloy specimen

# REFERENCES

1. Esslinger, P., Smarsly, W.: Intermetallische Phasen; Neue Werkstoffe für fortschrittliche Flugtriebwerke, MTU FOCUS 1(1991), p. 36- 42

2. König, W.: Fertigungsverfahren Band 1. Drehen, Fräsen, Bohren. 3. Auflage, VDI-Verlag, Düsseldorf, 1990

3. Erinski, D.: Metallkundliche Aspekte, technologische Grenzen und Perspektiven der Ultrapräzisionszerspanung von Titanwerkstoffen, Promotionsvortrag, RWTH Aachen, 1990

4. Kreis, W.: Verschleißursachen beim Drehen von Titanwerkstoffen, Dissertation, RWTH Aachen, 1973

5. Bömcke, A.: Ein Beitrag zur Ermittlung der Verschleißmechanismen beim Zerspanen mit hochharten polykristallinen Schneidstoffen, Dissertation, RWTH Aachen, 1989

6. Neises, A.: Einfluß von Aufbau und Eigenschaften hochharter nichtmetallischer Schneidstoffe auf Leistung und Verschleiß im Zerspanprozeß mit geometrisch definierter Schneide, Dissertation, RWTH Aachen, 1994

7. König, W., Neises, A.: Turning TiAl6V4 with PCD, IDR 2(1993), p. 85-88

8. Hartung, P. D., Kramer, B. M.: Tool Wear in Titanium Machining, Annals of the CIRP, Vol. 31/1(1982), p. 75-80

9. König, W., Gerschwiler, K.: Inconel 718 mit Keramik und CBN drehen, Industrie-Anzeiger 109(1987)13, p. 24/28

10. Lenk, E.: Bearbeitung von Titan- und Nickelbasislegierungen im Triebwerksbau, Vortrag anläßlich des DGM Symposiums "Schneidwerkstoffe, Spanen mit definierten Schneiden, Bad Nauheim, 1982

11. Vigneau, J.: Cutting Materials for Machining Superalloys, VDI-Berichte 762, p. 321-330, VDI-Verlag Düsseldorf, 1989

12. Narutaki, N., Yamane, Y., Hayashi, K., Kitagawa, T.: High-Speed Machining of Inconel 718 with Ceramic Tools, Annals of the CIRP, Vol. 42/1(1993)

# MACHINABILITY TESTING IN THE 21st CENTURY - INTEGRATED MACHINABILITY TESTING CONCEPT

**E. Kuljanic**

**University of Udine, Udine, Italy**

**KEYNOTE PAPER**

KEY WORDS: Trends, CIM, Machinability Testing, Artificial Intelligence, Intelligent Machine Tool

ABSTRACT: The trend in manufacturing is towards the intelligent machining system. The output of such a system depends significantly on machinability data of conventional and next-generation materials. This paper discusses trends in machinability testing: conventional machinability tests, short machinability tests, machinability index-rating and computerized machinability data system. A discussion of new technological conditions in manufacturing follows as well as a discussion of technological methodologies in the near future. Furthermore, the need for better understanding of machining is discussed. Finally, a proposal of an integrated manufacturing testing concept is proposed.

## 1. INTRODUCTION

In recent decades of our modern world the manufacturing technology and environment have undergone significant changes. However, the changes that will occur in the near future will be more dramatic. The trend in manufacturing is towards the intelligent machining system able to utilize experience, indispensable data and know-how accumulated during past operations, accumulates knowledge through learning and accommodates ambiguous inputs. The development of intelligent machine tool is given in Figure 1, [1]. How will the

Published in: E. Kuljanic (Ed.) *Advanced Manufacturing Systems and Technology*,
CISM Courses and Lectures No. 372, Springer Verlag, Wien New York, 1996.

manufacturing look like in the 21st century could be seen from M. E. Merchant's keynote paper presented at the AMST'93 [2].

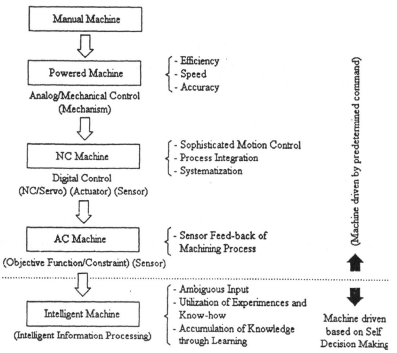

Figure 1. Development of the intelligent machine tool

What can we expect machinability testing to be like in the 21st century? To answer the question we will consider trends in machinability testing: conventional machinability tests, short machinability tests, machinability index-rating and computerized machinability data systems. A special attention is devoted to new technological conditions in manufacturing, technological methodologies and the need for better understanding of machining and other production processes. Finally, an integrated manufacturing testing concept is proposed.

## 2. TRENDS IN MACHINABILITY

It is known that the term "machinability" does not lend itself to an exact definition. Generally speaking, machinability signifies the "ease" of machining [3]. The general criteria for determining the "ease" of machining the material are as follows:

- tool life
- surface roughness
- surface integrity
- magnitude of cutting forces or energy (power) consumption, etc.

Which criterion or criteria will be chosen for determining machinability varies in accordance with the requirements of the particular operation or task to be performed.

F. W. Taylor was the first researcher who had done an extensive machinability testing. He was one of the most creative thinker according to M. E. Merchant [2]. In his well known work [4] presented at the ASME Winter Conference in New York exactly ninety years ago, Taylor raised three questions: **"What tool shall I use?, What cutting speed shall I use? and What feed shall I use?"**, with a following comment. "Our investigations, which were started 26 years ago with the definite purpose of **finding the true answer to these questions**, under all the varying conditions of machine shop practice, have been carried on up to the present time with this as the main object still in view."

It is significant to point out that, after so many years of new facility available such as an electron microscope, computer, machining systems, etc., we have had serious difficulties to find the right answers to the questions. Perhaps, we have to find new approaches in finding the answers to the above questions.

## 2.1. CONVENTIONAL MACHINABILITY TESTS

One of the result of the F. W. Taylor's machinability testing was the well known Taylor's equation:

$$T = K v_c^{k_v} \text{ , or} \tag{1}$$

$$v_c T^m = C \tag{2}$$

where $T$ is tool life in min, $C$ and $K$ are constants, $v_c$ is cutting speed in m/min, $k_v$ and $m$ are exponents, and $m = -1/k_v$. In order to obtain equations (1) and (2), first tool wear curves are to be determined using experimental tool wear data measured periodically after the effective cutting time, for example, after 5, 10, 15, etc. minutes. This procedure is given to point out that one tool wear curve is obtained with one tool. Yet, there are still misunderstandings in obtaining the tool wear curve.

The extended tool life equation is:

$$T = K v_c^{k_v} f^{k_f} a_p^{k_a} \tag{3}$$

where $f$ is feed in mm/rev, $a_p$ is depth of cut in mm, and $k_f$ and $k_a$ are exponents.

## 2.2. SHORT MACHINABILITY TESTS

Taylor's machinability tests are called conventional or long machinability tests. Since this procedure is material and time consuming, short or quick tests of machinability were introduced.

There are different criteria for quick machinability testing: drilling torque or thrust, drilling time or rate for penetration, energy absorbed in pendulum-type milling cut, temperature of cutting tool or chip, the degree of hardening of chip during removal, cutting ratio of chip, easy of chip disposal, etc.

In the 1950s and 1960s some new machinability testing methods came out. For example, quick tool life testing method was developed by applying face turning. The face turning is done at constant number of revolutions, starting cutting at a smaller diameter and moving to a greater diameter. In such a way the cutting speed is increased as the diameter increases according to:

$$v_c = \frac{D \pi n}{1000} \tag{4}$$

where $D$ is diameter in mm and $n$ is number of revolutions in rev/min.

The linear increase of the cutting speed increases the tool wear and decreases the tool life, thus making the test shorter. The reliability of such a method is small [5] due to different tool wear mechanisms, Figure 2, at a lower and at a higher cutting speeds. The tool wear is a sum of wear obtained by different tool wear mechanisms. Clearly, the obtained data can not be applied for different machining operations. In such cases not only the tool wear mechanisms are different but the chip formation is different in different machining operations.

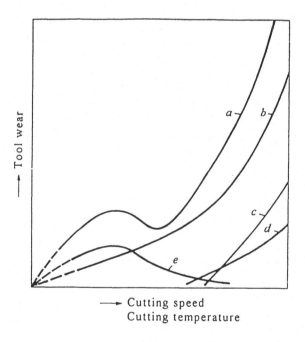

Figure 2. Tool wear mechanisms. $a$ - overall tool wear, $b$ - abrasive, $c$ - diffusion, $d$ - oxidation and $e$ - tool wear due to welding

Generally speaking, the reliability of machinability data obtained by quick machinability testing methods is low. Therefore, quick machinability testing could be applied only in some cases - in some machining operations as a preliminary test.

## 2.3. MACHINABILITY INDEX-RATING

Since the workpiece material properties are influencing tool life or machinability, the approach is to correlate tool life directly with case measured properties of the materials.
One of the first publications in machinability of steel rating was done by J. Sorenson and W. Gates [6]. They made a graphical representation of the general relation of machinability ratings - relative cutting speeds to hardness for hot-rolled SAE steels, Figure 3. A 100% rating was given to SAE 1112 steel cold rolled.

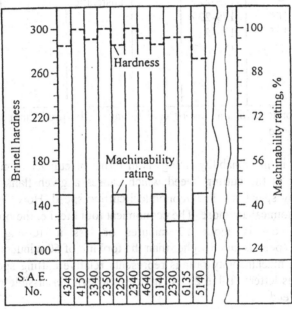

Figure 3. Machinability ratings - relative cutting speed for
hot-rolled SAE steels in relations to hardness

General machinability ratings - index for more common engineering metals and alloys were published by Boston et all [7] in 1943, Table 1. The ratings are expressed in terms of relative values. These figures are often called "percent machinability" or "relative machinability", and are representing the relative speed to be used with each given metal to obtain a given tool life. For example, a material whose rating is 50 should be machined at approximately half the speed used for the material rating 100, if equal tool life for both is desired.
It can be seen, that rating values in Table 1 are based on a rating of 100 for steel AISI B1112, cold-rolled or cold-drawn when machined with a suitable cutting fluid at cutting speed $v_c = 56$ m/min under normal cutting conditions using high-speed-steel tools. In Table 1 the ratings given for different classes of alloys, class I to class IV, represent their relative machinability within a given class, but the ratings for any class are not comparable with those for any other class.

Table 1. Machinability rating of various metals

| Class I | | | | Class IV | | |
| Ferrous 70% and higher | | | | Ferrous 40% and higher | | |
| AISI | Rating % | Brinell | | AISI | Rating % | Brinell |
| C1109 | 85 | 137-166 | | A2515[+] | 30 | 179-229 |
| C1115 | 85 | 147-179 | | E3310[+] | 40 | 170-229 |
| C1118 | 80 | 143-179 | | E9315[+] | 40 | 179-229 |
| C1132 | 75 | 187-229 | | Stainless | | |
| C1137 | 70 | 187-229 | | 18-8[+] | 25 | 150-160 |
| B1111 | 90 | 179-229 | | austenitic | | |
| B1112 | 100 | 179-229 | | | | |
| B1113 | 135 | 179-229 | | | | |
| A4023 | 70 | 156-207 | | | | |

Annealed prior to cold drawing or cold rolling in the production of the steel specially mentioned

The second approach in machinability ratings is in terms of equivalent cutting speed. The cutting speed number is the cutting speed which causes a given flank wear land in 60 minutes. Such a cutting speed is called economical cutting speed. However, the tool life of 60 minutes is not economical anymore. The economical tool life, i.e. the optimal tool life for minimum machining cost is about 10 minutes or less in turning. Therefore, the corresponding cutting speed is much higher than the tool life of 60 minutes.

The third approach in machinability ratings represents relative cutting speed values where the ratings are given as letters, Table 2. "A" indicates a high permissible cutting speed and "D" a lower cutting speed.

Table 2. Machinability rating of stainless steel (hot-
rolled, annealed)

| AISI No. | Hardness Brinell | Machinability rating[*] |
| --- | --- | --- |
| AISI 410 | 135-165 | C |
| AISI 416 | 145-185 | A |
| AISI 430 | 145-185 | C |
| AISI 446 | 140-185 | C |
| AISI 302 | 135-185 | D |
| AISI 303 | 130-150 | B |
| AISI 316 | 135-185 | D |

A - excellent, B - good, C - fair, D - poor

The forth approach is the correlation of tool life and the microstructure of the metal. Generally speaking hard constituents in the structure result in poor tool life, and vice versa. In addition, the tool life is usually better when the grain size of the metal is larger.

The correlation of microstructure of steels and tool life was studied by Woldman [8]. Averages relations of tool life and surface finish to microstructure of steel were reported as "good", "fair", "fair to good" and "poor", Table 3.

Table 3. Average relation of tool life and surface finish to microstructure of steels

| Class of steel | Structure | Tool life | Surface finish |
|---|---|---|---|
| Low - carbon steels | Cold - drawn, small grain size | Good | Good |
| | Normalized | Good | Fair |
| Mild medium - carbon steel | Perlitic, moderate grain size | Good | Good |
| | Perlitic, small grain size | Fair | Good |
| | Perlitic, large grain size | Good | Fair |
| | Spheroidized | Fair | Poor |

## 2.4. ACTIVITIES IN THE FIELD OF MACHINABILITY

Mostly the machinability data were obtained for turning. A research in machinability in milling was done at The Cincinnati Milling Machine Co., today Cincinnati - Milacron in Cincinnati, USA [9].

Perhaps most work has been done in machinability at Metcut Research Associates Inc. in Cincinnati. As a result of such a work is the Machinability Data Handbook [10], first edition published in 1966. The purpose of the Handbook is to provide condensed machining recommendation for significant workpiece materials and machining operations. It supplies the recommended speed, feed, tool material, tool geometry, and cutting fluid as well as data for determining power requirements for various machining operations. It also contains the most practical and readily accessible data for general shop application.

The CIRP (Colleáge international pour l' étude scientifique des tecniques de production mecanique) - Paris, and ISO (International Standard Organization) work in machinability was presented at the AMST'93 in [11]. In 1974 the CIRP Working Group "Milling" accepted the author's proposal to make a document on testing for face milling. The CIRP document "Testing for Face Milling" [12] was completed in 1977 and accepted by ISO as a basic document.

The Working Group 22 - Unification of Tool Life Cutting Test of the ISO Technical Committee 29 worked from 1977 to 1985 on three documents: ISO International Standard Tool Life Testing in Milling, Part 1 - Face Milling [13], Part 2 - End Milling [14] and updated the ISO International Standard Tool Life Testing with Single Point Turning Tools (ISO 3685, 1977) [15]. The CIRP document "Testing for Face Milling" and the ISO Standards are significant basic documents in machinability testing. A survey and discussion of the ISO Standards including the Volvo Machinability Test [16] is given in [11].

In 1986 J. Kahles proposed to conduct a survey in the CIRP, primarily in industry around the world, to ascertain the principal machinability data needs of industry. The results of the survey "Machinability Data Requirements for Advanced Machining Systems" were published in 1987 [17], and the discussion of the obtained results was presented by the author at the 1st International Conference AMST´87 [18]. The information were gathered from industries from twelve countries. The conclusion is that reliable machinability data on tool life, chip control, surface finish and surface integrity are indispensable for efficient computer integrated manufacturing.

## 2.5. COMPUTERIZED MACHINABILITY DATA SYSTEMS

The purpose of computerized machinability data systems is a systematic and rapid selection of machinability data for the user, i.e. to facilitate numerical control and conventional machining.

The development of computerized machinability data systems started in the United States and in other places at the beginning of 1970s. One of the first machinability data systems was developed at Metcut Research Associates Inc. - Machinability Data Center in Cincinnati. This system considers both the cost of operating the machine tool and the cost of tool and the tool reconditioning, and the total cost and operating time are calculated. In order to achieve this, tool life data related to cutting speed - tool life equations for various sets of machining parameters are required.

A series of programs were designed in different firms to help set time standards. Such a series of programs were developed at IBM under the name "Work Measurement Aids": Program 1 - Machinability Programming System, and Program 2 - Work Measurement Sampling. The basis for the Machinability Programming System is that machining operations can be related to a standard operation using a standard material and tool. Different machining operations are related to the standard operation using speed and feed adjustment factors. The factors are: base material, speed and feed, and adjustment factors for each operation and for conditions of cut. The Machinability Programming System allows for users specification of material data, machine group speeds and feeds, and operation factors.

ABEX Computerized System was developed using the Metcut tool life equations and the manipulative functions of the IBM Work Measurement Aids program to describe the part, store the available rates on a wide range of machine tools, and generate cutting conditions on short cycle operations [19]. In the ABEX system, shop-generated machining data are organized and selected on the basis of "operation family" concepts, while tool life is defined on the basis of the tool´s ability to hold a specified tolerances and surface finish. For example, in turning, workpiece diameter is often such a parameter, and all operations involved in obtaining a given range of diameters make up the "operation family". The ABEX system provides for the selection of metal cutting conditions based on strategies other than minimum cost or maximum production.

A mathematical model to determine cutting speed for carbide turning and face milling was applied in the General Electric "Computerized Machinability Program" [20]. The constants and exponents for tool life equations were developed from empirical data obtained in laboratory and shop tests. The considered factors affecting the tool life equations were: tool material and tool geometry, hardness of workpiece materials, surface conditions, feed, depth of cut, flank wear and machinability rating. The General Electric Computerized Machinability Program calculate speeds and feeds for minimum cost, maximum production, corresponding tool life, needed power, minimum part cost or maximum production rate.

At a same time the EXAPT Computerized Machinability Data System was also developed [19]. This system is an NC part programming system containing geometrical and technological features that select operation sequences, cutting tools and collision-free motions. The EXAPT processors determine optimum machining conditions from the economic, empirical and theoretical metal cutting conditions. The system stores information on materials, cutting tools and machine tools. An important and extensive work has been done by INFOS in machinability testing.

Thus, the basic data have to be obtained by testing for computerized machinability data systems or for conventional machinability data. The machinability testing methodology will depend on the predominant technologies and new conditions in the 21st century. Therefore, let us examine the nature and promise of the new technologies and conditions.

## 3. NEW TECHNOLOGICAL CONDITIONS IN MANUFACTURING

M. E. Merchant pointed out at the AMST'93 [2] that "the digital computer is an extremely powerful systems tool made us recognize that manufacturing is a system in which the operation can be optimized, and not just of those individual activities, but of the overall system as well. Thus today, as the world industry approaches the 21st century, it is engaged in striving toward accomplishment of computer integration, automation and optimized operation of the overall manufacturing enterprise. However, in pursuing that overall trend it is increasingly recognizing the dual nature of the concept of CIM enterprise, encompassing both in technological and managerial operations." We will discuss only some technological methodologies.

## 3.1. TECHNOLOGICAL METHODOLOGIES

According to M. E. Merchant [2] the first evolving methodology is **concurrent engineering** - concurrent engineering of the conception and design of product and of planning and execution of its manufacturing production and servicing. The second important evolving methodology is that of **artificial intelligence** in the manufacturing.

By applying concurrent engineering methodology the product costs are reduced and the industrial competitiveness is increased. About 70 percent of the cost of the manufacturing production of a product is fixed when its design is completed. Since the material of the components has to be chosen in the design, the "frozen" cost of the manufacturing can be reduced by selecting corresponding material with better machinability. Thus the purpose is to have more reliable machinability data.

Concurrent engineering also affects other cost savings and makes shorter the lead time between conceptual design of a product and its commercial production.

According to [2] "the **artificial intelligence** probably has greater potential to revolutionize manufacturing in the 21st century than any other methodology known to us today". Artificial intelligence has the potential to transform the non-deterministic system of manufacturing into an intelligent manufacturing system which is "capable of solving within certain limits, unprecedented, unforeseen problems on the basis even of incomplete and imprecise information" [21] - information characteristic of a non-deterministic system.

The realization of this tremendous potential of artificial intelligence to revolutionize manufacturing in the 21st century is surely the most challenging undertaking. It will require revolutionary developments in the technology of artificial intelligence and **massive research and development efforts in manufacturing** [2]. However, the rewards will be magnificent.

## 3.2. BETTER UNDERSTANDING OF MACHINING AND OTHER PRODUCTION PROCESSES

One part of the "massive research in manufacturing" will be carried out for better understanding of the fabrication processes of the removal and in machinability testing of conventional and new materials. The possibilities to do such a research are already enormous. The identification of the machining process can be easier and more reliable by applying mathematics of statistics, design of experiments, modeling, identification of the machining process by the energy quanta and the entropy [22], new software, new computer generations etc.

For example, this makes possible to determine more reliable models, as tool life equation which includes significant interactions proposed by the author in 1973 [23]. The equation could include: stiffness effect of machining system, the effect of number of teeth in the cutter in milling and interactions. Such tool life equation which has been determined from the experimental data [23] by applying new analysis facilities is:

$$T = 211.789 \cdot 10^5 \; v_c^{-4.0225} \; f_z^{-1.4538} \; z^{-10.2674} \; S^{-1.3292} \; \exp(2.3913 \ln v_c \ln z +$$

$$+ 0.3380 \ln v_c \ln S + 0.8384 \ln z \ln S + 0.0190 \ln v_c \ln f_z \ln S - 0.1972 \ln v_c \ln z \ln S) \tag{5}$$

where T is tool life in min, $v_c$ is cutting speed in m/min, $f_z$ is feed per tooth in mm, z is number of teeth and S is stiffness in N/mm. Multiple regression coefficient is $R = 0.91862$. The time needed to determine this equation is less than a second while thirty years ago it took days without computer. Thus it is easy to determine any equation. There is a need to obtain empirical surface roughness equations and better understanding of chip formation and chip control.

## 4. INTEGRATED MACHINABILITY TESTING CONCEPT

The integrated machinability testing is an approach in which tool life data and/or tool wear, tool wear images, machining conditions and significant output data as dimension changing of the machined workpiece, surface roughness, chip form etc. are registered and analyzed in an unmanned system, i.e., in intelligent machining system.

The analysis of the obtained data could be done for different purposes. From the tool wear or tool life data, the tool life equations can be determined and applied to optimize cutting conditions on the intelligent machining system. Secondly, an integrated machinability data bank could be built up by directly transferring machinability data from the intelligent machining system. The data obtained in this way could be used and analyzed for other purposes, for example, for process planning, for design, etc.

The integrated machinability testing can be used for roughing and finishing machining. It should be pointed out that the aim of the 26 years F. W. Taylor research work [4] was done only for roughing work. He emphasized in Part 1 of [4] "our principal object will be to describe the fundamental laws and principles which will enable us to do roughing work in the shortest time. Fine finishing cuts will not be dealt with." However, in integrated machinability testing the emphasis is on finishing or light roughing work due to the trend that dimensions of forgings and castings or workpieces produced by other methods are closer to the final dimensions of the part.

The main data that should be quoted to determine the conditions of machining are as follows:

- machine tool and fixturing data
- workpiece data
     material, heat treatment, geometry and dimensions
- tool characteristics
     tool material, tool geometry
- coolant data

These data will be registered automatically.

The cutting conditions: cutting speed, feed and depth of cut will be chosen by a computer applying the design of experiments. The advantage of applying the design of experiments in industrial conditions is to obtain data from the machining process in a shorter time. For example, the reduction of needed tests is from about fifteen tests to five or seven tests when Random Strategy Method [24] is applied using computer random number generator.

The tool wear and tool life will be determined by intelligent sensor systems with decision making capability [25]. The tool wear images and the dimensions of tool wear, like $VB$ - the average tool wear land, will be measured, analyzed and saved automatically in a computer.

The surface roughness, the hardness of the machined surface, perhaps the surface integrity, the dimensions of the machined workpiece will be analyzed and saved too. It will be possible to analyze the chip form and to classify it according to ISO numeric coding system [11]. From such data tool wear or the tool life equations (3) or (5), or some other relationships for the identification of machining process, will be easily determined by regression analysis.

The tool life equations obtained by integrated machinability testing can be used for different purposes. First for in-process optimization of machining conditions. Secondly, to building up the machinability data bank with more reliable data. Such a data bank would receive the machinability information, as tool life equations etc., directly from intelligent machining systems. Thus, the data in such a data bank would be more reliable. The machinability data bank would be self regenerative and a small factory could have a proper machinability data bank.

## 5. CONCLUSION

In accordance to the considerations presented in this paper, we may draw some conclusions about how the machinability testing might be in the 21st century. Based on what we already know the machinability testing will be possible to integrate with the machining in intelligent machining system in industrial conditions without a direct man involvement.

The integrated machinability testing (IMT) concept is an approach in which tool life data and/or tool wear, tool wear images, machining conditions, and the significant output data such as dimension changes of the machined workpiece, surface roughness, chip form etc., are registered and analyzed in an unmanned system, i.e. in intelligent machining system.

The machinability data and other information obtained from machining in new industrial conditions on such machining systems will be used for in-process optimization of machining conditions and self monitoring. For this purpose, quantitative machinability models for process physics description should be applied to intelligent machining systems.

An integrated machinability data bank could be built up even in a small factory by directly transferring machinability data, tool wear images and other relevant information from the intelligent machining system. Thus the reliability of the machinability data could be increased. The effect of stiffness of the machining system and other significant factors will be included. The data of the integrated machinability data bank will be useful in design for material selection to decrease the "frozen" cost, Chapter 3.1., for process planing, etc.

In order to improve intelligent machining systems and make them more reliable, machining and process physics researchers should be included more in control teams.

By applying the integrated machinability testing concept in the intelligent machining systems we could find more adequate answers to F. W. Taylor questions, "what speed shall I use" and "what feed shall I use", after hundred years.

## REFERENCES

1.  Moriwaki, T.: Intelligent Machine Tool: Perspective and Themes for Future Development, Manufacturing Science and Engineering, ASME, New York, 68(1994)2, 841-849

2.  Merchant, M.E.: Manufacturing in the 21st Century, Proc. 3rd Int. Conf. on Advanced Manufacturing Systems and Technology AMST'93, Udine, 1(1993), 1-12

3.  Tool Engineers Handbook, McGraw-Hill, New York, Toronto, London, 1950

4.  Taylor, F.W.: On the Art of Cutting Metals, Transactions of the ASME, 28(1906)

5.  Kuljanić, E.: Effect of Initial Cutting Speed on Tool Life in Short Tool Life Test, Strojniški Vestnik XIII, 1967, 92-95

6. Sorenson, J., Gates, W.: Machinability of Steels, Prod. Eng., 10(1929)

7. Boston, O.W., Oldacre, W.H., Moir, H.L., Slaughter, E.M.: Machinability Data from Cold-finished and Heat-treated, SAE 1045 Steel, Transactions of the ASME, 28(1906)1

8. Woldman, N.E.: Good and Bad Structures in Machining Steel, Materials and Methods, 25(1947)

9. A Treatise of Milling and Milling Machines, Sec. 2, The Cincinnati Milling Machine Co., 1946

10. Machinability Data Handbook, Metcut Research Associates Inc., Cincinnati, Ohio, First Ed. 1966, Second Ed. 1972, Second Printing 1973

11. Kuljanić, E.: Machinability Testing for Advanced Manufacturing Systems, Proc. 3rd Int. Conf. on Advanced Manufacturing Systems and Technology AMST'93, Udine, 1(1993), 78-89

12. Testing for Face Milling, Internal publication, CIRP, Paris, 1977

13. ISO International Standard: Tool Life Testing in Milling Part 1 - Face Milling, 8688/1, Stockholm, 1985

14. ISO International Standard: Tool Life Testing in Milling Part 2 - End Milling, 8688/2, Stockholm, 1985

15. ISO International Standard: Tool Life Testing with Single Point Turning Tools, Stockholm, 1977

16. Muhrén, C., Eriksson, U., Skysted, F., Ravenhorst, H., Gunarsson, S., Akerstom, G.: Machinability of Materials Applied in Volvo, Proc. 2nd Int. Conf. on Advanced Manufacturing Systems and Technology AMST'90, Trento, 1(1990), 208-219

17. Kahles, J.: Machinability Data Requirements for Advanced Machining Systems, Progress report No. 2, CIRP S.T.C. Cutting, Paris, 1987

18. Kuljanić, E.: Machining Data Requirements for Advanced Machining Systems, Proc. of the Int. Conf. on Advanced Manufacturing Systems and Technology AMST'87, Opatija, 1987, 1-8

19. N/C Machinability Data Systems - Numerical Control Series, SME, Dearborn, Michigan, 1971

20. Weller, E.J., Reitz, C.A.: Optimizing Machinability Parameters with a Computer, Paper No. MS66-179, Dearborn, Michigan, American Society of Tool and Manufacturing Engineers, 1966

21. Hatvany, J.: The Efficient Use of Deficient Information, Annals of the CIRP, 32(1983)1, 423-425

22. Peklenik, J., Dolinšek, S.: The Energy Quanta and the Entropy - New Parameters for Identification of the Machining Processes, Annals of the CIRP, 44(1995)1, 63-68

23. Kuljanić, E.: Effect of Stiffness on Tool Wear and New Tool Life Equation, Journal of Engineering for Industry, Transactions of the ASME, Ser. B, (1975)9, 939-944

24. Kuljanić, E.: Random Strategy Method for Determining Tool Life Equations, Annals of the CIRP, 29(1980)1, 351-356

25. Byrne, G., Dornfeld, D., Inasaki, I., Ketteler, G., König, W., Teti, R.: Tool Condition Monitoring (TCM) - The Status of Research and Industrial Application, Annals of the CIRP, 44(1995)2, 541-568

26. Kuljanić, E.: Materials Machinability for Computer Integrated Manufacturing, Proc. 2nd Int. Conf. on Advanced Manufacturing Systems and Technology AMST'90, Trento, 1(1990), 31-45

# HIGH-SPEED MACHINING IN DIE AND MOLD MANUFACTURING

H. Schulz

Technical University of Darmstadt, Darmstadt, Germany

**KEYNOTE PAPER**

KEYWORDS: High speed cutting, die and mold manufacturing.

ABSTRACT: For an economic application of high speed machining in die and mold manufacturing it is necessary to optimize the cutting process by specific cutting strategies. The influence of an optimized technology on the machine tool itself is very important and must be recognized for the development of new machines and their components.

## 1. INTRODUCTION

Molds and dies have to be manufactured at low cost in shorter and shorter time. For this reason, the entire product generation process must be investigated for opportunities of reducing the time from idea to final product as well as for simultaneously minimizing the costs [1-5]. As can be seen from Fig. 1, there are various approaches, with high speed cutting (HSC) of the mold contour being of special importance. Basically, high speed cutting of steel and cast-iron molds is however reasonable only for the finishing or pre-finishing operations (Fig. 2). Machining of steel or cast-iron molds must therefore be split up into

- rough-machining, on efficient standard NC machines as previously and
- finishing or pre-finishing on high speed machines.

Published in: E. Kuljanic (Ed.) *Advanced Manufacturing Systems and Technology*, CISM Courses and Lectures No. 372, Springer Verlag, Wien New York, 1996.

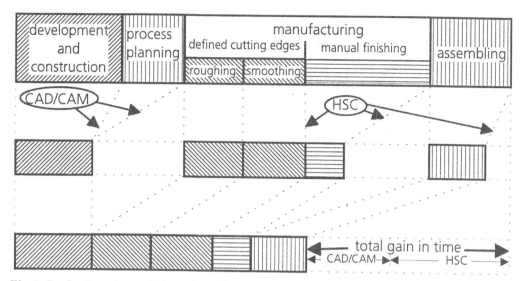

Fig 1: Production chain of die and mold manufacturing

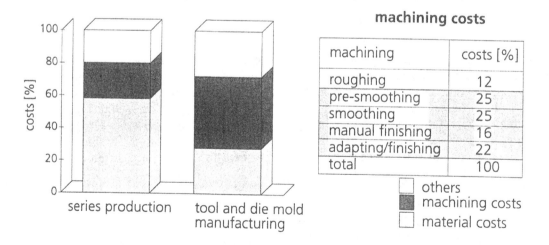

**machining costs**

| machining | costs [%] |
|---|---|
| roughing | 12 |
| pre-smoothing | 25 |
| smoothing | 25 |
| manual finishing | 16 |
| adapting/finishing | 22 |
| total | 100 |

□ others
■ machining costs
□ material costs

Fig 2: Cost structure in die and mold manufacturing

## 2. CUTTING STRATEGIES

The major objective in metal cutting is the closest possible approximation to the final contour, especially also for free-form surfaces. Since in high-speed cutting it is possible to use feeds five to eight times faster than usual, the cutter lines can for the same finishing time be made five to eight times closer than in conventional milling. As can be seen from Fig. 3, this results in a much better approximation to the final contour. For completion, only minor corrections of dimensions and surfaces are required which as a rule are made manually. This means that the manual rework time involved with smoothing and polishing is substantially reduced. This is the reason why for high-speed machining of molds and dies various strategies can basically be used:

1.  High speed cutting itself does not reduce the mechanical manufacturing time, but due to the close approximation to the final contour, manual rework is reduced substantially.

2.  HSC can also be used to reduce the manufacturing times in finishing and prefinishing operations, but in that case the potential of choosing very close line widths is not fully made use of. This will increase rework times to a certain extent.

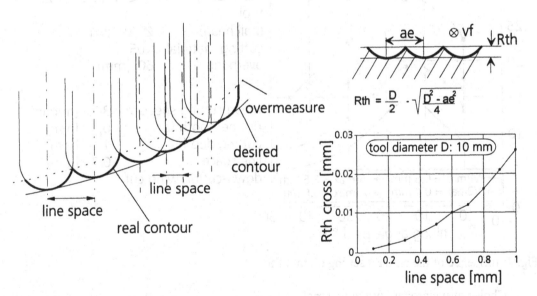

Fig. 3: Surface roughness - influence of line space

It is therefore necessary to consider in each case whether the cutting strategy 1 or 2 will in total bring about superior time and cost reductions.

High speed machining has many advantages, but one essential disadvantage, i.e tool wear which increases substantially with increasing cutting speed. It is therefore required to economically get this negative effect under control by using the following means:

1.  Optimization of technological parameters
    Tool life is a close function of feed, cutting speed and infeed depth, the tool life optimum being within a relatively small range of adjustment of the parameters which must be determined individually for each material/tool combination [6, 7].

2.  Improvement of setting conditions
    A decisive factor for wear behaviour of the tool is the choice of its lead angle in or across to the feed plane. For steel 40 CrMnMo 7, Fig. 4 clearly shows the influence exerted by the tilting angle on both tool life and surface quality in machining a mold. The tool life optimum is to be found at tilting angles of approx. 15 degrees [8]. This also results in the best surface quality characteristics. In this case, inclination in the feed plane is more useful, because a lead of the tool accross to the feed plane will cause higher impact loads to occur, thus reducing lool life substantially.

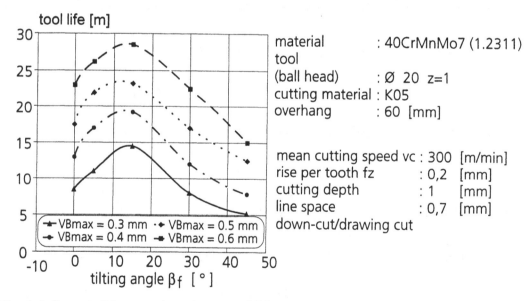

Fig. 4: Influence of the cutter's setting on tool life

3.  Choice of the proper cutting material
    The choice of the proper cutting material has decisive importance for tool life. It turns out that for cutting 40 CrMnMo 7 the use of Cermets (surface-treated) brings the best results (Fig. 5, left). On the other hand, CBN turns out to be a better choice for machining cast-iron GG25CrMo, a material also much used in die and mold making (Fig. 5, right).

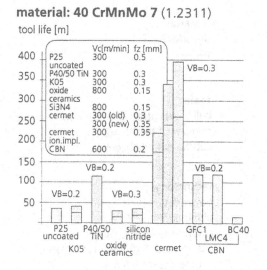

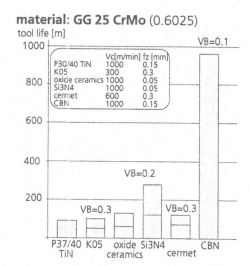

technology:
down-cut/drawing cut
tilting angle: + 15°
infeed: 1 mm
line space: 0.7 mm, 0.5 with CBN

Fig. 5: Wear of different cutting materials when machining

4. Choice of the proper tool

However, the tool life of the cutter to be chosen will not only depend on the cutting material but also on the geometry or the tool length possibly required for structural reasons. Fig. 6 shows that the feed per tooth has to be considerably decreased with increasing projection length of the cutter. However, it is important to consider all of the technological parameters simultaneously when tool life is to be optimized. Fig. 7 shows that, with the technological parameters mentioned, optimum values for machining 40 CrMnMo 7 can be achieved by a combination of cutting speed $v_c = 300$ m/min and feed per tooth $f_z = 0.3$ mm.

5. Use of the proper cutting strategies

A determining factor influencing both surface quality and dimensional and shape accuracy is the cutting strategy. The different effects of the various cutter guidance strategies can be seen from Fig. 8. Inclination in the feed plane combined with draw cut and down milling is the most appropriate strategy. Reciprocal milling for surface generation causes a substantial reduction of tool life. Boring cuts tilting angle exceeding +/- 15 degrees should be avoided as far as possible in order to keep dimensional deviations within reasonable limits (Fig.9).

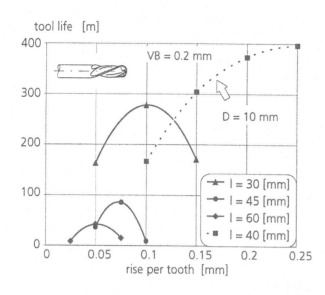

material:
40 CrMnMo7
(steel)

tool:
ball head cutter: Ø 6 (10) mm
number of teeth: z=2
cutting material: cermet

technology:
down-cut/drawing cut
tilting angle: + 15°
cutting speed: 300 m/min
line space: 0.5 (0.6 with Ø 10) mm
cutting depth: 0.5 mm

Fig. 6: Usage of cermet with different tool lengths

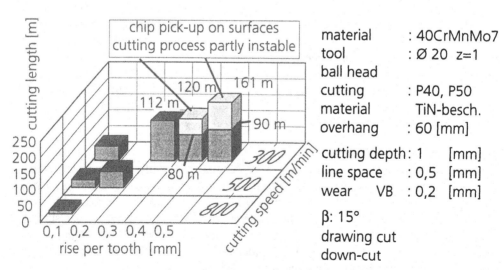

material        : 40CrMnMo7
tool            : Ø 20  z=1
ball head
cutting         : P40, P50
material          TiN-besch.
overhang        : 60 [mm]

cutting depth : 1    [mm]
line space    : 0,5  [mm]
wear    VB    : 0,2  [mm]

β: 15°
drawing cut
down-cut

Fig. 7. Cutting length depending on cutting speed and feed

tool life [m]

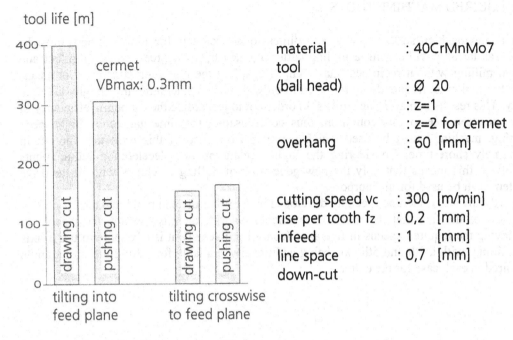

| | |
|---|---|
| material | : 40CrMnMo7 |
| tool | |
| (ball head) | : Ø 20 |
| | : z=1 |
| | : z=2 for cermet |
| overhang | : 60 [mm] |

| | | |
|---|---|---|
| cutting speed vc | : 300 | [m/min] |
| rise per tooth fz | : 0,2 | [mm] |
| infeed | : 1 | [mm] |
| line space | : 0,7 | [mm] |
| down-cut | | |

Fig. 8: Influence of different cutter settings

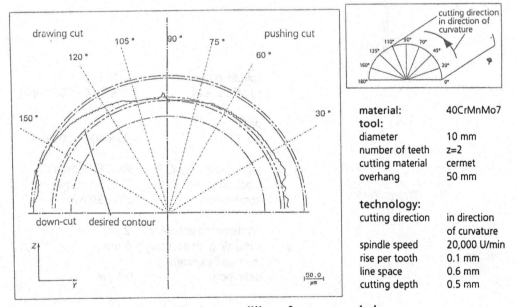

| | |
|---|---|
| material: | 40CrMnMo7 |
| tool: | |
| diameter | 10 mm |
| number of teeth | z=2 |
| cutting material | cermet |
| overhang | 50 mm |
| | |
| technology: | |
| cutting direction | in direction of curvature |
| spindle speed | 20,000 U/min |
| rise per tooth | 0.1 mm |
| line space | 0.6 mm |
| cutting depth | 0.5 mm |

Fig. 9: Tolerance when using the 3-axes milling of convex workpieces

## 3. REQUIRED MACHINE TOOLS

For achieving high surface quality and dimensional accuracy, the mold element must be made as far as possible with a milling cutter inclined by 15 degrees, using draw cut and down milling without reciprocal machining. Regarding the operating strategy, this means that the tool must frequently be retracted in order to permit machining in a single direction only. This results in major idle strokes. In order not to jeopardize the economic efficiency of the operation due to this condition, thus compensating the time gained by high speed cutting, machines must be used which permit performance of this retracting motion in extremely short time. Considering the high acceleration and deceleration requirements involved, this means that only the new generation of milling machines with linear drive systems can be used for this purpose.

Fig. 10 shows the machine specially developed by PTW for die and mold making.

In view of the fact that the machines available today frequently are not yet capable of achieving the required speeds in five axes (control problems!), it is advisable to use 3-axis machining, with a 4th and 5th axis being additionally available for setting the tilting angle required in each case for the cutter.

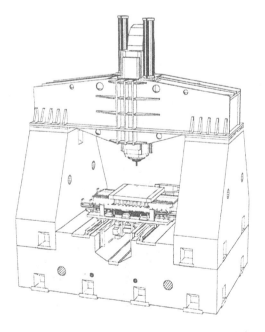

| | |
|---|---|
| spindle power | 7....12 kW |
| spindle speed | 30 000 - 40 000 1/min |
| path | |
| X x Y x Z | 400 x 400 x 300 mm |
| rapid motion X, Y, Z | 60 ... 80 m/min |
| machining feed X, Y, Z | 30 ... 50 m/min |
| acceleration | 20 ... 30 m/s$^2$ |
| positionning accuracy | 2 µm |
| dynamic path accuracy | 5 µm |
| measuring system definition | 0.1 µm |

Fig. 10: HSC-machine with linear drives

## 4. CONCLUSION

Application of high speed cutting in die and mold manufacturing results in substantial time reductions. In extreme cases, as is shown by a practical example, even finishing times slightly increased as compared with a conventional NC machine can permit obtaining overall time reductions. In the case of a mold rough-machined to 0.5 mm oversize, the finishing time on a conventional NC machine amounted to a total of 36 hours, the total manual reworking time was 70 hours, thus resulting in a total manufacturing time of 106 hours. As compared with this, application of a HSC machine involved a finishing time of 40 hours and manual reworking time of just 14 hours, i.e. a total of 54 hours. This resulted in a time reduction of almost 50 % due to the application of this new technology (Fig. 11).

For economic considerations it is therefore indispensable to include the production planning process as a whole.

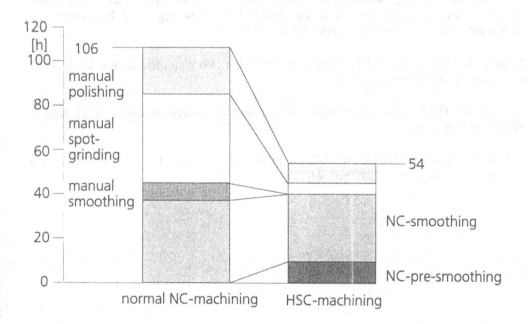

Fig. 11: Comparison of normal milling and high-speed milling

## 5. REFERENCES

1. Schulz, H.: Anwendung der Hochgeschwindigkeitsbearbeitung im Werkzeug- und Formenbau, Proceedings of 3. Internationaler Anwenderkongress, Dresden, 29./30.09.95

2. Hock, S.: Hochgeschwindigkeitsfräsen im Werkzeug- und Formenbau, Proceedings of Darmstädter Erfahrungsforum Werkzeug und Formenbau: „Schneller zu präzisen Werkzeugen", 6.12.95

3. Schmitt, T.: Schnelle Fräsmaschine für den Werkzeug- und Formenbau mit Linearantrieben, Proceedings of Darmstädter Erfahrungsforum Werkzeug und Formenbau: „Schneller zu präzisen Werkzeugen", 6.12.95

4. Hock, S.: Hochgeschwindigkeitsfräsen im Werkzeug- und Formenbau, Proceedings of Darmstädt Erfahrungsforum Werkzeug- und Formenbau:" Kostengünstigere Formen in kürzerer Zeit", 7.03.95

5. Schulz, H.: Hochgeschwindigkeitsbearbeitung im Werkzeug- und Formenbau, Proceedings: Fertigungstechnisches Symposium: "Der Werkzeug- und Formenbau im produktionstechnischen Umfeld der 90er Jahre", 17.06.93

6. Schulz, H.: Hochgeschwindigkeitsfräsen metallischer und nichtmetallischer Werkstoffe, Carl Hanser Verlag, München, 1989

7. Schulz, H.: Hochgeschwindigkeitsbearbeitung mit definierter Schneide, Carl Hanser Verlag, München 1996

8. Schulz, H.; Hoch, S.: High-Speed Milling of Dies and Molds - Cutting Conditions and Technology, Annals of the CIRP, Vol. 44/1, 1995

# FORMING PROCESSES DESIGN ORIENTED
# TO PREVENT DUCTILE FRACTURES

## N. Alberti and F. Micari
### University of Palermo, Palermo, Italy

## KEYNOTE PAPER

KEY WORDS: Ductile Fractures, Damage Mechanics, Numerical Simulations

ABSTRACT: During cold forming processes ductile fractures are sometimes encountered, depending on the operating parameters and on the material properties and determining the production of defective components to be discarded; for this reason the development of a general approach for the prediction of this type of defects is particularly suitable and this aim has been pursued by a large number of researchers in the last two decades. In the paper the most important and diffused models are described and analysed in order to outline the advantages offered by each of them. Some applications to typical metal forming processes are presented, carrying out a comparison between the numerical predictions and the experimental verifications.

## 1. INTRODUCTION

An always increasing emphasis on the production of components with "zero defects" is nowadays observed in the automotive and aerospace industries; "zero defects", in fact, means no discards and allows a significant cost reduction. These reasons have determined an increasing interest in developping a general approach for the design of processes aimed to prevent the occurrence of defects.

In the industries named above cold forming processes are largely used since they permit to obtain near-net shape or net shape forged parts characterised by shape and dimensions very close to the final desired ones, thus requiring little or no subsequent machining operations.

Published in: E. Kuljanic (Ed.) *Advanced Manufacturing Systems and Technology*, CISM Courses and Lectures No. 372, Springer Verlag, Wien New York, 1996.

During cold forming processes several classes of defects can arise; generally it could be observed that a defect occurs when the forged component do not conform to the design specifications, making it unsuitable for the purpose for which it was designed [1].

In particular the defects which most frequently are encountered in bulk metal forming processes can be classified as follows:

- ductile fractures (cracks), occurring inside or on the external surface of the forged component;
- flow defects, which include both imperfections such as buckling or folding, linked to plastic instability phenomena;
- shape and dimensional inaccuracies, caused by an uncomplete filling of the die cavity or by distorsions induced by residual stresses;
- surface imperfections, linked to an unappropriate quality of the manufactured surface, unsuitable for the purpose for which the component was designed;
- unacceptable modifications of the mechanical properties of the material, due to the microstructural transformations occurring during the forming process.

Among these classes of defects the former one is surely the most important, since its incidence in forged components can result in very serious consequences for the automotive and aerospace industries. Actually the production of a defective forged part could determine the occurrence of costs to the manufacturer at some different leves: first of all, if the defect is easily discernible to the naked eye, the costs are linked only to the discard of the defective component and, in a traditional industrial environment, to the development and experimental testing of a new forming sequence. The costs to the manufacturer increase if the defect is not detectable until all the subsequent processing has been carried out; finally the worst consequence level is reached when the forged component fails in service, causing the loss of the facility and sometimes of human lifes.

The above considerations show the importance of a new and advanced type of metal forming process design, namely a design to prevent process- and material-related defects. Such a design should be based on general methods able to assist the designer, suggesting the role of the process and the material parameters on the process mechanics in order to avoid the occurrence of defects.

Actually the causes of cracks insurgence in bulk forging processes are complicated and often there is not an unique cause for a defect. Despite these difficulties the most inportant causes of defects can be individuated in the metallurgical nature of the material, with particular reference to the distribution, geometry and volume fraction of second phase particles and inclusions, in the tribological variables, including die and workpiece lubrication, and finally in the forming process operating parameters, i.e. the forming sequence, the number of steps in the sequence and the geometry of the tools in each step.

Consequently in the last two decades the attention of most of the researchers has been focused on the understanding on the origin of ductile fractures; the mechanism which limits the ductility of a structural alloy has been explained by Gelin et al.[2] and by Tvergaard [3], taking into account the nucleation and the growth of a large number of small microvoids. The voids mainly nucleate at second phase particles by decohesion of the particle-matrix interface and grow due to the presence of a tensile hydrostatic stress; as a consequence the ligaments between them thin down until fracture occurs by coalescence.

These phenomena, generally called "damage" or "plastic ductile damage" [4] determine a progressive deterioration of the material, both as the elastic and the plastic properties are concerned, decreasing its capability to resist to subsequent loading.

Several criteria and theories have been developed and proposed in the literature which take into account the fracture mechanism described above and are aimed to follow the evolution of damage and consequently to evaluate the level of "soundness" of the forged material.

On the other hand other researchers [5,6] have observed that the limitations of the forming operations are linked mostly to plastic instability phenomena, which yield to the localization of the deformations in shear bands or strictions inside or on the surface of the forged component. Such a strain localization, in fact, favours the growth of damage and leads to ductile fracture. For this reason the attention of these researchers has been focused on the prediction of the plastic instability phenomena: the sufficient criteria for stability have been proposed, essentially based on the evolution of the incremental work in the neighbourhood of an equilibrium configuration.

In this paper the attention of the authors has been focused on the approaches based on the prediction of damage, which have been applied to a wide range of cold forging processes and have shown a well suitable predictive capability. In the next paragraph these approach will be discussed in detail and a proper classification will be carried out. Finally some applications to typical bulk cold metal forming processes will be described.

## 2. MODELLING OF DAMAGE

The aim to evaluate quantitatively the level of damage induced by a plastic forming process has been pursued by several researchers all over the world in the last two decades. Among the approaches published in the literature three main groups can be individuated, namely the one based on ductile fracture criteria, the one founded on the analysis of the damage mechanics, by means of specific yield functions for damaging materials and proper models able to analyse the evolution of damage in the stages of nucleation, growth and coalescence of microvoids, and finally the one based on the use of the porous materials formulation: in the latter case the yield conditions initially proposed for the analysis of forming processes on porous materials are in fact employed.

The three fundamental approaches are described in the next, highlighting the advantages offered by each of them and discussing their differences, which mainly rely on the capability to take into account the influence of damage occurrence on the plastic behaviour of the material and on the possibility to consider the nucleation of new microvoids.

### 2.1 Models based on the use of ductile fracture criteria.

These approaches have been the former to be proposed and still today they represent a powerful industrial tool to appreciate the "state" of the material, in terms of damage, by comparison with the critical value of the criterion at fracture.

The ductile fracture criteria proposed in the literature depend on the stress and strain conditions occurring in the workpiece during the forming process and have been formulated taking into account the role, described above, of the plastic straining and of the hydrostatic stress on the fracture mechanism. Moreover the ductile fracture criteria are generally characterised by an incremental form and require an integration procedure over the deformation path [7].

The first ductile fracture criterion was proposed by Freudenthal [8], who postulated that the plastic energy per unit volume is the critical parameter to be taken into account to predict ductile fracture:

$$\int_0^{\bar{\varepsilon}_f} \bar{\sigma}\, d\bar{\varepsilon} = C \tag{1}$$

Ductile fracture occurs when the plastic deformation energy reaches the critical value C. The validity of this criterion has been recently confirmed by Clift et al. [9], who performed several experiments to compare the results of several ductile fracture criteria and claimed that Freudenthal criterion predicted locations of cracks better than the others. However, Freudenthal criterion does not take into account the effect of hydrostatic stress, which is known to affect the ductile fracture significantly.

The mean stress is in fact detected as the main responsible of the fracture occurrence in the criterion proposed by Oyane [10], which is written in the following form:

$$\int_0^{\bar{\varepsilon}_f} \left(1 + \frac{\sigma_m}{A\bar{\sigma}}\right) d\bar{\varepsilon} = C. \tag{2}$$

The criterion has shown good capabilities to predict central bursting occurrence in drawing processes [11]. Cockroft and Latham [12] proposed that the maximum principal tensile stress, over the plastic strain path, is the main responsible of the fracture initiation. They postulated that fracture occurs when the integral of the largest principal tensile stress over the plastic strain path to fracture equals a critical value specific for the material:

$$\int_0^{\bar{\varepsilon}_f} \sigma_1\, d\bar{\varepsilon} = C. \tag{3}$$

An empirical modification of the Cockroft and Latham model has been proposed by Brozzo et al. [13] in order to explicitly include the dependence of ductile fracture on the hydrostatic stress. Brozzo model is shown in the following equation:

$$\int_0^{\bar{\varepsilon}_f} \frac{2\,\sigma_1}{3(\sigma_1 - \sigma_m)}\, d\bar{\varepsilon} = C. \tag{4}$$

In all the above equations:

$\bar{\varepsilon}$ and $\bar{\varepsilon}_f$  are the effective strain and the effective strain at fracture respectively;

$\bar{\sigma}$ and $\sigma_m$  are the effective stress and the mean stress respectively;

$\sigma_1, \sigma_2, \sigma_3$  are the principal stress components;

A and C    are material constants.

Several other ductile fracture criteria can be found in the literature: among them, the Ayada [14], McClintock [15], Rice & Tracey [16] and Osakada [17] ones should be noted.

The use of the ductile fracture criteria is very simple: the forming process is numerically simulated using a formulation based on the yield condition and the associated flow rule of the sound material, as it were undamaged; at each step of the deformation path the calculated stress and strain fields are introduced in the ductile fracture criterion, which allows to appreciate the level of "damage" of the material by means of the comparison with the critical value of the criterion at rupture. Consequently by integrating the ductile fracture criterion along the whole deformation path, it is possible to detect if and where fracture occurs. As regards the determination of the critical value at rupture, it is generally obtained by applying the criterion to the upsetting process of cylindrical specimens with sticking conditions at the punch-workpiece interface: the value reached by the criterion for a punch stroke corresponding to the comparison of the first naked eye discernible crack is assumed as the critical value.

The above considerations highlight that these criteria do not take into account the progressive degradating state of the material: the numerical simulation is carried out without that the plastic properties of the material are updated taking into account the influence of the accumulated damage. Consequently these criteria permit only a rough prediction of ductile fracture, very appreciated in the practical applications. For this reason other researchers have proposed to take into account damage occurrence and its evolution in the constitutive equations used for the analysis of metal forming processes.

### 2.2 Models based on the analysis of the damage mechanics.

The insurgence of damage determines a progressive deterioration both of the elastic and of the plastic properties of the material: in particular as regards the plastic properties the main effects of the nucleation and growth of microvoids are the insurgence of plastic dilatancy (irreversible volumic change), the consequent dependence of plastic yielding on the hydrostatic stress, and the occurrence of strain softening, i.e. a large reduction of the stress-carrying capability of the material, associated to an advanced evolution of damage. Finally the coalescence of micro voids yields to ductile fractures, either internal or external.

Consequently a suitable damage mechanics model must be able to take into account the insurgence of damage, its evolution, and the influence of the damage level on the evolutive degradation of the material. In order to pursue this aim,

- a new field variable able to describe the damage state must be selected;
- a new yield function for the damaging material must be defined;
- a proper model to analyse the evolution of damage in its various stages (nucleation, growth and coalescence) must be developed.

As regards the former the scalar variable $f$, void volume fraction, defined as the ratio between the volume of the cavities in an aggregate and the volume of the aggregate, is generally employed.

On the other hand, as far as the yield condition is concerned, the first set of constitutive equations including ductile damage was suggested by Gurson [18], who proposed a yield criterion depending on the first invariant of the stress tensor, on the second invariant of the deviatoric stress tensor and on the void volume fraction $f$. The Gurson yield criterion belongs to the so called "microscale-macroscale" approach [2], i.e. the macroscopic yield function has been derived starting from microscale considerations; Gurson, in fact, started from a rigid-perfectly plastic upper bound solution for spherically symmetric deformations around a single spherical void, and proposed his yield function for a solid with a randomly distributed volume fraction of voids $f$ in the form:

$$\Phi = \frac{\overline{\sigma}^2}{\sigma_0^2} + 2f \cosh\left(\frac{\sigma_{kk}}{2\sigma_0}\right) - \left(1 + f^2\right) = 0 \tag{5}$$

where

$$\overline{\sigma} = \left(\frac{3}{2}\sigma_{ij}'\sigma_{ij}'\right)^{\frac{1}{2}} \tag{6}$$

is the macroscopic effective stress, with $\sigma_{ij}$ macroscopic stress deviator, and $\sigma_0$ is the yield stress of the matrix (void-free) material. In eq.5 the effect of the mean stress on the plastic flow when the void volume fraction is non-zero can be easily distinguished, while for $f=0$ the Gurson criterion reduces to the von Mises one.

The components of the macroscopic plastic strain rate vector can be calculated applying the normality rule to the yield criterion above written.

Subsequently the Gurson yield criterion has been modified by Tvergaard [19] and by Tvergaard and Needleman [20], which introduced other parameters obtaining the following expression:

$$\Phi = \frac{\overline{\sigma}^2}{\sigma_0^2} + 2f^* q_1 \cosh\left(\frac{\sigma_{kk}}{2\sigma_0}\right) - \left[1 + \left(q_1 f^*\right)^2\right] = 0 \tag{7}$$

In the above expression the parameter $q_1$ allows to consider the interactions between neighbouring voids: Tvergaard in fact obtained the above criterion analysing the macroscopic behaviour of a doubly periodic array of voids using a model which takes into account the nonuniform stress field around each void. The value of $q_1$ was assumed equal to 1.5 by the same Tvergaard. On the other hand the parameter $f^*(f)$ was introduced in substitution of $f$ in order to describe in a more accurate way the rapid decrease of stress-carrying capability of the material associated to the coalescence of voids: in fact $f^*(f)$ is defined as:

$$f^*(f) = \begin{cases} f & \text{for } f \le f_c \\ f_c + \dfrac{f_u^* - f_c}{f_F - f_c} \cdot (f - f_c) & \text{for } f > f_c \end{cases} \tag{8}$$

where $f_u^* = 1/q_1$, $f_c$ is a critical value of the void volume fraction at which the material stress-carrying capability starts to decay very quickly (easily discernible in a tensile test curve) and finally $f_F$ is the void volume fraction value corresponding to the complete loss of stress-carrying capability. Again the constitutive equations associated to the Tvergaard and Needleman yield criterion can be determined by means of the normality rule.

Finally a model based on the analysis of the damage mechanics must be able to take into account the evolution of damage in its various stages (nucleation, growth, coalescence). At the end of each step of the deformation process the values of the void volume fraction calculated inside the workpiece under deformation should be updated: since the variation of the void volume fraction depends both on the growth or the closure of the existing voids and on the possible nucleation of new voids, the void volume fraction rate has to be written in the form:

$$\dot{f} = \dot{f}_{growth} + \dot{f}_{nucleation} \tag{9}$$

It is very simple to calculate the first term of eq.(9): the principle of the conservation of mass applied to the matrix material, in fact, permits to relate the rate of change of the void volume fraction to the volumetric strain rate $\dot{\varepsilon}_V$, i.e.:

$$\dot{f}_{growth} = (1 - f)\dot{\varepsilon}_V \tag{10}$$

On the other hand, several researchers have focused their attention on the nucleation of new voids. As regards this topic two main formulations have been proposed: the former is based on the assumption that nucleation is mainly controlled by plastic strain rate:

$$\dot{f}_{nucleation} = A\dot{\bar{\varepsilon}} \tag{11}$$

on the other hand in the latter formulation it is hypothised that the nucleation of new voids is governed by the maximum normal stress transmitted across the particle-matrix interface, i.e.:

$$\dot{f}_{nucleation} = B\left(\dot{\bar{\sigma}} + \dot{\sigma}_{kk}\right) \tag{12}$$

since, as suggested by Needleman and Rice [21] the sum of the effective and of the hydrostatic stress can be assumed as a good approximation of the maximum normal stress transmitted across the particle-matrix interface.

As concerns the parameter A and B, Chu and Needleman [22] have proposed that void nucleation follows a normal distribution about a mean equivalent plastic strain or a mean maximum normal stress: as an example, assuming the strain controlled nucleation model, eq.(11) can be rewritten in the form:

$$\dot{f}_{nucleation} = \frac{f_n}{s\sqrt{2\pi}} \exp\left[-\frac{1}{2}\left(\frac{\varepsilon_{eq} - \varepsilon_{mean}}{s}\right)^2\right]\dot{\varepsilon}_{eq} \tag{13}$$

where $s$ is the standard deviation which characterises the normal distribution, $\varepsilon_{mean}$ the mean strain for nucleation and finally $f_n$ is the volume fraction of voids which could nucleate if sufficiently high strains are reached. Voids are nucleated in the zones where tensile hydrostatic stresses occur.

The introduction of the yield condition for the damaging material and of the damage evolutive model in a finite element formulation permits to follow the evolution of the void volume fraction variable and consequently to evaluate the level of "soundness" of the forged components, by comparing the current value of the void volume fraction which the critical value $f_c$, which corresponds to the coalescence of the microvoids and to the steep decay of the stress-carrying capability of the material.

However the above considerations highlight that the use of a model based on the analysis of the damage mechanics requires a complete rewriting of the numerical code to be used to simulate the forming process. Moreover the formulation depends on several parameters, both as regards the yield condition and the damage evolution model, which must be properly selected in order to obtain a suitable prediction of fracture occurrence [23,24].

Actually this fact could represent a very important advantage offered by this type of approach, since it is possible to "calibrate" the model with respect to the actual state and properties of the material taken into account, but, on the other hand, it makes necessary the use of complex analytical tools in the stage of the parameters selection. This aim has been pursued by Fratini et al. [25], applying an inverse identification algorithm, based on an optimization technique which allows to determine the material parameters by comparing some numerical and experimental results and searching for the best matching between them. In particular the load vs. displacement curve during a tensile test on a sheet specimen has been employed to optimize the comparison between the numerical results and the experimental ones, and consequently to achieve the desired material characterisation.

The above considerations justify the observation that, even if the models based on the analysis of the damage mechanics are certainly the most correct from the theoretical point of view, their practical use in an industrial environment is limited to very few applications, while they encounter a larger interest in the academic and research fields.

### 2.3 Models based on the porous material formulation.

In some recent papers a further approach for the prediction of ductile fractures has been proposed, based on the formulation generally employed to simulate the forming processes on powder materials. Actually in these processes the powder material is compacted and consequently its relative density increases: on the contrary the application of the porous materials formulation to the analysis of ductile fractures insurgence is aimed to predict the reduction of the relative density (corresponding to the increment of the void volume fraction) associated to the occurrence of the defect.

The porous materials formulation is based on the yield condition initially proposed by Shima and Oyane [26], which depends on the first invariant of the stress tensor, on the second invariant of the deviatoric stress tensor and on a scalar parameter, which is, instead of the void volume fraction $f$ as in the previous models, the relative density R.

The Shima and Oyane yield function can be written in the form:

$$A\left\{\frac{1}{6}\left[\left(\sigma_x-\sigma_y\right)^2+\left(\sigma_y-\sigma_z\right)^2+\left(\sigma_z-\sigma_x\right)^2\right]+\left(\tau_{xy}^2+\tau_{yz}^2+\tau_{zx}^2\right)\right\}+$$

$$+\left(1-\frac{A}{3}\right)\left(\sigma_x+\sigma_y+\sigma_x\right)^2=\delta\sigma_0^2 \tag{14}$$

where $A=2+R^2$ and $\delta=2R^2-1$.

The introduction of this yield condition in a finite element formulation do not present particular problems [27]; moreover the possibility to simulate forming processes on porous material is supplied by most of the more diffused commercial codes.

Despite to these advantages with respect to the models based on the analysis of the damage mechanics, it must be outlined that the formulations used for the analysis of the compaction of porous materials take into account only the possibility of the growth or the closure of the existing porosity, while the nucleation mechanism is not considered.

At the end of each step of the deformation process, in fact, the value of the relative density, calculated at the integration points within each element is only updated taking into account that the relative density variation rate $\dot{R}$ and the volumetric strain rate $\dot{\varepsilon}_V$ are linked by:

$$\dot{\varepsilon}_v=-\frac{\dot{R}}{R} \tag{15}$$

By integrating equation (15) the updated relative density is calculated as:

$$R=R_0\exp\left(-\int\dot{\varepsilon}_v dt\right)\cong R_0\left(1-\Delta\varepsilon_v\right) \tag{16}$$

where $R_0$ is the relative density at the previous step and $\Delta\varepsilon_V$ the volumetric strain increment during the analysed step.

Following of the above considerations it derives that the porous materials formulation can be applied to the prediction of fractures only assuming the existence of an initial relative density lower than 100%. In other words it is necessary to hypothise that all the voids that could nucleate, nucleate once a material element enters the plastic zone: subsequently, depending on the stress conditions, they can grow or close. This assumption can be interpreted as a particular plastic strain controlled nucleation case, in which the plastic strain for nucleation is coincident with the tensile yield strain [28]. Furthermore such an assumption requires an appropriate choice of the initial relative density, depending on the purity of the considered material.

Nevertheless the application of this approach to the prediction of central bursting insurgence in the drawing process of Aluminum Alloy and Copper rods has allowed to obtain a very good overlapping of the numerical and the experimental results, as it will be shown in the next paragraph [23,29,30].

## 3. APPLICATIONS

A very large number of applications of numerical methods belonging to the three fundamental groups described in the previous paragraph can be found in the literature. Here a particular problem will be taken into account, namely the insurgence of the well known central bursting defect during extrusion and drawing.

Central bursts are internal defects which cause very serious problems to the quality control of the products, since it is impossible to detect them by means of a simple surface inspection of the workpiece. In the past some experimental and theoretical studies [31,32,33,34] have been performed in order to understand the origin of the defect and consequently to determine the influence of the main operating parameters (reduction in area, die cone angle, friction at the die-workpiece interface and material properties) on the occurrence of central bursts. The development of powerful numerical codes and suitable damage models has represented a new fundamental tool in order to pursue this attempt.

Some researchers [11,14,35] have used ductile fracture criteria, others have applied models based on the analysis of the damage mechanics [23,28], finally some researchers belonging to the group of the University of Palermo have used the model based on the porous materials formulation and in particular on the Shima and Oyane yield condition [29,30]. Some interesting results obtained by the latter group are presented below.

First of all the research has been focused on the drawing process of aluminum alloy (UNI-3571) specimens. The influence of the operating parameters (reduction in area and semicone die angle) on the insurgence of central bursting has been investigated both using a damage mechanics model founded on the Tvergaard and Needleman yield condition and the porous materials formulation, based on the Shima and Oyane plasticity condition. In order to compare the numerical predictions obtained with the two approaches, no nucleation of new microvoids has been considered in the former model and, taking into account the high percentage of inclusions and porosities of the material, the value of the initial void volume fraction has been fixed equal to 0.04 (i.e. the relative density has been fixed equal to 0.96).

A tension test has been performed, both to obtain the constitutive equation of the material to be used in the numerical simulation, and to determine the elongation ratio and the necking coefficient at fracture; these values in fact, are necessary in order to evaluate, by the comparison with the numerical simulation of the tensile test, the critical value of the void volume fraction (or of the relative density). In particular the numerical analysis has been stopped after a total displacement of the nodes of the specimen assumed clamped to the testing machine equal to the elongation ratio, and the maximum void volume fraction reached at this stage (or the minimum relative density) has been assumed as the critical value $f_C$, corresponding to the rapid decay of the stress-carrying capability of the material.

Subsequently the numerical analysis has been applied to several drawing processes, characterised by different reductions in area and die cone angles: for each of them the maximum achieved value of the void volume fraction has been calculated, and, by the comparison with the critical value, it has been evaluated if the coalescence of voids, i.e. the insurgence of the ductile fracture, should occur or not. The numerical results have shown a good agreement with the experimental ones.

Subsequently, in order to test the suitability of the employed models in an actual industrial problem, the research has been focused on the prediction of bursting occurrence in the drawing process of commercially pure copper specimens. This material, in fact, is typically used in industrial drawing operations, since it is characterised by a very low value of initial porosity and presents a small amount of inclusions; consequently, depending on the operating parameters (i.e. the reduction in area and the semicone die angle), the insurgence of defects could occur only after several drawing steps.

The prediction of defect insurgence has been carried out employing the model based on the porous materials formulation and assuming an initial value of the relative density equal to R=0.9998. Again this parameter has been selected by simulating the tensile test and choosing the value of R which allows a good overlapping between the numerical and the experimental results in terms of elongation ratio and of necking coefficient.

Several drawing sequences characterised by different values of operating parameters have been taken into account, according to the next Table 1:

|  | α=8° | α=12° | α=15° | α=20° | α=22° |
|---|---|---|---|---|---|
| RA=10% | X |  | X | X |  |
| RA=20% | X | X | X | X | X |

Table 1

For each sequence six subsequent reductions are carried out, maintaining constant the same reduction in area and semicone die angle.

The numerical simulations have supplied the relative density distributions inside the specimen during the deformation path. In particular, depending on the operating parameters, tensile mean stresses occur in the zone of the specimen close to the symmetry axis determining a reduction of the relative density.

The relative density trends for several different reduction in area - semicone die angle couples are reported in fig.1: for some operating conditions a slight variation of the relative density with respect to the initial value occurs, while for other conditions after few reductions a steep decay of the relative density arises. Moreover in the latter cases, in correspondence to the large reduction of the relative density, the numerical simulation stops due to the presence of a non positive definite stiffness matrix. This condition is the indication of the plastic collapse of the material i.e. of the insurgence of the defect.

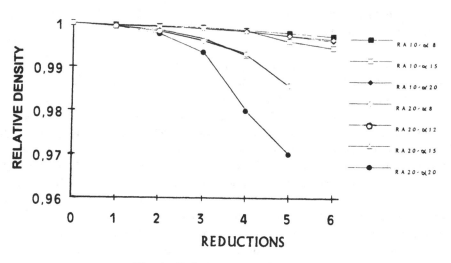

Fig. 1 - Relative density trends

The same combinations of the operating parameters, as reported in the previous Table 1, have been experimentally tested, repeating each sequence 10 times.

Table 2 reports the results of the experimental tests; in the table for each combination of the operating parameters, the number of defective specimens is reported.

| Drawing steps | 1st | 2nd | 3rd | 4th | 5th | 6th |
|---|---|---|---|---|---|---|
| RA=10% α = 8° | - | - | - | - | - | 0 |
| RA=10% α= 15° | - | - | - | - | - | 0 |
| RA=10% α= 20° | - | - | - | - | - | 0 |
| RA=20% α = 8° | - | - | - | - | - | 0 |
| RA=20% α= 12° | - | - | - | 0 | 6 | |
| RA=20% α= 15° | - | - | 0 | 4 | | |
| RA=20% α= 20° | - | - | - | 0 | 5 | |

Table 2. Number of defective specimens.

The results of Table 2 are in good agreement with the numerical predictions. After the drawing step for which a steep decay of the relative density had been predicted, an high number of defective specimens is always detected, while no defects are individuated for the "safe" sequences that were characterised by a smooth variation of the calculated relative density.

Very recently an approach based on the analysis of the damage mechanics has been applied by some researchers belonging to the group of the University of Palermo to the prediction of tearing in a typical sheet metal forming process, namely the deep drawing of square boxes [36]. The yield condition for damaging materials proposed by Tvergaard and Needleman and a strain controlled nucleation model have been introduced into a finite element explicit code in order to follow the evolution of the void volume fraction variable during the deep drawing operation.

First of all the model has been characterised with reference to the steel used in the experimental tests by means of the inverse identification approach described in reference [25]. Subsequently it hase been used to predict the insurgence of tearing in the deep drawing process, at varying the blank diameter and the lubricating conditions. In particular the void volume fraction trend has been analysed and it has been assumed that tearing occurs when the calculated void volume fraction reaches the critical value $f_c$ associated to the voids coalescence.

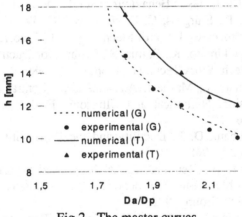

Fig.2 - The master curves

Fig. 2 shows the obtained results by means of master curves; in this diagram the drawing ratio, defined as the blank diameter $D_a$ vs. the punch side $D_p$ ratio is reported in the x-axis, while the box height h at which tearing is predicted, is reported in the y-axis. The master curves are referred to the two lubricating conditions investigated, namely using grease (G) or a thin film of teflon (T) at the die-sheet and blankholder-sheet interfaces.

In the same figure the experimental results are reported too: for a $D_a/D_p$ ratio lower than 1.7 no tearing has been observed, in agreement with the numerical prediction, neither using grease nor, obviously, teflon. For higher values of the blank diameter the box heights for which tearing has been experimentally observed are in good agreement with the predicted values again thus confirming the validity of the used formulation.

## 4. REFERENCES

[1]   Dodd, B., Defect in Cold Forging. Final Report the Materials and Defects Sub-group of the International Cold Forging Group, Osaka, 1993.

[2]   Gelin, J. C., Predelanu, M., Recent Advances in Damage Mechanics: Modelling and Computational Aspects, Proc. of Numiform '92, pp.89-98, 1992.

[3]   Tvergaard, V., Influence of Void Nucleation on Ductile Shear Fracture at a Free Surface, J. Mech. Phys. Solids, vol. 30, no. 6, pp.339-425, 1982.

[4]   Predeleanu, M., Finite Strain Plasticity Analysis of Damage Effects in Metal Forming Processes, Computational Methods for Predicting Material Processing Defects, pp.295-307, 1987.

[5]   Hill, R., A General Theory of Uniqueness and Stability for Inelastic Plastic Solids, J. Mech. Phys. Solids, vol.6, pp.236-249, 1958.

[6]   Storen, S., Rice, J.R., Localized Necking in Thin Sheets, J. Mech. Phys. Solids, vol.23, pp.421-441, 1975.

[7]   Barcellona, A., Prediction of Ductile Fracture in Cold Forging Processes by FE Simulations, Proc. of X CAPE, pp., 1994.

[8]   Freudenthal, A. M., The Inelastic Behaviour of solids, Wiley, NY, 1950.

[9].  Clift, S. E., Hartley, P., Sturgess, C. E. N., Rowe, G. W., Fracture Prediction in Plastic Deformation Processes, Int. J. of Mech. Sci., vol.32, no.1, pp.1-17, 1990.

[10]  Oyane, M., Sato, T., Okimoto, K., Shima, S., Criteria for Ductile Fracture and their Applications, J. of Mech. Work Tech., vol.4, pp.65-81, 1980.

[11]  Alberti, N., Barcellona, A., Masnata, A., Micari, F., Central Bursting Defects in Drawing and Extrusion: Numerical and Ultrasonic Evaluation, Annals of CIRP, Vol.42/1/1993, pp. 269-272.

[12]  Cockroft, M. G., Latham, D. J., Ductility and the Workability of Metals, J. Inst. Metals, vol.96, pp.33-39, 1968.

[13]  Brozzo, P., De Luca, B., Rendina, R., A New Method for the Prediction of Formability Limits in Metal Sheets, Proc. of the 7th Conference of the International Deep Drawing Research Group, 1972.

[14]  Ayada, M., Higashimo, T. Mori, K., Central Bursting in Extrusion of Inhomogeneous Materials, Advanced Technology of Plasticity, vol.1, pp.553-558, 1987.

[15]  McClintock, F., Kaplan, S. M., Berg, C. A., Ductile Fracture by the Hole Growth in Shear Bands, Int. J. of Mech. Sci., vol.2, p.614, 1966.

[16]  Rice, J., Tracey, D., On Ductile Enlargement of Voids in Triaxial Stress Fields, J. of Mech. Phys. Solids, vol.17, 1969.

[17]  Osakada, K., Mori, K., Kudo, H., Prediction of Ductile Fracture in Cold Forming, Annals of the CIRP, Vol. 27/1, pp.135-139, 1978.

[18]  Gurson, A. L., Continuum Theory of Ductile Rupture by Void-Nucleation and Growth: Yield Criteria and Flow Rules for Porous Ductile Media, J. of Eng. Mat. Tech., vol.99, pp.2-15, 1977.

[19]  Tvergaard, V., Ductile Fracture by Cavity Nucleation between larger voids, J. Mech. Phys. Solids, vol. 30, no. 4, pp.265-286, 1982.

[20] Needleman, A., Tvegaard, V., An Analysis of Ductile Rupture in Notched Bars, J. of Mech. Phys. Solids, vol. 32, No. 6: 461-490, 1984.

[21] Needleman, A., Rice, J.R., in Mechanics of Sheet Metal Forming, edited by D.P. Koistinen et al., Plenum Press, New York, p.237, 1978.

[22] Chu, C.C., Needleman, A., Jnl. Eng. Mat. Tech., vol.102, pp.249-262, 1980.

[23] Alberti, N., Barcellona, A., Cannizzaro, L., Micari, F., Predictions of Ductile Fractures in Metal Forming Processes: an Approach Based on the Damage Mechanics, Annals of CIRP, vol.43/1, pp.207-210, 1994.

[24] Alberti, N., Cannizzaro, L., Micari, F., Prediction of Ductile Fractures Occurrence in Metal Forming Processes, Proc. of the II AITEM Conference, pp.157-165, 1995.

[25] Fratini, L., Micari, F., Lombardo, A., Material characterization for the prediction of ductile fractures occurrence: an inverse approach, accepted for the publication on the Proceedings of the Metal Forming '96 Conference.

[26] Shima, S., Oyane, M., Plasticity Theory for Porous Metals, Int. J. of Mech. Sci., vol. 18, pp. 285, 1986.

[27] Kobayashi, S., Oh, S. I., Altan, T., Metal Forming and the Finite Element Method, edited by the Oxford University Press, 1989.

[28] Aravas, N., The Analysis of Void Growth that Leads to Central Bursts during Extrusion, J. of Mech. Phys. Solids, vol. 34, no.1, pp.55-79, 1986.

[29] Alberti, N., Borsellino, C., Micari, F., Ruisi, V.F., Central Bursting Defects in the Drawing of Copper Rods: Numerical Predictions and Experimental Tests, Transactions of NAMRI/SME, vol.23, pp.85-90, 1995.

[30] Borsellino, C., Micari, F., Ruisi, V.F., The Influence of Friction on Central Bursting in the Drawing Process of Copper Specimens, Proceedings of the International Conference on Advances in Materials and Processing Technologies (AMPT'95), pp.1230-1239, 1995.

[31] Avitzur, B., Analysis of Central Bursting Defects in Extrusion and Wire Drawing, Trans. ASME, ser. B, vol.90, pp.79-91, 1968.

[32] Orbegozo, J. I., Fracture in wire drawing, Annals of CIRP, vol.16/1, pp.319-322, 1968.

[33] Avitzur, B., Choi, C. C., Analysis of Central Bursting Defects in Plane Strain Drawing and Extrusion, Trans. ASME, ser. B, vol.108, pp.317-321, 1986.

[34] Moritoki, H., Central Bursting in Drawing and Extrusion under Plane Strain, Advanced Technology of Plasticity, vol.1, pp.441-446, 1990.

[35] Hingwe, A. K., Greczanik, R. C., Knoerr, M., Prediction of Internal Defects by Finite Elements Analysis, Proc. of the 9th International Cold Forging Congress, pp.209-216, 1995.

[36] Micari, F., Fratini, L., Lo Casto, S., Alberti, N., Prediction of Ductile Fracture Occurrence in Deep Drawing of Square Boxes, accepted for the publication on the Annals of CIRP, vol.45/1, 1996.

# WIRE DRAWING PROCESS MODELLIZATION:
## MAIN RESULTS AND IMPLICATIONS

**D. Antonelli, A. Bray, F. Franceschini, D. Romano, A. Zompì and R. Levi**

**Polytechnic of Turin, Turin, Italy**

**KEYNOTE PAPER**

KEY WORDS: Wire drawing, FEM, damage mechanism.

ABSTRACT: Numerical methods for description of processes involving large plastic deformation were exploited to model wire drawing, covering both single pass and sequences of consecutive passes. Results obtained from exploitation of models in terms of single and combined effects of process parameters on main responses, and of superimposed stress and strain patterns on material properties inclusive of defect evolution, are discussed in the light of theoretical prediction and experimental evidence.

## 1. INTRODUCTION

Progress in metal forming operations depends among other factors upon improved evaluation of single and combined effects of process parameters on product, entailing reliable estimation of material properties, both initial and as modified during production. The interplay between theory and experiment leads to enhanced modeling capability, and in turn to identification of areas with sizable potential for improvement in terms of process and product quality. Steel wire drawing is no exception, particularly in view of the ever increasing demands on production rate, product integrity, and process reliability. Some results obtained over a decade of applied research work performed on this subject at Politecnico di Torino are presented in this paper.

As a first step in process modeling, theoretical description of the basic wire drawing

Published in: E. Kuljanic (Ed.) *Advanced Manufacturing Systems and Technology*, CISM Courses and Lectures No. 372, Springer Verlag, Wien New York, 1996.

operation was relied upon to provide an assembly of single die functional models for simulation of an extended range of multipass production machines. Reduced drawing stress was evaluated (in terms of yield stress) according to either the classic slab method, or the upper-bound method, leading to fairly close results. Strain hardening, and related issues concerning central bursting, were also modeled according to established theory [Avitzur 1963, Chen et al. 1979, Godfrey 1942, MacLellan 1948, Majors 1955, Yang 1961].

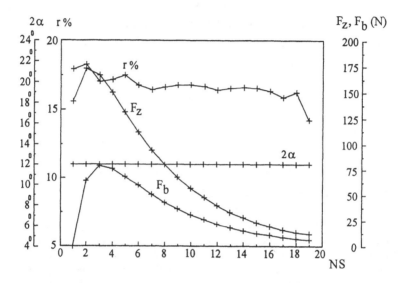

Fig. 1. Computed percentage reduction  r %, drawing force $F_z$ and back tension $F_b$ plotted (for a constant die angle $2\alpha$) versus step no. NS for a 19 pass drawing machine.

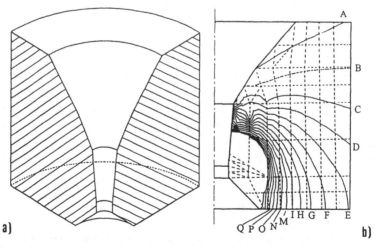

Fig. 2. Typical die configuration, a), and relevant hoop strain pattern, b), as evaluated with FEM. Strain levels shown range between -6 $\mu\epsilon$ (A) and 24 $\mu\epsilon$ (Q).

Software developed accordingly, catering for easy, user friendly comparative evaluation of drawing sequences on a what-if basis, see e.g. Fig. 1 [Zompì et al. 1990], proved practical enough to warrant regular exploitation in industrial environment for process planning and troubleshooting work.

Elastic stress distribution in the die was also modeled, typical results being shown in Fig. 2, and checked at selected locations against strain gage results, the latter involving some tricky undertaking as mechanical and thermal strains have by and large the same order of magnitude, not to mention low strain level and boundary condition problems. Quantitative information concerning scatter in drawn wire size, and drawing force, were also obtained, showing that while average drawing force increases slightly with drawing speed, scatter was found to be by and large constant over a 10:1 speed range [Zompì et al. 1991].

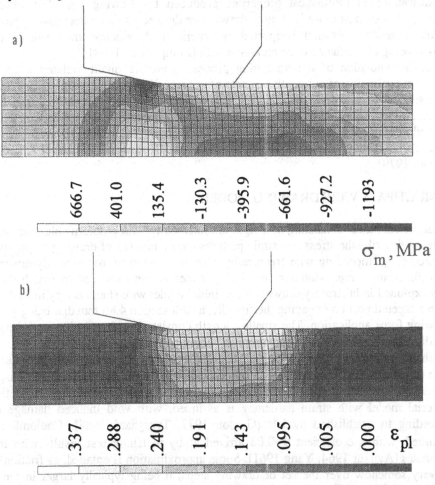

Fig. 3. Computed mean stress, a), and equivalent plastic strain pattern, b), corresponding to die angle $\alpha=12°$, reduction $r=16\%$, friction coefficient $\mu=0.05$ and strain hardening coefficient $\beta=0.32$.

As detailed analysis of effects of the drawing process on wire was sought, single pass operation was further modeled with FEM and run over a comprehensive array of combinations of operating parameters and material properties, namely die cone angle, percentage reduction, friction and strain hardening coefficients. Stress and strain patterns covering large plastic deformation were obtained, see Fig. 3, with features which enabled validation against established theoretical models; response variables included drawing force, peak normal stress and normal stress gradient.

Concise polynomial models were developed explaining with a handful of terms a range from over 95 to 99% of variation according to the response variable considered, a common feature being the predominant influence of quadratic and interaction terms, whose effects were frequently found to exceed those pertaining to first order ones. Detailed description of modifications in mechanical properties produced by drawing was thus obtained, and peculiarities such as those leading to drawn wire diameter smaller than that pertaining to die throat were demonstrated, supported by experimental evidence, the influence of contact length being also a factor to be reckoned with [Zompì et al. 1994].

Realistic simulation of manufacturing process, covering quality related aspects, entails however evaluation of cumulative effects of consecutive drawing passes on wire properties, inclusive of such defect evolution as may trigger either in process breakage or delayed failure. Since minor flaws may eventually cause sizable effects, refined material testing procedures are needed, particularly because some crucial information is available only in the neighborhood of plastic instability in tensile tests [Brown and Embury 1973, Goods and Brown 1979].

## 2. MULTIPASS WIRE DRAWING MODEL

A numerical model consisting of a rigid die interacting with a deformable wire was built in order to describe the stress and strain patterns over a number of drawing steps, also in view of predicting impending wire fracture by monitoring void propagation. Symmetry permits two dimensional representation; nominal parameters were selected to match those of die sets exploited in laboratory drawing tests. Initial model wire shape is a cylinder 5.5 mm dia. with a tapered portion engaging the first die, a stub section 4.86 mm dia. being provided for drawing force application. The smallest length consistent with reaching stationary drawing conditions was selected to keep computing time under control.

Mesh is made up by some 400 quadrangular axisymmetric linear isoparametric elements with full integration; boundary conditions include axial symmetry and constant displacement increments over the leading (right hand) edge of the model. An elasto-plastic flow curve material model with strain hardening is assumed, with void induced damage described according to established models [Gurson 1977, Tvergaard 1984]. Coulomb friction is assumed, with a coefficient $\mu=0.08$ arrived at by matching test results with theoretical estimates [Avitzur 1964, Yang 1961]. Some approximation is entailed, as friction is known to vary somehow over the set of drawing steps, it being typically larger in the first one owing mainly to coil surface conditions as affected by decarburization due to previous manufacturing operations.

The combined effects of reduced model length, rigid end displacement and finite number of elements induce some fluctuation in computed drawing force. Model size and discretization are a compromise between accuracy and computational time, the latter being fairly sizable (about 2 hours per pass on a IBM RISC/6000 platform) as the implicit Euler scheme resorted to for damage model integration requires rather small integration steps, say of the order of a few hundredths of the smallest element, if discretization errors are to be kept within reasonable bounds.

Model is remeshed whenever element distortion exceeds a given limit, and a tapered section matched to next die angle is introduced after every pass. Length increases due to drawing must be offset in order to avoid unnecessarily large run times after the first drawing steps. Ad hoc routines were developed to automate remeshing, a prerequisite for industrial application.

## 3. DRAWING AND TENSILE TESTS

Laboratory tests were performed on C70 steel wire specimens ranging from 5.5 to 2.75 mm diameter, the latter being obtained from the former in six steps of cold drawing. Main data concerning the dies used (throat dia., angle, reduction ratio and average true strain) are given in Table 1.

Table 1. Main die parameters and related values

| Die No. | $\varnothing_{out}$ [mm] | $\alpha$ [°] | r % | $\varepsilon = \ln(A_0/A)$ |
|---|---|---|---|---|
| 1 | 4.90 | 10.2 | 20.6 | 0.23 |
| 2 | 4.30 | 10.7 | 23.0 | 0.49 |
| 3 | 3.90 | 10.7 | 17.7 | 0.69 |
| 4 | 3.35 | 10.8 | 26.2 | 0.99 |
| 5 | 3.05 | 13.6 | 17.1 | 1.18 |
| 6 | 2.75 | 17.6 | 18.7 | 1.39 |

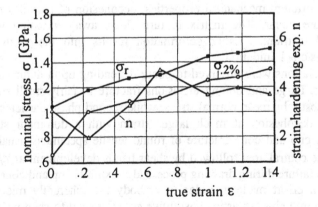

Fig. 4. Strain hardening exponent, yield and ultimate stress obtained from tensile tests at each drawing step.

Material properties were evaluated from tensile tests (after straightening specimens if required), at least three tests being performed at every drawing step. Tensile and drawing tests were carried out on a hydraulic material testing machine, with continuous monitoring of force, strain and crosshead displacement. Flow, yield (at 0.2% strain) and ultimate stresses were measured, and strain hardening exponent $n$ was evaluated, according to ASTM E 111 & 646-91, see Fig. 4.

The flow curve of the material was obtained by averaging several experimental tensile test curves; sizable prestraining of wire occurred before testing. As a consequence some scatter in Young's modulus was found, as underlined elsewhere [Hajare 1995]. An inherent limitation of the tensile test lies in the small value of uniform plastic strain obtained before the start of necking and then failure. After the onset of necking precious little data may be obtained with conventional techniques but for the local necking strain, leading to rather poor accuracy in the evaluation of flow curve. As simulation shows that the equivalent strain reached in proximity of die-wire interface exceeds by far what can be reached in tensile tests, extrapolation becomes necessary, entailing sizable uncertainties.

## 4. DUCTILE FAILURE MECHANISM

Since in metal forming processes such as extrusion and drawing ductility is progressively reduced by strain hardening, knowledge of material susceptibility to ductile failure was recognized of paramount importance for rupture prevention and accordingly investigated [Alberti et al. 1994]. Defect insurgence in highly deformed ductile materials occurs as a consequence of an evolutive process involving bulk material, although crack initiation is usually observed at selected locations only; as a matter of fact the amount of inclusion may be kept under control, as opposite to second phase particles [Benedens et al. 1994]. The underlying mechanisms were established in the course of a number of investigations, developed since the fifties [Puttick 1959, Rogers 1960].

Polycrystalline metals undergoing plastic deformation develop microvoids on sites of such potential nucleation elements as inclusions, second phase particles, precipitates, or grain boundaries. As these elements do not readily follow the plastic deformation of matrix material owing to different mechanical properties, decohesion of particles from the matrix and/or their fracture occur. The matrix in turn flows away initiating voids owing to reduction of local restraint. Further deformation results into the voids growing until coalescence triggers crack initiation.

Different void populations were observed to occur depending upon nature and size of void nucleating particles. Large and oblong inclusions fracture at fairly small strain, nucleation being controlled mostly by mean normal stress, while second phase particles, being smaller, produce voids by decohesion at much larger strain with a dominant strain controlled nucleation. Typical cup and cone fracture of round tensile specimens usually results from ductile failure in the central area followed by shear bands development at $\pi/4$ inclination in the outer rim. Wire failure during drawing process also exhibits cup and cone features.

Substantial research effort made in order to embody the inherently microscopic ductile damage mechanism into elasto-plastic constitutive equations led to such well known results as Gurson's model (1977), and Tvergaard's modification (1982). As far as macroscopic behavior is concerned, an equivalent "damaged" material can be defined to which the following yield criterion applies:

$$\sigma_{eq}^2 = \sigma_{eq}^2 \left( \overline{\sigma}^2, \sigma_m, f \right)$$

where $\sigma_{eq}$, $\overline{\sigma}$ and $\sigma_m$ are respectively von Mises equivalent stress for damaged material, yield stress of sound material, and mean normal stress, $f$ being a scalar state function representing the relative void volume fraction at any point of the specimen, thus indicating in a simple way the path towards ductile failure. Yield condition now depends on hydrostatic pressure and void volume; an increase of either mean normal stress and/or void volume fraction reduces equivalent stress and consequently stress-carrying capability. The state equation defining porosity rate has the form:

$$df = df_{nucleation} + df_{growth}, \quad \text{where}$$

$$df_{nucleation} = df\,(d\varepsilon_{eq}, d\sigma_m), \quad df_{growth} = df\,(f, d\varepsilon_v)$$

where $\varepsilon_{eq}$ is the equivalent plastic strain and $\varepsilon_v$ the volumetric strain. Change in porosity prior to failure is due to the combined effect of void nucleation and growth. Nucleation rate can be strain and/or stress controlled depending on the distribution of void nucleating

Table 2. Damage model parameters (Gurson-Tvergaard) used in multipass simulation.

| Initial void volume fraction | 0% |
|---|---|
| Mean equivalent strain for nucleation | 30% |
| St. dev. of equivalent strain for nucl. | 5% |
| Void volume fraction for coalescence | 15% |
| Void volume fraction for fracture | 20% |

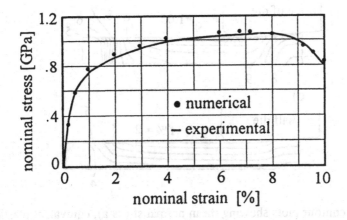

Fig. 5. Experimental and computed stress-strain data pertaining to tensile tests.

dilatancy while a negative one tends to close cavities. Eventually, coalescence occurs when porosity exceeds an established threshold, while upon reaching a higher ultimate level catastrophic failure is initiated.

Damage model was identified by fitting FEM simulation of wire tensile test results to experimental data with an iterative process; initial values were taken from literature [Tvergaard and Needleman 1984]. Model parameters were selected as corresponding to best fit to experimental stress-strain plots in the plastic range (see Fig. 5) and more closely matching reduction of area at fracture due to necking. MARC and ABAQUS codes were used, both catering for the damage model considered. Realistic values were obtained (see Table 2), well within the range of values reported in literature [Chu and Needleman 1980, Tvergaard 1982]. Strain controlled nucleation is adopted since the major ductile mechanism in a axisymmetric geometry is assumed to rely on voids formation by small particles-matrix decohesion and failure by coalescence [Brown and Embury, 1973].

## 5. SOME RESULTS AND IMPLICATIONS

Some results concerning simulation of six consecutive drawing steps are examined, covering material modeled both with and without damage. Fig. 6 shows for the fourth drawing step contour plots of mean normal stress, equivalent plastic strain and void volume density over the wire section.

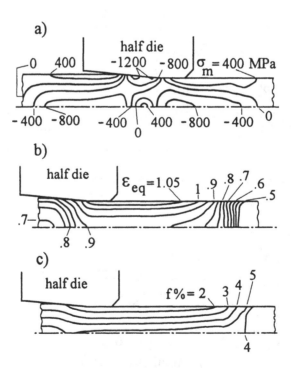

Fig. 6. Computed contour plots showing mean normal stress a), equivalent plastic strain b), and void volume density c), at the fourth drawing step.

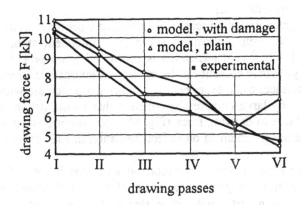

Fig. 7. Plot showing drawing forces computed (with and without damage) and measured over six consecutive passes.

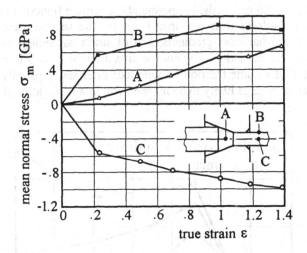

Fig. 8. Mean normal stress in wire at three characteristic locations versus true strain.

End effects due to finite length are apparent; steady state is however approached in the central part of the model towards the end of the drawing step, see plots b) and c). A plot showing over six steps computed and measured drawing forces (Fig. 7) indicates a substantial agreement of the latter with damage model results (average deviation of the order of 7%, with a maximum of 15%).

Scatter in test results and numerical fluctuation due to discretization would not justify seeking for a substantially better agreement. On the other hand, plain model exhibits not only larger differences (20% on the average) but also a marked discrepancy in trend, see step no. 6. Alternances between tension and compression along wire axis due to die's complex action are apparent in mean normal stress, accounting for incremental damage and its propagation over consecutive steps.

Process induced damage evolution presents some peculiar aspects susceptible of explanation in terms of interplay between hydrostatic stress and equivalent strain time and space histories. In a nutshell, while void nucleation is controlled (on a probabilistic basis) by a strain threshold, void growth is conditioned by the sign of hydrostatic stress. Therefore void nucleation near wire surface, while experiencing a rather sharp increase with the first step, is progressively curtailed by large compressive stresses due to die action which more than offset the tensile ones due to drawing force (Fig. 8), thus bringing about void closure. See for instance numerical results obtained on defect formation in rod drawing [Zavaliangos et al. 1991].

On the other hand the delay in void nucleation on wire axis is more than offset by the accumulated action due to subsequent steps, as the effect of hydrostatic tension occurring right under the die is only in part countered by steady state compressive stresses apparent after exit, see Fig. 8. Remark also that midway between surface and axis trends observed for void evolution duplicate closely those pertaining to surface, as opposed to what takes place along the axis after the fourth step (Fig. 9).

These findings point towards preferential void growth around wire axis, thus providing a mechanism in agreement with established experimental evidence [Goods and Brown 1979], and supported by microhardness tests performed across wire sections, see Fig. 10, where evidence of peculiar radial hardness gradients, and of strain hardening matching to some extent only flow stress increase induced by consecutive drawing passes is shown. It may be worth remarking that void volume fraction on wire axis after six steps only approaches the threshold for coalescence, a stage likely to be reached within a few additional steps.

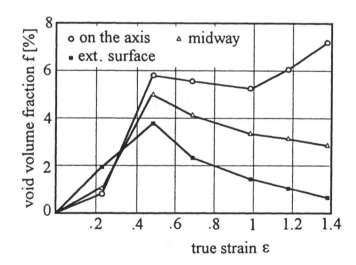

Fig. 9. Evolution of damage in multipass drawing on wire axis, outer surface and midway.
Remark crossover at second step and change of trend after fourth.

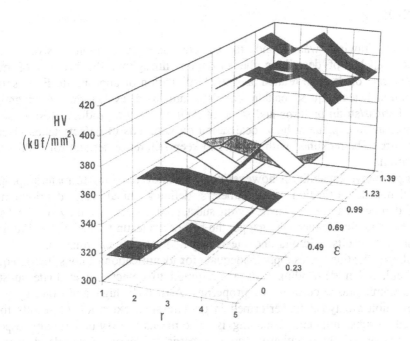

Fig. 10. HV microhardness values obtained at points equispaced along radius r over wire cross-sections (1=centre, 5=outer), corresponding to six drawing steps.

Risk of process disruption due to wire breakage may be readily estimated to a first approximation according to the well known Warner probabilistic model [Haugen 1968], taking ultimate stress as strength and flow stress as load. Main parameters of the relevant distributions are defined by material properties, inclusive of damage related terms, and strain hardening characteristics. Assuming for the sake of expediency both stresses to be independent, normally distributed random variables (a rash statement, to put it mildly), the maximum admissible ratio $k$ of flow stress to ultimate stress consistent with a given risk of tensile failure in process may be readily evaluated, and the feasibility of any given drawing sequence rated accordingly.

Taking e.g. for both stresses a coefficient of variation of 5% over a representative wire length, for risks at 1, 0.1, 0.01% level ratio $k$ would amount in turn to 0.85, 0.80, 0.77. However at 30 m/s drawing speed a 1% risk of failure per km of wire corresponds to a MTBF of the order of an hour, hardly appealing at factory floor level since it would entail a downtime almost as large as production time.

The argument is quite simple; the fly in the ointment is the requirement of reliable information about the main variance components pertaining to flow and ultimate stresses, and about departures from the normal form of the relevant tails of distributions involved. Extreme value distribution (which was found to fit rather well experimental data on yield stress) may be adopted to provide better approximation; a fairly substantial body of experimental evidence is however required if confidence intervals are not to become meaningless because of excessive width.

# 6. DISCUSSION

Multipass wire drawing was found to be amenable to comprehensive numerical modellization, capable of describing the evolution throughout the process of material properties inclusive of damage related aspects. Exploitation of appropriate FEM software, and computational techniques, enable detailed, full field evaluation of progressively superimposed stress/strain patterns and some of their effects on product. Results obtained provide a mechanism explaining how localization on wire axis of defects takes place, and which factors are to be reckoned with in order to keep failures under control within a given drawing program.

Unrestricted predictive capability is however not yet achieved, owing to material properties related problems. Identification of parameters defining initial void distribution, and evolution of damage may be performed with substantial uncertainty only on the basis of traditional tests; and, the smaller wire gauge, the higher are testing skills required to achieve adequate quality of results. Microhardness tests may add valuable information.

Conventional specifications are simply inadequate for identification of constitutive equation parameters; and, rather demanding tests are required to detect and evaluate substantial variation in dynamic plastic deformation properties. As tests at high strain rate up to large plastic deformation are typically performed in compression, extension of results to wire drawing, entailing substantial tensile loading, is by no means an easy undertaking, especially under consideration of the peculiarly non symmetric behavior of metals due to void presence. Time dependent defect evolution, liable to cause delayed rupture either at rest or under tension much lower than nominal load carrying capability, is another major issue to be addressed.

Effects observed were rather consistent; some however apply strictly to the sample examined, and to the relevant production lot. Coils from different heats, let alone from different steel mills, are routinely found to perform far from uniformly in multipass drawing machines on production floor, notwithstanding nominal material properties being well within tolerance limits. Close cooperation between steel mill and wire manufacturer is required in order to enhance final product quality without adversely affecting costs.

## REFERENCES

Alberti, N., Barcellona, A., Cannizzaro, L., Micari, F., 1994, Prediction of Ductile Fractures in Metal-Forming Processes: an Approach Based on the Damage Mechanics, Annals of CIRP, 43/1: 207-210

Anand, L., Zavaliangos, A., 1990, Hot Working - Constitutive Equations and Computational Procedures, Annals of CIRP, 39/1: 235-238

Avitzur, B., 1963, Analysis of Wire Drawing and Extrusion Through Conical Dies of Small Cone Angle, Trans. ASME, J. Eng. for Ind., Series B, 85: 89-96

Avitzur, B., 1964, Analysis of Wire Drawing and Extrusion Through Conical Dies of Large

Cone Angle, Trans. ASME, J. Eng. for Ind., Series B, 86: 305-315

Avitzur, B., 1967, Strain-Hardening and Strain-Rate Effects in Plastic Flow Through Conical Converging Dies, Trans. ASME, J. Eng. for Ind., Series B, 89: 556-562

Avitzur, B., 1983, Handbook of Metal Forming Processes, J. Wiley, New York

Benedens, K., Brand, W.D., Muesgen, B., Sieben, N., Weise, H.R., 1994, "Steel Cord: Demands placed on a High-Tech Product", Wire Journal International, April, pp. 146-151.

Bray, A., Forlin, G., Franceschini, F., Levi, R., Zompì, A., 1994, Messung der Mechanischen Eigenschaften von Feinstdrahten, Draht, v. 45, n.3: 181-188

Brown, L.M., Embury, J.D., 1973, The Initiation and Growth of Voids at Second Phase Particles, Proc. 3rd Int. Conf. on Strength of Metals and Alloys, Inst. of Metals: 164-169

Chen, C.C., Oh, S.I., Kobayashi, S., 1979, Ductile Fracture in Axisymmetric Extrusion and Drawing, Trans. ASME, , J. Eng. for Ind., Series B, 101:23

Chu, C.C., Needleman, A., 1980, Void Nucleation Effects in Biaxially Stretched Sheets, Journ. Eng. Materials and Technology, 102: 249-256

Godfrey, H.J., 1942, "The Physical Properties of Steel Wire as affected by Variations in the Drawing Operations", ASTM Trans., 42: 513-531.

Goods, S.H., Brown, L.M., 1979, The Nucleation of Cavities by Plastic Deformation, Acta Metallurgica, 27: 1-15

Gurson, A.L., 1977, Continuum Theory of Ductile Rupture by Void Nucleation and Growth: Part I - Yield Criteria and Flow Rules for Porous Ductile Media, Journ. Eng. Materials and Technology, 99: 2-15

Hajare, A.D., 1995, Elasticity in Wire and Wire Product Design, Wire Industry, 5: 271-3

Haugen, E.B., 1968, Probabilistic Approaches to Design, J. Wiley, New York

MacLellan, G.D.S., 1948, A Critical Survey of the Wire Drawing Theory, Journal of the Iron and Steel Institute, 158: 347-356

Majors, H., Jr., 1955, "Studies in Cold-Drawing - Part 1: Effect of Cold-Drawing on Steel", Trans. ASME, 72/1: 37-48.

Negroni, F.D., Thomsen, E.G., 1986, A Drawing Modulus for Multi-Pass Drawing, Annals of CIRP, 35/1: 181-183

Puttick, K.E., 1959, Ductile Fracture of Metals, Phil. Mag., 8th series, 4: 964-969

Rogers, H.C., 1960, The Tensile Fracture of Ductile Metals, Trans. Met. Soc. AIME, 218: 498-506

Siebel, E., 1947, Der derzeitige Stand der Erkentnisse über die mechanischen Vorgänge beim Drahtziehen, Stahl und Eisen, 66/67, 11/22: 171-180

Thomsen, E.G., Yang, C.T., Kobayashi, S., 1965, Mechanics of Plastic Deformation in Metal Processing, Macmillan, New York

Tvergaard, V., 1982, Ductile Fracture by Cavity Nucleation between Larger Voids, J. Mech. Phys. Solids, 30: 265-286

Tvergaard, V., 1984, Analysis of Material Failure by Nucleation, Growth and Coalescence of Voids, Constitutive Equations, Willam, K.J. ed., ASME, New York

Tvergaard, V., Needleman, A., 1984, Analysis of the Cup-Cone Fracture in a Round Tensile Bar, Acta Metallurgica, 32: 157-169

Yang, C.T., 1961, On the Mechanics of Wire Drawing, Trans. ASME, J. Eng. for Ind., Series B, 83: 523-530

Zavaliangos, A., Anand, L., von Turkovich, B.F., 1991, Towards a Capability for Predicting the Formation of Defects During Bulk Deformation Processing, Annals of CIRP, 40/1: 267-271

Zimerman, Z, Avitzur, B., 1970, Analysis of the Effect of Strain Hardening on Central Bursting Defects in Drawing and Extrusion, Trans. ASME, J. Eng. for Ind., Series B, 92: 135-145

Zompì, A., Levi, R., Bray, A., 1990, La misura dell'attrito nella trafilatura a freddo di fili in acciaio di piccolo diametro, Politecnico di Torino (unpublished report)

Zompì, A., Cipparrone, M., Levi, R., 1991, Computer Aided Wire Drawing, Annals of CIRP, 40/1: 319-332

Zompì, A., Romano, D., Levi, R., 1994, Numerical Simulation of the Basic Wire Drawing Process, Basic Metrology and Application, Barbato, G. et al ed., Levrotto & Bella, Torino, 187-192

# NEW METHODS OF MEASURING PLANNING
# FOR COORDINATE MEASURING MACHINES

**H.K. Tönshoff**
University of Hannover, Hannover, Germany

**C. Bode and G. Masan**
**IPH-Inst. f. Integrierte Produktion Hannover gGmbh, Hannover,**
Germany

## KEYNOTE PAPER

KEY WORDS: COORDINATE MEASURING MACHINE, MEASURING PLANNING,
PROBE CLUSTER, WORKPIECE ALIGNMENT, ACCESS AREA, PROBING GROUP

ABSTRACT: Measuring planning is a preparatory task ensuring the generation of efficient operation schedules for CMMs. This paper presents three consecutive and computerized methods to determine probe cluster and workpiece alignment. The first method determines the *access area* of a measuring point. The second method groups the access areas, thereby minimizing the number of required probes. The third method calculates the workpiece alignment and refines the probe cluster. Furthermore, the developed methods are applied to a car body part.

## 1. INTRODUCTION

Today, quality assurance is an important duty in industrial production. Measuring workpieces with conventional tools like gauges is an essential and well established part of quality assurance. But the time consumed by measuring processes can be reduced up to 75 % using coordinate measuring machines (CMM) [1].

Measurement must be as carefully planned as any other manufacturing process in order to utilize the full power of CMMs. Hence measuring planning is the preparatory task for generating an efficient and error-free measuring programme. During measuring planning the measuring technology, e.g. CMM type or additional devices, is chosen. Further subtasks of measuring planning are determination of workpiece alignment, probe cluster, fixtures, operation scheduling, etc.

Published in: E. Kuljanic (Ed.) *Advanced Manufacturing Systems and Technology*,
CISM Courses and Lectures No. 372, Springer Verlag, Wien New York, 1996.

## 2. MEASURING PLANNING

Measuring planning is based on the workpiece geometry and the inspection plan. The measuring plan guides the operator through all steps of the measuring process. If the desired CMM is integrated in a production line, measuring planning should be done offline.

AUGE did basic research on measuring planning [2]. GARBRECHT and EITZERT analyzed how the placing of measuring points affects the reproducibility of measuring results [3,4]. KRAUSE et al. developed a system for technological measuring planning in order to achieve reproducible measuring results [5]. This system takes into account fixed probes and standard features like planes, cylinders, circles, etc. Additional devices for CMMs (e.g. rotary tables or indexable probes), freeform surfaces, workpiece alignment and fixturing were not considered.

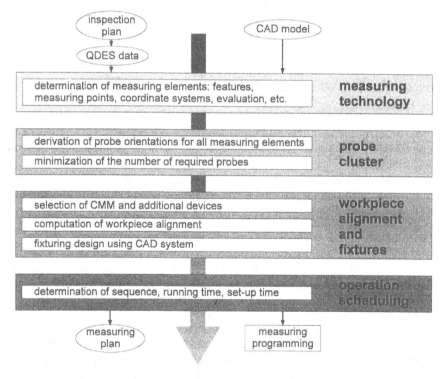

*Fig. 1: Sequence of tasks during measuring planning*

Measuring planning is a time consuming process. Depending on the complexity of the workpiece the process is multiply repeated. BODE showed that most repetitions are caused by the determination of probe cluster and workpiece alignment [6]. Further investigations of the measuring planning process showed a hierarchical dependency among the subtasks. Therefore the following sequence of subtasks is proposed: determination of measuring points and workpiece geometry, probe cluster, workpiece alignment, fixturing design, and operation scheduling. Based on this knowledge we developed a process plan for measuring planning and methods to compute probe cluster and workpiece alignment taking into account all kinds of geometry and CMM (Fig. 1).

## 3. COMPUTATION OF PROBE CLUSTER AND WORKPIECE ALIGNMENT

Determination of probe cluster and workpiece alignment ensures that at least one probe of the cluster can approach a specific measuring point of the workpiece. Therefore, the computation of the probe cluster depends on the measuring points and the workpiece geometry. Besides there is a close relationship between probe cluster design and workpiece alignment. A redesign of the probe cluster requires a change of the alignment and vice versa.

Especially the probe cluster has a significant impact on the efficiency of the measurement. For example, the set-up time required is proportional to the number of probes. If probes are changed during measure, each change lengthens the running time. Since a high number of probes decreases the quality of the entire probe cluster, one main objective of measuring planning is to reduce the number of required probes.

### 3.1 ACCESS AREAS OF A WORKPIECE

Each measuring feature has a distinct property called *access area*. Inside the access area at least one probe approaches the measuring element without colliding with the workpiece. An access area describes all possible probe orientations to touch the assigned measuring point.

The theoretical definition of an access area can be given by modelling a straight probe with infinitesimal diameter.

Definition 1:
An *access line* $\vec{R}_T$ of a measuring point is a half line starting from the measuring point and not intersecting with the workpiece.

Every access line represents one possible probe orientation. Consequently an access area includes all access lines of a measuring point (Def. 2). The so-called set-form of an access area allows any operations defined in set theory.

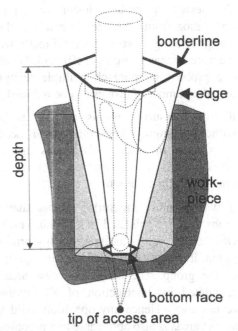

Fig. 2: Access area assigned to a measuring point

<u>Definition 2</u>:   An *access area* is a set consisting of all access lines of a
                measuring point.

Just as a probe has an extension the real access area is smaller than the theoretical. Typical access areas are cone-shaped solids. To represent the diameter of the probing sphere a bottom face is introduced, cutting off the tip of the access area (Fig. 2). This measure ensures the allocation of a suitable space around the measuring point. The depth limits the access area to the size of the entire probe. The borderline of access areas is of high interest. As several operations are carried out using this curve, the selection of an appropriate type is crucial. During our investigations convex polygons turned out to be most useful [6].

Access areas can be described conveniently on the surface of a unit sphere. Any probe orientation can also be represented as a point on the unit sphere. Therefore, a *normalized access area* is the intersection of a unit sphere and the given access area (Fig. 3). The normalized access area is described using the borderline curve. Normalization reduces the complexity of probe cluster computation.

## 3.2 DETERMINING A PROBE CLUSTER BY GROUPING ACCESS AREAS

To design a probe cluster one probe orientation from each access area is picked. But generating a probe cluster directly from the access areas is highly inefficient. Usually one probe can approach multiple features, thus the number of probes can be reduced.

If two measuring points can be probed without changing the probe, their access areas must intersect. This means that the access areas have one or more probe orientations in common.

If two or more measuring points share a probe they are collected in a *probing group*. One access area is assigned to any probing group. It is applicable to all measuring points of the group. The assigned access area is calculated by intersection of all involved access areas. Hence any operation valid for access areas is also applicable to a probing group.

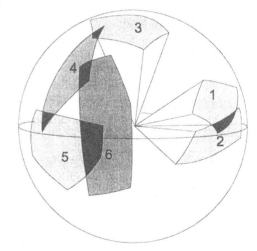

*Fig. 3: Normalized access areas.*
*Intersections are marked black.*

<u>Definition 3:</u> A probing group *G* is a set of n measuring points $M_i$ which access

areas $a_i$ satisfy the condition $\bigcap\limits_{i=1}^{n} a_i \neq \varnothing$.

To design an optimal probe cluster, the number of probing groups must be minimized. A capable algorithm to perform grouping was presented by CARPENTER/GROSSBERG and MOORE [3,5]. Probing groups are created by joining the 'closest' access areas. If two access areas intersect, they are close. We expressed the 'distance' A between a probing group G and an access area z as follows:

$$A(G,z) = A_1 - A_2 + \frac{A_3 + A_2 A_3}{16\pi} \qquad (1)$$

Primarily, the distance is determined by $A_1$. $A_1$ is the number of elements of the probing group $G$. Since the algorithm repeats grouping, $A_2=1$ must be subtracted if the access area $z$ is already an element of $G$. Secondarily, the surface measurement $A_3$ of the intersection of $G$ and $z$ is considered.

As MOORE showed the distance scale is crucial for the behaviour of the algorithm. We showed that the scaled described above satisfies the stability conditions proposed by MOORE [7]. Additionally, BODE proofed that the algorithm requires linear time [6].

## 3.3 PROBE ORIENTATIONS

In general, any type of probe or CMM has unusable probe orientations. A general approach to distinguish the unusable from the preferred probe orientations is a quality function. The quality function assigns to each probe orientation a value representing its usability. Since the value of a probe orientation is bound to the CMM configuration, the quality function is expressed in the machines coordinate system.

Especially when using indexable probes, unusable probe orientations are becoming important. An indexable probe cannot adjust to probe orientations inside a cone round the machine's z-axis. This cone contains the sleeve to which the probe body is mounted.

Preferred probe orientations are usually corresponding (or close) to the axes of the CMM. These probe orientations are easy to maintain during the measurement.

BODE developed quality functions for various types of inspection devices [2].

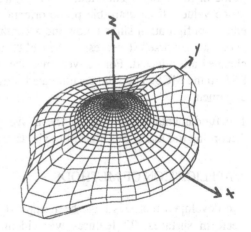

Quality functions are conveniently developed using the spherical coordinate system $K\{\varphi,\theta,r=1\}$. The following example was

*Fig. 4: Plot of quality function (cf. eq. 2)*

designed for a parallel system CMM equipped with indexable probes (Eq. 2, Fig. 4). The indexable probe imposes unavailable orientations as described above.

$$B_{IP}\left(\vec{R}_P(\varphi,\theta)\right) = \sin^4\varphi \cdot \sin^2\theta + \sin^4\theta \qquad (2)$$

## 3.4 DETERMINATION OF WORKPIECE ALIGNMENT

The list of probing groups for a specific workpiece constrains the set-up of the probe cluster to the most efficient ones. To select specific probe orientations from this list, the workpiece alignment must be taken in consideration. Therefore, probes and workpiece alignment are selected simultaneously.

Probing groups are bound to the workpiece, they are described within the part coordinate system (PCS). In oppositon, the probe cluster is described within the machine coordinate system (MCS). Both coordinate systems are linked by workpiece alignment. Hence, workpiece alignment is defined as follows:

Definition 4:   The workpiece alignment is the relative position of the part
                coordinate system in relation to the machine coordinate system.

Commonly the relation among two orthogonal coordinate systems is expressed in three tilt angles $\phi$, $\psi$, $\theta$. By convention these tilt angles are applied to the MCS.

Respecting these constraints, the idea of finding a workpiece alignment is as follows. At first, an initial workpiece alignment is randomly chosen. Each step of the algorithm slightly changes either one probe orientation within its access area or turns the whole probe cluster. Then, the quality of the new configuration is compared to the old one. Two configurations are compared by the value of their probe clusters. The quality function is applied to every probe of the cluster. The value of the entire cluster is computed to the sum of the single probe values. If an unusable probe orientation is included in the cluster, the value of the entire configuration sinks below the acceptance level. If the new configuration is better or „not much worse" it serves as the „old" configuration for the next step. Otherwise the changes are rejected. For convenience the initial workpiece alignment should equate the PCS to the MCS. Then the accumulated turns of the whole probe cluster are the workpiece alignment.

The fundamental algorithm for the above procedure was first presented by DUECK [9]. According to DUECK this algorithm yields results close to the absolute optimum [9].

## 4. APPLICATION OF METHODS

The developed methods have been applied to a car body part. This workpiece has many freeform surfaces, 22 features were identified for measuring. We compared the time consumption of different measuring planning procedures

The conventional offline planning process using a CAD system takes about 8.5 hours. The teach-in programming takes about 2 hours. Although teach-in programming requires less time, the necessity of a real workpiece devalues this procedure. Besides, teach-in programming hinders production taking place on the CMM.

Both conventional offline programming and teach-in programming set up a probe cluster consisting of seven probes (Fig. 5).

To utilize the new methods 28 access areas were directly derived from the CAD model. Six more access areas than features were created for cylindrical features. The construction of access areas and determination of workpiece alignment and probe cluster took about one hour. The computed workpiece alignment was similar to the previous one, but by a rotation of 180° the number of required probes was reduced to four. Thus, it is obvious that much time is saved applying the new methods (Table 1, Fig. 5).

| | teach-in | conventional offline progr. | supported offline progr. |
|---|---|---|---|
| entire measuring planning | 2 h | 10 h | 2 h |
| included time for determination of probe cluster, workpiece alignment, and detailed sequencing | 1.5 h | 8 h | 1 h |
| programming | 20 h | 6 h | 6 h |

Table 1: Comparison of time required by different methods of measuring planning

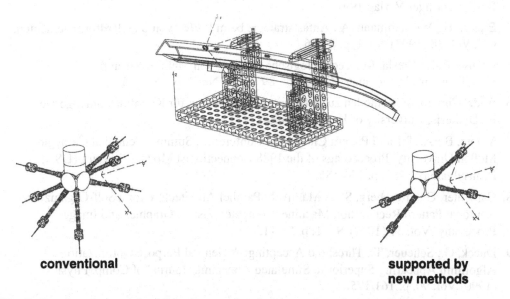

**conventional**　　　　　　　　　　　　　　　　**supported by new methods**

Fig. 5: Results of measuring planning: probe clusters and workpiece alignment

## 5. CONCLUSIONS

The examination of the currently performed measuring planning revealed a time-consuming procedure. This time-consumption is mainly caused by determination of probe cluster and workpiece alignment.

To shorten the expenditure of time for measuring planning we developed three consecutive methods. The efficiency of these methods has been proven in the automotive industry by applying these methods in the development of a car body part. The resulting effect was that a high proportion of time required during the measuring planning process has been saved.

The modified sequence outlined above will have an impact on the further development of measuring planning. After integration of the methods into CAD systems the designer is able to validate his work with respect to quality assurance [10]. This may lead to integrated Computer Aided Measuring Planning systems (CAMP).

## 6. REFERENCES

1. Hahn, H.:Wirtschaftlichkeit von Mehrkoordinaten-Meßgeräten mit unterschiedlicher Automatisation. Der Stahlformenbauer, Vol. 8 (1991) No. 6, p. 77/78

2. Auge, J. A.: Automatisierung der Off-line-Programmierung von Koordinatenmeßgeräten. Aachen: Ph.D. thesis, RWTH Aachen, 1988.

3. Garbrecht, T.: Ein Beitrag zur Meßdatenverarbeitung in der Koordinatenmeßtechnik. Berlin: Springer Verlag 1991

4. Eitzert, H.; Weckenmann, A.: Antaststrategie beim Prüfen von Standardformelementen. QZ, Vol. 38 (1993) No. 1, p. 41/46

5. Krause, F.-L.; Ciesla, M.: Technologische Planung von Meßprozessen für Koordinatenmeßmaschinen. ZWF, Vol. 89 (1994) No. 3, p. 133-135

6. Bode, Christoph: Methoden zur effizienten Meßplanung für Koordinatenmeßgeräte. Ph.D. thesis, University of Hannover, 1996.

7. Moore, B.: ART 1 and Pattern Clustering. Conference: Summer School at Carnegie Mellon University, Proceedings of the 1988 Connectionist Model, Pittsburg (USA, Pennsylvania), 1988, p. 174-185.

8. Carpenter, G.; Grossberg, S.: A Massively Parallel Architecture for a Self-Organizing Neuronal Pattern Recognition Machine. Computer Vision, Graphics and Image Processing, Vol. 37 (1987) No. 1, p. 54-115.

9. Dueck, G.; Scheuer, T.: Threshold Accepting: A General Purpose Optimization Algorithm Appearing Superior to Simulated Annealing. Journal of Comp. Physics, (1990) Vol. 90, p. 161/175.

10.Tönshoff, H.K.; et al.: A Workpiece Model for CAD/CAPP Applications. Production Engineering I/2 (1994) No. 1, p. 119-124

# ROLE AND INFLUENCE OF ECODESIGN ON NEW PRODUCTS CONCEPTION, MANUFACTURING AND ASSEMBLY

G.F. Micheletti
Polytechnic of Turin, Turin, Italy

## KEYNOTE PAPER

KEY WORDS: Ecodesign, Life Cycle Analysis, Life Cycle Engineering, L.C. Assessment, etc.

## ABSTRACT:

*The role and influence of eco-design on new product conception is investigated along four main guidelines as Life Cycle: Analysis, Engineering, Assessment, Development concept.*
*Friendly attitudes are the ethical and psycological basic factors. New design strategies require "ad hoc" methods and solutions, to reach the improvements. Which are the foreseable reactions of the entrepreneurs? Beside the CIM role, a more profitable approach could be envisaged; the concurrent design, supported by eco-auditors and eco-experts within the company, in order to face the "eco-labelling" prescriptions.*
*The total cost of the Life Cycle shall be included in the industrial costs in view of the heavy social costs that will be imposed for waste collection.*
*A prospect suggests a practical classification.*

Published in: E. Kuljanic (Ed.) *Advanced Manufacturing Systems and Technology*, CISM Courses and Lectures No. 372, Springer Verlag, Wien New York, 1996.

## 1. INTRODUCTION

I would like to start - in order to better investigate the role and the influence of eco-design on new product conception - by taking into consideration four main guidelines, along which the Life Cycle Research and Development should enhance the achievements:

- *the Life Cycle Analysis* that has been considered like a "balance *energy-environment"*, taking from the more conventional energy balance the basic study lines, due to the effect of the environment as an essential parameter, in each phase of the production process;

-   *the Life Cycle Engineering*, that has been designated not only as "engineering" but as "art of designing" a product, characterised by a proper Life Cycle derived from a combined choice of design, structure materials, treatments and transformation processes;

- *the Life Cycle Assessment* , that is the instrument able to visualise the consequences of the choices on the environment and on the resources, assuring the monitoring and the control not only on products and processes, but also on the environmental fall out;

- *the Life Cycle Development Concept*, that is centred on the product and summarises the techniques to be adopted in design phase, keeping in mind till from the first instant of the creative idea, the fall out that will influence the environment both during the period of the product's use, and in the following phase when the functional performance is no more existing, but the product still exists, to be dismantled or converted.

## A friendly attitude

Let me add one more ethical and psychological factor, where some basic friendly attitudes are emphasised:

- A product must become **a friend** of the environment;

- A process should respond of itself as **a friend** of the environment;

- A distribution to the market will act as **a friend** of the environment;

- A use of the product shall assure its behaviour as **a friend** of the environment;

- A re-use and a final dismantling of the components will finally validate all the previous steps as **friends** of the environment.

In contrast let me add the unfriendly attitudes that are enemies against which the environment asks to be defended:

- not bio-degradable and toxic materials;

- incompatible treatments;

- not destroyable components;

- not dismantables joints and groups.

All the above positive and negative factors are fitted as the basis for the increasing standardised prescription, starting from the famous "*State of the Art*" of Life Cycle Engineering design, that included the idea-design subsequently spread all over the world, in occasion of the "Health Summit - Agenda 21 - Action Program of the United Nations, held in Rio 1992, where the responsibilities especially for the more developed Countries where pointed out for a "sustainable global development"

As a direct consequence have been prepared the Standards ISO 14001, becoming operative during the present year 1996 as Environmental Management Standard, to which have to be added specific standards for industry, taking into account the more general ones to give space to an Environmental Management System (EMS).

For instance, within the frame of the Life Cycle Engineering/Design, several more relevant points have been identified:

- to reduce quantity of materials used for industrial production, through a rationalisation and simplification in design and dimensioning, responding to an analysis of the effects that every type of material induces onto the environment, at the same time saving at least or eventually including the functionality of the products;

- to centre the study not only on the production process (as it normally happens), but on the whole effects of the product's existence, by using in the best way the resources together with the protection of the environment.

Nevertheless there is still an evident gap between rules, expectations and realities.
I would introduce some concrete considerations

**New Design strategies**

It is clear that new design strategies still require to be invented and tested, as well as new "ad hoc" methods and instruments, properly addressed to the features, that could come from universities and from international initiatives.

Appropriate solutions to reach the improvements are to day considered as fundamental supports of the "Design for Engineering Saving" allowing to:
- optimise the materials handling system and the logistics in general;
- select the chemical elements and additives;
- stress the thermal and electric energy saving;
- reduce the overheads costs as heating, air conditioning, air filtering etc.

Consequently during the design phase the experts cannot any more disregard:

- to improve the modularity of the project in order to extend the possibility of derived solutions;
- to design by keeping in mind the evaluation of energy consumption, not only in manufacturing but in the use during the product life;
- at the same time of the assembly operations, to take into consideration also how simplify the subsequent dismantling phase (f.i. with the separation of different materials by group and the collection in function of the recycling);
- product volume and weight reduction (where possible), due to their negative influence in respect with logistical problems;
- to simplify the maintenance operations;
- evaluation and selection of packaging materials, trying to set up protective structures less fragile and prescribing as much as possible recycled materials.

**The reactions of the entrepreneurs**

As it is well known, the entrepreneurs don't like too much the theoretical debates, since they widely prefer practical indications.

How to manage the relationships towards environment ?
By which procedures ?

Two answers could firstly be addressed to their questions.

Every enterprise, even small-medium sized is aware to be like a system, in which all tasks are crossing the company context.

The CIM route for a global planning seems the one to be adopted, since it seemed - at least at the beginning - to manage the factory in a completely automatic way without the human intervention, automatically, even for a quite long period.

But some "errors" arose and should be recognised about this way of thinking:
- first of all with regard to men, not available to accept only secondary roles facing the machines and the process;
- the computers, very useful to integrate the distribution/diffusion of the information process are insufficient to accomplish the *functional integration*, required to realise all the factory functions simultaneously;
- small attention has been attributed to optimise the environmental resources;

- the environment policy, having reached in the last time a great importance, was not enough considered versus the standard already established within CIM strategy.

This is why the interpretation that seems to day more substantial within a CIM framework is as follows:
- CIM must be regarded as an important mean for the factory where the decision are confirmed as responsibility of the men, with the help of the computers;
- the organisation of an enterprise induces now a more flexible configuration enabling an easier dynamics in the companies;
- the management sets out to adopt engineering techniques able to improve the problems of the different factory areas in a punctual way, using the specialised co-operation coming from different/complementary competencies, included the environmental labelling;
- the product design, if and when Life Cycle criteria are adopted can be put in front in a CIM global way, towards the subsequent steps as manufacturing, marketing, use and, even beyond, can allow to identify the way of dismantling and reusing.

## A more profitable approach

Though a more profitable approach to the new environmental problems could be offered by the *"concurrent engineering strategy"* showing the better integration within the results and a fruitful stimulation, through the establishment in the company of groups involving experts of-and from-different areas, sustained by their personal experience and inclined to investigate together many issues of different nature.

The expression *"concurrent or simultaneous engineering"* in the actual case of designing should be properly indicated as *"concurrent design"*: a denomination that has been accepted at an international level since 1989 (being adopted in USA for the first time) to define a solution - very complex and ambitious - to support the competitiveness of industries, the improvement of the quality, the times reduction, the observance of the environmental rules.

This allows to put in action newly conceived Working Groups, including experts as designers, production engineers, quality responsibles, market operators, plus experts in the area of *eco-system*, with the possibility for them to be active since the first moment in which a new product is born, avoiding in this way the subsequent corrections that bring to time delays.

So the companies should be prompt to welcome their internal (and/or external) experts of environment in their teams as eco-auditors and eco-experts, in order to assure since the design phase and during the set up of the operation cycle, the criteria on which we are hereby dealing with.

### The green label

To adopt this methodology is and will be difficult in that industries that like to maintain a conventional organisation based on a hierarchical structure; in fact it is requested to consider how to establish Working Groups in both a horizontal and transversal configuration, with the task to operate from the idea to the end of the product and even farther.

These basic considerations are the guidelines for eco-design to which it is necessary to add some integrations, defined with specific and not only technological considerations, in the concurrent areas of marketing and economy.

That means, for a company "green labelled" to enhance its competitiveness versus other industrial producers; to highlight the image of the enterprise itself; to launch an ecological advertising campaign and commercial promotion; and, not as last point, to recover the additional costs that for sure the ecological accomplishment require.

The resonance on the market is acquiring a greater importance due to the criticisms coming from pollution claims.

It is probable that a deeper sensibilisation will be dramatically spread in favour of producers that appreciate the attention requested by the customers, now more and more aware of the ecology factors for a common benefit.

Some examples of "green company", very respectful of the environment have become a reality, that meets and earns the liking everywhere in the industrialised polluted world.

The producer can reasonably explain that a part of those costs, that in general are sustained by customers and users, also through a tax system, are up to him; this with reference also to the users costs and to the costs attributed to the social body (dismantling and discharge of materials after life cycle is ended).

## What specific solutions are involved in the production techniques?

It is clear that is a specific task of each company to define its own strategies. Some general elements can be indicated and adopted with advantage: f.i. the standardisation criteria to which has already been referred, facilitating the production, the assembly and maintenance; preliminary studies on producibility and on automation of operations and of assembly, flexibility of the production units and plants to facilitate the modification on the production level and the conversion between similar typologies of products or components (families); reliability and durability garanties, avoiding the limitation of product life.

It can be noticed that, till now the total cost of Life Cycle is not included in the industrial costs: but this shall be definitely considered also in view of the heavy cost that shall be imposed by public administrators on "waste" collection.

Coming back to terminology, several aspects should be cleared from the beginning. F.i. it's important to distinguish between:

*Life Cycle Analysis (LCA)* , that is - as already stated - the process of evaluation of the environmental "charges" connected with a product or an industrial activity; the analysis must identify and quantify the incidence of the involved amount of energy, of materials used, of scraps released within the environment;
the evaluation practically must take into account:
- the entire life cycle of the products;
- the treatments of raw materials;
- the manufacturing operations;
- the transportation;
- the distribution;
- the use;
- the eventual further re-use;
- the recycling modes;
- the final dismantling and disruption.
In addition to that, different meanings and contents have to be attributed with reference to:
*Eco-Balance,* that considers only:
- the cycle inside the plant, i.e.
- treatments;
- manufacturing;
- assembly;
- packaging.

In few words:
*Life Cycle Analysis* = product from cradle to grave
*Eco-Balance*          = from input to output inside the enterprises

Of course, both involve the management responsibility; in fact:
- new products' design and plant's control;
- the external and/or internal environmental rules;
- the expected precautions;
- the communication towards the subsequent clients and/or final consumers.

Which are the items, whose concrete impacts influence in a direct way the Company strategy?
An helpful Prospect (Source SPOLD*) is in the following page.

The Prospect suggests a good route towards a practical classification, the definition of the characteristics (i.e. to collect, aggregate, quantify the data); the operating evaluation for each component/subgroup/group/final product.

---

* SPOLD, a technical frame work for Life Cycle Assessment; the Society for Environmental Toxicology & Chemistry, Washington D.C. 1991

## Prospect:  Routes towards classification

| Category  Impact | Type of impact | Criteria of impact |
|---|---|---|
| | | |
| | Raw materials consumption | - Renewable Raw material consumption<br>- Not Renewable Raw mat. consumption |
| | Water consumption | - Water use, drying |
| Decay of resources and of the ecosystem | Consumption of energy resources | - Total consumption. of primary energies<br>- Consumption of renewable fossil sources |
| | Consumption of soil | - Erosion |
| | Damages to the natural environment | Damage to ecosystems, landscape etc. |
| | Human toxicity | - Toxic. effect (ingestion, inhalation, cancer risk). |
| Damage from emissions | Eco-toxicity | Eutrofiz. ecotox water ecotoxicity ground acidification |
| | Global effects on the atmosphere | Climate modifying gas emission;  ozone  danger; photochemical ozone emis. |
| | Other effects | -Noise;  smell; radiation heat dispersion, waste |

## CONCLUSION

Each Enterprise shall select the items that attains to its production; the basic recommendation is to start very seriously the analysis.
There are already available some examples, that give useful demonstrations on how some industrial sectors started their respective "Life Cycle Assessment" (some are reported in the Bibliography) other sectors will move the first steps.

One "must" is addressed to each Company: every delay in facing the environment obligations is not a benefit, but a penalisation and a lack of ethics (whose price will be paid later).

Life Cycling has already an important role in industrial management.
The *eco-management* is showing the importance of parameters not too much considered till yesterday, to day brought in evidence and to morrow belonging to the general sensitivity of the people.
Producers and users are together involved in respecting the ecology problems that with the finished product come back to the single components.
What the Life Cycle imposes, requires a co-operation involving "concurrent engineering", together with "eco-engineering", that goes far away from the actual laws with reference to the various emissions and stimulates study for modelling causes, behaviours and effects.
Till now the laws are prevailing on heavy risks, accidents, contaminations, but shall be necessary to prevent, with the design, in such a way to eliminate the causes of risks and accidents by means of the proper technologies; to this purpose shall be more and more necessary to create and update a "data-base" for materials, components, their chemical and physical properties, toxicological and eco-toxicological properties, bio-degradability etc. In the mean time many perspectives and hopes are connected with the introduction of *clean technology.*
At the same time, researches shall be developed to solve the maintenance problems in the light of eco-maintenance: chemical products (inside the machines), cooling fluids, disassembly with selection of materials.

A general monitoring, as a sum of a number of sub-monitoring and pre-monitoring, is the objective to reach.
Without the wish to criticise the strong discussion made in the last years on the "Industry of the Future", having recognised the outstanding value of the production technology and the organisation methodologies, it is necessary to recognise that nowaday some predictions (or at least projections) appear wrong, especially considering the protection of the environment, having all those considerations mostly the main task to increase the production rates, as a unique aim.

It seems appropriate to adopt a correct behaviour and an objective awareness, in order to conclude with Leo Alting: the influence of Life Cycle offers the desirable correction to foresee, more than a *"Factory of the Future"*, an important, stimulating position as the *"Factory with a Future"*

## Bibliography

1. *"Concurrent Engineering"*,  G. Sohlenius, Annals of CIRP, vol.41/2/1992

2. *"La metamorfosi ambientale"* , Carla Lanzavecchia, Ed. CELID, Torino, 1992

3. *"Ecodesign dei Componenti",* Domus Academy,  D.I.P.R..A. Facoltà di Architettura - Politecnico di Torino, *Progetto Riuso,* L. Bistagnino. Ed. Camera di Commercio di Torino 1993.

4. *"A Key Issue in Product Life-Cycle: Disassembly"*, L. Alting, A.Armillotta, W.Eversheim, K. Feldmann, F. Jovane, G. Seliger, n. Roth,  Annals of  CIRP  Vol. 42/2/1993.

4. *Management and Control of Complexity in Manufacturing"* , H.P. Wiendahl e P. Schoitissek, Annals of CIRP, pag. 533 vol.43/2/1994

5. *"Pour une intégration du facteur environnement dans l'entreprise: proposition d'une démarche de conception de produits respectant l'environnement"*, D. Millet, A.M. Cailleaux, R. Duchamp, ENSAM, Parigi. Ed. Rivista Design Recherche, n.6 Sett. 1994.

6. *"Co-Design"  "Attività progettativa fra stilisti e fornitori nell'ambito di ruote in lega leggera"* B. Corio, FIAT-AUTO, STILE-DESIGN,  1° Concurrent Engineering Award,  Milano, Settembre 1994

7. *"Life-Cycle Engineering and Design"*,  Leo Alting, Jens Brobech Legarth, Annals of CIRP, Vol. 44/2/1995

8. *"Designing for Economical Production"*, Truks H.E ",SME, Dearborn, USA, 1987

9. *"Tool and Manufacturing Engineers Handbook"*, Bakerjian R.  vol.6, SME, Dearborn, USA, 1992

10. *"Assembly Automation and Product Design"*, Boothroyd G.  Dekker, New York, 1992

11. *"Product Design for Manufacture and Assembly",* G. Boothroyd , P. Dewhurst, W. Knight Dekker, New York, 1994

12. *"Fertigungsverfahren"*, W. Koenig,  Band 4, VDI Verlag, Duesseldorf, 1990

13. *"Konstruktionselemente der Feinmechanik"*,W. Krause, Hanser, Muenchen, 1989

14. *"Plastic Part Technology"*, E.A.Muccio,  ASM, Materials Park, Ohio, USA, 1991.

15. *"Plastics waste-recovery of Economic value"*, J. Leidner ", Ed. K. Dekker, N.Y., 1981.

16. *"Le materie plastiche e l'ambiente"*, a cura dell'AIM, Ed. Grafis, Bologna 1990

17. *"Recupero post-consumo e riciclo delle materie plastiche"* F. Severini, M.G. Coccia Ed.IVR, Milano, 1990.

18. *"Recycling and Reclaiming of Municipal solid Wastes"* F.R. Jackson Noyes Data Corp., 1975.

19. *"Resource Recovery and Recycling Handbook of Industrial Waste"*, M. Sittig Noyes Data Corp., 1975.

20. *"Problematiche nel riciclo dei materiali plastici"*, P. La Mantia Macplas 116, p. 67, 1990.

21, *"Macromoleculs"*. H. G. Helias, Plenum Press, N.Y., 2°. Ed. 1994, vol. II, p. 858.

22., *"Recycling of plastics"*, W. Kaminsky, J. Menzel, H. Seim Conserv. Recycling 1, 91, 1976.

23. *"Sorting of Household waste"* A. Skordilis, Ed. M. Ferranti, G.Ferrero, Elsevier Appl. Sci. Pubbl., Londra 1985.

24. *"A technical framework for life cycle assessment "* SETAC ( The Society for Environmental Toxicology and Chemistry), Washington DC, 1991.

25. *"The LCA sourcebook, a European business guide to life cycle assessment SustainAbility"*, SPOLD Ltd., 1993.

26. *"The Fiat Auto recycling project: current developments"* S. Di Carlo, R. Serra, Fiat Auto, Auto Recycle Europe '94.

27. *"The life cycle concept as a basis for sustainable industrial production"* L. Alting, L. Jorgensen, Annals of the CIRP, 1993, vol.42, No.1.

28. *"Handbook of industrial energy analysis"*, I. Boustead, G.F. Hancock, Ellis Horwood Limited, Chicester, England, 1979.

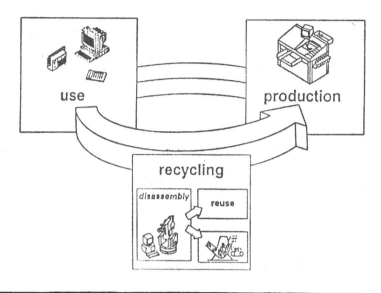

The life–cycle of products

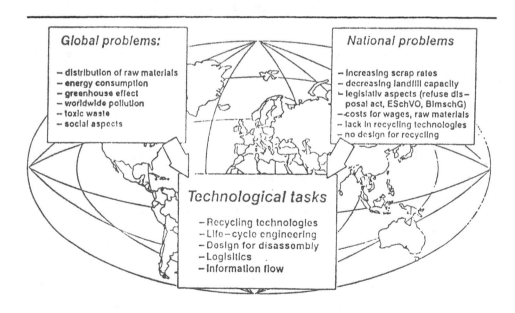

Global and national problems and
                                                    the resulting technological tasks

# PREDICTION OF FORCES, TORQUE AND POWER
## IN FACE MILLING OPERATIONS

**E.J.A. Armarego and J. Wang**
**University of Melbourne, Parkville, Victoria, Australia**

KEY WORDS: Face Milling, Force Modelling, Milling, Force Prediction.

ABSTRACT: The need for reliable quantitative predictions of the forces, torque and power in face milling operations is discussed and the alternative force prediction approaches reviewed and compared. It is shown that the traditional direct experimental or 'empirical' approach is expensive, laborious and only considers a few variables and average force components but the curve fitted 'empirical' equations are most suitable for economic optimisation. The semi-empirical 'mechanistic' approach is more comprehensive but does not consider all fluctuating and average force components or provide explicit force equations. By contrast the 'unified mechanics of cutting approach' is the most comprehensive and generic approach allowing for all the forces and most variables but results in complex equations. However extensive simulation studies of this proven approach have enabled comprehensive 'empirical-type' equations to be established without significant loss in predictive capability. These simpler equations which allow for numerous variables are most suitable for machine tool and cutter design purposes and for economic optimisation of these operations. The importance of fundamental cutting analyses in predictive modelling of machining operations is highlighted.

## 1. INTRODUCTION

Machining is one of the oldest process for shaping components and due to its versatility and precision, achieved through continual innovation, research and development, has become an indispensable process used in manufacturing industry. In more recent years machining has led the way towards the 'revolution' in modern computer based manufacturing through developments in computer controlled machining systems and

Published in: E. Kuljanic (Ed.) *Advanced Manufacturing Systems and Technology*, CISM Courses and Lectures No. 372, Springer Verlag, Wien New York, 1996.

flexible automation. These modern automated systems have been forecast to increase the total available production time a component spends being machined from 6% to 10% in conventional (manual) systems to 65% to 80% making machining more important than ever before[1, 2]. As a consequence the long recognised need for improved and reliable quantitative predictions of the various technological performance measures such as the forces, power and tool-life in order to optimise the economic performance of machining operations has also become more pressing than ever before if full economic benefits of the capital intensive modern systems are to be achieved [3]. This pressing need for reliable quantitative performance data has recently been re-emphasised in a CIRP survey [4].

The estimation or prediction of the various technological performance measures represents a formidable task when the wide spectrum of practical machining operations such as turning and milling as well as the numerous process variables such as the tool and cut geometrical variables, speed and material properties for each operation are considered [3]. Furthermore the establishment of comprehensive machining performance data, preferably in the form of equations, is an on-going task to allow for new developments in tool designs, materials and coatings as well as workpiece materials. Alternative approaches to machining performance prediction have been noted in the research literature and handbooks although the dearth of such data remains. It is interesting to note that a new CIRP Working Group on 'Modelling of Machining Operations' has been established to investigate the alternative modelling approaches to performance prediction [5, 6].

In this paper the alternative approaches to force, torque and power prediction for the important face milling operations will be reviewed and compared. Particular attention will be placed on the 'Unified Mechanics of Cutting Approach' developed in the author's laboratory [3] and the establishment of equations for these performance measures essential for developing constrained optimisation analysis for selecting economic cutting conditions in process planning [7, 8].

## 2. EMPIRICAL AND SEMI-EMPIRICAL APPROACHES

Traditionally the direct experimental or 'empirical' approach has been used for estimating and relating the various technological performance measures to the influencing variables for individual practical machining operations. Thus experiments have been planned to measure the required performance measures such as the force components, torque and power for one or more operation variables of interest, often varied at different levels on a one at a time basis. In view of the large number of influencing variables there has been a tendency to limit the number of variables studied to reduce the time and cost involved and to simplify the data processing. The performance data have often been presented in tabular form but have also been graphed and 'curve fitted' to establish 'empirical' equations. These equations enabled the performance measure to be estimated for any set of variables within the experimental testing domain. These equations would need to be re-established for each variable not included in the original testing programme such as the tool and work materials or even the tool geometrical features. In more recent times it has been found useful to statistically plan the testing programme and apply multi-variable regression analysis to curve fit the data to establish the 'empirical' equations.

These techniques enable a larger number of variables to be included for a given amount of testing and also provide estimates of the scatter or variability at defined levels of confidence [7, 9, 10].

Recent reviews of the forces in milling operations has shown that although empirical equations for some of the force components and the power were reported for peripheral milling very few equations were found for the popular face milling operations [11, 12]. Some earlier work by Roubik [13] using a planetary-gear torque-meter has shown that the tangential force $F_{tang}$ in face milling was dependent on the feed per tooth $f_t$ and the axial depth of cut $a_a$ when tested independently and given by eqs.(1) and (2) while Doolan et al [14] using a 'fly cutter' and a strain gauge mounted on the tooth flank used multi variable regression analysis to arrive at eqs.(3) for the tangential force $F_{tang}$ in terms of $f_t$, $a_a$ and the cutting speed V

$$F_{tang} = K_f \cdot f_t^{0.766}; \qquad F_{tang} = K_d \cdot a_a^{0.940} \qquad\qquad (1, 2)$$

$$F_{tang} = K \cdot f_t^{0.745} a_a^{0.664} V^{0.049} \qquad\qquad (3)$$

where $K_f$, $K_d$ and K were empirical constants.

Interestingly in a recent comprehensive Chinese handbook [15] (drawing on CIS data and handbooks) an empirical equation for the tangential force which allows for the majority of influencing variables in peripheral, end and face milling has been presented as shown below

$$F_{tang} = \frac{K \cdot a_a^{x_a} f_t^{x_f} a_r^{x_r} N_t}{D^{y_d} N^{y_n}} \qquad\qquad (4)$$

where K, $x_a$, $x_f$, $x_r$, $y_d$ and $y_n$ are 'empirical' constants and $a_a$, $a_r$, $f_t$ $N_t$, D and N are the axial and radial depths of cut, the feed per tooth, the number of teeth, the cutter diameter and the spindle speed. The values of the empirical constants have been given for a variety of common work materials and tool materials. Furthermore this equation enabled the torque $T_q$ and the power P to be evaluated from the known cutter radius (D/2) and peripheral cutting speed V, i.e.

$$T_q = F_{tang} \cdot D/2; \qquad P = F_{tang} \cdot V \qquad\qquad (5, 6)$$

Thus despite the very large number of variables considered comparable equations for the three practical force components, i.e. the average feed, side and axial force, were not available.

In general these empirical equations provided estimates for the average forces, torque and power but did not estimate the fluctuating forces in face or other milling operations. Furthermore few equations have been found for all the average force components so necessary for machine tool and cutting tool design, vibration stability analysis and constrained optimisation analyses for economic selection of milling conditions in process planning.

Despite the many disadvantages of the empirical approach noted above an important advantage is the form of empirical equations established which greatly enhance the development of constrained mathematical optimisation strategies of many operations [7, 8].

It should be noted that semi-empirical or 'mechanistic' approaches have been reported to predict the average and fluctuating forces in milling operations. These approaches involve a combination of theoretical modelling and 'matched' experimental milling tests assisted by computer programmes to numerically evaluate the forces after the 'empirical' constants have been established for the given milling cutter geometry and tool-workpiece material combination [16-18]. This approach generally considers one or two of the three force components in the milling tests to establish the required 'empirical' constants in the computer-aided mechanistic models. Furthermore the 'empirical' constants have to be experimentally established for each cutter tooth geometry. This approach is more comprehensive and complex than the traditional 'empirical' approach but not as generic as the mechanics of cutting approach discussed below.

## 3. MECHANICS OF CUTTING APPROACH

From a series of investigations the author and his co-workers have developed an alternative approach to force, torque and power prediction in practical machining operations such as turning, drilling and the different types of milling [3]. This 'Unified or Generalised Mechanics of Cutting Approach' is based on the modified mechanics of cutting analyses of 'classical' orthogonal and oblique cutting processes which incorporate the 'edge forces' due to rubbing or ploughing at the cutting edge [19]. It has been shown that a variety of practical machining operations such as face milling could be represented and mathematically modelled by one or more elemental 'classical' oblique cutting processes whose three elemental force components and torque could be resolved in the required practical directions (e.g. feed, side and axial) and integrated at each and every instant and cutter orientation to give the total instantaneous force, torque and power fluctuations as well as the corresponding average values per cycle [3, 11, 12, 20].

In this approach mathematical models for each practical machining operation have to be developed to allow for all the relevant tool and cut geometrical variables while a common data base of basic cutting quantities for each tool-workpiece material combination found from 'classical' orthogonal cutting tests can be used with each operation model to quantitatively predict the forces, torque and power. Additional data bases can be established for each new tool-work material combinations as can models for each new practical machining operation e.g. tapping. This approach has been shown to be ideally suited for integration into a modular software structure where independent modules for each operation model as well as independent data base modules can be developed and linked through the generic 'classical' oblique cutting module [3].

The predictive models for face milling operations allowing for tooth run-out have been developed and experimentally verified [12, 20]. It has also been shown from extensive simulation studies that while the tooth run-out can significantly affect the force components and torque fluctuations and peak values the average forces are not significantly affected when compared to the 'ideal case' of zero tooth run-out [12, 20].

Thus the explicit equations for the average forces and torque given below (i.e. eqs.(7) to (10)) can be used for prediction purposes and to study the effects of the many operation variables [12]. It is evident from these equations that although the effects of the number of teeth $N_t$, axial depth of cut $a_a$ and feed per tooth $f_t$ can be readily interpreted (i.e. linear increases in the forces with increases in these variables), the effects of the tooth cutting edge angle $\kappa_r$, radial depth of cut $a_r$, cutter axis 'offset' u (from the workpiece centre-line) and the cutter diameter D are not always obvious in view of the complex trigonometric functions involved. In addition, the effects of the tooth normal rake angle $\gamma_n$ and inclination angle $\lambda_s$ as well as the cutting speed V are not explicitly expressed since these are embedded in the modified mechanics of cutting analysis 'area of cut' and 'edge force' coefficients, i.e. the $K_c$'s and $K_e$'s in the equations. Thus despite the comprehensive nature of the predictive model and the average force and torque equations computer assistance is required to study the effects of a number of variables on the average forces. The effects of these operation variables on the average feed force $(F_x)_{avg}$, side force $(F_y)_{avg}$, axial force $(F_z)_{avg}$ and torque $(T_q)_{avg}$ are shown in Fig. 1 where the trends have been qualitatively and quantitatively verified [12, 20].

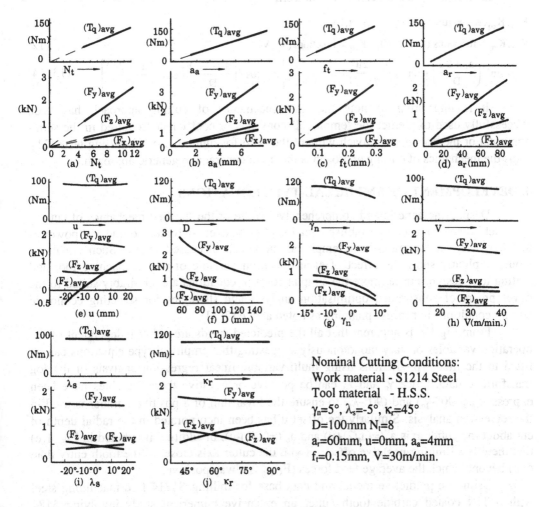

Nominal Cutting Conditions:
Work material - S1214 Steel
Tool material  - H.S.S.
$\gamma_n$=5°, $\lambda_s$=-5°, $\kappa_r$=45°
D=100mm $N_t$=8
$a_r$=60mm, u=0mm, $a_a$=4mm
$f_t$=0.15mm, V=30m/min.

Fig. 1 Typical average force and torque tredns in face milling (zero tooth run-out).

$$(F_x)_{avg} = \frac{N_t a_a}{4\pi}\left\{ f_t[K_{cp}\sin 2\theta_m \sin\theta_c + (K_{cq}\sin\kappa_r + K_{cr}\cos\kappa_r)(\theta_c - \cos 2\theta_m \sin\theta_c)] \right.$$
$$\left. + 4\left[\frac{K_{ep}}{\sin\kappa_r}\cos\theta_m \sin(\tfrac{1}{2}\theta_c) + \left(K_{eq} + \frac{K_{er}}{\tan\kappa_r}\right)\sin\theta_m \sin(\tfrac{1}{2}\theta_c)\right]\right\} \tag{7}$$

$$(F_y)_{avg} = \frac{N_t a_a}{4\pi}\left\{ f_t[K_{cm}(\theta_c + \cos 2\theta_m \sin\theta_c) - (K_{cq}\sin\kappa_r + K_{cr}\cos\kappa_r)\sin 2\theta_m \sin\theta_c] \right.$$
$$\left. + 4\left[\frac{K_{ep}}{\sin\kappa_r}\sin\theta_m \sin(\tfrac{1}{2}\theta_c) - \left(K_{eq} + \frac{K_{er}}{\tan\kappa_r}\right)\cos\theta_m \sin(\tfrac{1}{2}\theta_c)\right]\right\} \tag{8}$$

$$(F_z)_{avg} = \frac{N_t a_a}{2\pi}\left[2f_t(K_{cq}\cos\kappa_r - K_{cr}\sin\kappa_r)\sin\theta_m \sin(\tfrac{1}{2}\theta_c) + \left(\frac{K_{eq}}{\tan\kappa_r} - K_{er}\right)\theta_c\right] \tag{9}$$

$$(T_q)_{avg} = \frac{R \cdot N_t \cdot a_a}{2\pi}\left[2f_t K_{cp}\sin\theta_m \sin(\tfrac{1}{2}\theta_c) + \frac{K_{ep}}{\sin\kappa_r}\theta_c\right] \tag{10}$$

where $K_c$'s, $K_e$'s, $\theta_c$ and $\theta_m$ can be found from

$$K_{cp}, K_{cq}, K_{cr} = \text{functions }(\gamma_n, \lambda_s, r_1, \tau, \beta) \tag{11}$$

$$K_{ep}, K_{er} = \text{functions }(\gamma_n, \lambda_s, V); \quad K_{eq} = \text{functions}(\gamma_n, V) \tag{12}$$

$$\theta_c = \sin^{-1}\left(\frac{a_r + 2u}{D}\right) + \sin^{-1}\left(\frac{a_r - 2u}{D}\right); \quad \theta_m = \tfrac{1}{2}\left[\pi - \cos^{-1}\left(\frac{a_r - 2u}{D}\right) + \cos^{-1}\left(\frac{a_r + 2u}{D}\right)\right] \tag{13, 14}$$

It is interesting to note that this mechanics of cutting approach has been successfully used to predict the 'empirical' constants or coefficients required in the semi-empirical or mechanistic approach without the need for running special milling tests [21]. Thus the mechanics of cutting approach is the more general and generic approach.

## 4. DEVELOPMENT OF EMPIRICAL-TYPE EQUATIONS

Despite the generic and comprehensive nature of the 'unified mechanics of cutting approach' there is a need to establish equations for the average forces, torque and power of the type used in the traditional 'empirical' approach discussed above. Such equations would explicitly show the effect of each operation variable of use in machine tool and cutting tool design. Furthermore the simpler form of equations are admirably suited to the development of constrained optimisation analysis and strategies for selecting economic cutting conditions in process planning as noted above [7, 8].

From Fig.1 it is apparent that all the predicted trends are either independent of the operation variables or vary monotonously suggesting that empirical-type equations can be fitted to the model predictions using multi-variable linear regression analysis of the log transformed data. Since $\gamma_n$ and $\lambda_s$ can be positive or negative these variables have been expressed as $(90°-\gamma_n)$ and $(90°-\lambda_s)$ to ensure the logarithm of a positive number is used in the regression analysis. Similarly the offset u has been incorporated in the radial depth of cut about the cutter axis $a_i$ $(=(a_r/2)+u)$ and $a_o$ $(=(a_r/2)-u)$. In addition it is noted in Fig. 1(e) that there is a limiting negative offset u (with the cutter axis closer to the tooth entry than exit) beyond which the average feed forces $(F_x)_{avg}$ is alway positive.

Using the predictive model and data base for milling S1214 free machining steel with a TiN coated carbide tooth cutter an extensive numerical study involving 5184

$(2(\gamma_n) \times 2(\lambda_s) \times 3(\kappa_r) \times 3(D) \times 2(N_t) \times 3(a_r) \times 3(u) \times 2(a_a) \times 2(f_t) \times 2(V))$ different cutting conditions has been carried out and the predicted positive values curve fitted to yield the empirical-type equations below and the exponent values in Table 1. The corresponding equations for the limiting u for positive $(F_x)_{avg}$ have also been established which guarantee the validity of the equations below.

$$(F_x)_{avg} = K_x (90° - \gamma_n)^{eg_x} (90° - \lambda_s)^{el_x} \kappa_r^{ek_x} D^{ed_x} N_t^{en_x} a_i^{ei_x} a_o^{eo_x} a_a^{ea_x} f_t^{ef_x} V^{ev_x} \qquad (15)$$

$$(F_y)_{avg} = K_y (90° - \gamma_n)^{eg_y} (90° - \lambda_s)^{el_y} \kappa_r^{ek_y} D^{ed_y} N_t^{en_y} a_i^{ei_y} a_o^{eo_y} a_a^{ea_y} f_t^{ef_y} V^{ev_y} \qquad (16)$$

$$(F_z)_{avg} = K_z (90° - \gamma_n)^{eg_z} (90° - \lambda_s)^{el_z} \kappa_r^{ek_z} D^{ed_z} N_t^{en_z} a_i^{ei_z} a_o^{eo_z} a_a^{ea_z} f_t^{ef_z} V^{ev_z} \qquad (17)$$

$$(T_q)_{avg} = K_q (90° - \gamma_n)^{eg_q} (90° - \lambda_s)^{el_q} \kappa_r^{ek_q} D^{ed_q} N_t^{en_q} a_i^{ei_q} a_o^{eo_q} a_a^{ea_q} f_t^{ef_q} V^{ev_q} \qquad (18)$$

Table 1 The constant and exponents of the fitted empirical-type equations (S1214 work material and TiN coated carbide tool).

| | K | eg | el | ek | ed | en | ei | eo | ea | ef | ev |
|---|---|---|---|---|---|---|---|---|---|---|---|
| $(F_x)_{avg}$ | $1.24\times10^{-5}$ | 4.11 | -1.04 | 0.394 | -0.945 | 0.995 | 1.11 | -0.164 | 1.028 | 0.698 | 0.256 |
| $(F_y)_{avg}$ | 1.2946 | 1.328 | 0.069 | -0.078 | -0.913 | 0.999 | 0.376 | 0.537 | 0.988 | 0.849 | 0.071 |
| $(F_z)_{avg}$ | $9.29\times10^{-11}$ | 3.98 | 3.52 | -1.66 | -1.01 | 0.991 | 0.518 | 0.495 | 0.852 | 0.669 | 0.279 |
| $(T_q)_{avg}$ | $7.25\times10^{-4}$ | 1.35 | 0.064 | -0.077 | -.006 | 0.999 | 0.507 | 0.500 | 0.988 | 0.848 | 0.071 |

From a study of the exponents in Table 1 it can be deduced that the effects of the different operation variable on the average forces and torque are generally consistent with the trends in Fig. 1. In addition the predictive capability of the approximate empirical-type equations with respect to the rigorous model predictions has been assessed in terms of the percentage deviation (e.g. %dev=100×(Empirical pred.-Model Pred.)/Model Pred.)). The histograms of the percentage deviations in Fig.2 show that the average %dev. is very close to zero for all force components and torque with the largest scatter from -10.8% to 13.5% occurring for the average feed force $(F_x)_{avg}$. Thus the predictive capability of the simpler empirical-type equations can be considered to be very good.

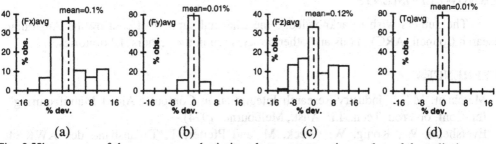

(a)      (b)      (c)      (d)

Fig. 2 Histograms of the percentage deviations between equation and model predictions.

The predictive capability of these empirical-type equations with respect to the experimentally measured average forces are shown in Fig. 3 where again good correlation is evident. The result in Fig. 3 have been obtained from 150 different milling conditions.

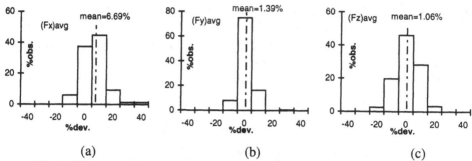

(a)                                    (b)                                    (c)

Fig. 3 Histograms for percentage deviations between equation predictions and
measured forces (TiN coated carbide machining S1214 steel).

## 5. CONCLUSIONS

There is a pressing need for reliable quantitative predictions of the forces, torque, power and tool-life in face milling, and machining in general, to optimise machining conditions and gain maximum economic benefits of the increased productive times in modern computer based manufacturing.

The traditional 'empirical' approach to force prediction is laborious, expensive and primarily considers the tangential force and power for some variables although the 'empirical' equations are most suitable for economic optimisation. The semi-empirical 'mechanistic' approach is more comprehensive but does not consider all the fluctuating and average force components and relies on some special milling tests and computer assistance for quantitative predictions. The 'unified mechanics of cutting approach' is the most comprehensive and generic approach which allows for all the forces, torque and power, encompasses the 'mechanistic' approach, but results in complex equations. Nevertheless this generic approach can be used to develop comprehensive 'empirical-type' equations for use in CAD/CAM applications and economic optimisation.

The importance of fundamental cutting analyses and data bases in predictive modelling of machining operations is highlighted in this work.

## ACKNOWLEDGMENTS

The authors wish to acknowledge the financial support offered by the Australian Research Council (ARC) in this and other projects run in the authors' laboratories.

## REFERENCES

1.  Merchant, M.E., "Industry-Research Integration in Computer-Aided Manufacturing", Int. Conf. on Prod. Tech., I.E. Aust., Melbourne (1974).
2.  Eversheim, W., König, W., Weck, M. and Pfeifer, T. Tagunsband des AWK'84, Aachener Werkzeugmaschinen-Kolloquim, (1984).
3.  Armarego, E.J.A., "Machining Performance Prediction for Modern Manufacturing", 7th Int. Conf. Prod./Precision Engineering and 4th Int. Conf. High Tech., Chiba, Japan, K52, Keynote paper, (1994).

4. Kahles, J. F., "Machinability Data Requirements for Advanced Machining Systems", CIRP. Report, Annals of the CIRP, 36/2, (1987), p523.

5. Van Luttervelt, C.A., Working Paper, STC.'C' Working Group, "Modelling of Machining Operations", 1995.

6. Armarego, E.J.A., Jawahir, L.S., Ostafiev, V.A. and Patri, K. Venuvinod, Working Paper, STC.'C' Working Group, "Modelling of Machining Operations", (1996).

7. Armarego, E.J.A. and Brown, R.H., "The Machining of Metals", Prentice Hall Inc., New Jersey, 1969.

8. Armarego, E.J.A., Smith, A.J.R. and Wang, Jun., "Constrained Optimisation Strategies and CAM Software for Single Pass Peripheral Milling", Int. J. Prod. Res., Vol.31, No.9, (1993), p2139.

9. Leslie, R.T. and Lorenz, G., "Tool-Life Exponents in the Light of Regression Analysis", Annals of CIRP, 12/1/(1963-1964), p226.

10. Wu, S.M., "Tool-life Testing by Response Surface Methodology", Trans. Amer. Soc. Mech. Engrs., J. of Eng. for Industry, Vol.86(2), (1964), p49.

11. Deshpande, N. P., "Computer-Aided Mechanics of cutting Models for Force Predictions in End Milling, Peripheral Milling and Slotting", Ph. D. Thesis, The University of Melbourne, (1990).

12. Wang, Jiping., "Computer-Aided Predictive Models For Forces, Torque and Power in Face Milling Operations ", Ph. D. Thesis, The University of Melbourne, (1995).

13. Roubik, J.R., "Milling Forces Measured With a Planetary-Gear Torquemeter", J. of Eng. for Industry, Trans. Amer. Soc. Mech. Engrs., Nov. (1961), p579.

14. Doolan, P., Phadke, M.S. and Wu, S.M., "Computer Design of a Minimum Vibration Face Milling Cutter Using an Improved Cutting Force Model", J. of Eng. for Industry, Trans. Amer. Soc. Mech. Engrs., Aug. (1976), p807.

15. AI, X. and XIAO, S., "Metal Cutting Conditions Handbook", Mechanical Industry Press, P. R. China, 1985.

16. Zhou Ruzhong and Wang, K.K., "Modelling of Cutting Force Pulsation in Face-Milling ", Annals of the CIRP, 32/1, (1983), p21.

17. Kline, W.A., Devor, R.E. and Lindberg, J.R., "The Prediction of Cutting Forces in End Milling With Application to Cornering Cuts", Int. J. Mach. Tool Des. & Res. 22/1, (1982), p7.

18. Fu, H. J., Devor, R. E. and Kapoor, S. G., "A Mechanistic Model for the Prediction of the Force System in Face Milling Operations", J. of Eng. for Industry, Trans. Amer. Soc. Mech. Engrs., Vol. 106, (1984), p81.

19. Armarego, E.J.A., "Practical Implications of Classical Thin Shear Zone Analysis", UNESCO/CIRP Seminar on Manufacturing Technology, Singapore, (1983), p167.

20. Armarego, E.J.A., Wang, Jiping and Deshpande, N.P., "Computer-Aided Predictive Cutting Model For Forces in Face Milling Operations Allowing For Tooth Run-Out. ", Annals of the CIRP Vol. 44/1, (1995), p43.

21. Budak, E., Altintas, Y., Armarego, E.J.A., "Prediction of Milling Force Coefficients From Orthogonal Cutting Data", J. of Eng. for Industry, Trans. Amer. Soc. Mech. Engrs., (Accepted in Nov. 1994).

# EVALUATION OF THRUST FORCE AND CUTTING TORQUE IN REAMING

R. Narimani and P. Mathew

University of New South Wales, Sydney, Australia

**KEYWORDS:**    Reaming, Metal Cutting, Thrust and Torque, Modelling

**ABSTRACT:**    Reaming is a process that is widely used in industry with very little theoretical modelling being carried out. In this paper the cutting action of the reaming operation is presented by explaining the thrust and torque involved. A model based on an orthogonal theory of machining and variable flow stress is presented in order to predict the thrust and torque involved in the cutting process. A comparison of the predicted and experimental results give good correlation and thus indicates that the procedure used is viable.

## 1.0   INTRODUCTION

Reaming is an internal machining operation which is normally performed after drilling to produce holes with better surface finish and high dimensional accuracy. A reamer consists of two major parts. The first part is the chamfer length for material removal and the second part, the helical flute section, carries out the sizing operation of the hole. During a reaming operation the chamfer will first remove the excess material left from the drilling operation which is then followed by the helical flutes which size the hole precisely and produce a good surface finish. In analysing the thrust force and cutting torque in a reaming operation the first step will consider the action of the chamfer length. To study the forces acting on the chamfer, it is necessary to investigate the cutting action of a single tooth in the reamer (Figure 1). The investigation of the single tooth showed that the cutting edge represented a

Published in: E. Kuljanic (Ed.) *Advanced Manufacturing Systems and Technology*,
CISM Courses and Lectures No. 372, Springer Verlag, Wien New York, 1996.

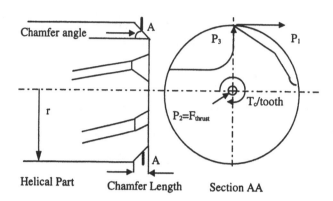

Figure 1: Reamer geometry and cutting forces

similar geometry to a single point turning tool involved in an external turning operation. Thus it is expected that the reaming process can be modelled by assuming that a single tooth is a single point tool with zero or low normal rake angle, low inclination angle and a negative side cutting edge angle (chamfer angle). Using this assumption then it is possible to use a predictive theory of machining developed by Oxley and his co-workers [1] to predict the thrust and torque in reaming.

In the following sections, an experimental investigation of the reaming process is given followed by a brief description of the machining theory and the extension to the reaming operation. The experimental results are compared with the predicted results to verify the theoretical development. Finally conclusions are presented.

## 2.0   INITIAL THRUST AND TORQUE OBSERVATIONS

An experimental investigation was carried out to observe the thrust, the torque and the chip formation during the reaming operation. The experiments involved reamer sizes of $\phi 10.0$mm to $\phi 16.0$mm with varying drill sizes ($\phi 9.5$mm to $\phi 15.5$mm) for each reamer. A total of 18 experimental observations were obtained for a plain carbon steel with a chemical composition of 0.48%C, 0.021%P, 0.89%Mn, 0.317%Si, 0.024%S, 0.07%Ni, 0.17%Cr, 0.02%Mo, 0.04%Cu, 0.03%Al. An example of the experimental condition is as follows: drilled hole size = $\phi 9.5$mm or $\phi 9.65$mm; reamer = $\phi 10.0$mm; rotational speed = 140 rpm; feed-rate = 0.252mm/r. The experimental results were measured using a Kistler 9257 two component force/torque dynamometer and Kistler 5001 charge amplifiers connected to a PC-based data acquisition system using a RTI815 A/D card and a 80386 computer with accompanying software. A total of 1000 data points for each component were collected over a 30 second period and an example of the results obtained is shown in Figure 2. From this figure it is clear that the thrust force increases quickly to its maximum value when the chamfer length is in full contact with the workpiece. The thrust force remains fairly constant throughout the cutting period until the reamer exits the hole at the

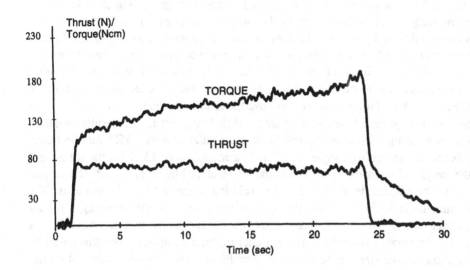

Figure 2: Variations of thrust and torque in reaming

bottom end when the thrust force drops down to zero in a very short time. The torque increases rapidly at the beginning of the operation as soon as the cutting commences and then gradually increases as the helical component of the reamer increases its contact with the newly created surface. This trend continues until the reamer exits the hole when there is a sudden decrease in the torque. However there is still a component of torque present due to the rubbing of the helical component of the reamer with the newly formed surface. This torque is seen to gradually decrease as the hole reaches its required size.

The following observations can be made in relation to the results obtained. The time period for the rapid increase in the thrust and torque is dependent on the time taken for the full contact of the chamfer length of the reamer (ie the feed rate) while at the exit end the rapid decrease is dependent on the time taken for the chamfer length to disengage. From the results it can be said that the chamfer length performs the cutting action or material removal while the helical part carries out a rubbing action to clean out the hole to create good surface finish and correct hole size[2]. The torque generated during the material removal process then is going to be a summation of the two components ie cutting torque and rubbing torque. It is expected that the rubbing torque will keep on increasing as the length of engagement of the reamer with the hole increases till the full depth is reached. Thus the torque can be distinguished into two components as given by the following equation:

$$T_m = T_c + T_b \qquad\qquad (1)$$

where $T_m$ is the total torque in Ncm, $T_c$ is the cutting torque in Ncm due the chamfer length and $T_b$ is the rubbing torque in Ncm due to the rubbing effect of the helical part of the reamer. It is expected that $T_c$ will be constant during the reaming process while $T_b$ will change with specimen length or hole depth and the frictional condition at the interface for a given cutting condition. In order to verify the experimental observations another experiment was carried out to compare the thrust and torque obtained when reaming a φ9.65mm hole created by a larger drill. In this case the amount of material removed is less than the previous cutting condition due to a larger drill being used. The results obtained indicate a reduction in the thrust and torque with the results showing 55N for the thrust force and 85 Ncm for the cutting torque, Tc. This is a reduction of 15 N for the thrust and 20Ncm for the torque which is expected due to the less material being removed. There was also difference in the magnitude of the maximum rubbing torque attained by the tests. The difference in magnitude is 53Ncm with the test for the smaller width showing a higher rubbing component of 135Ncm compared to 82Ncm for the larger width. This result is interesting in that the smaller width indicates a higher rubbing component and this could be due to the variation in hole size due to the smaller width and less material removed during cutting. So the helical part carries out the extra material removal.

Since the material removal in reaming is seen to be similar to the turning operation a model is developed to predict the thrust and torque in reaming taking into account the variable flow stress orthogonal machining theory developed by Oxley and co-workers [1].

## 3.0   DEVELOPMENT OF A THEORY OF REAMING

To predict the thrust and cutting torque it is essential to know the actual geometry of the cutting edge of a reamer. The cutting edge can be modelled as an oblique tool with a specific geometry[3]. Once this geometry is known, then for a single straight cutting edge in oblique machining, the method uses the experimental observations (i) that for a given normal rake angle and other cutting conditions, the force component in the direction of cutting, $F_C$, and the force component normal to the direction of cutting and machined surface, $F_T$, are nearly independent of the cutting edge inclination angle, $i$, and (ii) that the chip flow direction, $\eta_c$, satisfies the well known Stabler's flow rule ($\eta_c = i$) over a wide range of conditions. It is assumed that $F_C$ and $F_T$ can be determined from a variable flow stress orthogonal machining theory by assuming zero inclination angle irrespective of its actual value and with the rake angle in the orthogonal theory taken as the normal rake, $\alpha_n$, of the cutting edge. The tool angles associated with the cutting edge of the reamer together with the predicted values of $F_C$ and $F_T$ and the values of $\eta_c$ and $i$ are then used to determine $F_R$ the force normal to $F_C$ and $F_T$ which results from a non-zero inclination angle, from the relation

$$F_R = \frac{F_C(\sin i - \cos i \sin\alpha_n \tan\eta_c) - F_T \cos\alpha_n \tan\eta_c'}{\sin i \sin\alpha_n \tan\eta_c + \cos i} \qquad (2)$$

For a tool with a non-zero side cutting edge angle, $C_s$, the force components $F_T$ and $F_R$ no longer act in the feed and radial directions. Therefore, the force components are redefined as $P_1$, $P_2$ and $P_3$ of which the positive directions are taken as the velocity, negative feed and radially outward directions as shown in Fig.1. For the equivalent cutting edge these are given by the following equations

$$
\begin{aligned}
P_1 &= F_C \\
P_2 &= F_T \cos C_s + F_R \sin C_s \\
P_3 &= F_T \sin C_s - F_R \cos C_s
\end{aligned}
\tag{3}
$$

In predicting the force components $F_C$ and $F_T$, the orthogonal machining theory as described by Oxley[1] is used. The chip formation model used in predicting the forces is given in Figure 3.

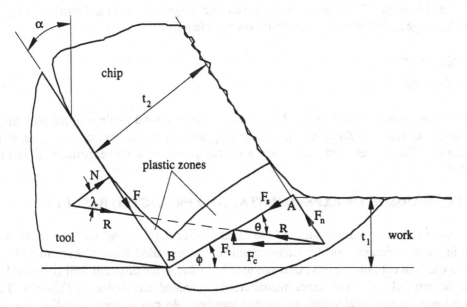

Figure 3: Chip formation model

In order to calculate forces, temperatures etc. it is first necessary to determine the shear angle $\phi$ which defines the chip geometry. In order to find $\phi$ for given cutting conditions and work material properties the method used is briefly as follows. For a suitable range of shear angle values, the resolved shear stress at the tool chip interface is calculated from the resultant cutting force determined from the stresses on the plane AB (Figure 3) within the chip formation zone. For the same range of values of $\phi$ the temperature and strain-rate in the interface plastic zone are then calculated and used to determine the shear flow stress in this zone. The solution for $\phi$ is taken as the value which gives a shear flow stress in the chip material at the interface equal to the resolved shear stress, as the assumed model of chip formation is then in equilibrium. Once the shear angle is known then $F_C$, $F_T$ etc. can be determined. Details of the theory and its applications have been given by Oxley [1]. In addition the thermal properties of the work material used in the experiments which are also needed in applying the machining theory are found from relations given by Oxley.

In order to work out the equivalence of the reamer cutting edge to a single point tool the following is carried out. The feed per revolution is converted into a feed per tooth or cutting edge, $f_t$, and the cutting velocity is worked out. Since the chamfer has a negative 45° chamfer angle ($C_s$) it is necessary to convert the feed per tooth into an equivalent undeformed chip thickness, $t_1$, for the principal cutting edge by using $t_1 = f_t \times \cos C_s$ and the width of cut, w = radial difference in the reamer and hole size $\div \cos C_s$. These values are then inputted into the orthogonal theory to determine values of $F_C$ and $F_T$. These values are then used with the inclination angle of 10° to determine $F_R$ and then the values of $P_1$, $P_2$ and $P_3$ are determined. From these values the cutting torque, $T_C$, and the thrust force $F_{thrust}$, per cutting edge is determined using the following relations

$$F_{thrust} = P_2$$
$$T_c = P_1 r \tag{4}$$

where r is the radius of the reamer. From these values the total cutting torque and thrust force are calculated by multiplying the values in equation by the number of cutting edges in the reamer. These predicted values are now compared with the experimental results obtained.

## 4.0 COMPARISON OF EXPERIMENTAL AND PREDICTED RESULTS

The experiments carried out used the same rotational speed and feed per revolution as the initial experiments but the reamer diameters were varied from $\phi$10mm to $\phi$15mm with the width of cut ranging from 0.07mm to 0.5mm due to the different drill sizes used to create the original hole. The experimental results obtained are shown in Figure 4. The predicted and experimental values are plotted together. As can be seen from Figure 4 the correlation between the experimental and predicted results is good given that the model is based on the orthogonal machining theory for single point tools. The maximum difference seen is between the predicted and experimental thrust forces (Figure 4a) with the biggest difference being 105% for the $\phi$10mm reamer and the 0.31mm width of cut. However it must be noted that this attempt at predicting the thrust force has given values of the same magnitude. The differences in the data can be explained by the inconsistency in the generation of the drilled hole as the drill could have inconsistent cutting action along its flutes and thus creating an uneven surface for the reamer to follow. The results for the thrust of the larger reamer ($\phi$15.0mm) are in good agreement with the maximum variation being only 20%.

The results in Figure 4b indicate excellent agreement between the predicted and experimental cutting torques. The biggest variation observed between the results is approximately 35% when the width of cut is very small. This seemingly large variation could be again due to the drilling action of the previous tool causing variations in width of cut and thus material removal. Overall the results presented here indicate that the variable flow stress theory of machining is capable for predicting the thrust and torque in reaming.

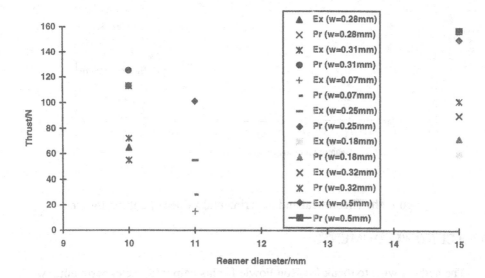

Figure 4a: Theoretical and experimental values of thrust force

## 5.0 CONCLUSION

The work in this paper indicates that the reaming operation is not a simple operation but involves two types of operations for material removal and sizing of the hole. The reaming operation involves a cutting action by the chamfer length and then this is followed by a rubbing action of the helical part to create the precise hole. The thrust force is fairly constant during the operation with the torque being made up of two components as given by equation (1). The rubbing component increases as the length of the reamer in contact with hole increases. This will remain until the hole is fully cleaned out by the reamer as indicated by the results in Figure 2. The prediction of the thrust and torque using the variable stress machining theory has been successful however further work is required to improve the correlation. Finally the rubbing component needs further investigation to fully understand the reaming operation and this is currently being carried out.

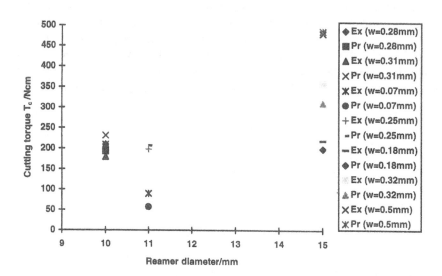

Figure 4b: Theoretical and experimental values of cutting torque

## 6.0  ACKNOWLEDGMENTS

The authors wish to thank Mr Ron Fowle for his help with the experimental work.

## 7.0  REFERENCES

1.    Oxley, PLB, The Mechanics of Machining: An Analytical Approach to Assessing
      Machinability, Ellis Horwood, Chichester, 1989.

2.    Collwell, LV, and Branders, H, Behaviour of Cutting Fluids in Reaming Steels,
      Trans ASME, (July 1958), 1073-1078.

3.    Lin, GCI, Mathew, P, Oxley, PLB, and Watson, AR, Predicting Cutting Forces for
      Oblique Machining Conditions, Proc Instn Mech Engrs, 196, No 11 (1982), 141-148.

# WEAR OF CERAMIC TOOLS
# WHEN WORKING NICKEL BASED ALLOYS

S. Lo Casto, E. Lo Valvo, M. Piacentini and V.F. Ruisi
University of Palermo, Palermo, Italy

KEY WORDS: Ceramics, Cutting, Nickel-based alloy

ABSTRACT: In order to improve the toughness of alumina materials, various trials have recently been made. These include toughening by the addition of zirconia and of significant amounts of titanium carbide to ceramic oxide $Al_2O_3$ and the more recent use of nitride based ceramics, which have resulted in an increase of fracture toughness and in a significant improvement of ceramic tool performance. Another very recent way of improving ceramic materials consists in adding SiC whiskers to $Al_2O_3$ matrix. This composite material is also suitable for machining nickel based alloys. In order to evaluate and to qualify these materials some test cycles have been carried out in continuous cutting conditions, employing cutting parameters chosen following an experimental design and suitable testing. The wear tests have been carried out on an AISI 310 steel tube, with cutting speeds ranging from 1m/s to 4m/s, using as cutting tools silicon nitride, alumina based and mixed-base alumina, alumina reinforced with SiC whiskers and sintered carbide. Their chemical stability, together with their good mechanical properties can explain the appreciable results obtained with sintered carbide and alumina reinforced with SiC whisker tools for the cutting parameters and tool geometry utilised.

## 1. INTRODUCTION

The principal properties required of modern cutting tool materials for a high production rate and high precision machining include good wear resistance, toughness and chemical stability under high temperatures.

Published in: E. Kuljanic (Ed.) *Advanced Manufacturing Systems and Technology*,
CISM Courses and Lectures No. 372, Springer Verlag, Wien New York, 1996.

Tool failure is also usually attributed to excessive wear on tool flank and rake face, where the tool is in close contact with the workpiece and the chip, respectively.

For this reason many tool materials have been developed in the last ten years, such as ceramics. For a long time alumina ceramics have held great promise as cutting tool materials because of their hardness and chemical inertness, even at high temperatures. However, their inability to withstand mechanical and thermal shock loads make them unpredictable for most cutting operations.

In order to improve the toughness of alumina ceramics, various research has recently been done [1,2,3,4,5]. These include transformation toughening by the addition of zirconia, the incorporation of significant amounts of titanium carbide, and reinforcement with silicon carbide whiskers [6,7,8,9,10]. Recently this group has been supplemented by silicon nitride. In our previous papers [11,12] we report the performances and wear mechanisms of some ceramic materials when cutting AISI 1040. In the light of the results obtained after machining new materials for special uses, it become very relevant to study the effect of Nickel and Chromium on tool life. These latter are present in special refractory steels.

The AISI 310 nickel based alloy is one of the most frequently employed materials for equipment subjected to high chemical wear at working temperatures of up to 1100°C. Generally the AISI 310 steel belongs to the group of "hard to machine" materials. Its low specific heat, thermal conductivity and hardness as compared to AISI 1040 steel, its pronounced tendency to form a built-up-edge, strain hardening and the abrasive effect of intermetallic phases result in exceptionally high mechanical and thermal stresses on the cutting edge during machining. Due to the high cutting temperatures reached, sintered carbide inserts can be used only at relatively low cutting speeds. Because of these difficulties, recent improvements have made some ceramic tool grades suitable for machining nickel-based alloys.

For these reasons the purpose of this paper is to refer to the performance of some ceramic materials when cutting AISI 310 steel, nickel based alloy, in continuous cutting with cutting speeds ranging from 1m/s to 4m/s.

## 2. EXPERIMENTAL SECTION

A set of tool life tests, in continuous dry turning with three-dimensional cutting conditions, was performed on AISI 310 steel whose characteristics are reported in Tab. I

Table I  Characteristics of AISI 310 steel

Chemical composition: C=0.084%; Cr=24.89%; Ni=20.72%; Mn=1.29%; Si=0.92%;
P=0.021%; S=0.015%
Tensile strength:      R = 620 N/mm$^2$
Hardness:              $HBN_{(2.5/187.5)}$ = 160

The material worked was a commercial tube with an outer diameter of 250mm and an inner one of 120mm. A piece of this tube, approximately 750mm long, was fixed between chuck and tail stock. The commercially available ceramic materials selected for the tests, according to the insert number SNG 453, were as follows:
- Zirconia-toughened Alumina ($Al_2O_3$-$ZrO_2$ 7%vol.), in the following called "F";
- Mixed-based Alumina ($Al_2O_3$-TiN-TiC-$ZrO_2$), in the following "Z";
- Alumina reinforced with SiC whiskers ($Al_2O_3$ - $SiC_w$), in the following "W";
- Silicon nitride ($Si_3N_4$), in the following "S";
- Sintered carbide grade P10 (WC-TiC-Co), in the following "C".
The inserts was mounted on a commercial tool holder having the following geometry:
- rake angle                $\gamma = -6°$
- clearance angle           $\alpha = 6°$
- side cutting edge angle    $\psi = 15°$
- inclination angle         $\lambda = -6°$
The tests were carried out with the following parameters:
- depth of cut: d=2.0mm;
- feed: f=0.18mm/rev;
-speeds: $v_1$=1.3m/s; $v_2$=2.1m/s; $v_3$=3.3m/s.
In each test the cutting tool wear level was periodically submitted first to a classical control by profilometer and then to observation of rake face and flank by computer vision system. Each image of the cutting tool observed was digitized by a real time video digitizer board. Finally the image thus obtained was stored in an optical worm disk. With this technique one can always measure the flank wear and observe and check the crater dimensions.


## 3. ANALYSIS OF THE RESULTS

The most important observations is that in all tests carried out with AISI 310 a large groove at the end of the depth of cut immediately begins and rapidly grows, according to the cutting tool material. It could be thought as due to the abrasive effect of intermetallic phases on all the tool materials used.

During the tests it has been observed that the only material which has shown a crater and flank wear was type S, while the tool materials type C, F, W and Z have shown a large groove at the end of the cut. For this reason it has been decided to stop the tests just when the height of the primary groove reaches approximately 2.0mm.

Type S tool materials show a high level of flank wear, Vb, and a low level of groove, Vbn, at speed of 1.3m/s, Fig.1. With the increase of speed the flank wear increases. At 3,3 m/s the flank wear reaches a level of 2.1mm after 30" of cutting.

Type W tool material show a slow increase of groove and a very low flank wear at a lower speed, Fig.2. With the increase of speed, flank wear and groove grow more quickly. At the speed of 3.3m/s tool life is reduced to 210", with very little crater wear even at a higher speed.

Type F and Z tool materials show a very short life also at a lower speed. At 1.3m/s the groove was 2.2mm for F and 2.4mm for Z after 240" of cutting. The flank wear was very low at all speeds. At a higher speed both tool materials reached the maximum groove level after 30" .

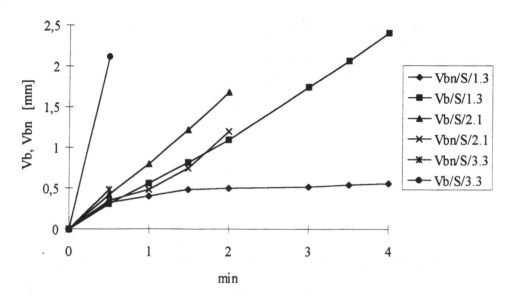

Fig. 1 - Wear of S-type tool vs cutting time.

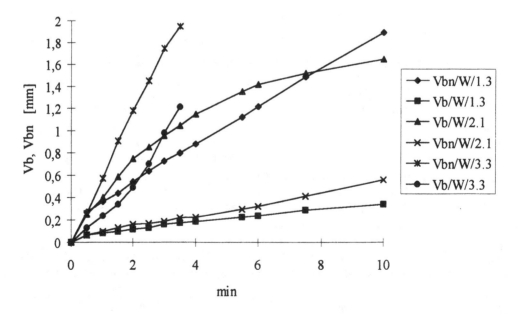

Fig. 2 - Wear of W-type tool vs cutting time.

Type C tool material, Fig.3, was very interesting at low and medium speed. At 1.3m/s the flank wear was 0.2mm after 2400" of cutting. After the same time the groove reached the level of 1.9mm. At the speed of 2.1m/s after 2400" of cutting the flank wear reached the level of 0.8mm and the groove the level of 1.5mm. At the speed of 3.3m/s the flank wear was predominant and after 400" of cutting reached the level of 1.1mm and the groove after the same time the level of 0.3mm.

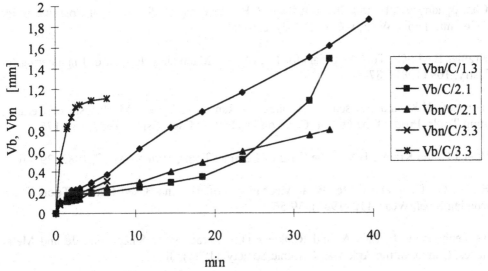

Fig. 3 - Wear of C-type tool vs cutting time.

During cutting the chip became hotter at its external extremity. This was probably due to the abrasive effect of intermetallic phases.
In the tests the limit imposed by regulations regarding flank wear and groove were exceeded. The work was continued until the workpiece was well finished.


## 4. CONCLUSIONS

After the tests carried out with ceramic tool materials on cutting AISI 310 steel we can conclude:
- type S tool material wears very quickly and is the only one which displays crater and flank wear;
- types F and Z tool materials have a very short life and only displays groove wear;
- type W tool material was very interesting at low and medium speeds;
- type C tool material was the most interesting because of its long life, 40min at.low speed and 35min at medium speed. At high speed the life shortens at 9min.
The tool materials type C, F, W and Z do not displays crater wear.

ACKNOWLEDGEMENTS

This work has been undertaken with financial support of Italian Ministry of University and Scientific and Technological Research.

REFERENCES

1. Chattopadhyay A.K. and Chattopadhyay A.B.: Wear and Performance of Coated Carbide and Ceramic Tools, Wear, vol. 80 (1982), 239-258.

2. Kramer B.M.: On Tool Materials for High Speed Machining, Journal of Engineering for Industry, 109 (1987), 87-91.

3. Tönshoff H.K. and Bartsch S.: Application Ranges and Wear Mechanism of Ceramic Cutting Tools, Proc. of the 6th Int. Conf. on Production Eng., Osaka, 1987, 167-175.

4. Huet J.F. and Kramer B.M.: The Wear of Ceramic Tools, 10th NAMRC, 1982, 297-304.

5. Brandt G.: Flank and Crater Wear Mechanisms of Alumina-Based Cutting Tools When Machining Steel, Wear, 112 (1986), 39-56.

6. Tennenhouse G.J., Ezis A. and Runkle F.D.: Interaction of Silicon Nitride and Metal Surfaces, Comm. of the American Ceramic Society, (1985), 30-31.

7. Billman E.R., Mehrotra P.K., Shuster A.F. and Beeghly C.W.: Machining with $Al_2O_3$-SiC Whiskers Cutting Tools, Ceram. Bull., 67 (1980) 6, 1016-1019.

8. Exner E.L., Jun C.K. and Moravansky L.L.: SiC Whisker Reinforced $Al_2O_3$-$ZrO_2$ Composites, Ceram. Eng. Sci. Proc., 9 (1988) 7-8, 597-602.

9. Greenleaf Corporation: WG-70 Phase Transformation Toughened Ceramic Inserts, Applications (1989), 2-3.

10. Wertheim R.: Introduction of $Si_3N_4$ (Silicon Nitride) and Cutting Materials Based on it, Meeting C-Group of CIRP, Palermo, (1985).

11. Lo Casto S., Lo Valvo E., Lucchini E., Maschio S., Micari F. and Ruisi V.F.: Wear Performance of Ceramic Cutting Tool Materials When Cutting Steel, Proc. of 7th Int. Conf. on Computer - Aided Production Engineering, Cookeville (U.S.A.), 1991, 25-36.

12. Lo Casto S., Lo Valvo E., Ruisi V.F., Lucchini E. and Maschio S.: Wear Mechanism of Ceramic Tools, Wear, 160 (1993), 227-235.

# TITANIUM ALLOY TURBINE BLADES MILLING
## WITH PCD CUTTER

**M. Beltrame**

**P. Rosa TBM, Maniago, Italy**

**E. Kuljanic**

**University of Udine, Udine, Italy**

**M. Fioretti**

**P. Rosa TBM, Maniago, Italy**

**F. Miani**

**University of Udine, Udine, Italy**

KEY WORDS: Diamond Machining, Titanium Alloys, Milling Blades PCD, Gas Turbine

ABSTRACT: Is milling of titanium alloys turbine blades possible with PCD (polycrystalline diamond) cutter and what surface roughness can be expected? In order to answer the question a basic consideration of diamond tools machining titanium alloys, chip formation and experimental results in milling of titanium alloy TiAl6V4 turbine blades are presented. The milling results of a "slim" turbine blade prove that milling with PCD cutter is possible. The tool wear could not be registered after more than 100 minutes of milling. The minimum surface roughness of the machined blade was $R_a = 0.89$ μm. Better results are obtained when wet milling has been performed. Therefore, finishing milling of titanium alloy TiAl6V4 turbine blades with PCD cutter is promising.

## 1. INTRODUCTION

Contemporary technology relies much on the exploitation of new and advanced materials. Progress in Materials Science and Technology yields year by year new applications for new materials. The field of gas turbine materials has experienced the introduction of several advanced materials [1] for both the compressor and the turbine blades: respectively titanium and nickel based alloys have met thorough industrial success. Compressor blades are used with high rotational speeds; materials with high Young modulus $E$ and low density are required to obtain a high specific modulus, which is the ratio of the two and is one of the

Published in: E. Kuljanic (Ed.) *Advanced Manufacturing Systems and Technology*,
CISM Courses and Lectures No. 372, Springer Verlag, Wien New York, 1996.

key factors in controlling the rotational resonance. TiAl6V4 (IMI 318), an alloy with a mixed structure α (hexagonal close packed) and β (body centered cubic) with a room temperature proof stress [2] of 925 MPa and a relative density of 4.46 kg/dm$^3$ is now almost universally used for blades operating up to 350 °C.

Titanium alloys are generally machined with uncoated carbides tools at speeds that have been increased in the last decades much less than the ones employed in steel cutting. A possibility to apply PCD tools in turning for titanium – based alloys is presented in [3]. As far as the authors know there are no publications on titanium alloys turbine blades milling with PCD cutter.

Is milling of titanium alloys turbine blades possible with PCD (polycrystalline diamond) cutter and what surface roughness can be obtained? To answer the question we will present some basic considerations of diamond tools machining titanium alloys, chip formation and experimental results in milling of titanium alloy TiAl6V4 turbine blades, obtained in Pietro Rosa T.B.M., a leader in manufacturing compressor gas and steam turbine blades.

## 2. BASIC CONSIDERATIONS OF DIAMOND TOOLS MACHINING TITANIUM

Cutting forces in titanium machining are comparable to those required for steels with similar mechanical strength [4]; however, the thermal conductivity, comparing to the same class of materials, is just one sixth. A disadvantage is that the typical shape of the chip allows only a small surface area contact. These conditions cause an increase in the tool edge temperature. Relative machining times increase more than proportionally than Brinell hardness in shifting from the pure metal to α alloy to α/β to β alloys as in the following table [5]:

Table 1. Ratio of machining times for various titanium alloys

| Titanium alloy | Brinell Hardness | Turning WC Tools | Face Milling WC Tools | Drilling HSS Tools |
|---|---|---|---|---|
| Pure metal Ti | 175 | 0.7 | 1.4 | 0.7 |
| Near α TiAl8Mo1V1 | 300 | 1.4 | 2.5 | 1 |
| α/β TiAl6V4 | 350 | 2.5 | 3.3 | 1.7 |
| β TiV13Cr11Al3 | 400 | 5 | 10 | 10 |

In roughing of titanium alloys with a 4 mm depth of cut and feed of 0.2 mm/rev, the cutting speeds are influenced not only by the hardness but also by the workpiece material structure, as seen in the Figure 1.

Kramer et al. [6, 7] have made an extensive analysis of the possible requirements for improved tool materials that should be considered in titanium machining. In such an interesting analysis a tool material should:

- promote a strong interfacial bonding between the tool and the chip to create seizure conditions at the chip–tool interface,

- have low chemical solubility in titanium to reduce the diffusion flux of tool constituents into the chip,
- have sufficient hardness and mechanical strength to maintain its physical integrity.

Polycrystalline diamond (PCD) [8] possesses all these requirements. The heat of formation of TiC is among the highest of all the carbides [9] (185 kJ/mol), the chemical solubility is low, even if not negligible (1.1 atomic percent in α Ti and 0.6 atomic percent in β Ti), and comparing it with single crystal diamond, has indeed enough hardness, along with a superior mechanical toughness. PCD is thus a material worth of being considered for machining titanium alloys, if correct cutting conditions are chosen. The correct cutting conditions can be found out only by experiments.

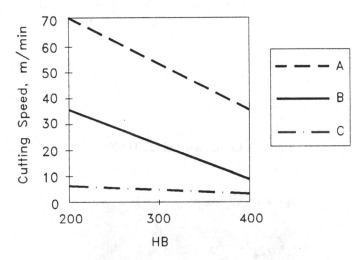

Figure 1. Titanium roughing with a 4 mm depth of cut and feed of 0.2 mm/rev.
A – α WC tools, B – α and α+β HSS tools, C – β HSS tools

## 3. EXPERIMENTAL APPARATUS AND PROCEDURE

### 3.1. Workpiece and Workpiece Material

The workpiece is a compressor blade of a gas turbine, Figure 2. Such a "slim" blade was chosen on purpose to have an extreme low stiffness of the machining system. The effect of stiffness of machining system on tool wear in milling was considered in [10].
The material of the workpiece is TiAl6V4 titanium alloy heat treated HB400 usually used for turbine construction.

### 3.2. Machine Tool and Tool

The milling experiments were performed on a CNC five axis milling machine, at Pietro Rosa facilities in Maniago, $P = 16$ kW, with Walter end milling cutter, 32 mm diameter and 3 inserts PCD (Figure 3).

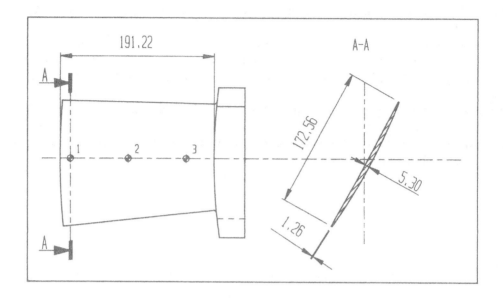

Figure 2. Compressor blade TiAl6V4

Figure 3. End Milling Cutter – 3 inserts PCD

## 3.3. Experimental Conditions

The pilot tests were performed to determine the adequate experimental cutting conditions. Finishing milling was done at constant cutting conditions: cutting speed $v_c$ = 110 m/min, feed per tooth $f_z$ = 0.135 mm, and depth of cut $a_p$ = 0.2 mm. The experiments were performed dry and wet. The coolant was the solution of 7% Cincinnati Milacron NB 602 and water.

## 4. EXPERIMENTAL RESULTS AND DISCUSSION

The stiffness of the workpiece, as well as of the machining system was extremely low in order to be able to answer the former question.

### 4.1. Chip Formation

A typical characteristic of chip formation in machining of titanium alloys is the formation of lamellar chip. This can be seen in Figure 4.

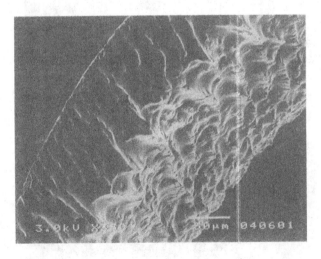

Figure 4. SEM chip micrograph of TiAl6V4 in milling

Such lamellar chip formation causes the change of cutting force and thermal stress periodically as a function of cutting time. This holds true for continuous and interrupted cutting, for example, for turning and milling respectively. However, there is a sudden increase of cutting force and temperature in milling, when the tooth enters the workpiece. A sudden decrease of the cutting force occurs at the tooth exit. Furthermore, a strong thermal stress is present when cooling is applied. Therefore, it is hard to find a tool material to meet the requirements for low tool wear and cutting edge chipping.

## 4.2. Tool Wear

An investigation of tool wear in milling was done in [11]. The characteristics of diamond tools are high hardness and wear resistance, low friction coefficient, low thermal expansion and good thermal conductivity [12].

In these experiments the crater or flank wear was not observed after 108 minutes of dry milling, Figure 5. The same results were obtained, Figure 6, in wet milling at the same cutting conditions.

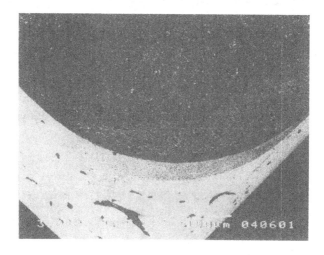

Figure 5. Cutting edge after 108 minutes of dry milling

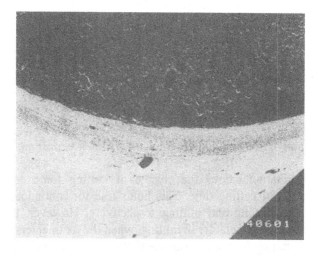

Figure 6. Cutting edge after 108 minutes of wet milling

There is no difference between new cutting edge and even after 108 minutes of dry or wet milling. There is an explanation for such a behavior of PCD tool when turning titanium alloys [3]. The formation of titanium carbide reaction film on the diamond tool surface protects the tool particularly of the crater wear. Further work should be done for better understanding of this phenomenon.

## 4.3. Surface Roughness

Surface roughness is one of the main features in finishing operations. The surface roughness was measured at three points: 1, 2 and 3 on both sides of the blade, Figure 2. The minimum value of surface roughness was $R_a = 0.89$ µm measured in feed direction, and the average value was $R_a = 1.3$ µm in both dry and wet milling. It can be seen that the obtained surface roughness is low for such a "slim" workpiece and for milling operation.

## 5. CONCLUSION

Based on the results and considerations presented in this paper, we may draw some conclusions about milling of titanium alloy turbine blades with PCD cutter. The answer to the question raised at the beginning, whether milling of titanium alloys turbine blades may be performed with PCD (polycrystalline diamond) cutter, is positive.
The crater of flank wear of PCD cutter does not occur after 108 minutes of milling.
The minimum surface roughness of the machined surface is $R_a = 0.89$ µm, and an average value is $R_a = 1.3$ µm measured in feed direction.
Milling of TiAl6V4 with PCD cutter could be done dry or wet. However, it is better to apply a coolant.
In accordance with the presented results, milling of titanium based alloy TiAl6V4 blade with PCD cutter is suitable for finishing operation. This research is to be continued.

## ACKNOWLEDGMENTS

The authors would like to express their gratitude to Mr. S. Villa, Technical Manager of WALTER – Italy. This work was performed under sponsorship of WALTER Company.

## REFERENCES

1. Duncan, R.M., Blenkinsop, P.A., Goosey, R.E.: Titanium Alloys in Meetham, G.W. (editor): The Development of Gas Turbine Materials, Applied Science Publishers, London,1981

2. Polmear, I.J.: Light Alloys, Metallurgy of the Light Metals, Arnold, London, 1995

3.   Klocke, F., König, W., Gerschwiler, K.: Advanced Machining of Titanium and Nickel–Based Alloys, Proc. 4th Int. Conf. on Advanced Manufacturing Systems and Technology AMST'96, Udine, Springer Verlag, Wien, N.Y., 1996

4.   Chandler, H.E.: Machining of Reactive Metals, Metals Handbook, Ninth Edition, ASM, Metals Park Ohio, 16(1983)

5.   Zlatin, N., Field, M.: Titanium Science and Technology, Jaffee, R.I., Burte, H.M., Editors, Plenum Press, New York, 1973

6.   Kramer, B.M., Viens, D., Chin, S.: Theoretical Considerations of Rare Earth Compounds as Tool Materials for Titanium Machining, Annals of the CIRP, 42(1993)1, 111-114

7.   Hartung, P.D., Kramer, B.M.: Tool Wear in Titanium Machining, Annals of the CIRP, 31(1982)1, 75-79

8.   Wilks, J., Wilks, E.: Properties and Applications of Diamond, Butterworth Heinemann, Oxford, 1994

9.   Toth, L.E.: Transition Metal Carbides and Nitrides, Academic Press, New York, 1971

10.  Kuljanic, E.: Effect of Stiffness on Tool Wear and New Tool Life Equation, Journal of Engineering for Industry, Transaction of the ASME, Ser. B, (1975)9, 939-944

11.  Kuljanic, E.: An Investigation of Wear in Single-tooth and Multi-tooth Milling, Int. J. Mach. Tool Des. Res., Pergamon Press, 14(1974), 95-109

12.  König, W., Neise, A.: Turning TiAl6V4 with PCD, IDR 2(1993), 85-88

# TRANSFER FUNCTION OF CUTTING PROCESS BASED ON OUTPUT INPUT ENERGY RELATIONS

**S. Dolinsek**
University of Ljubljana, Ljubljana, Slovenia

KEY WORDS: Cutting process, Identification, Transfer Function, Tool wear

ABSTRACT: In the following paper some results of the on-line identification of the cutting process in the macro level of orthogonal turning, are presented. The process is described by the estimation of the transfer function, defined by output-input energy ratios. The estimated parameters of the transfer function (gain, damping) vary significantly with different tool wears and provide a possibility for effective and reliable adaptive control.

## 1. INTRODUCTION

Demands on machining cost reduction (minimization of the operators assistance and production times) and improvements in product quality are closely connected with the successful monitoring of the cutting process. Thus, building up an efficient method for on-line tool condition monitoring is no doubt an important issue and of great interest in the development of fully automated machining systems. In detail, we describe a reliable and continuous diagnosis of the machining process (tool failures, different tool wears and chip shapes), observed under different machining conditions and applied in practical manufacturing environments. A great effort has been was spent during the last decade in researching and introducing different applications of tool monitoring techniques [1]. Numerous research works have addressed these questions, related to the complexity of the

Published in: E. Kuljanic (Ed.) *Advanced Manufacturing Systems and Technology*, CISM Courses and Lectures No. 372, Springer Verlag, Wien New York, 1996.

cutting process, but marketable monitoring applications are still too expensive and unreliable. They are more useful in tool condition monitoring techniques. The completed monitoring system usually consists of sensing, signal processing and decision making. According to different approaches to monitoring problems, methods can be divided into two categories: model based and feature based methods. A comprehensive description of different methods is depicted in Fig 1. [2]. The most widely used are features based methods, where we can observe some features, extracted from sensor signals to identify different process conditions. In model based methods sensor signals are outputs of the process, which is modeled as a complex dynamic system. These methods consider the physics and complexity of the system and they are the only alternative in modeling a machining system as a part of the complex manufacturing system [3]. However, they have some limitations, real processes are nonlinear, time invariant and difficult for modeling.

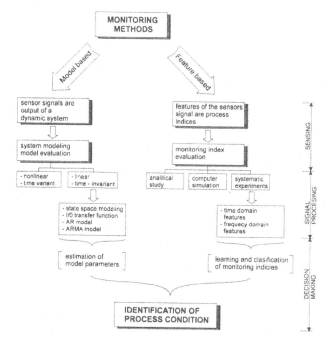

Fig.1.: Monitoring methods divided according to different research approaches.

One of the model based methods is the energy based model, proposed by Peklenik and Mosedale [4]. They introduced the stochastic time series energy model, which treats the cutting process as a closed loop. An analytical description was made on the basis of estimates of the energy average values, variances, autocorrelation functions and Furrier transformations to power spectra. The identification of dynamic structure of the cutting process can be determined by on-line estimation of the output-input energy time series and their transfer function [5]. The latest research was possible due to fast developments of sensor and signal processing techniques, so new results at the micro [6] and macro level [7] of the cutting process were obtained.

## 2. ENERGY MODEL AND TRANSFER FUNCTION OF THE CUTTING PROCESS

The classic energy model for orthogonal cutting, proposed by Merchant [8], expresses the input energy of the process as a sum of transformation and output energies. Input energy is used for the transformation of workpiece material to chip, to overcome the chip and tool friction and also for the chip acceleration and new surface formation. Regarding the letter the kinetic and surface energies are normally neglected, but they should be considered in high speed cutting. According to the presentation of stochastic character of energies, the extended energy model was proposed in the form of a time series [4]:

$$U_i(t) = U_t(t) + U_o(t) \tag{1}$$

For the orthogonal cutting model, presented in Fig.2., the input and output energies are expressed in the form of their time series parameters [5]:

$$U_i(t) = F_z(t)\left[v_i(t) \pm \dot{z}(t)\right] \tag{2}$$

$$U_o(t) = F_y(t)\left[v_o(t) \pm \dot{y}(t)\right] \tag{3}$$

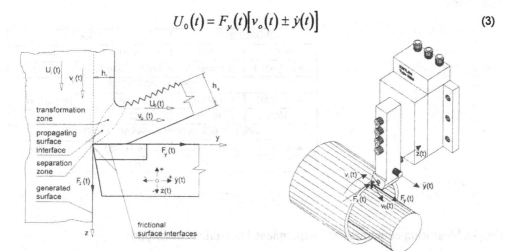

Fig 2.: Energy model of cutting process [6] and practical orthogonal implementation [7].

Fig.2. also shows the practical solution of orthogonal cutting in the case of side turning of the tube and the necessary measuring points to access input-output parameters in energy equations. The cutting process can be described in this way by on-line estimation of the input output energies and their spectral estimations. The transfer function is defined as follows [8,6]:

$$\hat{H}(f) = \frac{\hat{G}_{U_i U_o}(f)}{\hat{G}_{U_i U_i}(f)} = \left|\hat{H}(f)\right| e^{-j\phi(f)} \tag{4}$$

At the transfer function equation, $Gu_iu_o$ represents the cross power spectrum estimate between the input and output energies and $Gu_iu_i$ the input energy power spectrum estimate. An estimated transfer function could be described in the form of its parameters, gain (amplitude relationship) and damping factors ( impulse response).

## 3. EXPERIMENTAL SET UP DESCRIPTION AND TIME SERIES ENERGY ASSESSMENTS

For the verification of the presented model, it is necessary to build-up a proper machining and measuring system. The sensing system for accessing the parameters in energy equations consists of a force sensor, cutting edge acceleration ( velocity displacement ) sensors and a cutting speed sensor. With their characteristics, they do not interfere within the studied frequency range of the cutting process. The greatest problem exists in measuring the chip flow speed. On-line possibilities have so far not been materialized so that the speeds had to be defined from interrelations between chip thickness and cutting speeds. To record all measured parameters simultaneously in real time, a sophisticated measuring system was used. Fig. 3. shows the basic parts of equipment for signal processing and also a description of workpiece material, cutting tool geometry and the range of selection in the cutting conditions.

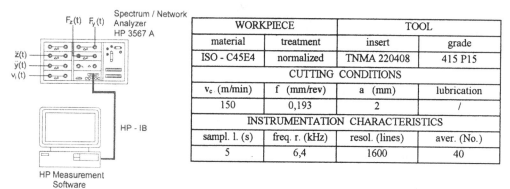

| WORKPIECE | | TOOL | |
|-----------|-----------|-----------|-----------|
| material | treatment | insert | grade |
| ISO - C45E4 | normalized | TNMA 220408 | 415 P15 |
| CUTTING  CONDITIONS | | | |
| $v_c$ (m/min) | f  (mm/rev) | a  (mm) | lubrication |
| 150 | 0,193 | 2 | / |
| INSTRUMENTATION  CHARACTERISTICS | | | |
| sampl. l. (s) | freq. r. (kHz) | resol. (lines) | aver. (No.) |
| 5 | 6,4 | 1600 | 40 |

Fig. 3.: Measuring equipment and experimental conditions description.

The machining system (machine tool and tool holder) should ensure well-known and unchangeable characteristics throughout a whole range of applied, realistically selected cutting parameters. Suitable static and dynamic characteristics of the machine tool need not be defined since for the verification of the cutting model on the macro level it is enough if the process is observed only at the cutting point and all the necessary characteristics are defined in accordance with this point. For a clear explanation and presentation of the structure dynamics of the cutting tip, the dynamic characteristics of the particular parts and responses of the assembled cutting tip (tool holder-dynamometer-machine tool) were first defined using widely known model testing methods [10]. A comparison between the frequency responses at the cutting tip with the frequency analysis of the measured parameters of the process ( power spectrum of the cutting force and displacement speed ) is shown in Fig. 4. From the above study we can conclude that the energy of the cutting process is, in the case of the real turning process mainly, distributed in the range of the natural frequencies of the cutting tool tip.

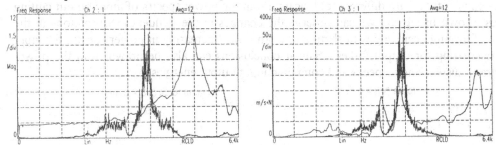

Fig 4: Power spectra of force and displacement velocity plotted in comparison of tool-tip modal characteristics in input direction.

The fluctuations of input-output energies are determined from real time series records of measuring parameters in energy equations. Fig. 5. shows an example of time series records of the input parameters ( cutting speed, acceleration, computed displacement speed, the difference between the cutting and displacement speeds and the input force) and a calculated time series record of the input energy. Similar results have also been obtained for the output energies. An analysis of stochastic time series records of cutting forces and displacement velocities signals shows stability, normality and sufficient reproducibility of measuring results. Changes in cutting conditions significantly influence the static and dynamic characteristics of the parameters in the energy equations.

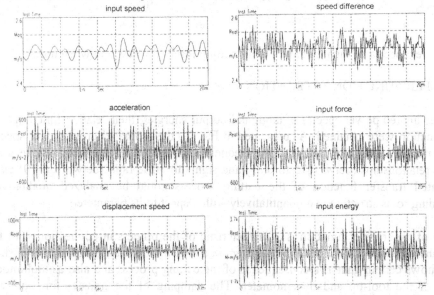

Fig. 5.: Time series of the measured parameters and input energy evaluation.

A power spectrum analysis of energies shows the distribution of the spectrum corresponding to input or output forces and velocity of tool displacements. Also the autocorrelation analysis expresses a certain periodicity in energies, detailed results have already been presented [11]. Interesting conclusions can be drawn from these results, obtained in cutting with different tool wears. From the power spectra of input output cutting forces, shown in Fig. 6., we can observe the changes in the frequency distributions in the event of turning occurring with sharp and worn tools (increasing in power spectra for input and decreasing for output forces). Similar conclusions can also be drawn from the power spectra of displacement speeds and computed input output energies.

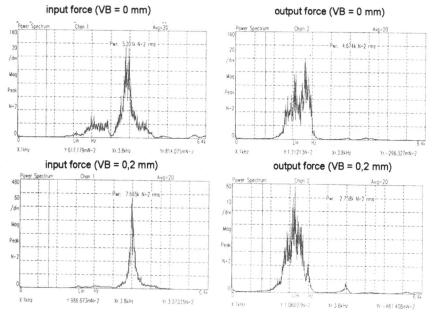

Fig. 6.: Power spectra of the input and output cutting forces for new and worn tool.

## 4. TRANSFER FUNCTION ESTIMATION

From the estimated power spectra of input and output energies, their relation functions and transfer function were obtained. As presented in Fig. 7., the estimated cross power spectra show a common signal component in the frequency range of 2 to 2,5 kHz, where good coherence relationships (betw. 0,75 to 0,85) and signal to noise ratio (betw. 5 to 10) exist. An estimated transfer function of the cutting process could be analyzed qualitatively corresponding to its structure and quantitatively with respect to its parameters.

The shape of the transfer function is a characteristic of the process in connection with the structure characteristics of the machining system in a closed loop. In amplitude relationships its shape shows certain gain as a consequence of the cutting process and also as multimodal responses of the tool-tip and dynamometer. The damping of the transfer function was obtained from its impulse response, which is a damped one - sided sine wave.

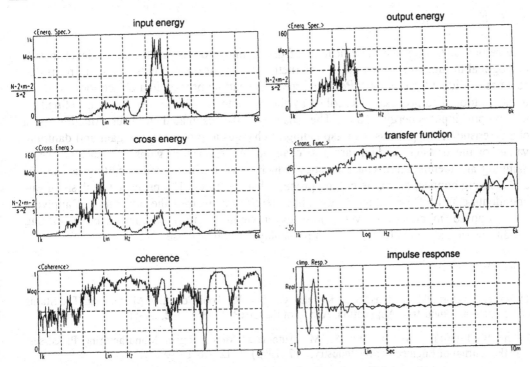

Fig. 7.: Estimations of the power spectra and transfer function of the cutting process.

In the region of a strong connection between energies we can locate changes in the gain and damping of the transfer function as an influence of cutting conditions [11]. These parameters could therefore be a basis for the identification of the process and a criterion for on-line adaptation of cutting conditions according to optimal cutting circumstances[12]. Tool wear is certainly one of the most unfavorable phenomena in cutting processes. Research on turning with a worn tool indicated an increase at power spectra of input energy and decrease of output energy. The most significant changes are in the transfer function parameters, while the shape remains unchangeable. Fig. 8. indicates the influence of different tool wears on the parameters of the estimated transfer function. Decreasing in gain and increasing in damping are the identification characteristics which confirm unfavorable cutting conditions and accuracy of the identification process.

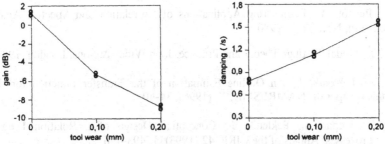

Fig. 8. Gain and damping values of the transfer function for different tool wears.

## 5. CONCLUSIONS

The results of the on-line identification of the cutting process on the macro level in side orthogonal turning are presented. A cybernetic concept of the machining system, proposed as a basis of identification by J. Peklenik, treats the cutting process in a closed loop with the machine tool. The process is described by the estimate of the transfer function defined by the output input energy ratios. The shape of the estimated transfer function is a characteristic of the process. Corresponding to changes in tool wear, the gain and damping values of the transfer function are also changed. A decrease in gains and an increase in damping are identification characteristics, which show the unfavorable cutting conditions and a strong connection to the cutting process characteristics. The proposed analysis of on-line identification of the cutting process has therefore confirmed the possibility of applying the proposed models, however, more experimental verifications in different cutting conditions should be made to provide a possibility for practical use.

## 6. REFERENCES

1. Byrne, G., Dornfeld, D., Inasaki, I., Ketteler, G., Teti, R.: Tool Condition Monitoring, The Status of Research and Industrial Application, Annals of the CIRP, 44 (1995), 2, 24-41

2. Du, R., Elbestawi, M., A., Wu, S., M.: Automated Monitoring of Manufacturing Processes, ASME, Journal of Engineering for Industry, 117 (1995), 3, 121- 141

3. Serra, R., Zanarini, G.: Complex Systems and Cognitive Processes, Sprin. - Verlag, Berlin, 1990

4. Peklenik, J., Mosedale, T.: A Statistical Analysis of the Cutting System Based on an Energy Principle, Proc. of the 8th Intern. MTDR Conference, Manchester, 1967, 209-231

5. Mosedale, T.W., Peklenik, J.: An Analysis of the Transient Cutting Energies and the Behavior of the Metal-Cutting System using Correlation Techniques, Adv. in Manuf. Sys., 19, 1971, 111-141

6. Peklenik, J., Jerele, A.: Some Basic Relationship for Identification of the Machining Process, Annals of the CIRP,41(1992), 1, 129 -136

7. Dolinsek, S.: On-line Cutting Process Identification on Macro Level, Ph.D.thesis, University of Ljubljana, 1995

8. Merchant, M., E.: Mechanics of the Metal Cutting Process, Journal of Applied Physics, 16 (1945), 3, 267 - 275

9. Bendat, J., Piersol, A.: Engineering Applications of Correlation and Spectral Analysis, John Willey and Sons Ltd, New York, 1980

10. Ewins, D., J.: Modal Testing- Theory and Practice, John Willey & Sons, London, 1984

11. Dolinsek, S., Peklenik, J.: An On-line Estimation of the Transfer Function for the Cutting Process, Technical paper of NAMRI/SME, 27 (1996), 34-40

12. Kastelic, S., Kopac, J., Peklenik, J.: Conceptual Design of a Relation Data Base for Manufacturing Processes, Annals of the CIRP, 42 (1993), 1, 493-496

# A NEURAL NETWORK ARCHITECTURE FOR TOOL WEAR DETECTION THROUGH DIGITAL CAMERA OBSERVATIONS

C. Giardini, E. Ceretti and G. Maccarini
University of Brescia, Brescia, Italy

KEY WORDS : Cutting operations, Tool replacement, Image recognition, Neural Network.

ABSTRACT : In flexible manufacturing systems with unmanned machining operations, one of the most important issues is to control tool wear growth in order to identify when the tool needs to be replaced. Tool monitoring systems can be divided into on-line and off-line methods. The authors have already conducted both on-line and off-line analyses. The simplest way to check the tool status is to measure either the flank and the crater wear levels or the presence of a cutter breakage. This task, which can be easily performed off-line by the operator, gives a lot of problems when it is conducted automatically on-line. In the present paper a neural network for image recognition is applied for the wear level detection. The network receives as input an image of the tool, acquired by a digital camera mounted near the machine tool storage, and provides a binary output which indicates whether the tool can continue to work.

## 1. INTRODUCTION

In industrial applications, where flexible manufacturing systems are employed, one of the most important tasks is to control the tool status in order to replace it as it loses its cutting capability. Amongst the various methods of controlling the tool status, a subdivision can be made between *on-line* and *off-line* methods.

Published in: E. Kuljanic (Ed.) *Advanced Manufacturing Systems and Technology*, CISM Courses and Lectures No. 372, Springer Verlag, Wien New York, 1996.

The Authors have already conducted both on-line (related to machine tool vibrations) [1] and off-line (related to direct measurement of the flank wear by means of a microscope) analyses showing how, sometimes, it is rather difficult to correctly correlate the actual tool status with the physical variables chosen to monitor the system. The simplest way to conduct off-line check of tool status is to measure either the flank and the crater wear levels (see Figure 2) or to evidence the presence of a cutter breakage. This operation gives a lot of problems when it is conducted automatically on-line [2, 3, 4, 5].

A suitable mechanism for automatic recognition of the tool wear level can be found applying a neural network trained to perform image recognition. The neural network proposed by the authors is a multi-layer one where both the input nodes (518) and the second layer (37) perform non-linear operations.

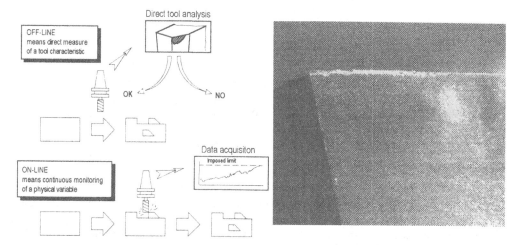

Figure 1 - On-line and off-line tool controls          *Figure 2 - Flank wear on tool edge*

## 2. NEURAL NETWORKS

The field of Artificial Neural Networks has received a great deal of attention in recent years from the academic community and industry. Nowadays A.N.N. applications are found in a wide range of areas, including Control Systems, Medicine, Communications, Cognitive Sciences, Linguistic, Robotics, Physics and Economics [6].

As a definition we can say that A.N.N. is a parallel, distributed information processing structure consisting of processing elements (which can possess a local memory and can carry out localised information processing operations) interconnected via unidirectional signal channels called *connections*. Each processing element (called *neuron*) has a single output connection that branches into as many collateral connections as desired. Each of them carries the same signal, which is the processing element output signal. The processing element output signal can be of any mathematical type desired. The information within each element is treated with a restriction that must be completely local : this means that it depends only on the current values of the input signal arriving at the processing element via incoming connections and on the values stored in the processing element's local memory.

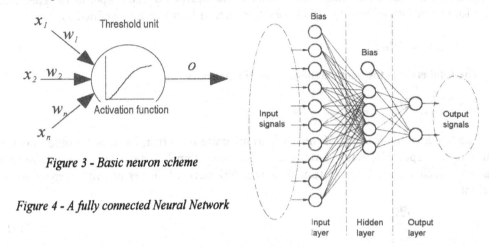

Figure 3 - Basic neuron scheme

Figure 4 - A fully connected Neural Network

As opposed to conventional programmed computing, where a series of explicit commands (a *programme*) processes information on an associated database, A.N.N. develops information processing capabilities by learning from examples. Learning techniques can be roughly divided into two categories :

⇒ supervised learning

⇒ self-organising or unsupervised learning

Both of them require a set of examples for which the desired response is known. The learning process consists in adapting the network in such a way that it will produce the correct response for the whole set of examples. The resulting network should be able to generalise (i.e. give a good response) even for cases not found in the set of examples. In the case of supervised learning A.N.N. the network topology is selected before the learning algorithm starts and remains fixed. On the contrary, in the second the network topology is modified as the learning process continues adding new elements (neurons).

The A.N.N. we refer to is a supervised learning type. The adopted algorithm is the Back-propagation one.

## 3. THE BACK-PROPAGATION ALGORITHM

The activation function for the basic processing unit (neuron) in a B.P. network is the *sigmoid function* (please refer to Figure 3):

$$o = f_s(net) = \frac{1}{1+e^{-net}} \text{ where } net = \sum_{i=1}^{n} w_i \cdot x_i - \theta \tag{1}$$

where $\theta$ is the *bias unit* input and $w_i$ are normally called *weights*.

The unknowns in this expression are represented by $w_i$ and $\theta$ for each neuron in each layer. The error of the network can be evaluated comparing the outputs obtained by the net ($o_{po}$) with the desired ones ($d_{po}$) extended to all the neurons of the output layer. This means that for one of the *pattern* (p) used to teach the net, the error can be defined as:

$$E_p = \sum_o \left(d_{po} - o_{po}\right)^2 \tag{2}$$

The total error is the sum extended to all the patterns :

$$E = \sum_p \frac{1}{2} \sum_o \left(d_{po} - o_{po}\right)^2 = \sum_p \frac{1}{2} E_p \tag{3}$$

The target of the learning of the net is to minimise this error, i.e. it is possible to correct the weights repeatedly until convergence. The algorithm [7] states that the change to be made to each weight is proportional to the derivative of the error with respect to that weight :

$$\Delta_p w_{ij} = -\eta \cdot \frac{\partial E_p}{\partial w_{ij}} \tag{4}$$

where $w_{ij}$ is the weight between the output of unit $i$ and unit $j$ in the next layer. $\eta$ represents the *learning rate* of the process.

It is our intent to use such a network (once adequately taught) to determine automatic recognition of the tool status.

## 4. THE ARCHITECTURE OF THE SYSTEM DEVELOPED

The system developed is schematically represented in Figure 5.

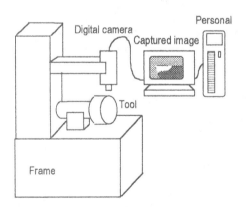

*Figure 5 - Schematics of the system developed*

In this system a digital camera, placed near the working machine (mill), takes the tool edge image and stores it in a binary file. The tool is carried onto a suitable frame in order to easily reproduce the cutter placement with respect to the camera.

The procedure is subdivided into two different moments: In the first the images saved are collected together with the tool status (received by the operator) in order to prepare a suitable set of patterns to teach the net. In the second phase the net automatically decides if the tool can continue to work or must be substitued.

## 5.  THE STEPS FOLLOWED

The digital images are processed according to the following steps (the software has been developed by the authors) :

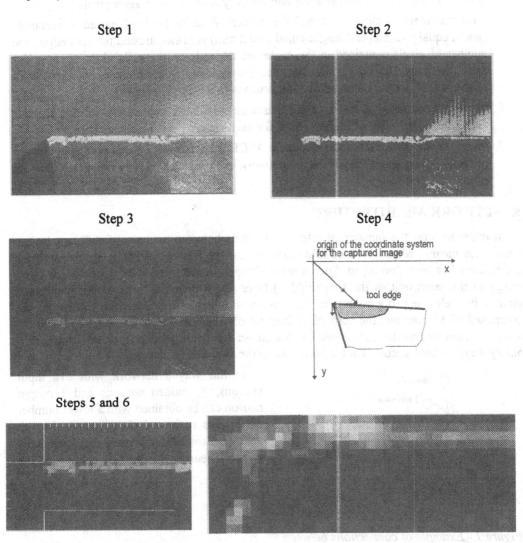

Step 1                                                    Step 2

Step 3                                                    Step 4

origin of the coordinate system
for the captured image

tool edge

Steps 5 and 6

*Figure 6 - Different steps of the image treatment*

1. the original digital image is recorded on the hard disk of the personal computer ; the images are stored using 256 grey levels ;

2. the maximum value of the grey level of the background is used in order to obtain a black background ; the mean grey level is also calculated ;

3. all the non-black pixel modify their grey level in order to obtain a mean value equal to 127 (256/2 - 1) ;

4. a suitable procedure (starting from the origin of the co-ordinate system) finds the tool edge and then the tool corner analysing the grey level of the image's pixels ;

5. once the image is cleaned up and the corner of the cutter is identified, a rectangle (width equal to 2 mm and height equal to 0.2 mm) is drawn around the tool edge ; the number of pixels contained in this rectangle depends on the scale along the x and y axes : there are 185 x 70 pixels; this means that, using all these pixels, the input layer of the A.N.N. would consist of 12950 neurons (a too large number) ;

6. the pixels contained in this area are therefore tessellated into squares of 5 x 5 pixels taking the mean value of the grey level for each square ; the tassels are then 37 x 14 ;

7. these grey levels are stored in a suitable ASCII file which represents one of the inputs of the A.N.N. ; using these files the network can be taught or can recognise the tool status.

## 6. NETWORK ARCHITECTURE

In order to limit the number of inter-connections between neurons of different layers, which also means reducing the calculus time during the training phase, a special network architecture has been developed. Such a network operates on small sub regions of the input image so that each unit of the first (hidden) layer is connected to a restricted number of input units, belonging to a column of the tessellated image. This means the hidden layer is composed of 37 neurons, the activation function of which is a sigmoid. All these neurons are then connected to the output one (with sigmoid activation function) which furnishes (in binary way) the tool status : 1 for a good tool, 0 for a tool to be changed.

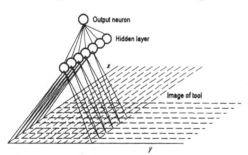

In this way a network with 518 input neurons, 37 hidden neurons and 1 output neuron can be obtained with a lower number of weights to be calculated, since the $i$-th neuron of the hidden layer is only connected to the $j$ neurons of the $i$-th column of the input grid.

*Figure 7 - Example of connections between the various layers*

## 7. RESULTS OBTAINED

Once the network architecture was decided, it was tested with actual cases taken from sampled images. Some of them have been used in the training phase and others in the validation phase. In particular the network has been trained using *518* input neurons, *37* hidden neurons, *1* output neuron, *24* samples of both good and worn tools, a learning rate equal to *0.5* and a maximum acceptable error equal to *0.01*. In order to facilitate the network convergence a suitable procedure was adopted. More specifically the network is trained using only one set of the samples (5 at the beginning) and then (once convergence occurred) a new image is put in the sample set which is used for a new convergence, and so on until all the images are considered. The convergence of the network was obtained within 20 minutes. Figure 8 shows as an example one of the input images and the weights distribution between the input and the hidden layers after the training phase.

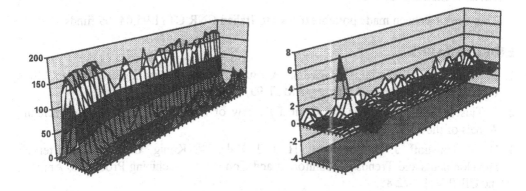

*Figure 8 - An example of the grey level for one of the sampled images and the weight distribution for the trained network*

To test the network's capacity for recognising tool status, 28 other images were experimented. The answers of the network are shown in Table 1. Only in one case did the network give a wrong answer and in two cases it was not sure of the tool status.

| | Tool status | A N N output | which means |
|---|---|---|---|
| 1 | GOOD | 0.455 | NOT SURE |
| 2 | GOOD | 0.999 | GOOD |
| 3 | GOOD | 0.996 | GOOD |
| 4 | GOOD | 0.989 | GOOD |
| 5 | GOOD | 0.536 | NOT SURE |
| 6 | WORN | 0.001 | WORN |
| 7 | WORN | 0.171 | WORN |
| 8 | GOOD | 0.999 | GOOD |
| 9 | GOOD | 0.999 | GOOD |
| 10 | GOOD | 1.000 | GOOD |
| 11 | GOOD | 1.000 | GOOD |
| 12 | GOOD | 0.999 | GOOD |
| 13 | GOOD | 0.999 | GOOD |
| 14 | WORN | 0.013 | WORN |

| | Tool status | A N N output | which means |
|---|---|---|---|
| 15 | WORN | 0.000 | WORN |
| 16 | GOOD | 0.999 | GOOD |
| 17 | GOOD | 0.999 | GOOD |
| 18 | GOOD | 0.999 | GOOD |
| 19 | GOOD | 0.999 | GOOD |
| 20 | GOOD | 0.957 | GOOD |
| 21 | GOOD | 0.995 | GOOD |
| 22 | WORN | 0.841 | GOOD |
| 23 | GOOD | 0.703 | GOOD |
| 24 | GOOD | 0.999 | GOOD |
| 25 | GOOD | 0.999 | GOOD |
| 26 | GOOD | 0.999 | GOOD |
| 27 | GOOD | 0.903 | GOOD |
| 28 | GOOD | 0.979 | GOOD |

*Table 1 - Output of the trained network during the test phase*

## 8. CONCLUSIONS

The present work has shown how it is possible to identify the tool status by means of off-line analysis using a suitably trained neural network. The architecture proposed here has lead to a small number of interconnections (weights) even though the input layer has a large number of neurons. The procedure of increasing the number of images in the training set has also facilitated the network convergence reducing the calculus time. Once the network is trained, the answer to a new image is given in real time.

## ACKNOWLEDGEMENTS

This work has been made possible thanks to Italian CNR CT11 95.04109 funds.

## BIBLIOGRAPHY

1. E. Ceretti, G. Maccarini, C. Giardini, A. Bugini: Tool Monitoring Systems in Milling Operations : Experimental Results, AMST 93, Udine April 1993.

2. J. Tlusty, G. C. Andrrews : A Critical Review of Sensors for Unmanned Machining, Annals of the CIRP Vol. 32/2/1983.

3. H. K. Tonshoff, J. P. Wulsfberg, H. J. J. Kals, W. Konig, C. A. Van Luttervelt : Developments and Trends in Monitoring and Control of Machining Processes, Annals of the CIRP Vol 37/2/88.

4. J. H. Tarn, M. Tomizuka : On-line Monitoring of Tool and Cutting Conditions in Milling, Transaction of the ASME Vol. 111, August 1989.

5. C. Harris, C. Crede : Shock and Vibration Handbook, Vol. 1 Mc Grow Hill , England 1961.

6. Toshio Teshima, Toshiroh Shibasaka, Masanori Takuma, Akio Yamamoto, "Estimation of Cutting Tool Life by Processing Tool Image Data with Neural Network", annals of CIRP, vol. 41/1/1993, 1993.

7. Rumelhart D.E., Hinton G.E., Williams R.J., "Learning Internal Representations by Error Propagation", in Parallel Distributed Processing : Explorations in the Microstructures of Cognition, Runelhart D.E., McClelland J.L. and the PDP Research Group Editors, vol. 1, Cambridge, MA. MIT Press, 1986.

# INFLUENCE OF PRE-DRILLING ON LIFE OF TAPS

**G.M. Lo Nostro**
University of Genoa, Genoa, Italy

**G.E. D'Errico and M. Bruno**
C.N.R., Orbassano, Turin, Italy

KEY WORDS: Tapping, Pre-Drilling, Optimisation.

ABSTRACT: An investigation of pre-drilling effects on tap performance is developed in a statistical framework. Results of extensive steel machining experiments are presented in terms of observations on drill wear and tap lifetime. Work-hardening is also discussed. A tool life criterion for worn drill replacement aimed at internal threading process optimisation is proposed and illustrated by a case-study.

## 1. INTRODUCTION

Lifetime of high speed steel (HSS) tools for steel machining, usually exhibits coefficients of variation in the range 0.30-0.45 [1-3]. Nevertheless, analyses of cutting performance of taps point out that tool life coefficients of variation may have higher values, in the range 0.6-0.8. Since lifetime of taps is affected by the pre-drilling process, it is interesting to investigate if a relationship exists between wear of drills and life obtained by taps.

In order to contribute to a deeper insight, the present paper is focused on an array of experiments designed and performed as follows.

A set of twist drills is partitioned into subsets such that elements in a subset are characterised by values of wear belonging to a given range. A lot of steel workpieces is also sub-divided into groups such that each group is composed of workpieces machined using drills belonging to a single subset only. Each group is further machined, until tap breakage,

Published in: E. Kuljanic (Ed.) *Advanced Manufacturing Systems and Technology*, CISM Courses and Lectures No. 372, Springer Verlag, Wien New York, 1996.

using nominally equal taps. During the experiments, observations are made in terms of number of holes obtained by a tap before its breakage.

A statistical treatment of experimental results allows to find out a relationship between wear of twist drills and life of taps. This result translates also into optimisation criteria for the ·complete tapping process.

## 2. EXPERIMENTAL DETAILS

Discs (thickness 16 mm) of steel 39NiCrMo3 UNI 7845 (HV50 = 283-303) are used for workpieces prepared according to UNI 10238/4. On each workpiece 53 through holes are obtained by wet drilling operations at cutting speed $v_c$ = 20 m/min and feed $f$ = 0.08 mm/rev.

The workpieces are subdivided into 36 groups. Each group contains 8 specimens denoted by numerals $i$ ($i$=1,...,8). More, 36 twist drills UNI 5620 (diameter 6.80 mm) are used to perform drilling operations such that a single drill per group is used. Hence 424 holes are drilled by each drill (53 per specimen). The mean value of flank wear $V_B$ is measured on the drill after 53 holes are completed on a single specimen. Therefore a basic experiment consisting in drilling 8 specimens in the sequence from 1 to 8 is iterated 36 times (i.e. the number of the specimen groups

After the drilling operations, the above specimens are tapped using taps M8 (UNI-ISO 54519). Tapping is performed at a cutting speed $v_c$ = 10 m/min. The same machine is used for both wet drilling and wet tapping operations.

Tapping experiments are performed as follows. The drilled workpieces are grouped into 8 groups such that each group $j$ ($j$=1,...,8) contains the 36 specimens denoted by $i$=$j$. All the 36 specimens of each single group are tapped by a single new tap: the tap allocated to group $j$ is denoted by $j$ ($j$=1,...,8). The total number of holes on a complete set of 36 specimens is 1908. A tap is used until a catastrophic failure occurs. If a tap happens to fail before a set of specimens is completely machined, the worn tap is replaced by a new tap, giving rise to a second run of experiments. Such a situation occurs when tapping specimens belonging to the groups from no. 4 to no. 8. In the case of group no. 1, 7 extra specimens (for a total of 43 specimens) are tapped before the tap breakage. Results of tapping experiments are summarised in Table 1 in terms of number $m$ of holes threaded during a tap lifetime.

**Table 1** *Results of tapping experiments.*

| group no. | $m$ ($1^{st}$ run) | $m$ ($2^{nd}$ run) |
|:---:|:---:|:---:|
| 1 | 2120 | - |
| 2 | 1621 | - |
| 3 | 1101 | - |
| 4 | 735 | 957 |
| 5 | 653 | 899 |
| 6 | 645 | 789 |
| 7 | 517 | 823 |
| 8 | 566 | 844 |

## 3. DISCUSSION OF RESULTS

Individual values of flank wear measured on drills are collected in a set, such that 8 sets of 36 data each are obtained. Each set is processed in order to check if relevant data fit a normal distribution, using a Kolmogorov-Smirnov test (K-S). Table 2 reports the mean values of $V_B$ (mm) along with standard deviations $\sigma$ and parameter D2 relevant to the distributions obtained.

**Table 2** *A synopsis of statistical data.*

| group no. | mean $V_B$, mm | $\sigma$ | D2 |
|-----------|----------------|----------|-------|
| 1 | 0.076 | 0.0152 | 0.157 |
| 2 | 0.121 | 0.0283 | 0.175 |
| 3 | 0.164 | 0.0299 | 0.191 |
| 4 | 0.204 | 0.0285 | 0.134 |
| 5 | 0.228 | 0.0295 | 0.144 |
| 6 | 0.259 | 0.0334 | 0.181 |
| 7 | 0.291 | 0.0341 | 0.097 |
| 8 | 0.330 | 0.0514 | 0.173 |

Since all D2 values are less than 0.27, the critical value at the significance level $1-\alpha=99\%$, results of the (K-S) test are always positive.
The mean flank wear $V_B$ (mm) observed on the drills may be plotted versus the number of drilled holes $n$ on the base of a 3rd order polynomial [4]:

$$V_B = 5.2 \cdot 10^{-9} \cdot n^3 - 3.95 \cdot 10^{-6} \cdot n^2 - 0.157 \cdot 10^{-2} \cdot n - 0.1618 \cdot 10^{-2} \quad (1)$$

Equation (1) has a correlation coefficient $r=0.913$, a quite high value given the sample size.

**Table 3** *List of $n_{25}$ values and correlation coefficients of relevant 3rd order polynomials.*

| Drill no. | $n_{25}$ | $r$ | Drill no. | $n_{25}$ | $r$ | Drill no. | $n_{25}$ | $r$ |
|-----------|----------|-------|-----------|----------|-------|-----------|----------|-------|
| 1 | 327 | 0.958 | 13 | 332 | 0.978 | 25 | 196 | 0.983 |
| 2 | 286 | 0.972 | 14 | 297 | 0.963 | 26 | 291 | 0.986 |
| 3 | 355 | 0.941 | 15 | 274 | 0.972 | 27 | 243 | 0.943 |
| 4 | 295 | 0.966 | 16 | 247 | 0.960 | 28 | 197 | 0.967 |
| 5 | 325 | 0.948 | 17 | 376 | 0.952 | 29 | 242 | 0.974 |
| 6 | 341 | 0.979 | 18 | 301 | 0.984 | 30 | 283 | 0.977 |
| 7 | 270 | 0.974 | 19 | 301 | 0.924 | 31 | 290 | 0.957 |
| 8 | 363 | 0.989 | 20 | 308 | 0.971 | 32 | 292 | 0.961 |
| 9 | 338 | 0.980 | 21 | 310 | 0.987 | 33 | 305 | 0.972 |
| 10 | 231 | 0.966 | 22 | 231 | 0.956 | 34 | 246 | 0.983 |
| 11 | 212 | 0.962 | 23 | 270 | 0.989 | 35 | 327 | 0.997 |
| 12 | 270 | 0.943 | 24 | 257 | 0.971 | 36 | 278 | 0.965 |

Following the above treatment, if also individual curves $V_B = V_B(n)$ pertaining to each single drill are interpolated by 3rd order polynomials, the mean number of holes $n_{25}$ obtained until reaching a $V_B = 0.25$ mm can be estimated. The set of these 36 values of $n_{25}$ fits a Weibull distribution, characterised by the parameters $\alpha = 7.33$ and $\beta = 305.0$. The significance of this distribution according to a (K-S) test, results in a confidence level higher than 99% (D2=0.987>0.27, the critical value). Estimated values of $n_{25}$ are listed in Table 3 along with correlation coefficients.

If data of tapping experiments are cross-correlated with data of drilling experiments, the following equation can be obtained with a regression coefficient $r = 0.897$:

$$m = \frac{21750.673}{n^{0.596}} \tag{2}$$

A correlation between $V_B$, mm, (UNI ISO 3685) and $m$ can be obtained by the following equation with a regression coefficient $r = 0.896$:

$$m = \frac{233.884}{V_B^{0.852}} \tag{3}$$

Equations (2) and (3) have prediction limits for the individual values respectively given by the following equations (4) and (5):

$$\exp\left[\ln m \pm t_{\alpha/2}\sqrt{0.003075 + 0.00777(\ln n - 5.464)^2}\right] \tag{4}$$

$$\exp\left[\ln m \pm t_{\alpha/2}\sqrt{0.00312 + 0.01619(\ln V_B + 1.5404)^2}\right] \tag{5}$$

where $t_{\alpha/2}$ is the $\alpha/2$ point of the $t$-distribution for $\nu = 11$.

Equations (2-3) are respectively plotted in Figures 1-2, with their confidence intervals ($\alpha = 0.05$).

It is likely that tap life decrease with increase of twist drill wear may be due to the greater work hardening that the worn drills produce on the hole walls. Also the hole diameter reduction due to drill wear might be supposed to be a cause of this situation, but such a conjecture is discarded on the base of the following procedure performed (before tapping) on 3 holes in each test specimen. Actually, a statistical analysis of measurements of pairs of diameters, taken at 1/3 and 2/3 of distance from the hole bottom respectively, points out that the max difference between the mean diameters of the holes made by new drill and by worn drill is maximum 0.0207 mm only. This reduction corresponds to a removed-material increase of less than 3%. Accordingly, it is deemed that the effect of hole diameter reduction on tap life is negligible, at a first level of approximation.

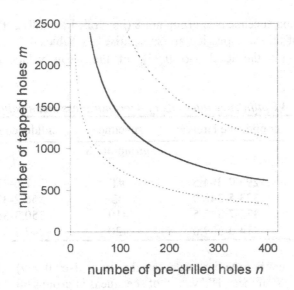

Figure 1: Number of tapped holes *vs.* number of pre-drilled holes.

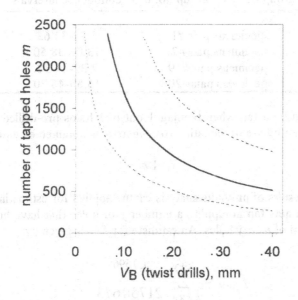

Figure 2: Number of tapped holes *vs.* drill's flank wear.

In order to verify the extent of work hardening on the hole walls, the 3 above mentioned pre-drilled holes of eight specimens are sectioned. In these sections, 40 Vickers microhardness measurements (loading mass 50 g) are performed at a distance of 0.1 mm from the wall. Four specimens in group no. 1 (i.e. drilled by new twist drills) are compared with four corresponding specimens in group no. 8 (i.e. drilled by drills at the maximum wear condition). Results of this investigation are summarised in Tables 4-5. Table 4 reports the confidence intervals (at the level $1-\alpha=0.95$), of the mean values of microhardness measurements.

**Table 4** *Confidence intervals of microhardness mean values.*

| specimen | confidence intervals | specimen | confidence intervals |
|---|---|---|---|
| group no. 1 | | group no. 8 | |
| #1 | 293.0-304.5 | #1 | 312.0-324.4 |
| #2 | 333.5-348.9 | #2 | 356.2-377.8 |
| #19 | 332.7-345.5 | #19 | 350.3-362.6 |
| #20 | 314.0-339.0 | #20 | 347.3-367.6 |

Table 5 reports the confidence intervals (at the level $1-\alpha=0.999$) of the differences (according to UNI 6806) of mean HV values of specimens in groups no. 1 and no. 8.

**Table 5** *Confidence intervals of mean microhardness differences.*

| group no. 1 vs. group no. 8 | confidence intervals |
|---|---|
| specimens pair #1 | 5.37-33.63 |
| specimens pair #2 | 13.00-38.50 |
| specimens pair #19 | 2.32-32.38 |
| specimens pair #20 | 15.60-46.20 |

The life $f_1$ obtained by a tap when tapping 1 out of $n$ holes pre-drilled by the same single drill can be estimated by use of equation (6), where $m$ is obtained by equation (2):

$$f_1 = 1/m \tag{6}$$

In general terms, results of previous analysis can be applied for estimating the tool life ratio $f_{x+y}$ attainable by a new tap in tapping a number $y$ of holes that have been pre-drilled by a drill used for a total of $n=x+y$ holes. An estimate for $f_{x+y}$ is given by:

$$f_{x+y} = \sum_{k=x+1}^{x+y} \frac{k^{0.596}}{21750.673} \tag{7a}$$

subject to

$$f_{x+y} \leq 1 \tag{7b}$$

In particular, when a new (or re-sharpen) drill is used to pre-drill a total of $y$ holes, the following conditions apply:

$$x=0, \text{ and } y=n=m \tag{7c}$$

and equation (7a) becomes:

$$f_y = \sum_{k=1}^{y} \frac{k^{0.596}}{21750.673} \tag{7d}$$

From equation (7d), the quantity $y=700$ is obtained such that $f_y=1$: this value of $y$ represents the maximum number of holes attainable by a tap during its whole lifetime.

An optimal value for $y$ can be derived on the base of economical considerations. If costs related to accidental tap breakage are not taken into account, the cost $C$ for tapping $y$ holes already pre-drilled by a single drill can be estimated. Assuming that a single new tap is used for tapping $y$ holes, the following equation can be derived:

$$C = \frac{c_d + n_r \cdot c_r}{(n_r +1) \cdot y} + \frac{T_s \cdot M}{y} + f_y \cdot \frac{c_t}{y} \tag{8}$$

where:

$c_d$ is the cost of a new drill;
$c_r$ is the cost for re-sharpening a worn drill,
$c_t$ is the cost of a tap;
$n_r$ is the umber of expected re-sharpening operations;
$T_s$ is the time required for substitution of a worn drill;
$M$ is the machine-tool cost per time unit;
$f_y$ is the exploited ratio of tap life.

It is worth noting that if a tap fails before the drill, then $n_r$ represents the number of re-sharpening operations performed on the drill before the tap failure occurs: in this case is $(n_r+1) \cdot y=1$, and the last term in equation (8) should be replaced by the quantity $c_t/(n_r+1)$.

If costs are expressed in Italian liras and time in minutes, the following cases can be developed by introducing in equation (8) the values $c_r=500$, $c_t=15000$, $T_s=0.5$, and:

(a) $c_d=5000$, $M=1000$;
(b) $c_d=3250$, $M=1000$;
(c) $c_d=5000$, $M=50$;
(d) $c_d=5000$, $M=1500$.

The relationships $C=C(y)$ obtained by equation (8) are plotted in Figure 3, where curves relevant to cases (b) and (c) appear overlapped.

A minimum in a plot in Figure 3 indicates the number of holes over which it is convenient to replace a worn twist drill. It should be noticed that minima in such plots may become even more important if a term related to cost of expected damage due to tool failure is added in equation (8).

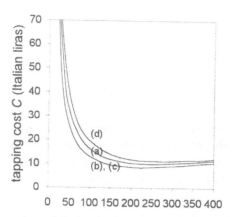

number of holes before drill replacement *y*

Figure 3: Tapping cost plots.

## 4. CONCLUSIONS

The investigation performed propose a comprehensive a treatment of tool life variability which affects performance of tools for internal threading. In the light of the experimental results presented and discussed, the following conclusions may be drawn.

1. A statistical model may be developed which points out a possible relationship between wear of twist drills and life of taps.
2. It is likely that tap life decrease with increase of twist drill wear may be due to the greater work hardening that the worn drills produce on the hole walls.
3. A minimum cost criterion provides the basis for an optimal replacement strategy of worn drills.

## REFERENCES

1. Wager, J.G., M.M. Barash: Study of the Distribution of the Life of HHS Tools, Transactions of the ASME, 11 (1971), 1044 - 1050.

2. Ramalingam, S., Tool-life Distribution. Part 2: Multiple Injury Tool-life Model, Journal of Engineering for Industry, 8 (1977), 523 - 531.

3. Lo Nostro, G.M., E.P. Barlocco, P.M. Lonardo, Effetti della Riaffilatura di Punte Elicoidali in HSS ed in ASP, nude o rivestite, La Meccanica Italiana, 12 (1987), 38-45.

4. De Vor, R.E., D.R. Anderson, W.J. Zdebelik, Tool Life Variation and its Influence on the Development of Tool Life Models, Transaction of the ASME, 99 (1977), 578-584.

# INFLUENCES OF NEW CUTTING FLUIDS
# ON THE TAPPING PROCESS

**J. Kopac, M. Sokovic and K. Mijanovic**
**University of Ljubljana, Ljubljana, Slovenia**

KEY WORDS: Cutting Fluids, Tapping Process, Ecological Aspects

ABSTRACT: The development of additives for cutting fluids has led to new products which are used in specific machining procedures such as tapping. The improved quality shown in smaller roughness of threads and considerably smaller cutting force moment in the process of tapping. In the research study, some new cutting fluids were tested in cutting threads into steel and Al-alloys. Serving as a basis in all experiments were the preliminary hole and the tool (tap) with a given cutting geometry adopted for tapping into steel or Al-alloys. The parameters of cutting were chosen from the technological database with respect to particular machined material / cutting tool combinations.

## 1. INTRODUCTION

For reasons of rationalization of the process high costs of purchasing some of the newest cutting fluids have to be compensated, by a suitable increase in production rate or considering smaller tool wear, and by the environmental effects: the new cutting fluids have to be environment-friendly with a possibility of recycling. The results of the research are presented in diagrams and tables serving as guidelines where by the pondered values method the suitability and economy of particular cutting fluids in cutting threads were assessed.

Published in: E. Kuljanic (Ed.) *Advanced Manufacturing Systems and Technology,*
CISM Courses and Lectures No. 372, Springer Verlag, Wien New York, 1996.

## 2. ECOLOGICAL ASPECTS OF CUTTING FLUIDS

Data analysis of scientific research works in the fields of ecology and science of residence quality shows that most problems are due to technological carelessness, unsolved problems of manufacturing processes and human behavior. Manufacturing processes produce substances which pollute air, water and soil. These processes are part of business systems and they represent the basic reasons for polluting the living and working environment. They are also a source disturbances of the natural system balance. When this system becomes incapable of renewing it own resources, then disturbances occur having an mankind influence and the next generations. Most of the dust deposited in the earth crust will undergo some process of biodegradation or some other changes that will make it less harmful. Problems arise when dusts deposits grow faster than is the natural ability of their neutralization, and the results are disturbances of natural balance. The process is just the opposite with natural resources. We consume natural goods faster than is their renewal ability. Business systems are in both examples the largest dust producer and the largest consumer of resources at the same time, Figure 1.

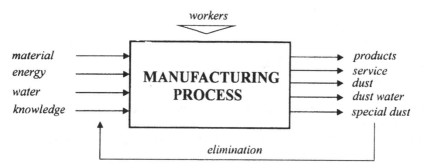

Fig. 1. Manufacturing processes polluting the living environment

We are looking for cutting fluids which would better meet the ecological demands. Emulsion Teolin AIK is a mixture of alkiestres of phosphorus acid with condensation products of high grease acid. This emulsion has a good solubility with water in all proportions. It is used first of all for machining of aluminium and other color metals. The surface after machining is smooth and shinning. Clinical investigations of Teolin AIK influences on human respiratory system show it is benign in solutions and when concentrated. Investigations of biological decomposition show that Teolin AIK decomposes in dust water.

Table 1. Data on the cutting fluids used

| cutting oil with chlorinated hydrocarbons | semysynthetic fluid | cutting oil |
|---|---|---|
| KUTEOL CSN 5 Teol Ljubljana | TEOLIN AIK Teol Ljubljana (20% emulsion) | SKF |

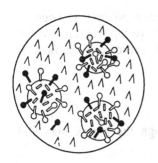

○ - oil phase
∧ - molecule of water
═ - molecule of oil
⊂═ - surface active molecules
●━ - aditive in water
o - mulecules of soluble material

Fig. 2. Schematic representation of polysynthetical cutting fluid structure

## 3. PROPERTIES OF WORK PIECE MATERIALS AND TOOLS

Table 2. Chemical structure and mechanical properties of steel Ck 45

| Chemical structure % | | | | | Flow stress N/mm$^2$ | Tensile stress N/mm$^2$ | elongation % |
|---|---|---|---|---|---|---|---|
| C | Si | Mn | P | S | | | |
| 0.42 | 0.15 | 0.50 | max. | max. | min. | | min. |
| 0.50 | 0.35 | 0.80 | 0.035 | 0.035 | 420 | 670 | 16 |

Table 3. Chemical structure and mechanical properties of alloy AlMgSiPbBi (design. T8)

| Chemical structure % | | | | | Flow stress N/mm$^2$ | Tensile stress N/mm$^2$ | elongation % |
|---|---|---|---|---|---|---|---|
| Si | Mg | Pb | Fe | Bi | | | |
| 1.11 | 1.00 | 0.53 | 0.30 | 0.56 | 365 | 387 | 11.4 |

Another aim of this paper was to establish suitable combinations of cutting speeds and cutting fluids from the point of view of machinability. Experiments were planned with one-factor plan by Boks-Wilson method inside cutting speed limits. A mathematical model of machinability function, which is used for searching the response inside of experimental space, is a potential function:

$$M = C \cdot v_c^{\,p} \tag{1}$$

where are:    C, p    machinability parameters,
              $v_c$     cutting speed as control factor and
              M        moment as process state function.

The moment is the output which results from measurements of tapping process. The cutting speed varied from $v_{c,min} = 12$ m/min to $v_{c,max} = 25$ m/min. SKF M10 tap was used with a shape for common holes deep $l = 10$ mm, which were selected considering the plan of experiments and optimal cutting geometry for Al-alloy or steel.

## 4. MEASUREMENT RESULTS

Measurement results are directly shown in Table 4 and 5 through the matrix plan and statistic values.

Table 4. Tapping into Al - alloy AlMgSiPbBi with 20 % AIK emulsion

| No. | Matrix plan | | Variable value | Measured value | $y = \ln M$ | $y^2$ | $y_0^2$ |
|---|---|---|---|---|---|---|---|
| | $x_0$ | $x_1$ | $v_e$ m/min | M Nm | | | |
| 1 | 1 | +1 | 25.13 | 3.740 | 1.319 | 1.74 | - |
| 2 | 1 | -1 | 12.57 | 3.325 | 1.202 | 1.44 | - |
| 3 | 1 | 0 | 17.59 | 3.618 | 1.286 | 1.65 | 1.65 |
| 4 | 1 | 0 | 17.59 | 3.692 | 1.306 | 1.71 | 1.71 |
| 5 | 1 | 0 | 17.59 | 3.740 | 1.319 | 1.74 | 1.74 |
| 6 | 1 | 0 | 17.59 | 3.723 | 1.315 | 1.73 | 1.73 |
| | | | | | $\Sigma$ | 10.01 | 6.83 |

Table 5. Tapping into steel Ck 45 with 20 % AIK emulsion

| No. | Matrix plan | | Variable value | Measured value | $y = \ln M$ | $y^2$ | $y_0^2$ |
|---|---|---|---|---|---|---|---|
| | $x_0$ | $x_1$ | $v_e$ m/min | M Nm | | | |
| 1 | 1 | +1 | 12.57 | 10.787 | 2.378 | 5.65 | - |
| 2 | 1 | -1 | 6.28 | 9.581 | 2.260 | 5.11 | - |
| 3 | 1 | 0 | 8.79 | 11.569 | 2.448 | 5.99 | 5.99 |
| 4 | 1 | 0 | 8.79 | 10.522 | 2.353 | 5.37 | 5.37 |
| 5 | 1 | 0 | 8.79 | 11.182 | 2.414 | 5.83 | 5.83 |
| 6 | 1 | 0 | 8.79 | 10.697 | 2.370 | 5.62 | 5.62 |
| | | | | | $\Sigma$ | 33.57 | 22.81 |

Equations which represent the relationship between the cutting force moment and cutting speed are:

- for steel using a 20 % Teolin AIK emulsion:

$$M = 7.385 \cdot v_c^{0.17} \tag{2}$$

- for steel using KUTEOL oil

$$M = 10.34 \cdot v_c^{-0.029} \tag{3}$$

- for Al-alloy using 20 % Teolin AIK emulsion:

$$M = 2.23 \cdot v_c^{0.17} \tag{4}$$

- for Al-alloy using KUTEOL oil

$$M = 5.323 \cdot v_c^{-0.182} \tag{5}$$

The model requires a variation of 24 tests carried in four repetitions in the zero point and distributed according to the matrix. The coefficients of regression were considered and the results verified by the Fischer criterion $F_{rLF}$, the dispersion of the experimental results mean values ($S_{LF}^2$) being compared with respect to the regression line ($S_R^2$):

$$F_{rLF} = \frac{S_{LF}^2}{S_R^2} \tag{6}$$

The condition of suitability is fulfilled if $F_{rLF} < F_t$ : $F_t = 9.55$ and is obtained from the table of the degrees of freedom. The calculation of $F_{rLF}$ yield the following results:

- Ck 45 / 20 % emulsion       $F_{1LF} = 1.6$   $< F_t$,
- Ck 45 / oil                 $F_{1LF} = 7.03$   $< F_t$,

- Al-alloy / 20 % emulsion    $F_{1LF} = 1.57$   $< F_t$,
- Al-alloy / oil              $F_{1LF} = 0.84$   $< F_t$,

which show adequacy of these models for each <u>workpiece material</u> / <u>cutting fluid</u> combination used.

Diagrams which following from equations (2) to (5) are shown below in Figs. 3 and 4.

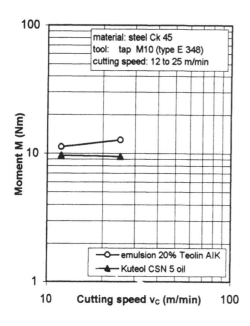

Fig. 3 Moment vs. cutting speed in tapping steel Ck 45

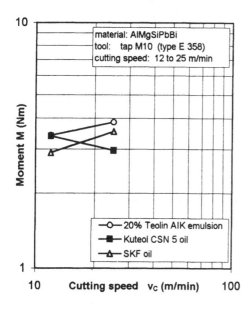

Fig. 4 Moment vs. cutting speed in tapping Al - alloy AlMgSiPbBi

## 5. CONCLUSIONS

Cutting fluid has, besides of cutting speed, a big influence on forces and moments in tapping thread into steel Ck 45 and Al-alloy AlMgSiPbBi. An appropriate combination of cutting speed and cutting fluid is important for the efficiency and required quality of machining.

Table 6. Recommended combinations of tool, cutting fluid and cutting parameters in tapping thread into steel and Al-alloy

| Workpiece material | Used tap | Used cooling fluid | Recommended cutting speed $v_c$ m/min |
|---|---|---|---|
| steel<br><br>Ck 45 | M10, DIN 317<br>type E 348<br>(catalogue SKF 1992)<br>For thread length < 2,5 D | 20% emulsion<br>Teolin AIK with water<br>Q = 0.5 l/min | 6 |
| | | Cutting oil<br>Kuteol CSN 5<br>Q = 0.5 l/min | 12 |
| AlMgSiPbBi<br><br>(T8) | M10, DIN 317<br>type E 358<br>(catalogue SKF 1992)<br>For thread length < 2,5 D | 20% emulsion<br>Teolin AIK with water<br>Q = 0.5 l/min | 12 |
| | | Cutting oil<br>Kuteol CSN 5<br>Q = 0.5 l/min | 25 |

We can see that the moment is decreasing with cutting speed Kuteol CSN 5 oil is used, but its usage is limited because of chlorinated hydrocarbons contents. Because of that, we must reconcile ourselves with the primary goal of this paper stating ecological effects as the main factor. The achievement of goals which are related to clean nature and improvement of working and living environment definitely require more costly production.

## REFERENCES

1. E.M. Trent: Metal Cutting, Edition, Butterworths, London, 1984

2. K. Mijanović: Research of quality of TiN coated taps, Master thesis, Faculty of mechanical engineering, Ljubljana, 1992

3. L. Morawska, N. Bafinger, M. Maroni: Indoor Air Integrated Approach ES1[st], Elsevier, Oxford, 1995

4. G. Douglas, W.R. Herguth: Physical and Chemical Properties of Industrial Mineral Oils affecting Lubrication, Lubrication Engineering, Vol. 52, No. 2, January 1996, 145 - 148

5. W.D. Hewson, G.K. Gerow: Development of New Metal Cutting Oils With Quantifiable Performance Characteristics, Lubrication Engineering, Vol. 52. No. 1, January 1996, 31-38

6. K. Heitges: Metallwerkerlehre, Band 2, Umformen mit Maschinen Arbeiten an Werkzeugmaschinen, Lehrmittelverlag Wilhlem Hagemann - Dusseldorf, Dusseldorf, 1972

7. A. Salmič: Measurement of Tapping Forces, Diploma thesis, 1996, Ljubljana, Slovenia

8. J. Kopač, M. Soković and F. Čuš: QM and costs optimization in machining of Al - alloys, Total Quality, Creating individual and corporate success, Institute of Directors, Excel Books, First Edition, New Delhi, 1996, 150 - 156

# INTENSIFICATION OF DRILLING PROCESS

**A. Koziarski and B.W. Kruszynski**
Technical University of Łødz, Łødz, Poland

KEY WORDS: Drilling, Coatings, Geometry Modification

ABSTRACT: Two methods of intensification of the drilling process are described in the paper. The first one - by application of different deposited or chemically treated layers for twist drills made from high speed steels. Three coatings were applied: titanium carbide, titanium nitride and diamond-like thin layers. Also, a vacuum nitriding was applied. Tool life was investigated when cutting carbon steel, laminated glass fibre and hardboard. In the second part of the paper the possibility of minimising of cutting forces and torque in drilling is discussed. The influence of different modifications of chisel edge was investigated experimentally. The results of these investigations and conclusions are presented in the paper.

## 1. INTRODUCTION

Drilling is one of the most popular manufacturing methods of making holes. Twist drills are the most common tools used in mass production as well as in small batch production. They are made more and more frequently from cemented carbides, but mainly for economical reasons, drills made from high speed steels are still in use. It concerns, for instance, drilling holes of small diameters, where the risk of destruction of brittle cemented carbide drills is very high.

Published in: E. Kuljanic (Ed.) *Advanced Manufacturing Systems and Technology*, CISM Courses and Lectures No. 372, Springer Verlag, Wien New York, 1996.

Commercially available HSS twist drills have some disadvantages. Their wear resistance is often insufficient, which makes their life too short for economical application. It sometimes can be observed in cutting non-metallic materials which can consist of highly abrasive particles or have poor thermal conductivity. In consequence, the intensity of tool wear is very high. Also geometry of commercial twist drills is not optimal for effective drilling. For example, relatively long chisel edge makes cutting forces high, which increases the risk of tool breakage, increases tool wear and decreases drilling accuracy.

All of these makes investigations of HSS twist drills with modifications increasing their life and decreasing cutting forces important for economical application of these tools.

## 2. INVESTIGATION OF TOOL LIFE OF THERMOCHEMICALLY TREATED TWIST DRILLS

One of the ways of tool wear resistance improvement is application of thermochemical treatment to their working surfaces. There are a lot of methods of treatment existing. This paper deals with methods in which temperature of treatment is below tempering temperature for high speed steels. In such cases there are no structural changes, thermal stresses, cutting edge deformations, etc. in tool material.

Commercially available twist drills made of SW7M HSS, $\varnothing$ 3.8 mm in diameter were investigated. Drills were divided into five groups. Four groups were additionally treated - according to information given in table 1. The fifth group of drills remained with no additional treatment for comparison. For statistical reasons each group consisted of nine drills.

Table 1 Thermochemical treatment methods applied to twist drills

| Kind of treatment | References |
|-------------------|:----------:|
| Vacuum nitriding - NITROVAC'79 | [1] |
| TiC layer deposition | [2] |
| Diamond-like layer deposition | [3] |
| TiN layer deposition | [4] |

The following workmaterials were cut:
- laminate of glass fibre, 1 mm thick, applied for printed circuits,
- particle hardboard, 20 mm thick, applied for furniture,
- mild steel, 2 mm thick, for comparison.

Cutting speed and feed rate were constant - $v_c$=28 m min$^{-1}$ and f=0.1 mm/rev, respectively.

Measurementsof tool wear for each drill were performed after making every 50 holes and drilling was continued until tool wear limit was reached. In these cutting conditions flank wear dominated, cf. fig. 1, and was taken as tool wear indicator. Measurements were carried out on a microscope using a special device.

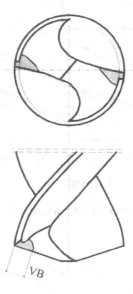

Fig. 1  Flank wear of
twist drill.

The following results of experiments, carried out according to [5], were obtained:

- comparison of drill wear characteristics for different workmaterials

$$VB = f(l_w) \qquad (1)$$

- comparison of tool wear intensity coefficient

$$K_{VB} = VB/l_w \qquad (2)$$

where: $l_w$ is length of drilling [mm].

Tool wear characteristics, average for each group of twist drills in drilling laminate of glass fibre are shown in fig. 2. Flank wear VB=0.25 mm was taken as limiting value for these cutting conditions because higher values causes burr appearance which is unacceptable in this finishing operation.

Experimental average tool wear characteristics for hardboard drilling are shown in fig. 3. Tests were stopped at $l_w$=800 mm because of drill flutes loading with workmaterial and appearance of tool wear trend shown in fig.3.

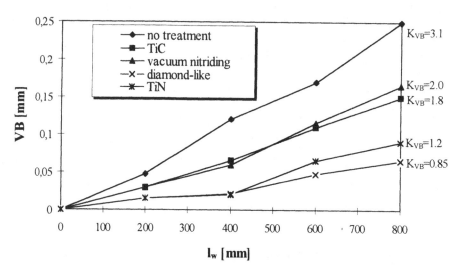

Fig. 2 Twist drill wear vs. drilling length for drilling hardboard

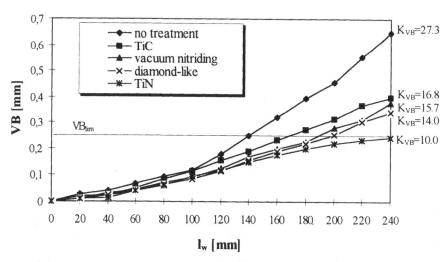

Fig. 3 Twist drill wear vs. drilling length for drilling hardboard

Experimental, average tool wear characteristics for mild steel drilling are shown in fig. 4. Tool wear limit $VB_{x\ lim}=0.4$ was assumed.

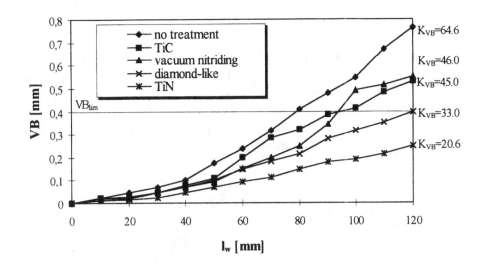

Fig. 4 Twist drill wear vs. drilling length for drilling mild steel

## 3. INFLUENCE OF CHISEL EDGE MODIFICATION ON CUTTING FORCES AND TORQUE

The second part of investigations was devoted to modifications of twist drill geometry and its influence on cutting forces and torque. Modifications of drill geometry concern chisel edge in the manner shown in fig. 5. The chisel edge was partly (fig. 5b) or entirely (fig. 5c) removed. For comparison also twist drills without modifications (fig. 5a) were investigated. For quantitative estimation of changes the chisel edge modification coefficient (CEMC) was introduced, see also fig. 5:

$$CEMC = \frac{l_s - l_o}{l_s} \ 100\% \tag{6}$$

where: $l_s$ is unmodified chisel edge length and $l_o$ is chisel edge length after modification.

Investigations were performed on a stand shown in fig. 6. Feed force $F_f$ and torque $M_s$ were measured by means of 4-component Kistler 9272 dynamometer. The following cutting conditions were applied:

- workmaterial:         carbon steel .45%C, 210HB,
- twist drills diameters:    Ø7.4 mm, Ø13.5 mm, Ø21 mm,
- cutting speed:        0.4 ms$^{-1}$ (constant),
- feed rates:          0.1 mm/rev, 0.15 mm/rev, 0.24 mm/rev,
- cutting fluid:        emulsion.

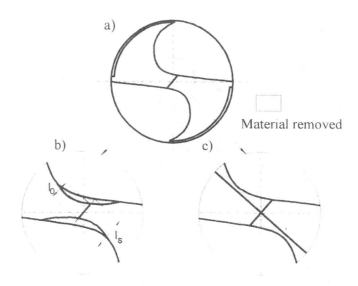

Fig. 4 Chisel edge modifications
a) chisel edge without modification, b) chisel edge shortened,
c) chisel edge entirely removed (cross-cut)

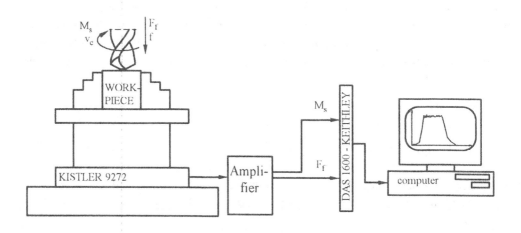

Fig. 6 Experimental set-up

Results of investigations are shown in figures 7-12. In these figures changes of feed force and torque versus CEMC are presented.

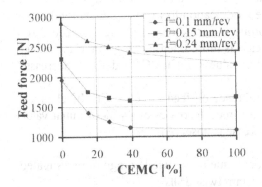

Fig. 7 Feed force vs. CEMC for ⌀13.5 mm
twist drill

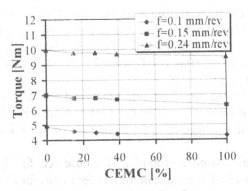

Fig. 8 Torque vs. CEMC for ⌀13.5 mm
twist drill

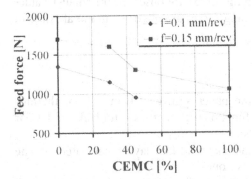

Fig. 9 Feed force vs. CEMC for ⌀7.4 mm
twist drill

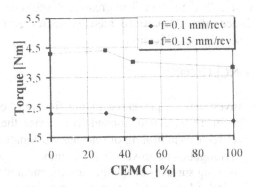

Fig. 10 Torque vs. CEMC for ⌀7.4 mm
twist drill

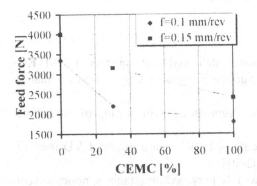

Fig. 11 Feed force vs. CEMC for ⌀21 mm
twist drill

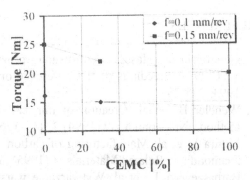

Fig. 12 Torque vs. CEMC for ⌀21 mm
twist drill

The most detailed investigations were carried out for $\varnothing 13.5$ mm twist drills, figs. 7 and 8. It is clearly seen from these figures that chisel edge modification has very significant influence on feed force $F_f$, fig. 7, and its influence on torque is hardly observed, fig. 8. Feed force decreases with increase of CEMC and this decrease is more intensive in a range of smaller values of this coefficient (0%-40%). For higher values of CEMC feed force decreases slightly.

It is also seen from fig. 7 that feed rate has an influence on feed force vs. CEMC changes. For small value of feed rate (f=0.1 mm/rev.) feed forces decreases of 44 % of initial value while for feed rate of 0.24 mm/rev. only 24% decrease is observed.

The maximum drop of 12% is observed, fig. 8, for torque in the whole range of investigated chisel edge modifications and feed rates for $\varnothing 13.5$ mm twist drills.

Similar trends are observed for $\varnothing 7.4$ mm, figs. 9 and 10, and $\varnothing 21$ mm, figs. 11 and 12, twist drills diameters. The greatest changes of cutting forces are observed for smaller values of feed rates. Also, more intensive decrease of cutting forces are observed in initial range of chisel edge modification.

## CONCLUSIONS

1. Investigations presented in the first part of the paper showed advantages of thermo-chemical treatment applied to increase their functional properties - tool life and wear resistance - in drilling of selected workmaterials.
2. Investigations showed that titanium nitride and diamond-like layers deposited on the working surfaces of the drill are the most effective ones.
3. Chisel edge modification has important influence on feed force and relatively small influence on cutting torque.
4. Chisel edge modification of about 40% is sufficient to reduce significantly cutting forces. Further chisel edge reduction has practically no influence on cutting forces.

## REFERENCES

1. Gawroński Z., Haś Z.: Azotowanie próżniowe stali szybkotnącej metodą NITRO-VAC'79. Proceedings of II Conference on Surface Treatment, Częstochowa (Poland), 1993, 255-259.
2. Wendler B., et al: Creation of thin carbide layers on steel by means of an indirect method., Nukleonika, 39 (1994) 3, 119-126.
3. Mitura S. et al: Manufacturing of carbon coatings by RF dense plasma CVD method, Diamond and Related Materials, 4 (1995), 302-305.
4. Barbaszewski T., et al: Wytwarzanie warstw TiN przy wykorzystaniu silnoprądowego wyładowania łukowego, Proceedings of Conference on Technology of Surface Layers, 1988, Rzeszów, 88-98.

# MILLING STEEL WITH COATED CERMET INSERTS

G.E. D'Errico and E. Guglielmi

C.N.R., Orbassano, Turin, Italy

KEY WORDS: Cermet Insert, PVD Coatings, Milling, Cutting Performance

ABSTRACT: A commercial cermet grade SPKN 1203 ED-TR is PVD coated. The following thin layers are deposited by a ion-plating technique: TiN, TiCN, TiAlN, and TiN+TiCN. Dry face milling experiments are performed with application to steel AISI-SAE 1045. A comparative performance evaluation of uncoated and coated inserts is provided on the base of flank wear measurements.

## 1. INTRODUCTION

Cermets, like hardmetals, are composed of a fairly high amount of hard phases, namely Ti(C,N) bonded by a metallic binder that in most cases contains at least one out of Co and Ni [1]. The carbonitride phase, usually alloyed with other carbides including WC, Mo$_2$C, TaC, NbC and VC, is responsible for the hardness and the abrasive wear resistance of the materials. On the other hand, the metal binder represents a tough, ductile, thermally conducting phase which helps in mitigating the inherently brittle nature of the ceramic fraction and supplies the liquid phase required for the sintering process.
Cermet inserts for cutting applications can conveniently machine a variety of work materials such as carbon steels, alloy steels, austenitic steels and grey cast iron [2-11].
A question of recent interest is to assess if the resistance of cermet cutting tools to wear mechanisms may be improved by use of appropriate hard coatings.
Controversial conclusions are available in the relevant technical literature, since positive results are obtained in some cutting conditions but unsatisfactory results may also be found, especially with application to interrupted cutting processes [12-16].
The present paper deals with the influence of some Physical Vapour Deposition (PVD) coatings on the performance of a cermet tool when milling blocks of normalised carbon

Published in: E. Kuljanic (Ed.) *Advanced Manufacturing Systems and Technology*, CISM Courses and Lectures No. 372, Springer Verlag, Wien New York, 1996.

steel AISI-SAE 1045. The following thin layers are deposited by a ion-plating technique [17]: TiN, TiCN, TiAlN, and TiN+TiCN.

Dry face milling tests are performed on a vertical CNC machine tool. The cutting performance of the uncoated and coated inserts is presented and compared in terms of tool life obtained until reaching a threshold on mean flank wear.

## 2 EXPERIMENTAL CONDITIONS

The cermet used for substrate in the present work is a commercial square insert with a chamfered cutting edge preparation (SPKN 1203 ED-TR) for milling applications (ISO grade P25-40, M40). The insert micro-geometry is illustrated in Figure 1. This insert is the most reliable found in a previous comparative work [4] among a set of similar commercial cermet inserts for milling applications tested. This cermet has the following percentage volume composition [3, 4]: 52.04 Ti(C,N), 9.23 Co, 5.11 Ni, 9.41 TaC, 18.40 WC, and 5.80 Mo$_2$C, and a hardness of 91 HRA.

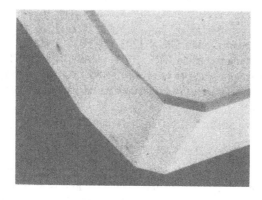

**Figure 1** *Micro-geometry of the cermet insert used for substrate.*

As regards PVD coatings, three mono-layers (thickness 3μm) of TiN, TiCN, and TiAlN and a multy-layer (thickness 6μm) of TiN+TiCN are deposited by an industrial ion-plating process [17].

**Table I** *Nominal values of main characteristics of PVD coatings (industrial data).*

| Characteristic | TiN | TiCN | TiAlN |
|---|---|---|---|
| thickness (μm) | 1-4 | 1-4 | 1-4 |
| hardness (HV 0.05) | 2300 | 3000 | 2700 |
| friction coefficient *vs* steel (dry) | 0.4 | 0.4 | 0.4 |
| operating temperature (°C) | 600 | 450 | 800 |
| thermal expansion coefficient ($10^{-3}$/°K) | 9.4 | 9.4 | - |

The deposition temperature is around 480 C which allows adhesion to substrates avoiding also deformation and hardness decay. The nominal values of some main characteristics of these coatings are given in Table I, according to industrial data. Ti-based coatings are used as a thermal barrier against the temperature raise during the cutting process, they reduce friction between cutting edge and workpiece, chemical-physical interactions between insert and chip, crater and abrasive wear, and built-up-edge formation. Further, TiCN is valuable in application to difficult-to-cut materials, and TiAlN provides a raised resistance to high temperature and to oxidation.

As far as machining trials are concerned, dry face milling tests are performed on a vertical CNC machine tool (nominal power 28 kW). Workpieces of normalised carbon steel AISI-SAE 1045 (HB 190±5) were used in form of blocks 100·250·400 mm$^3$. In such conditions, the length of a pass is $L = 400$ mm, and the time per pass $T = 31.4$ s).

The cutting parameters and the geometry of the milling cutter (for six inserts) are shown in Table II.

Table II *Machining conditions.*

| Cutting parameters | | Milling cutter geometry | |
|---|---|---|---|
| cutting speed $v_c$ | 250 m/min | cutter diameter $\phi$ | 130 mm |
| feed $f_z$ | 0.20 mm/tooth | corner angle $\kappa_r$ | 75° |
| axial $a_a$ , radial $a_r$ | | orthogonal $\gamma_o$, axial $\gamma_p$, radial $\gamma_f$ | |
| depth of cut | 2 mm, 100 mm | rake angles | 2°, 7°, 0° |

## 3. RESULTS AND DISCUSSION

Mean values of six $VB_B$ data (and standard deviations) measured after 22.05 minutes (i.e. 42 cuts) on each insert type are reported in Table 2 along with tool lives calculated from intersections of the straight line $VB_B$=0.20 mm with each wear curve plotted in Figure 2.

Table·III *A Synopsis of Experimental Results.*

| Coating | $VB_B$max: Mean, mm (Std.Dev) | Tool life, min | % Tool life variation |
|---|---|---|---|
| uncoated | 0.218  (0.0259) | 19.7 | base line (100) |
| TiN | 0.189  (0.0107) | 23.9 | 121.5 |
| TiCN | 0.241  (0.0119) | 18.3 | 93.2 |
| TiAlN | 0.189  (0.0141) | 23.4 | 119.0 |
| TiN+TiCN | 0.193  (0.0252) | 23.3 | 118.3 |

Percentage variations of tool life obtained by use of coatings are also shown in Table III, with reference to the uncoated insert. Evolution of maximum flank wear $VB_B$ measured every 3.15 minutes (i.e. every 6 cuts) during the milling operations is plotted in Figure 2.

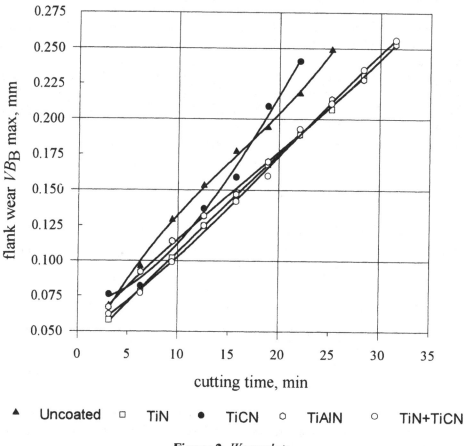

Figure 2 *Wear plots.*

In general, the PVD coatings used give increments in tool life obtained in the experimented conditions with respect to the cermet substrate, but the TiCN coating gives a tool life decrement around 7 per cent (Table III). Initially the wear behaviour of the TiCN coated insert is better than the substrate's behaviour, but after 33 cuts this coating is outperformed by the uncoated insert (Figure 2).

As far as wear morphology is concerned, no important difference can be observed among the coated inserts. Photographs in Figures 3-4 (low magnifications) point out the existence of both flank and crater wear, and also of microcracks on the cutting edge of: a TiN coated insert (Fig.3a) after 28.35 minutes ($VB_B$=0.243 mm); a TiCN coated insert (Fig.3b) after 22.05 minutes ($VB_B$=0.253 mm); a TiAlN coated insert (Fig.4a) after 31.50 minutes ($VB_B$=0.257 mm); a TiN+TiCN coated insert (Fig.4b) after 31.50 minutes ($VB_B$=0.254 mm).

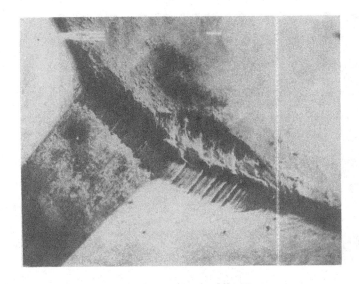

(a): a TiN coated insert

(b): a TiCN coated insert

**Figure 3** *Examples of wear morphology.*

(a): a TiAlN coated insert

(b): a TiN+TiCN coated insert

**Figure 4** *Examples of wear morphology.*

Focusing on positive results, since the wear plots in Figure 2 relevant to TiN, TiAlN, and TiN+TiCN show similar behaviours, it can be deduced that the multy-layer does not improve the cutting performance of the mono-layers.

It is worth noting that the TiN and TiAlN mono-layers exhibit quite high operating temperatures, respectively 600 °C, and 800 °C (Table I). In interrupted cutting, this characteristic is more important than hardness: actually TiCN has the highest hardness value (3000 HV 0.05), but also the lowest operating temperature, 400 °C (Table 1). In dry milling operations the cutting insert is subject to high cutting temperature, and to thermal shock as well as to mechanical impacts. Using a cermet substrate, the resistance to mechanical impacts (which relates to toughness) is controlled mainly by molybdenum carbide in the substrate composition (volume percentage of $Mo_2C$ is 5.80% in the cermet used for substrate). The resistance to thermal shock is controlled both by the substrate (particularly by tantalum carbide and niobium carbide) and by the coating: in this case, cermet substrate has 9.41% by volume of TaC, while NbC is not included in the composition.

## 4. CONCLUSIONS

In the light of the experimental results obtained in dry face milling tests using diverse PVD coatings of a cermet insert, the following conclusions can be drawn.

1. Results of cutting performance are almost scattered: if the uncoated cermet insert is referred to as the baseline, tool lives vary from ~−7% to ~22%.
2. The best performing coating is TiN, while the worst one is TiCN.
3. Efficiency of PVD coatings of cermet inserts for interrupted cutting is related to the possibility of raising the substrate's resistance to temperature-controlled wear mechanisms.

## REFERENCES

1.  Ettmayer, P., W. Lengauer: The Story of Cermets, Powder Metallurgy International, 21 (1989) 2, 37-38.
2.  Amato, I, N. Cantoro, R. Chiara, A. Ferrari: Microstructure, Mechanical Properties and Cutting Efficiency in Cermet System, Proceedings of 8th CIMTEC-World Ceramics Congress and Forum on New Materials, Firenze (Italy), 28 June- 4 July, 1994, Vol. 3, Part D, (P. Vincenzini ed.), 2319-2326.
3.  Bugliosi, S., R. Chiara, R. Calzavarini, G.E. D'Errico, E. Guglielmi: Performance of Cermet Cutting Tools in Milling Steel, Proceedings 14th International Conference on Advanced Materials and Technologies AMT'95, Zakopane (Poland), May 17-21, 1995, (L.A. Dobrzanski ed.), 67-73.
4.  D'Errico, G.E., S. Bugliosi, E. Guglielmi: Tool-Life Reliability of Cermet Inserts in Milling Tests, Proceedings 14th International Conference on Advances in Materials and Processing Technologies AMPT'95, Dublin (Ireland), 8-12 August, 1995, Vol. III, (M.S.J. Hashmi ed.) 1278-1287.

5.  Destefani, J.D.: Take Another Look at Cermets, Tooling and Production, 59 (1994) 10, 59-62.
6.  Doi, H.: Advanced TiC and TiC-TiN Base Cermets, Proceedings 1$^{st}$ International Conference on the Science of Hard Materials, Rhodes, 23-28 September, 1984, (E.A. Almond, C.A. Brookes, R. Warren eds.), 489-523.
7.  Porat, R., A. Ber: New Approach of Cutting Tool Materials, Annals of CIRP, 39 (1990) 1, 71-75.
8.  Thoors, H., H. Chadrasekaran, P. Olund: Study of Some Active Wear Mechanism in a Titanium-based Cermet when Machining Steels, Wear, 162-164 (1993), 1-11.
9.  Tönshoff, H.K., H.-G. Wobker, C. Cassel: Wear Characteristics of Cermet Cutting Tools, Annals of CIRP, 43 (1994) 1, 89-92.
10. Wick, C.: Cermet Cutting Tools, Manufacturing Engineering, December (1987), 35-40.
11. D'Errico, G.E, E. Gugliemi: Anti-Wear Properties of Cermet Cutting Tools, presented to International Conference on Trybology (Balkantrib'96), Thessaloniki (Greece), 4-8 June, 1996.
12. König, W., R. Fritsch: Physically Vapor Deposited Coatings on Cermets: Performance and Wear Phenomena in Interrupted Cutting, Surface and Coatings Technology, 68-69 (1994), 747-754.
13. Novak, S., M.S. Sokovic, B. Navinsek, M. Komac, B. Pracek: On the Wear of TiN (PVD) Coated Cermet Cutting Tools, Proceedings International Conference on Advances in Materials and Processing Technologies AMPT'95, Dublin (Ireland), 8-12 August, 1995, Vol. III, (M.S.J. Hashmi ed.), 1414-1422.
14. D'Errico, G.E., R. Chiara, E. Guglielmi, F. Rabezzana: PVD Coatings of Cermet Inserts for Milling Applications, presented to International Conference on Metallurgical Coatings and Thin Films ICMCTF '96, San Diego-CA (USA), April 22-26, 1996.
15. D'Errico, G.E., E. Guglielmi: Potential of Physical Vapour Deposited Coatings of a Cermet for Interrupted Cutting, presented to 4$^{th}$ International Conference on Advances in Surface Engineering, Newcastle upon Tyne (UK), May 14-17, 1996.
16. D'Errico, G.E., R. Calzavarini, B. Vicenzi: Performance of Physical Vapour Deposited Coatings on a Cermet Insert in Turning Operations, presented to 4$^{th}$ International Conference on Advances in Surface Engineering, Newcastle upon Tyne (UK), May 14-17, 1996.
17. Schulz, H., G. Faruffini: New PVD Coatings for Cutting Tools, Proceedings of International Conference on Innovative Metalcutting Processes and Materials (ICIM'91), Torino (Italy), October 2-4, 1991, 217-222.

# SAFE MACHINING OF MAGNESIUM

**N. Tomac**
**HIN Narvik Institute of Technology, Norway**

**K. Tønnessen**
**SINTEF, Norway**

**F.O. Rasch**
**NTNU, Trondheim, Norway**

KEY WORDS: Machining, Magnesium, Hydrogen

## ABSTRACT

In machining of magnesium alloys, water-base cutting fluids can be effectively used to eliminate the build-up formation and minimize the possibility of chip ignition. Water reacts with magnesium to form hydrogen which is flammable and potentially explosive when mixed with air. This study was carried out to estimate the quantity of hydrogen gas formed in typical machining processes. Air containing more than 4 vol% of hydrogen is possibly flammable and can be ignited by sparks or static electricity. To facilitate this study, a test method which can measure the quantity of hydrogen formation was designed. The results of the study show that the amount of hydrogen generated is relatively small. Part of the research was carried out in order to determine a safe method for storage and transport of wet magnesium chips.

## 1. INTRODUCTION

Magnesium alloys have found a growing use in transportation applications because of their low weight combined with a good dimensional stability, damping capacity, impact resistance and machinability.

Magnesium alloys can be machined rapidly and economically. Because of their hexagonal metallurgical microstructure, their machining characteristics are superior to those of other structural materials: tool life and limiting rate of removed material are very high, cutting forces are low, the surface finish is very good and the chips are well broken. Magnesium dissipates heat rapidly, and it is therefore frequently machined without a cutting fluid.

Published in: E. Kuljanic (Ed.) *Advanced Manufacturing Systems and Technology*, CISM Courses and Lectures No. 372, Springer Verlag, Wien New York, 1996.

The high thermal conductivity of magnesium and the low requirements in cutting power result in a low temperature in the cutting zone and chip.

Literature indicates that magnesium is a pyrophoric material. Magnesium chips and fines will burn in air when their temperature approaches the melting point of magnesium (650 °C). However, magnesium sheet, plate, bar, tube, and ingot can be heated to high temperatures without burning [1].

It has been described [2] that when cutting speed increases to over 500 m/min, a build-up material may occur on the flank surface of the tool. This phenomenon may lead to a high deterioration of surface finish, an increase in cutting forces and a higher fire hazard. When machining with a firmly adhered flank build-up (FBU), sparks and flashes are often observed. It has been reported that FBU formation is essentially a temperature-related phenomenon [3].

Water-base cutting fluids having the best cooling capabilities, they have been used in cutting magnesium alloys at very high cutting speeds to reduce the temperature of the workpiece, tool and chip [4]. In addition, the coolant always plays a major role in keeping the machine tool at ambient temperature and in decreasing the dimensional errors resulting from thermal expansion. Water-base cutting fluids are cheaper and better coolants, and also easier to handle in comparison with other cutting fluids.

Despite these excellent properties, water-base cutting fluids are historically not recommended in the machining of magnesium due to the fact that water reacts with magnesium to form hydrogen gas, which is flammable and explosive when mixed with air. Our research shows that magnesium alloys can be machined safely with appropriate water-base cutting fluids. The FBU problem was completely eliminated and the ignition risk presented by the magnesium chips was minimized. Frequent inspection of the machine tool area with a gas detector did not show any dangerous hydrogen level. The engineers concluded that the prohibition of magnesium machining with water-base cutting fluids is no longer justified [5].

In this paper, further attempts have been made to examine hydrogen formation from wet chips. The major outcome of the present work is the determination of a safe method for storage and transport of wet magnesium chips.

## 2. FORMATION OF HYDROGEN GAS

Magnesium reacts with water to form hydrogen gas. Hydrogen is generated by corrosion in cutting fluid [6]. Corrosion is due to electrochemical reactions, which are strongly affected by factors such as acidity and temperature of the solution. Electrochemical corrosion, the most common form of attack of metals, occurs when metal atoms lose electrons and become ions. This occurs most frequently in an aqueous medium, in which ions are present in water or moist air [7]. If magnesium is placed in such an environment, we will find that the overall reaction is:

$$Mg \rightarrow Mg^{2+} + 2e^- \text{ (anode reaction)}$$
$$2H^+ + 2e^- \rightarrow H_2 \uparrow \text{ (cathode reaction)}$$
$$Mg + 2H^+ \rightarrow Mg^{2+} + H_2 \uparrow \text{(overall reaction)}$$

The magnesium anode gradually dissolves and hydrogen bubbles evolve at the cathode.

Magnesium alloys rapidly develop a protective film of magnesium hydroxide which restricts further action [8].

$$Mg + 2H_2O \rightarrow Mg(OH)_2 + H_2 \uparrow$$

Hydrogen is the lightest of all gases. Its weight is only about 1/15 that of air, and it rises rapidly in the atmosphere. Hydrogen presents both a combustion explosion and a fire hazard. Some of the properties of hydrogen which are of interest in safety considerations are shown in Table 1. However, when hydrogen is released at low pressures, self ignition is unlikely. On the contrary, hydrogen combustion explosions occur which are characterized by very rapid pressure rises. It is important to emphasize that open air or space explosions have occurred due to large releases of gaseous hydrogen [8].

**Table 1: Properties of Hydrogen of Interest in Safety Considerations [8]**

| Property | Value |
|---|---|
| Flammability limits in air, vol% | 4.0-75.0 |
| Ignition temperature, $^0C$ | 585 |
| Flame temperature, $^0C$ | 2045 |

The flammability limits are expressing the dependence of the gas concentration. If the concentration of the gas in air is beyond the limits, the mixture will not ignite and burn.

## 3. AREAS OF HYDROGEN FORMATION

The formation of hydrogen takes place in three different locations of the machine tool system: in the working area of the machine, in the external part of the cutting fluid system and in the chip transportation and storage unit (Figure 1).

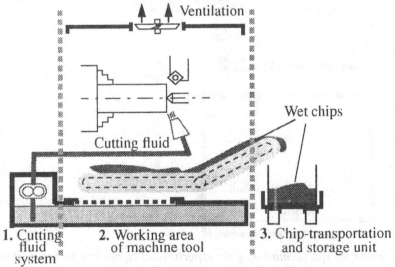

*Figure 1. Schematic representation of areas where hydrogen is generated*

To minimize the generation of hydrogen, accumulation of chips, turnings, and fine particles in the machine tool should be avoided. Chips should be stored in nonflammable, ventilated containers.

It was demonstrated that magnesium can be safely machined with the use of appropriate water-base cutting fluids. Large quantities of cutting fluid are needed to keep the workpiece, cutting tool and chips cool during high-speed machining operations.

Frequent safety inspection of the working area of the machine tool and the area of the cutting fluid system with a portable gas detector did not reveal any dangerous concentration of hydrogen. Especially, the portable instrument was used to detect the presence of hydrogen gas in confined areas of the machine tool [4].

## 4. EXPERIMENTAL

Several commercial Mg alloys were used as test materials. The magnesium chips used in this study were produced in continuous fine turning. The depth of cut 0.4 mm and the feed per revolution 0.1 mm were held constant. The chips formed when cutting magnesium are easily broken into short lengths with small curvatures. The hexagonal crystal structure of magnesium is mainly responsible for the low ductility and results in segmented chips. The ratio between the volume and weight of chips is low due to the very short chips lengths and the low density of magnesium.

For continuously monitoring the levels of hydrogen gas generated, the EXOTOX 40, portable gas detector was used. It is also designed to register when the lower flammable limit is reached such that attention is drawn to this fact. The development of hydrogen concentration in a container completely filled with wet magnesium chips, was measured by the experimental equipment illustrated in Figure 2.

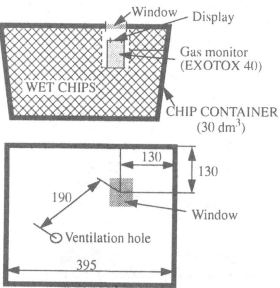

*Figure 2. Schematic representation of the experimental equipment for measuring the concentration of hydrogen generated from wet magnesium chips*

## 5. RESULTS

### 5.1 RATE OF HYDROGEN FORMATION

The hydrogen gas formation can be estimated based on the test results given in Figure 3, which shows the percentage of hydrogen generation for distilled water and two cutting fluids. Of these three fluids, the cutting fluid with pH 9.5 gave the best results.

The rate of corrosion of magnesium and consequently the generation of hydrogen gas in aqueous solutions is severely affected by the hydrogen ion concentration or pH value. The generation of hydrogen gas is several times higher at low pH values. Therefore the pH value of the cutting fluid used in the machining of magnesium should be chosen as high as possible.

It is important to note that, to avoid health hazard the pH-value should however not exceed about 9.5.

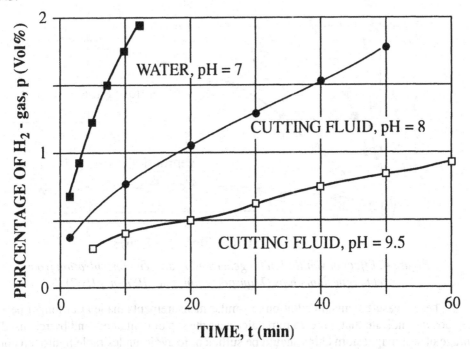

*Figure 3. Measured values of generated hydrogen concentration in the closed container filled with wet magnesium chips*
*(Transportation time from cutting zone to container: 10 min)*

The generation of hydrogen from wet magnesium chips is also highly affected by the temperature of the applied cutting fluid. The hydrogen gas formation is about 25 times higher at 65 °C than at 25 °C. An increase in temperature affects the chemical composition and physical properties of the fluid. The composition of the cutting fluid is affected by changes in solubility of the dissolved magnesium, which cause an increase of pH value. Simultaneously, the volume of generated hydrogen rises [4].

## 5.2 EFFECT OF VENTILATION ON HYDROGEN CONCENTRATION

Although hydrogen gas has a strong tendency to escape through thin walls and openings, the measuring results indicate that containers used for storage of wet magnesium chips need to be well ventilated. As illustrated in Figure 4, the lower flammable limit was reached within 7 hours even when the experimental container had a 10 mm ventilation hole in the top cover. When the ventilation hole had a diameter of 25 mm, the percentage of generated hydrogen remained well below the lower flammable limit.

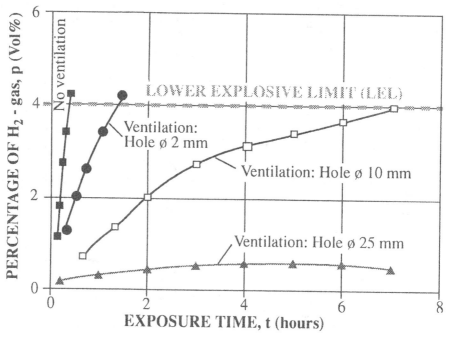

*Figure 4. Effect of ventilation on generated hydrogen concentration from wet magnesium chips (Transportation time: 10 min, pH=7)*

These measurements, in addition to similar measurements made after longer periods of exposure, indicate that three 25 mm holes near the top of containers and barrels used for storage of wet magnesium chips should be sufficient to avoid undesirable hydrogen concentrations. It is recommended that the holes are positioned just below the top cover (to avoid the entry of rain water on the magnesium chips).

## 5.3 EFFECT OF "DRYING TIME" ON HYDROGEN FORMATION

Figure 5 clearly illustrates a substantially reduced generation of hydrogen from chips dried in air for 4 hours at 25 °C in atmospheric conditions, than from wet chips stored immediately after machining.

Newly generated magnesium chips exposed to indoor or outdoor atmospheres will develop a gray film which protects the metal from corrosion while generating a negligible amount of hydrogen. X-ray diffraction analysis of corrosion films indicates that the primary reaction in corrosion of magnesium is the formation of magnesium hydroxide ($Mg(OH)_2$), which terminates the generation of hydrogen [8].

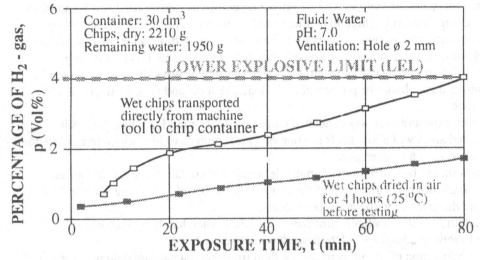

*Figure 5. Effect of drying time of wet magnesium chips on hydrogen formation, before storing in a slightly vented container*

Figure 6 shows the concentrations of hydrogen in a barrel filled with magnesium chips was first dried in air for two days. As illustrated, the generation of hydrogen decreases from day to day and was at any time much lower than the lower flammable limit of hydrogen. Experimental results presented in Figure 5 and 6 verify that the wet magnesium chips have to be dried before storage in containers and barrels, e.g. with the aid of chip centrifuges. The container used for storing magnesium chips had three 25 mm ventilation holes. The containers and barrels should be stored in a safe and readily accessible location and the area should be effectively ventilated.

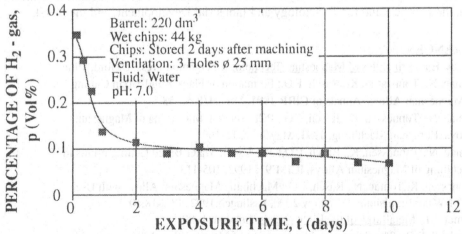

*Figure 6. Concentration of hydrogen in a barrel relative to exposure time*

## 6. CONCLUSION

The analysis of experimental results, presented above, provides some useful information on the formation of hydrogen gas from wet magnesium chips. It has been found that:

1.  A cutting fluid with a high pH-value must be selected. The generation of hydrogen gas in aqueous solutions is strongly influenced by the hydrogen ion concentration or pH-value. Cutting fluids with higher pH-values will generate less hydrogen gas.
2.  To maintain safe machining operations and handling of wet chips, accumulation of hydrogen gas should be prevented in the machine tools and the containers for chip storage.
3.  A 50 h exposure of newly generated chips to indoor or outdoor atmospheres in isolated areas will allow the formation of a gray, protective film which terminates the generation of hydrogen gas.
4.  Chips should be stored in nonflammable, ventilated containers in isolated areas. Three 25 mm holes at the top of the containers and barrels should be sufficient to avoid dangerous hydrogen concentrations.
5.  In addition to the available information on safety when handling magnesium, the following additional safety procedures are suggested:
    -   Separation of the adhered cutting fluid from wet magnesium chip with the aid of chip centrifuges.
    -   Turnings and chips should be dried before being placed in containers.
    -   Wet magnesium chips should be stored and transported in ventilated containers and vehicles.

Further investigation is desirable in order to determine the design of the ventilation systems in relation to the quantity of metal and water. There is also a need of additional investigation of the influence of environmental temperature and ventilation conditions on hydrogen formation.

## 7. ACKNOWLEDGMENT

This research was sponsored by the Royal Norwegian Council for Scientific and Industrial Research, the Nordic Fund for Technology and Industrial Development, and the Norsk Hydro.

## 8.REFERENCES

1.  Emley, E.F.: Principles of Magnesium Technology, Pergamon Press, Oxford, 1966
2.  Tomac, N., Tønnessen, K., Rasch, F.O.: Formation of Flank Build-Up in Cutting of Magnesium Alloys, Annals of CIRP, 1991, Vol. 41/1, 55-58
3.  Tomac, N., Tønnessen, K., Rasch, F.O.: PCD Tools in Machining of Magnesium Alloys, European Machining, 1991, May/June, 12-16
4.  Tomac, N., Tønnessen, K., Rasch, F.O.: The Use of Water Based Cutting Fluids in Machining of Magnesium Alloys, ICIM'91, 1992, 105-113
5.  Tønnessen, K.,Tomac, N., Rasch, F.O.: Machining Magnesium Alloys with Use of Oil-Water Emulsions, Tribology 2000, Esslingen,1992, 18.7-18.10
6.  Palmer, I.J.: Metallurgy of the Light Metals, London, 1989
7.  Askeland, D.R.: The Science and Engineering of Metals, PWS, 1994
8.  Herman, F.M.: Encyclopedia of Chemical Technology, John Wiley&Sons, Vol 12, 1978

# FAST SENSOR SYSTEMS FOR THE MONITORING
# OF WORKPIECE AND TOOL IN GRINDING

**H.K. Tönshoff, B. Karpuschewski and C. Regent**
**University of Hannover, Hannover, Germany**

KEY WORDS:  Grinding, Sensors, Residual Stress, Barkhausen Noise, Laser
Triangulation Sensor

ABSTRACT: In grinding a thermal damage of the workpiece has to be avoided reliably. For this aim various techniques are used to check the integrity state of a ground workpiece, to monitor the process itself or to evaluate the grinding wheel wear. In this paper the principles and measuring results of different suitable sensor systems are presented. A micromagnetic system is used to characterise the surface integrity state of case-hardened workpieces. A laser triangulation sensor has been developed, which can be installed in the workspace of a grinding machine. The evaluation of the sensor parameters allows to describe the micro- and macrogeometrical wear state of the grinding wheel. Examples of industrial application demonstrate the efficiency of these technologies to avoid thermal damage.

## 1. INTRODUCTION

In grinding a thermal damage of the workpiece has to be avoided reliably. The result of the grinding process strongly depends on the choice of input variables and on interferences. During grinding this interferences are usually vibrations and temperature variations. Besides the machine input variables and the initial state of the workpiece the topography state of the

Published in: E. Kuljanic (Ed.) *Advanced Manufacturing Systems and Technology*,
CISM Courses and Lectures No. 372, Springer Verlag, Wien New York, 1996.

grinding wheel is of substantial importance for the achievable machining result. The quality of the workpiece is mainly determined by geometrical characteristics. Furthermore the integrity state of the surface and sub-surface is of importance [1-3].

In industrial applications various techniques are used to check the surface integrity state of ground workpieces. Laboratory techniques like X-ray diffraction, hardness testing or metallographical inspection are time consuming and expensive, while etching tests are critical concerning environmental pollution. Visual tests and crack inspection are used extremely rarely because of their low sensitivity. The direct quality control of ground workpieces is possible with the micromagnetic analysing system. Based on the generation of Blochwall-motions in ferromagnetic materials a Barkhausennoise-signal is investigated to describe the complex surface integritiy state after grinding. White etching areas, annealing zones and tensile stresses can be separated by using different quantities. Acoustic Emissions (AE) systems are installed in the workspace of the grinding machine suitable to detect deviations from a perfect grinding process. The AE-signal is influenced by the acoustic behaviour of workpiece and grinding wheel. Furthermore the monitoring of the grinding processes is possible evaluating forces and power consumption. Laser triangulation systems mounted in the workspace allow to describe the micro- and macrogeometrical wear state of grinding wheels.

## 2. OPTICAL MONITORING OF GRINDING WHEEL

The quality of ground workpieces depends to a great extent on the grinding wheel topography, which changes during the tool life. In most cases dressing cycles or wheel changes are carried out without any information about the actual wheel wear. Commonly, grinding wheels are dressed or replaced without reaching their end of tool life in order to prevent workpiece damages, e.g. workpiece burning. As a rule the different types of wheel wear are divided in macroscopic and microscopic features. Macroscopic features describe the grinding wheel shape. Radial run out caused by unbalance or non uniform wear is the most important one. But also ellipticity and wheel waviness belong to the group of macroscopic features. Apart from the shape of the grinding tool the microscopic quality features of the wheel play an important role. The cutting capacity of the tool is described by the roughness and is influenced by grain breakage, grain pullout, flat wear and wheel loading, which may damage the workpiece surfaces due to higher thermal load. Up to now there are no measuring techniques available for monitoring the state of wheel wear in order to assure a high quality level. Therefore a new sensor has been developed at the Institute for Production Engineering and Machine Tools (IFW) based on the principle of triangulation [6]. Fig. 1 shows the basic principle of the triangulation sensor and the characteristic quantities of the bearing ratio curve (b.r.c.) that are used for the evaluation of microgeometrical grinding wheel wear. A laser diode emits monochromatic laser light on the grinding wheel surface. The scattered reflected light is focused on a position sensitive detector (PSD). If the distance to the sensor changes, the position of the reflected and focused light on the PSD also changes. The b.r.c is calculated from surface roughness profiles recorded along the grinding wheel circumference. The point of interest is the variation in slope of Abott's b.r.c.. For this purpose a new German standard has been developed with special parameters for describing the shape of curve. These parameters divide the total peak-to-valley height of a surface profile into three portions corresponding to three heights: $R_K$, the kernel or core roughness, $R_{PK}$, the reduced peak height, and $R_{VK}$, the reduced valley depth.

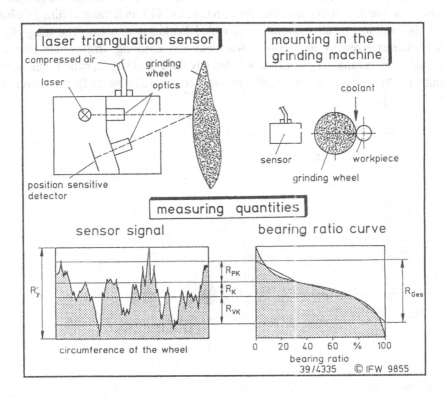

Fig. 1: Triangulation sensor for monitoring of grinding wheel wear

# 3. MICROMAGNETIC EVALUATION OF GROUND WORKPIECES

As mentioned above the surface quality can significantly be influenced by the grinding process. Until now these characteristics are determined with measuring methods, which are time consuming and cannot be used for real time testing. Most of the industrial measuring methods are only suitable for detecting occurred damages especially on the workpiece surface. The measuring principle is based on the fact, that the magnetic domain structure of ferromagnetic materials is influenced by residual stresses, hardness values and metallurgical parameters in sub-surface zones. Adjacent ferromagnetic domains with different local magnetization directions are separated by Bloch-walls. There are two kinds of Bloch-walls to be distinguished; the 180°-walls with comparatively large wall-thickness and the 90°-walls having a small wall-thickness. An exciting magnetic field causes Bloch-wall motions and rotations. As a result the total magnetization of the workpiece is changing. With a small coil of conductive wire at the surface of the workpiece the change of the magnetization due to the Bloch wall movements can be registered as an electrical pulse, fig. 2 . This magnetization is not a continuous process, rather the Bloch walls move in a single sudden jumps. Prof. Barkhausen was the first to observe this phenomenon in 1919. In honour of him the obtained signal of the addition of all movements is called Barkhausen noise. The magnetization process is characterized by the well-known hysteresis shearing. Irreversible Bloch wall motions lead to remaining magnetization without any field intensity H, called remanence $B_r$. For eliminating this remanence, the application of a certain field intensity, the coercivity $H_C$ is necessary [4,5].

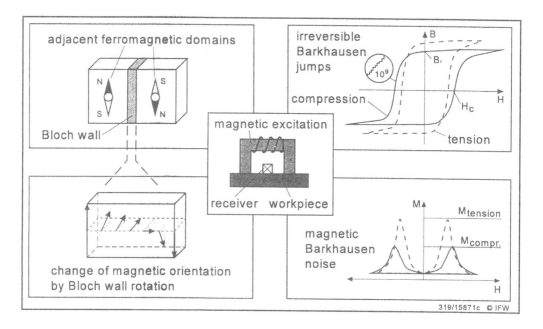

Fig. 2: Micromagnetic structure and measuring quantities

The Barkhausen noise is damped in the material due to the depth it has to pass. The main reason is the eddy current damping effect that influences the electromagnetic fields of the moving Bloch walls. The presence and the distribution of elastic stresses in the material influence the Bloch wall to find the direction of easiest orientation to the lines of magnetic flux. Subsequently the existence of compressive stresses in ferromagnetic materials reduces the intensity of the Barkhausen noise whereas tensile stresses will increase the signal. In addition to these stress sensitive properties also the hardness and structure state of the workpiece influences the Barkhausen noise. Fig. 3 shows a comparison of micromagnetic, X-ray analysis and hardness testing.

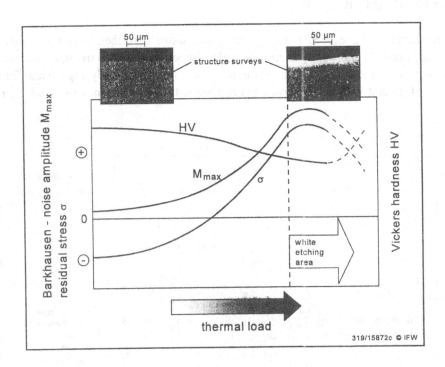

Fig. 3: Comparison of micromagnetic analysis to laboratory methods

An increasing thermal load leads to a reduction of compressive stresses and furthermore to tensile stresses at ground workpieces. As mentioned above is it frequently caused by grinding wheel wear or increasing specific material removal rate provoked by heat treating distortion. The amplitude of the Barkhausen noise is increasing with an increase of the thermal load. Negative values of the Barkhausen noise are not possible due to the principle of signal generation. So low values of $M_{max}$ correspond to compressive residual stresses, while tensile stresses correlate to high amplitudes. The Barkhausen noise amplitude is obviously a stress sensitive quantity [7]. Besides the thermal workpiece damage influences the measuring result of hardness testing, however the sensitivity is low in comparison to the residual stresses. Increasing thermal load leads to decreasing hardness of the ground surface due to annealing effects. Extreme thermal load causes a rapid increase of hardness in the surface layers because a rehardening of the material is obtained, which is visible in structure surveys as a white etching area. These significant structural changes have also an influence on residual stresses and $M_{max}$. If the thickness of the rehardening zone is exceeding 15 µm, a reduction of tensile stresses and Barkhausen noise amplitudes will be observed. The Vickers hardness measurement is reacting with a delay, because the thickness of this white layer has to reach a specific height. Otherwise the diamond is penetrating through this rehardening zone.

## 4. INDUSTRIAL APPLICATION

Many investigations have been made adapting the above described sensor systems for industrial applications and testing their potentials. Especially the laser triangulation sensor for the monitoring of grinding wheel wear and the micromagnetic analysing system for the monitoring of ground workpieces have been applied with high efficiency to avoid thermal damage.

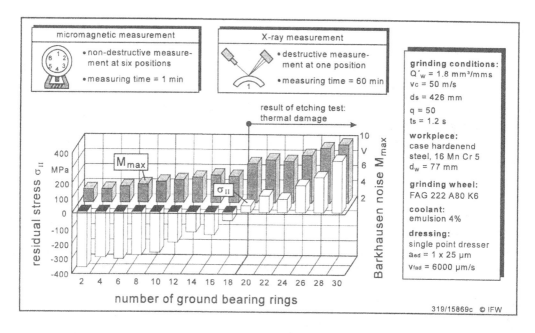

Fig. 4: Barkhausen noise as a function of grinding wheel wear

Results of micromagnetic analysis that were made with an industrial application of the above presented sensor system are shown in fig. 4. The comparison of micromagnetic and X-ray measurements represent the capability of the sensor monitoring workpiece surface integrity to detect and avoid thermal damage. For these investigations the cutting speed was set at 50 m/s and the specific material removal rate at 1.8 mm³/mms while grinding case hardened steel 16 Mn Cr 5 with a corundum wheel. With the increasing wear of the grinding wheel due to the increasing number of ground workpieces without dressing the wheel the energy in the zone of contact is significantly influenced. The increasing amount of heat penetration into the workpiece is leading to an increase of residual stresses in the workpieces. The highest X-ray measured residual tensile stresses are found for the highest workpiece number. A decrease of the compressive stresses leads to a corresponding increase of the maximum of the Barkhausen amplitude. In this investigated combination of grinding wheel and workpiece a threshold of 6 V for the Barkhausen noise amplitude can be determined. If this threshold is not reached, a thermal damage of the workpiece can be excluded. Between 5 and 6 V a control limit can be defined. Exceeding the 6 V threshold the workpiece is without any doubt thermally damaged and has to be rejected due to high tensile residual stresses and likely structural changes. Corresponding to tensile stresses and Barkhausen noise amplitudes of 6 V the bearing rings are rejected due to thermal damage. The micromagnetic analysis has been carried out on six positions spaced regularly on the circumference of the workpiece, whereas the results of the X-ray analysis base on the measurement at one position. The Barkhausen evaluation takes about one minute whereas the results of the X-ray analysis are available after 60 minutes. Therefore, the advantages of the Barkhausen noise measurement consist in a short measuring time. Also the destruction of the workpiece for measuring is not necessary as for X-ray analysis because of the limited space in the diffractometer .

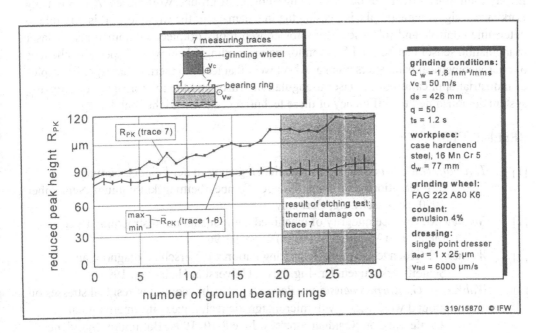

Fig. 5: Reduced peak height $R_{PK}$ as a function of grinding wheel wear

Furthermore the optical monitoring of the grinding wheel has been carried out during the investigations described above. Fig. 5 presents the microscopic parameter reduced peak height $R_{PK}$ and the result of the workpiece etching test in dependence upon grinding wheel wear. It is distinguished between traces one and six and trace seven, placed near to an edge of the grinding wheel. Regarding the low increase of $R_{PK}$ on traces one to six no significant change can be recognized. On the other hand $R_{PK}$ measured on trace seven increases significantly. It is obvious that wheel wear was not steadily on the whole width of the grinding wheel. The corresponding trace on the workpiece was proved by the etching test to be thermally damaged. Evidently the measured increase of reduced peak height $R_{PK}$ leads to thermal damage of the bearing rings. In comparison with a simulation which has been done the increase of $R_{PK}$ can be interpreted as grain obstruction. Monitoring the grinding wheel topography makes it possible to detect wheel wear and find the optimum time for dressing respectively change of grinding wheel.

## 5. CONCLUSIONS AND OUTLOOK

This paper describes different sensor systems for the monitoring of the tool, the workpiece and the process during grinding. A micromagnetic analysing systems based on the Barkhausen noise effect can be used for monitoring of ground workpieces. An increasing Barkhausen signal indicates the increasing thermal damage of the workpiece. It is possible to determine control- and tolerance limits for chosen tool/workpiece combinations. A laser triangulation sensor can be used for monitoring of grinding wheel wear. Especially changes of microgeometrical parameters point at wheel wear that leads to thermal damage. Examples of industrial application of the laser triangulation sensor and the micromagnetic analyzing system demonstrated the efficiency of these technologies to avoid thermal damage.

## REFERENCES

[1]     *Tönshoff, H.K., Brinksmeier, E.:* Qualitätsregelung in der Feinbearbeitung,
               6. Internationales Braunschweiger Feinbearbeitungskolloquium, September
               1990, Braunschweig, Germany
[2]     *Neailey, K.:* Surface integrity of machined components-microstructural aspects,
               Metals and Materials, 4 (1988) 2, pp. 93-96
[3]     *Werner, F.:* Hochgeschwindigkeitstriangulation zur Verschleißdiagnose an
               Schleifwerkzeugen, Dr.-Ing. Diss., Universität Hannover, 1994
[4]     *Wobker, H.-G., Karpuschewski, B., Regent, C.:* Quality control of residual stresses on
               ground workpieces with micromagnetic techniques, 4th International
               Conference on Residual Stresses, June 8-10, 1994, Baltimore, Maryland
[5]     *Birkett, A.J.:* The use of Barkhausen noise to investigate residual stresses in machined
               components, International Conference on applied stress analysis, Universitiy
               of Nottingham, 30-31 August, 1990
[6]     *Karpuschewski, B.:* Mikromagnetische Randzonenanalyse geschliffener
               einsatzgehärteter Bauteile, Dr.-Ing. Diss., Universität Hannover, 1995
[7]     *Hill, R., Geng, R.S., Cowking, A.:* Acoustic and electromagnetic Barkhausen emission
               from iron and steel: application of the techniques to weld quality assessment
               and nde, British Journal of NDT, Vol. 35, No. 5, May 1993

# TRUING OF VITRIFIED-BOND CBN GRINDING WHEELS
## FOR HIGH-SPEED OPERATIONS

**M. Jakobuss and J.A. Webster**
**University of Connecticut, Storrs, CT, U.S.A.**

KEY WORDS: Grinding, Superabrasives, Conditioning

ABSTRACT:

This study investigated the truing of vitrified-bond CBN grinding wheels. A rotary cup-type diamond truer was used to condition grinding wheels under positive as well as negative dressing speed ratios. Cutting speeds between 30 and 120 m/s were used in an external cylindrical grinding mode. The grinding wheel surface was recorded during all stages of truing and grinding using 2D surface topography measurements. The performance of each truing condition on wheel topography, grinding behavior, and on workpiece topography and integrity were compared. Dressing speed ratio has been found to be a suitable parameter for appropriate CBN grinding wheel preparation.

## 1. INTRODUCTION

Grinding is perhaps the most popular finishing operation for engineering materials which require smooth surfaces and fine tolerances. However, in order to consolidate and extend the position of grinding technology as a quality-defining finishing method, improvements in the efficiency of grinding processes have to be made. Powerful machinery and, especially,

Published in: E. Kuljanic (Ed.) *Advanced Manufacturing Systems and Technology*,
CISM Courses and Lectures No. 372, Springer Verlag, Wien New York, 1996.

extremely wear resistant superabrasives, such as diamond and cubic crystalline boron nitride (CBN), are available and are opening up economical advantages.

CBN has excellent thermal stability and thermal conductivity and its extreme hardness and wear resistance has the potential for high performance grinding of ferrous materials, hardened steels, and alloys of hardness Rc 50 and higher. Four basic types of bond are used for CBN grinding wheels: resinoid, vitreous, metal, and electroplated. The vitreous bond material, also known as glass or ceramic bond material, is regarded as the most versatile of bonds. It provides high bonding strength, while allowing modifications to the strength and chip clearance capability by altering the wheel porosity and structure. But the most important advantage of vitrified-bond CBN grinding wheels is the ease of conditioning.

The conditioning of a wheel surface is a process comprised of three stages: truing, sharpening and cleaning. Truing establishes roundness and concentricity to the spindle axis and produces the desired grinding wheel profile. Sharpening creates the cutting ability and finally, cleaning removes chip, grit, and bond residues from the pores of the grinding wheel [1]. The variations in the conditioning process can lead to different grinding behavior even with the same grinding parameters [2]. By selecting an appropriate conditioning process and by varying the conditioning parameters, the grinding wheel topography can be adapted to grinding process requirements. Higher removal rates require sharp grit cutting edges and, especially, adequate chip space, whereas good workpiece surface qualities require a larger number of grit cutting edges [3]. The grinding wheel behavior during the machining process is influenced strongly by the grinding wheel topography, which is dictated by the conditioning operations, as shown by Pahlitzsch for conventional abrasives[4].

Therefore, an investigation was conducted, to examine the link between grinding wheel topography and grinding performance, as an initial step towards the development of an in-process grinding wheel topography monitoring system for adaptive process control.

## 2. EXPERIMENTAL PROCEDURE

Experimental plunge grinding experiments have been performed on an external cylindrical grinder. Three vitrified-bond CBN grinding wheels with the following specifications were used:
- CB 00080 M 200 VN1 (diameter $d_s$ = 100 mm, width $b_s$ = 13 mm),
- CB 00170 M 200 VN1 (diameter $d_s$ = 100 mm, width $b_s$ = 13 mm),
- B 126-49-R0200-110-BI (diameter $d_s$ = 125 mm, width $b_s$ = 10 mm)

Prior to grinding, conditioning of the grinding wheel surface was conducted through the use of a rotary cup-type diamond truer (diameter $d_d$ = 50.8 mm, contact width $b_d$ = 1.75 mm), which was mounted on a liquid-cooled high-frequency spindle The orientation of the truing device was set up for cross-axis truing. By turning the truer spindle through 20°, the circumferential velocities in the wheel-truer-contact could be uni-directional (downdressing,

$q_d > 0$) or counter-directional (updressing, $q_d < 0$), see Figure 1. Three dressing speed ratios were used to true the grinding wheels:
- uni-directional, downdressing:     $q_d = + 0.75$,
- uni-directional, downdressing:     $q_d = + 0.35$,
- counter-directional, updressing:  $q_d = - 0.5$.

The traverse truing lead of the grinding wheel across the truing cup was set at $f_{ad} = 0.23$ mm/U and a radial truing compensation of $a_{ed} = 1.27$ µm incremented every pass.

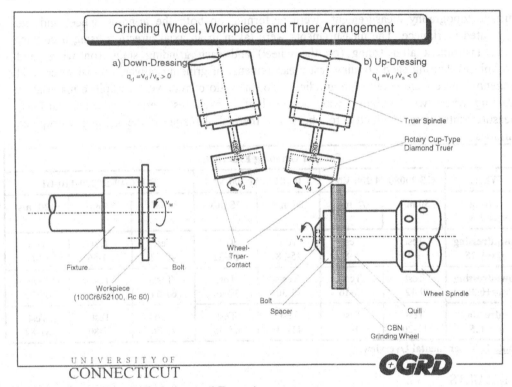

**Figure 1: Grinding Wheel, Workpiece and Truer Arrangement.**

Three cutting speeds, $v_c$, were used throughout the experiments:
- cutting speed: $v_c = 30$ m/s,
- cutting speed: $v_c = 75$ m/s,
- cutting speed: $v_c = 120$ m/s.

Two wheels were tested with the low and the medium cutting speed. The third wheel could be used in all three cutting speeds, due to the higher speed rating.

Both truing and grinding were performed using a water-based metalworking fluid in a concentration range between 6.5% and 7.5%. Upon completion of the truing operation the grinding wheels were self-sharpened during machining of the workpieces. The workpieces were of 100Cr6 (AISI-52100) bearing steel, heat treated to Rc 60 with a diameter of

$d_w = 92.1$ mm and a width of $b_w = 6.35$ mm. A total of 23.5 mm was machined off the diameter in four separate stages. First, a self-sharpening grind with a diametric stock removal of 0.5 mm, then a grinding sequence with a diametric stock removal of 3 mm was followed by two diametric stock removals of 10 mm. The rotations of workpiece and grinding wheel were in opposite directions, with a fixed workpiece speed of $v_w = 2$ m/s. The specific material removal rate was fixed at $Q'_w = 2.5$ mm$^3$/(mm ·s), without a sparkout at the end of each grind, for the 84 grinding tests outlined in Table 1.

Surface topography measurements were performed on both the grinding wheel and the associated workpiece. The measurements were taken in an axial direction using a contact stylus instrument after truing (on the wheel) and each grinding stage (on wheel and workpiece). During each grinding test measurements of grinding power, normal force, and tangential force were taken. The grinding ratio, the ratio of removed workpiece material to grinding wheel wear volume, was measured after each test series and microhardness measurements were conducted to check for thermal damage of the ground workpiece surfaces.

| Table of Grinding Experiments | | | | | | |
|---|---|---|---|---|---|---|
| **Wheel** | **CB 00080 M 200 VN1** | | **CB 00170 M 200 VN1** | | **B 126-49-R0200-110-BI** | | |
| cutting speed $v_c$ | 30 m/s | 75 m/s | 30 m/s | 75 m/s | 30 m/s | 75 m/s | 120 m/s |
| downdressing $q_d$=+0.75 | Test 1-4 | Test 5-8 | Test 25-28 | Test 29-32 | Test 49-52 | Test 53-56 | Test 57-60 |
| downdressing $q_d$=+0.35 | Test 9-12 | Test 13-16 | Test 33-36 | Test 37-40 | Test 61-64 | Test 65-68 | Test 69-72 |
| updressing $q_d$=-0.5 | Test 17-20 | Test 21-24 | Test 41-44 | Test 45-48 | Test 73-76 | Test 77-80 | Test 81-84 |

**Table 1: Experimental Overview.**

## 3. RESULTS

This experimental work was conducted to examine the influence of vitrified-bond CBN grinding wheel topography on grinding behavior and on ground workpiece topography and integrity.

In Figure 2 the specific normal force, $F'_n$, is plotted versus specific material removal, $V'_w$, for cutting speeds $v_c = 30$ m/s and $v_c = 75$ m/s. Initial normal forces were high, as the wheels were used directly after truing without a separate sharpening process, since required adequate chip space between the CBN grits did not exist. Further grinding gradually opened up the chip space and grinding forces reduced accordingly.

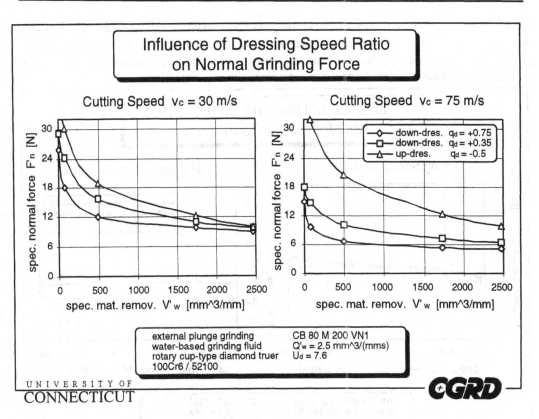

**Figure 2: Influence of Dressing Speed Ratio on Normal Grinding Force.**

The influence of dressing speed ratio, $q_d$ , on specific normal grinding force, $F'_n$ , was well pronounced, see Figure 2. Significantly higher normal forces were measured for the updressed wheels, leading to dull grinding due to the smooth wheel surface generated during truing. The circumferential velocities of wheel and truer were directionally opposite during truing and therefore led to increased shearing and decreased splitting of the CBN grits. Bonding material covered the CBN grains and no cutting edges protruded out of the matrix. The high normal forces for grinding with updressed wheels can result in the danger of thermal workpiece damage. The downdressed wheels, especially, with a dressing speed ratio of $q_d = +0.75$, machined the workpieces with lower forces. The truing operations with a positive dressing speed ratio, $q_d$ , subjected the CBN grits to higher compressive loads and increased the likelihood of grit and bond material splitting. The rougher grinding wheel surfaces result in higher specific material removal rates $Q'_w$ , without danger of thermal workpiece damage.

During the tests it was observed that increasing cutting speeds led to lower grinding forces. This was expected, since the average ground chip thickness decreased for higher cutting speeds and the work done by each single cutting edge became smaller.

As stated earlier, surface roughness measurements were taken on the ground workpieces in the axial direction. Examination of surface finish, $R_a$ , showed influences of grinding wheel grit size, cutting speed, $v_c$ , specific material removal, $V'_w$ , and dressing speed ratio, $q_d$ , on workpiece roughness values. A graphical representation is given in Figure 3.

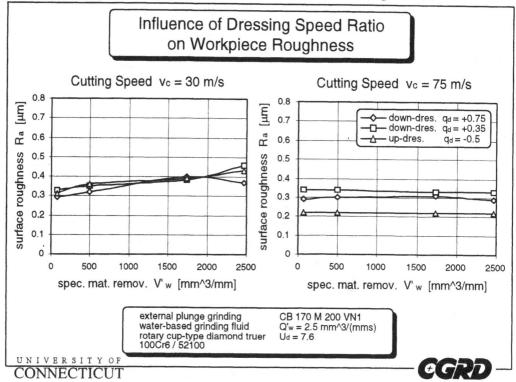

**Figure 3:** Influence of Dressing Speed Ratio on Workpiece Roughness.

The surface finish, $R_a$ , of workpieces ground with low cutting speeds, $v_c$ , deteriorated with increasing specific material removal, $V'_w$ . An explanation might be that low speed grinding leads to an increase of chip thickness, consequently subjecting the abrasive grits to higher loads. The high loads crush the vitrified bond material forcing the loosely bonded grits to be released. This rough grinding wheel surface produced the worst workpiece finish values. In contrast, the higher cutting speed of $v_c = 75$ m/s led to better workpiece surface finish values. The values remained at the same level throughout the whole experiment but were significantly affected by the dressing speed ratio, $q_d$ . The workpiece $R_a$ values were always best for samples ground with updressed wheels. The smooth grinding wheel surface generated during truing with negative dressing speed ratio, which caused the highest normal force values earlier (see Figure 2) machined the best workpiece surfaces.

Figure 3 proves that there is a relationship between the dressing speed ratio, $q_d$ , and the ground workpiece surface roughness. In addition, Figure 2 clearly shows the influence of dressing speed ratio, $q_d$ , on normal grinding force, $F'_n$ , i.e. wheel sharpness. Tests were conducted to find a correlation between grinding wheel topography and produced workpiece topography. But examined topography values, including wheel surface finish, $R_a$ , average peak-to-valley-height, $R_z$ , wheel waviness, $W_t$ , and bearing ratio, $t_p$ , unfortunately, did not lead to correlations of sufficient confidence.

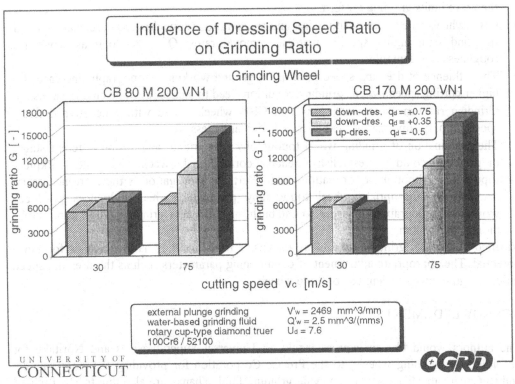

**Figure 4: Influence of Dressing Speed Ratio on Grinding Ratio.**

After the finishing grind of each experiment a workpiece was lightly ground to imprint both the worn and the unworn part of the grinding wheel surface in the workpiece. This wear sample was traced by the contact stylus instrument, to measure the radial wear depth of the wheel, from which the final G-ratio was calculated. Figure 4 shows the results. A considerable improvement in tool life could be noted for the updressed wheels when used with $v_c = 75$ m/s. Conversely, the downdressed wheels wore up to twice as fast as the updressed wheels. For the low cutting speed, $v_c = 30$ m/s, the influence of dressing speed ratio ,$q_d$ , could be neglected. The higher loads caused by larger average chip thicknesses seemed to be dominant, giving nearly constant G-ratios.

## 4. CONCLUSIONS

This research has shown that the dressing speed ratio, $q_d$, a parameter describing the ratio of circumferential velocities in the contact zone between grinding wheel and truing tool, influences grinding behavior and tool life of vitrified bond CBN grinding wheels as well as the produced workpiece topography. These influences are as follows:

- Truing with negative dressing speed ratio (updressing) generates a grinding wheel surface which produces best workpiece roughness, but grinds with high forces, and therefore limits grinding performance.
- CBN wheels trued with positive dressing speed ratios (downdressing) are sharp enough to grind with higher specific material removal rates, $Q'_w$ , as long as workpiece roughness is acceptable.
- The influence of dressing speed ratio $q_d$ on ground workpiece topography increases for higher cutting speeds $v_c$ , as grinding conditions lead to smaller average chip thicknesses.
- Grinding ratio measurements show that CBN wheels, trued with a negative dressing speed ratio, $q_d$ (updressing), experience significantly less wear.
- The examination of grinding wheel topography after truing, and during different stages of grinding, could not establish a reliable correlation between wheel and workpiece topography, although, a correlation exists. Further investigation is therefore necessary, to find the appropriate wheel topography parameter, which influences produced workpiece topography, in order to establish an in-process grinding wheel topography monitoring system.

Vitrified-bond CBN grinding wheels are expensive tools but they have a high performance potential. The appropriate adjustment of conditioning parameters such as the dressing speed ratio, $q_d$ , guarantee efficient use of these wheels.

ACKNOWLEDGMENTS

The authors would like to express thanks to Universal Superabrasives and Noritake for providing the grinding wheels, to the Precise Corporation for providing the truer spindle, and to Cincinnati Milacron for supplying grinding fluid. Thanks are also due to Dr. Richard P. Lindsay for his advice during this project.

REFERENCES

1. Saljé, E., Harbs, U.: Wirkungsweise und Anwendungen von Konditionierverfahren, Annals of the CIRP, 39 (1990) 1, pp. 337-340
2. Rowe, W.B., Chen, X., Mills, B.: Towards an Adaptive Strategy for Dressing in Grinding Operations, Proceedings of the 31st International MATADOR Conference, 1995
3. Klocke, F., König, W.: Appropriate Conditioning Strategies Increase the Performance Capabilities of Vitrified-Bond CBN Grinding Wheels, Annals of the CIRP, 44 (1995), pp. 305-310
4. Pahlitzsch, G., Appun, J.: Einfluss der Abrichtbedingungen auf Schleifvorgang und Schleifergebnis beim Rundschleifen, Werkstattechnik und Maschinenbau, 43 (1953) 9, pp. 369-403

# FORCES IN GEAR GRINDING - THEORETICAL
# AND EXPERIMENTAL APPROACH

**B.W. Kruszynski and S. Midera**
**Technical University of Łódz, Łódz, Poland**

KEY WORDS: Generating Gear Grinding, Grinding Forces

ABSTRACT: Forces in generating gear grinding were investigated. On the basis of detailed analysis of generation of a tooth profile in this manufacturing operation theoretical approach was developed to calculate grinding forces in each generating stroke of grinding wheel. Variations of such parameters like: cross-section area and shape of the material layer being removed, wheelspeed, radius of curvature of conical grinding wheel, etc. were taken into consideration. Measurements of grinding forces were also carried out and experimental results were compared with the calculated ones.

## 1. INTRODUCTION

Generating gear grinding with a trapezoidal grinding wheel, the so-called Niles method, is often used in industrial practice as finishing operation for high quality, hardened gears. It has many advantages like high machining accuracy and independence of tool geometry from geometry of gear being machined. It is also relatively easy to perform any modification of a tooth profile.

The choice of grinding parameters in this processes should be a compromise between high productivity rates and high quality of gears being ground. High productivity rates usually lead to the increase of grinding forces and grinding power which in turn leads to deterioration of gear accuracy and quality of a surface layer of ground gear teeth. It means that pre-

Published in: E. Kuljanic (Ed.) *Advanced Manufacturing Systems and Technology*,
CISM Courses and Lectures No. 372, Springer Verlag, Wien New York, 1996.

diction and control of grinding forces in this complex manufacturing method is a key problem for making high quality gears effectively. Also possibility to calculate grinding forces
on the basis of grinding conditions is a basic problem in modelling of generating gear grinding process [1].

## 2. CREATION OF TOOTH PROFILE IN GENERATING GEAR GRINDING

Generating gear grinding is a complex manufacturing method. To obtain an involute shape
of tooth profile the following kinematical movements have to be performed, cf. fig.1:
- strictly related linear ($v_{st}$) and rotational ($n_{fa}$) movements of the workpiece which
  both give a generating movement,
- workspeed $v_w$ which is performed by reciprocating movement of grinding wheel,
- rotational movement of grinding wheel.

After completing grinding of one tooth space an indexing movement of workpiece is
performed to start machining of the subsequent tooth profile.

Due to such kinematics grinding allowance along the tooth profile is removed in subsequent
generating strokes of grinding wheel in the manner shown in fig. 2. Layers being removed in
these material layers being ground in subsequent generating strokes have complex shapes
which changes from one stroke to another. Grinding depth, wheel/workpiece contact area,

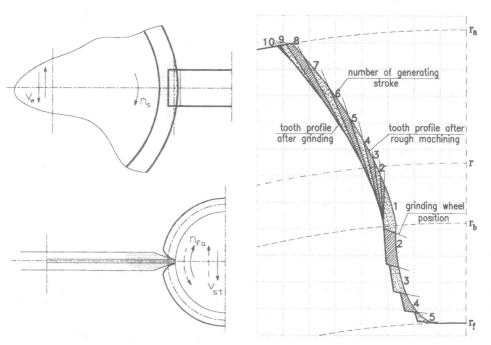

Fig.1 Kinematical movements of the gear        Fig.2 Division of the grinding allowance
             grinding process

grinding wheel curvature, etc. also change which makes calculations of grinding forces much more difficult than in surface or cylindrical grinding. In fig. 3 an example of changes of cross-section area is shown for tooth profile shaped in 10 generating strokes.

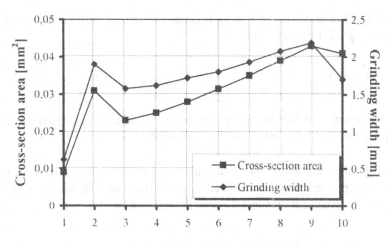

**Generating stroke number**

Fig. 3   Changes of cross-section area and grinding width
in subsequent generating strokes
m=5, z=22, $a_e$=0.03mm, 10 generating strokes

Because of changes of grinding conditions in subsequent generating strokes grinding forces also change from one stroke to another. An example of such changes is shown in fig. 4.

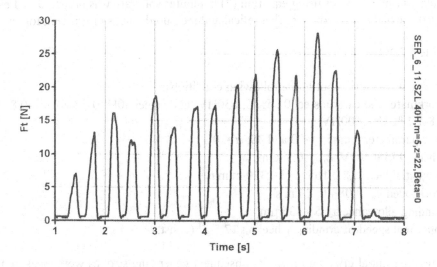

Fig. 4 Changes of tangential grinding force in shaping of one tooth profile

It is visible from this figure that in some initial generating strokes (five in this case) relatively higher grinding forces are observed, which is due to additional material removal in the root fillet area.

To calculate tangential grinding force in each generating stroke of grinding wheel the well-known equation (eg. [2]) was adapted:

$$F_t' = F_0 \, h_{eq}^f \tag{1}$$

where: $F_t'$ is tangential grinding force per unit grinding width, $h_{eq}$ is equivalent chip thickness, $F_0$ and $f$ are constants dependent on work material.

Assuming that grinding conditions in particular stroke are similar to surface grinding with variable grinding depth and that a workspeed $v_w$ is constant the following equation was developed to calculate tangential grinding force in the generating stroke „i":

$$F_{ti} = F_0 \, v_w^f \int_0^{b_{Di}} a_i^f \, v_{si}^{-f} \, dx \tag{2}$$

where: $b_{Di}$ is grinding width (variable in subsequent strokes), $a_i$ is grinding depth (variable over grinding width), $v_s$ is wheelspeed (variable over grinding width due to a conical shape of grinding wheel).

To calculate grinding forces using equation (2) computer software was prepared and experiments were carried out to compare theoretical and measured values of grinding forces.

## 3. EXPERIMENTS

Experiments were carried out in the following conditions:
- workmaterials: carbon steel 0.55%C, 650HV; alloy steel 40H (0.4 %C, 0.5-0.8 %Mn, 0.8-1.1 % Cr), 600HV,
- gear parameters: $m=5$; $z=20$ and $30$; $\alpha=20°$,
- wheel: 99A80M8V, $\phi_{max}=0.330$ m;
- tangential feed $v_{st}$: 0.165 and 0.330 m min$^{-1}$;
- workspeed $v_w$: 0,08 - 0, 24 m s$^{-1}$
- grinding allowance $a_e$: 5 - 135 μm;
- rotational speed of grinding wheel $n_s$: 27.5 s$^{-1}$ (const.)

In each test tangential grinding forces in subsequent generating strokes were measured with KISTLER 9272 dynamometer.

Preliminary tests were carried out to evaluate model parameters - constants $F_o$ and f in equation (3). Values of these constants are presented in table 1. The exponent f is of the same value for both workmaterials but constant $F_o$ is higher for alloy steel.

Table 1 Model parameters for different workmaterials

| Workmaterial | $F_o$ | f |
|---|---|---|
| Carbon steel .55%C | 17 | 0,3 |
| Alloy steel 40H | 20 | 0,3 |

Having model parameters determined, it was possible to check model validity. Results of measurements of grinding forces obtained for the wide range of grinding conditions were compared with calculated values.

In figures 5 and 6 experimental and calculated tangential forces are shown for different grinding conditions. It is visible from these figures that tangential grinding force changes significantly from one generating stroke to another. It can also be seen that calculated values of grinding forces coincide with those obtained from experiments. The coefficient of correlation of measured and calculated values in all tests was never lower than 0.9. It indicates that the developed model is correct and calculations based on equation (2) are accurate enough to determine grinding forces in generating gear grinding process.

Differences of measured values of grinding forces observed in every two consecutive strokes of grinding wheel are due to the fact that in reciprocating movement of grinding wheel one of them is performed as up-grinding and the other one as down-grinding. It is possible to include this phenomena into the model by evaluating two sets of constants in equation (3) - one for up-grinding and the other for down-grinding.

## 4. CONCLUSIONS

The model of generating gear grinding process described above makes calculations of forces in generating gear grinding possible. Correctness of the model was proved in a wide range of grinding conditions. This model will allow theoretical analysis of surface layer creation, grinding wheel load and wear processes, etc. in this complicated manufacturing process.

ACKNOWLEDGEMENTS: The research was funded by the Polish Committee for Research (Komitet Badań Naukowych), Grant No 7 S102 026 06.

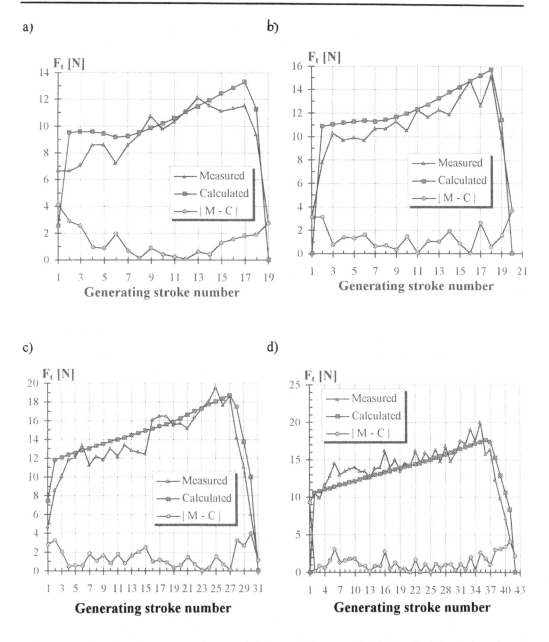

Fig. 4 Comparison of calculated and measured tangential forces in subsequent
generating strokes. Work material 0.55%C carbon steel
a) $v_w$=0.08 ms$^{-1}$; DH=75 min$^{-1}$; $a_e$=0.037 mm
b) $v_w$=0.08 ms$^{-1}$; DH=75 min$^{-1}$; $a_e$=0.051 mm
c) $v_w$=0.06 ms$^{-1}$; DH=75 min$^{-1}$; $a_e$=0.095 mm
d) $v_w$=0.06 ms$^{-1}$; DH=100 min$^{-1}$; $a_e$=0.120 mm

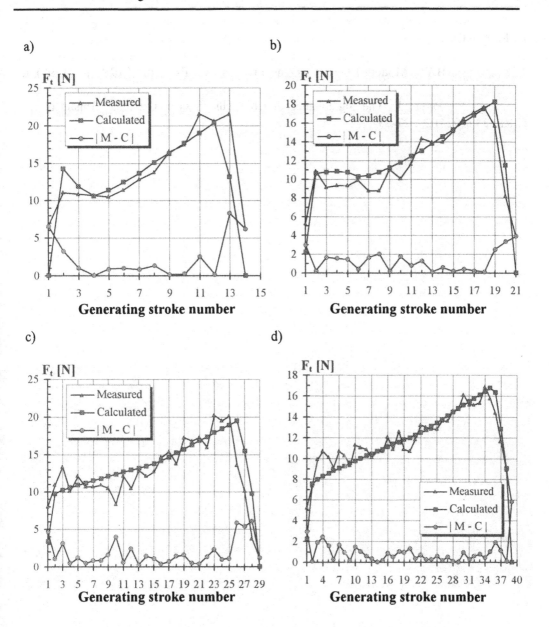

Fig. 5 Comparison of calculated and measured tangential forces in subsequent generating strokes. Work material: 40H alloy steel
a) $v_w$=0.11 ms$^{-1}$; DH=50 min$^{-1}$; $a_e$=0.040 mm
b) $v_w$=0.08 ms$^{-1}$; DH=75 min$^{-1}$; $a_e$=0.050 mm
c) $v_w$=0.08 ms$^{-1}$; DH=100 min$^{-1}$; $a_e$=0.080 mm
d) $v_w$=0.06 ms$^{-1}$; DH=100 min$^{-1}$; $a_e$=0.075 mm

REFERENCES

1. Kruszyński B.W.: Model of Gear Grinding Process. Annals of the CIRP, 44 (1995) 1, 321-324.
2. Snoeys R., Peters J.: The Significance of Chip Thickness in Grinding. Annals of the CIRP, 23 (1974) 2, 227-237.

# OPTIMUM CUTTING CONDITION DETERMINATION
## FOR PRECISION TURNING

**V. Ostafiev**

**City University of Hong Kong, Hong Kong**

**D. Ostafiev**

**University of Melbourne, Melbourne, Australia**

KEY WORDS: Precision Turning, Optimization, Multi-Sensor System

ABSTRACT: One of the biggest problems for precision turning is to find out cutting conditions for the smallest surface roughness and the biggest tool life to satisfy the high accuracy requirement. A new approach for the optimum condition determination has been developed by using multiple sensor machining system. The system sensors mounted on the CNC lathe have measured tool vibrations, cutting forces, motor power and e.m.f. signal. This paper presents the method leading to the determination of the optimal feed value for given cutting conditions. The experimental results indicate that, under gradually changing cutting tool feed, the all turning parameters also alter emphasizing their close correlation. Under this feed variation some of the system measuring parameters have minimum or maximum points that relieve optimum feed determination taking into account all machining conditions. Also, the maximum approach of the tool life has been defined under the optimum feed. The cutting speed increasing leads to increasing optimum feed and its value changes for every pair of cutting tool and machining materials. The results show that the turning under optimum feed would improve surface roughness and tool life at least of 25 percent.

## 1. INTRODUCTION

The main purpose for precision turning is to get high accuracy symmetrical surface with the smallest roughness. The solution depends very much on cutting conditions. Usually the depth of cut is predicted by a part accuracy but cutting speed and feed should be properly selected. It was found out the minimum surface roughness could be determined under some optimum feed rate. [1,2]. The optimum feed proposition for fine machining had been made by Pahlitzsch and Semmer [1] who established minimum undeformed chip thickness

Published in: E. Kuljanic (Ed.) *Advanced Manufacturing Systems and Technology,* CISM Courses and Lectures No. 372, Springer Verlag, Wien New York, 1996.

with the set of gradually changing feed for different cutting speeds. The feed changing had been led by means of the scanning frequency generator (1).The generator frequency monitoring has been provided by the frequency counter (4) and connected to the CNC lathe (2) for the feed step motor control (3). The frequency impulse has been recorded at the same time as all other turning parameters. The feed rate changing was checked by repeated cutting tool idle movement after workpiece turning and the necessary surface marks were made for different feed rate values.

The dynamometer (5) for three cutting force component measurements . has been connected through the amplifier (6) to the same recorder (7) as the frequency counter (4) for obtaining their signal comparisons more precisely.

E.m.f. measurement was made by using slip ring (8) , preamplifier (9), amplifier (10) and oscilograf (7). The same process was used for vibration measurements by accelerometer (14) working in frequency range of 5 Hz...10 kHz. through amplifier (15). The machining power signals from a motor (11) were measured by the wattmeter (12) and through the amplifier (13) send to the recorder (7). This kind of setup has permitted to record all turning parameters simultaneously at one recorder to make analysis of their interrelations while gradually changing the cutting conditions.

The machining surface roughness has been measured by the stylus along the workpiece and their meanings were evaluated according to the surface marks made for different feed rate values. Additional experiments were made for tool life determination under fixed cutting conditions in the range of feed rate and cutting speed changing.

The machining conditions were chosen as follows: feed rate 0.010- 0.100 mm/r.; cutting speed 1.96-2.70 m./ s., depth cut 0.5 mm. The workpiece materials were carbon steel S45, chromium alloy steel S40X and nickel-chromium-titanium alloy steel 1X18H9T.The standard Sandvik Coromant cutting tip from cemented carbide had been used with nose radii 0.6 mm.

## 3. RESULTS AND DISCUSSIONS

The experimental results of turning changing parameters at different feed rates are shown in Fig.2. It can be observed from the figure that all parameters except for e.m.f. have minimum and maximum values for feed rate increasing from 0.010 to 0.095 mm /r.

The vibration signal G has changed twice : at the beginning it was increasing up to feed rate 0.020 mm/r. than it started decreasing to the minimum signal at the feed rate 0.050 mm./r. and after that it increased gradually. The surface roughness, in spite of the increasing feed rate from 0.010 to 0.050 mm./r., has decreased from Rmax = 5.4 μm to Rmax = 3.6 μm. and than increase to Rmax = 7.0 μm with feed rate increasing to 0.090 mm./r. The power P has a small minimum at the feed rate close to 0.050 mm /r. indicating less sensitive correlation for feed rate changing. The tool life has a maximum T = 72 min. at the feed rate 0.045 mm./r. However, at smaller feed rates 0.006 and 0.080 mm./r. the tool life decreased to 54- 58 min. ( in the amount of 23% ).

Analysis of these results pointed out that all turning parameters have minimum value for surface roughness, cutting forces, power and maximum tool life for feed rate in the range of 0.045-0.50 mm./r. and with cutting speed of 1.96 m/s. The surface roughness has more close correlation with vibration signal and cutting force that on line monitoring could be used for cutting condition determination.

for cutting. But the optimum feed determination had been described only by the kinematics solution. The properties of cutting tool-workpiece materials under cutting, machining vibrations, workpiece design should certainly be taken into account in real machining for the feed determination.
The micro turning research [2] also shows the change in Rmax is described by the folder line whose breakpoint depends on the cutting conditions, tool geometry and cutting liquids.

But now it takes a lot of time to find out, on-line, this optimum cutting speed and feed values for precision turning of different workpiece designs and materials. Because there is a gradually increasing precision of operation performance the new express methods for optimum precision cutting condition determination need to be investigated and developed for industry application. According industry demands in achieving the optimum machining performance for precision turning many are concerned with the smallest surface roughness and highest workpiece accuracy .

## 2. EXPERIMENTAL EQUIPMENT AND PROCEDURES

According many investigations [3,4] there is a close correlation between machining vibrations, cutting forces, tool life and surface roughness etc. while turning. Thus the correlation coefficient between vibrations and surface roughness has been determined as 0.41-0.57.[5]. Also it has been shown the nonmonotonic dependencies of cutting forces, surface roughness, tool life, vibrations, e.m.f. when cutting feed and speed are changing gradually. There are some minimum and maximum values for almost every turning parameter while cutting condition changes. But to find out the optimum cutting conditions for precision turning all parameters should be studied simultaneously to take into account their complex interrelations and nonmonotonic changing. The CNC lathe advantage to gradually change feed and speed by their programming and measuring all the necessary turning parameters simultaneously opens a new opportunity for the express method of optimum condition determination. To solve the problem the special experimental setup had been designed as shown in Fig.1. The setup has been mounted on a precision CNC lathe

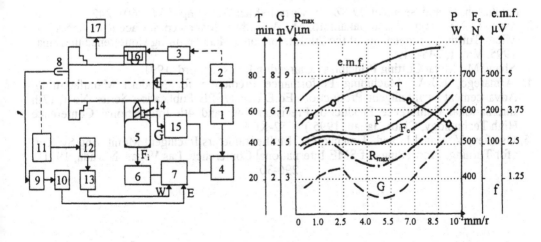

Fig. 1 Experimental setup          Fig. 2. Dependence turning parameters on feed rate

To investigate cutting speed influence on the vibration signals, some experiments have been conducted under gradually feed rate changing for different cutting speeds .The results show a general trend of increasing feed rate for vibration signal minimums with increasing cutting speed. Thus the minimum vibration signal for cutting speed 1.96 m/s. is 0.050 mm/r., for cutting speed 2.3 m/s. is 0.060 mm/r. and for cutting speed 2.7 m/s. is 0.068 mm/r. Also the minimum pick is much more brightly expressed for a smaller cutting speed than for the higher speed. The surface roughness has not changed much under this cutting speed increasing ( from Rmax = 3.6 µm to 3.1 µm.) but tool life decreased significantly in 1.5-2 times. The tools wear investigation has found out any relationship with surface roughness under this conditions.

Another experiment has been conducted to determine the interrelations for turning parameters for different kinds of workpiece materials. The minimum vibration signals have been received at feed rate 0.038mm/r. for steel 1H18H9T, at 0.43mm/r. for the chromium alloy steel S20X and at 0.45mm/r. for carbide steel S45. The latest two closer to each other because the corresponding materials have similar properties. Therefore, there is a positive correlation between vibration signal and surface roughness indicating the smallest vibration signal as well as surface roughness for carbon steel S45 (Rmax =3.6 µm) and the bigger their values (Rmax = 5.6 µm) obtained for nickel-chromium-titanium alloy steel 1X18H9T.

## 4. CONCLUSION

The new method for determining optimum cutting conditions for precision turning has been developed. The method uses of the high level correlation between precision turning parameters for their monitoring. The optimum cutting conditions could be more precisely specified by the determination of vibration signal minimum while feed rate is gradually changing. The vibration signal could easily be taken on line cutting and all machining conditions would be taken into account for the optimum feed rate determination.

## REFERENCES

1. Pahiltzsch, G. and Semmler, D.: Z. fur wirtschaftich Fertigung. 55 (1960), 242.
2. Asao, T., Mizugaki, Y. and Sakamoto, M.: A Study of Machined Surface Roughness in Micro Turning, Proceedings of the 7th International Manufacturing Conference in China 1995, Vol. 1, 245-249
3. Shaw, M.: Metal Cutting Principles, Clarendon Press. Oxford. 1991
4. Armarego, E.J.A.: Machining Performance Prediction for Modern Manufacturing, Advancement of Intelligent Production, Ed.E.Usui, JSPE Publication Series No.1, (7th International Conference Production/Precision Eng. and 4th International Conference High Technology, Chiba, Japan, 1995 ): K52-K61
5. Ostafiev, V. Masol I. and Timchik, G. : Multiparameters Intelligent Monitoring System for Turning, Proceedings of SME International Conference, Las Vegas, Nevada, 1991, 296-301

# COMPUTER INTEGRATED AND OPTIMISED TURNING

**R. Mesquita**
**Instituto Nacional de Engenharia e Tecnologia Industrial (INETI),
Lisboa, Portugal**

**E. Henriques**
**Instituto Superior Tecnico, Lisboa, Portugal**

**P.S. Ferreira**
**Instituto Tecnologico para a Europa Comunitaria (ITEC), Lisboa,
Portugal**

**P. Pinto**
**Instituto Nacional de Engenharia e Tecnologia Industrial (INETI),
Lisboa, Portugal**

KEY WORDS: Computer Integrated Manufacturing, Process Planning, Machining

ABSTRACT: The reduction of lead time is a vital factor to improve the competitiveness of the Portuguese manufacturing SMEs. This paper presents an integrated and optimised system for turning operations, aiming the reduction of process planning and machine setup time, through the generation of consistent manufacturing information, technological process optimisation, manufacturing functions integration and synchronisation. The integrated system can strongly contribute to decrease the unpredicted events at the shop floor and increase the manufacturing productivity without affecting its flexibility.

## 1. INTRODUCTION

New patterns of consumer behaviour together with the extended competition in a global market call for the use of new manufacturing philosophies, simultaneous engineering processes, flexible manufacturing, advanced technologies and quality engineering techniques. The reduction of both the product development time and the manufacturing lead time are key objectives in the competitiveness of industrial companies.
The reduction of lead time can be achieved through the use of integrated software applications, numerical control systems and a dynamic process, production and capacity

Published in: E. Kuljanic (Ed.) *Advanced Manufacturing Systems and Technology*,
CISM Courses and Lectures No. 372, Springer Verlag, Wien New York, 1996.

planning environment, where resources allocation can be supported by real-time information on the shop floor behaviour.

Setup time and process planning time, as components of lead time, should be minimised (figure 1). The use of numerical control equipment, multi-axis machining, computer aided fixturing systems, automated and integrated process planning and tool management systems can contribute to the reduction of lead time. The required information for setup, the suitable fixturing and cutting tools should be selected and made available at the machine tool just-in-time. The minimisation of unexpected events, rework cycles, non conformities and process variability should be the final goal. Product delivery time, quality and cost determine the customer order and job execution.

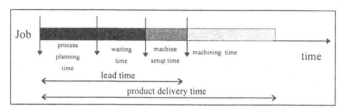

Fig. 1 - Components of product delivery time

Within the different components of product delivery time (figure 1), machining and waiting time could also be reduced, by adjusting the machining parameters for maximum production rate and using an appropriate planning technique. However, it should be noted that these components of delivery time will be considered invariable in this study.

## 2. ARCHITECTURE

In this context, a cooperative research project under development at INETI, IST and ITEC aims the integration of software applications in the areas of design, process planning, computer aided manufacturing, tool management, production planning and control, manufacturing information management and distributed numerical control.

This paper presents the architecture of a computer integrated and optimised system for turning operations in numerical control machine-tools. The system includes a CAD (Computer Aided Design) interface module, a CAPP (Computer Aided Process Planning) and tool management module, a CAM (Computer Aided Manufacturing) package and a DNC (Distributed Numerical Control) sub-system. A manufacturing information management sub-system controls the flow of information to and from the modules. The information required to drive each module is made available through a job folder. Figure 2 presents the sequence of manufacturing functions together with the information flow. One of the key objectives of our integrated system is to reduce manufacturing lead time. The manufacturing information should be generated with promptness, the machine setup time should be reduced and the unexpected events at the shop floor (tools not available, impossible turret positions, incorrect cutting parameters, etc.) should be minimised.

With the CAPP system under developement [1,2] it is possible to generated the manufacturing information in a short time and assure its quality and consistency.

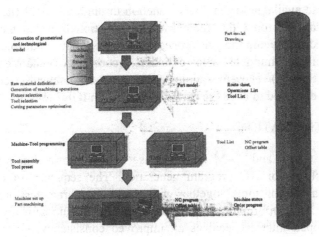

Fig. 2 - Manufacturing functions and information flow

Starting from a geometrical and technological part model, the CAPP function identifies the sequence of the required processes, selects the machines and fixturing devices and defines the elementary operations, in conformity with a built-in machining strategy. For each operation, the "best" tool is selected, automatically, from a database of available tools. Tool selection is developed in three phases [3,4]:

- a preliminary tool selection, by which a limited set of tools among those able to perform the operation are selected from the database, using an heuristic search method;
- a cutting parameters optimisation, in a constrained environment, considering a compromise solution of three objective functions - machining cost and time and number of passes;
- a final selection, constrained by the availability of machine turret positions, produces the "best" set of tools (minimum cost tool for each operation).

## 3. INTEGRATION AND OPTIMISATION OF FUNCTIONS

As mentioned, the prompt generation of high quality data for manufacturing is the key objective. Stand alone applications supporting some manufacturing functions are being considered as components of the product and process-oriented information processing system. The development of interfaces between commercial applications, the design of shared databases and the development of new methodologies for information generation and information flow is a possible approach to reach the integration of design and manufacturing systems and processes, aiming the reduction of manufacturing lead time.

## 3.1 CAD/CAPP INTEGRATION

In order to allow a full CAD/CAPP integration, part model is generated by the CAD function and is made available to the CAPP function through an IGES file. Actually, this model should contain not only the geometrical information but also the technological information required to develop the part process planning: dimensional tolerances, surface quality and type of machining features (cylinders, cones, faces, chamfers, axial holes, torus, grooves and threads). These machining features are automatically recognised and a methodology was developed to code the part model using an IGES format.

## 3.2 CAPP/CAM INTEGRATION

The integration of CAPP and CAM aims the use of the information provided by the process planning phase for NC program generation. The sequence of operations, the required and optimal cutting tools, together with machining parameters are generated by the CAPP system. This information has to be transmitted automatically to the CAM system. The developed interface enables an improved consistency, since toolpaths are generated in a computer-assisted sequence, in accordance with the operations list provided by the CAPP system. The CAM system being used is Mastercam, a PC / MS Windows based application that offers a programmatic interface. The ability to customise and extend the functionality of a commercial CAM with user code was found to be very important for our integration.

New functions were added to this system, associated with menu items, giving the operator some shortcuts to perform his/her work faster and more reliably, assisting him/her with the tasks involving machining process knowledge. The machining strategies offered by the CAM system were enhanced with the introduction of the operation sequence contained in the previously generated operations list.

At the beginning of the process, the system reads the operations list and loads the part model from the IGES file. This file contains not only the part model but also auxiliary data defining the raw material geometry and the elementary operations boundaries, to be used for toolpath generation. In order to allow the automatic recognition of this information some rules must be followed, concerning the distribution of entities by layers. Tool data (shape, dimensions, turret position and offset table position) and machining parameters are loaded. A new menu item is used to start toolpath definition: the system displays the operation description and auxiliary operation contour, so that the CAM operator can easily pick the right entities to perform the contour chaining. The cutting tool and the machining operation previously defined in the CAPP function are associated and a tool sketch is made on the fly for the cutting simulation. This process is repeated for every operation.

## 3.3 CAPP/TMS INTEGRATION

Tool management functions include, together with the selection of the suitable set of tools, tool list distribution, tool order and inventory control, tool assembly and pre-setting, tool delivery at the workstation and tool tables administration for part program generation and machine tool controller setup.

The integration of process planning and tool management can only be accomplished if a central tool database exists. This calls for a tool database standard which is not yet implemented in commercial systems. Consequently, it was decided to implement a solution aiming to integrate to some extent the process planning and tool management functions, using independent software applications [5].

It is used a tool management system (Corotas from Sandvik Automation) which supports all tool related sub-functions - tool identification, tool assembly, tool measuring, inventory control. A Zoller V420 Magnum tool pre-setter is used for tool measuring. The proprietary measuring programme - Multivision was interfaced with the Corotas package.

The tool lists generated by the CAPP system convey all the data required for tool management in the Corotas system through a file formatted as required by the TM system, for data import. Since a common database does not exist, tool data exchange between CAPP system and TM is performed through import and export functions. A file export facility, specifically designed for Corotas, was built.

Corotas operates at the tool room level together with the pre-setter system. Once Corotas is fed with the tool lists produced by the CAPP system (one tool list per workstation), the toolkits (a set of tools required to machine a part at a workstation) are defined, the tool items are identified and the number of sister tools is calculated. The tool items are allocated to the job, removed from its stock location and assembled. The assemblies are labelled (with the tool label code) and delivered for measurement and pre-setting at the tool pre-setter workstation.

The link between Corotas and Multivision is already implemented in Corotas. Tool nominal values are send to the tool pre-setter and the tool measured dimensions are returned. After tool measuring, the actual dimensions of the tools are written to a file, which is post-processed for the particular CNC controller and stored in the corresponding job folder. The TM function supplies the job folder with all NC formatted tool offset tables and tool drawings, and provides the workstation with the toolkits, properly identified, for machine setup.

## 4. INFORMATION SYSTEM

Considering the target users, composed by SME's, the underlying informatic base system is a low cost system, easily operated and maintained, widely spread and open. A PC network with a Windows NT server and Windows for Workgroups and DOS clients is used. Industrial PCs (from DLoG) are attached to the machine-tools controllers for DNC (Direct Numerical Control) and monitoring functions. Complementary ways of communication are used - data sharing through a job central database and a message system for event signalling, both windows based, and file sharing for the DOS clients.

In the proposed architecture, a job manager is at a central position and plays a major role as far as the information flow and the synchronisation are concerned (Figure 3). Assisted by a scheduler, the manager defines the start time of each task, attributes priorities, gathers the documents and pushes the input documents to the corresponding workstation, using a job management application. This application presents a view of the factory, keeps track of all I/O documents and promotes document update by tracing its dependency chain. The system manages a hierarchical structure of information composed by objects, such as, job folders, workstation folders, documents, workstations and queues.

At each workstation, the operator uses a task manager to select a job from the job queue, to identify the required input documents, to run the application and to file the output documents.

At the workstation located in the shop floor level, the operator uses his local task manager (built with the DLOG development kit). After selecting the job from the queue, all the required manufacturing information about the job and machine setup is available. Batch size, due date, operations list, tool data, drawings, fixturing data, NC program, tool offsets and instructions can be presented on the screen. If a CNC part program has to be modified at this workstation, the required modifications are registered for evaluation and updating at the process planning workstation.

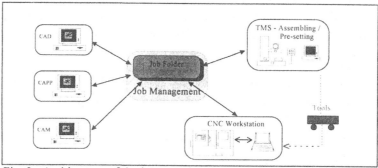

Fig. 3 - Architecture of the integrated system and distributed workstations

A snapshot of the on-going production and of the use of productive capacity is built out of the information collected at the workstations. Machine signals are fed into DLOG IPCs and filtered so that relevant status changes will be detected and sent to the monitoring station. Additional software indicate order completion and messages for process and job diagnosis.

## 5. PROCEDURE DESCRIPTION

Typical Portuguese job shops manufacture small batches of medium to high complexity parts, being critical the product delivery time. The system described in this paper is particularly suitable to the following company profile.

There is a limited number of qualified staff. Computer systems were selected and are used to assist a particular function. Usually, there is no formal process plan, with detailed operations lists, cutting tools list and optimised machining parameters (records of previous experiences are non-existing or are out of date). All this information is selected empirically, on the fly, at the CAM station. The generated NC program is always modified at the machine controller. There is no tool database and the amount and type of cutting tools is limited by part programmer experience and knowledge. Tool management function is rather limited and prone to errors or delays. Tool holders are located near the machine-tools and there is no tool pre-setting device (dry cutting and tool offset changes are compulsatory). The comments included, by the part programmer, in the NC listing provide the operator with the information required to select the needed tools (no availability checking is made). Machine-tool operator has to assemble, mount (in an undefined turret position) and measure all the cutting tools. The machining parameters are adjusted during the operation (machining cost and time is not a constraint). The number of required tools (sister tools) is unknown. Tool replacement is determined by the observed surface quality. As a result of this procedure, machine setup, even for very simple parts, takes a long time, being the whole process highly dependent on the planner expertise who is also the CAM system operator.

The integrated and optimised procedure that can be used with the proposed system is described in this section. A geometrical and technological model of the part is created in a CAD system and exported through an IGES file. CAD files delivered by the customer can also be used as input data for the process planning phase. The CAPP system interprets this model and automatically selects the sequence of operations and required machine-tools (workstations) and generates the elemental sequence of operations, as presented in figure 4. Tools are selected from a tool database containing only existing tools. A different set of tools will be found by the system if a larger database of existing tools is available, since a minimum time and cost criteria is used. Machining parameters are determined by the optimisation process.

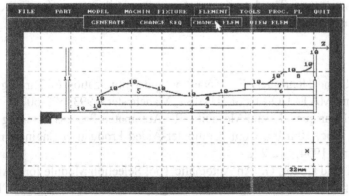

Fig. 4 - Generation of operations by the CAPP system.

Concurrently with part programme generation, all the tool management functions are performed, based in the same set of manufacturing information provided by the CAPP system. This procedure enabled the generation of the documents presented in figure 5, as shown at the DNC terminals. Optimal and existing cutting tools together with tool setting data and ready to use part programmes are made available just-in-time at the workstation for minimum setup time.

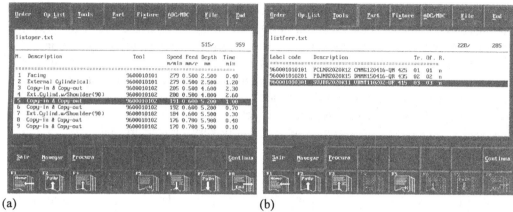

(a)                                                                                      (b)

Fig. 5 - Operations list (a) and tool list (b), as viewed at the machine workstation

## 6. CONCLUSIONS

The implementation of an integrated system as described in this paper can contribute to the competitiveness of SME job shops. Particularly, a higher machine productivity and shorter product delivery time can be achieved. The consistency of the manufacturing information can be improved and the process can be technologically optimised. The synchronisation of tool related functions and the quality of the numerical control program allow the reduction of the machine setup time. The described system can contribute to maintain the consistency of planned and achieved machining costs and time and to reduce the lead time.

## REFERENCES

1. Mesquita, R. and Henriques, E.: Modelling and Optimization of Turning Operations. Proc. 30th Int. MATADOR Conf., UMIST, Manchester, 1993, 599-607

2. Nunes, M., Henriques, E. and Mesquita, R.: Automated Process Planning for Turning Operations, Proc. 10th Int. Conf. Computer-Aided Production Engineering, Palermo, Univ. Palermo, 1994, 262-272

3. Mesquita, R. and Cukor, G.: An Automatic Tool Selection Module for CAPP Systems, Proc. 3rd Int. Conf. Advanced Manufacturing Systems and Technology, CISM, Udine, 1993, 155-165

4. Mesquita, R., Krasteva, E. and Doytchinov S.: Computer-Aided Selection of Optimum Machining Parameters in Multipass Turning, Int. Journal of Advanced Manufacturing Technology, 10 (1995) 1, 19-26

5. Mesquita, R., Henriques, E., Ferreira, P.S. and Pinto, P.: Architecture of an Integrated Process Planning and Tool Management System, Proc. Basys'96, Lisboa, 1996

# MULTI-CRITERIA OPTIMIZATION IN MULTI-PASS TURNING

**G. Cukor**
University of Rijeka, Rijeka, Croatia

**E. Kuljanic**
University of Udine, Udine, Italy

KEY WORDS: Multi-Criteria Optimization, Optimization Methods, Multi-Pass Turning

ABSTRACT: The basic idea of the paper is to overcome the barrier of cutting conditions optimization estimated either on the basis of minimum unit production cost or minimum unit production time. In order to evaluate both criteria simultaneously and their mutual dependence, the concept of double-criteria objective function is proposed. The approach adopted in solving the constrained nonlinear minimization problem involved a combination of theoretical economic trends and optimization search techniques. Finally, popular multi-pass rough turning optimization strategy of using equal cutting conditions for all passes is shown to be useful approximation but more rigorous computer-aided optimization analysis yielded unequal cutting conditions per pass.

## 1. INTRODUCTION

The trend of present manufacturing industry toward the concept of *Computer Integrated Manufacturing* (CIM), imposes as an imperative, the *Computer-Aided Process Planning* (CAPP). The development of a cutting conditions optimization module is essential to the quality of CAPP software.

Optimization of multi-pass rough turning operation means determination of optimal set of cutting conditions to satisfy an economic objective within the operation constraints. The solution to the problem of selecting optimum cutting conditions requires sophisticated

Published in: E. Kuljanic (Ed.) *Advanced Manufacturing Systems and Technology*,
CISM Courses and Lectures No. 372, Springer Verlag, Wien New York, 1996.

computer-aided optimization search techniques. However, the results obtained from the optimization search will depend on the mathematical models of the process, and to a greater extent on the optimization method used.

In principle, the cutting conditions are usually selected either from the viewpoint of minimizing unit production cost or from the viewpoint of minimizing unit production time if cost is neglected. It has also been recognized that between these two criteria there is a range of cutting conditions from which an optimum point could also be selected in order to increase profits in the long run. Furthermore, maximum profit is in reality the major goal of industry.

This work has been developed under the assumption that an optimum economic balance between the criteria both the minimum unit production cost and the minimum unit production time will theoretically result in maximum profit rate criterion. A solution using the *multi-criteria optimization procedure* is introduced as the basis of further discussion.

## 2. MODELLING OF MULTI-PASS ROUGH TURNING OPERATIONS

The economic mathematical models of multi-pass rough turning operations have been formulated by many investigators [1, 2, 3]. However, optimized cutting conditions in view of a given economic objective are in general assessed on the basis of some form of *tool life equation* [4]. In the multi-pass turning each pass is denoted by the corresponding workpiece diameter $D_j$ [mm], depth of cut $a_{pj}$ [mm], feed $f_j$ [mm] and cutting speed $v_{cj}$ [m/min]. The unit production cost and the unit production time for $i_p$ passes can then respectively be expressed as:

$$y_1(x) = c_1 = c_o t_1 + c_t \sum_{j=1}^{i_p} \frac{\pi l D_j}{1000 v_{cj} f_j} \frac{1}{T_j} \tag{1}$$

$$y_2(x) = t_1 = t_{np} + \sum_{j=1}^{i_p} \frac{\pi l D_j}{1000 v_{cj} f_j} \left(1 + \frac{t_c}{T_j}\right) + i_p t_r \tag{2}$$

with

$$D_j = D_0 - 2\sum_{j=1}^{i_p} a_{pj-1} \tag{3}$$

$$T_j = T_j\left(v_{cj}, f_j, a_{pj}\right) \tag{4}$$

and

$$\sum_{j=1}^{i_p} a_{pj} = a_w \tag{5}$$

where $c_1$ = unit production cost [\$/min], $t_1$ = unit production time [min], $c_t$ = tool cost per cutting edge [\$], $l$ = length to be machined [mm], $T_j$ = tool life [min], $t_{np}$ = non-productive time [min], $t_c$ = tool changing time [min], $t_r$ = tool reset time [min], $D_0$ = initial workpiece diameter [mm], and $a_w$ = total depth to be machined [mm].

In order to find a compromise solution the above mentioned criteria can be aggregated into a global one using the following *two-criteria objective function* [5]:

$$y(x) = \frac{w}{y_1^*(x)} y_1(x) + \frac{1-w}{y_2^*(x)} y_2(x) \tag{6}$$

where $w$ is the weight coefficient, $y_1^*(x)$ and $y_2^*(x)$ represent minimum values of the corresponding criterion when considered separately.

It should be noted that two-criteria optimization becomes one-criterion for $w = 1$, from the viewpoint of minimum unit production cost. If $w = 0$, then this is the optimization from the viewpoint of minimum unit production time. For $0 < w < 1$ yields a compromise solution to obtain an optimum economic balance between the unit production cost and time. Hence, the cutting conditions, at which this occurs, will theoretically result in maximum profit rate.

The objective function (6) should be minimized while satisfying a number of process constraints which limit the permissible values of the cutting conditions $v_{cj}$, $f_j$ and $a_{pj}$ for each pass. Most of the constraints are quite obvious, such as:

- min., max. depth of cut
- min., max. feed
- min., max. cutting speed
- chip shape
- min., max. spindle speed
- available power of the machine
- allowable cutting force
- available/allowable chucking effect
- rigidity of machining system
- min. tool life, etc.

These constraints need not to be discussed further. Some other variable bounds and operation constraints is given in [6]. However, the following constrained nonlinear minimization problem has to be solved:

$$\text{minimize } y(x)$$
$$\text{subject to } g_j(x) \{\geq ; =\} \, 0, \, j = 1, ..., m \tag{7}$$

where $m$ is the total number of nonlinear inequality and equality constraints. Since the mathematical model of the multi-pass rough turning optimization problem has an explicit, multivariable, nonlinear objective function constrained with nonlinear inequality constraints and some equality constraints, the optimization method selected should fulfil the model features.

## 3. OPTIMIZATION METHOD

The optimization method described herein is based on optimization procedure of multi-pass turning operations previously developed [2], but is coupled with *general nonlinear programming methods.*
For the successful operation of advanced and more sophisticated flexible machining systems, it is essential that the cutting conditions selected are such that easily disposable chips are produced. Since cutting speed has only a secondary effect on chip-breaking, when compared with feed and depth of cut, the possible cutting region to satisfy the chip-breaking requirement can be represented by the $a_p$-$f$ diagram. This diagram is defined as combinations of depth of cut $a_p$ and feed $f$ which produce easily disposable chips. Such diagrams are generally available from cutting tool manufacturers and an example is shown in Figure 1.

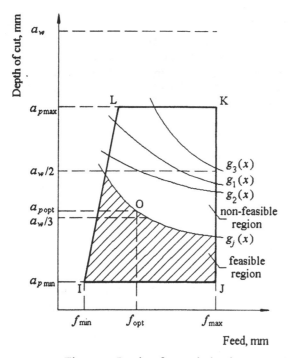

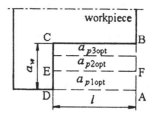

Figure 1. Region for optimization          Figure 2. Multi-pass turning

Let us assume that, in the workpiece shown in Figure 2, area ABCD has to be machined. To start with, the feasible and non-feasible regions are separated in the $a_p$-$f$ plane by a concave curve of the most significant constraint acting on the process, as illustrated in Figure 1. The point O at which the objective function (6) is a minimum, is selected as the optimum point. This point always lies on the boundary separating the feasible and non-feasible regions. In order to ensure finding global optimum, a combination of techniques both the *Direct Search of Hooke and Jeeves* and *Random Search* can be successfully applied [5].

Direct search is performed three times at different starting points chosen with respect to feasibility and criterion values out of randomly generated points. For handling the constraints, the *modified objective function method* is implemented. The constraints are incorporated into the objective function (6) which produces an unconstrained problem:

$$\text{minimize } y(x) + C_F \sum_{j=1}^{m} k\big(g_j(x)\big) \tag{8}$$

where $C_F$ is correction factor and $k\big(g_j(x)\big)$ is the *exterior penalty function* of the form:

$$k\big(g_j(x)\big) = \begin{cases} 0 & g_j(x) \geq 0 \\ g_j(x) & g_j(x) < 0 \end{cases} \tag{9}$$

Penalty function is used in order to apply a penalty to the objective function at non-feasible points, thus forcing the search process back into the feasible region.

The resulting shape of the workpiece after the first pass would be given by DEFB. This is now considered to calculate the optimum cutting conditions for the second pass. An identical $a_p$-$f$ diagram is again assumed, but the optimization search techniques may result different cutting conditions. This could be due to several reasons, one of which is the change in the workpiece flexibility and another, the reduced workpiece radius at which the cutting forces are applied. This procedure of determining the optimum cutting conditions for each pass is repeated until the sum of the optimum depths of cut equals the total depth $a_w$ to be machined.

In addition, from Figure 1, it is obvious that the assumption of equal depths of cut is not generally valid, especially if a considerable amount of material has to be removed with roughing passes. Namely, when the total depth $a_w$ is outside the feasible region, as usual, then the number of passes has to be increased until the plausible depth of cut inside the feasible region is found. Only a limited number of $a_p$ values can be obtained. For instance, in Figure 1, the first feasible depth of cut $a_p = a_w/3$ lies below the optimum point. It is evident, that by choosing this value, a small error is made. However, by multiplying this small error with the total number of passes we obtain a bigger error which significantly affects the reliability of achieved optimum. Therefore, the assumption of equal depths of cut for all passes is not expected to yield optimal solution.

## 4. NUMERICAL STUDY

Applying the above optimization method, the interactive program system PIVOT for multi-criteria computer-aided optimization of cutting conditions in multi-pass rough turning operations, applicable in CAPP environment, is developed. Detailed computer flow diagrams are given in [5, 6].

The application of the PIVOT program system has been tested in industrial work conditions. More elaborate description of the industrial setup in Figure 3 is given in [5]. The optimization of each single pass (case a) was compared with the popular optimization strategy of using equal cutting conditions for all passes and hence ignoring the decrease of workpiece diameter $D_j$ (case b).

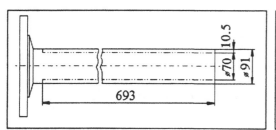

machine tool:        CNC lathe (33 kW)
workpiece material: 42 CrMo 4
tool:               CNMM 190612
tool material:        P25 (ISO)

case a) optimization of each single pass
case b) passes of equal depths

Table 1. Sample results of multi-pass turning optimization

| Case | Pass | $v_c$ [m/min] | $f$ [mm] | $a_p$ [mm] | $c_1$ [$] | $t_1$ [min] |
|------|------|------|------|------|------|------|
| a | 1 | 84.200 | .401 | 3.755 | 6.414 | 22.567 |
|   | 2 | 85.570 | .376 | 3.984 |  |  |
|   | 3 | 85.409 | .490 | 2.761 |  |  |
| b | 1 | 80.667 | .456 | 3.500 | 6.975 | 23.583 |
|   | 2 | 80.667 | .456 | 3.500 |  |  |
|   | 3 | 80.667 | .456 | 3.500 |  |  |

Conditions for Cases a and b: $t_{np}$ = 5 min, $t_c$ = 0.5 min, $t_r$ = 0.25 min, $c_o$ = 0.04 $/min, $c_t$ = 1 $, $T$ = 1836.3 $v_c^{-1.34} f^{-0.76} a_p^{-0.35}$

reduction in  ⇒  cost per unit: 8.7 %
time per unit: 4.5 %

Figure 3. Example for an economical comparison between optimization of each single pass and passes of equal depths

In Table 1 the computed results for cases (a) and (b) are presented. These results could be used to check the validity of the proposed optimization method and the percentage difference of production cost (i.e. % $c_1$) and time (i.e. % $t_1$) per unit when comparing cases (a) and (b), i.e. the penalty of using the equal cutting conditions for all passes instead of the superior optimization of each single pass.

While a total saving of about 8.7 % of production cost could be achieved, the production time per unit could be reduced about 4.5 % with respect to case (a). Therefore, it is evident that any optimization method which has forced equal cutting conditions for all passes can not be considered valid in CNC machining.

## 5. CONCLUSIONS

In this study, the mathematical model of the multi-pass rough turning optimization problem has been formulated. This model has the explicit, multivariable, nonlinear objective function constrained with nonlinear inequality constraints and some equality constraints. Also, a method for cutting conditions optimization has been proposed, based on which the PIVOT program system was developed.

The approach adopted in solving the nonlinear constrained minimization problem involved a combination of mathematical analysis of the theoretical economic trends and optimization search techniques. The implemented optimization method is based on the multi-criteria optimization under the assumption that an optimum economic balance between the unit production cost and time yields a maximum profit rate. The estimation contains the weight coefficient so that the level of the profit rate may be modelled.

The numerical study and results have supported the proposed optimization method and highlighted the superiority of optimization of each single pass over the optimization strategy of using equal cutting conditions for all passes. Therefore, any optimization method which has forced equal cutting conditions for all passes can not be considered valid or applicable for both the multi-pass rough turning operations on CNC lathes and the constraints considered in this paper, except in conventional machining.

## REFERENCES

1. Kals, H.J.J., Hijink, J.A.W., Wolf, A.C.H. van der: A Computer Aid in the Optimization of Turning Conditions in Multi-Cut Operations, Annals of the CIRP, 27(1978)1, 465-469

2. Hinduja, S., Petty, D.J., Tester, M., Barrow, G.: Calculation of Optimum Cutting Conditions for Turning Operations, Proceedings of the Institution of Mechanical Engineers, 199(1985)B2, 81-92

3. Chua, M.S., Loh, H.T., Wong, Y.S., Rahman, M.: Optimization of Cutting Conditions for Multi-Pass Turning Operations Using Sequential Quadratic Programming, Journal of Materials Processing Technology, 28(1991), 253-262

4. Kuljanić, E.: Machining Data Requirements for Advanced Machining Systems, Proceedings of the International Conference on Advanced Manufacturing Systems and Technology AMST'87, Opatija, 1987, 1-8

5. Cukor, G.: Optimization of Machining Process, Master of Science Thesis, Rijeka: Technical Faculty, (1994)

6. Cukor, G.: Computer-Aided Optimization of Cutting Conditions for Multi-Pass Turning Operations, Proceedings of 3rd International Conference on Production Engineering CIM'95, Zagreb, 1995, D27-D34

# ON-LINE CONTROL TECHNIQUES: OPTIMISATION OF TOOL SUBSTITUTION INTERVAL

**E. Ceretti, C. Giardini and G. Maccarini**
**University of Brescia, Brescia, Italy**

**KEY WORDS** : Cutting operations, Tool substitution interval, ARIMA Models.

**ABSTRACT** :

The present paper describes an on-line method to control the tool wear during machining. The variable detected is the tool vibration, which is related to the tool status. The dependence between the two variables is studied by the authors with a mathematical model able to adapt itself to the stochastic nature of the phenomenon analysed.

## 1. INTRODUCTION

Useful cutting tool life strongly influences the economy of the operations. Referring to this aspect, it is important to monitor tool working conditions and to identify when the tool needs to be replaced.

In the past, the machine tool operator decided when to change the tool on the basis of practical experience. The rise of automatic working centre, characterised by very high productivity and by an automatic tool storage, have caused the cutting tool to be replace at convenient intervals (i.e. the end of the working shift, or after a fixed number of shifts).

The diffusion of FMS (Flexible Manufacturing Systems) and the use of unmanned systems of production, requires a different approach for tool substitution.

In this environment, the operator can not decide the tool substitution interval ; in addition, the cutting conditions may change frequently and the identification of a planned tool substitution policy is difficult.

Published in: E. Kuljanic (Ed.) *Advanced Manufacturing Systems and Technology*, CISM Courses and Lectures No. 372, Springer Verlag, Wien New York, 1996.

A tool monitoring system is required to optimise the cutting operations. A cutting tool needs to be changed for excessive tool wear or for a sudden tool breakage. Tool monitoring systems have to adapt themselves to identify these main faults, that is tool breakage and excessive tool wear. So, it is necessary to develop sensitive, accurate and reliable techniques.

## 2. TOOL MONITORING SYSTEMS

Tool monitoring systems can be divided into direct and indirect methods [1, 2, 3].

The *direct methods* are based on the direct measure of the tool volumetric material loss. These techniques are mainly *off-line*; that is, the information is acquired once the cutting operation is finished or interrupted. In particular, the *direct off-line* methods measure directly the cutting edge wear by means of probes or microscopes, while between the devices used to detect *direct on-line* methods there are thermocouples, electrical resistance or optical instruments.

The *indirect methods* utilise variables characterising the cutting process, which can be acquired in real time and directly related to tool wear. The main advantage is that these are *on-line* techniques and the information is acquired without halting the working operations. The physical variables utilised by *on-line indirect* methods are vibrations, forces, torques, acoustic emission, temperatures and so on. The *off-line indirect* methods are based on the measurement of the workpiece dimensional variations or on the workpiece roughness.

The drawbacks of the *off-line* techniques are:

1. a loss of time due to the interruption of the cutting process to detect the tool wear;

2. tool breakage can not be detected or forecast.

The *on-line* techniques can solve these limitations, but they are usually expensive due to the electronic devices and instruments required. In addition, direct on-line methods present problems because the sensors need to be positioned close to the zone where the cut is performed. Since this area is lubricated and subjected to very high temperature these techniques are not widely diffused.

## 3. TOOL SUBSTITUTION INTERVAL

Tool substitution interval is strictly related to the dimensional accuracy of the workpiece, to the time required to perform the cutting operation and to the cutting speed.

The optimisation of cutting operations depends on the cutting speed chosen. Several parameters influence the choice, such as workpiece material properties, tool material properties, tool geometry, lubrication conditions at the tool workpiece interface, operation performed and dimensional accuracy of the workpiece.

The choice between high cutting speed and short tool life, and low cutting speed and long tool life is an economic decision. It depends on the tool cost, maintenance time, manpower costs, machine amortisation costs.

As Known, the following formula defines an economic cutting speed, which corresponds to minimum productive cost as a function of tool substitution policy (the tool is considered a stochastic variable), dead times, manpower costs etc..[4]

$$C_{tot} = E(n) \cdot [t \cdot C_0 + t_a \cdot C_0] + C_0 \cdot t_s + C_c + [1 - R(\alpha)] \cdot [P_t \cdot C_0 + P_c]$$

$$h_p = h_\alpha \cdot R(\alpha) + h_b \cdot (1 - R(\alpha)) \qquad E(n) = \frac{h_p}{t}$$

where

$E(n)$ is the expected number of produced parts, $C_0$ is manpower cost per minute,

$t$ is the working time per part, $t_s$ is the time for tool substitution,

$t_a$ is the passive auxiliary time, $C_c$ is the cutter cost,

$P_t$ is the penalty time, $P_c$ is the penalty cost,

$R(\alpha)$ is the tool reliability, $h_p$ is the productive time,

$h_\alpha$ is the tool life with a reliability of $R(\alpha)$, $h_b$ is premature tool breakage life,

$P_c$ is the penalty cost, $[1-R(\alpha)]$ premature tool breakage probability.

For a cutting speed it is possible to determine reliability and productive time and to calculate the associated cost. Varying the cutting speed the relative cost curve presents a minimum which represents the optimal working conditions.

## 4. VIBRATION AND TOOL WEAR DETECTION

The present paper aims to identify a model able, on the basis of acquired information, to forecast the tool condition. That is, the model should say whether it is possible for a tool to perform another operation or if the critical wear value has been reached.

The operation analysed is dry face milling, during the tests a single cutter was mounted on the mill, the material machined is C40 steel, the tool material is hard metal (P25) and the cutting parameters are: cutting speed 484 rev/min and cutting feed 59 mm/min [5, 6].

The input variables for the model are vibrations and the tool wear. Vibrations are detected on the piece holder by means of an inductive accelerometer. The acquired signal is filtered and processed. The RMS (Root Mean Square) of the signal for each pass is recorded, figure 1 shows the typical behaviour of an RMS for a cutter as the number of passes increases.

At the end of each pass the cutting operation is interrupted, the tool is positioned under a microscope and the value of the flank wear is measured.

## 5. THE MATHEMATICAL MODEL

Once the signals of vibrations and tool flank wear are detected there is the problem of interpreting these signals as historical series by means of a mathematical model.

The mathematical model is the ARIMA (Auto Regressive Integrated Moving Average). This model was created to explain and forecast the behaviour of economic parameters (Box & Jenkins) [7, 8]. By means of the ARIMA model the last data of a series $(z_t)$, which represents a time dependent phenomenon, can be expressed as a function of the previous data plus a linear combination of random factors and of white noise $(a_t)$:

$$z_t = \phi_1 z_{t-1} + \phi_2 z_{t-2} + \ldots + \phi_p z_{t-p} + a_t + \theta_1 a_{t-1} + \theta_2 a_{t-2} + \ldots + \theta_q a_{t-q}$$

where $\phi_i$ and $\theta_i$ are the parameters of the model to be determined.

The models reported above express the relation between the last data of the series and the previous data and random factors. If there are two series of data (vibrations and wear), the bivariate models can represent the relation between the two variables. The input data is a statistical variable, called cross autocorrelation, which expresses the link between data of a series and the previous data of the other series. In particular, a bivariate ARIMA model is used to create the vibration model. The last data of the series depends on the history of the series itself and on external factors related to that series. That is, vibrations are an ARIMA function of the previous vibrations and of tool wear level. The bivariate model can be expressed as :

$$X_t = \phi_{11}X_{t-1} + \phi_{12}Y_{t-1} + \phi'_{11}X_{t-2} + \ldots \theta_{11}a_{t-1,1} + \theta_{12}a_{t-1,2} + \theta'_{11}a_{t-2,1} + \ldots + a_{t,1}$$

$$Y_t = \phi_{21}X_{t-1} + \phi_{22}Y_{t-1} + \phi'_{21}X_{t-2} + \ldots \theta_{21}a_{t-1,1} + \theta_{22}a_{t-1,2} + \theta'_{21}a_{t-2,1} + \ldots + a_{t,2} \; ;$$

Both in monovariate and bivariate models it is necessary to define the model order, that is, the coefficient $p$ of the autoregressive part and the coefficient $q$ of the moving average part (model identification phase), and to estimate the model coefficients $\phi$ and $\theta$ (parameters estimation phase).

The vibration model (ARIMA) enables the vibration level for the next operation to be forecast. If the forecast level exceeds the limit level (defined by the correlation between vibrations and wear) the tool has to be changed before performing the next operation. The equation of the forecast ARIMA model is :

$$z_t(l) = (\phi_1 z_{t+l-1} + \ldots + \phi_{p+d} z_{t+l-p-d}) + (a_{t+l} - \theta_1 a_{t+l-1} - \ldots - \theta_q a_{t+l-q})$$

Figure 1 shows the comparison between the experimental (RMS) and the forecast vibration curves.

## 6. SIMULATIONS

The assumptions described above have been verified by means of a simulation of the actual process. The series of the average wear data, detected in the University Labs, is used to obtain the simulated wear series, under the hypothesis of a normal wear distribution around the average values and with an increasing variance as the number of passes increases (Figure 2).

The vibration series is generated by an autoregressive bivariate model previously defined by means of the two starting series, using wear numerical values derived from the wear series previously generated, as described above. So, for each simulated series the correlated vibration series is obtained.

The equations of the ARIMA model, which best fit the vibrations and wear data collected during the experiments at the University Labs, have been included in a simulation programme. The results of the ARIMA model, in terms of definition of tool substitution, have been used for the calculation of the cost function.

The simulation programme works as follows:

1.     Wear and vibration curves calculation for each working pass using the models previously defined;

2.     For each pass the wear and vibration curves are compared with limit values "a priori" defined: if the result is negative the next pass is performed.

3.     If the comparison is positive there are two possibilities:

A) The wear curve exceeds the limit value before vibrations. This is the case in a sudden tool breakage, the last part produced is scrapped.

B) The vibration curve exceeds the limit value before wear. The tool is changed and the last part produced is good.

1. Through the cost function, described above, and on the basis of the total number of produced parts the average unitary cost of the operation is calculated.

The process is repeated for a fixed number of cycles; the average values of the unitary cost, obtained for the different parameters (wear limits accepted), are calculated. Then the vibration level is changed and other unitary cost curves are calculated.

The cost curve becomes more significant as the number of cycles increases and the minimum position is more precise as a smaller increment of the limit vibration level is used. Figures 3, 4, 5 show the comparison of the results obtained with the simulative and forecast methods at different wear limit levels. Results obtained with forecast method are always slightly better than the simulative ones.

## 7. RESULTS

To check the validity of the method the cost curves, obtained with variable tool substitution interval (ARIMA), Figure 6, have been compared with the cost curves obtained with a fixed tool substitution interval (fixed number of parts), Figure 7.

The results show that the model based on the comparison between the vibration values and the relative imposed limits gave results which are slightly better than the results obtained with a fixed number of parts. A big advantage can be obtained using a tool substitution method based both on the comparison of the vibration level for the pass $x$ with the imposed limit and on the comparison of the forecast value for the $x+1$ pass with the same imposed limit. In more detail, the comparison with the forecast vibration value is performed first, then the comparison with the simulated vibration is calculated (in this way the effects of a wrong forecast are ignored).

## 8. CONCLUSIONS

Figures 6 and 7 show that the ARIMA model gives a better optimisation of the residual tool life and lower costs than the strategy of changing the tool after a fixed number of parts produced.

In conclusion, the model proposed by the authors seems to be a good methods for the definition of the tool substitution interval in unmanned machining operations.

## ACKNOWLEDGEMENTS

This work has been made possible thanks to Italian CNR CT11 95.04109 funds.

## REFERENCES

1. J. Tlusty, G. C. Andrrews : A Critical Review of Sensors for Unmanned Machining, Annals of the CIRP Vol. 32/2/1983.

2. H. K. Tonshoff, J. P. Wulsfberg, H. J. J. Kals, W. Konig, C. A. Van Luttervelt : Developments and Trends in Monitoring and Control of Machining Processes, Annals of the CIRP Vol 37/2/88.

3. J. H. Tarn, M. Tomizuka : On-line Monitoring of Tool and Cutting Conditions in Milling, Transaction of the ASME Vol. 111, August 1989.

4. S.S. Sekulic : Cost of Cutting Tools and Total Machining Cost as a Function of teh Cutting Tool Reliability in Automatic Flow Lines, Int. J. Prod. Res. Vol. 20 N. 2 1982.

5. C. Harris, C. Crede : Shock and Vibration Handbook, Vol. 1 Mc Grow Hill , England 1961.

6. E. Ceretti, G. Maccarini, C. Giardini, A. Bugini: Tool Monitoring Systems in Milling Operations : Experimental Results, AMST 93, Udine April 1993.

7. G. E. P. Box, G. M. Jenkins : Time Series Analysis : Forecasting and Controls, Holden-Day, San Francisco 1970.

8. L. Vajani : Analisi Statistica delle Serie Temporali, Vol. I & II CLEUP, Padova 1980.

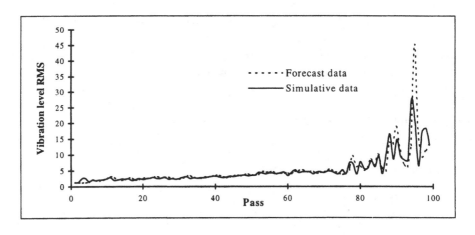

Fig. 1 - Slop of RMS vibration level detected and forecast.

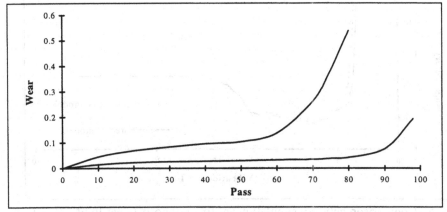

Fig. 2 - Upper and lower limits of simulated wear series.

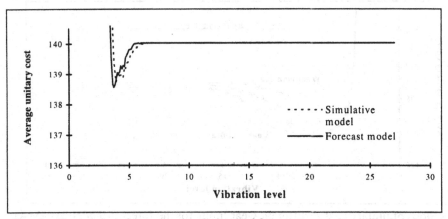

Fig. 3 - Comparison between simulation anf forecast costs (wear limit = 0.1).

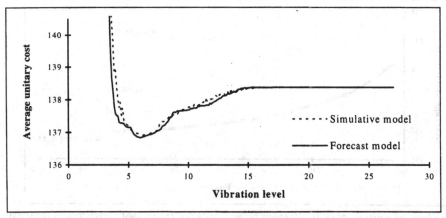

Fig. 4 - Comparison between simulation anf forecast costs (wear limit = 0.25).

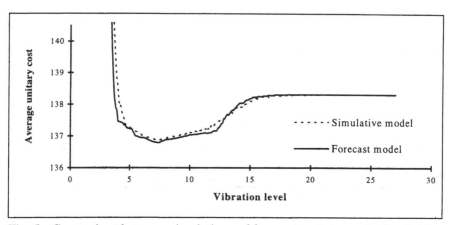

Fig. 5 - Comparison between simulation anf forecast costs (wear limit = 0.35).

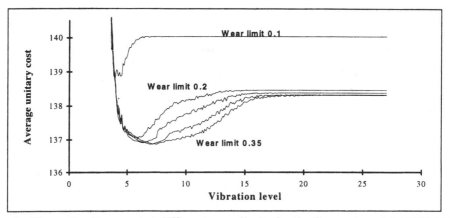

Fig. 6 - Slops of unitary cost at different wear limits for the forecast model (ARIMA).

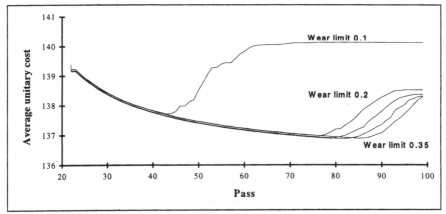

Fig. 7 - Slops of unitary cost at different wear limits for fixed tool substitution interval.

# STEP-ORIENTED DATA AND KNOWLEDGE SHARING FOR CONCURRENT CAD/CAPP INTEGRATION

**W. Eversheim and P.Y. Jiang**

**Technical University of Aachen, Aachen, Germany**

KEY WORDS: Data and Knowledge Sharing, CE, CAD/CAPP Intergation, STEP

ABSTRACT: This paper deals with a methodology approach to object-oriented data and knowledge sharing for concurrent part design and process planning. CE server, a special application, interfaces, dynamic data and knowledge records, and logic model for expressing part/process-planning data and knoweledges are presented. Here, activity-based solving logic is the keypoint of realizing data and knowledge sharing. ISO 10303 is referred to and EXPRESS language is used for modelling a logic model which is composed of four units of functionality. With the help of mapping mechanism of ONTOS database system, physical schemata are generated. The correspondent interfaces can be realized by means of using clear text encoding of exchange structure specified by ISO 10303 Part 21. CE server is just used for generating a special application, scheduling it, and sharing the dynamic data and knowledges related to a designed part during the whole procedure of concurrent part design and process planning. At last, conclusions are given.

## 1. INTRODUCTION

Concurrent engineering (CE), also called as simultaneous engineering, is considered to be feasible and practical product development mode in the industrial applications of today and future. Since 1988, researchers have been focusing on the CE issues and have also provided many available methodologies and application prototypes. Some large companies such as GE have used some of CE methodologies for their industrial practices[1]. However, CE is a very complex systemized technique and deals with a quantity of technical, social, and cognitive problems to have been found and to be discovered. It is still under development.

Published in: E. Kuljanic (Ed.) *Advanced Manufacturing Systems and Technology*, CISM Courses and Lectures No. 372, Springer Verlag, Wien New York, 1996.

So a research project, named „Models and Methods for an Integrated Product and Process Development" (SFB361), supported by Deutsche Forschungsgemeinschaft, and being researched at RWTH Aachen led by the Laboratory for Machine Tools and Production Engineering(WZL), is just for solving the issues on CE application. This paper focuses only on one aspect of sub-project in this project, that is, dynamic data and knowledge sharing for concurrent CAD/CAPP integration.

From the angle of global viewpoints, CE researches can be particularly emphasized from one of three aspects which are listed as follows, although some overlaps exist:
- high-level configurating, controlling, and scheduling mechanism (such as activity-based modelling, activity planning, overall structural configuration and scheduling, etc),
- data and knowledge sharing (D&K sharing), and
- specific application tools or systems in low level (such as DFM, CAD, CAPP,etc).

The typical presentations for the first keypoint research include a methodology and PiKA tool developed at WZL[2], structured activity modelling by K.Iwata[3], etc. The second keypoint research just deals with SHARE and SHADE project[4][5], PDES/STEP-based data preparation for CAPP by Qiao[6], etc. As to the researches on application tools in low level, we can find many cases about design for manufacturability[7][8][9], environmental conscious design[10], modified process planning, etc[9][11].

Mainly from the angle of dynamic data and knowledge sharing, together with activity-based modelling, we will discuss the integrating issues for concurrent part design and process planning procedure in this paper. Here, developing a logic model for unified representations of dynamic part/process-planning data and knowledges, achieving a kind of physical realization, realizing the correspondent interface functions, and creating a CE data server and further extending it into activity-based scheduling tool are just main goals in this research. In the following sections, the detailed contents will be described. Here, section 2 is for the analysis of data and knowledge sharing procedure, section 3 for the overall logic architecture of data and knowledge sharing, section 4 for the unified part/process-planning integrated model, section 5 for the methodology of interface development, and section 6 for the realization of CE server. At last, conclusions are given.

## 2. ANALYSIS FOR PROCEDURE OF DATA AND KNOWLEDGE SHARING

In order to analyze the procedure of data and knowledge sharing, first of all, we explain the following concepts:

**Dynamic Data and Knowledges:** So-called dynamic data and knowledges are defined as ones that are generated during the procedure of concurrent part design and process planning. They include initial, intermediate and final data and knowledges related to a designed part, for example, part geometric and technical information, activity-based management data, resource data, etc. In this paper, we formalize four types of data and knowledges listed as follows:
- feature-based part/process-planning data,
- dynamic supporting resource data,
- dynamic knowledges, and
- activity-based management data.

**Activity and Meta-activity:** In this paper, activity is defiined as a procedure to finish a specific basic design task under concurrent integrated CAD/CAPP environment, such as designing structural shapes of part. While an activity can be decomposed further into three meta-activities, that is, designing a basic task, analyzing and evaluating a basic task, and redesigning a basic task. Realizing a meta-activity needs a functional tool. In fact, each functional tool may have a specific data structure of itself. It means that sometimes it is not avoidable to exchange data between meta-activities or functional tools with the help of database.

**Scenario for Realizing Concurrent Part Design and Process Planning:** A Scenario specifies the static procedure of problem solving for concurrent part design and process planning, which does not consider adding the stochastical activity. So scenario, in fact, is a collection of activities.

According to above concepts, we may analyze how to generate and reuse dynamic data and knowledges inside an activity. As shown in Fig.1, when a meta-activity is finished, the correspondent functional tool generates dynamic data and knowledges and proceed them into the database. When a meta-activity begins, the functional tool reuses dynamic data and knowledges from database. It should be pointed out that no exchange exists except the generating-operations when sever-al functional tools are realized inside a special application tool/system. Based on above analysis, the procedure of dynamic data and knowledge sharing can be illustrated in Fig.2. Here, the relationships among activities, scenario, interfaces, records, and database are indicated. Using this mode, concurrent design history and designers'intents are stored if necessary. Further, these records can be used as cases for retrieval or fundamental data for example-based learning.

Fig.1 General D&K Exchange Procedure for Functional Tools inside An Activity

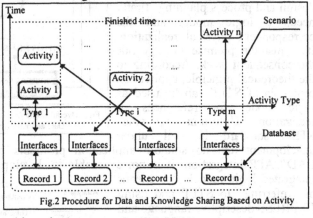

Fig.2 Procedure for Data and Knowledge Sharing Based on Activity

3.OVERALL LOGIC ARCHITECTURE FOR DYNAMIC DATA AND KNOWLEDGE SHARING

As shown in Fig.3, a logic architecture for dynamic data and knowledge sharing is composed of a special application, CE server for the integration of CAD/CAPP, dynamic data and knowledge base, interface library, application tool library, and EXPRESS schemata for logic data model. The correspondent supporting resources include ISO 10303, ONTOS DBMS, and a generator for mapping EXPRESS to ONTOS physical schemata. Here, CE server is a defining, planning, and scheduling mechanism for carrying out a special application. Application tool library is used for describing possible application systems which contain the data structures of themselves, such as feature-based modeller, process planning system, manufacturability evaluation tool, etc. According to the same reasons, interface library is used. In Fig.3, further, the main functions of CE server are listed and the information correlationships among functional modules are also indicated with arrow lines.

## 4.DEVELOPING UNIFIED PART/ PROCESS-PLANNING INTEGRATED MODEL

It is obvious that realizing STEP-based data and knowledge sharing which is based on object-oriented database system and supports the concurrent solving activities for part design and process planning firstly needs a logic model and the correspondent physical realization. It is just the premise to carry out the consequent work. According to the theoretical principle, application protocols of STEP can be used as logic model for references or realization. Unfortunately, since STEP standard is still under

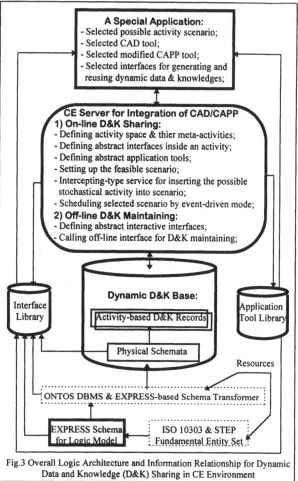

Fig.3 Overall Logic Architecture and Information Relationship for Dynamic Data and Knowledge (D&K) Sharing in CE Environment

development and no suitable application protocols can be used for our concurrent CAD/CAPP integration, it is not avoidable to develop an unified part/process-planning

integrated logic model by means of referring to STEP and using EXPRESS descriptive language and its graphical version, EXPRESS-G, which are specified in ISO 10303.

Based on above ideas, the research on modelling a basic framework of STEP-based unified part/process-planning integrated logic model has been finished. The overall EXPRESS-G diagram can be seen in Fig.4. Where, four units of functionality, that is, feature-based

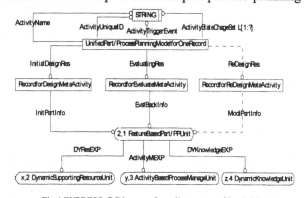

Fig.4 EXPRESS-G Diagram: Overall structure of Logic Model

part/process-planning unit, dynamic supporting resource unit, dynamic knowledge unit, and activity-based process management unit, are determined for describing the correspondent logic model which meets the needs of concurrent CAD/ CAPP procedure. Feature-based part/process-planning unit shown in Fig.5 is just a core of this logic model. As soon as obtaining a logic model, we can use the following steps for finishing its physical realization based on selected object-oriented database ONTOS:
- transforming EXPRESS-G into EXPRESS-L by directly mapping or appending some additional items;
- generating a test scenario and the correspondent data and knowledges;
- realizing soft-test, in which possible dynamic data and knowledges for a designed part are generated by manual mode and are filled in the correspondent data structure specified by logic entity group or schema, for verifying the correctness of the logic model;
- modifying the logic model if there are mistakes;
- using schema generator in ONTOS for automatically transfering EXPRESS-L into ONTOS physical schema.

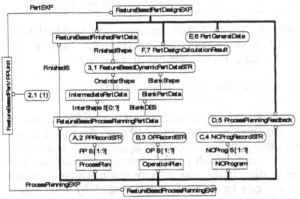

Fig.5 EXPRESS-G Diagram: Feature-based Part/Process-Planning Unit

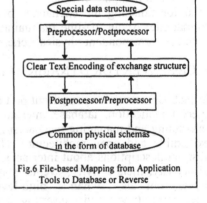

Fig.6 File-based Mapping from Application Tools to Database or Reverse

## 5. METHODOLOGY FOR DEVELOPMENT OF INTERFACES

The realization of an interface depends on the requirement of data transfer, the related entity group in physical schemata, the data structure of used application tool and its openness, etc. Here, the first item indicates what data need to be exchanged, the second declares which aspects of physical schemata are dealt with, and the third, on the one hand, means what relationship can be obtained between the entity group in physical schemata and the data structure of used application tool, on the other hand, either interactive or automatical interface types are needed according to the openness of application tool. As shown in Fig.6, typically, an interface for generation and reuses of dynamic data and knowledges can be created by using clear text encoding of exchange structure specified in ISO 10303 Part 21. Here, mapping between physical schema of database and clear text encoding of exchange structure can be realized by combining the hierarchical modulized constructing-blocks which deal with the correspondent classified entity or entity group. The smallest combinating-blocks are entity-related. In addition, some of combinating-blocks are linked to either the sub-unit of functionality or a group of entities between which there are inheritant relationships. In our research, these constructing blocks are being programmed in C++ language based on ONTOS. According to the logic model for concurrent part design and process planning, a hierarchical classification for constructing blocks is illustrated in Fig.7.

It must be pointed out the types of interfaces for on-line generation and reuses of data and knowledges depend on the openness of specific data structure of used application tool. If this specific data structure is unreadable and closed, the standard text exchange file will be generated or reused by interactive mode. If the specific data structure is open and readable, an automatical mapping will be

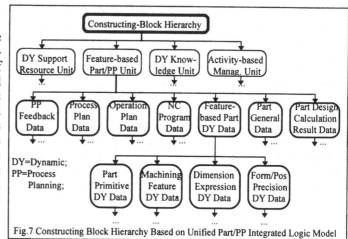

Fig.7 Constructing Block Hierarchy Based on Unified Part/PP Integrated Logic Model

used for finishing the exchanges between the specific data structure and standard text format file. About off-line maintainment of dynamic data and knowledges, it only deals with the correspondent physical schemata. No standard text format files are needed.

## 6. CE-SERVER FOR DYNAMIC DATA AND KNOWLEDGE SHARING

In fact, CE server for concurrent part design and process planning builds a bridge among a special application, database, interface library, and application tool library. As a high-level scheduling and managing mechanism, it is independent to specific applications, and has a extendible and open architecture. Not only static high-level information, such as the abstract descriptions about interfaces, application tools, and activity space which consists of activity, meta-activity, events, states, etc can be defined and managed by using this server, but also dynamic high-level information, for example, scenario planning, intercepting-type activity inserting, event-driven scenario scheduling, etc, can be operated and managed. Seeing CE server from the angle of functional realization, on-line sharing and off-line maintaining for dynamic data and knowledges must be included. The correspondent software prototype development just bases on above two keypoints.

**On-line Dynamic Data and Knowledge Sharing:** On-line sharing occurs during concurrent part design and process planning. In order to realize it, definition, planning, and scheduling are three basic operations. In the aspect of definition, activity space, abstract feasible application tools, abstract feasible interfaces are dealt with. Here, activity space hierarchy which deals with activities, the correspondent meta-activities, triggering events, state set, etc, can be illustrated by using Fig.8. Similar to this, definitions for abstract application tools and interfaces can also be obtained. In addition, we also provide a managing function so as to maintain the correctness of above definitions. In the aspect of planning, a series of activities can be selected with the correspondent mechanism of CE server in order to generate a scenoario which specifies the static procedure for executing concurrent part design and process planning. In the aspect of scheduling, an event-driven intelligent scheduler based on rule-based reasoning is used. The formalization of the correspondent scheduling knowledges depends on the current executed event and state changes. The following is a rule for generating a new event of meta-activity:

IF: (Current event of activity from scenario is Activity (Structural-Design-Stage, $W, Structural-Shapes)) .&.

(Last used event of meta-activity is Evaluate (Structural-Design-Stage, $W, Structural-Shapes) ) .&.
(Current state is IS (Data-GEN, $X, Yes)) .&.
(Current state is IS (Solving, $X, Yes)) .&.
(Current state is IS (Intercept, No)) .&.
THEN: (Next executical event of meta-activity from scenario is Redesign (Structural-Design-Stage, $W, Structural-Shapes))

Here, $W and &X mean variables. From it, we can know that dynamic states play the important roles in the procedure of formalizing knowledges. As shown in Fig.9, dynamic states during executing an event or a meta-activity can be generated by initial state input, intercepting demands for inserting an additional activity, and state generator. The state generator is main source of generating dynamic states and will run after calling the application tool and finishing the correspondent operations.

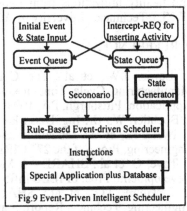

Fig.9 Event-Driven Intelligent Scheduler

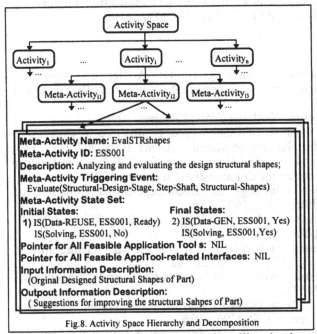

Fig.8. Activity Space Hierarchy and Decomposition

**Off-line Dynamic Data and Knowledge Maintaining:** In the same way, off-line dynamic data and knowledge maintaining deals with interactive interface definition and running. Here, interactive interfaces are only related to physical schemata and their calling is also simple. So no detailed contents are discussed further.

**Structured Software Devewlopment for CE Server:** Structured control (SC) diagrams plus Input/Process/Output (IPO) diagrams, as structured software design means, are being used for developing this CE server. Here, SC diagrams are utilized for describing the module hierarchy and calling relationships between modules. While IPO diagrams are just used for describing the variable list, process functions, etc. It must be noted that the same method is also applied to developing necessary interfaces.

## 7. CONCLUSIONS

According to the description mentioned above, a methodological framework for CE application based on dynamic data and knowledge sharing is provided.

To sum up, we can know that using STEP-based part model is correspondent with the needs of utilizing the unified product model, that exchanging data and knowledges by interfaces oriented from the standard text format of STEP can integrate some application tools which contain the data structure of themselves with database, and that using the functions of intelligent CE server, such as defining, planning, scheduling, and managing, assures a flexibility to reach the goal of CE application.

## 8. ACKNOWLEDGEMENTS

Dr.Ping-Yu Jiang would like to express his thanks for the financial supports from Alexander von Humboldt Foundation and especially thanks to Prof. Dr.-Ing. W. Eversheim, Dr.-Ing. Matthias Baumann, and Mr. Richard Gräßler for academic discussions and facility utilizations from Laboratory of Machine Tools and Production Engineering (WZL), Technical University of Aachen, The Federal Republic of Germany.

## 9. REFERENCES

1. Lewis, J.W., et al.: The Concurrent Enfineering Toolkit: A Network Agent for Manufacturing Cycle Time Reduction, Proc. of Concurrent Engineering Research and Application, Pittsburgh, PA, 1994
2. Bochtler, W.: Dr.-Ing. Diss.Manuscript, WZL der RWTH Aachen, 1996
3. Iwata, K., et al.: Modelling and Analysis of Design Process Structure in Concurrent Engineering, Proc. Of the 27th CIRP Int. Seminar on Manuf. Sytsems, 1995, 207-215
4. Toye, G. et al.: SHARE—A Methodology and Environment for Collaborative Product Developmemt, Proc. of IEEE Infrastructure for Collaborative Enterprise, 1993
5. McGuire, J.G., et al.: SHADE—Technology for Knowledge-Based Collaborative Engineering, Technical Report, Standford Univ., 1992, 17p
6. Qiao, L., et al.: A PDES/STEP-based Product Data Preparation Procedure for CAPP, Computer in Industry, 21(1993) 1, 11-22
7. Gupta, S.K. and Nau, D.S.: Systematic Approach to Analysing the Manufacturability of Machined Parts, Computer-Aided Design, 27(1995) 5, 323-342
8.Mill, F.G., et al.: Design for Machining with a Simultaneous Engineering Workstation, Computer-Aided Design, 26 (1994) 7, 521-527
9. Ping-Yu JIANG and W.Eversheim: Methodology for Part Manufacturability Evaluation under CE Environment, to appear in Proc. of CESA'96, 1996
10. Johnson, M, et al.: Environmental Conscious Manufacturing--A Life-cycle Analysis, Proc. of 10th Int. Conf. on CAD/CAM, Robotics, and Factories of the Future, 1994
11.Herman, A., et al.: An Opportunistic Approach to Process Planning within a Concurrent Engineering Environment, Annals of the CIRP, 42 (1993) 1, 545-548
12. ISO 10303 Part 1: Overview and fundamental Principles. 1992
13. ISO 10303 Part 11: The EXPRESS Language Reference Manual, 1992
14. ISO 10303 Part 21: Clear Text Encoding of the Exchange Structure, 1993.
15.Marczinski, G: Verteilte Modellierung von NC-Planungsdaten: Entwicklung eines Datenmodells für NC-Verfahrenskette auf Basis von STEP, WZL der RWTH Aachen. Dr-Ing. Diss., 1993, 135Seite
16. Chan, S, et al.: Product Data Sharing with STEP, in Concurrent Engineering: Methodology and Application, Edited by P.Gu and A.Kusiak, Elsevier, The Netherland, 1993, 277-298
17. Gu, P and Norrie, D.H.: Intelligent Manufacturing Planning, Chapman & Hall, London, 1995
18.Schützer, K: Integrierte Konstruktionsumgebung auf der Basis von Fertigungsfeatures, Dr.-Ing. Dissertation, TH Darmstadt, 1995, 202 Seite

# AUTOMATIC CLAMPING SELECTION IN PROCESS PLANNING USING TOLERANCE VERIFICATION ALGORITHMS

**L.M. Galantucci and L. Tricarico**
**Polytechnic of Bari, Bari, Italy**

**A. Dersha**
**Universiteti Politeknik i Tiranes, Tirana, Albania**

KEY WORDS: Process Planning, Clamping Selection, Precision Requirements, Tolerance Verification, Machining Errors.

ABSTRACT: In this paper is presented a module developed inside a Computer Aided system for automatic programming of NC lathes, that integrates Process Planning functions with algorithms which control the compatibility of precision design requirements with machining accuracy. Algorithms has been implemented for radial tolerance verification, that takes in consideration the uncertainties created by the elastic part's deflection under the clamping forces, the influence of clamping devices and tool wear errors. Probabilistic methods are used for axial tolerance verification, based on reduced chains. The system is implemented in C++ language, creating graphical user interfaces (GUI) in Windows environment.

## 1. INTRODUCTION

Nowadays, Computer Aided precision planning functions have to be included in CAPP systems, especially for precision manufacturing. Although many authors have discussed this issue, few existing CAPP systems offers the functions of automated tolerance assignment or tolerance verification, and furthermore, very little has been discussed from the point of view of NC machines specification [1]. Application of NC machines in manufacturing production makes precision dimensional requirements of every surface dependent (apart from the class of the machine tool) only by the tool and workpiece position, respects the absolute machine tool co-ordinate system. Therefore, the tolerance

Published in: E. Kuljanic (Ed.) *Advanced Manufacturing Systems and Technology*, CISM Courses and Lectures No. 372, Springer Verlag, Wien New York, 1996.

chains reflected in the part drawing in designing phase, normally, has a minor effect in the accumulation of tolerances in manufacturing design phase.

The aim of this work is to develop an interactive Computer Aided Clamping Selection and Tolerance Verification Module, inside an integrated CAD-CAPP-CAM system [2], proposing the integration of Process Planning decisions, as surface clamping selection and operational sequence determination, with algorithms that controls and guarantees part's precision requirements.

## 2. MODULE'S DESCRIPTION

At present, in modern production, most NC lathes are equipped with a positioning resolution of 1 μm. Various machining errors in the finishing turning, however, degrades the accuracy to a level of approximately 10 μm [3]. But, such accuracy (that can be considered fully acceptable for most of parts with normal precision requirements), can degenerate if we do not consider in the manufacturing design phase the influence of other factors, like part's deformation under cutting forces, set-up errors, tool changes errors, tool wear, etc.

Substantial research has been carried out regarding optimal tolerance allocation; the majority of them are for tolerances design purpose, and only a few of them deal with manufacturing tolerances [1][4].

However, in the latter ones, most of the studies consider the manufacturing process as a fixed one, proposing the distribution of the tolerances on each phase of the machining process. Such approaches, assure the precision requirements of part's design, without considering the role of precision requirements in manufacturing process design (process planning).

Furthermore, in most of tolerance distribution approaches, the dimension chains are constructed grouping all dimensions (tolerances) that form the sum dimension (tolerance), for a determined operational sequence. That is undoubted the right solution for manual conventional machine tools, but may cause an unjustifiable accumulation of tolerances, and consequently, very closed tolerances for individual operational dimensions in manufacturing phase.

In this work, refereed to symmetrical rotational parts, the authors propose to consider only errors that derive by the change of the workpiece co-ordinate system (set-up errors), and "real" tool position (tool change errors and tool wear), respect to the absolute co-ordinate system (machine tool co-ordinate system). The modules for these functions are:

- *clamping selection module*, which analyse all possible set-up conditions, searching the best one, which can guarantees the part's precision requirements, and
- *tolerance control modules*, called by clamping surface selection module, any time when a precision requirement is violated.

## 2.1 CAD AND CAM INTERFACES

The implementation of a Computer Aided Module for Clamping Selection and Tolerance

Verification, certainly pretends, the existence of automatic interfaces with CAD and CAM systems. An interface with CAD system is implemented [2], extracting by 2D CAD wireframe models (IGES files), the geometrical, technological and precision information (*Extract* function of *Data* pull-down menu, fig.1).

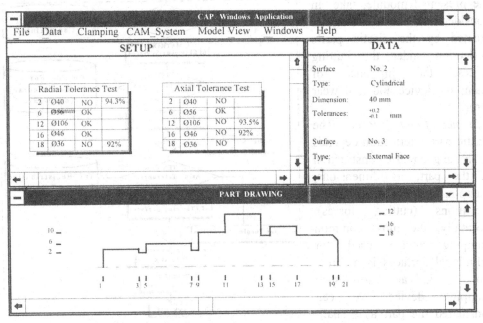

*Fig. 1 - Graphical User Interface (three view model)*

In the study, AutoCAD 12 is used as CAD systems. The CAM modules can also generate script files for NC Programmer of Lathe Prod. Package V4.60 (NC Microproducts) [2].
The system is implemented in Visual C++ language using graphical user interfaces (GUI) based on Microsoft Windows 3.11 (fig.1 and fig.4), permitting an easily user interaction with the system, by means of pull-down menu's functions, and windows dialog-box interfaces. The interface with the user is realised in different model data views, depending on the user demands and the data quantity of the information.

## 3. CLAMPING SELECTION MODULE

The choice of the clamping surface is one of the most important task in process planning, due to the great influence of the set-up position, in the determination of operational sequence. The data flow in clamping selection module is presented in fig.2 . All part surfaces are scanned by algorithm, constructing a clamping matrix containing the surfaces which geometrically can be potentially clamping surfaces (cylindrical surfaces which satisfy a set of geometrical constraints, depending on the type chuck used). Then checking their capability, these surfaces are tested to guarantees all radial precision requirements,

## 4. RADIAL TOLERANCE VERIFICATION MODULE

The implemented algorithms in the presented module, takes in consideration the uncertainties created by the elastic part's deflection under the cutting forces, the influence of clamping device and tool wear errors.

*Influence of cutting forces:* The elastic part's deflection depends upon the geometric construction of the part, its dimensions, clamping system and machining conditions (cutting forces). Normally, the most common clamping device used for cylindrical surfaces is a three-jaw chuck, and for such clamping device we can considered the part as a shaft, fitting at the supposed clamping surface, under the cutting forces at the right hand of the machining surface (Fig.3).

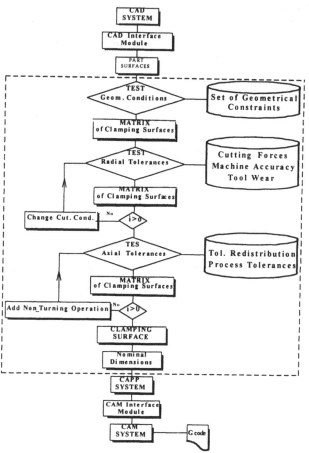

*Fig. 2 - Data flow in the clamping selection module*

Considering that the moment M(z) vary lineament in z-axis as:
the angle of the elastic deflection line $\delta\alpha_i$, between sections "i" and "i-1", can be calculated as:
and the relative displacement of the deflection line $\delta f_i$, respectively the "i-1" section of the shaft is:
For the supposed clamping position, the different sections "i" represents a displacement:

$$M_{(z)} = M_{i-1} + \delta M * z / l_i$$

$$\delta\alpha_i = ( M_{i-1} + M_i ) / 2 * ( l_i / E_{ji} )$$

$$\delta f_i = (2 * M_{i-1} + M_i)/3 * (l_{i2} / 2 * E_{ji})$$

$$f_i = f_{i-1} + l_i * \alpha_{i-1} + \delta f_i$$

These values represent for the tolerance control module, the pre- clamping selection determined error by the influence of cutting forces.

The algorithm calculates these values for all examined potentially clamping surfaces, considering as "i"-section, the right vertical extremity of the part surface, related with a radial precision requirements (fig.3).

*Influence of clamping device and machining accuracy:* A lot of work has been done for the mathematical prediction of machining errors, and various models have been proposed to represent the cause-and-effect relationship between the error and the source [3]. Others researches in precise manufacturing, offers a large number of experimental results for the prevention of clamping device and machine accuracy errors [5].

The implemented clamping selection module offers as alternatives for the predetermination of machine accuracy errors:

a - the calculation of these errors based in error functions reported by T.Asao [3]:

$$f(x) = C_1 * [1 - \exp(C_2 * x)]$$

where:
f (x) - error function;
x - length of cutting,
$(C_1, C_2) = f(V, a, f, \text{tool material})$.

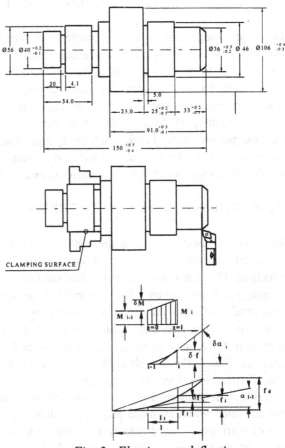

Fig. 3 - Elastic part deflection

The respective data are stored in the database of clamping selection module.

b - the consideration of clamping device's influence and machining accuracy as fixed values, using predetermined errors values (experimental results represented in the literature), or allowing the user to impose his values, according to the real accuracy of a specified NC machine tool (an user interface is implemented to change system's default values).

If none of clamping surfaces guarantees the precision requirements, the system advises the user, and imposes a reduction of the cutting force, altering feed rate and depth of cut, or permits the user to impose his decisions by means of a dialog-box.

In the worst case, evaluating the impossibility to guarantees the radial precision requirements with finishing operations, the clamping selection module adds in the process plan finishing non-turning operation (e.g. grinding), determining the stock removal amount for such operations.

## 5 AXIAL TOLERANCE VERIFICATION MODULE

In the second stage, the system evaluates the possibilities of each possible clamping surface to realise the axial precision requirements. If an axial tolerances is violated, the system starts the tolerance verification (redistribution) module.

The implemented CAPP system, consider the part's drawing requirements as constraints; this fact emphases the necessity to ensure the correctness of part's drawing, because any incompleteness in part dimensions input, can bring wrong decisions of the tolerance distribution module or can break down the system. These controls are carried out applying the graph theory [6]. The incorrectness in part drawing are reflected in the created oriented graphs as close loops (superfluous dimensions) or as unlinked nodes (missing dimensions).

### 5.1 ALGORITHMS FOR TOLERANCES REDISTRIBUTION

The base concept used to evaluate the quality of manufactured product is the conformance to specifications expressed by the allowable variances (tolerances), defined in design specifications. But, since tolerances implies randomness, probabilistic methods have been used, being advantageous over the deterministic approaches (handling only the 100% in-spec case, and resulting with tolerances which may be more conservative than necessary).

The objective of the tolerance control (redistribution) module is the determination and verification of tolerances for individual dimensions, based on a specified confidence level (the user can change the default confidence level) and on a datum tolerance for the "sum dimension" (axial tolerances to be guarantee).

The algorithms implemented in the proposed system, use the concept of tolerances chains, and are activated when an axial precision requirement could be violated., depending on changes of workpiece co-ordinate system relative to absolute co-ordinate system (machine tool co-ordinate system). In the proposed solution, tolerances chains are created with the minimal number of components, taking into account that in NC machines, during manufacturing process, the tool positions relative different part's surfaces, depends only by the position of workpiece and machine tool co-ordinate systems.

If we consider Ai-j, the dimension between the surfaces i and j, related with an axial precision requirement, it can be presented using two different situations:

- both surfaces i and j can be realised in the same set-up (which depends by the position of clamping surface). In such case, Ai-j represent for the system, both design and machining dimension;

- surfaces i and j must be realised in different set-ups; in that case this dimension is not guarantee and the system compose Ai-j dimension (tolerance), creating a tolerance chain with four components:

Ai-j = Ai-s + Aj-s + As + At

where:
- Ai-s - between surface i and set-up surface s;
- Aj-s - between surface j and set-up surface s;
- As  - mean of set-up errors;
- At  - mean tool change errors.

The procedure for tolerance distribution is based in the unified model developed by Bjorke [7]. The real determination of individual tolerances uses the following formula:

$$\sum_{i=1}^{n_t} A_i^2 Tx_i^2 \, \text{var} \, z_i = \left\{ \frac{Tx_\Sigma}{Tw_\Sigma} \right\}^2 - \sum_{i=1}^{n_s} A_i^2 Tx_i^2 \, \text{var} \, z_i$$

where:

$n_t$ - number of dimensions with determinable tolerances;

$n_s$ - number of dimensions with predetermined tolerances.

In the approach have been determined the normalised tolerance value $Tw_\Sigma$ for beta distribution [7], based in the selected confidence level, and are considered as predetermined tolerances the set-up errors and the tool change errors, using the experimental results available literature.

If the same dimension take place in different dimension chains, the system determines the smallest value as the tolerance for that dimension, and consider that value as a predetermined tolerance for the other dimension chains.

After the examination of all axial precision requirements of the part drawing, the system determine as clamping surface, the cylindrical part surface that allows largest tolerances (if all tolerances are larger than process tolerances). Contrarily, the system advice the user and add a finishing non-turning operations (e.g. grinding) to guarantee the violated tolerances.

In every phase of the test, the clamping module gives to the operator the testing results, showing its decisions.

In fig.1 is presented the user interface of Tolerance Control Module. The fourth column in the matrices of potentially possible clamping surfaces, represent the confidence level for the clamping surfaces that can violates axial precision requirements of part drawing. It generates alternative process plans for every possible clamping surface (fig.4), helping the operator in the decisions, if any possible set-up position can not guarantee the precision requirements of the part's drawing.

## 5. CONCLUSIONS

The implemented system realise the integration of Process Planning functions like surface clamping selection and operational sequence determination with algorithms that controls and guarantees the part's precision requirements.

It composes an alternative tool to surpass some critical problems apparent in existing CAPP systems as:

- the reliability of existing CAPP system in the definition of optimal process plans in conformance with quality drawing requirements;
- the possibility that the system decisions are controlled and influenced by alternative user decisions, in every phase of process planning.

The step by step method used (pull-down menu functions situated in a GUI Interface), permits not only a good system's visibility, but also the evaluation of different user decisions (generation of alternative process plans, evaluation of confidence levels, etc.).

Although the created system doesn't interfere in product design phase, it can evaluate the correctness of input information and in case of incompleteness or incorrectness, automatically make the necessary changes in the input part drawing.

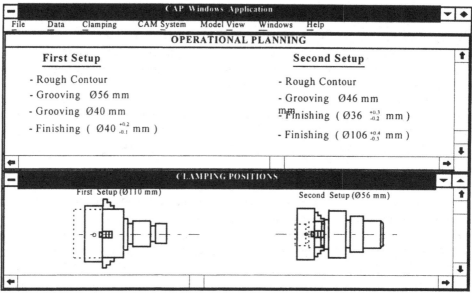

*Fig.4 - GUI of process plans for the examined clamping surface (two view model)*

ACKNOWLEDGEMENTS

This research has been supported with Italian MURST funds. The authors thank Prof. Attilio ALTO, for the precious suggestions and the enthusiastic support during the work.

REFERENCES

[1] H.C.Zhang, J.Mei, R.A.Dudek: Operational Dimensioning and Tolerancing in CAPP, Annals of the CIRP, Vol.40/1/1991, 419÷422.

[2] L.Galantucci, M.Picciallo, L.Tricarico: CAD-CAPP-CAM Integration for Turned Parts based on Feature Extraction from IGES files, CAPE 10, Palermo, 1994.

[3] T.Asao, Y.Mizugaki, M.Sakomoto: Precision Turning by Means of a Simplified Predictive Function of Machining Error, Annals of the CIRP, Vol.41/1/1992, pp.447÷450.

[4] B. Anselmetti, P.Bourdet: Optimisation of a workpiece considering production requirements, Computers in Industry, 21, 1993, pp.23-34.

[5] M. Rahman, V.Naranayan: Optimization of Error-of-Roundness in Turning Processes, Annals of the CIRP, Vol.31/1/1989, 81÷85.

[6] A.Dersha: Computer Aided Process Planning for Symmetrical Rotational Parts, Research Report for Post-Dott. Research Activity at Politecnico di Bari, Italy, 1996.

[7] Ø. Bjørke: "Computer Aided Tolerancing", Tapir Publishers, Trondheim,1978.

# OPTIMIZATION OF TECHNOLOGICAL PROCESS
# BY EXPERIMENTAL DESIGN

**N. Sakic and N. Stefanic**
**University of Zagreb, Zagreb, Croatia**

KEY WORDS: Experimental Design, Optimization, Response Surface, Canonical Function, Welding, Experiment Plans

ABSTRACT: In the design of technological processes, i.e. in determining of the process parameters, selection of optimal parameters is important factor, regardless of the optimization criterion. If the first order experimental plans are used for finding the response function of measured variable (criterion function), problem is reduced to the determination of boundary values of individual parameters, which correspond to the extreme value of criterion function. Application of the second order experimental plans, however, makes certain difficulties, since, in this case, criterion functions are nonlinear (they usually consist of second order elements and interaction elements, besides linear elements), namely the saddle surface. Considering of many parameters (three, four and more), and many optimization criteria, creates even greater problem. The paper deals with the optimization procedure for such functions by analyzing different cases regarding the number of parameters and complexity of the model. The algorithm is tested on suitable practical examples. The article is accompanied with suitable graphs and diagrams.

## 1. INTRODUCTION

In the process of planing of experiment it is important to choose the correct model for calculation of criterion function. Selection of model is defined by the nature of analyzed technological process. The application and elaboration of the model of welding processes

Published in: E. Kuljanic (Ed.) *Advanced Manufacturing Systems and Technology*, CISM Courses and Lectures No. 372, Springer Verlag, Wien New York, 1996.

results in data about linear and interactive influence as well as the square contribution of each factor.

The paper presents the mathematical model of the square function with two independent variables, generally formulated as:

$$Y(x) = b_0 + b_1 \cdot x_1 + b_2 \cdot x_2 + b_{11} \cdot x_1^2 + b_{22} \cdot x_2^2 + b_{12} \cdot x_1 \cdot x_2 \qquad (1)$$

where: $b_0$ is a constant; $b_1$ and $b_2$ are parameters describing the linear relationship between $x_1$ and $Y(x)$ or $x_2$ and $Y(x)$, respectively; $b_{11}$ and $b_{22}$ are parameters describing the square relationship between $x_1$ and $Y(x)$ or $x_2$ and $Y(x)$, respectively; $b_{12}$ is a parameter describing linear interaction between $x_1$ and $x_2$ in their relation to $Y(x)$. If there are more factors in the design, equation (1) is expanded accordingly.

After the experimental data processing and determination of coefficients in equation (1), it is very important to select such a graphical presentation of criterion function, that would enable relatively fast perceiving of the optimal solution of the problem. Since the three-dimensional presentation of results will not always make this possible, it is necessary to convert the equation (1) to a suitable form in plane coordinate system.

The paper develops a method for translation of criterion functions given in (1), into the canonical form, suitable for the presentation in plane coordinate system. The accuracy of this approach is verified and confirmed on a practical example.

## 2. DEVELOPMENT OF OPTIMIZATION METHOD

### 2.1. Method description

Square approximation of real function, given by the expression (1), can present different types of response surfaces, i.e. those with maximum, minimum or saddle point. It is dependent on the values of its coefficients.

To calculate the optimum of given function (1), it is, in the first instance, necessary to determine its first partial derivatives:

$$\frac{\partial Y}{\partial x_1} = b_1 + 2b_{11} \cdot x_1 + b_{12} \cdot x_2 \qquad (2)$$

$$\frac{\partial Y}{\partial x_2} = b_2 + 2b_{22} \cdot x_2 + b_{12} \cdot x_1 \qquad (3)$$

If we equalize the expressions (2) and (3) with zero, the solution of the obtained equation system will produce values of the stationary points. To determine the type of extreme of response function in stationary point, it is necessary to calculate second derivatives of given function:

$$\frac{\partial^2 Y}{\partial x_1^2} = 2b_{11} \qquad (4)$$

$$\frac{\partial^2 Y}{\partial x_2^2} = 2b_{22} \tag{5}$$

$$\frac{\partial^2 Y}{\partial x_1 \partial x_2} = b_{12} \tag{6}$$

The type of extreme can now be determined from solution of the square equation (7).

$$(2b_{11} - b')\cdot(2b_{22} - b') - b_{12}{}^2 = 0 \tag{7}$$

There are three possible solutions:
    a) $b_1'$ and $b_2'$ are both positive - given function has minimum
    b) $b_1'$ and $b_2'$ are both negative - given function has maximum
    c) $b_1'$ and $b_2'$ have opposite signs - given function has saddle point
To present the response function in two dimensions, it is necessary to translate the expression (1) into canonical form:

$$Y(x) - Y(x^*) = b_{11}' \cdot x_1'^2 + b_{22}' \cdot x_2'^2 \tag{8}$$

where $Y(x^*)$ is value of function $Y(x)$ in the middle of response surface; $b_{11}'$, $b_{22}'$ are transformed coefficients; $x_1'$, $x_2'$ are translated axes

Interpretation of canonical function types (8) is given in Table 1.

| Coefficient | | Signs | | Contour | Geometrical | Middle |
|---|---|---|---|---|---|---|
| Case | Relations | $b'_{11}$ | $b'_{22}$ | type | interpretation | point |
| 1 | $b'_{11} = b'_{22}$ | - | - | circles | circular prominence | max. |
| 2 | $b'_{11} = b'_{22}$ | + | + | circles | circular subsidence | min. |
| 3 | $b'_{11} > b'_{22}$ | - | - | ellipses | elliptical prominence | max. |
| 4 | $b'_{11} > b'_{22}$ | + | + | ellipses | elliptical subsidence | min. |
| 5 | $b'_{11} = b'_{22}$ | + | - | hyperbolae | symmetrical saddle | saddle point |
| 6 | $b'_{11} = b'_{22}$ | - | + | hyperbolae | symmetrical saddle | saddle point |
| 7 | $b'_{11} > b'_{22}$ | + | - | hyperbolae | stretched saddle | saddle point |
| 8 | $b'_{22} = 0$ | - | | straight lines | stationary ridge | none |

Table 1. Interpretation of canonical function

Before it can be transformed into canonical form, linear and interactive members of the function (1) must be eliminated. The elimination of linear members is obtained by translation of coordinate system to the extreme point of the function, while the rotation of translated coordinate system eliminates interactive members (Fig. 1).

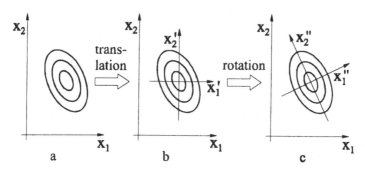

Fig. 1. Elementary transformations

## 2.2. Simulation results

The results of experimental research [4] of welding process and obtained optimal parameters of the impulse MIG welding of aluminum alloy AlMg$_3$, with wire ($\Phi = 1.2$ mm) made of AlMg$_5$, were used to verify the selected model.

The equation of response function for the width of penetration, given in paper [4], is the following:

$$Y = 9.456 + 0.211 \cdot x_1 - 0.223 \cdot x_1 \cdot x_2 + 1.106 \cdot x_1^2 \tag{9}$$

According to proposed method, the response function must be transformed to canonical form by calculating the stationary points:

$$\frac{\partial Y}{\partial x_1} = 0.211 + 2.212 \cdot x_1 - 0.223 \cdot x_2 = 0 \tag{10}$$

$$\frac{\partial Y}{\partial x_2} = -0.223 \cdot x_1 = 0 \tag{11}$$

Solution of this equation system determines the value of the stationary point:

$$x_1 = 0, \ x_2 = -0.946$$

The values of second derivatives of response function in stationary point are the following:

$$\frac{\partial^2 Y}{\partial x_1^2} = 2.212, \ \frac{\partial^2 Y}{\partial x_2^2} = 0, \ \frac{\partial^2 Y}{\partial x_1 \partial x_2} = -0.223$$

Expanded form of the square equation (7) for this example is expressed by:

$$b'^2 + 2.212 \cdot b' + 0.0497 = 0,$$

and solutions are of equal signs:

$b_1' = -2.189$, $b_2' = -0.0227$ ,

which leads to conclusion that the response surface has the form of stationary ridge. Value of response function in stationary point is:

$Y(0, 0.946) = 9.456$

The equation of response surface in canonical form, for the case of penetration, is represented as:

$$Y' - 9.456 = -2.189 \cdot x_1'^2 \tag{12}$$

Canonical formulation, described by (12), is presented graphically by Statgraphics software, as showed in figures 2 and 3.

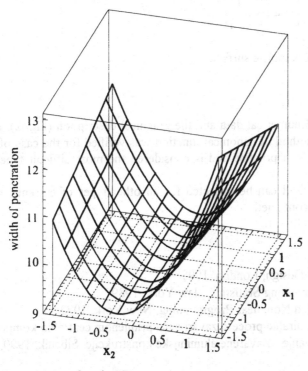

Fig. 2. Response surface in 3D

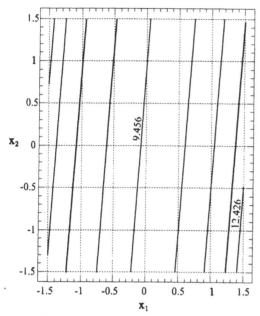

Fig. 3. Canonical form of response surface

## 3. CONCLUSION

Comparison of the obtained numerical data and the graphical description (Fig. 2), justifies the selected method. The method of canonical function was verified for the case of saddle surface, which often appears in practice and is considered the most difficult type of the square function extreme.

Validity of the selected method can be expected for all other types of extreme, and the research in this area will be continued.

## REFERENCES

1. Morgan E.: Experimental Design, London, 1995
2. Hogg R., Leddter J.: Engineering Statistics, New York, 1989
3. Himmelblau D. M.: Applied Nonlinear Programming, New York, 1972
4. Šakić N., Grubić K.: Optimiranje procesa zavarivanja primjenom centralno kompozitnih planova pokusa, Savjetovanje - Zavarene aluminijske konstrukcije, Šibenik, 1990, 105-112

# ROBUST DESIGN OF AUTOMATED GUIDED VEHICLES
# SYSTEM IN AN FMS

**A. Plaia, A. Lombardo and G. Lo Nigro**
**University of Palermo, Palermo, Italy**

KEY WORDS: AGV, Robust Design, Noise Factors, Control Factors

ABSTRACT: Automated Guided Vehicles (AGV), as material handling systems, are widely diffused in FMS environment. The design of such a system involves the selection of the most suitable lay-out on one hand, and the choice of the "optimal" level of some parameters such as the number of vehicles. machine buffer capacity, the number of pallets, vehicle and part dispatching rules, etc. on the other. The optimal combination of these factor levels, that maximises a certain output variable, could be uncovered by Response Surface Methodology (RSM). But two are the problems that immediately arise in the application of such a technique:

- how to consider the qualitative variables. like dispatching or loading rules;
- how to think about some uncontrollable factors, like production mix or machine Mean Time Between Failure (MTBF), whose level can be controlled only during simulations (Noise Factors - NF).

The Japanese researcher Taguchi suggests to study the variability in the system performance induced by these NFs in order to select the best setting (the least sensitive to the induced variability) of the controllable ones (CFs).

In this research we propose a new approach to get the best solution to our problem, considering (in contrast with what Taguchi advises) a single matrix for both CFs and NFs, potentially fractionable, using different responses for the analysis of simulation results, including Taguchi's S/N (signal to noise) ratio.

Research supported by M.U.R.S.T. grant

Published in: E. Kuljanic (Ed.) *Advanced Manufacturing Systems and Technology*, CISM Courses and Lectures No. 372, Springer Verlag, Wien New York, 1996.

## 1. INTRODUCTION

Design and analysis of Flexible Manufacturing Systems (FMS) is most of all based on simulation, but, as everybody knows, this methodology does not provide optimal solutions. Usually, by optimal solution, we mean the optimum setting of system parameters, related to a performance measure of the FMS. But, in many cases, some parameters, while influencing the system performance, are not controllable "in process". Anyway, their value can be set during simulation, according to a plan of experiments. In order to obviate this drawback J. S. Shang [1] proposed to use a methodology developed by the Japanese researcher Taguchi [2]: he suggests to study the variability in the system performance induced by these uncontrollable (Noise) factors (NF) in order to select the best setting (the least sensitive to that variability) of the controllable ones (CF).

While Taguchi method considers just qualitative or discrete variables, the statistical methodology or Response Surface (RSM) is used for quantitative and continuous input variables. Actually these two methods have been also applied sequentially [1]: first Taguchi to set qualitative variables and to consider NFs, then RSM to fine-tune the solution. But at least two problems arise immediately. First, the best setting of qualitative factors by Taguchi method is left unchanged while applying RSM: but this can be considered correct only after proving that no interaction exists between qualitative and quantitative variables. Second, the NFs in the first step are changed into CFs in the RSM, but this is absolutely conflicting. In this research we propose a different approach to get the best solution to our problem, basing on the belief that both the interactions between CFs and NFs and inside CFs need to be studied.

## 2. SYSTEM CHARACTERISTICS AND SIMULATION MODEL

The layout of the FMS [3] under study is shown in fig. 1.

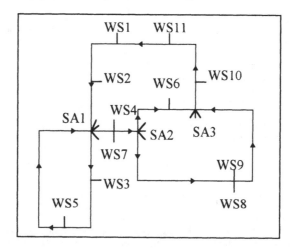

**Fig. 1 FMS lay-out**

The system consists of 11 WorkStations (WSs) including a receiving department (WS1) and a storage/shipping department (WS11). All jobs enter the system through WS1 and leave it through WS11. Parts are handled inside the system by means of Automated Guided Vehicles (AGVs). Each AGV moves a part between the WSs along a unidirectional path layout. Along this path layout three staging area are strategically positioned, each of which consists of three links able to accept one vehicle each. Each WS has limited input (BIj j=1, 2, ..., 10) and output (BOj j=2, 3, ..., 11) buffers where parts can wait before and after an operation. Five part-types are processed concurrently. Parts arrive at the BI1 (WS1 input buffer) with interarrival of 4 minutes (the order of entry is established according to a dynamic loading rule LR) and then they are mounted to fixtures at the WS1 and held in the BO1 (WS1 output buffer); finally they enter the system on a FCFS in BO1 basis. In order to avoid system congestion, the parts residing in the BOj are transported when both the requested transport is feasible and a queue space at the destination WS is available. Parts visit the WSs according to their routing (TAB.1) If no machine in the current WS is free, the part remains in its BI; as soon as a resource becomes idle a part is chosen from the BI queue basing on a Shortest Processing Time (SPT) dispatching rule. After being processed it waits in the BO until the required transport is feasible. A transport of a part is feasible if:
- an AGV is free;
- a place is available in the BI of the next WS according to the part routing (considering also the parts already enrouted to it);
- the part has the highest priority in the queue of the parts awaiting for a transport.
Usually central buffer areas are provided in order to prevent system locking or blocking; if a capacitated system, as our system, is liable to blocking and locking problems, will depend on the operating policies used to run it.
Upon the completion of a part transfer, the idle AGV sends its availability to the AGV_DR (AGV dispatching rule) controller. The AGV_DR controller, basing on the AGV_DR used, selects a new part, if no part can be transported or if the AGV queue is empty, it singles out the nearest staging area; if no place is free, the AGV will go around the path layout until a task is assigned to it by the controller or a place in a staging area becomes free.
A SIMAN [4] discrete simulation model was developed to represent the FMS.
As observed by Garetti et. al. [5] in the actual production the plant will be found to operate both under steady state conditions and in conditions of fulling and emptying (transient) because the FMS is seen by the production planning system as a "machine" to which is assigned "a job" to be finished within a certain available time.
Moreover, the effectiveness of a loading rule is revealed not only in its steady state performance, but also in the rapidity with which the steady state performance is achieved; the loading rule is also responsible for the mix to be realised when the transient is disappeared. Therefore simulations were performed considering the production horizon required to produce the entire volume (V=300 units).
O'keefe and Kasirajan [6] suggest to use a steady state approach to overcome the bias introduced in the results by the starting and ending conditions, while they argue that most manufacturing systems never reach a steady state.

**Table 1  Job-type production mix and Process route**

| Job type | part-mix level 1 | part-mix level 2 | Routing | Processing times (min/unit load) |
|---|---|---|---|---|
| 1 | .10 | .25 | 1-2-4-8-9-10-11 | 8.0-6.0-22.8-8.0-9.2-4.0-2.0 |
| 2 | .20 | .30 | 1.2.4.7.9.6.10.11 | 8.0-6.0-9.2-12.4-8.0-7.6-14.0-14.0-2.0 |
| 3 | .30 | .10 | 1-2-7-9-6-10-11 | 6.0-4.5-23.4-7.2-9.6-3.0-2.0 |
| 4 | .20 | .10 | 1-2-3-5-9-6-11 | 8.0-6.0-17.2-20.4-6.8-26.8-2.0 |
| 5 | .20 | .25 | 1-2-4-8-10-11 | 8.0-6.0-26.4-9.2-4.0-2.0 |

**Table 2  Machine center capacities**

| Workstation | 1 | 2 | 3 | 4 | 5 | 6 | 7 | 8 | 9 | 10 | 11 |
|---|---|---|---|---|---|---|---|---|---|---|---|
| Number of resources | 5 | 3 | 1 | 7 | 1 | 3 | 5 | 2 | 4 | 2 | 1 |

## 2.1 Experimental conditions and assumptions

The following hypotheses characterise our system:
I.   Pre-emption is not allowed and set up time is included in the operation time;
II.  Whenever a machine and an AGV become idle, the next job in the queue is processed immediately (i.e. no delay scheduling);
III. Part transfer time by an AGV is directly proportional to the distance travelled (AGV speed is constant);
IV. Upon job completion at any workcenter, if there is more than one AGV available to transfer the part to next station, the one closest to the workcenter which is demanding service is selected;
V.  Machines in a workcenter are subject in turn to breakdown: when such an event occurs the processing part is positioned in the BI of the WS during the repairing time (TTR).
VI. Control zones are provided at the intersections of tracks to avoid vehicle collision. At the intersections in the AGV path network, FCFS rule govern the node assignment.

We have selected 4 controllable factors and two uncontrollable factors that affect the overall system performance, as listed below:

Controllable factors:

a) number of AGV (AGV#), examined at 3 levels:
   1. 4 vehicles
   2. 6 vehicles
   3. 8 vehicles
b) AGV dispatching rules (AGV_DR) are used when multiple tasks are awaiting pick up simultaneously at different locations. That is, they are vehicle initiated task assignment rules [7]. They are examined at 3 levels:
   1. FCFS rule assigns vehicles to demands sequentially, as requests for AGVS are received from different machines;
   2. STD (shortest travelling distance) rule minimises the time to satisfy the request (time

vehicle travels empty + time vehicle travels busy).
  3. LQS (largest queue size) rule selects the WS that has the longest output queue size.
c) loading rule (LR), which consists in the sequencing of the jobs at the entrance of the system in order to make them accessible for production. It is examined at 2 levels:
  1. MB (mix based) rule selects the job with the maximum value of the VR (volume rate)

$$VR = \frac{M_i^*(t)}{M_i}$$

where $M_i \cdot V$ is the volume of part i to be realised (V is the total production) and $M_i^*(t)$ indicates the percentage of parts-type i produced.
  2. BNB (bottleneck based) rule: assuming $WS_k$ as the bottleneck WS, the part with the shortest operation time on it will enter the system.
d) buffer capacity (BUF_DIM), considered the same for all the WSs, with 3 levels:
  1. 3 places
  2. 4 places
  3. 5 places
Noise factors:
e) mean time between failure (MTBF), with 2 levels:
  1. 150 minutes
  2. 300 minutes
f) part production mix (MIX), which depends on the market condition, with 2 levels, as reported in table 1.

## 3. OUR APPROACH TO PARAMETER DESIGN

According to Taguchi's approach, we need to distinguish CFs from NFs: the latter levels can be set only during simulation. Basing on the concept that we essentially need to consider interactions between the two types of factors, Taguchi proposes to select two detached orthogonal arrays, one for each type of factor, and to replicate each level of the CF array for the complete array of the NFs. By this way we are able to study all level of interactions between CFs and NFs, but usually we are not interested to the higher levels; on the contrary, we may be interested in studying some interactions inside CFs , but, as in order to reduce the number of runs fractional design are used for the CFs, sometimes we may not.
We propose to consider a single matrix, including both CFs and NFs, potentially fractionable. Moreover, we use different methodologies to analyse simulation results, including Taguchi's S/N (signal to noise) ratio, where "signal" stands for the desirable components, while "noise" represents the uncontrollable variability.

### 3.1 The design of experiments

As explained in section 1, we aim at studying the performance of an FMS under varying input conditions. As performance measure, we have considered both part mean flow time and the WIP (work in process), while six are the input variables.

Among the CFs, two are qualitative and two quantitative, while inside the NFs we distinguish a qualitative variable and a quantitative one. Table 3 summarizes the level assumed by each of these factors.

**Table 3 Factor Levels**

| LR | AGV_DR | AGV# | BUF_DIM | MIX | MTBF |
|----|--------|------|---------|-----|------|
| MB | FCFS | 4 | 3 | Level 1 | 150 |
| BNB | STD | 6 | 4 | Level 2 | 300 |
| | LQS | 8 | 5 | | |

## 4. RESULTS

According to the levels of each variable, a factorial design of 216 ($2^3 \times 3^3$) runs has been considered. On both the output variables we performed an Analysis of Variance, one of which is shown in table 4.

**Table 4. Analysis of Variance on flowtime**

| Source | DF | SS | MS | F | P |
|--------|----|----|----|----|----|
| LR | 1 | 62152 | 62152 | 174.34 | 0.000 |
| AGV_DR | 2 | 227110 | 113555 | 318.53 | 0.000 |
| AGV# | 2 | 798553 | 399276 | 1119.98 | 0.000 |
| BUF_DIM | 2 | 304 | 152 | 0.43 | 0.654 |
| LR* AGV_DR | 2 | 115 | 57 | 0.16 | 0.852 |
| LR* AGV# | 2 | 2909 | 1454 | 4.08 | 0.019 |
| LR* BUF_DIM | 2 | 3095 | 1548 | 4.34 | 0.015 |
| AGV_DR * AGV# | 4 | 369238 | 92310 | 258.93 | 0.000 |
| AGV_DR * BUF_DIM | 4 | 1647 | 412 | 1.16 | 0.333 |
| AGV#* BUF_DIM | 4 | 11816 | 2954 | 8.29 | 0.000 |
| LR* AGV_DR * AGV# | 4 | 2486 | 621 | 1.74 | 0.143 |
| LR* AGV_DR * BUF_DIM | 4 | 188 | 47 | 0.13 | 0.971 |
| LR* AGV#* BUF_DIM | 4 | 1347 | 337 | 0.94 | 0.440 |
| AGV_DR * AGV#* BUF_DIM | 8 | 6798 | 850 | 2.38 | 0.019 |
| LR*AGV_DR*AGV#*BUF_DIM | 8 | 824 | 103 | 0.29 | 0.969 |
| Error | 162 | 57753 | 357 | | |
| Total | 215 | 1546335 | | | |

Of course, the analysis has been performed only on the CFs, while NFs are used to create replicates: in fact we aim at selecting the best level of the input setting corresponding not only to the minimum in the performance variable, but also the least sensitive to the variability induced by NFs. As we want to keep under control this variability, the only table of the ANOVA cannot help us if we do not associate to it an analysis of means performed on significative interactions. This analysis of means for the highest interaction AGV_DR*AGV# is shown in Table 5.

**Table 5 Analysis of means**

| AGV_DR | AGV# | | | |
|--------|------|------|------|------|
|        | 4 | 6 | 8 | ALL |
| FCFS | 120.14 | 116.09 | 112.40 | 116.21 |
|      | 24.88 | 25.63 | 25.08 | 25.05 |
| STD | 297.90 | 132.01 | 113.22 | 181.05 |
|     | 26.66 | 33.80 | 25.70 | 88.31 |
| LQS | 324.64 | 127.69 | 112.76 | 188.36 |
|     | 24.90 | 29.99 | 25.42 | 100.77 |
| ALL | 247.56 | 125.26 | 112.80 | 161.87 |
|     | 94.79 | 30.33 | 25.05 | 84.81 |

where in each cell we find the mean and the standard deviation (SD) of flowtime.
If we consider the columns of this table, we can immediately see that we cannot consider 4 vehicles in the system, as we have great values both for the mean and for the SD (except for cell FCFS-4); in the other two columns we do not find a high variability, especially in the last one, corresponding to 8 vehicles in the system. If we repeat the analysis omitting the observations corresponding to 4 AGVs, we get that AGV_DR is no more significative, while the only significative interaction is the LR*BUF_DIM, on which we perform an analysis of means. Examining the corresponding table 6

**Table 6 Analysis of means**

| LR | BUF_DIM | | | |
|----|---------|------|------|------|
|    | 3 | 4 | 5 | ALL |
| MB | 99.87 | 99.85 | 99.90 | 99.87 |
|    | 8.24 | 8.21 | 8.25 | 8.12 |
| BNB | 151.48 | 135.74 | 127.35 | 138.19 |
|     | 31.42 | 28.19 | 20.58 | 28.56 |
| ALL | 125.68 | 117.79 | 113.62 | 119.03 |
|     | 34.59 | 27.40 | 20.80 | 28.41 |

we immediately see that we cannot use the second loading rule, to whom both higher means and SDs correspond. Repeating the analysis with just the last three CFs and only the first loading rule we get no more significative interactions and just the AGV# as significative factor. This means that, setting the LR at its first level and considering just the highest two levels for AGV#, we can select arbitrarily the vehicle dispatching rule, considering that we get better performance (lower flowtime and less induced variability) with 8 vehicles. The same results we get performing the analysis on the other output variable.

In order to strengthen our analysis, we applied Minimax and S/N ratio to the columns we get separating the four replicates (2 X 2) given by the NFs (the S/N ratio for a smaller the better output variable is "$-10\log(\frac{1}{n}\sum_{i=1}^{n}y_i^2)$" ).

## MINIMAX

We considered the maximum value of the output variables among the 4 replicates corresponding to each of the $2\times3^3$ level set of the CFs (said column MAX); we chose, as optimum input level set, the one corresponding to the minimum in the above mentioned column MAX. According to this methodology, the optimum input level sets are MB - FCFS - 8 - 4 (or 5), that agrees perfectly with the analysis performed in the above section.

## S/N ratio

On the replicates described in the Minimax section we computed S/N ratios, getting a column of $2\times3^3$ values. We chose, as optimum input level set, the one corresponding to the maximum in the above mentioned column. According to this methodology, the optimum input level sets are the same obtained in the MINIMAX analysis.

## 5. CONCLUSION

With all the methodology applied in this research we have got a univocal solution to our problem. But it may happen that some interaction cannot be cut off. Anyway, its stressing help us to deal with it adequately.

## REFERENCES

1. Shang, J. S.: Robust design and optimization of material handling in an FMS, Int. J. Prod. Res. (1995), 33 (9), 2437-2454.
2. Taguchi G.: System of Experimental Design, (1987), Vol. 1 & 2. Quality Resources - Kraus & American Supplier Institute.
3. Taghaboni-Dutta, F. & Tanchoco, J. M. A.: Comparison of dynamic routeing techniques for automated guided vehicle systems. Int. J. Prod. Res. (1995), 33 (10), 2653-2669.
4. SIMAN IV Ref. Guide 1989. System Modeling Corp.
5. Garetti, M., Pozzetti, A., Bareggi, A.: An On-line Loading and Dispatching in Flexible Manufacturing Systems, Int. J. Prod. Res. (1990), 28 (7), 1271-1292.
6. O'Keefe, R. and Kasirajan, T.: Interaction between dispatching and next station selection rule in a dedicated Flexible Manufacturing System, Int. J. Prod. Res. (1992), 30 (8), 1753-1772.
7. Egbelu, P. J. & Tanchoco, J. M. A.: Characterization of automated guided vehicles dispatching rules, Int. J. Prod. Res. (1984), 22 (3), 359-374.

# CONSTRAINT PROGRAMMING APPROACH FOR THE OPTIMIZATION OF TURNING CUTTING PARAMETERS

**L.M. Galantucci, R. Spina and L. Tricarico**
**Polytechnic of Bari, Bari, Italy**

KEYWORS: CSP, Turning Optimization

ABSTRACT: In this work the authors tackle the problem of the optimal parameters definition in chip removal processes using a constraint programming approach; the reference process is the multiphase turning. The developed models allow to find the optimal values of the feed and cutting speed; the analysis has been extended in order to consider the influence of the tool reliability. The paper highlights the schemes that has been followed in the formulation of the problem; the results gained with the developed models are then compared with those available in literature obtained with other programming techniques. Models implementation has been realized in C++ using an high level constraint programming tool.

## 1 INTRODUCTION

One of the first fundamental step in the planning of the machining process and in the choice of the cutting parameters consists in the analysis of the feature requirements of the part and of the manufacturing system capability (the limits imposed by the available machine tools and tools). An example is the turning process, where it is necessary to respect the macro and the micro-geometric specifications of the piece, and the constraints imposed by the power and the rigidity of the selected machine tool. The determination of the optimum cutting conditions in the process can be defined as a Constraint Satisfaction Problem (CSP), because it is necessary to find those operative parameters values which satisfy the constraints imposed by the system, searching in the same time the optimal values which minimize the working cost or the working time.

The optimization of operative parameters in the chip removal processes has been approached with different techniques [1,2,3,4]; as concern the turning process,

Published in: E. Kuljanic (Ed.) *Advanced Manufacturing Systems and Technology*, CISM Courses and Lectures No. 372. Springer Verlag, Wien New York, 1996.

different models have been proposed for the optimization of single or multi-pass operations. The geometric programming allows to analyze the case of two passes turning operations [1,5]; in this case the functions non linear that describe the optimization model can be linearized to build a more simple model. This latter can be afterwards analyzed with the classical linear programming techniques. Another approach uses the multi-objective programming to extend the optimization to multi-pass turning operations [2]. In this case the problem is approached without the linearization of the functions, but easily transforming the constraint inequalities in constraint equations, that are afterward solved assigning a well defined precedence. These two examples represent two different applications in the solution of a Constraint Satisfaction Problem.

The constraint programming technique represents an alternative and more efficient way to solve a CSP, because it allows to save specific knowledge of the problem without the necessity to separate the problem and its model of representation. Furthermore, this technique is characterized by some features which accelerate the search of the solutions; these are the constraint propagation (that allows the reduction of the domain of analysis), and solution search algorithms, such as the Tree Traversal and the Backtracking. In this paper the concepts of constraint programming have been used for the parameter optimization of two-pass turning operations; the first part describes the constraint programming method and the fundamental equations that control the strategies for the selecting of the optimum process conditions. Afterwards two examples are presented, in order to highlights the characteristics of the proposed method.

## 2 PROBLEM POSITION

Manufacturing is an activity which combines productive factors and products features. In removal chip processes, the manufacturing cost $C_{TOT}$ of a part is composed by various items, that include the cost of cutting phase, unproductive times, tool and tool change ($C_L$, $C_I$, $C_U$ $C_{CU}$). The total cost for piece can be expressed with the following equation [1,6]:

$$C_{TOT} = C_I + C_L + C_{CU} + C_U / N_P \qquad (1)$$

where $N_P$ is the number of pieces that the tool can work before it is sharped or replaced ($N_P = T/t_L$). The expression (1) can be explicited in function of the work place unit-cost $C_P$, the unproductive times $t_I$, the tool change time $t_{CU}$, the cutting time $t_L$ and the tool life T.

$$C_{TOT} = C_P \cdot (t_I + t_L + t_{CU} / N_P) + C_U / N_P \qquad (2)$$

The generalized Taylor equation can be used in the estimation of T:

$$V_c \cdot T^n \cdot f^m \cdot d^x = C_T \qquad (3)$$

where f, $V_c$ and d are the cutting parameters (feed, cutting speed and depth of cut) and n, m, x and $C_T$ are constant values. The choice of cutting parameters influences the cost equation: higher cutting speeds give for example lower working times, that mean lower cutting phase cost; moreover there is a lower tool life and this increases the tool costs for the wear and the tool breakage. In additional to these economic considerations, the final product features are also characterized by quality and functional specifications; these features depend from the process capability and from the market requirements which limit the productive factors with constraints. In turning, constraints are represented, for example, by the available machine spindle power P, the maximum radial deflection $\delta_{max}$, the surface roughness requirements Ra. The addition of the constraints modifies the initial problem, because optimal values are obtained applying only economic considerations could not satisfy the limitations imposed by the system. The operative parameters that effect the analysis, the type of constraints that parameters have to respect, and the function cost to optimize, arrange the chip removal process as a Constraint Satisfaction Problem.

## 3 THE CONSTRAINT PROGRAMMING

The constraint programming can analyze and solve problems which belong to the category of the CSP [7]. In general a CSP is represented with a set of variables, their variability field and a list of constraints; the search of the solution consists in the determination of the variable values which satisfy a whole of imposed limitations. This CSP category is sometimes defined as a finite resources problem because its goal is to find an optimal distribution of the resources (values of the variable which belong to the variability fields and satisfy the constraints). In the more complex case in which the CSP has also an objective function, the CSP aim is the optimization of this function in the respect of the operative limitations tied to the nature of the problem.

## 3.1 CONSTRAINT PROPAGATION AND REDUCTION OF ANALYSIS DOMAIN

One of the fundamental operations in constraint programming techniques is to consider, through the specific knowledge of the problem, the variables which have a sensitive influence on the solution; for a completely definition of a variable of a problem it is necessary to assign the type (integer, real, boolean) and its variability range, that is the set of values that the variable can assume. The choice of each variable needs to consider different elements as for example the required level of the problem complexity and the importance of that variable in the model. It is evident that an increase of the variable number increases the complexity, but at the same time extends the field of the analysis.

The constraints are an other important element of the model; they define the existing relations among the variables. In the traditional programming the constraints represent control relations, because firstly it is necessary to assign a value to the variable (included in its variability range), and then to verify if this value satisfies the constraints. This method requires recursive phases of assignment and control of the variables, is done also for those values which don't respect constraints. In the constraint programming technique instead, the assignment reduces variability ranges automatically without any preliminary values assignment, and before the solution phase. If for example two integer variable x and y have the same variability field (0,10), and must satisfy the relation $x<y$, the values 10 of x variable and 1 of y variable are not solutions of the given problem, because they don't respect the imposed constraints and thus they don't need to be considered.

The domain reduction of a variable, which is the exclusion of all the values that don't satisfy the constraints, occurs through the constraint propagation, which is the extension of the domain reduction of a variable to the domains of the other ones. Let us consider for example three integer variables x, y and z, with variability range (0,10), and two constraint equation x<y and x+ y<z: the application of the first constraint reduces x and y intervals respectively to (0,9) and to (2,10), without effect on z variable. The second constraint works on new domains and involves a successive reduction of x, y and z domains, respectively to (0,8), (1,9) and (2,10). This effect of the propagation represents an important factor to decrease the complexity of the CSP, because the domain reduction allows to start the solution phase using one variable values which respect the constraints.

## 3.2 TECHNIQUES TO SEARCH THE OPTIMAL SOLUTIONS

The reduction of the variable domains and the constraints propagation are useful means to reduce the complexity of the model, but they are not still sufficient to solve it completely; many values of the variable could be infact potentially solutions of the problem; moreover not always the propagation of the constraints is able to exclude all the variable which do not completely satisfy the problem. This indetermination is resolved with a following phase which uses opportune solving techniques.

The Tree Traversal technique [7,8,9] realizes for example a non deterministic search of the solution, in function of the available variables and the breadth of variability domain; the process starts with the building of a tree structure, where the root is the select variable for the beginning of the search, the nodes represent the other variables of the problem and the branches characterize the values assigned to the variables. Using the constraint propagation, the choice of the beginning branch determines a reduction of the domains of the residual variables ; if all the constraints are respected, the whole of the actual values of the variables represents one of the solutions. In the hypothesis that during the constraint propagation appears an inconsistency in the solution phase (the values of the variable don't succeed to satisfy the constraints), it follows the phase of Backtracking [7,8,9]. In this case the original domains are restored, excluding the value of the variable that has determined the inconsistency, and it is selected a following value of the new domain to continue the search phase. If this search has success, the solution is obtained, while in contrary case it is activated a new phase of backtracking.

## 4 CSP IN CHIP REMOVAL PROCESSES

The optimization of chip removal processes could be treated as a CSP and then approached and solved with a constraint programming technique. Two examples are reported in this paper.

## 4.1 OPTIMIZATION OF TWO PASSES TURNING PROCESS

The goal is the optimization of the cutting speed and the feed in two passes turning operation (roughing and finishing type) with a constant depth of cut in roughing ($d_R$) and in finishing ($d_F$). The elements of the problem are the variables ($V_c$ and f), the constraints and the objective function cost. The hypothesized $V_c$ and f variability fields are respectively (0,2) [mm/rev] and (0,500) [m/min]. The objective function cost is obtained from the equations (2) and (3).

$$C_{TOT} = K_{O1} \cdot V_c^{\alpha_{o1}} \cdot f^{\beta_{o1}} + K_{O2} \cdot V^{\alpha_{o2}} \cdot f^{\beta_{o2}} + C_p \cdot t_I \qquad (4)$$

The constraints are related to the operation type and to the process (turning). In the roughing operation the feed is limited to a maximum value $f_{max}$.

$$f \leq f_{max} \tag{5}$$

In the finishing operation the constraints are linked to the required $R_a$:

$$K_1 \cdot V_c^{\alpha_1} \cdot f^{\beta_1} \leq R_a \qquad \text{for } V_c \leq 207 \text{ [m/min.]} \tag{6}$$

$$K_2 \cdot f^{\beta_2} \leq R_a \qquad \text{for } V_c > 207 \text{ [m/min.]} \tag{7}$$

The constraints relative to the turning process concern the available machine spindle power P and the max radial flexion $\delta_{max}$ (half of the dimensional tolerance on D diameter), in the hypothesis of turning between pointed center and counterpart center; they are described with the following equations:

$$K_3 \cdot V_c^{\alpha_3} \cdot f^{\beta_3} \cdot d^{\gamma_3} \leq P \tag{8}$$

$$K'_4 \cdot V^{\alpha_4} \cdot f^{\beta_4} \cdot (d+\Delta)^{\gamma_4} \cdot \left[ G_c + G_P \cdot \left(\frac{L-X}{L}\right)^2 + G_{CP} \cdot \left(\frac{X}{L}\right)^2 + \frac{64000 \cdot (L-X)^2 \cdot X^2}{3 \cdot \pi \cdot E \cdot D^4 \cdot L} \right] \leq \delta_{max} \tag{9}$$

where L is the turning length of the semifinished part, X is the distance of the tool from the pointed center (hypothesized equal to L/2) and $G_C$, $G_P$ and $G_{CP}$ are respectively the compliances of the tool holder, of the pointed center and of the counterpart center [1]. The values of the constants are reported in the tables 1 and 2.

| | | | | | |
|---|---|---|---|---|---|
| D=50 [mm] | L=500[mm] X=250[mm] | x=0.35 | m=0.29 | n=0.25 | |
| $G_C$=1.3 [mm/daN] $G_P$=0.15 [mm/daN] | | $G_{CP}$=0.75 [mm/daN] | | $\alpha_1$=-1.52 | |
| $\alpha_3$=0.91 $\alpha_4$=-0.3 | $\beta_1$=1.004 $\beta_2$=1.54 | $\beta_3$=0.78 $\beta_4$=0.6 | $\gamma_3$=0.75 | $\gamma_4$=0.9 | |
| $K_1$=2.22×10$^4$ | $K_2$=12.796 | $K_3$=5×10$^{-2}$ | $K'_4$=240 | P = 10 [HP] | |

Table 1. Constant values of the constraints

| | | | |
|---|---|---|---|
| $d_R$ = 4.15 [mm] | $d_F$=.85[mm] $d_{TOT}$=5[mm] | $C_P$=400[Lire/min] | |
| $C_U$=2000[Lire/cut. edge] | $C_T$=300 | $t_{PS}$=1.2[min/piece] $t_{PF}$=1.0[min/piece] | |
| $t_{CU}$=0.5[min.] $\alpha_{O1}$=1 $\alpha_{O2}$=1/n-1 | $\beta_{O1}$=1 | $\beta_{O2}$=m/n-1 | |
| $K_{O1} = \pi \cdot L \cdot D \cdot C_P /1000$ | $K_{O2} = \pi \cdot L \cdot D \cdot \left(C_p \cdot t_{CU} + C_U\right) \cdot d^{x/n} /1000 \cdot C_T^{1/n}$ | | |

Table 2. Constant values of the objective function

The solution of the CSP is realized searching the minimum of the function cost and then afterward determining the values of $V_c$ and f which allow to obtain this value. In the first phase the constraints of the problem are imposed; using the Tree Traversal and the Backtracking technique, the constraints act on the variables $V_c$ and f, propagating the effects of the domain reduction to the variable $C_{TOT}$. This method allows to obtain not only the variable ranges that respect the constraints, but also a first approximation solution of the minimum cost (in practice the lower bound $C_{TOT}^{lower}$ of the $C_{TOT}$ variable domain). Because of the numerical approximations of the solution phase, the absolute minimum of the cost ($C_{TOT}^{min}$) is close to $C_{TOT}^{lower}$ value; the objective of the second phase is to approximate

$C_{TOT}{}^{min}$ through a search in a range of $C_{TOT}{}^{lower}$. This is realized with an additional constraint, which limits $C_{TOT}$ variable to the upper bound of $C_{TOT}{}^{lower}$ range; the constraint propagation proceeds this time from $C_{TOT}$ to the $V_c$ and f variables, and has the goal to reduce $C_{TOT}{}^{lower}$ range values.

The proposed approach has been realized using an high constraint programming language [9]. In the first phase the constraint assignment (5)-(9) and the domain reduction of the variable is realized using the functions *IlcTell(expression 1 operator expression 2)* and *IlcSolveBounds (variables, precision)*, where **expression 1** and **expression 2** are respectively the analyzed function and the assigned constraint functions, while *operator* represents the relation among the two expressions. The term *precision* represents the numerical approximation during the Domain Reduction. The second phase is realized through an iterative procedure, which has the goal of converging toward a minimum of $C_{TOT}{}^{lower}$ range value. The results obtained with this approach are reported in table 3; in the same table are reported the solutions gained with the geometric programming technique [1]:

| Constraint programming | | | | | | |
|---|---|---|---|---|---|---|
| $(C_{TQT})_{Rough}$ | = | [2293.52 - 2294.27] | [Lire] | $(C_{TOT})_{Finish}$ = | [921.676 - 922.036] | [Lire] |
| $(V_c)_{Rough}$ | = | [171.412 - 173.973] | [m/min] | $(V_c)_{Finish}$ = | [218.855 - 228.140] | [m/min] |
| $(f)_{Rough}$ | = | [0.14857 - 0.14968] | [mm/rev] | $(f)_{Finish}$ = | [0.29936 - 0.29964] | [mm/rev] |
| Geometric programming | | | | | | |
| $(C_{TOT})_{Rough}$ | = | 2729.26 | [Lire] | $(C_{TOT})_{Finish}$ = | 929.017 | [Lire] |
| $(V_c)_{Rough}$ | = | 188 | [m/min] | $(V_c)_{Finish}$ = | 246 | [m/min] |
| $(f)_{Rough}$ | = | 0.11 | [mm/rev] | $(f)_{Finish}$ = | 0.3 | [mm/rev] |

Table 3. Solution obtained

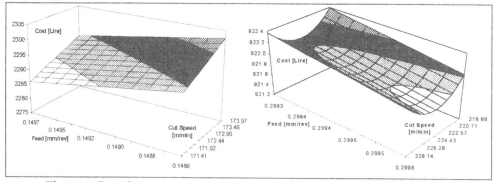

Figure 1. Rough cost                              Figure 2. Finish cost

Some considerations are necessary to explain the differences between the two methods: the solutions of the constraint programming are variable domains, because the search has been addressed toward the extraction of the least breadth range that contains $C_{TOT}{}^{min}$, this is represented by the figures 1 and 2; they highlight the limited variability of the cost in the calculated domain of speed and feed.

All the values of $V_c$ and f included in these intervals allow to obtain a cost that belongs to the domain solution of $C_{TOT}$, and this gives to the process planner a certain flexibility in the choice of cutting parameters, avoiding the sensibility analysis of the solution in the case of small variation of the parameters. The domain breadth highlights the weight of the variable on the objective function $C_{TOT}$: where the domain breadth is larger, the sensibility of solution to its changes is smaller. Another consideration is that the values obtained with the geometric programming are more higher than those ones obtained with the constraint programming; this fact highlights that the geometrical programming finds more conservative solutions to the problem.

## 4.2 TOOL RELIABILITY

An important feature of the constraint programming technique is that is easy to extend the model together with the problem evolution: the addition of new variables and constraints is infact translated in terms of input of new code lines to the original structure of the program. In the following example is introduced the tool reliability R(T) in the model of cost optimization of two passes turning. The problem is resolved describing the constraints and the objective function using the three variables $(V_c, f, R(T))$ without the need to reduce the complexity transforming the model in a two variables one (for example speed and reliability for fixed values of feed [3]). The new $C_{TOT}$ function takes in account the risks connected to the premature tool failure using the time and cost penalties $P_T$ and $P_C$ [3,6].

$$C_{TOT} = C_P \cdot t_I + C_P \cdot \left(t_L + t_{cu}/N_P\right) + C_U/N_P + P_R \qquad (10)$$

$$P_R = \left(C_P \cdot P_T + P_c\right) \cdot \left(1 - R(T)\right)/N_P \qquad (11)$$

The input of the reliability variable R(T) presupposes the knowledge of the statistic model which better describe the tool failure behavior; in the example the Weibull distribution has been used, with null threshold parameter, and $\beta$ and $\mu$ as shape and position parameters $(R(T)=\exp(-(T/\mu)^\beta))$. On the base of these hypothesis it is possible to express the expected $N_P$ value and the expected tool life $t_R$ with a definite reliability:

$$t_R = T \cdot \exp\left[\ln\left(\ln(1/R)\right)/\beta\right]/\Gamma(1/\beta + 1) \qquad (12)$$

$$t_U = \left(t_R + \left[t_R\right]_{R=0.999}\right)/2 \qquad (13)$$

$$N_P = \left(t_R \cdot R + t_U \cdot (1 - R)\right)/t_L \qquad (14)$$

where the cutting time $t_L$ in a turning operation is:

$$t_L = (\pi \cdot L \cdot D)/\left(f \cdot V_c \cdot 1000\right) \qquad (15)$$

The equations (12)-(15) and (3) allow to define completely the solution model. It has been solved for two different values of penalty cost $P_C$ and using a shape parameter of the Weibull distribution of 2.5.

The obtained results have been reported in Table 4; the results highlight the congruence of the optimization model; for example for an increasing of the penalty cost $P_C$, there is an increase of the optimal cost and reliability values [6], and a decrease of the cutting speed in rough and in finish operation. With respect to the previous model, the interaction among a wider number of variables shows an increase of the variable breadth.

| Penalty cost: $10C_0$ | |
|---|---|
| $(C_{TOT})_{Rough} = [2381.46\text{-}2587.82]$ [Lire] | $(C_{TOT})_{Finish} = [930.748\text{-}1017.32]$ [Lire] |
| $(V_c)_{Rough} = [157.064\text{-}174.695]$ [m/min] | $(V_c)_{Finish} = [207.000\text{-}235.280]$ [m/min] |
| $(f)_{Rough} = [0.13076\text{-}0.14495]$ [mm/rev] | $(f)_{Finish} = [0.25867\text{-}0.29964]$ [mm/rev] |
| $(R(T))_{Rough} = [0.50000\text{-}0.81018]$ | $(R(T))_{Finish} = [0.63101\text{-}0.75101]$ |
| Penalty cost: $50C_0$ | |
| $(C_{TOT})_{Rough} = [2724.79\text{-}3196.83]$ [Lire] | $(C_{TOT})_{Finish} = [1122.21\text{-}1229.04]$ [Lire] |
| $(V_c)_{Rough} = [127.362\text{-}141.764]$ [m/min] | $(V_c)_{Finish} = [207.000\text{-}220.744]$ [m/min] |
| $(f)_{Rough} = [0.10975\text{-}0.13285]$ [mm/rev] | $(f)_{Finish} = [0.27071\text{-}0.29964]$ [mm/rev] |
| $(R(T))_{Rough} = [0.80049\text{-}0.97185]$ | $(R(T))_{Finish} = [0.88734\text{-}0.95162]$ |

Table 4. Optimal solutions for different value of penalty cost

## 5 CONCLUSIONS

The optimization problems of cutting parameters in the chip removal processes have been approached and solved analyzing the model as a CSP, and applying a constraint programming technique. The use of this method has highlighted some advantages respect to other techniques, such as the coincidence of the problem with the representation model and its simplicity of implementation and extension. Peculiar characteristics of the proposed method is the reduction of the analysis domain using constraint propagation and backtracking; these techniques have allowed to find not only the solution intervals of the variables which optimize the objective function cost, but also to weight the solution sensibility to each variable.

## REFERENCES

1. Galante G., Grasso V., Piacentini: M. Ottimizzazione di una lavorazione di tornitura in presenza di vincoli, La Meccanica Italiana, 179 (1984), 45-49
2. Galante G., Grasso V., Piacentini: Approccio all'ottimizzazione dei parametri di taglio con la programmazione multiobiettivo, La Meccanica Italiana, 181 (1984), 48-54
3. Dassisti M., Galantucci L.M.: Metodo per la determinazione della condizione di ottimo per lavorazioni ad asportazione di truciolo, La Meccanica Italiana, 237 (1990), 42-51
4. Kee P.K.: Development of constrained optimisation analysis and strategies for multi-pass rough turning operations, Int. J. Mach. Tools Manufact., 36 (1996) 1, 115-127
5. Hough C. L., Goforth R. E.: Optimization of the second order logarithmic machining economics problem extended geometric programming, AIIE Transactions, 1981, 151-15
6. Bugini A., Pacagnella R., Giardini C., Restelli G.: Tecnologia Meccanica - Lavorazioni per asportazione di truciolo, Città Studi Edizione, Torino, 1995
7. Smith B.M., Brailsford S.C., Hubbard P.M., Williams H.P.: The progressive Party Problem: integer linear programming and constraint programming compared, School of Computer Studies Reserach Report Series - University of Leeds (UK) - Report 95.8, 1995
8. Puget J.F., A C++ Implementation of Constraint Logic Programming, Ilog Solver Collected papers, Ilog tech report, 1994.
9. ILOG Reference Manual - Version 3.0, Ilog, 1995

# APPLICATION OF CUSTOMER/SUPPLIER RELATIONS ON DECENTRALIZED SHORT TERM PRODUCTION PLANNING AND CONTROL IN AUTOMOTIVE INDUSTRIES

**K. Mertins, R. Albrecht, O. Bahns, S. Beck and B. La Pierre**

**(IPK Berlin), Berlin, Germany**

KEY WORDS: automotive industries, sequential production, production control, pull principle, decentralization

ABSTRACT:

Existing production and control systems of automotive industries are based on traditional „push - oriented" principles. These principles are characterized by a central production planning procedure which creates a sequence of cars to be produced, the input of the planned production sequence into the body shop as the first shop of the production line and a production control which follows the philosophy to keep the original sequence by any means throughout the complete production. The strict plan orientation of these traditional systems restricts the flexibility to react on short-term disturbances of the production processes.

The paper describes a new, „pull - oriented" concept for the production management and control of automotive industries which is implemented at an asian automotive manufacturer. By establishing customer /supplier relations between the different shops production planning and control tasks are decentralized to the production segments. Based on the „customer orders" each shop optimizes its production.

Published in: E. Kuljanic (Ed.) *Advanced Manufacturing Systems and Technology,*
CISM Courses and Lectures No. 372, Springer Verlag, Wien New York, 1996.

## 1. INTRODUCTION

Today's market of automotive industries is characterized by a global competition. This situation forces car manufacturers around the world to redesign their corporate structure and to go new ways in performing all enterprise functions.

In the field of production conceptual headlines like „Lean Production", „Agile Manufacturing" and „Market - In Orientation" reflect the requirements for new production management and control systems. Reducing the costs of production, increasing the stability of the production against disturbances and in - time delivery to the customer are the top goals to be reached. The high complexity of the production processes and logistics of automotive production requires extremely flexible mechanisms for production planning and control tasks to fulfill these requests.

During the last years IPK has developed new strategies for production planning and control which meet the special requirements of different types of production. Conducting industrial projects as well as international research projects the IPK is able to combine public funded research with experiences from industrial cooperation. Thus research is driven by actual needs of industrial partners and industrial projects are based on latest research results.

In 1995 IPK started a project with a car manufacturer from Asia. The target of the project was to develop a future oriented production management system for a new plant in Asia. Based on an actual state analysis of „state of the art" production management systems in automotive industries and a detailed goal determination for the new system in the first step the project partners designed a concept for a „pull - oriented" production management system which includes the planning of the production sequence, the production control as well as the material handling tasks for body, paint and trim shop. In the next step a system specification was done as a basis for the system realization which was completed beginning of this year. During the system design and realization phase a prototype and simulation system was implemented to support the development process and to evaluate the benefits of the new system [1].

## 2. THE TRADITIONAL APPROACH: PUSH - ORIENTATION

In today's automotive industries the common way is to look upon the production system as **one continuous production line** from body to trim shop. This philosophy finds its expression in one central production planning procedure for all shops (figure 1).

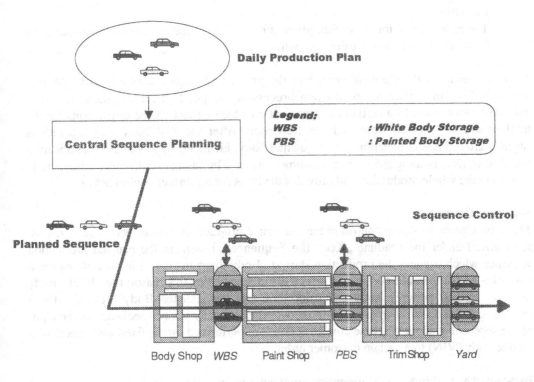

*figure 1: Push-oriented Production Management System*

Focusing one the two basic production management tasks of sequence planning and sequence control the current strategy can be characterized as follows:

### Sequence Planning

Based on a daily production plan, which includes the customer orders to be produced during a certain day, the sequence planning creates the production sequence. This sequence determines the order how the customer orders should be processed in the production line. The goal of this planning procedure is to calculate the most efficient sequence for the total production process. So the sequence planning performs a global optimization of the daily production based on the criteria of the different production areas:

Trim Shop
For the assembly process at the trim shop the sequence planning has to consider the aspects of

- workload leveling (spreading of heavy option cars within in sequence) and
- smoothing of part consumption to minimize the line side buffer inventory.

Paint Shop
The main criteria for the optimization of the painting process is to minimize the costs for color changes in the painting booths.

As a first result of the planning procedure the production sequence as a list of customer orders is fixed to initiate the production process at the point of „Body Shop In". Each customer order refers to a certain car specification which represents the instructions for the total production process from body to trim shop. After the first body shop operations (usually after „floor completion") each unique body, identified by a Vehicle Identification Number (VIN), is assigned to one customer order. This assignment, called christening, applies to the whole production and provides the basis for all further control actions.

## Sequence Control

The main idea of sequence control in the traditional production management system can be summarized under the headline „**Keep the Sequence**". Based on the planned production sequence which triggers the production start at „Body Shop In" the main task of sequence control is to compensate possible sequence disturbances. For this reason the „White Body Storage" (WBS) between body and paint shop and the „Painted Body Storage" (PBS) between paint and trim shop are used to recover the originally planned sequence as the input of the succeeding shop. All these control actions are performed on the fixed assignment of a unique body (VIN) to a certain customer order.

Basically the „traditional, push-oriented" approach can be characterized by:

- determination of one common production sequence for all shops
- fixed assignment of a certain body to a customer order at the beginning of production at the body shop
- pushing of the planned production through the line (from body through paint to trim) using the philosophy of „Keep the Sequence" at sequence control

## 3. THE NEW APPROACH: PULL - ORIENTATION

Because of the extremely high complexity of the car building process in terms of production technology as well as material logistic processes the experience at automotive companies all over the world has shown, that a centralized, push-oriented production management philosophy can not fulfill the requirements of today's marketplace. The necessity of a reliable on-time delivery to the customer and considerable cost reduction requires a production management of high flexibility to handle unexpected events during the production process.

## Basic Strategy

For this reason the IPK Berlin developed a concept for production management in automotive industries which focuses on decentralized, short planning and control loops. The initial idea was to transform new approaches of pull-oriented production management between autonomous production units, which have been successfully implemented at other industries in the last years, into the automotive industries. Based on this vision the IPK designed in cooperation with a car manufacturer from Asian a production management system for a new plant in Asia [2]. The following actions provide the basis for the new system:

- establishment of separate production units for planning and control tasks (**segmentation**)
- classification of product types in order to define products for each production unit („**shop-products**")
- abolition of the strict connection between a certain body and a customer order (**customer anonymous production**)
- introduction of a **multiple christening** system
- implementation of **customer/supplier relations** between the different production units

The initial step of the development was to change the point of view on the car building process. Instead of looking upon the production as one continuous line the project team defined the different shops as autonomous production units. As a result of this **segmentation** the new system includes separate planning and control procedures for body shop, paint shop and trim shop. To unlink the different shops to independent planning and control units „**shop-products**" were defined for each segment. In the new system the objects of body shop's planning and control actions are the different body types to be produced, the paint shop is dealing with paint types as a combination of a body type and color and the trim shop products are defined by a paint type plus trim shop options. As a further step to achieve more autonomous and flexible units a combined system of **multiple christening** and **customer anonymous production** has been designed. In order to keep the flexibility of production management as high as possible, a basic philosophy of the new system is to deal with product types instead of unique customized cars as long as possible. For example the body shop produces bodies of different body types. For the production management tasks it is of no meaning if a certain body will be painted in red ore blue at the paint shop. The body shop only needs the information which body types it has to produce. So the first step of the multiple christening is the identification of a body type. After the paint shop production management has decided the color for a certain body the next christening point determines the paint type of the car. The last christening point is the point where the „classic" christening takes place. Instead of at the begin of body shop the connection between customer (buyer) and a certain vehicle can be established at the „sign-off" as the last station at the trim shop. The implementation of **customer/supplier relations** between the different shops represents the main philosophy for the „shop-to-shop" communication within the new system . Each shop orders the preceding shop to deliver a certain amount of its different shop products at a certain point of time (figure 2).

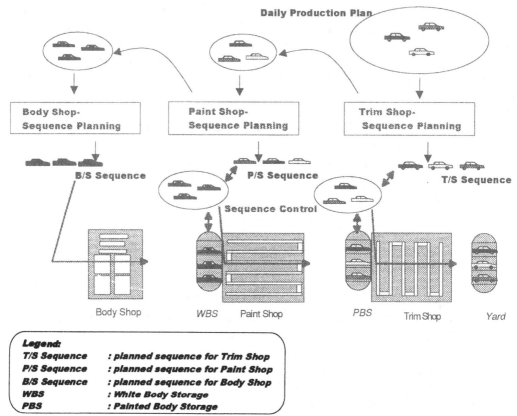

*figure 2: Pull-oriented Production Management System*

**Sequence Planning**

Based on the above described basic strategies for the new production management system the appropriate logic for sequence planning and control was developed. The sequence planning generates optimal production sequences for trim, paint and body shop based on the demand determined in the daily production plan (figure 3). The sequence planning procedure is performed in the following main steps:

1. Based on the daily production plan, which includes the trim type of the ordered car for each customer order to be processed at the certain day, the optimal trim shop production sequence is created by taking all relevant trim shop criteria for workload leveling and smoothing of part consumption into consideration.

2. The generated trim shop sequence is devided into several time buckets. Each bucket contains a defined amount of bodies, which depends on the cycle time at the assembly lines and the capacity of the PBS. Considering the paint type of one car only each bucket represents one order for the paint shop.

3. After the paint shop has received the orders from trim shop. The paint shop production sequence of paint types is calculated. The sequence planning of paint job production management now rearranges the order of bodies corresponding to paint criteria for color grouping. In order to fulfill the demand of the trim shop the paint shop sequence planning is only applied within the buckets.

4. After the paint shop production sequence is created the orders for body shop (containing the needed body types within a body shop time bucket) is derived.

5. The body shop sequence planning calculates its optimal sequence by rearranging the order of bodies within the bucket borders. This sequence initiates the production at the body shop.

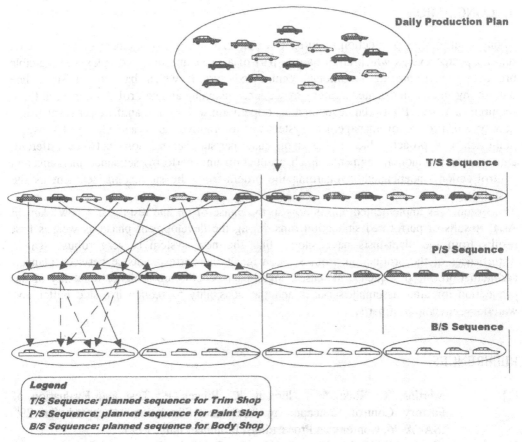

*figure 3: Basic Logic of Sequence Planning*

## Sequence Control

The main application fields of sequence control are the buffers between the different production units (White Body Storage (WBS) between body and paint and Painted Body

Storage (PBS) between paint and trim). The sequence control compensates disturbances of the production process in the preceding shop to prevent negative effects on the succeeding shop production. Corresponding to the concept of multiple christening and shop-products the basic control logic can be described as follows. Based one the actual inventory of bodies in the buffer (e.g. WBS) the control algorithm selects a body of a product type (e.g. body type) which is required to be introduced next in the succeeding shop (e.g. paint shop) to fulfill the planned production sequence of this shop. While determine the next body to be processed the according step of christening takes place automatically. To react on major disturbances of production the sequence control provides supporting functionality for decentralized re-planning of the affected part of the planned sequence.

## 4. CONCLUSION

Today's automotive production requires a flexible production management system to react on unexpected events which are a natural part of a more and more complex and unstable production environment. Traditional, centralized system which base on a strict line philosophy and a pull oriented strategy of sequence planning and control can not fulfill these requirements. The IPK Berlin developed in cooperation with a car manufacturer from Asia a new type of production management system for automotive industries. The system uses a pull-oriented approach which is based on customer/supplier relations between different, autonomous production segments. Each production unit performs sequence planning and control actions independently regarding the orders from the succeeding segment as the customer.

The system was implemented and is now in the phase of on-line testing at a new plant in Asia. Results of performed simulation runs during the development phase as well as first results from the plant-tests have shown that the new system is very robust against disturbances of the production process. As a result the correspondence between planned and actual produced sequence increases. Significant benefits are a higher reliability of the production for after assembly services and the possibility to reduce line-side buffer and warehouse inventory of parts.

## REFERENCES

[1]        Mertins, K.; Rabe, M.; Albrecht, R.; Rieger, P.: Test and Evaluation of Factory Control Concepts by Integrated Simulation. Accepted for 29th ISATA`96, Conference Proceedings, Florence, Italy, 1996.

[2]        Mertins, K.; Rabe, M.; Albrecht, R.; Beck, S.; Bahns, O.; La Pierre, B.; Rieger, P.; Sauer, O.: Gaining certainty while planning factories and appropriate order control systems - a case study. Proceedings Seminar CAD/CAM`95, Bandung, 1995, p. 8B1 - 8B20.

# ARTIFICIAL INTELLIGENCE SUPPORT SYSTEM FOR MANUFACTURING PLANNING AND CONTROL

**D. Benic**
**University of Zagreb, Zagreb, Croatia**

KEYWORDS: intelligent manufacturing, production planning and control, neural networks, constraint-based reasoning, genetic algorithms, production systems

ABSTRACT: Modern approach to manufacturing management requests quick decisions and appropriate real-time solutions. Paper presents the framework for the integrated solution of the scheduling and transportation problem that uses the AI production system supported by neural networks, constraint-based reasoning and genetic algorithms. The framework is superior to the traditional MS/OR approach in real-time manufacturing planning and control and produces quick and reasonable (sometimes optimal) solutions.

## 1. INTRODUCTION

Production planning and control (PPC) is the essence of the manufacturing management. Traditional view at the manufacturing considers: man, machine, material and money. The purpose of the management is to determine the quantities and terms when some resource must or could be available. It is the base for the material resource planning (MRP) as the base for planning the operations and business activities. Some of resources for some activities will be common (material) and some activities' shares or attaches to the same resources (man or machine). The consequence of sharing resources will be in overflowing the production lead time, because of having not enough free resources that enable continuous job-flow through the system. Traditional manufacturing management anticipates that problem, but does not provide appropriate solution. It results with the consequence that

Published in: E. Kuljanic (Ed.) *Advanced Manufacturing Systems and Technology*, CISM Courses and Lectures No. 372, Springer Verlag, Wien New York, 1996.

most of MRP systems are not completely implemented in manufacturing which leads to various strategies of the manufacturing management (MRP II, OPT, JIT, etc.).

Changes in manufacturing request appropriate conceptual and practical answers that enable all advantages of increasing performances of modern technologies. The paper considers some solutions related to intelligent-based framework for solving some tasks in PPC. The emphasis is at the production planning and scheduling in a taught methodological way that enables efficient decision support for use in the concept of intelligent manufacturing. The concept of intelligent manufacturing is the frame that couples process planning and manufacturing management with recent MS/OR methods and advances in AI. Due to the traditional view at the management, new factors are considered: method and time (figure 1). Some of the systems that enable some aspects of 'intelligent' manufacturing are given in [1], [2], [3] and [4]. This paper presents some results of the research that focuses to conceptual and methodological aspect of modelling 'intelligent' decision support.

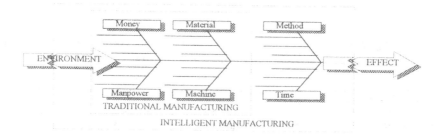

Fig. 1.  The management of manufacturing: Generic Cause-and-Effect Diagram

## 2.  THE INTELLIGENCE AND MANUFACTURING

What does it mean to be an intelligent? Capabilities of receiving, analysing, presenting and using information are certainly attributes of intelligence. The intelligence in manufacturing can be fit in many ways but the emphasis is at the new attitude to management. It means not only use of AI but also the way that to build and manage with the system. The goal is to build a system that consists of 'intelligent units' that solve problems in its domain by they own. The 'intelligence' is possible as a problem-oriented specific designed system that 'acts' the 'intelligence' (AI terminology: production system).

The framework that enables intelligence basis upon holism. It contains as much as possible abilities and knowledge that has to be implemented in the manufacturing and means serious and bright conceptual-based management (figure 2). The holism implemented in a system expresses in using a set of different methods capable to solve specific tasks. As in our case methods came from AI, the selection of the proper method expresses in the usage of 'the best' from each method and combine them in the decision support.

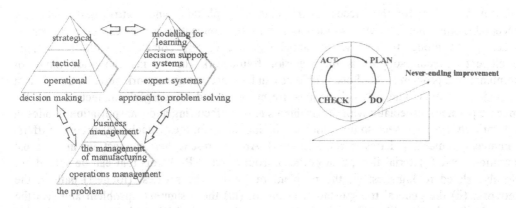

Fig. 2.  Decision making in manufacturing          Fig. 3.  The improvement principle

The intelligent framework in manufacturing management that enables continuous and never-ending process improvement (Deming cycle, figure 3) is similar to the natural process of genetic adaptation. Such framework, that came from the biological sciences, can be present as a never-ending loop that is 'the wheel' of the natural selection. Such kind of optimisation calls the algorithm of genetic adaptation. It is an optimisation program that starts with some encoded procedure, mutates it stohastically, and uses a selection process to prefer the mutants with high fitness and perhaps a recombination process to combine properties of (preferably) the successful mutants.

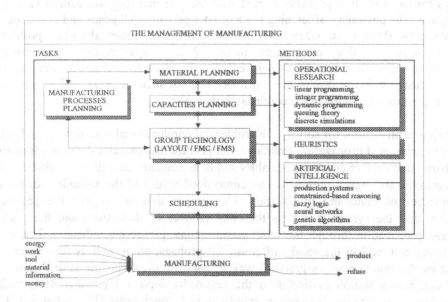

Fig. 4.  The management of manufacturing: Tasks and Methods

Holism is the key for the success in manufacturing planning and control that guarantees acceptable solutions. The clear modelling is how the conceptual model can be observed by focusing the model to narrow sub-areas that are closely connected with the system. Intelligent decision support is the superior framework for solving the wide range of combinatorial problems. All those problems can be represent as network systems that consist of nodes and arcs. The network entities are in direct and meaningful connection with the specific problem, especially with the problems in manufacturing and transportation. Nodes in network always represent some stationary or mobile resources (machines, robots, buffers, conveyers, vehicles, palettes, parts, jobs, etc.). Arcs represent their connections in the most common sense (material-flow, transportation routes, etc.). Problems that can be solved are closely related to logistics: (i) the problem of finding the shortest (longest) path in the network, (ii) the general transportation problem, (iii) the assignment problem and (iv) the general scheduling problem. In compare to the traditional MS/OR and the usage of related methods (figure 4), approach that uses AI is superior in solving combinatorial problems.

## 3. THE FRAMEWORK OF INTELLIGENT DECISION SUPPORT

The large scale of AI methods is probably the most promising tool for solving the problems of manufacturing planning and control. In such a sense artificial neural networks, constraint-based reasoning (CBR) and genetic algorithms (GA) are applicable.

Artificial neural networks are computing systems that process information by their dynamic state in response to the external inputs that have been proven as an effective approach for solving problems. One reason of their effectiveness is in their surprising ability of being well excuses for heurism. It is probably related with the fact that they are natural base for the intelligence. The possibility of learning is a natural aspect of intelligence and in the such way of neural networks too. But when using them in cases where practically every problem is a new one, it is more than unreasonable to use all their capabilities. By preventing their capability to learn, they become a well resource for algorithms that can give reasonable initial solutions. It finally results as a capability that stimulates a great deal of interest that results in their usage in wide range combinatorial problems.

CBR is powerful AI tool for solving the problem of combinatorial optimisation and is based upon the constrained propagation technique as a process that requests deductive framework. The clauses that declare constraint variables and their domains, as well as the clause defining the constrains, are logical premises. The constrained values of the variables comprise the consequences of constraint propagation. Changes in domains trigger the propagation procedure and the system maintains the consequences of deductions and the results of constraint propagation through the same premise-consequence dependencies. This scheme can retrieve and explain the results of constraint propagation as well as the consequences derived by the rule system. A practical way of applying the technique is to formulate the knowledge representation particular to the first-order logic - logic that deals with the relationship of implication between assumptions and conclusions (IF assumption THEN conclusion). When solving the problem by the top-down mechanism, reasoning is backward from the conclusion and repeatedly reduces goals to subgoals, until eventually all subgoals are solved directly by the original assertions. The bottom-up mechanism reasons forward from the hypotheses, repeatedly deriving new assertions from old ones, until eventually the original goal is solved directly by derived assertions. Such kind of problem solving approaches enables that system can 'act' some basic aspects of intelligence. The system is capable to explain why the specific solution is selected. Also, the system can give an explanation how the specific solution was found.

Applied to manufacturing, the algorithm of genetic adaptation is closely related to a system that learns and solves tasks due to the build-in decision procedures. Even the level of the 'intelligence' in today manufacturing systems is close to 'intelligence' of the most primitive organisms in nature, many tasks can be automated by appropriate procedures that contents of one or several methods that can enable the 'intelligence'. The possibility of a real-time (optimal) control seems important because it enables to system to quickly react to unpredictable disturbances. Another aspect that enables intelligent manufacturing is the hierarchy of the system functions and decision procedures as instances of a complex scheme of a decision support. Also, only the concept of distributed management and control enables true simplicity and efficiency that result with increased effects to the system as an entire and system units as instances that solve some specific tasks. The key for the success is in simplicity and is placed in the triangle of 'intelligent' unit, 'intelligent' method and holism.

## 4. THE SCHEDULING PROBLEM

The scheduling is the problem of appropriate operation sequencing for the each product that enables successful completition of the production time-phased plan. It implies assigning specific operations to specific operating facilities with specific start- and end-time indicated with high percentage of orders completed on time, high utilisation of facilities, low in-process inventory, low overtime and low stockouts of manufactured items. The new attitude to product and component designing that concerns concurrent engineering is in idea of schedulability as the key that enables good schedule [5]. The design rules that enable good shedulability are: (i) minimising the number of machines involved in process, (ii) assigning parts/products to the machining/assembly cells, (iii) maximising the number of parallel operations, (iv) maximising the number of batches assigned to parallel machines and (v) allowing for the usage of alternative manufacturing resources.

The solution that we describe is based upon the usage of neural networks and CBR. Bi-directional Hopefield neural network (BHNN) identifies possible bottlenecks in work sequences and facilities. It produces the solution for the simple scheduling problem where the set of $i$ jobs must be distributed to $j$ facilities. Every $j$-th facility can be assumed to every of $i$-th job and, generally, there are $ij$ possible alternatives. Each alternative represents one possible weighted connection between input and output layer. Neurones in both levels are directly and recurrent connected between output and input layer. Each output layer neurone computes the weighted sum of its inputs and produces output $v_i$ signal that is then operated on by the simple threshold function to yield the input $v_j$ signal. To produce the best possible solutions, system enables several strategies for determine the output node ( min $e(j)$, max $e(j)$, min-max $e(j)$ and max-min $e(j)$ ) and only one strategy - min $e(i)$, to determine the input node. The $e(j)$ represents the network energy function:

$$e(j) = \sum_i v_i w_i \tag{1}$$

The method guarantees efficiency in finding the initial solution (practically at one moment) that is then improved by the IF-THEN production rules that gives the initial sequences:

**IF** *last job i at workplace j determines lead time* **THEN**
  **IF** *job i has an alternative k at some other workplace and k≠j* **THEN**
    **IF** *rearranging of i-th job at workplace k minimises lead time* **THEN** (2)
        *rearrange the job and calculate new lead time*

The purpose of the procedure (2) is in balancing the utilisation of the facilities at each sequence and to reduce the sequence lead time as much as possible.

The production system as an intelligent simulation allows description of the specific knowledge and represents objects being reasoned about but rules as well. Such kind of simulations enables assigning additional preferences such are activities connected with the specified facility and fixing it start-, duration- and/or end-time. Two classes of rules take actions that reduce the lead time. First one rearrange the orders of the activities:

**IF** (*R-th facility is bottleneck it the (k+1)-th sequence*) **.and.** (*activities of R-th facility in k-th sequence are distributed at other various facilities*) **THEN** (*rearrange order of activities in R-th facility by the end-time of activities in previous (k+1)-th sequence where l = 0 ∧ 1 ∧ 2 ∧ ... ∧ N* )

Second one is consist of rules that enables reducing of the slack times between sequences:

$$\text{IF } t_{RT_k} < t_{RO_{k+1}} \quad \text{THEN } t_{RP_{k+1}} = \max\left\{ t_{RT_k}, t_{RO_k} + d_{RO_k} \right\} \tag{3}$$

$$\text{IF } t_{RT_k} < t_{RO_k} + d_{RO_k} \quad \text{THEN } t_{RT_k} = t_{RO_k} + d_{RO_k} \tag{4}$$

where $t_{RT_k}$ is the maximal time $T$ in $k$-th sequence at $R$-th facility, $t_{RP_k}$ is the start-time $P$ $k$-th sequence at $R$-th facility, $t_{RP_{k+1}}$ is the start-time $P$ $(k+1)$-th sequence at $R$-th facility, $t_{RO_k}$ is the start-time of $O$-th activity in $k$-th sequence at $R$-th facility and $d_{RO_k}$ is duration of $O$-th activity in $k$-th sequence at $R$-th facility. The main system procedure is the event-driven program. Rule (3) causes event that triggers following rules:

**IF** $t_{RO_k}$ *has changed* **THEN** *end-time of O-th activity in k-th sequence at R-th facility is changed.*

**IF** *R-th facility at k-th sequence has more than one activity* **THEN** *start-times and end-times of all other activities in k-th sequence at R-th facility are changed using* (3) *and* (4)

The main system procedure successive from the end-sequence to the begin-sequence rearranges orders of activities at each facility by previous rules and constructs the feasible solution that guarantees (at least) good solution of the scheduling problem. Because of using the best possible solutions from the first phase, system saves the computer time by reducing the number of simulation experiments. Traditional system that consists four sequences, four jobs where each job in each sequence can be performed by the every of four facilities requests up to 1024 experiments for each priority strategy or their combination. Our system requests only one experiment with concerning no priorities.

## 5. THE TRANSPORTATION PROBLEM

The transportation problem is the important aspect that must be seriously considered when developing practical solutions for 'intelligent' manufacturing. In such a sense the task of finding the shortest path in the network is of great importance because the claim is in real-time control.

The solution we propose uses AI production system with BHNN and CBR in a tough methodological frame capable to quickly identify well (in most cases optimal) solution. The networks that represent the transportation problems consist of nodes and arc where nodes represent the stages and arcs represent the possible transportation routes. Each node represents one stage of the transportation route and is with weightings' $w_{ij}$ (transportation flows) connected with all other $J$ nodes including node as itself ($i=j$). Node at $k$-th stage is connected with each of $j$ nodes in ($k+1$)-st stage and with each of $i$ nodes in ($k-1$)-st stage. In such way there are $ij$ alternatives through $k$-th stage. Complete network representation requests coefficient that represents the current state of node activities. For input signal $v_i$ this state is fixed at level -- $-1$ and for output signal $v_j$ at level -- $1$. The calculation of the energy level for each neurone follows (1) and corresponds with the energy activation function of neural network where $v_i$ connects $i$-th and $j$-th neurone. The $v_i$ value is -- $|1|$ if connection exists or -- $0$ if connection does not exist. To produce the initial solution, system enables four strategies as it was explained in section that describe the solution for the scheduling problem. Backpropagation starts with the node that represents the output layer and the exit from system. In such sense, the number of network layers depends about the number of nodes (transportation stages) and the transportation routes. Input network layer is the node that represents the input in a system. The procedure guarantees high efficiency in finding the initial solution for the system with one input and output. The initial solution is then easy to improve by IF-THEN rules:

**IF** (*two nodes at the transportation route that are not directly connected
have direct connection*) **.and.** (*the value of transportation intensity function
is less then those in the transportation route*) **THEN**                            (5)
*rearrange transportation route by using this network connection*

**IF** (*two nodes at the transportation route are directly connected*) **.and.** (*the value of
transportation intensity function is less when using indirect route through one or
several nodes that are not on the transportation route*) **THEN**                    (6)
*rearrange transportation route by using this network connection*

Iterative improvement for all initial solutions prevents the system to 'stuck in' some local
optimum and maximises probability that the selected solution is the global optimum. Even
only the formal mathematical programming can validate the optimality of the solution,
system guarantees that in the most cases optimal solutions can be achieved very quickly.

## 6. CONCLUSION

The paper presents some ideas and results that concern the possibility of implementing the
AI in solving some problems of manufacturing planning and control. The results point to a
more than usable ideas, principals, methods and practical solutions that can be easily fit into
the manufacturing management decision support. The framework with such methods is
superior to traditional MS/OR one, because quickly produces reasonable (in most cases
optimal) solutions. That is why it is applicable in real-time control of the manufacturing
systems and can be easily fit into a computer-assisted decision support. The structure of the
solution that enables such intelligent-based support must be hierarchical, distributed and
modular to provide efficient decision support for all manufacturing units. Also, the
framework of CBR enables some basic aspects of intelligence as it is the capability to
explain why and how the specific solution is selected.

## REFERENCES

1.  Kusiak A.: Intelligent Manufacturing Systems, Prentice-Hall, 1990
2.  Chao-Chiang Meng, Sullivan M.: LOGOS - A Constraint-directed Reasoning Shell for
Operations Management, IEEE Expert, 6 (1991) 1, 20-28
3.  Karni R., Gal-Tzur A.: Frame-based Architectures for Manufacturing Planning and
Control, AI in Eng., 7 (1992) 3, 63-92
4.  Bugnon B., Stoffel K., Widmer M.: FUN: A dynamic method for scheduling problems,
EJOR, 83 (1995) 2, 271-282
5.  Kusiak A., He W.: Design of components for schedulability, EJOR, 76 (1994) 1, 49-59
6.  Underwood, L.: Intelligent Manufacturing, Addison-Wesley, 1994
7.  Freeman J.A.: Simulating Neural Network with Mathematica, Addison-Wesley, 1994
8.  Freuder E.C., Mackworth A.K.: Constraint-based Reasoning, The MIT Press, 1994
9.  Benić, D.: An Contribution to Methods of Manufacturing Planning and Control by
Artificial Intelligence, Ph.D. Thesis in manuscript, University of Zagreb

# COMPUTER AIDED MANUFACTURING SYSTEMS PLANNING

**T. Mikac**

**University of Rijeka, Rijeka, Croatia**

KEY WORDS: Manufacturing system concept, Manufacturing system planning, Computer aid.

ABSTRACT: The process of manufacturing systems planning is a very complex activity, where the great quantity of data, the planning frequency, the complexity of the manufacturing system models, the need for shorter time of project's elaboration and other factors of influence, create a need to develop a computer programming aid to planning. A particularly important phase of the planning process is the early phase of the project's elaboration, which represents the basic support of further detailed elaboration of the project. In this phase, for reasons of speed, as well as quality of the project's concept, it is necessary to use an organized programming aid. In this paper, a software program for manufacturing system planning (PPS) is described. This was developed and tested on examples, representing an efficient mean of more rationale, creative and quality planning activity.

## 1. INTRODUCTION

The present moment is characterized by an intensive development of science, and in the field of industrial production the market changes are strong and frequent, so there's the problem of planning the adequate manufacturing systems (MS) which would satisfy those demands [1,2,3].

Published in: E. Kuljanic (Ed.) *Advanced Manufacturing Systems and Technology*,
CISM Courses and Lectures No. 372, Springer Verlag, Wien New York, 1996.

The complexity of the planning process itself is increased by the bigger planning frequency, by manufacturing systems which are getting more complex, the need for a shorter time of project's elaboration and the demand for high quality of project's solution that reflects itself in the choice of objective criteria and the choice of an optimal solution with minimal differences between planned and real characteristics.

What is particularly important here, is the early phase of planning, which refers to the definition of an approximate global project's concept [4], therefore the realization of an adequate program aid to the activities related to this planning phase contributes to the efforts of finding adequate answers to the solution described.

## 2. PPS SOFTWARE PROGRAM CONCEPT

A developed methodology for MS planning is used to achieve the following effects:
- the increase in project's solutions quality
- the shortening of project's elaboration time
- the possibility to increase consistency of data and information, and the possibility to generate more variants of the solutions for manufacturing program assortment segmenting
- joining the adequate model of a basic manufacturing system (BMS) to generated groups of products with regard to the assortment and the quantity of applied manufacturing equipment and its layout, already in the early phase of planning
- cutting down the planning costs through increase of planning speed, productivity and quality in the early phase of planning
- the improvement of information flow and relieving the planner of routine activities

The concept of a PPS software program aid is based upon:
- adequate data-basis
- numerical data processing
- interactive activity of the planner

Managing the activities related to the MS concept definition with a PPS program implies the use of commands that proceed partly automatically or by interactive activity of the planner. The structure of data-base is such that it is possible to have their simple and continual additional construction, and their application on smaller computer structures (PCs).

The interactive activity enables the planner to use his creativity and experience by choosing those global input parameters of the project's task which are not automatically comprised in the program's algorithm. Thus it is possible to enable the changing of project's tasks and parameters and to realize a number of alternative solutions of groups of workpieces as a segment of the manufacturing program total assortment, and an adequate choice of an associate BMS model as a basic module for a complex MS.

The computer aid is developed in the Pascal program language and the fundamental information flow of the PPS software program is represented in figure 1.

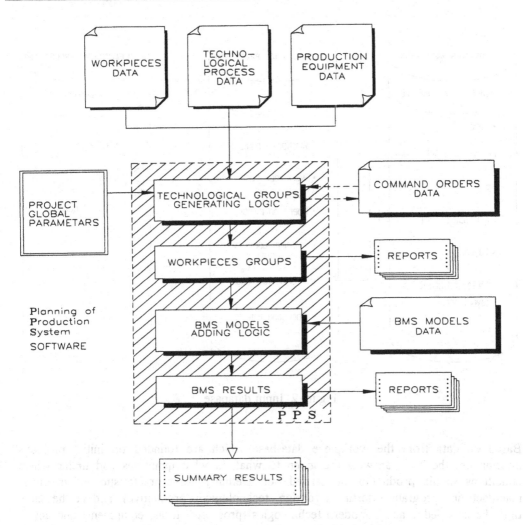

Figure 1. PPS software program information flow

## 3. PROGRAM ACTIVITIES DESCRIPTION

### 3.1. FORMATION OF THE INPUT DATA-BASE

The complexity of activities that need to be done within the frame of the task of MS concept formation, implies their gradual and multilevel process functioning under a determined sequence and mode. The first step in MS planning, and at the same time in application of the proposed PPS software program, is represented by formation of an input data-base. This is the most extensive routine part of the work. The data-base is structured as in figure 2.

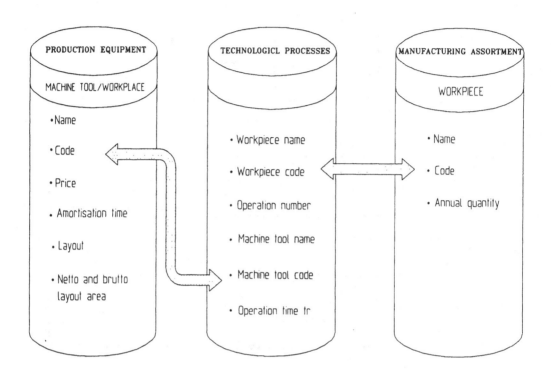

Figure 2. Input database

Based on data from the workpiece data-base which are founded on initial project's information, the basic answers are given to what, in what quantities and under which conditions should production be carried out. Defining the characteristics of the total manufacturing program assortment (design, technological, productive), and at the same time the knowledge about modern technologies (processes, tools, equipment) and actual manufacturing policy, the technologist - planner does the correlation of those influence variables. Based on data about available manufacturing equipment from the equipment data-base planner forms approximate, preliminary technological processes and fills the technological processes data-base as a base for defining the project's policy expressed through the attitudes on characteristics that must be acquired by BMS.

Based on that knowledge the planner defines certain global parameters as input data like number of annual working days, number of working hours in a shift, the level of time utilization of available working time, which are being entered in the program forming thus a data-base for global parameters of the project's task.

Those data are necessary to define the available capacities that, in relation to the real load of applied manufacturing equipment, represent one of the basic criteria for segmenting the manufacturing program total assortment into homogen groups of similar workpieces in the continuation of the application of PPS software program.

## 3.2. GROUPING OF SIMILAR WORKPIECES

By grouping workpieces into homogen groups by design, technological and productive characteristics the segmenting of the total assortment is being achieved. Thus the division of total CMS into larger number of different groups of manufacturing equipment within the frame of BMS on which a determined workpieces group will be completely processed is possible.

The goal of adequate workpieces assortment segmenting and forming a BMS model is:
- to shorten the manufacturing cycle
- to simplify the material and information flows
- to achieve a better conduction of the process
- to cut down the stock of material in the process, and thus the production costs
- to increase manufacturing quality
- to specialize the staff

A base for workpieces grouping is represented by the use of homogen manufacturing equipment based on correlation of workpieces and manufacturing equipment through manufacturing operations. As a criterion for decisions about purposeful formation of workpieces groups and related BMS, besides the homogeneity of applied manufacturing equipment a satisfactory level of its exploitation is also used.

The grouping process based on Cluster methodology is done automatically by the computer together with the defining of grouping strategy by the planner. As a similarity function, beside technical (time) level of equipment exploitation, an economic (value) level of exploitation is taken, which takes into account the value of single manufacturing equipment expressed through size and amortization time.

The procedure is lead by the computer forming a matrix of relations between equipment and workpieces through technological operations $M_{trij}$, calculates the matrix of manufacturing equipment annual load $M_{tgrij}$, serving as a base upon which the equipment (tool machine) quantity is established, and which is expressed using a matrix of necessary manufacturing capacities $M_{srij}$, as in expressions (1), (2), (3). Since equipment quantity must be expressed by an integer and not by a decimal, comparing those two values the total level of equipment exploitation $\eta_i$ (4) is received, which is a base for calculation of total system exploitation level $\eta_s$ (5) for a single proposed solution.

$$M_{trij} = i \left| \overset{j}{t_{rij}} \right| \tag{1}$$

$$M_{tgrij} = \left| M_{trij} \right| \cdot \left| q_{gj} \right| \tag{2}$$

$$M_{srij} = \frac{\left| M_{tgrij} \right|}{\left| K_i \right|} \tag{3}$$

$$\eta_i = \frac{\sum\limits_{j=1}^{n} S_{rij}}{S_{ic}} \tag{4}$$

$$\eta_s = \frac{\sum\limits_{i=1}^{m} S_{ic} \cdot \eta_i}{S_s} = \frac{\sum\limits_{i=1}^{m} \sum\limits_{j=1}^{n} S_{rij}}{\sum\limits_{i=1}^{m} S_{ic}} \tag{5}$$

The levels of exploitation, regarding the necessity of optimal grouping of adequate BMS models, differ by detailed criteria with regard to production control organization that is the workpieces flow through the system, into line and return levels. Thus four levels of manufacturing equipment exploitation are identified, which serve not only as a workpiece grouping criterion but also as a basis for future joining of BMS to so formed groups.

## 3.3. JOINING OF A BMS MODEL

To each of these workpieces groups formed by segmenting of assortment with similarities of determined design, technological and productive characteristics, suitable BMS models adequately formulated for this planning phase and characterized by a limited number of input parameters of the project's task are joined automatically by the computer.

The models are systematized with regard to a larger number of criteria, so as to allow the difference between line and return models based on the way of manufacturing process controlling, expressed by a possible number of operations on one manufacturing capacity, on the quantity of manufacturing equipment in the BMS, on the number of workpieces processed on the BMS, on the characteristics of the workpieces' flow through the system expressed indirectly with the correlation coefficient and the proposed equipment layout in the BMS and as a lower border on one of the equipment exploitation levels.

At the PPS program such modules are entered into the model's data-base with the interactive work of the planner, but the data-base can be completed or reshaped also for the purpose of later planning phase or final MS planning.

After the model has been entered in the model's data-base, the computer automatically joins adequate BMS models to single workpieces groups based on previously formulated and explained criteria, allowing the final listing of single BMS concept solutions as in figure 3. On it, a workpieces group and a joined BMS model, the quantity and the type of applied manufacturing equipment with the price, annual load, total investment in equipment, and the levels of its exploitation (time and value) can be seen. In the solution's listing the range of correlation coefficients, as a characteristic for leading the manufacturing process is also expressed. Based on this listing with regard to the informations from manufacturing equipment data-base, it is also possible to make a calculation of the total BMS surface required, as another criteria for optimizing the concept project's solution.

| 6   PSS Path model | | | | | Util. fact. | | Corr.co. | | Sum |
|---|---|---|---|---|---|---|---|---|---|
| #  | Group name | Wcnt | Mcnt | | Tech | Econ | min | max | EtaMax |
| 24 | GROUP 22 | 3 | 6 | | 0,623 | 0,705 | -3 | 4 | - |

| PRIRUBNICA-29 | VILJUSKA-32 | | VILJSKLOPKE-35 | | | | |
|---|---|---|---|---|---|---|---|
| # | Machine types | Price | Wrk hours | Sr | Sp | Sr/Sp | Value |
| 1 | LATHE-TOK5 | 280000 | 3420,00 | 0,967 | 1 | 0,967 | 280000 |
| 2 | RADIAL DRILLING MACHINE-BUR1 | 75000 | 3460,00 | 0,979 | 1 | 0,979 | 75000 |
| 3 | RO-TABLE-RST1 | 5000 | 1420,00 | 0,402 | 1 | 0,402 | 5000 |
| 4 | MILLING MACHINE-GLH1 | 130000 | 2240,00 | 0,633 | 1 | 0,633 | 130000 |
| 5 | DRILLING MACHINE-BUJ2 | 40000 | 1440,00 | 0,407 | 1 | 0,407 | 40000 |
| 6 | IND HARDENING MACHINE-INK1 | 200000 | 1240,00 | 0,351 | 1 | 0,351 | 200000 |
| | | | 13220,00 | | 6 | | 730000 |

Figure 3. BMS concept report

## 4. CONCLUSION

The program aid developed on the basis of the described procedure allows a quick, efficient and scientifically based elaboration of MS project concept, as a basic ground for further detailed continuation of MS planning.

A developed program comprises:

- the generation of a data-base
- segmenting of a CMS into BMS, based on suitable grouping of similar workpieces (by design, technological and productive characteristics)
- dimensioning of the system by capacity
- defining and joining of BMS models suitable in the concept's implementation phase, by which the manufacturing equipment group structure is determinated, mode of manufacturing process control, surface and price.

The quality is represented also by interactivity, specially in the case when planner's intervention is needed in order to include influence characteristics or project's limitation that could not have been quantified of algorithmically processed.
Generally, by application of such program the total planning time is shortened and the quality of project's solution is increased.


List of symbols:

$t_{rij}$ —  r operation time when j workpiece is producing on the i equipment
$q_{gj}$ —  j workpiece annual quantity
$K_i$ —  annual available time capacity
$S_{rij}$ —  i equipment quantity necessary for r operation of j workpiece
$S_{ic}$ —  responsible i production equipment quantity (integer)

REFERENCES

1. Vranješ, B.: Jerbić, B.; Kunica, Z.:   Programska podrška projektiranju proizvodnih sistema, Zbornik radova Strojarstvo i brodogradnja u novim tehnološkim uvjetima, FSB-Zagreb, Zagreb, 1989, 159-164.

2. Muftić, O.; Sivončuk, K.:   Ekspertni sistemi i njihova primjena, Strojarstvo 31(1), 1989, 37-43.

3. Tonshoff, H.K.; Barfels, L.; Lange, V.:   Integrated Computer Aided Planning of Flexible Manufacturing Systems, AMST'90, Vol. I, Trento, 1990, 85-97.

4. Mikac, T.:   Grupiranje izradaka i izbor modela proizvodnog sustava, Zbornik radova 2. međunarodnog savjetovanja proizvodnog strojarstva CIM'93, Zagreb, 1993, I79-I89.

# LINEAR PROGRAMMING MODEL
# FOR OPTIMAL EXPLOITATION OF SPACE

**N. Stefanic and N. Sakic**
**University of Zagreb, Zagreb, Croatia**

KEY WORDS: Optimization, Linear Programming, Arrangement of Elements, Palette, Integer Programming

ABSTRACT: Optimal utilization of given space is a significant problem which occurs in design of the elements of transporting and storage systems. Optimal space utilization figures as a factor which can influence the unit selling price of a product. Different possible cases of space arrangement are analyzed in the paper. The selected of linear programming proved to give satisfying results in determining of the optimal space arrangement. The results will be compared with the results of some practical methods. Finally, the direction of possible further work in this area will be given at the end of the article.

## 1. INTRODUCTION

The problem of arrangement of elements is a very complex combinatorial, present in many industrial branches. These are:
- metal industry,  textile industry, paper industry, ladder industry, wood industry, glass industry,  food and drugs industry and plastic industry.
The range of fields, in which the arrangement of elements is the significant problem, is very wide:

Published in: E. Kuljanic (Ed.) *Advanced Manufacturing Systems and Technology*,
CISM Courses and Lectures No. 372, Springer Verlag, Wien New York, 1996.

- microminiaturization, crystallography, complex design approach, modeling of various economic, physical and chemical processes, typesetting printing industry, transport and storage

The problem of arrangement of elements is complex in both theoretical and practical sense. Theoretical complexity of the problem is manifested in selection of appropriate mathematical model, which is used in optimal arrangement of elements. Combinatorial nature of the problem, which imposes limitations on optimal arranging of elements by an expert, or a team, presents its practical complexity.

The arrangement of elements can be performed by the following approaches:

a)heuristic approach - arrangement of the elements is based on the experience of an expert or a team, without any mathematical modeling. This approach will not lead to optimal solution and will therefore have degree of utilization.

b)heuristic-exact approach - arrangement of the elements is based on both heuristic and exact approach. Only some variants of arrangement are considered, i.e. those that are carried out manually or by some procedure. Although these solutions have certain practical value, they are not optimal.

c)exact approach - presents the highest level in solving the problem of element arrangement. The arrangement of elements is performed by strictly formalized procedure (according to mathematical model), using computer. Consequently, the solutions are optimal.

## 2. METHOD DEVELOPMENT

Firstly, it is necessary to adopt certain terminology in the field of arrangement of elements. There are three basic concepts in arranging of elements:

a)element of arrangement - is a geometrical object (triangle, quadrangle, polygon, circle, ellipse, or any plane surface originating from them), which is a part of a product or a semi-finished product. The element of arrangement can be defined with two or three data, which relate to its dimensions.

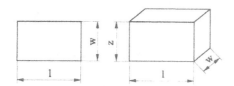

Fig. 1. Element of arrangement

b)region of arrangement - presents an object on which the elements are arranged. Region of arrangement can be defined with two (two-dimensional arrangement) or three (three-dimensional arrangement) data which relate to dimensions. These parameters are length (L) and width (W) for 2D arrangements, or length (L), width (W) and height (Z) for 3D element arrangements.

c)refuse - remainder of material after cutting out the elements.

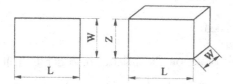

Fig. 2. Region of arrangement

Basic spaces of arrangement of elements according to Fig. 3, are:

a)one-dimensional real space R,

b)two-dimensional real space $R^2$ and

c)three-dimensional real space $R^3$.

Fig. 3. Basic spaces of arrangement

## 2.1. Mathematical model

Mathematical model for 3D problems of space arrangements is often applied in industry, especially in the area or transportation and storage, where the problem of optimal cutting of three-dimensional objects occurs.

Of all possible cases of arrangement, this paper deals with the arrangement of handling units on a palette. handling units can have:

a)prismatic form

b)cylindrical form

Fig. 4. Arrangement of cylindrical forms on palette

For case b), shown in Fig. 4, if the cylindrical handling units with diameter 2R are arranged on a palette with dimensions 6R × 4R, the obtained arrangement efficiency is at the same time maximum and equals:

$$\eta = \frac{6 \cdot R^2 \cdot \pi}{6 \cdot R \cdot 4 \cdot R} = \frac{\pi}{4} = 0.7854$$

From the preceding section follows that, considering maximum efficiency, case a) is more interesting than case b). The following are some specific features that will occur, and which should be considered:

a) shape of the cutting object is not standardized,

b) dimensions of the palette are standardized,

c) handling units will be folded over each other on the palette, in order to achieve compact and solid cargo.

These specific features of prismatic units arrangement on a palette are shown in Fig. 5.

Fig. 5. Arrangement of prismatic forms on palette

The problem can be formulated mathematically, as follows:

Region of arrangement has dimensions $L \times W \times Z$. The set of prismatic handling units, with dimensions $l_i \times w_i \times z_i$ ($i = 1 \dots m$) has to be arranged on the region $L \times W \times Z$, in such a way to achieve the optimal utilization of the region of arrangement. Each of the elements is assigned particular priority. On the basis of these data, specific algorithm is generated. Obviously, the three-dimensional problem can be reduced to a two-dimensional, due to the arrangement technology itself, as well as mathematical simplicity of the problem. Let $B_z$ be the requirement vector which results from given demands, and let B be the vector of arrangement of elements on a palette. The condition is following:

$$\sum_i B_i = B_z \tag{1}$$

We need to define the minimal coefficient of efficiency. According to given algorithm, valid variants of arrangements are determined by varying the priorities of the elements. The obtained variants are forming the matrix of coefficients A. Now, the problem can be solved as a 0-1 programming problem:

min (C·X)

with conditions:    $A \cdot X \geq B_z$

$\rho \cdot V_i \cdot g \leq N$

$x_i = 0, 1 \dots n$

where: C is one-dimensional vector which components represent refuse of the region of arrangement for each of the generated arrangement variants; X is n-dimensional vector of variables which represent the unknown number of application of single arrangement variant; A is matrix (m×n) comprised of column vectors $A_j$ which are connected to generated arrangement variants; ρ is density of material; $V_i$ is volume of the handling unit; N is carrying capacity of palette (kN).

The condition of non-negativity and integer condition are applied to vector X, which provides an interpretation of obtained solutions.

## 2.2. Testing of the mathematical model

At the Faculty of Mechanical Engineering and Naval Architecture, the OPTIMA computer program for calculation of optimal arrangement is developed.
For practical example NSHR1, handling units data are given in table 1. The height of each handling unit is 150 mm, and used material is steel. The carrying capacity of palette is 10 kN.

| Handling unit | Length (mm) | Width (mm) | Name | Rotation | Quantity |
|---|---|---|---|---|---|
| 1 | 200 | 100 | PL1 | Y | 1000 |
| 2 | 450 | 150 | PL2 | Y | 1000 |
| 3 | 500 | 350 | PL3 | Y | 1000 |
| 4 | 600 | 500 | PL4 | Y | 1000 |
| 5 | 300 | 600 | PL5 | Y | 1000 |
| 6 | 400 | 700 | PL6 | Y | 1000 |

Table 1. Basic data for example   NSHR1

Results obtained with OPTIMA software are given in tables 2 and 3. Table 2 contains data of appropriate schemes of arrangements per block, as well as total efficiency for each block.

| Block number | Scheme number | | | | | | Efficiency (%) |
|---|---|---|---|---|---|---|---|
| 1 | 1 | 2 | 3 | 4 | 5 | 6 | 73.24 |
| 2 | 1 | 2 | 4 | 5 | 6 | 7 | 88.45 |
| 3 | 1 | 2 | 4 | 5 | 7 | 8 | 88.65 |
| 4 | 1 | 4 | 5 | 7 | 8 | 9 | 91.69 |
| 5 | 4 | 5 | 7 | 8 | 9 | 10 | 92.94 |
| 6 | 5 | 7 | 8 | 9 | 10 | 11 | 96.09 |

Table 2. Block data for example NSHR1

Optimal solution of the problem is given in table 3. It can be concluded, from presented results, that required quantity of handling units has been achieved. Also, table 3 gives comparison of number of schemes needed for transport, and data about connection between handling units and appropriate scheme of arrangement.

| Scheme number | Scheme quantity | PL1 | PL2 | PL3 | PL4 | PL5 | PL6 |
|:---:|:---:|:---:|:---:|:---:|:---:|:---:|:---:|
| 5 | 184 | 0 | 0 | 0 | 0 | 0 | 3 |
| 7 | 500 | 0 | 0 | 0 | 2 | 2 | 0 |
| 9 | 125 | 2 | 8 | 2 | 0 | 0 | 0 |
| 10 | 75 | 10 | 0 | 4 | 0 | 0 | 0 |
| 11 | 225 | 0 | 0 | 2 | 0 | 0 | 2 |
| Required | - | 1000 | 1000 | 1000 | 1000 | 1000 | 1000 |
| Achieved | - | 1000 | 1000 | 1000 | 1000 | 1000 | 1002 |

Table 3. Optimal solution for NSHR1

Graphical presentation of the example NSHR1 is given in Fig. 6.

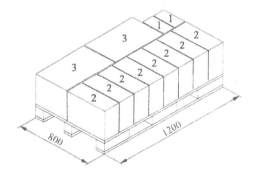

Fig. 6. Graphical presentation of the optimization results

## 3. CONCLUSION

When modular principle of arrangement is not applied, the linear programming model proved to be very suitable for the problems of optimal space utilization on a palette.

## REFERENCES

1. B. Mađarević: Rukovanje materijala, Tehnička knjiga, Zagreb, 1972.
2. P. Bauer: Planung ind Auslegung von Palettenlagern, Springer-Verlag, Berlin, 1985.
3. N. Štefanić: Matematički modeli kod krojenja materijala, Magistarski rad, FSB Zagreb, 1990.

# A FAMILY OF DISCRETE EVENT MODELS
## FOR FLOW LINE PRODUCTION

M. Lucertini
University "Tor Vergata", Rome, Italy
F. Nicolò
Terza Università, Rome, Italy
W. Ukovich
University of Trieste, Trieste, Italy
A. Villa
Polytechnic of Turin, Turin, Italy

KEY WORDS: production systems, flow lines, push systems, pull systems, perturbations.

ABSTRACT: general, abstract model is proposed for a production system consisting of a flow line, made of several machines, which can produce different products. Based on this model, a formal definition of push and pull concepts is provided. Different operating conditions are considered, and an abstract characterization of the set of all fesible production plans is proposed. Some of its general properties are then investigated: in particular, a minimum wip production plan is characterized, apt to tackle bounded perturbations of processing times.

# 1 INTRODUCTION

After the first enthusiasm that hailed the advent of *Just–In–Time (JIT)* and *pull* systems, many practitioners realized that "the techniques used to implement JIT are, in many ways, identical to those found in the "out–dated" reorder–point and/or *push*" (cf. for instance [7]). Is there any "real" difference? And, if so, what is it? Several authors tried to give their answers, but it is quite possible that a completely general and satisfactory answer cannot be given, for the following reasons.
One reason could be that many of the exciting performances attributed to Kanban are mainly due to the manufacturing environment where it is applied (see [5]). Another

Published in: E. Kuljanic (Ed.) *Advanced Manufacturing Systems and Technology*, CISM Courses and Lectures No. 372, Springer Verlag, Wien New York, 1996.

reason could be that performances of Kanban systems depend on the use of appropriate modelling and/or optimization techniques (see [1], [4], [7]). As a third reason, maybe the most important one, the differences between push and pull are not so sharp as they may appear at a first glance, because the two concepts are still not well defined: in fact, a traditional *push* system embedded in an MRP system computes release dates *from* the due dates by subtracting the *leadtime* of the line, or of the more general production system: in a way, this is an application of the *pull* concept.

The last simple remark suggests that the real difference is that *push* systems have large or very large closed loops, while in *pull* systems loops are as small as possible. Being this statement true or not, it is clear that we need well defined *basic* quantitative models, which allow to analyze the system structure and to evaluate the opportunities it provides. The purpose of this paper is to give a first very basic deterministic model in the most simple situation of a line of single server stations (a chain of machines) that must process a given sequence of parts.

When we have a chain of machines with intermediate buffers, it is clear that, in some sense, *push* implies to process jobs at the earliest time, while *pull* implies processing jobs at the latest possible time. Does this mean that push always implies a forward computation of the schedule from the release dates, and pull a backward computation from the due dates? The simple model presented in this paper provides a motivated answer to such a question. Furthermore, it also helps to understand at a very basic level a few classical problems, such as the role of buffer sizes (see for instance [8]), and the effect of Kanban on the Work–In–Process (see [3]).

# 2   DESCRIPTION OF THE SYSTEM

We consider a production system consisting of a simple flow line, made of several machines, which can produce different products. The basic ingredients of the system are: $n$ machines, denoted as $j = 1, 2, \ldots, n$, that can process parts to provide finished products; $m$ parts denoted as $i = 1, 2, \ldots, m$, that have to be processed by the machines. Each part $i$ has a release time $a(i)$, and a due date $d_i$. Each pair $(i, j)$ part–machine has a processing time $d(i, j)$. Furthermore, there are $n$ buffers (one before each machine), denoted as $j = 1, 2, \ldots, n$ in the same order as machines. Buffer $j$ has a capacity $k(j)$. A machine can be: busy, when there is a part on it, or idle. A waiting part can be either in buffer $j$, waiting to be processed by machine $j$, or it can be on machine $j$ still waiting to be processed, or after it has been completed. For the sake of simplicity, we always assume zero transportation times.

The state of a buffer at a given time instant is specified by the number of parts it contains. However, for our purposes it matters to discriminate between two relevant conditions only: full, when the buffer accommodates a number of parts equal to its capacity. Otherwise, the buffer is non–full.

The system operates as follows. Each part always retains its identity. No assembling operations are performed between parts of the system, nor a part can be disassembled into different sub–parts. Each part is an indivisible unit, so it cannot be shared by different machines. Each machine can process at most one part at any time instant. Processing of a part $i$ by a machine $j$ must progress, once it has started, and terminates exactly after $d(i,j)$ time instants from the beginning. Each part can be processed as soon as the machine waiting for it is idle: no setup operations are required. However, we don't assume that processing begins as soon as a part enters a machine. Each part must be processed by all machines, in a well definite order, which is the same for all parts. For ease of presentation, we assume that machine names $j = 1, 2, \ldots, n$ are ordered according to the sequence in which they are visited by parts. To simplify presentation, and without loss of generality, we also assume that the processing order corresponds to the name order of parts: $i = 1, 2, \ldots, m$. In other words, we assume that $i_1 < i_2$ implies $a(i_1) \le a(i_2) \ \forall \ 0 < i_1, i_2 \le m$. This is also the release order for parts. Parts leave each buffer in the same order as they enter it on the ground of a FIFO buffer discipline. All parts are processed by each machine always in the same order.

# 3  A GENERAL MODEL

The data for our model are the following:

$d(i,j)$ duration of the operation of part $i$ on machine $j$ (processing times)

$a(i)$ release date of part $i$, i.e. the time from which part $i$ is available to the first machine to start processing

$b(i)$ due date of part $i$, i.e. the time at which part $i$ must have left the last machine

$k(j)$ capacity of buffer $j$.

We choose the following variables to represent the way our system operates:

$x(i,j)$ time instant at which part $i$ enters machine $j$

$y(i,j)$ time at which part $i$ leaves machine $j$.

Then the problem of determining these variables can be formulated as

## Production Planning Problem

given $a, b \in \Re^m$, $d \in \Re^{m+n}$, $k \in \Re^n$, find $x, y \in \Re^{m+n}$ such that

$$x(i,1) \geq a(i) \quad i = 1, 2, \ldots, m \tag{1}$$
$$y(i,n) \leq b(i) \quad i = 1, 2, \ldots, m \tag{2}$$
$$x(i,j) \geq y(i-1,j) \quad i = 2, \ldots, m, \ j = 1, 2, \ldots, n \tag{3}$$
$$x(i,j) \geq y(i,j-1) \quad i = 1, 2, \ldots, m, \ j = 2, \ldots, n \tag{4}$$
$$y(i,j) \geq x(i,j) + d(i,j) \quad i = 1, 2, \ldots, m, \ j = 1, 2, \ldots, n \tag{5}$$
$$y(i,j) \geq x(i - k(j+1), j+1)$$
$$j = 1, 2, \ldots, n-1, \ i = k(j+1) + 1, k(j+1) + 2, \ldots, m. \tag{6}$$

Eq. 1 says that each part cannot enter the first machine before it is made available. Eq. 2 requires that each part must have left the last machine before its due date. Then any part (except the first one) cannot enter a machine before the previous part has left the same machine, according to Eq. 3. Also, any part cannot enter a machine (except the first one) before it has left the previous machine, by Eq. 4. Furthermore, Eq. 5 states that each part cannot leave a machine before the processing time has elapsed, since the moment in which it entered that machine. Finally, each part $i$ cannot leave a machine $j$ if the output buffer $j+1$ (i.e. the buffer following that machine, with $k(j+1)$ places) has no empty positions. Considering part $i$, buffer $j+1$ is full if parts $i-1, i-2, \ldots, i-k(j+1)$ lie in that buffer. The first of those parts leaving that buffer is $i - k(j+1)$, according to the FIFO discipline. This is expressed by Eq. 6.

# 4   OPERATING STRATEGIES

Due to its staircase structure, the Production Planning Problem can be solved in a relatively easy way using iterative procedures.

We first consider two basic solution methods, which correspond to popular operating strategies for our system: push and pull.

## 4.1   PUSH SYSTEMS

**Definition 4.1** *In a push system, input and output times are determined, starting from the release dates, in the ascending order of $i$ and $j$. More precisely, for each piece (in the ascending order), all times are determined for each machine (again, in the ascending order).*

According to the above definition, in a push system input and output times are determined from what happened *in the past*. Past events "push" the system forward from the past. So push systems are causal systems, in the sense that each event is completely determined by what happened in its past.

Note that, in determining input and output times in a push system, it is essential to scan pieces in an outer loop, and machines in an inner loop. If the opposite sequence would be adopted (scanning each piece for each machine), it could be impossible to determine some output times, since they depend from input times of previous pieces (i.e. with lower index $i$) to the following machine, due to the finite capacity of buffers (see Eq. 6).

Once all times have been determined, their feasibility with respect to the due dates is verified. It is therefore natural to set times *as early as possible*. Such an **earliest time assumption** turns out to be an essential feature of the push systems.

As a consequence, the constraints of Eqs. 3 and 4 of our model give the:

**push constraints**

$$x(i,j) = \max\{y(i-1,j), y(i,j-1)\} \quad i = 1, 2, \ldots, m, \; j = 1, 2, \ldots, n. \qquad (7)$$

Eq. 7 says that part $i$ enters machine $j$ when it has left the *previous* machine $(j-1)$ *and* the *previous* part $(i-1)$ has left the same machine. They are called *push* constraints since they express how what happens *upstream* in the line affects downstream operations.

The earliest time assumption transforms the constraints of Eqs. 4 and 5 in the

**pull constraints**

$$y(i,j) \;=\; \max\{x(i,j) + d(i,j), x(i-k(j+1), j+1)\}$$
$$j = 1, 2, \ldots, n-1, \; i = k(j+1)+1, k(j+1)+2, \ldots, m. \quad (8)$$

Eq. 8 says that part $i$ leaves machine $j$ when it has finished processing *and* the previous $k(j+1)$th part has entered the *following* machine (thus making available a place in the $j+1$th buffer). Those constraints are called *pull* constraints, since they express how what happens *downstream* in the line affects upstream operations.

Note that the push strategy is described by both push and pull constraints. The presence of pull constraints in this case is due to the fact that in a push system, input and output times may be affected not only by what happens upstream, but also by what happens downstream, due to the finite capacity of buffers. In the case of buffers with an infinite capacity, the influence of downstream operations disappears.

So in general the push concept signifies that current events depend from previous events. "Previous" must be intended in strict chronological sense in the definition of push systems, whereas in the definition of push constraints it refers to the order in which machines are visited. Such an ambiguity disappears with infinite buffers.

## 4.2   PULL SYSTEMS

Pull systems are defined in a completely symmetric way with respect to push systems.

**Definition 4.2** *In a pull system, input and output times are determined, starting from the due dates, in the descending order of i and j. Like for push systems, also in this case for each piece (outer loop), all times are determined for each machine (inner loop).*

So in a pull system, event times are determined backward in time in order to guarantee that specified conditions (event times and due dates) will be met *in the future*. Future events "pull" the system forward from the future. In this sense, *kanban* can be considered as a smart way to implement pull systems using feedback.

Once all times have been determined, their feasibility with respect to the release dates is verified. It is therefore sensible to set times *as late as possible*. Such a **latest time assumption** is the key feature of pull systems.

The conditions ruling a pull system may be derived from Eqs. 1 ÷ 6 using a reasoning which parallels the one used for push systems. They are symmetric to the conditions of Eqs. 7 and 8, with $x$ and $y$ swapped, *max* replaced by *min*, and sum and difference swapped. The only asymmetry concerns the index of $k$ in the pull constraints, which becomes just $j$ for pull systems (by the way, the role of push and pull constraints is also swapped). This is due to the fact that buffers are named after the machine they precede.

# 5   FEASIBLE PRODUCTION PLANS

Given $a$, $b$, $d$ and $k$, the set of all feasible production plans $(x, y)$, that is the set of all the solutions of the system of inequalities of Eqs. 1 ÷ 6 is a polyhedron $\Omega(a, b, d, k)$ in the space $\Re^{m+n} \times \Re^{m+n}$.

## 5.1   EXTREMAL PRODUCTION PLANS

Now we present some interesting properties of the set $\Omega$. The reader is referred to [6] for formal proofs.

Let $(x^e, y^e)$ denote input and output times obtained according to the earliest time assumption for given $a$, $b$, $d$ and $k$, and $(x^l, y^l)$ the corresponding times obtained according to the latest time assumption. Then

**Theorem 5.1** $\Omega \subseteq B \doteq \{(x, y) : \ x^e \leq x \leq x^l, \ y^e \leq y \leq y^l\}.$

**Corollary 5.1** $\Omega \neq \emptyset \iff B \neq \emptyset.$

## 5.2   INFLUENCE OF PROCESSING TIMES

The following results show the relation between the schedules of two sets of parts with different processing times.

**Theorem 5.2** *Consider two sets of parts that have to be processed by the same system: suppose they have identical release and due dates, and processing times $d$ and $\bar{d}$, respectively, such that $d(i,j) \le \bar{d}(i,j) \; \forall \, i,j$. Then $\Omega \doteq \Omega(a,b,d,k) \supseteq \Omega(a,b,\bar{d},k) \doteq \bar{\Omega}$.*

As a special case, we have the following result

**Corollary 5.2** *For the same conditions of Theorem 5.2, if $\Omega$ and $\bar{\Omega}$ are both nonempty, then*

$$x^e \le \bar{x}^e \le \bar{x}^l \le x^l, \quad y^e \le \bar{y}^e \le \bar{y}^l \le y^l.$$

## 5.3   MIN WIP PRODUCTION PLANS WITH PERTURBED PROCESSING TIMES

Now for each production plan we consider the WIP $w(t)$ it produces at any given time $t$, i.e. the number of parts that are on a machine or in an intermediate buffer. Note that

$$w(t) = |W(t)| \quad \text{with} \quad W(t) = \{i \,|\, x(i,1) \le t \le b(i)\}, \tag{9}$$

i.e. $W(t)$ is the set of parts within the system at time $t$.

Then the above results can be used for dealing with the following problem, where processing times are subject to nonstochastic uncertainty, i.e. they are only known to lie anywhere within given ranges (for other problems tackling such a kind of uncertainty, see for instance [2]).

**Problem 1** *Given due dates $\bar{b}$ and upper bounds on processing times $\bar{d}$, find a production plan $(x^0, y^0)$ such that:*

1. *$(x^0, y^0)$ solves the Production Planning Problem with given due dates $\bar{b}$ and buffer sizes $k$, suitable release dates $a$, and any processing times $d \in D \doteq \{d \,|\, d \le \bar{d}\}$*

2. *the WIP $w^0(t)$ produced by $(x^0, y^0)$ at any time $t$ is not larger, for all $t$, than the WIP $w(t)$ produced by any other production plan satisfying 1.*

Then we have the following result.

**Theorem 5.3** *For any given $a$ and $k$, $\Omega(a, \bar{b}, \bar{d}, k)$ is the largest set of production plans which are feasible for $a$, $\bar{b}$, $k$ and any processing time $d \in D$.*

Now the solution of Problem 1 is immediate.

**Corollary 5.3** *$(\bar{x}^l, \bar{y}^l)$ solves Problem 1.*

It is worth to point out that for Problem 1 upper bounds $\bar{d}(i,j)$ for the the processing times give the worst case for all processing times in $D$, in the sense that any production plan which is feasible for $\bar{d}$ is also feasible for all $d \in D$. This is a remarkable property of Problem 1, which is not always true in other situations with nonstochastic uncertainty (see [2]).

# 6 CONCLUSIONS

The chain model presented in the paper allows us to say that only in special situations the earliest time schedule (*push*) implies forward computation of the start times (*pure push*) and the latest time schedule implies backward computation (*pure pull*). In the chain model these two situations correspond to infinite and zero buffer capacities, respectively.

In the simple context of this paper, all the possible schedules (feasible production plans) lie within a polyhedron in the space of the times at which parts enter and leave machines. In such a situation, *pull* produces the lowest WIP.

Of course, the model presented in this paper is just a simplified model, since in practical cases several parallel machines may be available at each stage. Our results apply to this case only if part sequences do not change on each machine, which is a rather strong assumption in practice. Nevertheless, such a simple model turned out to be convenient, for instance, to gain a deeper insight into relevant concepts such as *push* and *pull*.

# REFERENCES

1. Bitran, G.R., and Chang, L., "A mathematical programming approach to a deterministic Kanban system," *Management Science* 33, 1987, 427–441.

2. Blanchini, F., Rinaldi, F., and Ukovich, W., "A network design problem for a distribution system with uncertain demands," *SIAM Journal on Optimization*, accepted for publication.

3. Karmarkar, U.S., "Kanban systems," *Paper Series N. QM8612*, Center for Manufacturing and Operation Management. The Graduate School of Management, University of Rochester, Rochester, N.Y., 1986.

4. Kimura, O., and Terada, H., "Design and analysis of pull systems: A method of multi–stage production control," *International Journal of Production Research* 19, 1981, 241–253.

5. Krajewski, L.J., King, B.E., Ritzman, L.P., and Wong, D.S., "Kanban, MRP, and shaping the manufacturing environment," *Management Science* 33, 1987, 39–57.

6. Lucertini, M., Nicolò, F., Ukovich, W., and Villa, A., "Models for Flow Management", *International Journal of Operations and Quantitative Management*, to be published.

7. Spearman, M.L., and Zazanis, M.A., "Push and pull production systems: Issues and comparisons," *Operations Research* 40, 1990, 521–532.

8. Villa, A., Fassino, B., and Rossetto, S., "Buffer size planning versus transfer line efficiency," *Journal of Engineering for Industry* 108, 1986, 105–112.

# CHARACTERIZATION OF Ti AND Ni ALLOYS
# FOR HOT FORGING: SETTING-UP
# OF AN EXPERIMENTAL PROCEDURE

**P.F. Bariani, G.A. Berti, L. D'Angelo and R. Guggia**
**University of Padua, Padua, Italy**

KEYWORDS :Physical Simulation, Testing, Superalloys, Hot Forging

ABSTRACT : There is a growing interest in the application of Titanium and Nickel alloys forged components in the aerospace and Hi-Tech industry.
Hot forging of these alloys requires a precise definition and control of operating parameters, such as forging temperature, punch speed and lubricant.
A procedure is presented to obtain true stress true strain curves for Ti and Ni alloys at hot forging conditions, that is especially designed to guarantee homogeneity at deformation and temperature in the specimen during the overall compression test.

## 1. INTRODUCTION

Materials sensitivity to the temperature and strain rate should be carefully considered when forming processes are designed, due to the high influence of these parameters on a successful and correct production. From an industrial point of view, forming processes require, as much as possible,
- high rates of working, which are desirable for reason of economy and to minimise heat transfer ;
- high temperature, in order to reduce material's resistance to deformation.

Published in: E. Kuljanic (Ed.) *Advanced Manufacturing Systems and Technology*,
CISM Courses and Lectures No. 372, Springer Verlag, Wien New York, 1996.

The knowledge of the behaviour of metals as a function of temperature, strain and strain rate is a basic step for (i) optimising the industrial process and (ii) dimensioning the forming machines tooling-set. To obtain data on the characteristics of the materials it is important that temperature, strain and strain rate replicate the parameters of the real process.

The precise characterisation of the material is a fundamental step when the forming process is designed for Ti and Ni alloys components, which have a growing applications in hi-tech industry (aerospace and aeronautics). These alloys present a narrow range of temperature, strain and strain rate for the formability. For these reasons a particular care should be taken in fine controlling these parameters during the test.

This paper is focused on the optimisation of a procedure for determining the behaviour of these alloys at forging condition. The procedure should be set-up in order to :

- assure homogeneity of deformation and temperature in the specimen during the whole compression test ;
- evaluate true stress-true strain curves of Titanium and Nickel alloys in hot forging condition.

## 2. THE EQUIPMENT IN PHYSICAL SIMULATION

Methods and equipment used in physical simulation depend upon the process to be studied. The physical simulator should offers a wide range for thermal and mechanical parameters, in order to replicate the operating condition of the real process. When these requirements are satisfied, it is possible to replicate the thermal mechanical history on the specimen and to determine the effects of thermal (temperature and heating/cooling rate) and mechanical parameters (strain and strain rate) on material and process.

There are different kinds of testing machine [1] :

- *servohydraulic load frame connected to a furnace or an induction heater* . This system is satisfactory for process applications where the temperature is changing slowly during the process ;
- *cam plastometer* . The simulation takes place at one defined temperature. It can be considered a single 'hit' device. Multiple 'hit' programs are possible in principle, but the time to change strain rates and temperature, usually make the replication of multy stage forming processes very difficult or impossible ;
- *torsion testing machine* . The tester provides shear data for simulation in a wide range of strain rates (up to 100/second) and strains. The most notable benefits of torsion testing are the large amount of strain possible without necking or barrelling, typically up to strain of 5.0. The temperature is usually hold constant or may be changed at slow rates.

A GLEEBLE 2000® System [2], installed at DIMEG's lab in Padua, has been utilised to conduct the tests presented in the paper. It is an electronically controlled, hydraulically operated testing machine used for :

- thermal and mechanical analysis of materials for research,
- quality control,
- process simulation, and

– a wide variety of metallurgical studies.

Accurate temperature control is the significant characteristic of the Gleeble machine. Specimen temperature is monitored by a thermocouple spot welded on the specimen surface and the heat input and rate are controlled according to a predetermined programmed cycle chosen by the researcher. The system can be consider as a sophisticated *universal testing machine* capable of heating rates of 10.000 °C/s, speed of 2000 mm/s with a maximum load of 20 tons. The two servo hydraulic systems can assure precise strain and strain rate even in a multiple-hit test. Strain can be measured by means of either a crosswise gauge or a lengthwise gauge ; in both cases stress evaluation is based on the assumption of constant volume and section area. To satisfy this condition no barrelling should develop during the test. At this regard, it should be noted that barrelling is influenced by axial gradient of temperature, as well as by friction at the interface specimen-anvils. In order to reduce the barrelling effect [3] friction at the interfaces and temperature axial gradient should be minimised, as well as slenderness ratio (l/d) should be optimised.

## 3. THE TESTS

Most metalworking processes involve compressive deformation : for this reason the uniaxial compression test [4,5] has been widely used for studying metals workability and characteristics.
In the present work, Titanium and Nickel alloys behaviour, in forging conditions, is studied. Due to the high sensitivity of these alloys to the working temperature, strain and strain rate, some difficulties may arise in the hot upsetting [6] of a cylindrical specimen :
– axial temperature gradient ;
– specimen end surfaces cooler than mid section ;
– non uniform deformation during the test.

Fig. 1                                                        Fig. 2

Fig. 1 shows a Ti-6Al-4V specimen that presented a non uniform temperature distribution along the axis ; in Fig. 2 it is evident the final deformation of a specimen (same alloy) with the end surfaces cooler than mid section.

The Gleeble heating systems is based on Joule effect : the current passes through the specimen and heat it up. With this system isothermal planes are obtained in the specimen, but an axial gradient is introduced due to the cooling effect induced by the punches. In order to reduce the axial gradient the following measures should be adopted:
– reducing heat loss from the end surfaces with a thermal barrier ;
– increasing electrical resistance of the surfaces in order to increase temperature ;
– reducing mass ratio between punches and specimen.

The system is equipped with several adjustments, which allow to operate on a wide range of specimen, as concerns size, shape and resistivity. Thermal power should be selected according to the requirements relevant to specimen size, heating rate and thermal distribution. The combination of nine transformer taps with a switch for four different specimen sizes provides the most suitable power range depending on the specimen characteristics; the best switches combination gives the minimum thermal gradient along the specimen axis.
The following three kinds of test [7] have been conducted to reduce barrelling :
     *(i)* heating tests for selecting the right thermal power of the system;
     *(ii)* tests with conventional lubricant ;
     *(iii)* tests with different layers of lubricant.

### *(i) heating tests for selecting the right thermal power of the system*

Several tests have been conducted on $\phi$12x14 mm long specimens heated in the range 600-900 °C without lubricant (scheme in Fig. 3). The measurement of the specimen axial gradient has been done in steady conditions (60 seconds after reaching the programmed temperature).
Four thermocouples were spot welded inside four drilled holes to place them at the specimen core.
Best combination of switch positions and taps have been investigated. Fig. 4 (a) and (b) show respec-

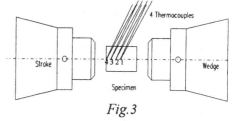

*Fig. 3*

tively the result of measurements at 645 °C in the best and in the worst condition. It is evident that even in the best situation (a) the maximum Δt, close to 35 °C, is not acceptable.

### *(ii) tests with conventional lubricant*

Isothermal conditions are important in flow stress measurement. When a thermal gradient exist along the axis of the specimen, barrelling occurs during deformation, no matter what lubricant is used (Fig. 5).

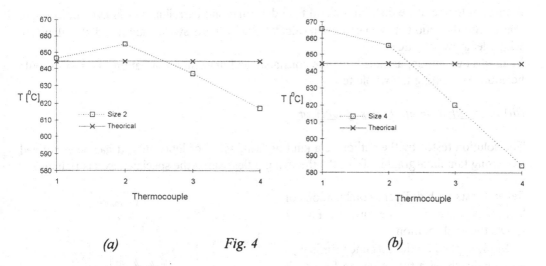

(a)                    Fig. 4                    (b)

Three different lubricants, MoS$_2$ powder, graphite foil and tantalum foil have been tested [8] with the twofold aim to reduce:
– friction at high temperature ;
– heat loss introducing a thermal barrier.

MoS$_2$ powder is relatively simple to use when mixed with alcohol, but some difficulties may arise during heating because of the bad electric contact between specimen and punch surfaces. It is usually applied at a temperature below 600 °C, above which it breaks into the Mo oxide.

Fig. 5

Graphite foil can be used above 600 °C, but only if the diffusion does not become a problem. During tests at high temperature a piece of tantalum foil can be utilised between the specimen and the graphite foil as a diffusion barrier and to protect graphite from high temperature, avoiding self burning.

Tantalum foil is used for two reasons : it is a good thermal barrier and, due to its high resistivity, increases the electrical resistance at the punch-specimen interface.

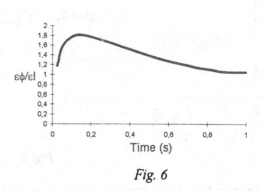

Fig. 6

The best results have been obtained with tantalum-graphite foils, but they aren't acceptable

in term of temperature distribution and final deformation : barrelling is still evident in Fig. 6 that shows the ratio between εφ (strain calculated with a crosswise gauge) and εl (calculated with a lengthwise gauge).
In an ideal condition, assumed volume constancy and uniform deformation, the ratio should be equal to 1 during the whole test.

*(iii) tests with different layers of lubricant.*

The solution tested by the authors is a kind of "sandwich" of lubricants. It has been noticed that using tantalum-graphite foils, the thermal gradient along the specimen axis is reduced.

Several tests with different combinations of lubricants have been performed. Fig. 7 shows the final solution :
– MoS₂ applied on the specimen surface ;
– a sandwich with two graphite foils with two tantalum foils on each side of the specimen.

This configuration gives the best results as concerns temperature and deformation uniformity.

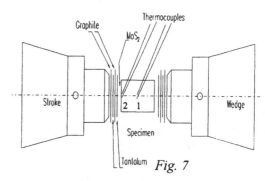

Fig. 7

As concerns temperature, the gradient is highly reduced : Fig. 8 shows the difference of the measured temperature between thermocouple 1 and 2 (see Fig. 7) after 200 s from the beginning of test : (a) is relevant to the test with both graphite and tantalum foils, (b) the same test with the multy-layered "sandwich". It can be noticed that in the second configuration the temperature gradient is reduced to 5 degrees.

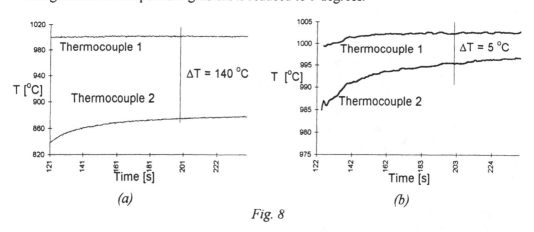

Fig. 8

Good results have been obtained even in terms of deformation using the "sandwich": Fig. 9 presents the εφ / εl ratio during the test, which is almost constant and close to 1 (ideal condition). Fig. 10 shows the deformation of the specimen during the test where barrelling can be neglected.

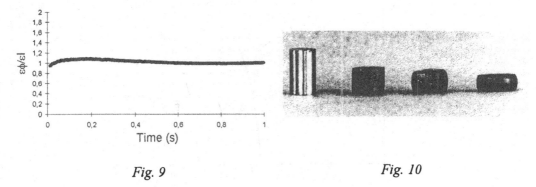

Time (s)

Fig. 9

Fig. 10

## 4. CHARACTERISATION OF Ti AND Ni ALLOYS

The developed test configuration allowed the characterisation of materials which present high sensitivity to temperature and strain rate. [8]
True stress - true strain curves have been calculated for Ti-6Al-4V in the following conditions :
- Temperature :    850, 880, 910 °C
- Strain rate :    1, 3, 5, 7, 9   s$^{-1}$

As example of this characterisation two true-stress vs. true-strain curves at 850 °C are presented in Fig. 11.

Operating conditions relevant to a complete forging sequence have been simulated for Nimonic 80A alloy. The test has been focused on the evaluation of the material response to forming operations where small amount of deformation per step has been performed at strain rates in the range of 4-10 s$^{-1}$ :

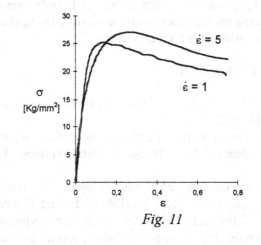

Fig. 11

- heating up to 1150 °C in 1 minute;
- soak time at temperature for 30 s ;
- deformation at 4 s$^{-1}$ with 2.3 as amount strain ;
- holding specimen for 15 s;
- re-heating up to 1150 °c for the next deformation.

Four stages as described above have been performed at strain rate of 4, 5, 7, 9 s$^{-1}$. Fig 12 shows the four curves at different stages of strain.

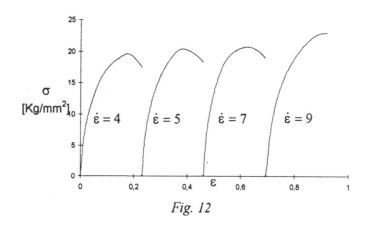

*Fig. 12*

## 5. CONCLUSIONS

Some progress in the setting-up of the procedure to characterise materials such as Ti and Ni alloys have been presented. Effects of different lubricants as concern temperature uniformity and barrelling have been investigated. A multi-layered lubricant, tantalum, graphite and $MoS_2$, has been tested giving good results due to its effects (thermal and diffusion barrier, lubricant, local increase of resistivity). Using this sandwich the Ti-6Al-4V and Nimonic 80A materials have been characterised.

REFERENCES

1. H. E. Davis, G.E.Troxell, G.F.W.Hauch , The testing of engineering materials, McGraw-Hill 1892
2. Ferguson, H.F., Fundamentals of Physical Simulation, DELFT Symposium, Dec. 1992
3. Schey, J.A., Tribology in Metal Working : Friction, Lubrication and Wear, ASM, June 1984
4. K.Pöhlandt : 'Materials Testing for the Metal Forming Industry'. Springer-Verlag 1989
5. George E. Dieter, Workability Testing Techniques, American Society for Metals, 1984
6. K.Pöhlandt, S.N. Rasmussen, Improving the Accuracy of the Upsetting test for determining Stress-Strain Curves, Advanced Technology of Plasticity 1984
7. Dal Negro T., De Vivo D., Ottimizzazione del Ciclo di Stampaggio a Caldo di Superleghe di Nickel e Titanio Mediante Simulatore Fisico di Processi Termomeccanici, thesis degree in Italian, DIMEG June 1996
8. E. Doeghe, R. Schneider, Wear, Friction and Lubricants in Hot Forging, Advanced technology of Plasticity 1984
9. H.G.Suzuki, H.Fujii, N. Takano K.Kaku, Hot Workability of Titanium Alloys, Int. Symposium on Physical simulation of welding, hot forming and C. casting, Ottawa 1988

# AN ANALYSIS OF FEMALE SUPERPLASTIC FORMING PROCESSES

A. Forcellese, F. Gabrielli and A. Mancini
University of Ancona, Ancona, Italy

KEY WORDS: Superplastic forming, FEM.

ABSTRACT: An analytical model of a sheet superplastic forming process is proposed. It provides the pressure versus time relationship that guarantees the optimum superplastic flow condition by taking into account also the entry radius of the die. For simplicity, the process was studied in two separated stages: in the former, the superplastic flow of the membrane occurs in a free bulging condition whilst, in the latter, the membrane is in contact with the die wall. The analytical model has shown an excellent agreement with numerical predictions.

## 1. INTRODUCTION

The industrial development of new superplastic materials provides new chances for manufacturing complex components by superplastic forming techniques (SPF) that are not suitable for conventional alloys [1]. Several industrial sectors, among which the aerospace industry, are interested in SPF processes for manufacturing light weight-highly complex components [2]. The very high ductility exhibited by superplastic materials is one or two order of magnitude higher than the one of conventional materials. As a consequence, processes taking advantage of such extremely high ductilities can be used to obtain complex shape components that, if produced in a traditional way, would require expensive assembly operations. Several SPF methods and techniques have been developed [3]. In spite of the remarkable attention paid to SPF of sheets, the relation between material behaviour and process parameters is not yet well defined. The effective industrial application of SPF processes requires the selection of the technological parameters that

Published in: E. Kuljanic (Ed.) *Advanced Manufacturing Systems and Technology*, CISM Courses and Lectures No. 372, Springer Verlag, Wien New York, 1996.

preserve the superplastic condition throughout of the process. In hot blow forming, the rate of pressurisation is normally established so that the strain rates induced in the sheet are maintained within the superplastic range of the material. In the practice, the pressurisation rate is determined by trial and error techniques or by analytical methods.

The first engineering model was proposed by Jovane [4] that, on the basis of the membrane theory and elementary geometrical considerations, gave the pressure versus time relationship. Basically, this model assumes that: i) the material is isotropic and incompressible, ii) the regime is membranous, iii) the elastic strains are negligible, iv) the material is not strain-hardenable, and v) at any instant, the membrane is part of a thin sphere subjected to internal pressure, i.e. the curvature and thickness are uniform and a uniform biaxial tension state exists. Unfortunately, the hypothesis of a uniform sheet thickness of the dome is not consistent with the experimental results that show a thinning of the sheet during forming owing to stress, strain and strain rate gradients. In practice, the indeformability of the die in proximity of periphery makes the circumferential strain negligible and lower than the meridian strain: then, due to the continuity law, the strain component along the thickness direction increases. In order to reduce such inconsistency, Quan and Jun [5] considered both the non-uniformity in sheet thickness and the anisotropy of the sheet along the thickness direction. They, starting from a basic theory of plastic mechanics for continuous media and using both the Rosserd's viscoplastic equation [6] and the constitutive equation with a strain rate sensitivity varying with strain rate [5], provided a theoretical analysis of the free bulging that is more consistent with the real processes. In this model, the previous basic hypotheses [4] were assumed and, beyond the anisotropy in the thickness direction, the geometrical shape of the mid layer of the sheet as part of a spherical surface during the bulging process was also considered.

Although the complex analytical models and those based on the finite element method (FEM) [7,8] provide a better accuracy, they require longer computation times. Therefore, from an engineering point of view and for its simplicity, the most interesting approach remains the model proposed in [4] that, however, needs to be improved.

Recently, a model that, beside the basic assumptions, assumes that the median part of the formed dome is spherical at any instant and that each meridian passing to the dome apex is uniformly stretched, was developed [9]. However, this model leads to results that are very similar to the ones proposed by Jovane [4].

All the models proposed neglect bending of the sheet at the clamps; moreover, during deformation the rotation of the sheet is allowed at the extremities similarly to a frictionless hinge. In this way, at any instant, the membrane maintains a spherical shape, also in the region next to the periphery. This does not correspond to real SPF processes since it means to build a zero entry radius die. Such a die causes strong local stress concentrations and the corresponding stress gradients can lead to a dramatic thinning and rupture of the sheet. Therefore, in order to obtain a realistic model, the die entry radius cannot be neglected. The effect of the die entry radius is considered in the present paper and a new model incorporating such aspect is proposed. In particular, the sticking friction condition at the sheet-die interface is assumed since such condition, although represents a limit situation of the real process, provides a better realistic starting point than the frictionless one.

## 2. THE MODEL

The model, aimed at obtaining the pressurisation rate that maintains the superplastic condition during the process, was developed according to the scheme in fig. 1.

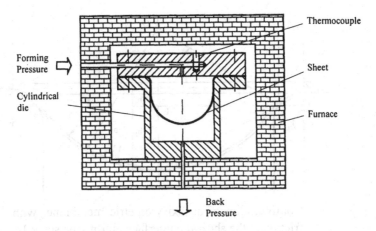

Fig. 1    Schematic representation of a female SPF process.

The pressure versus time relation was obtained by using the equation of equilibrium of the forces applied to the membrane. The state of equilibrium depends on the applied pressure, material properties, strain levels and boundary conditions. The basic assumptions of the model are the following: a) the material is isotropic and governed by the $\bar{\sigma} = k\,\bar{\varepsilon}^m$ law, b) the volume is constant, c) the elastic strains are negligible, d) the material is not strain-hardenable, with very low yield strength, and e) the regime is membranous. Such hypotheses are similar to those used in previous models [4,5]. Beside these hypotheses, the present model assumes that: f) the sheet is rigidly clamped at the periphery where bending is allowed around the entry die profile, and g) no sliding at the die-sheet interface occurs. The last condition causes a reduction in sheet thickness from the initial value to the uniform value in the spherical part of the membrane.

For sake of simplicity, the process was divided in two stages. The stage 1 deals with the study of the optimum conditions that guarantee the superplastic flow of the membrane from the initial configuration to the instant when the free bulging region assumes the hemispherical shape. In the stage 2, the membrane is in contact with the cylindrical die.

### 2.1 Stage 1 (X<R)
The initial step is to calculate the membrane thickness (s) versus geometrical parameters by using the constant volume law. In order to find a more general solution, two dimensionless geometrical parameters were defined:

$$X = \frac{x}{a} \qquad\qquad R = \frac{r}{a} \qquad\qquad (1)$$

where x is the abscissa of B that indicates the end point of the contact region between sheet and die, r is the curvature radius of the die entry and a is the radius of the initial undeformed blank (Fig. 2).

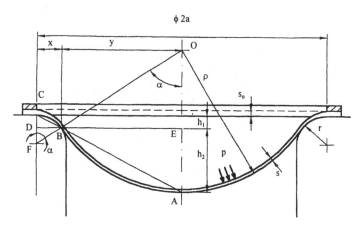

Fig. 2    Configuration of an axisymmetric membrane, with sticking friction at the sheet-die interface, during the stage 1.

For the constant volume law, the volume of the initial blank ($V_o$) is equal to the volume of the formed blank ($V_{AB} + V_{BC}$) that is calculated by means of the Guldino theorem:

$$V_o = \pi\, a^2 s_o \qquad\qquad V_{AB} = 2\pi\, \rho\, h_2\, s \qquad\qquad V_{BC} = \alpha\, r\, s\, 2\pi\, (a - x_G) \qquad (2)$$

Therefore:

$$\pi\, a^2 s_o = 2\pi\, \rho\, h_2 s + \alpha\, r\, s\, 2\pi\, (a - x_G) \qquad (3)$$

where $s_o$ is the thickness of the undeformed blank, $\rho$, $h_2$ and s are, respectively, the radius, height and thickness of the part of the sphere representing the membrane in the region where no sheet-die contact occurs, $x_G$ is the position of the centre of gravity of the part of the sheet between B and C. In order to obtain an equation relating the thickness s to the parameters X and R, $\rho$ and $h_2$ must be expressed as a function of X and R. Since:

$$x = r \sin\alpha \qquad\qquad y = \rho \sin\alpha \qquad\qquad x + y = a \qquad (4)$$

the value of $\rho$ versus the dimensionless parameters is:

$$\frac{\rho}{a} = R\,((1 - X)/X) \qquad (5)$$

Furthermore, being the triangles $\hat{BEA}$ and $\hat{CDB}$ similar, the height $h_2$ is given by:

$$h_2 = (y/x)\, h_1 \qquad (6)$$

By substituting eqns. (4) into eqn. (6), the height $h_2$ becomes:

$$h_2 = h_1\,((a - x)/x) \qquad (7)$$

The term $h_1$, that depends on $\alpha$ (Fig. 2), results:

$$h_1 = r(1 - \cos\alpha) \tag{8}$$

where $\cos\alpha$, that can be obtained versus x by considering the triangle $F\hat{D}B$, is equal to:

$$\cos\alpha = \frac{\left(r^2 - x^2\right)^{1/2}}{r} \tag{9}$$

Putting eqns. (8) and (9) into eqn. (7) and using the parameters X and R, one obtains:

$$h_2 = a\left[R - \left(R^2 - X^2\right)^{1/2}\right]\left(\frac{1-X}{X}\right) \tag{10}$$

The position of centre of gravity of the arc CB, whose length is $\alpha r$, is determined by means of the definition of gravity centre:

$$x_G = \frac{a}{\alpha}\left[R - \left(R^2 - X^2\right)^{1/2}\right] \tag{11}$$

By substituting eqns. (5), (10), and (11) into the constant volume law (Eqn. (3)), the thickness s can be expressed versus the dimensionless parameters X and R:

$$s = s_0 \frac{\left[1 - R\left(\arcsin\left(\frac{X}{R}\right) - R + (R^2 - X^2)^{1/2}\right)\right]}{\left[2R\left(\frac{1-X}{X}\right)^2\left(R - \left(R^2 - X^2\right)^{1/2}\right) + R\left(\arcsin\left(\frac{X}{R}\right) - R + (R^2 - X^2)^{1/2}\right)\right]} \tag{12}$$

In order to get the optimum superplastic condition, the equivalent stress ($\bar{\sigma}$) and the equivalent strain rate ($\bar{\varepsilon}$) must be kept equal to the values for which the material exhibits the highest elongation. Therefore, the forming pressure must vary versus time so that:

$$\bar{\sigma} = \frac{p\,\rho}{2\,s} = \bar{\sigma}_0 = \text{constant} \tag{13}$$

where $\bar{\sigma}_0$ is the optimum flow stress for SPF. By substituting the eqns. (5) and (12) into eqn. (13) and solving it, the dimensionless pressure ($p$) results:

$$p = \frac{p\,a}{\bar{\sigma}_0\,s_0} = \frac{2\,X\left[1 - R\left(\arcsin\left(\frac{X}{R}\right) - R + (R^2 - X^2)^{1/2}\right)\right]}{R^2(1-X)\left[2\left(\frac{1-X}{X}\right)^2\left(R - \left(R^2 - X^2\right)^{1/2}\right) + \arcsin\left(\frac{X}{R}\right) - R + (R^2 - X^2)^{1/2}\right]} \tag{14}$$

The calculation of the dimensionless pressure versus time is difficult since it requires the solution of the following differential equation:

$$\bar{\varepsilon} = \frac{d\bar{\varepsilon}}{dt} = -\frac{1}{s}\frac{ds}{dt} = \bar{\varepsilon}_0 = \text{constant} \tag{15}$$

where $\bar{\dot{\varepsilon}}_o$ is the optimum strain rate for SPF. Eqn. (15) expresses the dependence of the dimensionless parameters on time. By substituting eqn. (13) into eqn. (15), it results:

$$\tau = \int_o^t \bar{\dot{\varepsilon}}_o dt = \int_{X_o}^X b(R, X)\, dX \tag{16}$$

where $\tau$ is the dimensionless time, $X_o(\neq 0)$ is the value of X at t=0, and:

$$b(R, X) = [2R(1-X)^2 / (X(R^2 - X^2)^{1/2}) - 4R(1-X)^2(R - (R^2 - X^2)^{1/2}) / X^3 +$$
$$-4R(1-X)(R - (R^2 - X^2)^{1/2}) / X^2 + R(1-X) / (R^2 - X^2)^{1/2}] /$$
$$[2R((1-X)/X)^2[R - (R^2 - X^2)^{1/2}] + R(arcsen(X/R) - R + (R^2 - X^2)^{1/2})] + \tag{17}$$
$$+R((1-X) / (R^2 - X^2)^{1/2}) / [1 - R(arcsen(X/R) - R + (R^2 - X^2)^{1/2})]$$

Due to mathematical complexity, the $\bar{p} = \bar{p}(\tau)$ cannot be described by an elementary function. In order to give a representation, the curves $\bar{p} = \bar{p}(X)$ and $\tau = \tau(X)$ obtained by eqns. (14) and (16), respectively, can be plotted: in the range $X_o \leq X < R$, for each value of X a pair of values $\bar{p}$ and $\tau$ can be obtained and plotted.

### 2.2 Stage 2 (W≥0)

The deforming sheet is divided in two parts (Fig. 3): AB (no sheet-die contact) and BD (sheet-die contact). The thickness of the part identified by AB is s whilst an average thickness $\bar{s} = (s_o + s)/2$ was assumed for the part BD.

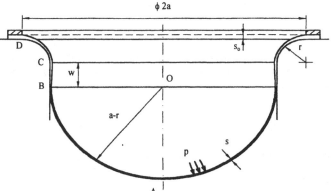

Fig. 3 Configuration of the axisymmetric membrane, with sticking friction at the die-sheet interface, during the stage 2.

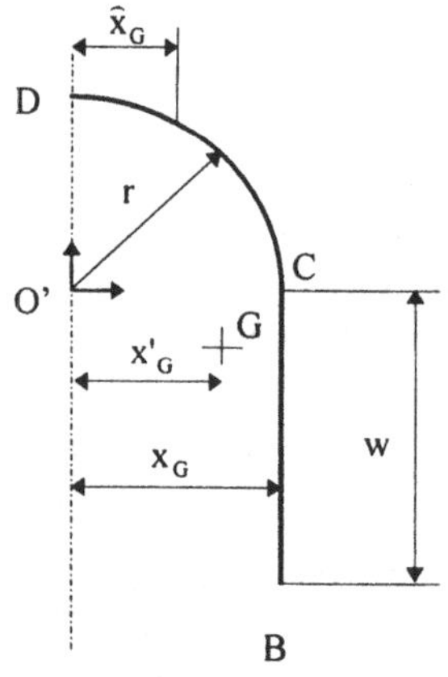

Fig. 4 Schematic representation of the sheet-die contact zone.

From the constant volume law, the equality of the initial volume with the volume of the deforming sheet, calculated by the Guldino's theorem, provides:

$$\pi a^2 s_o = 2\pi(a-r)^2 s + \left(\frac{\pi}{2}r + w\right)\left(\frac{s + s_o}{2}\right) 2\pi(a - x'_G) \tag{18}$$

where $x'_G$ is the distance from the centre of gravity of the part BCD to the axis O'D. It can be determined by fig. 4 where $\hat{x}_G$ and $\bar{x}_G$ represent the distance from the centre of gravity of the sections identified by the arc CD and by the straight line BC to the axis O'D, respectively. Therefore, if $l_{BC}$ and $l_{CD}$ are the lengths of BC and CD, respectively, it is:

$$x'_G = \frac{\sum x_i l_i}{\sum l_i} = \frac{\hat{x}_G l_{CD} + \bar{x}_G l_{BC}}{l_{CD} + l_{BC}} \tag{19}$$

where:

$$\hat{x}_G = \frac{2r}{\pi} \qquad \bar{x}_G = r \qquad l_{CD} = \frac{\pi}{2} r \qquad l_{BC} = w \tag{20}$$

By substituting eqn. (20) into eqn. (19), it results:

$$x'_G = \frac{2r(r+w)}{\pi r + 2w} \tag{21}$$

By inserting eqn. (21) into eqn. (18) and solving with respect to the thickness s:

$$s = s_o \frac{\left[1 - \left(\frac{\pi}{2} R + W\right)\left(1 - \frac{2R(R+W)}{\pi R + 2W}\right)\right]}{\left[\left(\frac{\pi}{2} R + W\right)\left(1 - \frac{2R(R+W)}{\pi R + 2W}\right) + 2(1-R)^2\right]} \tag{22}$$

where W=w/a is the dimensionless length of the straight line between the two concavities of the membrane. Since:

$$\rho = a(1-R) = \text{constant} \tag{23}$$

by substituting eqns. (22) and (23) into eqns. (13) and (15), the following relationships are obtained:

$$p = \left(\frac{2}{1-R}\right) \frac{\left[1 - \left(\frac{\pi}{2} R + W\right)\left(1 - \frac{2R(R+W)}{\pi R + 2W}\right)\right]}{\left[\left(\frac{\pi}{2} R + W\right)\left(1 - \frac{2R(R+W)}{\pi R + 2W}\right) + 2(1-R)^2\right]} \tag{24}$$

$$\begin{aligned}
\dot{\varepsilon}_o dt = &\{[1 - 2R(R+W)/(\pi R + 2W) + (\pi R/2 + W)(4R(R+W)/(\pi R + 2W)^2 + \\
&-2R/(\pi R + 2W))]/[2(1-R)^2 + (\pi R/2 + W)(1 - 2R(R+W)/(\pi R + 2W))] + \\
&-[-1 + 2R(R+W)/(\pi R + 2W) - (\pi R/2 + W)(4R(R+W)/(\pi R + 2W)^2 - 2R/ \\
&(\pi R + 2W))]/[1 - (\pi R/2 + W)(1 - 2R(R+W)/(\pi R + 2W))]\} \, dW
\end{aligned} \tag{25}$$

The integration of eqn. (25) from $t_o$ to t provides the dimensionless time versus W:

$$\tau - \tau_o = \dot{\varepsilon}_o(t - t_o) = \ln\left[\frac{(-2 + \pi R - 2R^2)(4 - 8R + \pi R + 2R^2 + 2W - 2RW)}{(4 - 8R + \pi R + 2R^2)(-2 + \pi R - 2R^2 + 2W - 2RW)}\right] \tag{26}$$

where $\tau_o$ is the dimensionless time after that the membrane shape does not change anymore

### 2.3 Geometrical representation of the model

The manufacturing of very complex components by SPF requires an accurate control of the process parameters; in particular, the control can be made more effectively by an accurate analysis of the dependency of the height H on thickness s. In this framework, the analytical models are very helpful. Actually, even if achieved by simplifying assumptions, the models provide the guidance to define the optimum conditions for superplastic flow. Therefore, the height H of the pole (figs. 2 and 3) in the stage 1 can be expressed as:

$$H = H_1 + H_2 = \frac{1}{a}\left(h_1 + h_2\right) = \frac{1}{X(\tau)}\left[R - \left(R^2 - X^2(\tau)\right)^{1/2}\right] \tag{27}$$

and in the stage 2 as:

$$H(\tau) = 1 + W(\tau) \tag{28}$$

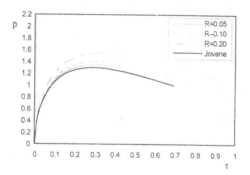

Fig. 5    Influence of die entry radius on dimensionless pressure.

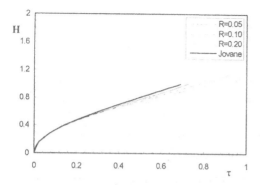

Fig. 6    Influence of the die entry radius on the dimensionless polar height.

The results obtained from by the proposed model, in terms of pressurisation curve, variation of the height of the pole and thinning ratio $s/s_o$, are plotted in figs. 5-7. The analysis of the effects produced by an increase in the die entry radius shows that when the radius increases the pressure increases and the polar height decreases; the effect on pressure becomes significant after short times from the beginning of the process ($\tau\sim0.1$) whereas the effect on the polar height becomes significant after longer times ($\tau\sim0.3$). As far as the sheet thinning is concerned, it appears that the thinning factor $s/s_o$ decreases with increasing $\tau$ and decreasing die entry radius. It can be observed that for all the parameters considered, the results of the proposed model tend to the ones of the model reported in [4] that can be assumed as the limit values of pressure for the die entry radius tending to zero.

## 3. VALIDATION OF THE MODEL

In order to assess the validity of the model proposed, an attempt was carried out by

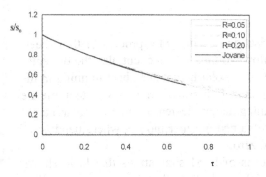

Fig. 7    Influence of the die entry radius
          on the thinning factor.

means of a FEM analysis of the SPF process. The numerical simulations were performed by applying the pressurisation rates for the two stages of the process given by the eqs. (14), (16), (24) and (26). The material was a titanium alloy (Ti-6Al-4V), with optimum superplastic flow conditions at 927°C and $2 \cdot 10^{-4}$ s$^{-1}$, at which correspond a strain rate sensitivity (m) equal to 0.6, a strength coefficient (k) of 1160 MPa$\cdot$s$^{-1}$, and an optimum flow stress ($\bar{\sigma}_0$) of 7 MPa [10]. Only one half of the sheet and die were considered in the simulations owing to the axisymmetry of both sheet and die.

A sheet blank with thickness and diameter of 2 mm and 220 mm, respectively, was used. The internal diameter of the cylindrical die was 160 mm, and the curvature radius of the die entry was 20 mm. The sheet was modelled by using 480 isoparametric bilinear quadrilateral elements whilst the die was modelled by a rigid surface. In order to model sticking conditions, a shear friction factor equal to 1 was used at the sheet-die interface.

The numerical results, obtained by a rigid-plastic FEM analysis, show a non-uniform sheet thickness throughout the process: the thickness is the lowest in the pole and increases moving towards the die. As a consequence, also the strain rate distribution in the deforming sheet is non-uniform and $\dot{\varepsilon}$ is the highest in the pole of the membrane. The strain rate versus time data predicted by the FEM analysis in the pole are shown in fig. 8.a; it can be observed that they are lightly higher than the optimum one obtained without taking into account the non uniform sheet thinning. However, they are within the superplastic flow range of Ti-6Al-4V ($10^{-3} \div 10^{-4}$ s$^{-1}$) [3,10]. Even more so, in the other parts of the deforming sheet the flow behaviour is superplastic. Also the predicted polar height versus time data by FEM techniques are in excellent agreement with the analytical data.

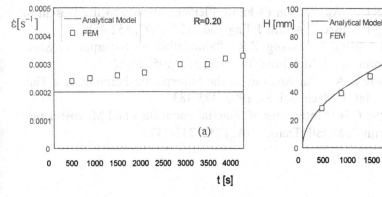

Fig. 8    Comparison between FEM and analytical results.

## 4. CONCLUSION

An analytical model for predicting the pressurisation rate in a SPF process of thin sheets was proposed. As to the previous models, it allows to take into account also the die entry radius. The process parameters, in terms of pressure, polar height and sheet thinning in the pole of the dome, vary with time and die entry radius. In particular if, at any instant, the die entry radius decreases, the pressure and thinning factor decrease and the polar height increases: these parameters tend to the values obtained in other models where bending of the sheet at clamps is neglected when R tends to zero.

The validity of the model was verified by means of FEM simulations that have shown results that, in terms of strain rate, polar height and sheet thinning, are in excellent agreement with the analytical results provided by the model.

## ACKNOWLEDGEMENTS

The authors express their warm thanks to Dr. Valeri Berdine for the helpful suggestions in the preparation of this paper. The financial support of MURST (40%) is acknowledged.

## REFERENCES

[1]    Padmanabhan K.A., Davies G.J.: "Superplasticity", Springer-Verlag, Berlin, 1980.
[2]    Baudelet B.: "Industrial Aspects of Superplasticity", Mat. Sci. Eng., A137, 1991, 41-55.
[3]    Pilling J., Ridley N.: "Superplasticity in Cristalline Solids", The Institute of Metals, London, 1989.
[4]    Jovane F.: "An Approssimate Analysis of the Superplastic Forming of a Thin Circular Diaphragm: Theory and Experiments", Int. J. Mech. Sci. 10, 1968, 403-427.
[5]    Quan S.Y., Jun Z.: "A Mechanical Analysis of the Superplastic Free Bulging of Metal Sheet", Mat. Sci. Eng., 84, 1986, 111-125.
[6]    Odqvist F.K.: "Mathematical Theory of Creep and Creep Rupture", Clarendon, Oxford, 1966.
[7]    Chandra N., Rama S.C.: "Application of Finite Element Method to the Design of Superplastic Forming Processes", ASME J. Eng. Ind., 114, 1992, 452-458.
[8]    Zhang K., Zhao Q., Wang C., Wang Z.R.: "Simulation of Superplastic Sheet Forming and Bulk Forming", J. Mat. Proc. Technol., 55, 1995, 24-27.
[9]    Enikeev F.U., Kruglov A.A.: "An Analysis of the Superplastic Forming of a Thin Circular Diaphragm", Int. J. Mech. Sci. 37, 1995, 473-483.
[10]  Ghosh A.K., Hamilton C.H.: "Influences of Material Parameters and Microstructure on Superplastic Forming", Metall. Trans., 13A, 1982, 733-743.

# OVERVIEW OF PROESTAMP: AN INTEGRATED SYSTEM FOR THE DESIGN, OPTIMIZATION AND PERFORMANCE EVALUATION OF DEEP DRAWING TOOLS

**R.M.S.O. Baptista**

Instituto Superior Técnico, Lisboa, Portugal

**P.M.C. Custódio**

Instituto Politécnico de Leiria, Leiria, Portugal

KEY WORDS: Sheet Metal Forming, Deep Drawing, Tool Design, Computer Aided Process Planing CAPP

ABSTRACT: The paper discusses the philosophy of PROESTAMP - a Computer Aided Process Planning (CAPP) system for the design of deep drawing tools. PROESTAMP will interface with a finite element code and a material database that includes the mechanical, anisotropic and formability properties of the material experimentally determined in order to enable the preliminary and final design of the deep drawing tools from the geometrical or physical model of the part. The CAPP system will include the capability to store in a multimedia environment the record of all of the above-mentioned procedures including the experimental analysis of the strains imposed on the part and the try-out results. This integrated system will lead to the capability of producing sound parts at first and in a shorter period of time. Simultaneously, an improvement in the design and product quality will be achieved.

## 1. INTRODUCTION

The designers of deep drawing tools are facing new and greater challenges. To start with, there is a need for a quick and reliable preliminary plan of the deep drawing tool so that the increasing number of budget requests can be met. This is a very critical phase because the company needs to meet budget requests within a short amount of time. Notice that budget deadlines are continuously being shortened and only a small number of budget requests become orders. So, it is clear the need to develop a quick and reliable tool for the design of

Published in: E. Kuljanic (Ed.) *Advanced Manufacturing Systems and Technology*,
CISM Courses and Lectures No. 372, Springer Verlag, Wien New York, 1996.

deep drawing tools that integrates the empirical knowledge of the designers and craftsmen with a more scientific understanding achieved by a theoretical and experimental analysis of the deep drawing process.

However, the challenges do not end with the order. After that phase, it is necessary to design the final plan, making use of modelling capabilities, for optimisation of the operative parameters, linked with the knowledge of the mechanical, anisotropic and formability properties of the material.

Finally, there is the phase of building and testing the deep drawing tool. That phase usually includes the production of acceptance parts or even a pre-series run. During the try-out of the tools some corrections are frequently made, usually based on the craftsmen empirical knowledge and accumulated skill. No record is kept of the arising problems and of the solutions found.

In the past years a large effort has been put on the development and application of finite element codes to the simulation of the deep drawing process. Despite all of these developments for the most part, the design is still done by process engineers in the traditional way, as well as the tools try-out. In order to take advantage of these "two worlds" the more scientific knowledge of the process, achieved by the theoretical modeling, and the know how acquired by the technicians a friendly user system that integrates all of these benefits was envisaged.

The development of an integrated tool/methodology to face these challenges are the main objective of a three years project recently launched by the authors and financially supported by JNICT. A Computer Aided Process Planning (CAPP) system that interfaces with a finite element code and a database with the mechanical, anisotropic and formability properties of the material experimentally determined will enable the preliminary and final design of the deep drawing tools from the geometrical or physical model of the part. The CAPP system will include the capability to store in a multimedia environment the record of all of the above-mentioned procedures including the experimental analysis of the strains imposed on the part and the try-out results. This integrated system will lead to the capability of producing sound parts at first and in a shorter period of time. Simultaneously, an improvement in the design and product quality will be achieved.

In process design, the goal is to define the most adequate process sequence that will transform the original blank geometry into the final part. This target is very up-to-date and requires a large and continuous effort in developing time as it can be noticed in the continuous works of Tisza [1-3] and Karima [4, 5], for example. All of these works have a main objective that is to transform the present "art" of tooling development according to the technological skills developed by the artisans to an experience-enhanced science based technology to be used by the practitioners, [6].

## 2. GENERAL CONCEPT OF PROESTAMP

The main idea is not to develop anything completely new but only to integrate the know how already available in the industry and their methodologies, beginning with the most easy cases, in a computer aided process planing for the design of the deep drawing process as a whole, i.e., from the blank development to the try-out of the tools and data management.

The originality came from the fact that using PROESTAMP the designer can develop the whole process at the computer and all the required information is available as needed. Finally all the design and try-out steps will be kept on a multimedia archive in order to be used in future works.

In Figure 1, the general architecture of PROESTAMP is shown. In a synthetic way the system has a main driving sequence of modules that interfaces, since the very beginning until the end, with a multimedia archive and documentation module and with a data base module divided into several relational data bases.

It is evident that the main effort is to develop an assisted system as close as possible to the usual way things are done industrially. Simultaneously trying to make available and easy to apply by the practitioners the new facilities as, for example, the FEM analysis, the circle grid analysis and the multimedia facilities.

All programs are being developed on the PC-basis and using Microsoft Visual C++ under the Microsoft Windows environment. The main factors for this option are:
- The large diffusion and acceptance of this operating system now and in the near future, at least;
- The possibility of generating familiar interfaces for the user, in order to minimize the learning time and to be user friendly;
- Once the system has to be an open one, for continuous improvement, it was decided to use an object oriented programming language;
- Compiler with the capability for an easy introduction of menus, dialog boxes, icons, bitmaps, etc.

All of the development has being carried out taking into account that the program interactively with the user shall be as close as possible to the processes in use by the Portuguese industry in order to minimize eventual rejecting attitudes of the addressed people. Another very important factor is that all the suggestions offered by the program must be validated by the user and if he does not agree he can easily change the suggestion.

So, the target is to develop a cheap system specific tailored for the Portuguese industry that enables the beginning of a move for a more scientific design of the deep drawing process, keeping the possibility of being used and applied by the industrialists.

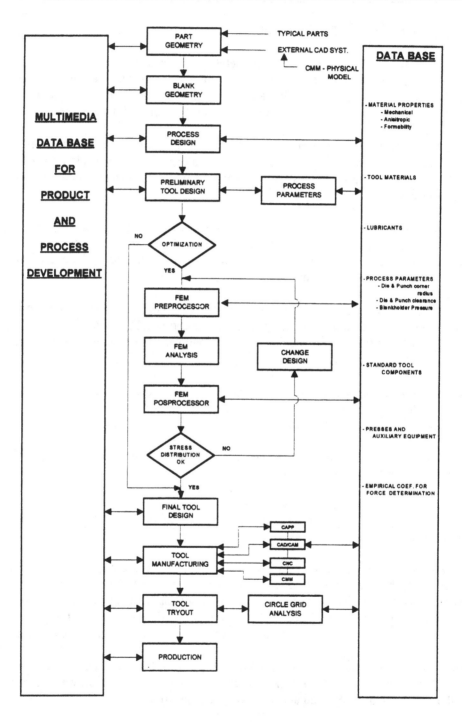

Fig. 1 - Flow Chart of PROESTAMP

The process design begins with the definition of the geometry of the part to be produced. This task can be done in three different ways:

- For typical parts, Fig. 2, the user select the desired shape and gives the required dimensions;
- Import the part geometry from external CAD systems, using neutral interfaces, DXF and IGES at the moment;
- To draw directly on PROESTAMP. Nevertheless this capability does not intend to substitute a CAD system but only to take advantage of the drawing capabilities implemented in the Edit functionality;
- From a physical model a Coordinate Measuring Machine, CMM; will scan the model and export the data to a CAD system.

At any stage the user can change the part geometry and the system automatically update all of the computations already done.

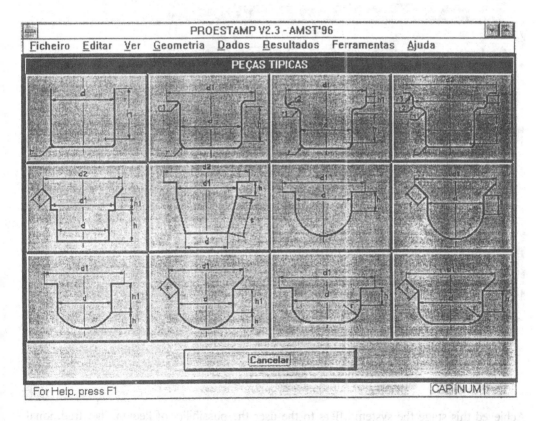

Fig. 2 - Typical parts

The next step is to compute the initial blank geometry, Fig. 3. This is done assuming volume constancy in plastic deformation and average thickness constancy in deep drawing. So, the initial blank is obtained by surface constancy.

At this moment the user shall indicate the material type. All the relevant data should be available on the material data base, if not, the user can update the material data base.

At this moment the system is able to propose the sequence of operations necessary for producing the part. The main criteria applied in this module is the use of the Limit Drawing Ratio, LDR, of the material. At the same time all the relevant process parameters are taken from the data base and a forecast of the forces involved is made. These tasks are performed under the preliminary tool design module.

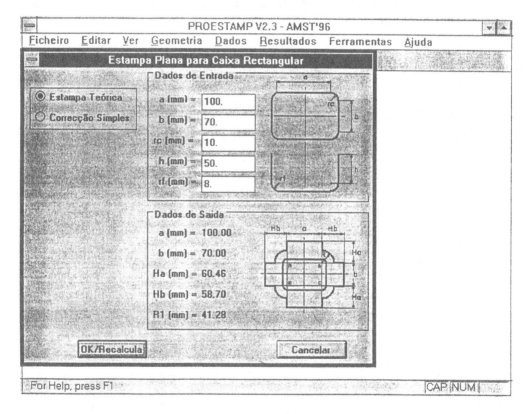

Fig 3 - Blank Geometry

Achieved this stage the system offers to the user the possibility of keeping her traditional way, move directly to the final tool design, or the possibility of optimizing the sequence of operations and the values of the process parameters. Once, more and more FE programs come available for analysis of the deep drawing process the main idea is to implement a pre and a postprocessor module for a FE program, ABAQUS for example. The main criteria for the optimization procedure is the comparison between the predicted strain distribution on the part versus the Forming Limit Curve, FLC, of the material.

Once the preliminary tool design is finished with this optimization procedure the final design consists mainly on the interfacing of the available geometric and operating data with a CAD system for the final design and a CAPP/CAM system for planning the manufacturing of the tool.

Finally the circle grid analysis, and the FLC, will be used during the try-out of the tool.

All of these procedures will be registered, documented, using multimedia facilities, including the video register of tool details and of the main occurrences during the try-out.

## 3. CONCLUSIONS

A very ambitious three years project, actually at the first half-year, for the development of a computer assisted system for the design of deep drawing tools was presented.

These type of tools are very attractive because the industrialist has to cut in the time to market and they need to make very realistic quotations in a short time.

Once the system has so many modules each of them can be of large or small degree of difficulty. We decided to emphasize the development of the total philosophy, from the very beginning until the end, with the easiest possible cases and after to include the more complexes cases as the complex shapes, for example.

The system will always offer to the user the possibility of working in the traditional way he is used to, and to incorporate all of theirs know how, as well as the possibilities of gradually introduce a more scientific approach.

The most interesting particularity is that the designer will be able of designing the whole process at the computer without needing to look for recommended values on manuals.

## ACKNOWLEDGES

This project is sponsored by JNICT "Junta Nacional de Investigação Científica e Tecnológica" which is gratefully acknowledged.

## REFERENCES

1. Tisza, M.: A CAD System for Deep-drawing, 2nd ICTP, Stuttgart, 1987, 145.
2. Tisza, M.:An Expert System for Process Planning of Deep-drawing, 4th ICTP Conference, Beijing, 1993, 1667.
3. Tisza, M.; Rácz, P.: Computer Aided Process Planning of Sheet Metal Forming Processes, 18th Biennial Congress of IDDRG, Lisboa, 1994, 283.
4. Karima, M.: From Stamping Engineering to an Alternative Computer-Assisted Environment, Autobody Stamping technology Progress SP-865, USA, 1991, 97.
5. Karima, M.: Practical Application of Process Simulation in Stamping, Journal of Materials Processing Technology 46, 1994, 309.
6. Keeler, S.P.: Sheet Metal Stamping Technology - Need for Fundamental Understanding, Mechanics of Sheet Metal Forming, Plenum Press, N.Y., 1978, 3.

# LAYERED TOOL STRUCTURE FOR IMPROVED FLEXIBILITY OF METAL FORMING PROCESSES

**T. Pepelnjak, Z. Kampus and K. Kuzman**
University of Ljubljana, Ljubljana, Slovenia

KEY WORDS: Metal Forming , Rapid Tooling, Layered Tools, FEM Simulation

ABSTRACT: The  paper presents the concept of layered-construction metal forming tools. Such tool design allows the adaptation of its geometry where a set of products have to be manufactured. The composing of a forming tool from existing tool steel plates with different thicknesses enables quick tool geometry changes. The decrease of tool production times and costs is possible with computer-aided decision systems and through the use of existing tool plates.

The concept of layered forming tools is shown with the deep drawing of a set of thick rotational cups. Determined real tool geometry has been estimated with FEM simulations and compared with experiments to justify the FEM model and to verify the proper choice of tool geometry.

## 1. INTRODUCTION

Nowadays industrial production tends towards the shortening of production times and towards the ability of fast reactions to market demands. Modern manufacturing technologies and computer-aided systems offer manufacturers increased production flexibility, which is still very low in metal forming processes. Forming tools are mainly designed for one product, which restricts manufacturing adaptability due to high tool costs. To fulfil the demands for flexibility in metal forming, special tooling concepts have been developed.

Published in: E. Kuljanic (Ed.) *Advanced Manufacturing Systems and Technology*, CISM Courses and Lectures No. 372, Springer Verlag, Wien New York, 1996.

These tools are based on a layered or laminated structure, which enables fast and inexpensive manufacturing of tools (e.g. with laser beam cutting)[1,2,3] and increases their geometrical adaptability [4]. They also can be designed in segmented structure to enable changing of tool geometry with the setting of tool segments into appropriate positions [5]. Another possibility is geometrical adjustable pin-structure [6] or tools with elastic elements. All above mentioned tools are for use mainly in prototyping and small quantity production where large demands for fast, inexpensive and/or adaptable tooling systems are presented.

The paper presents a possibility of increasing the geometrical adaptability of forming tools based on layered tool structure and analysed using the deep drawing process. The conventional tools used in this forming process are clearly unsuitable for any geometrical adaptations caused by changes in product geometry.

## 2. FLEXIBILITY OF DEEP DRAWING TOOLS

The adaptation of tool geometry in deep drawing tools was realised with the development of laminated tools in the Forming Laboratory of the Faculty of Mechanical Engineering, University of Ljubljana, Slovenia. The lamellas for the tools were laser beam cut from 1 mm thick sheet steel. Manufacturing of these tools is very fast. The manufacturing times were five times shorter than for tools produced with conventional manufacturing technologies [4]. A forming analysis of deep drawing with laminated tools without blankholder for rotational and non-rotational cups from thick sheet steel was performed. It showed that laminated structure allows the changing of a particular tool part (wear or fracture-damaged part of drawing die) or even the use of different materials in the same die.

Laminated tools also enable the adaptation of tool geometry according to the demands of the forming process. Without larger design or manufacturing efforts some lamellas can be changed or removed and new tool parts can be added between existing lamellas [4]. The removal of some lamellas makes possible the manufacture of a limited set of geometrically similar products. On the other hand it allows one to control and optimise the drawing process by changing the punch-die clearance.

The active die geometry of the designed laminated tool is prescribed with a drawing curve which has in the analysed tool the form of a tractrix curve in order to achieve the minimal drawing forces. The tool geometry reduces the flexibility in terms of the production of defined product geometry which does not match to existing die openings achieved with the removal of some lamellas. An additional problem to tool flexibility is the stiffness of lamellas, which is in comparison to hardened tool steel very low. Low stiffness of lamellas requires a thicker calibration part which receives increased local strains at the end of the forming process and has to be changed on each variation of the bottom die opening.

The products chosen for the process analysis are rotational cups with outer diameter of 55, 60 and 65 mm from 5 mm low-carbon deep drawing steel (DIN RSt13). FEM simulations were performed for all three cup dimensions and experiments for the first two dimensions. In order to increase the tool's flexibility and stiffness and to achieve the defined cup diameters, a set of tool plates was developed to replace a part of the laminated tool. For shortening the design, the establishing of a computer-aided CAD-CAPP-CAM system is recommended [7,8]. The system, based on decision-making criteria such as the technological parameters of the forming process, the difference between real and ideal tool geometry, tool materials, prescribed minimal and maximal thickness of the tool plates etc., can support the design of the forming tool, enable the implementation of existing tool plates and shorten the necessary design and manufacturing times. A diagram of the decision-making system is shown in *fig. 1.*

The tool geometry, made up of lamellas in the form of tractrix curves and tool plates in the form of broken cones for the inner die profile has been checked using defined forming parameters with FEM simulation.

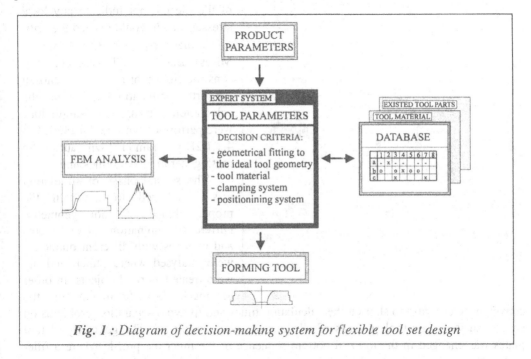

*Fig. 1 : Diagram of decision-making system for flexible tool set design*

## 3. FEM ANALYSIS OF DEEP DRAWING

The research work on deep drawing without blankholder [9] has shown that with FEM simulation it is possible to predict with good accuracy the course of the forming process. Analyses of possible reductions of forming force with a redesign of the drawing tool [4] and achievement of maximal drawing ratios through optimisation of tool geometry [10] have shown that both criteria can be fulfilled only with an increase of die height.

In the redesign phase of the laminated die, the boundary conditions of maximal tool thickness have been stated. This should not exceed the thickness of the laminated die with drawing profile in the form of a tractrix curve. This inlet die shape offers one of the lowest forming forces in deep drawing without blankholder. It is to be expected that a maximal drawing ratio of $\beta_{max}$=2.8 (in special cases up to 3.25 [10]) and forming forces achieved with tractrix inlet profile could not be reached.

The FEM simulations of deep drawing were performed in two phases. In the first phase the reference die with inner die profile in the form of a tractrix curve was simulated with two different die models, rigid and elastic (see *fig. 2*). Comparison of both models has shown nearly identical results for forming forces, cup geometry and stress conditions. The stress condition in the elastic model of the die did not indicate any local stresses which would exceed the critical values ($\sigma_{die,max}$< 320 N/mm$^2$ - Mises). Based on all these comparisons the rigid tool model was chosen for further simulations to shorten the calculation times. All simulations were performed with the DEFORM$^{TM}$ 4.1.1 2D program [11] with automatic mesh generation and remeshing.

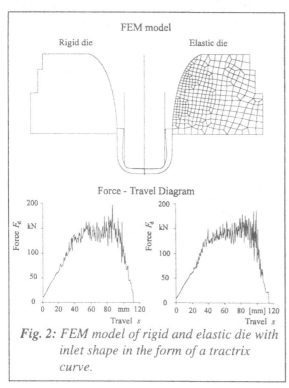

*Fig. 2: FEM model of rigid and elastic die with inlet shape in the form of a tractrix curve.*

The second phase of simulation was established to ascertain the proper choice of tool geometry. Different combinations of tool plates and punches with different diameters were analysed where punch and die were treated as rigid objects. In order to reach fast fulfilment of the convergence criterion to shorten the calculation times and to avoid separation problems on contact surfaces, the tool surfaces of all die parts were simplified. The geometry of tool plates was changed in the top and bottom sequence of the inner die profile where a fillet with 5 mm radius has been chosen to bind the cone geometry of neighbouring plates (see *fig. 3*). These simplifications have to be considered through comparison of FEM simulation with the experimental results of deep drawing.

## 4. TOOL-SET DESIGN

The tooling set is designed with the following boundary conditions:

- die plate opening is defined through outer cup geometry,
- the geometry of the broken cone should best fit on an ideal drawing profile in the form of a tractrix curve,

- the dimensions of tool plates should fit the tractrix die profile (see above condition) in their input and output diameters ($\phi$70,$\phi$65, $\phi$60, $\phi$55 mm),
- the geometry change from tractrix profile to broken cone profile should not increase the die thickness,
- the replacement of each part of the die has to be fast and simple to perform.

The designed set of tool plates is based on partial replacement of lamellas in laminated drawing dies to achieve better production accuracy and/or stiffness on particular parts of the die. As simple as possible a combination of tool plates was achieved, starting from the smallest outer cup diameter ($\phi$55 mm); each next bigger cup diameter in the set can be produced by removing a single tool plate. In *fig. 4* the tool flexibility by removing the tool plates is shown with the corresponding product geometry simulated with the FEM program for each combination punch-die and with constant blank diameter of $D_0 = 100$ mm.

The stiffness of tool plates which result from their material (plates are made of hardened tool steel and lamellas from steel sheets with 0.61% C), heat treatment and thickness which is 10-20 times larger in comparison with lamellas allow the use of these plates also as bottom clamping plates.

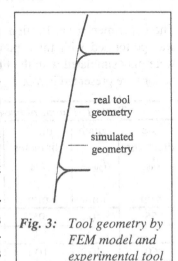

real tool geometry

simulated geometry

Fig. 3: Tool geometry by FEM model and experimental tool

The top and bottom geometry of the drawing profile of each plate is made with transitional radii to decrease the local stresses in them (see *fig. 3*). The inlet and output radius in the cone profile influences the forming process, which is evident from experimentally determined forming forces.

The production of cup sets requires in addi-

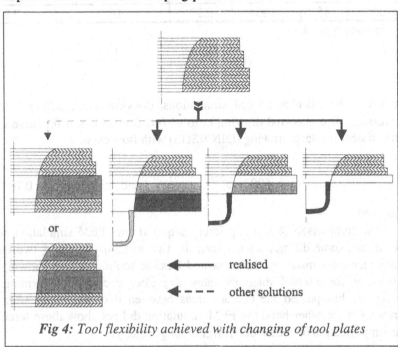

or

realised

other solutions

Fig 4: Tool flexibility achieved with changing of tool plates

tion to the changing of the die opening also the changing of the punch diameter. In the presented work the increase of tool flexibility was limited to the die design and conventional drawing punches were used. The research of tooling sets for fast and inexpensive punch geometry changes using special front plates is in progress.

## 5. EXPERIMENTAL VERIFICATION

The experimental verification of the flexibility of the newly designed deep drawing tool was performed with three punch-die combinations whereas another three combinations were only simulated with the FEM program. The chosen die-punch-blank diameter combinations are presented in *tab. 1*.

| *Table 1: Punch-blank-die combinations for flexible production of cup sets* | | | | | | |
|---|---|---|---|---|---|---|
| die diameter $d_{die}$ | punch $d_{punch}$ | blank diameter $D_0$ | drawing ratio $\beta$ | maximal drawing force (simulation) $F_{d,FEM}$ ($\mu$=0.1) | maximal drawing force (experiment) $F_d$ | one-sided clearance $z$ |
| mm | mm | mm | / | kN | kN | mm |
| 55 | 40 | 95 | 2.375 | 145.7 | 169 | 7.5 |
| | | 100 | 2.5 | 173 | 188 | |
| 60 | 40 | 100 | 2.5 | 179.5 | 184 | 10 |
| | | 110 | 2.75 | -- | 202 * | |
| 60 | 45 | 100 | 2.22 | 169 | 186 | 7.5 |

* cup bottom breakage

$$\beta = \frac{D_0}{d_{punch}} \quad , \quad z = \frac{d_{die} - d_{punch}}{2}$$

To verify the calculated FEM simulations, drawing with different drawing ratios was performed. The maximal drawing ratio of $\beta$=2.5 was chosen for these experiments. Low-carbon steel for deep drawing (DIN RSt13) with flow curve

$$\sigma_f = 690.1 \cdot \varphi_e^{0.221} \left[ \frac{N}{mm^2} \right] \quad \text{and anisotropy } \bar{r} = 0.78$$

was used.

The comparison of forming forces acquired with FEM simulation and experimentally has shown some differences between the two methods. In *fig. 5* two cups with different outer diameters made by the presented flexible tooling set are shown. The experimental results of force-travel diagram show a greater decrease of forming force when the workpiece has passed the contact areas between two tool plates (*fig. 6* - indicated by arrow). On the other hand the FEM simulation did not show these force decreases due to the simplification of the model of rigid tool geometry.

## 6. CONCLUSIONS

The aim of this paper is to illustrate the possibility of flexible tooling in metal forming production. The demands on tool adaptability and flexible tool design are strongly presented in prototyping and small quantity production where tool optimisation is often not required.

To achieve faster and more efficient decisions about proper tool set geometry, an expert system will be developed. It will be based on decision criteria to fulfil pre-defined technological parameters of forming processes, demands on tool materials, allowed geometrical difference between real and ideal tool geometry etc. The use of a database for already existing tool parts and their implementation into newly designed tools will decrease tool costs and shorten its production times.

The chosen geometry of the new tool is to be verified with FEM simulation. It has been shown that FEM simulation of deep drawing processes are accurate enough to predict the course of the forming process, its forming forces, stress conditions and product geometry.

The adding or removing of particular tool plates enables fast and simple changes of tool geometry. This allows the production of a set of cups with different outer diameter and represent a new approach to the design of deep drawing

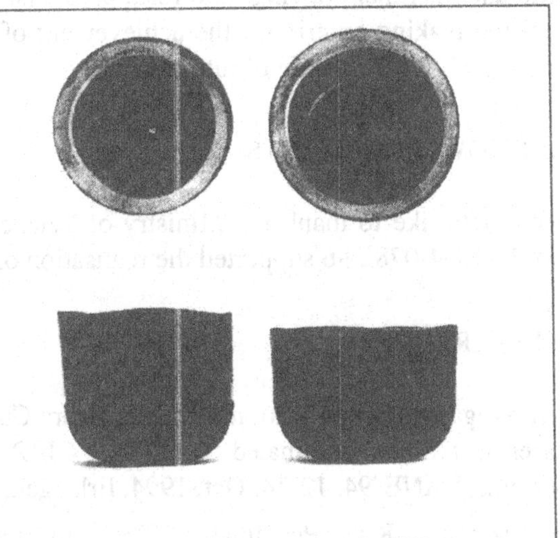

Fig. 5:  Deep drawn cups with different outer diameters

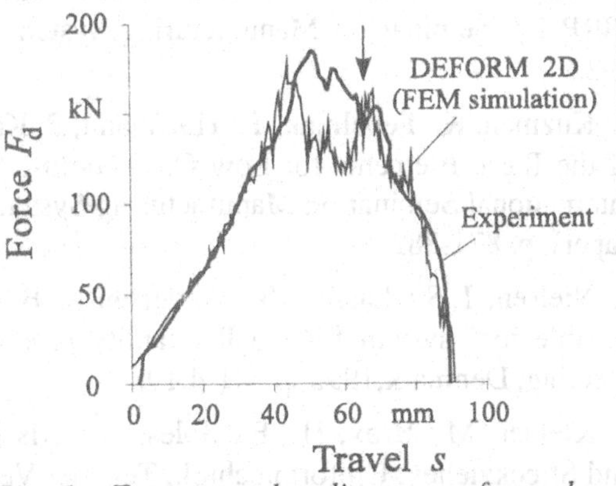

Fig. 6:  Force-travel diagram for deep drawing with laminated die with two tool plates

tools. The changes of die opening can also be used to control the production accuracy of deep drawing processes using different die-punch clearances.

The use of flexible tooling sets of combined laminated parts and tool plates has an influence on the optimal parameters of forming processes. On one side, with existing tool parts the die can be designed quickly and inexpensively, and on the other side tool geometry optimisation can be performed in order to achieve optimal drawing conditions. With the removal of one or more tool plates the inner tool geometry does not fit the optimal drawing profile - tractrix curve - which offers one of the best forming parameters in deep drawing without a blankholder. The correlation between geometrical optimisation ratio and maximal drawing ratio for a particular tool geometry in order to build up reliable decision-making criteria for the achievement of the maximal drawing ratio with a chosen tool set will be the theme of future work.

## ACKNOWLEDGEMENTS

We would like to thank the Ministry of Science and Technology, which through project No. J2-7064-0782-96 supported the realisation of the presented work.

## 7. REFERENCES

1. Nakagawa T.: Applications of Laser Beam Cutting to Manufacturing of Forming Tools-Laser Cut Sheet Laminated Tool, 26th CIRP International Seminar on Manufacturing Systems, LANE '94, 12-14. Oct. 1994, Erlangen, Germany, p. 63-80.

2. Kuzman, K.; Pepelnjak, T.; Hoffmann P.: Flexible Herstellung von Lamellenwerkzeugen mittels Laserstrahlschneiden, Blech Rohre Profile, 41(1994) 4, p. 241-245 (in German).

3. Franke, V.; Greska, W.; Geiger, M.: Laminated Tool System for Press Brakes, 26th CIRP Int. Seminar on Manufacturing Systems, LANE '94 , Erlangen, Germany, p. 883-892.

4. Kuzman, K.; Pepelnjak, T.; Hoffmann, P; Kampuš, Z.; Rogelj V.: Laser-cut Sheets - one of the Basic Elements for Low Cost Tooling System in Sheet Metal Forming, 26th CIRP International Seminar on Manufacturing Systems, LANE '94 , Erlangen, Germany, (invited paper), p. 871-882.

5. Nielsen, L.S.; Lassen, S.; Andersen, C.B.; Grønbæk, J.; Bay, N.: Development of a flexible tool system for small quantity production in cold forming, 28th ICFG Plenary Meeting, Denmark,1995, p. 4.1-4.19.

6. Kleiner, M.; Brox, H.: Flexibles, numerisch einstellbares Werkzeugsystem zum Tief- und Streckziehen, Umformtechnik, Teubner Verlag, Stuttgart 1992, p. 71-85 (in German).

7. Balič, J.; Kuzman, K.: CIM Implementation in Forming Tools Production, Proc. of 2nd Int. Conf. on Manufacturing Technology, Hong Kong, 1993, p. 361-366.

8. Brezočnik, M.; Balič, J.: Design of an intelligent design-technological interface and its influence on integrational processes in the production, Master Thesis, University of Maribor, 1995, 91 p.

9. Kampuš, Z.; Kuzman, K.: Experimental and numerical (FEM) analysis of deep drawing of relatively thick sheet metal, J. Mat. Proc. Tech., 34 (1992), p. 133-140.

10. Kampuš, Z.: Optimisation of dies and analysis of longitudinal cracks in cups made by deep drawing without blankholder, 5th ICTP, October 1996, Ohio, USA, (accepted paper).

11. Scientific Forming Coop.: DEFORM 2D - Ver. 4.1.1., Users Manual, 1995.

# PHYSICAL SIMULATION USING MODEL MATERIAL FOR THE INVESTIGATION OF HOT-FORGING OPERATIONS

P.F. Bariani, G.A. Berti, L. D'Angelo and R. Meneghello
University of Padua, Padua, Italy

KEY WORDS : Physical Simulation, Hot Forging, Model Material

ABSTRACT : Physical simulation using model materials is an effective technique to investigate hot forging operations of complex shapes and can be a significant alternative to a numerical approach (F.E.M) in the preliminary phases of process design. This approach is suitable to evaluate die filling, material flow, flow defects and to predict forging load system. The paper is focused on i) presentation of the equipment, developed by the Authors, and ii) its application to model the hot forging of a crane link.
The developed equipment is suitable to replicate forging operations using wax and lead as model materials, as well as to reconstruct the system of forces acting on the dies.
Characterisation of model material and forged steel has allowed to obtain an estimation of real forging load at each stage of the sequence.

## 1. INTRODUCTION

Physical simulation of forging operations consists of different techniques aimed to i)reproducing operating conditions using real materials on simple geometry specimen [1,2], ii) simulating the forming process using either real geometry and model materials (waxes, plasticine, lead) or visioplasticity techniques [3] and analysing flow behaviour.
The paper presents some progresses in the investigation of forming processes model materials and its application to the study of an hot forged crane link. The model materials present i) a lower load for deformation, if compared with real forged materials, and ii) the

Published in: E. Kuljanic (Ed.) *Advanced Manufacturing Systems and Technology*,
CISM Courses and Lectures No. 372, Springer Verlag, Wien New York, 1996.

deformation can be performed at room temperature, instead of hot forging temperature. For these reasons the laboratory tests are faster, easier and less expensive than a sub scale production process. Furthermore, the dies utilised in the test and reproducing the geometry of real dies can be manufactured in resin, aluminium, Plexiglas (in the case of waxes and plasticine) or carbon steels (in the case of lead).

Investigations based on model material can be focused on different aspects, such as flow behaviour (die filling, defects recognition), forming load requirements [4], parting line location, flash design, die attitude optimisation, billet location, etc.

## 2. PHYSICAL SIMULATION TECHNIQUE

A new facility has been developed [5] and installed at DIMEG, University of Padova, devoted to physical simulation using model materials. It consists of a 2000 kN lab press (named Toy Press and shown in Fig. 1) equipped with a multi-axes force- and moment-transducer. The transducer, a 3-plate die set with 3 piezo-electric three axial load cells, connected to a PC-based acquisition system, provides, during the forming cycle, the history of the three components of force and moment, as well as the attitude of the resultant of the forming forces, reconstructing them from the nine load cells signals. Different plots can be obtained, such as force and moment versus time/die stroke, attitude of resultant of the forming force versus time/die stroke, as well as application point of resulting force mapped over the cycle [6].

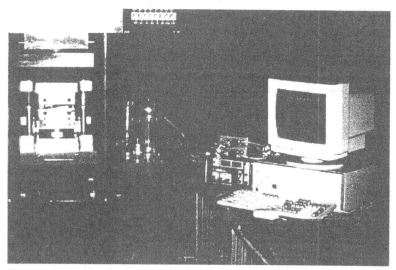

Fig. 1  The equipment for the reconstruction of forces and moment over the forging cycle

The direct analysis of the attitude of resultant and mapping of its application point can suggest modification of die attitude, parting line location and billet positioning, in order to reduce lateral forces and moment acting on the dies.

Force and moment history over the forging cycle allow the recognition of symmetries/asymmetries in the dies and in the flow.

A comparison among simulations obtained using alternative forming dies gives information on effectiveness of solutions adopted in die design [7, 8] relevant to die filling, forming load reducing, defects eliminating and flash minimisation. Defects in material flow and in the die filling can be recognised by visual inspection of model material preforms and using a multi-colour layered billet. This approach results to be particularly useful in the preliminary phases of process design, when alternative solutions should be rapidly evaluated in order to determine the optimal one, without manufacturing expensive die sets and testing them at operative conditions.

An extreme care should be taken when the load of real forming operation has to be predicted. The following rules should be applied :
- plastic behaviour of model material should reproduce as close as possible, in reduced scale, the real material behaviour at forging conditions,
- an equivalent effect due to the lubrication should be reproduced at the die-material interface using oil, solid soap, plastic films, etc.

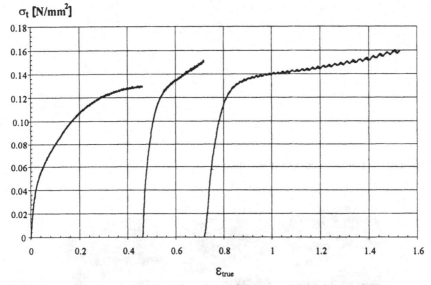

Fig. 2  Multi-step true stress - true strain curve of the wax (model material)

As concerns model material characterisation, cylindrical specimens have been upset; good lubrication condition should be assured in order to minimise the barrelling, otherwise the state of stresses is three-axial. When true strain ($\varepsilon$) is above 0.4, lubrication becomes not effective ; therefore, the characterisation test should be splitted into a number of steps, each one performed to a strain less then  0.4 and reconstructing the lubricating film before each step. The resulting multi-step true stress-true strain curve is presented in Fig. 2.

3.    APPLICATION EXAMPLE

Physical simulation using model material is applied to the study of a hot forged crane link. This new-design large crane link for earth moving machines (see Fig. 3) will be produced on a three stage vertical hot former. Main difficulties in forging this crane link are relevant to die attitude and too high forming forces compared with the press loading capacity.
Material is 35MnCr5 steel forged in the range 1200-1250 °C. Forging sequence dies used in physical simulation are shown in Fig. 4 ; a flash trimming stage, not shown, ends the sequence. The starting billet is 100x100 mm (square section), 350 mm long.

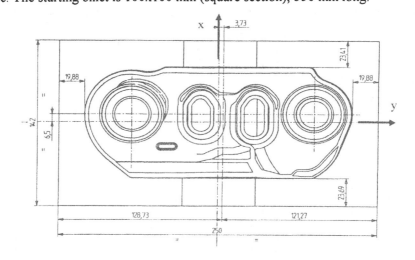

Fig. 3  Top view of the new-design crane link

Fig. 4  Dies for the simulation of hot forged crane link (3 forming steps : preforming, blocking and finishing)

In order to investigate both die attitude and required forging force, dies for physical simulation have been NC manufactured using a resin in halfsize scale respect to designed dies for real process.

As concerns model material, a wax has been used which offers a behaviour similar to the real material. In Fig. 5 the true stress-true strain curve of wax is compared with the true stress-true strain curve of 35MnCr5 steel (T=1200°C, $\dot{\epsilon}$=11 s$^{-1}$ ). The curves of this steel have been obtained using the Gleeble 2000™ thermo-mechanical simulator in the range of temperature, strain and strain rate present in the process.

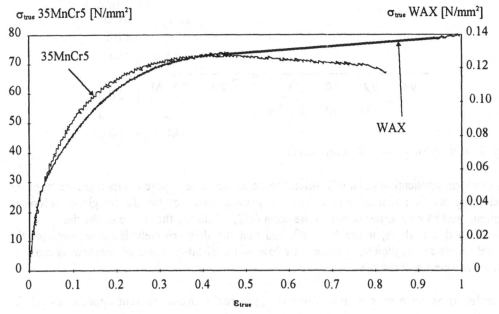

Fig. 5  Comparison of true stress - true strain curve of 35MnCr5 (T=1200°C, $\dot{\epsilon}$=11 s$^{-1}$ ) steel with wax curve

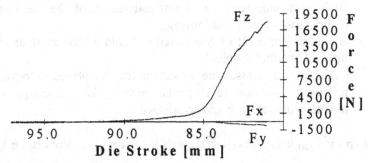

Fig. 6  History of 3 force components

A direct analysis of forces, torque and application point plots gives information on correctness of partition line definition, dies orienting and billet positioning. The lateral forces (Fx, Fy) in the finishing die, shown in Fig. 6, are negligible and main contribution to resultant force is due to the Fz component. The moment My and Mz (se Fig. 7) are low due to the facts that the die is symmetric respect to the y axis (My → 0) and the lateral forces (Fx, Fy) are negligible (Mz → 0). Presence of moment Mx can be explained by the fact that dies parting line is not in a single plane.

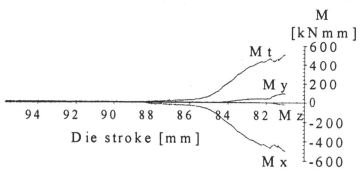

Fig. 7  History of 3 moment components

As concerns application point of resultant force, it can be recognised in its mapping over the cycle (Fig. 8), that i) it is located near the gravity centre of the die (origin of reference system), and ii) it becomes closer to the point (-12,20) during the stroke of the die.
Based on this analysis, it can be confirmed that the dies are well designed, because the lateral forces are negligible, moments are low and application point of resultant is close to the gravity centre of the die.

In order to estimate maximum of force ($F_{max}$) in real process different approaches can be chosen.
The first one [3], based on similitude,  requires the satisfaction of different conditions depending  on type of processes :

  plastostatic : strain-hardening exponent of real material should be the same as the model material one (cold forming), or
  strain-rate exponent of real material should be the same as the model material one (hot forming)

  dynamic : (when inertia stresses are important for the plastic deformation) the following condition (equivalence between kinetic energy ratio and internal work ratio) should be satisfied

$$(\tfrac{1}{2} \bullet \text{Vol} \bullet \rho \bullet v^2)_{model} / (\tfrac{1}{2} \bullet \text{Vol} \bullet \rho \bullet v^2)_{real} = (\text{Vol} \bullet \sigma_0 \bullet \varepsilon)_{model} / (\text{Vol} \bullet \sigma_0 \bullet \varepsilon)_{real}$$

where $\rho$ is the density, $v$ is the speed, Vol is the volume of the workpiece, $\sigma_0$ is the flow stress, $\varepsilon$ is the strain.

The fulfilment of similitude conditions allows the determination of load for real process (F$_{real}$) on the basis of

$$F_{real}(\varepsilon) = F_{model}(\varepsilon) \bullet \sigma_{0real}(\varepsilon)/\sigma_{0model}(\varepsilon)$$

The second approach [9], which gives a very approximate estimation, is based on the following assumption : *the maximum of force is reached at the end of the forming process, when the dies are filled and flow of material is essentially located in the flash.* In this case the following relation can be used

$$F_{max} = K_f \bullet A \bullet \sigma_f$$

where  A is the projection area in the forming direction including the flash,
      $K_f$ is the complexity factor,
      $\sigma_f$ is the material flow stress.

The complexity factor mainly depends on geometry of dies and preforms. Taking in account this simplification, the $K_f$ factor can be considered independent from material and it can determined as

$$K_f = F_{maxm} /( A_m \bullet \sigma_{fm}) = 5.8$$

using the F$_{maxm}$ obtained from physical simulation and $\sigma_{fm}$ as the flow stress of the model material. The maximum of force in the real process (F$_{maxr}$) results to be

$$F_{maxr} = K_f \bullet A_r \bullet \sigma_{fr} = 28900 \text{ kN}$$

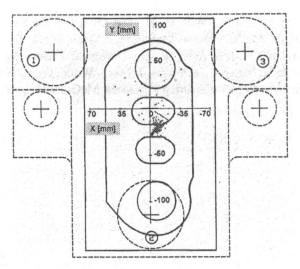

Fig. 8  Application point of resultant force

## CONCLUSIONS

A recent application of physical simulation technique to the analysis and modelling of hot forging operations has been presented. The paper has illustrated the developed equipment and the procedure utilised in the investigation of hot forming by physical modelling, which includes the monitoring of force-and-moment history over the forging cycle of complex parts. The application of this technique to the forging of a crane link has been presented, demonstrating the power of this approach in the preliminary phases of process design if compared with expensive trial sub scale production or time consuming F.E. simulations.

## REFERENCES

1. Altan, T. and Lahoti, G.D : Limitations, Applicability and Usefulness of Different Methods in Analysing Forming Problems, Trans. of ASME, May 1970
2. Ferguson, H.F. : Fundamentals of Physical Simulation, DELFT Symposium, December 1992
3. Wanheim, T. : Physical Modelling of Metalprocessing, Procesteknisk Institut, Laboratoriet for Mekaniske Materialeprocesser, Danmarks Teknisk Højskole, Denmark, 1988
4. Altan, T., Henning, H.J. and Sabroff, A.M., The Use of Model Materials in Predicting Loads in Metalworking, J. Eng. for Ind., 1970
5. Bariani, P.F., Berti, G., D'Angelo, L., et al. : Some Progress in Physical Simulation of Forging Operations, II AITEM National Conference, Padova, September 1995
6. Bariani, P.F., Berti, G., D'Angelo, L. and Guggia, R. : Complementary Aspects in Physical Simulation of Hot Forging Operations, submitted to 5th ICTP Conference, Ohio, USA, 1996
7. Pihlainen, H., Kivivouri, S. and Kleemola, H., Die Design in the Extrusion of Hollow Copper Sections Using the Model Material Technique, J. Mech. Work. Tech., 1985
8. Myrberg, K., Kivivouri, S. and Ahlskog, B., Designing Preforming Dies for Drop Forging by Using the Model Material Technique, J. Mech. Work. Tech., 1985
9. Schey, J.A., Introduction to Manufacturing Processes, McGraw-Hill, 1988

# TEXTURE EVOLUTION DURING FORMING
# OF AN Ag 835 ALLOY FOR COIN PRODUCTION

**L. Francesconi**

**University of Ancona, Ancona, Italy**

**A. Grottesi, R. Montanari and V. Sergi**

**University "Tor Vergata", Rome, Italy**

KEY WORDS: Forming, Coining, Texture evolution, Ag alloy

ABSTRACT: The Ag 835 alloy is currently used for the production of coins by the Italian Mint Service. After continuos casting, ingots are formed into sheets following two different cycles (normal and proof). Blanks are cut from the sheet and then subjected to coining. The texture evolution of the material has been characterized after each stage of the two cycles. The results of this work show that remarkable grain re-orientations take place during processing. Before coining, the normal cycle gives rise to a {111} texture whereas a mixed texture with a strong {110} component is produced by the proof one.

## 1. INTRODUCTION

The Italian Mint Service (Istituto Poligrafico e Zecca dello Stato) has developed two different cycles for the production of commemorative coins: the normal cycle and the proof one (Fig. 1). The proof cycle is used when a mirror-like surface is requested for the final product.

Ingots (70 mm wide, 15 mm high and 300 mm long) are produced by continuos casting. The microstructure of the ingots is not homogeneous and shows evident segregation of Cu at the surface. In order to reduce segregation phenomena the ingots are annealed 6 hrs at

Published in: E. Kuljanic (Ed.) *Advanced Manufacturing Systems and Technology*,
CISM Courses and Lectures No. 372, Springer Verlag, Wien New York, 1996.

600°C in air (stage 2 - normal cycle) or their surface is milled removing 0.5 mm from each side (proof cycle).

Following the normal cycle, ingots are then cold rolled (total height reduction $\Delta h = 12.5$ mm), annealed 6 hrs at 600°C in air and again cold rolled into sheets of 1.6 mm in thickness. Blanks are cut from the sheets, annealed 6 hrs at 600°C in air, pickled with $H_2SO_4$ and then hemmed.

The stages of the proof cycle are: cold rolling ($\Delta h = 12.4$ mm), annealing 1.5 hrs at 600°C under inert atmosphere, pickling with $H_2SO_4$ and hemming.

The total height reduction was obtained by multi-pass 2-high non reversing mill; the stock is returned to the entrance of the rolls for further reduction by means of a platform.

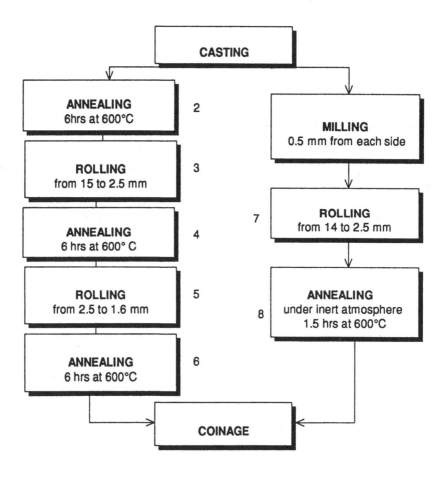

Figure 1. Flow-chart showing the stages of the normal and proof cycles.

The following rolling schedule was employed:

| Stage | Number of passes | Average reduction % per pass |
|-------|------------------|------------------------------|
| 3 | 16-18 | 11 |
| 5 | 10 | 5 |
| 7 | 20 | 10 |

Before coining the blank surface results to be enriched in Ag with respect to the mean chemical composition due to the milling (proof cycle) or to the pickling, which remove the Cu oxide scale formed during thermal treatments in air (normal cycle).

The microstructure evolution and the mechanical characterization during each stage of the two cycles have been object of a previous work [1]. It was found that after the normal cycle, the hardness of the material is slightly higher and the microstructure less homogeneous but despite these differences, blanks show a better formability which results in a longer life of the dies.

This work aims to evaluate the grain orientation after continuos casting and its evolution during processing.

## 2. EXPERIMENTAL

The studied alloy has the following chemical composition (wt. %): Ag 83.5 %, Cu 16.44 %, P 0.06 %.

The alloy consists of Ag-Cu eutectic grains in an Ag-rich matrix [1].

For X-ray diffraction (XRD) measurements a diffractometer SIEMENS D5000 equipped with Euler cradle has been employed with Ni-filtered Mo-K$\alpha$ radiation ($\lambda = 0.71$ Å). XRD patterns of bulk specimens were collected by step scanning with $2\theta$ steps of 0.005° and counting time of 20 s per step. The texture has been evaluated from the (111), (200) and (220) pole figures of the Ag- and Cu-rich phases by the reflection method. The intensities measurements have been performed in the ranges $0 < \chi < 70°$ and $0 < \phi < 355°$ and with a step size of 5° and a counting time of 5 s for each step. Data have been then corrected for background and defocusing. The experimental data were elaborated by a series expansion method in order to have computed pole figures covering the whole $\chi$ range up to 90°.

## 3. RESULTS AND DISCUSSION

XRD patterns of the samples after each stage of the two cycles are shown in Fig. 2.

The ingots can be considered texture-free being the relative intensities of the reflexions of the Ag-rich and Cu-rich phases similar to those of a sample with random oriented grains.

The texture development and evolution after each stage of the two cycles may be explained as the results of two different processes: the cold-rolling deformation and the following recrystallization due to thermal treatments.

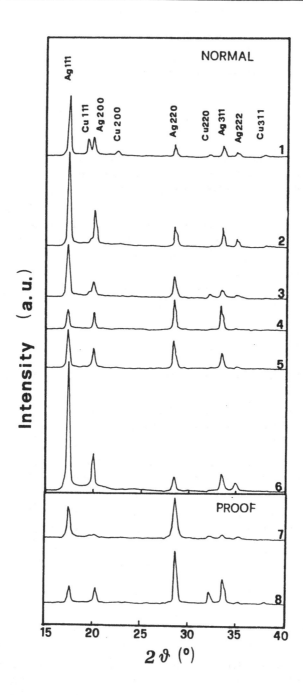

Figure 2. XRD patterns of the samples after each stage of the normal and the proof cycles.

## 3.1 Normal cycle

The initial homogeneization treatment (stage 2) does not induce an appreciable texture change in the material (Fig. 2). After the first cold-rolling (stage 3) some poles appear in the Ag {111} pole figure (Fig. 3.a). The texture change corresponds to a development of a mixed texture showing a {110} component as indicated by the markers in the figure. The {110} texture forms after cold rolling in low stacking fault energy F.C.C. metals as Ag [2-5]. The presence of Cu in solid solution and in the eutectic may be a factor responsible for the weakening of this type of texture. The texture component {110} is further reinforced by the following heat treatment (stage 4) (Fig. 3.b) during which recrystallization occurred [1]. The second cold rolling (stage 5) induces a texture which is quite similar to that visible after stage 3 (Fig. 3.c). After the final thermal treatment (stage 6), i.e. before coining, the material exhibits a {111} texture (Fig. 3.d).

## 3.2 Proof cycle

In this case the texture of the blanks exhibits after stage 8, i.e. before coining, strong {110} components (Fig. 4).

Recrystallization was found to occur after the intermediate annealing of the normal cycle and the final annealing of both cycles [1]; nevertheless, the resulting texture are much more different. After stage 4 and 8, a reinforcement of the {110} components, already present after cold rolling, has been observed while a {111} fibre texture develops after stage 6. Further investigations are underway to study in particular by TEM observations the reasons of this behaviour.

Figure 5 shows the pole figures of the Cu-rich phase after stage 8 which confirm the crystallographic relationships in the Ag-Cu eutectic found by Cantor & Chadwick [6] and Davidson & Smith [7]: {100}Ag // {100}Cu and <010>Ag // <010>Cu.

## 4. CONCLUSIONS

The results of the present investigation on the texture of the Ag 835 alloy can be summarized as follow:

1. The crystalline texture evolves after each stage of the two cycles.

2. The final grain orientation is different in the two cases. Before coining, the normal cycle gives rise to a {111} texture whereas a mixed texture with a strong {110} component is produced by the proof one.

3. The {111} texture of the blanks obtained by the normal cycle is preferable for the subsequent coining.

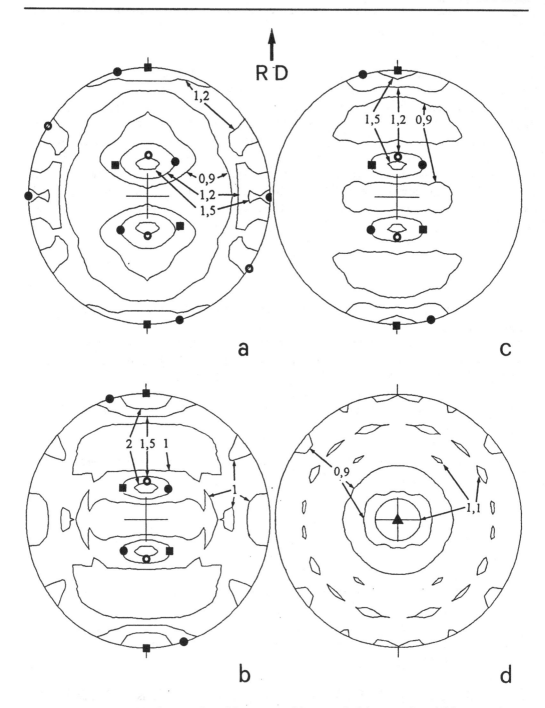

Figure 3. Ag {111} pole figure after: (a) stage 3, (b) stage 4, (c) stage 5 and (d) stage 6.
●(110)[1$\bar{1}$2], ○(110)[001], ■(110)[$\bar{1}$11], ▲{111} pole (the other poles lie on the circle
at 70° from the central pole).

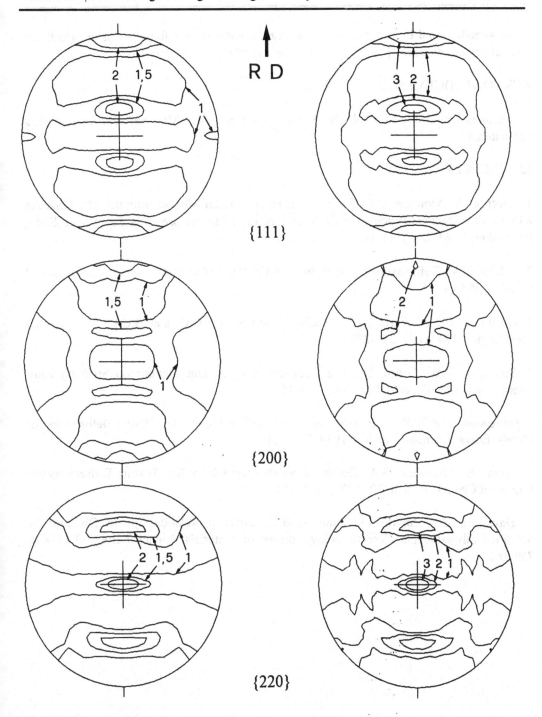

RD

{111}

{200}

{220}

Figure 4. Pole figures of the Ag-rich phase after stage 8.

Figure 5. Pole figures of the Cu-rich phase after stage 8.

4. As already found by other authors, the Ag-Cu eutectic was found to have orientation relationships close to: {100}Ag // {100}Cu and <010>Ag // <010>Cu.

ACKNOWLEDGEMENTS

The authors would like to thank Dr. A. Tulli (I.P.Z.S. Italian Mint Service) for providing the material.

REFERENCES

1. Grottesi, A., Montanari, R., Sergi, V., Tulli, A.: Studio Microstrutturale della Lega Ag 835 nei Vari Stadi del Processo di Fabbricazione delle Monete presso la Zecca dello Stato, Proceeding 3° AIMAT, in press

2. Calnan, E.A.: Deformation Texture on Face-Centred Cubic Metals, Acta Metallurgica, 2 (1954), 865-874

3. Dillamore, I.L., Roberts, W.T.: Rolling Textures in F.C.C. and B.C.C. Metals, Acta Metallurgica, 12 (1964), 281-293

4. Smallman, R.E., Green, D.: The Dependence of Rolling Texture on Stacking Fault Energy, Acta Metallurgica, 12 (1964), 145-154

5. Honeycombe, R.W.K.: Anisotropy in Polycrystalline Metals, The Plastic Deformation of Metals, Edward Arnold, London, 1984, 326-341

6. Cantor, B., Chadwick, G.A.: Eutectic Crystallography by X-Ray Texture Diffractometry, Journal of Crystal Growth, 30 (1975), 109-112

7. Davidson, C.J., Smith, I.O.: Interphase Orientation Relationships in Directionally Solidified Silver-Copper Eutectic Alloy, Journal of Materials Science Letters, 3 (1984), 759-762

# DETERMINATION OF THE HEAT TRANSFER COEFFICIENT IN THE DIE-BILLET ZONE FOR NONISOTHERMAL UPSET FORGING CONDITIONS

**P.F. Bariani, G.A. Berti, L. D'Angelo and R. Guggia**
University of Padua, Padua, Italy

KEY WORDS : forging, heat transfer, inverse analysis

ABSTRACT : A number of tests were conducted to measure the heat transfer coefficient between billet and tool in non-isothermal upset forging conditions. Elastic deformation of cylindrical specimens between flat punches was used to verify the proposed procedure, whose applicability to large plastic deformation was verified in plane strain tests. This paper mainly insists on theoretical aspects and on presentation of axisymmetrical tests.
A fixture consisting of two flat 304 stainless steel dies was instrumented with type-K thermocouples to elastically deform 1100 Al cylindrical specimens at forging temperature, also instrumented with thermocouples.
A procedure was designed to measure heat transfer coefficient. Inverse analysis based on comparison between numerical and experimental data was applied. A fitting technique, based on adjustment of input data, was aimed at (i) reaching a satisfactory agreement between experimental and analytical results and (ii) determining a mean value for the heat transfer coefficient. Different operating conditions were tested.

## 1. INTRODUCTION

Some recent trends in metal forming as, for instance, forging of complex geometries and/or with close tolerances highlight the importance of studying interface phenomena: friction, wear, heat transfer. Recently at DIMEG a research work has been started on such topics.
Heat transfer is a process which heavily influences tool behaviour at work. Tool hardness

Published in: E. Kuljanic (Ed.) *Advanced Manufacturing Systems and Technology*,
CISM Courses and Lectures No. 372, Springer Verlag, Wien New York, 1996.

rapidly decays when temperature is higher than the selected recovery temperature. For this reason it is desirable that heat transfer between workpiece and tool is low in forging processes. According to lubricant producers desiderata, even a qualitative standard test could be useful to compare behaviour of different lubricants as concerns thermal barrier effect. In spite of these considerations and of remarkable works already developed [1÷4], reliable data on heat transfer between tools and workpiece in forming processes are not available.

Direct measurement of the heat transfer coefficient is not possible, therefore non-direct ways have been developed, usually combining different techniques. When experimental tests are used, the starting point is measurement of temperature through thermocouples in selected points inside tools and workpiece. Location of measurement points inside the workpiece is difficult during forging tests because hot junctions shift from the original positions due to material deformation. As a consequence, a temperature Vs time diagram obtained from a test does not refer to a point but to a point path.

The present work relates to a first set of experiments developed to evaluate the heat transfer coefficient at tool-specimen interface at forging conditions. Some simplifications were used. (i) 1100 Aluminium was chosen as the specimen material to reduce test temperature and to use low cost and easy-to-work tool materials, in this case AISI 304 stainless steel. (ii) Specimens are elastically deformed, then point shifts inside the specimen are so small that the position of measuring points inside the specimen can be considered known throughout the test.

## 2. PROCEDURE

To evaluate the heat transfer coefficient between punch and specimen an inverse analysis technique is used, based on combination of experimental tests and numerical simulations. The test chosen to this aim is upsetting of a round cylindrical specimen between flat punches.

In this test the heat transfer is supposed to be one-dimensional, from the specimen to the punches [1,2,3], that is to say the temperature is uniform on each plane parallel to the punch faces (cross sections). With this assumption two phenomena are neglected: lateral cooling and the distribution of the heat transfer coefficient at interface. Lateral cooling determines a radial temperature gradient in the components (punches and specimen), its influence increasing in slow tests. Pressure distribution is not uniform in this kind of test. It has been shown [5] that higher pressures determines higher values for the heat transfer coefficient. As a consequence, also the temperature distribution close to the specimen end surfaces is not uniform.

Locations for hot junctions were chosen taking into account the following requirements :
• thermocouples should not be too close to each other to reduce mutual influence,
• thermocouples should be far from lateral surface and close to interface to reduce influence of lateral cooling.

In addition, all thermocouples were set at the same distance from the axis, to reduce the influence of pressure distribution at interface.

Four thermocouples are used, two inside the specimen and two in one punch.
The procedure is made up of 5 steps, listed below.
1. conduction of the experimental test, whose output are the four temperature Vs time diagrams given by thermocouples.
2. extraction of a "first attempt value" for the heat transfer coefficient, to be used for the numerical simulation. The "first attempt value" is based on the hypothesis of one-dimension heat flow and constant axial temperature gradient inside components.
3. development of the numerical simulation and derivation of temperature Vs time diagrams in "control points" corresponding to the thermocouple locations.
4. comparison between experimental and numerical diagrams to decide if the heat transfer coefficient should be increased or reduced in the numerical simulation.
5. development of a new simulation using the new value for the heat transfer coefficient.
The procedure ends when a good agreement between experimental and numerical results is reached.

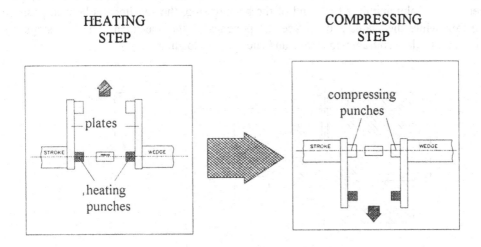

Fig. 1 - Main steps of the procedure

## 3. EXPERIMENTAL SET UP

A Gleeble 2000 high temperature testing system was used to conduct experimental tests. Characteristics of the system useful for this work are (i) possibility to conduct both mechanical and thermal tests, (ii) possibility to choose stroke and temperature control, (iii) an integrated acquisition and registration system for thermal and mechanical data, and (iv) its control software, suitable to command movement of non-standard devices.
Specimens are heated through Joule effect, the resulting temperature distribution being made up of isothermal cross sections inside the specimen. Two independently controlled hydraulic pistons (stroke and wedge) are used to deform specimens, the maximum load

being 20 tons. Up to six thermocouples may be connected to the acquisition system, four type-K (chromel-alumel) and two type-S (platinum-platinum rhodium). In the presented tests four type-K were used.

To assure good test reproducibility a special equipment was used, designed by the authors. A feeding ram is used for correct specimen positioning. Two sliding stainless steel plates are inserted between hydraulic pistons and punches. Their movement is possible through a pneumatic piston, software controlled. Plates are water cooled for high temperature tests. The resulting assembly allows use of two different couples of punches, one copper couple (electrodes) for heating specimens and one stainless steel for deforming them. This solution was chosen because heating by Joule effect causes increase of temperature in punches as well as in specimen. In that case punches would reach much higher temperatures and operating conditions during deformation would be farther from industrial forging conditions.

The testing procedure is shown in Fig. 1. The specimen is first carried by the feeding ram between the electrodes (heating step). Here it is heated till 430°C, an usual forging temperature for aluminium. At the end of the heating step, the specimen is kept in position by the ram while the plates slid to face the punches to the specimen. The last steps are compression to the programmed stroke and specimen unloading.

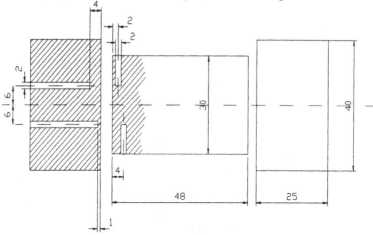

Fig. 2 - Test components geometry

Fig. 2 shows the test geometry. The specimen is a $\phi 30 \times 48$ mm aluminium cylinder in which two holes were drilled at different distances from the end surface. The two punches are $\phi 40 \times 25$ mm, of which one was drilled to insert thermocouples. All hole bottoms are 6 mm far from the axis, on different directions to reduce mutual influence. The distortion of the thermal field due to (i) heat conduction along the thermocouple wires and to (ii) holes for the thermocouples was calculated according to Attia and Kops theory [6]. It results that the main distortion is due to thermocouples farther from interface, but still acceptable.

For the selected thermocouples $\phi 0.25$ mm wires were used, having a time constant of 0.016 seconds. The time constant was evaluated moving the hot junction of a thermocouple from environment to boiling water. To connect thermocouples to components, the hot junction results from (i) spot soldering thermocouple wires together at one end, (ii) setting it in position and (iii) soldering it at the bottom of holes. Both ceramic and Teflon sheaths are used for thermocouples, ceramic inside holes because it resists high temperatures and flexibility is not required, Teflon is used outside holes. Connection between hot junction and hole bottom was strengthened by means of a special cement which also contributes to thermal insulation.

## 4.  NUMERICAL SIMULATION

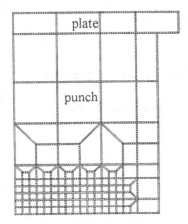

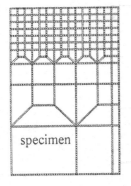

Fig.3 - Simulation mesh

A coupled thermal-mechanical simulation of the test was developed using ABAQUS. True stress-true strain curves for both punch and workpiece materials were developed using the Gleeble system. Punch material was modelled as simply elastic, the elastic-plastic model was considered the most suitable for the specimen. It was taken advantage of existing symmetries and one quarter geometry was analysed. The utilised mesh is shown in Fig. 3. The simulation is made up of two steps: (i) elastic compression of the specimen at the starting temperature of 420°C (to take into account specimen cooling during transfer from heating to the compressing punches), stroke and speed being the same as in the real test, and (ii) simple contact between specimen and punches for 30 seconds. Four control points were selected corresponding to the position of the four thermocouples. Some experimental tests were conducted to establish starting temperature distribution on the boundary of both punch and specimen. A plate at constant temperature is used to simulate punch water cooling. To select a value for the heat transfer coefficient for the first run, a specific procedure was used, as presented below:

- calculation of the heat flow in the specimen and in the punch:

$$q_s = \frac{\lambda_s}{l_s}(t_{s1} - t_{s2})$$

$$q_p = \frac{\lambda_p}{l_p}(t_{p1} - t_{p2})$$

where:    $q_s$ and $q_p$      are the specific heat flow inside, respectively, the specimen and the punch,

$\lambda_s$ and $\lambda_p$     are the thermal conductivity of specimen and punch materials,

$l_s$ and $l_p$     are the axial distance between thermocouples, respectively inside specimen and punch,

$t_{s1}, t_{s2}, t_{p1}, t_{p2}$     are the temperatures measured by thermocouples at the end of the test.

- extrapolation of surface temperatures at interface:

$$t_{ss} = t_{s2} - \frac{q_s}{\lambda_s} l_{s2}$$

$$t_{sp} = t_{p1} - \frac{q_p}{\lambda_p} l_{p1}$$

where:     $t_{ss}$ and $t_{sp}$     are the interface temperature of, respectively, the specimen and the punch,

$l_{s2}$ and $l_{p1}$     are the axial distance of thermocouples s2 and p1 from the interface.

- evaluation of the specific heat transferred at the interface:

$$q = (q_s + q_p)/2$$

- evaluation of the heat transfer coefficient:

$$K = \frac{q}{t_{ss} - t_{sp}}$$

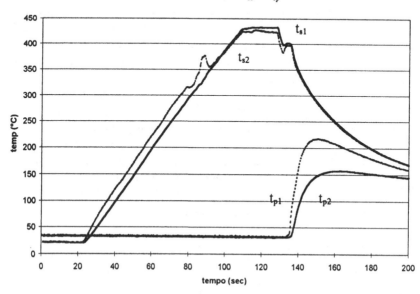

Fig. 4 - Temperature Vs. time diagrams from a test

temp.

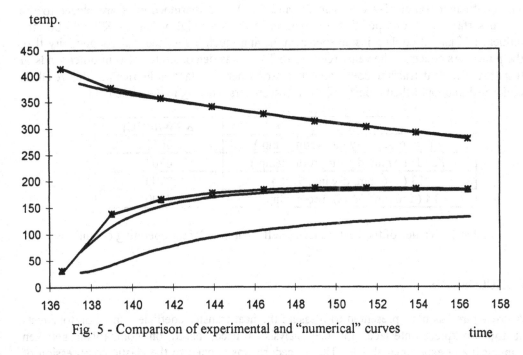

Fig. 5 - Comparison of experimental and "numerical" curves          time

## 5. RESULTS

The results of experimental tests are sets of four temperature Vs time diagrams relevant to thermocouple positions, an example is in Fig. 4. Each diagram is made up of four steps: (i) specimen heating, (ii) temperature homogenisation for 20 seconds, (iii) elastic compression, and (iv) keeping in position for 30 seconds.

Operating conditions of the tests were, as follows:

- hot junctions 6 mm far from the axis, dry interface, die initially at room temperature (T1);
- hot junctions 10 mm far from the axis, dry interface, die initially at room temperature (T2);
- hot junctions 6 mm far from the axis, dry interface, die starting from 75°C (T3);
- hot junctions 6 mm far from the axis, interface lubricated with $MoS_2$, die starting from room temperature (T4);

Each test was conducted three times; data dispersion is within 9%. FEM was used to simulate the compression step, so temperature-time diagrams of control points derived from simulations must be compared with one part of experimental diagrams. One comparison is shown in Fig. 5, percentage temperature difference being in all cases within 5%. Due to high thermal conductivity of aluminium, curves relevant to the specimen overlap. Details of results can be found in [7].

Table 1 summarises values of the heat transfer coefficient calculated for the different operating conditions.

The coefficient has similar values in T1 and T3. In T2 thermocouples are closer to the lateral surface, then one possible explanation of the very high value for K could be the influence of lateral cooling inside specimen, having much higher thermal conductivity than the punch. As concerns the value relevant to T4, it was demonstrated also in other kinds of test that the heat transfer coefficient increases when interface is lubricated; this result is confirmed in unpublished reports of other researchers working in the same field.

| Test | $K (W/m^2 \, ^\circ C)$ |
|---|---|
| T1 (d=6 mm, dry int., room temp.) | 3050 |
| T2 (d=10 mm, dry int., room temp.) | .6700 |
| T3 (d=6 mm, dry int., 75°C) | 2700 |
| T4 (d=6 mm, MoS$_2$, room temp.) | 4500 |

Table 1 - Values of the heat transfer coefficient in different operating conditions

## CONCLUSIONS

A procedure has been presented to measure the heat transfer coefficient in laboratory tests at tool-workpiece interface. Inverse analysis is used, based on comparison between numerical and experimental data. The procedure was applied to the elastic compression of cylindrical specimens between flat punches. Results have been presented relevant to application of the procedure to tests with different operating conditions.

## REFERENCES

1. Burte, Y.-T. Im, T. Altan, S.L. Semiatin, *Measurement and Analysis of Heat Transfer and Friction During Hot Forging*, Transactions of the ASME, Vol. 112, Nov. 1990
2. J.G. Lenard, M.E. Davis, *An experimental Study of Heat Transfer in Metal-Forming Process*, Annals of the CIRP, Vol. 41/1/1992
3. Burte, T. Altan, S.L. Semiatin, *An Investigation of the Heat Transfer and Friction in Hot Forging of 304 Stainless and Ti-6Al-4V*, Proceedings of Symposium of Advances in Hot Deformation Textures and Microstructures, 1993
4. Z. Malinowski, Lenard and M.E. Davis, *A study of the heat-transfer coefficient as a function of temperature and pressure*, J. Mat. Proc. Tech., Vol. 41, 1994, pp. 125-142
5. B.K. Chen, P.F. Thomson and S.K. Choi, *Temperature Distribution in the Roll-Gap during Hot Flat Rolling*, J. Mat. Proc. Tech., Vol. 30, 1992, pp.115-130
6. M.H. Attia, L. Kops, *Distortion in the Thermal Field Around Inserted Thermocouples in Experimental Interfacial Studies-Part 2*, J. Eng. For Ind., Vol. 110
7. D. Zardo, *Characterisation of interface between workmaterial and tool in hot forging ; heat transfer coefficient determination*, Graduation Thesis, DIMEG-University of Padua, 1995 (in Italian)

# DETERMINING THE COMPONENTS FORCES
# BY ROTARY DRAWING OF CONICAL PARTS

**D.B. Lazarevic, V. Stoiljkovic and M.R. Radovanovic**
University of Nis, Nis, Yugoslavia

KEY WORDS: Rotary Drawing, Conical parts, Maximal Components Forces

ABSTRACT: In this paper the procedure is presented of determining maximal components forces by the conical parts rotary drawing with respect to the "sine law" $\left(s_1 = s_0 \cdot sin\alpha_0\right)$ and with respect to a deviation from it. In addition, the positions of the maximal components forces during the rotary drawing process are determined.

## 1. INTRODUCTION

This paper gives an analysis of the components forces which appear in the drawing of conical parts. In view of the complexity of the deforming process itself, the existing literature gives numerous solutions for the forces, that are the solutions based upon various approximations [1,2].

In this paper the procedure is presented of determining maximal component forces by the conical parts rotary drawing with respect to the "sine law" $\left(s_1 = s_0 \cdot sin\alpha_0\right)$ and with respect to a deviation from it.

The component forces are determined on the basis of the stresses found out in close neighborhood of stressing and of areas involved in the transmission of the respective

Published in: E. Kuljanic (Ed.) *Advanced Manufacturing Systems and Technology*, CISM Courses and Lectures No. 372, Springer Verlag, Wien New York, 1996.

component forces [3,4]. In addition, the positions of the maximal component forces during the rotary drawing process are determined.

## 2. FORCES AT $\left(s_1 \neq s_0 \cdot sin\alpha_0\right)$

A detailed analysis of the stress / deformation state in the rotary drawing is given in the Ref. [3]. Fig.1. shows the meridian stresses during the deforming process which has been used as the basis for determining the component forces. The most immediate deforming zone for the case $s_0 > s_1 \neq s_0 \cdot sin\alpha_0$ consists of two zones, namely the first (I) zone where the reduction is done with respect to the diameter, and the second (II) zone in which the reduction is done with respect to thickness.

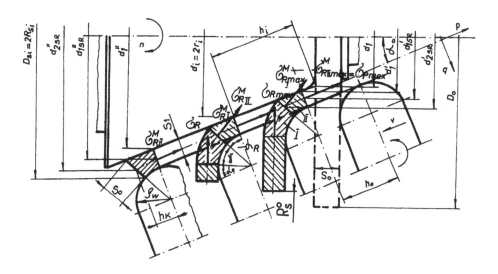

Fig. 1. Meridian stresses during the deforming process

The component forces $(F_P, F_Q, F_T)$ in the direction of the axes $p$, $q$ and $t$ are shown in Fig. 2. for the case $s_0 > s_1 \neq s_0 \cdot sin\alpha_0$ Fig. 2. also shows areas involved in the transmission of the respective component forces $\left(A_P, A_Q, A_T\right)$.

The maximal component forces in the direction of the axis $p$ $\left(F_P\right)$ appears at the moment of the total grasp of the shaping arbor radius $(R)$ which is experimentally proved [3]. The pressure roll path along the cone generating line starting from the moment of touching a workpiece till an achievement of the maximal force in the direction of the axis $p$ is equal

to:     $$h_0 = \frac{R}{cos\alpha_0} - \left(\rho_w + s_0 + R\right) \cdot tg\alpha_0 + \rho_w \cdot sin\alpha_0 \cdot tg\alpha_0 + \frac{s_0}{cos\alpha_0} + \rho_w \cdot cos\alpha_0 \qquad (1)$$

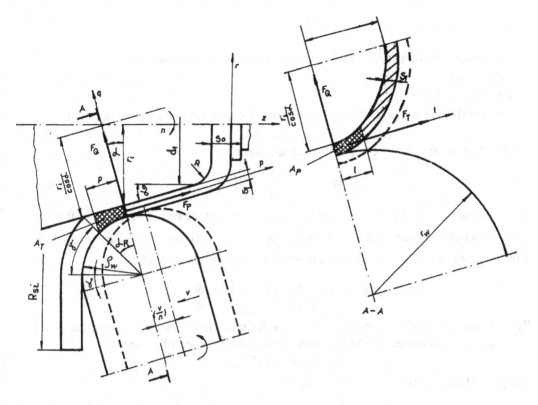

Fig.2.Components forces

The maximal stress in the direction of the axis $p$ at the exit from the zone II is given by the equation (reduction with respect to a workpiece diameter and thickness) [3,4]:

$$\sigma^{M}_{Rl\,lmax} = 1,15 \cdot K_{llsr} \left\{ \left[ 1 + \frac{\mu}{sin\alpha} \left( 1 - \frac{\sigma^{M}_{Rlmax}}{1,15 K_{llsr}} \right) - \frac{\mu}{sin\alpha} \cdot ln\frac{s_o}{s_1} \right] \cdot ln\frac{s_o}{s_1} + \frac{\sigma^{M}_{Rlmax}}{1,15 K_{llsr}} + \frac{sin\alpha}{2} \right\} \quad (2)$$

where:

$\sigma^{M}_{Rl\,max}$ - maximal meridian stress at the exit from the zone I (Fig. 1,2)

$$\sigma^{M}_{Rl\,max} = \left( 1,1 K_{lsr} \cdot ln\frac{R^{o}_{s}}{r_1} + K_{lsr} \frac{s_o}{2\rho_w + s_o} \right) \cdot (1 + \mu\gamma) \quad (3)$$

$K_{lsr}, K_{llsr}$ - specific deformation resistance,

$R^{o}_{s}$ - radius of the cone collar at the maximal force moment,

$d'_1 = 2r'_1 = d_1 + 2R \cdot tg(90^0 - \alpha_0) \cdot sin\alpha_0$ -diameter of the cone at the maximal force moment,

$s_o$ - starting thickness of a workpiece,

$$\gamma_0 = 90^0 - \alpha_R - \alpha_0, \qquad \alpha_R = arccos\frac{\rho_w + s_1}{\rho_W + s_0}$$

$\mu$ - connecting friction coefficient between material and the pressure roll,

$d_w = 2r_w$ - pressure roll diameter,

$\rho_w$ - pressure roll radius,

The maximal-force component in the direction of the axis p is given by the equation:

$$F_{P\,max} = \sigma_{P\,max} \cdot A_P \tag{4}$$

Where $A_P$ is an area involved in the force transmission.

$$A_P = \sqrt{\frac{d_1' \cdot d_w}{d_1' + d_w} \cdot \left(\frac{v}{n}\right)} \cdot s_1 \tag{5}$$

The maximal force in the direction of the axis $q$ appears at the end of the process, that is at the moment of disappearance of the zone I $\left[h_k = (\rho_w + s_0/2) \cdot sin\alpha_0\right]$[3]. The stress in the direction of the axis $p$ at the observed moment is given by the equation:

$$\sigma_P = 1{,}15K_{llsr}\left\{\left[1 + \frac{\mu}{sin\alpha}\left(1 + ln\frac{s_0}{s_1}\right)\right] \cdot ln\frac{s_0}{s_1} + \frac{sin\alpha}{2}\right\} \tag{6}$$

By using an approximate plasticity condition at the moment of disappearance of the zone I, the expression is obtained for the maximal stress in the direction of the axis q:

$$\sigma_{Q\,max} = \sigma_P + 1{,}15K_{sr} \tag{7}$$

The pressed surface in the direction of the axis $q$:

$$A_Q = \sqrt{\frac{d_1'' \cdot d_x}{d_1'' + d_w} \cdot \left(\frac{v}{n}\right)} \cdot tg\frac{\alpha_R}{2} \cdot \rho_w \cdot sin\alpha_R \tag{8}$$

where: $d_1''$ - diameter of the cone at the end of the process, $\left(\frac{v}{n}\right)$ - pressure roll path along the cone generating line, then the maximal force component in the direction $q$ is:

$$F_{Q\,max} = \sigma_{Q\,max} \cdot A_Q \tag{9}$$

The tangential force component:

$$F_T = \sigma_T \cdot A_T \tag{10}$$

where for the case of the plane deformation state:

$$\sigma_T = \frac{\sigma_T + \sigma_Q}{2} \tag{11}$$

$$A_T = \frac{1}{2}(s_0 + s_1) \cdot \rho_w \cdot sin\alpha_R \tag{12}$$

For the case $s_1 = s_0$ the maximal force component in the direction of the axis $p(F_P)$ appears at a distance $h_0$ (Equ.1) from the moment when a workpiece is touched:

$$F_{P\,max} = \sigma_{P\,max} \cdot A_P \tag{13}$$

where the maximal stress in the direction of the axis $p$ (reduction with respect to diameter only) is

$$\sigma_{Pmax} = \left(1.1 K_{lsr} \cdot ln \frac{R_s^o}{r_1} + K_{lsr} \frac{S_o}{2\rho_w + S_o}\right) \cdot (1 + \mu\gamma) \tag{14}$$

whereas an area involved in the force transmission (Fig. 2):

$$A_P = \sqrt{\frac{d_w \cdot d_1'}{d_w + d_1'} \cdot \left(\frac{v}{n}\right)} \cdot S_0 \tag{15}$$

The maximal force in the direction of the axis $q$ appears immediately before the end of the process [3]. At that moment the stress is $\sigma_P \approx 0$. From an approximate plasticity condition the maximal stress is obtained in the direction q (Fig. 2):

$$\sigma_{Qmax} = \sigma_P + 1.15 K_{sr} = 1.15 K_{sr} \tag{16}$$

If the pressed surface in the direction of the axis $q$ taken to be:

$$A_Q = \sqrt{\frac{d_1'' \cdot d_w}{d_1' + d_w} \cdot \left(\frac{v}{n}\right)} \cdot \sqrt{2\rho_w \cdot \left(\frac{v}{n}\right)} \tag{17}$$

The maximal component force in the direction of the axis $q$:

$$F_{Qmax} = \sigma_{Qmax} \cdot A_Q \tag{18}$$

If we take into consideration that in deforming a plane deformation state appears then the expression for the tangential stress is:

$$\sigma_T = \frac{\sigma_T + \sigma_Q}{2} \tag{19}$$

Whereas the tangential stress component is:

$$F_T = \sigma_T \cdot A_T \tag{20}$$

where:

$$A_T = \sqrt{2 \cdot \rho_w \cdot \left(\frac{v}{n}\right)} \cdot S_0 \tag{21}$$

## 3. FORCES WITH RESPECT TO THE CONDITION $s_1 = s_0 \cdot sin\alpha_0$ ("sine law")

During the rotary drawing of conical parts with respect to the "sine law" that is the condition is that the wall thickness of a cone part is $s_1 = s_0 \cdot sin\alpha_0$, then the collar diameter retains its constant value throughout the deforming process and it is equal to the initial workpiece diameter $(D_0 = D_{si} = const)$ Accordingly, the reduction is not done with respect to diameter; it is only done with respect to the workpiece thickness so that the stress in the direction of the axis $p$ is given by the equation [3,4]:

$$\sigma_P = 1.15 K_{lsr} \left\{\left[1 + \frac{\mu}{sin\alpha}\left(1 + ln\frac{S_0}{S_1}\right)\right] \cdot ln\frac{S_0}{S_1} + \frac{sin\alpha}{2}\right\} \tag{22}$$

According to the equation (22) the stress during the process has a constant value. The experimental research performed [Ref. 3] shows that the component forces gradually increase during the deforming process. The increases of the component forces appear due to an increase of the connecting surface during the process $\left(d_1' \div d_1''\right)$ The greatest contacting surface is immediately before the end of the process and consequently, the components forces are the greatest.

The maximal components force in the direction of the axis $p$ :

$$F_{P\,max} = \sigma_P \cdot A_P \tag{23}$$

Area involved the force transmission in the direction of the axis $p$ for the very end of the process $\left[ h_k = \left(\rho_w + {}^{s_0}\!/_2\right) \cdot \sin\alpha_0 \right]$ :

$$A_P = \sqrt{\frac{d_1'' \cdot d_w}{d_1'' + d_w} \cdot \left(\frac{v}{n}\right) \cdot \frac{1}{2}(s_1 + s_0)} \tag{24}$$

The components forces in the direction of the axes $q\left(F_Q\right)$ and $t\left(F_T\right)$ can be determined with respect to the same given equations (7,8,9,10,11,12) along with taking consideration about the stress constancy as well as about a small force increase due to an increase of the contacting surface between the pressure roll and the working cone.

The above-given expression can be used for the rotary drawing with pressure rolls having a radius $\left(\rho_w\right)$ or a cone on its top (angle of clearance $\alpha$). The connection between the pressure roll radius $\left(\rho_w\right)$, that is the angle $\alpha_R$ and the angle $\alpha$ of clearance is given by the equation:

$$\alpha = \frac{\alpha_R}{2} = \frac{1}{2} arccos \frac{\rho_w + s_1}{\rho_w + s_0} \tag{25}$$

## 4. EXPERIMENTAL RESEARCH (EXPERIMENT)

In order to verify the correctness of the obtained expressions, respective experimental examinations have been carried out upon rotary drawing machine HYCOFORM of the firm BOKÖ. In order to register the components forces a special three-components dynamometer has been used on the basis of measuring tapes. The recording of the pressure roll stroke (path) has been performed by means of an inductive path recorder of the W100 type; for amplification of the measuring signal a six-channelled amplifier KWS/6A has been used. The signals emitted are transmitted from an amplifier to a computer. Recorded values are obtained with the aid of computer by using corresponding programmes; they are sorted out and drawn on the plotter [3].

Experimental research has been done upon the following materials: C0148, CuZn37, CuZn63, Al99,5; ZnSn30; copper, duralumin. The nominal thicknesses of the used sheet metal are 0,8; 1; 1,5; 2; 2,5 /mm/. The main arbor rotations number n = 500o/min. Whereas the pressure roll paths are 0,204; 0,086; 0,364 mm/o. The pressure roll had an external diameter of $d_w = 250 / mm /$ and the radius around the top is $\rho_w = 12 / mm /$.

During the experiment the diagrams are recorded of the application of the component forces $F_P, F_Q$ and $F_T$ as well as the path $(h)$ of the pressure roll are time-dependent. The diagrams are recorded for the conical parts whose thickness deviates from the "sine law" and especially for various materials and process parameters [ 3].

In order to view more clearly the flow of changes of the components forces the diagrams are given of the pressure roll path. Fig. 3. gives diagrams components forces for a conical part when $s_1 = s_0 \cdot sin\alpha_0$ whereas in Fig.4. diagrams are given of the forces for the rotary drawing with respect to the "sine law" $(s_1 = s_0 \cdot sin\alpha_0)$.

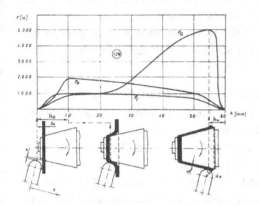
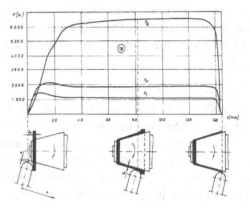

Fig.3.The component forces diagrams when $s_0 > s_1 \neq s_0 \cdot sin\alpha_0$

Fig.4.The component forces diagrams in the deforming process with respect to the "sine law" $(s_1 = s_0 \cdot sin\alpha_0)$

## 5. CONCLUSION

On the basis of the results obtained by theoretical elaboration as well as by experimental research the following conclusions can be drawn:

- values of the maximal components forces ($F_P, F_Q$ and $F_T$) determined by theoretical analysis and experimentally measured upon the recorded diagrams agree very well,

- positions of the component forces' maximum arrived at by theoretical analysis agree with the experimental values since the theoretical assumptions are achieved on the basis of the component forces' analysis during the experiment,

- a flow of the components forces changes upon the recorded diagrams agrees with the theoretical assumptions (in view of the fact that sudden changes for particular processes could not be involved),

- maximal component forces in the direction of the axis $p$ for the case of a deviation from the "sine law" appears at distance:

$$h_0 = \frac{R}{cos\alpha_0} - (\rho_w + s_0 + R) \cdot tg\alpha_0 + \rho_w \cdot sin\alpha_0 \cdot tg\alpha_0 + \frac{s_0}{cos\alpha_0} + \rho_w \cdot cos\alpha_0 \text{ from the moment}$$

when a workpiece touches the pressure roll,

- maximal value of the component force in the direction of the axis $q$ appears immediately before the end of the process for $h_k = \left(\rho_w + \frac{s_0}{2}\right) \cdot sin\alpha_0$,

- tangential force component during the deforming process has an approximately constant value,

- in the rotary drawing of the conical parts with respect to the "sine law" and in the very beginning of the process there is a sudden increase of the component forces which retain an approximately constant value after the process is firmly established,

- a negligible increase of the component forces in the rotary drawing with respect to the "sine law" appears due an increase of the contacting surface during the deforming process $\left(d_1' \div d_1''\right)$,

- if the rotary drawing process at $s_1 \neq s_0 \cdot sin\alpha_0$ approximates the process at $s_1 = s_0 \cdot sin\alpha_0$ (from upper $s_0 > s_0 \cdot sin\alpha_0$ or from lower $s_1 < s_0 \cdot sin\alpha_0$) then it should be stressed that the maximums of the component forces are less and less expressed,

- in the rotary process with respect to the "sine law" the elements' collar is all the time perpendicular to the rotation axis in the form of a shaping arbor.

On the basis of the above-stated considerations and conclusions the suggested expressions can be used in solving problems of rotary drawing of conical parts with respect to the "sine law" and with respect to a deviation from it $\left(s_1 \neq s_0 \cdot sin\alpha_0\right)$.

REFERENCES

1.Kalpakcioglu,S.:An Experimental Study of Plastic Deformation in Power Spinning, CIRP Annalen, 10, $N^0$ 1,1962,58-64

2.Kobayashi,S. and K.Hall.A Theory of She ar Spinning of Cones, Trans.,ASME.(1967)

3..Lazarevic,D.: Master's Work, Mechanical Engineering Faculty,Nis,1983

4.Velev,S.A.;Kombiniravanaja glubokaja vitjazka listavih materialov, "Masinostroenie", Moskva,1973

# DETERMINATION OF THE OPTIMAL PARAMETERS OF CASTIN A COPPER WIRE BY THE APPLICATION OF NEURAL NETWORKS

V. Stoiljkovic, M. Arsenovic, Lj. Stoiljkovic and N. Stojanovic
University of Nis, Nis, Yugoslavia

KEY WORDS: Neural Networks, Casting, Plastic Properties, Elongation

ABSTRACT: Considering all shortcomings of the existing procedures for continual copper wire casting a new process has been developed based on the "Upcast" casting system principles as well as on those for casting by continual hardening immediately from a molten material. The aim is to produce a copper wire of 8 mm in diameter that can be directly subjected to cold treatment without any previous hot treatment processing. In order to create the conditions for such manufacture a great number of experiments have been performed with casting parameters and the effect they have upon the obtained wire's quality. This paper presents an analysis of the data acquired by using neural networks that have provided for a relatively easy determination of input parameters for manufacturing copper wire with desired characteristics.

## 1. INTRODUCTION

The Copper Institute of Bor has been trying for several years to develop technology for continual wire casting as well as that for profiles of small cross-sections made of pure metals and their alloys by crystallization above the molten metal. The first aim of developing such continual casting technology is to meet the demands of the lacquer wire factory of Bor, that is to manufacture a cast copper wire of 8 mm in diameter that can be directly subjected to cold plastic treatment without any previous hot treatment operations. In addition to the fact that it does not need to be treated in hot state before drawing, the wire has to have good plastic properties so that it can undergo a high degree of reduction.

Published in: E. Kuljanic (Ed.) *Advanced Manufacturing Systems and Technology*, CISM Courses and Lectures No. 372, Springer Verlag, Wien New York, 1996.

In this way it can be used for fine drawing for diameters below 0.1 mm. In order to develop such technology a great deal of experimental research is needed for the sake of defining the effect of great many factors upon casting velocity and copper wire quality. Part of this research is presented in this paper.

What has been tested is the effect of many casting parameters upon the process stability and the cast copper wire's quality for the sake of separating the parameters that most affect the casting velocity and wire plastic properties. The determination of these factors would present the basis for new research aiming at increasing casting velocity and cast wire quality. The results of the research presented in this paper can also serve as the basis for choosing optimal parameters for getting the best quality wires as well as for manufacturing a new structure of cooler. Besides, a new solution can be obtained for a wire drawing device in order to obtain much greater capacity and quality.

## 2. CASTING PROCEDURE

The continual copper wire drawing procedure developed at Bor Copper Institute is one of the continual casting procedures by crystallization above the molten metal (1).

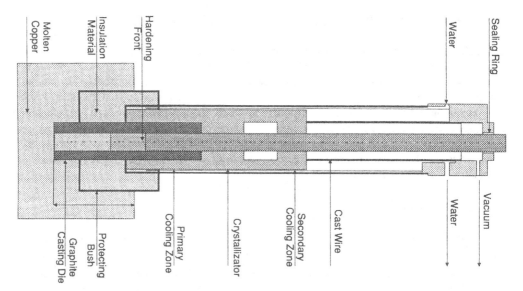

Fig. 1 Diagram of Continual Copper Casting

The principle of this casting procedure is schematically presented in Fig.1. The cooler used for casting copper wires is immersed in molten copper until the depth $h$. The protecting bush made of water-resistant material that does not react with molten copper protects the cooler from the effect of both molten copper and from high temperatures. The same function is performed by a layer of thermal insulation material. The hydrostatic pressure of the surrounding molten copper presses the molten metal into the graphite die. The molten metal hardens in the die since heat is carried away *via* the crystallizator's primary part cooled by water. The hardened wire leaves the graphite die at high temperature. Within the cooler, and in order to prevent the oxidation of the cast wire surface due to high temperature vacuum is used. In addition to this function, vacuum is used for providing a necessary pressure differential within the cooler thus enabling the molten metal to enter the graphite die. In order to prevent the cast wire oxidation after its leaving the cooler, the temperature on its surface must be below 60C (2). This is provided by cooling the cast wire in the crystallizator's secondary part. The cast wire drawing is done according to the motion-pause pattern.

The process stability is provided by adjusting the wire drawing velocity to the process leading away heat from its side surface.

## 2. DEFINING CASTING PROCESS AND WIRE QUALITY FACTORS

The process of the continual copper wire casting by crystallization over the molten metal is based on agreement between thermodynamic parameters and those of cast wire motion (3). Both hardening character and cast wire quality are directly influenced by the degree of the acquired agreement. The determination of the most optimal continual casting regime by calculation is very difficult due to the fact that a series of unknown values appear, namely those that cannot be contained within one formula. Their definition requires experimental research.

The factors affecting the process stability and the cast wire quality are divided into five following groups:

1. Parameters for Crystallizator Cooling Water,
2. Parameters of Cast Wire Motion,
3. Structural Parameters,
4. Vacuum in Casting Cooler, and,
5. Parameters of Molten Metal.

For experimental needs six measuring places have been determined at which the process input parameters are recorded (11) as well as the cast wire quality parameters (4). Samples have been taken for every change of any casting parameter as shown in Table 1.

Table 1

| CASTING PARAMETERS | Number of Samples (Total 250) | | | | |
|---|---|---|---|---|---|
| | 1 | 2 | 3 | 4 | 5 |
| P1 Water Flow for Cooling the Crystallizator in dm³/h | 80 | 100 | 160 | 200 | 300 |
| P2 Water Velocity for Cooling Crystallizator in dm³/h | 0.617 | 0.772 | 1.233 | 1.544 | 2.13 |
| P3 Casting Velocity in m/min | 0.57 | 0,57 | 0.68 | 0.68 | 0.80 |
| P4 Motion Parameter 1 | 5 | 5 | 2 | 1 | 5 |
| P5 Motion Parameter 2 | 1 | 1 | 1 | 1 | 1 |
| P6 Crystallizator Parameter A | 8 | 8 | 8 | 8 | 8 |
| P7 Crystallizator Parameter B | 1 | 1 | 1 | 1 | 1 |
| P8 Crystallizator Parameter C | 0 | 0 | 0 | 0 | 0 |
| P9 Vacuum Parameter | 290 | 290 | 310 | 330 | 330 |
| P10 Casting Temperature in °C | 1160 | 1160 | 1160 | 1160 | 1160 |
| P11 Drawing Rolls Diameter in mm | 75 | 45 | 40 | 75 | 75 |

| CHARACTERISTICS OF THE OBTAINED COPPER WIRE | | | | | |
|---|---|---|---|---|---|
| I1 Tensile Strength in N/mm$^2$ | 158.9 | 167.1 | 174.1 | 177.5 | 183 |
| I2 Relative Elongation A5 u % | 38.82 | 44.47 | 47.42 | 50.1 | 41.32 |
| I3 Results of Bending Test | 22 | 20 | 41 | 65 | 46 |
| I4 Results of Alternate Bending | 3 | 3 | 2 | 2 | 4 |

The casting parameters that cannot be shown in their real form for the sake of protecting the technology are denoted as Motion Parameters 1 and 2, Crystallizator Parameter A, B and C and Vacuum Parameters (such as the time for drawing wire in one cycle, the time for wire' s rest in one cycle, the value of cross-section for crystallization water-cooling, structural parameter that is supposed to increase the crystallization front lifting, the structural parameter that increases crystallizator's length as well as vacuum intensity).The feed value that cannot be shown in its real value is expressed through the drawing rolls diameter. The shown values for this group of parameters are given in this way so as not to change the function flow of the obtained dependencies between the cast wire plastic properties and these values. The overall number of the performed experiments was 250; some of them are given in Table 2.

## 3. NEURAL NETWORKS

The basic aim of the obtained experimental results is to model the observed wire casting system. Since the system is described by means of many concrete input-output values, neural networks impose themselves as an ideal modelling method primarily because of their abilities for machine-learning.

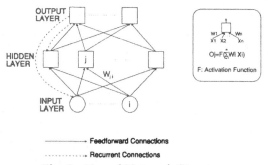

Fig. 2   Neural Network Structure

Neural networks are artificial intelligence systems and they represent a simplified model of human nervous system (4). They consist of a large number of process elements (neurons) that are similar to biological neurons and that can be hierarchically ordered at layers (input, hidden, output) mutually linked by the connections of given weights (Fig.2). The signals are distributed throughout the network in the input-output direction and each neuron performs a particular processing over the signals with respect to its activation function. The network knowledge is encoded in the connections' weights as well as in the very network structure.

The network acquires knowledge (learns) by adjusting the connections' weights in the network training process that is taking place according to the determined learning algorithm. In the learning process so-called training samples are used, that is pre-determined pairs of input and output values as the basis for weights' adjustment.

The neural network structure involves the following tasks:
- determination of the network structure: network topology (number of layers, number of neurons and their connection) and activation function
- choice of learning algorithm
- choice of training samples

## 3.1. NEURAL NETWORK STRUCTURE

The basic task of the neural structure is to build in the way the tested characteristics of the cast wire depend on the variable casting parameters. On the basis of the Kolmogorov convergence theorem a two-layer neural network has been applied (in addition to input layer) - that is, multilayer perceptron. The network has 11 neurons in the input layer (11 input parameters P1 - P11), 4 neurons in the output layer (4 output parameters I1 - 14). The number of neurons in the hidden layer is experimentally determined. The neuron' activation function is sigmoidal (...). The network is trained by the same set of samples for varying number of neurons in the hidden layer and the best results (the least error) are achieved in the case of 12 neurons in the hidden layer. With a lesser number of neurons in the hidden layer the network is not able to converge. With a greater number of neurons it only memorized the samples (input-output pairs) and it loses its generalization ability.

## 3.2. NEURAL NETWORK TRAINING ALGORITHM

Since neural networks acquire their knowledge about the problem by learning, the task of the training process itself is to enable the networks to detect dependencies among the processed data while simultaneously bridging the gap between particular examples and general relations and dependencies. The trained network models the mapping of a set of input data vectors (process parameters) into a set of output vectors (copper wire's characteristics) and thus it represents the model of the copper wire generation.

For the network training a backpropagation algorithm has been used (4) based on the error backwards propagation as the most commonly used network training algorithm for mapping networks training.

As for the algorithm parameters (4) that cannot be discussed here due to the lack of space, it is important to note that the initial network weights values are initialized at the value 1/(2.. number of neurons in the previous layer) while the learning rate is set at 0.3.

## 3.3. CHOICE OF SAMPLE TRAINING

In order to obtain necessary training samples numerous experiments have been carried out in real conditions with the following parameters:

a) Casting Parameters (system input values and at the same time neural network input values) shown in Table 1 in the first 11 columns

b) Output characteristics of the manufactured wire (system output values and at the same time neural network output values) shown in the last 4 columns in Table 1.

The overall number of the performed measurements is 250, namely 250 various parameters' combinations have been observed as the ones affecting the copper wire quality. The test results are only points in multidimensional space (11 input variables) though it is necessary to define the whole space. However, the values obtained as the test results are not able to lead us to any conclusion about the way particular parameters as well as their mutual effects influence the copper wire quality. Not even the experts with great experience in this field are able to determine what manufacturing conditions are needed for obtaining copper wire of particular quality. In practice the trial method is applied as a non-efficient and non-economical procedure since it wastes both material and precious time. Therefore, the obtained data processing by regression analysis as well as the acquisition of analytical expressions are not sufficient to make any reliable conclusion about the effect of the discussed factors. It is for this reason that the knowledge acquisition from the known data is found in the application of artificial intelligence, that is in the use of neural networks.

The average number of iterations needed for the neural network to learn the given mapping is 100000. It is not pre-determined; instead, the network is trained until the overall error is obtained as less than 10.

In other words, the trained neural network represents a model of the cooper wire acquisition described by empirical connections between the system input and output. The neural network behaves as an adaptive system since it learns by self-adjustment; when the learning process is over, it identifies and simulates the system for acquiring copper wire. This procedure is known as forward modelling - mapping of the direct system dynamics.

The result of the forward modelling is a system model that identifies the observed non-linear system in an extremely accurate and effective way. This means that this model is extremely suitable for various experiments with the system behavior; for instance, this includes the experiments aimed at discovering what particular wire characteristics will be obtained in the case of arbitrarily chosen casting parameters. Likewise, this prediction can be output-input oriented, that is it can be oriented towards the determination of the system input parameters' values that would give optimal output values - quality. This problem can also be formulated in another way: if the output system values are given, then the values to be led to the system outputs should be defined. This problem belongs to the so-called inverse modelling.

## 4. INVERSE MODELLING

The inverse system modelling plays an important part in a wide class of control structure. There are many procedures for inverse modelling and they are discussed briefly (5).

1. The simplest and, at the same time, the most tiresome procedure is manual adjustment of input values as well as output values control. Thus fine input adjustment is performed in order to obtain the desired output. This procedure is time-consuming especially if the input is multidimensional.

2. A better procedure is the so-called direct inverse modelling based on the identical training procedure as in the forward modelling, though the roles of the system inputs and outputs are changed. Namely, training samples are thus synthesized that the outputs of the original training samples are brought to the network input. The system outputs are measured, whereas the system inputs await at the output. Clearly, such structure tends to an effective mapping of the system inverse model. This procedure, however, has some shortcomings, primarily in the case when the system-defining mapping is not one-one since for the two identical combinations of the input signals different output will be obtained.

3. The third procedure of the inverse modelling that can overcome all the above-listed shortcomings is known as a specialized inverse learning.

The idea is quite clear from the mathematical standpoint. Let X and Y be inverse and direct system models, while $a$, $b$ and $c$ are system inputs or outputs (models) and let:

Since the inverse and direct models are mutually symmetrical, that is since the number and character of the inverse network inputs correspond to the number and character of the direct network outputs and *vice versa*, the number and character of the inverse network outputs correspond to the number and characters of the direct network inputs (number of hidden neurons cannot be identical since they depend on training samples), it is possible to couple them in such a way that the one set of outputs is brought to the other's inputs, that is, it is possible to obtain the following combination:

If these are really complementary models (direct and inverse), then it is going to stand that:

Namely, what is brought to the inverse model input appears at the direct model output.

Thus this network combination has to provide for identical mapping and in this sense it should be trained under supervision. This does not present any special difficulty.

It is important to note that, since the direct network has already been trained, only weights belonging to the inverse modelling network should be adjusted while the error from the input, namely, the one used for adjusting weights, is normally distributed through the direct model as well.


## 5. OPTIMIZATION OF CAST COPPER WIRE PLASTIC PROPERTIES

Since the forward modelling for the wire casting system is described in the section 3, the previously presented solutions can be used for obtaining the inverse system model that would provide for generating quality in development. They can also be used to predict the above-listed process parameters in order to acquire the desired copper wire characteristics.

The determination of the desired input parameters' values is experience-based during the development of technology and characteristics of a high-quality copper wire:

Relative Elongation $A_5 \geq 40\%$

Results of Bending Test $\geq 70$

Results of Alternate Bending $\geq 10$.

The procedure 3. from the previous section has been applied for inverse modelling; the procedure 1 has been used for testing. The obtained results matched each other till satisfactory accuracy is obtained.

The used neural network(s) structure is shown in Fig. 3.

Namely, the inverse neural network outputs are brought to the trained direct neural networks. The direct network outputs are compared with the inverse networks inputs and by the above-described backpropagation algorithm the training is performed until the unit mapping is completed.

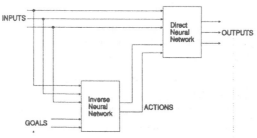

Fig. 3 Network Structure for Inverse Modelling

Already-prepared samples () are used as training samples; the training has been performed only with output values from the original training samples.

The inverse network structure that has satisfied the convergence conditions for the shown training samples has contained 8 neurons (number of input and output neurons determined by inputs and outputs from the direct network).

Thus obtained structure has enabled the determination of the input values (casting parameters) in order to acquire optimal characteristics of copper wiring. Some results are shown in Table 2 :

| CASTING PARAMETERS | | | | | | | | | | | CHARACTERISTICS | | | |
|---|---|---|---|---|---|---|---|---|---|---|---|---|---|---|
| P1 | P2 | P3 | P4 | P5 | P6 | P7 | P8 | P9 | P10 | P11 | I1 | I2 | I3 | I4 |
| 401 | 4.939 | 1.14 | 1.20 | 1.17 | 5 | 1 | 0 | 350 | 1160 | 50.3 | 180.0 | 42 | 75 | 10 |
| 450 | 3.968 | 1.18 | 1.12 | 1.20 | 7 | 1 | 0 | 350 | 1160 | 50.1 | 187.0 | 47 | 75 | 11 |
| 400 | 4.939 | 0.90 | 1.05 | 1.02 | 5 | 1 | 0 | 350 | 1160 | 40.0 | 184.0 | 45 | 70 | 10 |

## 6. CONCLUSION

The formulated new knowledge contained in neural networks enables a further development of new copper wires with pre-determined characteristics. Thus important results have been undoubtedly obtained for applying this technology without which there is no high-quality drawn wire. They have confirmed, in the best possible way, both importance and possibilities that exist in the generation of the product quality in development.

REFERENCES

[1] Murty, V.Y., Mollard, R.F.: Continuous casting of small cross sections, AIME, New York, 1981

[2] Hobbs, L., Ghosh, N.K.: Manufacture of copper rod, Wire Industry, 1 (1986), 42-44

[3] Bahtiarov, R.A.: Vlijanie temperaturi i skorosti litja na strukturu i svojstva slitkov splavov na mednoj osnove, Cvetnie metalli, 1 (1974), 68-71

[4] Fu, L.: Neural Networks in computer intelligence, McGRAW-HILL, New York, 1994

[5] Miller III, W.T., Sutton, R.S., Werbos, P.J.: Neural Networks for Control, MIT Press, Cambridge, MA, 1991

# FLEXIBLE MACHINING SYSTEM FOR PROFILE AND WIRE COLD ROLLING

**M. Jurkovic**

**University of Rijeka, Rijeka, Croatia**

KEY WORDS: Flexible Machining System, Profile Rolling, Drawing

ABSTRACT: The paper deals with the research and development flexible machining systems (FMS) in one relative wide area of profiles and wire production. The designed technology is based on stress state optimization in the zone of deformation, where are three different technological methods were applied: rolling, roll drawing and matrix drawing. The described concept and realized FMS in industry, is efficient means for more rational cold forming of profiles, particularly for small and medium batches of part production.

## 1. INTRODUCTION

Flexible automated manufacturing recently obtains more importance, particularly for small and medium batches of part production. We are faced with stronger requirements for a fast adaptation to marketing conditions, higher level of accuracy and reliability of machining systems and for raising of small production batches to a higher and better quality level. The contemporary machining systems are strongly required to provide simultaneously a flexibility and productivity and those conditions, at present, are met only by flexible machining systems (FMS). Flexible machining systems of profiles and wires manufacturing are still insufficiently researched, developed and applied in practice [1, 2, 3, 4, 5]. On that account, it is necessary to dedicate particular attention to determination of optimal

Published in: E. Kuljanic (Ed.) *Advanced Manufacturing Systems and Technology*, CISM Courses and Lectures No. 372, Springer Verlag, Wien New York, 1996.

technology and developed FMS for complex profiles and wire manufacturing. The first FMS FLEXIPROF 40-7 presented in this work is designed at Mechanical Faculty and was put in trial operation on 1988, [6, 7]. This work deals with a structural construction, technology, principle and control model of FMS.

## 2. ELEMENTARY TEHNOLOGICAL BASE FOR DEVELOPMENT AND DESIGN OF FMS

### 2.1. Group technology and classification systems for rolling and drawing profiles

Realize a sufficient level of productivity, together with flexibility, it is necessary to put the manufacturing program assortment into manufacturing homogenic groups of profiles having technological and manufacturing similarity (Fig. 1).

The classification system and profile classifier are elementary technological base for design of technologies and flexible technological modules (FTM). A classifier is defined by the numerical code characters (Fig. 2), which enables the using computer material selection, tools and technological methods (CAPP). Application of integral classification systems in describing of profiles production has a special importance in the technological preparation of production and products design (CAD).

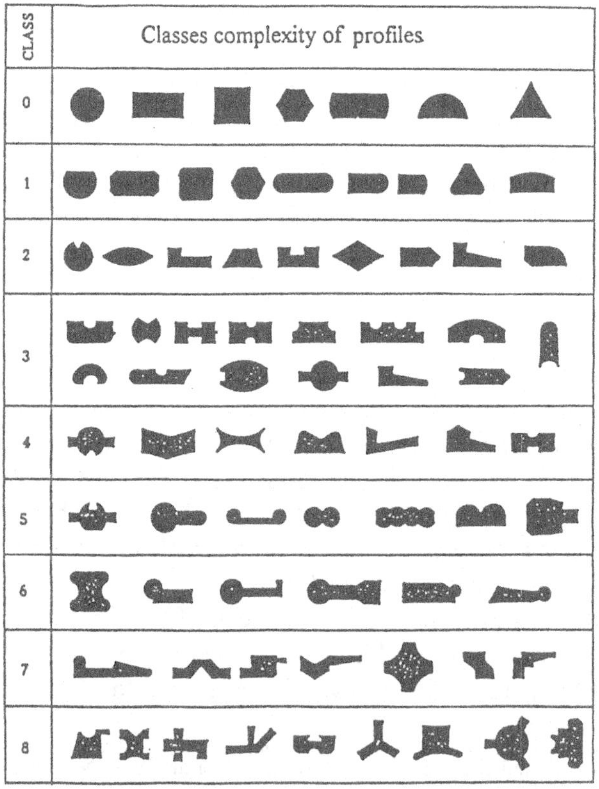

Figure 1. Classes complexity of some profiles

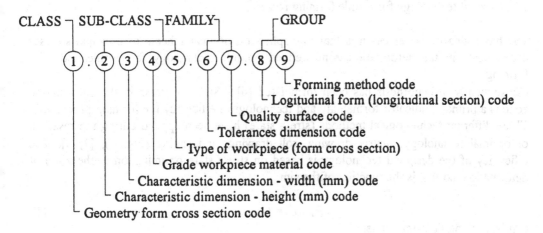

CLASS ⌐ SUB-CLASS ⌐FAMILY⌐          ⌐GROUP

①.②③④⑤.⑥⑦.⑧⑨
                    └ Forming method code
                 └ Logitudinal form (longitudinal section) code
              └ Quality surface code
            └ Tolerances dimension code
          └ Type of workpiece (form cross section)
        └ Grade workpiece material code
      └ Characteristic dimension - width (mm) code
    └ Characteristic dimension - height (mm) code
  └ Geometry form cross section code

Figure 2. Classification system - coding system [1, 2]

## 2.2. Flexible technology for profile forming

The profile forming process for production of small and medium batches of parts is developed, for which only the flexible technology and flexible technological modules are satisfying (Fig. 3).

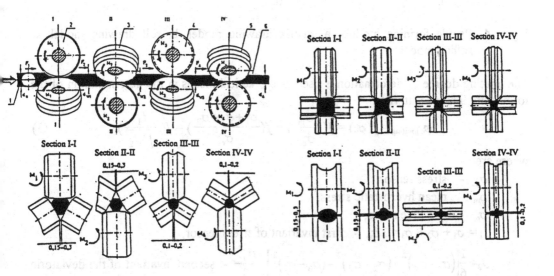

Figure 3. Illustration of flexible technological models (forming station) for rolling and for roll drawing with 2, 3 and 4 tools

## 2.3. Optimal technology for profile forming research

The basic idea of the research is that by application optimal scheme of principal stresses components in the deformation zone defined optimal technology for profile and wire forming.

Consequently, a hypothesis has been formulated [6]: "State of stress in the deformation zone is a primary indicator formability and technological efficiency the forming processes". Three different technological methods (stress state) have been applied during the research of optimal technology: matrix drawing, roll drawing, and rolling (Fig. 4), [7, 8]. The efficiency of the designed technology is based on stress state optimization in the zone of deformation, so that is the function of the aim:

$$F_a = \sigma_{ij} = (\sigma_{ij})_{opt}, \tag{1}$$

that is, limiting deformation is:

$$\varphi_{e\ limiting} = \varphi_{max} \tag{2}$$

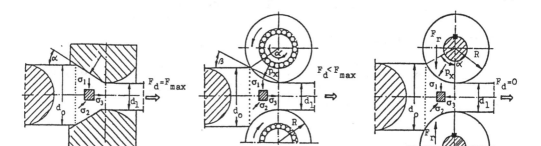

Figure 4. Profile forming models: a. matrix drawing model, b. roll drawing model, c. rolling model

The limiting degree of deformation is used as a practical means in the determination of the formability limit of materials:

$$\varphi_{e\ limiting} = f(\sigma_{ij}) = f(\frac{3\sigma_m}{\sigma_e}) = f(\frac{\sigma_1 + \sigma_2 + \sigma_3}{\sigma_e}) = f(\frac{I_1}{\sqrt{3J_2}}). \tag{3}$$

where are:

$\sigma_{ij}$  -  stress tensor,

$\sigma_m$  -  mean hydrostatic stress,

$\sigma_e$  -  effective stress,

$I_1 = \sigma_1 + \sigma_2 + \sigma_3 = 3\sigma_m$  -  first invariant of stress tensor,

$$J_2 = \frac{1}{6}\left[(\sigma_1 - \sigma_2)^2 + (\sigma_2 - \sigma_3)^2 + (\sigma_3 - \sigma_1)^2\right] = \frac{\sigma_e^2}{3}$$  -  second invariant of the deviatoric

stress tensor.

As at cold forming processes of axial - symmetrical profiles $\varphi_1 = \varphi_2$ and $\varphi_3 = -2\varphi_1$, that is $\varphi_1 = \varphi_2 = -0,5\varphi_3$, this equivalent strain can be represented form is:

$$\varphi_e = \varphi_{e\ limiting} = \varphi_3 = \ln\frac{A_0}{A_1} \qquad (4)$$

A similar pressed state of stress as that of rolling, showed a maximum formability in metal forming processes:

$$\varphi_{e\ max} = \varphi_{e\ rolling} = 2,810 > \varphi_{e\ roll\ drawing} = 1,650 > \varphi_{e\ matrix\ drawing} = 1,520 = \varphi_{e\ min}$$

The experimental values for material steel R St 42-1 (DIN 17006) are shown in Table 1.

Table 1. Results of experimental investigation of optimal technology [1]

| Num | FACTORS OF OPTIMIZATION | EXPERIMENTAL VALUES, % | | | AIM FUNC. $F_a$ |
|---|---|---|---|---|---|
| | | Rolling | Roll drawing | Matrix drawing | |
| 1 | Maximal degree of the deformation for machining station, $\varphi_{ms}$ | 100 | 54,0 | 41,5 | $F_a = \varphi_{msmax}$ |
| 2 | Total degree of the deformation, $\varphi = \varphi_{e\ limiting}$ | 100 | 58,7 | 54,1 | $F_a = \varphi_{max}$ |
| 3 | Velocity of profiles, $v$ | 100 | 85 | 60,5 | $F_a = v_{max}$ |
| 4 | Number of machining stations, $n$ | 100 | 160 | 185 | $F_a = n_{min}$ |
| 5 | Strengthening of workpiece material, $\Delta\sigma$ | 100 | 108 | 130 | $F_a = \Delta\sigma_{min}$ |
| 6 | Temperature of tool contact surface, $t$ | 100 | 107 | 153,8 | $F_a = t_{min}$ |
| 7 | Process productivity, $q$ | 100 | 68 | 35 | $F_a = q_{max}$ |
| 8 | Energy consumption, $E$ | 100 | 96 | 138 | $F_a = E_{min}$ |
| 9 | Tool life, $T$ | 100 | 93 | 30 | $F_a = T_{max}$ |

Optimization of designed technology resulted with 250% higher productivity in comparison to matrix drawing technology.

2.4. Design of forming process and number FTM

During design of forming process and number FTM important is of knowledge: geometry and characteristic dimensions cross section of profiles, initial workpiece material and number of phase plastic forming. Number of phase forming, that is, number FTM is defined by the expression:

$$n = \frac{\ln\lambda_t}{\ln\lambda_m} = \ln\frac{A_0}{A_n}\frac{1}{\ln\lambda_m} \qquad (5)$$

Total coefficient elongation of materials:

$$\lambda_t = \lambda_1,\ \lambda_2,\ \dots,\ \lambda_n = \lambda\ {}^n_m = \left(\frac{1}{1-\varepsilon_m}\right)^n \tag{6}$$

where are:

    $A_0,\ A_n$ - initial, that is, finite cross section of workpiece,

    $\lambda_1,\ \lambda_2,\ \dots,\ \lambda_n$ - coefficient elongation from phase of forming,

    $\varepsilon_m$ - mean strain of workpiece,

    $\lambda_m$ - mean coefficient elongation of workpiece,

    1, 2, ..., n - phase forming.

Some possibilities of the forming process of cross section of profiles and wire in Fig. 5 are shown.

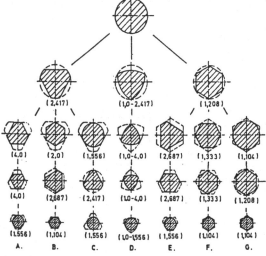

Figure 5. Phase forming of cross section of profiles

## 3. FLEXIBLE MACHINING SYSTEM FOR PROFILE AND WIRE COLD ROLLING

Function of FMS FLEXIPROF 40-7 is forming assortment wide of profiles with various cross section, and material quality [2, 4]. In Fig. 6 the layout and brief description of technical components of FLEXIPROF 40-7 with added equipment are given:

1. Workpieces accumulation module
2. Preparation material (mechanical pickling and/or brushing) module
3. Transport system
4. Flexible technological modules (forming stations) for rolling
5. Flexible technological module (forming stations) for roll drawing
6. Controlling CNC module
7. Module for electric - motor drivers
8. Lubricant module
9. Module for cooling
10. Measurement, control and monitoring (process measurement) module
11. Module for products accumulation
12. Module for a calibration
13. Module for electricity and controlling
14. Module (console) for controlling (tool station control)
15. Regal technological modules (tools) store - pallets and modular clamping devices

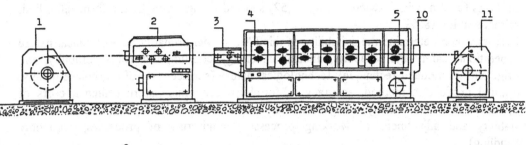

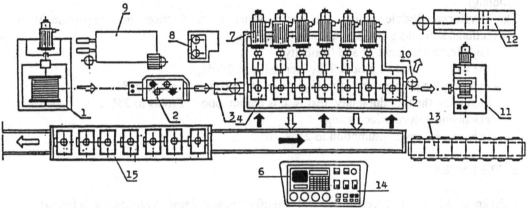

Figure 6. Technical structure of FMS for cold profile rolling

Main constructional-technological features of CNC-FMS are:
- Number of technological modules: 7
- Cross section area of a product: 3 mm² to 100 mm² (maximum 150 mm²)
- Tolerances of finished product: ±0,1 to ±0,02 mm
- Number of rolls per technological module: 2, 3, 4, and 6
- Production speeds: 100 m/min to 1000 m/min (maximum 1.800 m/min)
- Roller diameter: 160 to 250 mm
- Electric motor power: 40 kW

## 5. CONCLUSIONS

Produced prototype and realized CNC FMS in industry offer great possibilities for further development of machinery in this area of metalworking with the adverse main stress schemes (matrix drawing) and with machining systems which are characteristic with rigid conventional automation.

This research carried out shows that by using an optional scheme of principal stress components the potential of metal materials formability substantially increases, which has an exceptional importance in projecting technology, tools and machining systems. Thus, a similar pressed state of stress (cold rolling $\varphi_{e\ limiting}=2,81$) related to the heterogeneous state of stress (cold matrix drawing $\varphi_{e\ limiting}$=1,52) showed in increasing of the formability limit of material for 185%.

Factors which determine the suitability of this FMS model primarily have technical nature (operational safety and reliability, unification, energy consumption, tools durability, and originality of modular technique); technological nature (surface state of workpiece quality, optimal production method, accuracy of forming); economic nature (minimal costs of operation and profitability); organizational and information nature (process control, stability and adjustment of working processes, monitoring of processes, ergonomy handling).

The developed flexible rolling line has a series of preferences in comparison with conventional automatic line, i.e., automatic rigid line:

- high degree automation and flexibility,
- the working process productivity increasing to 250%,
- reducing the preparatory - finishing times and auxiliary times to 50%,
- increasing the tool durability, that is, less consumption of tools to 350%,
- machining accuracy increasing,
- reducing energy consumption to 25%, etc.

## REFERENCES

1. Jurković, M.: Flexible technology and manufacturing systems in process of deformation, Proceeding 1/91, University of Banja Luka, 1991, 67-92

2. Jurković, M., Popović, P.: Die verwendete Technologie und flexibles Bearbeutngssystem für Profilherstellung, Proceedings FOSIP '88, Bihać, 1988, 5-61

3. Reuter, R.C,: The rod rolling revolution of the '88s, Continuus Spa Milano, Wire Industry, 4(1983), 203-205

4. Esipov, V.D., Iljuković, B.M.: Prokatka specialnyh profilei složnoi formy, Tehnika, Kiev, 1985

5. Jurković, M., Mečanin, V.: Development and itroduction of new flexible computer controlled line for wire and full profiles rolling, Proceedings 8th International Conference BIAM '86, Zagreb, 1986, 226-229

6. Jurković, M.: The state of stress in the zone of deformation is a fundamental pointer of the deformability of material and effectiveness of cold work, Proceedings 4. International Symposium on the Plasticity and Resistance to Metal Deformation 1984, H. Novi, 1984, 354-356

7. Jurković, M., Ćurtović, K.: Erwägung der Wirksamkeit der verwendung von Bearbeitungsverfahren zur Ausarbeitung der Aschensymetriscen und ähnlichen umrisse, Technology of Plasticity, 17(1983), 41-61

8. Jurković, M.: Determination of the Formability Limit of Metals in Processes of Plastic Forming, Engineering Review, 15(1995)1, 9-22

# INFLUENCE OF TOOL CLAMPING INTERFACE
# IN HIGH PERFORMANCE MACHINING

**E. Lenz and J. Rotberg**

Technion - Israel Institute of Technology, Haifa, Israel

KEY WORDS: High PerformanceMachining, Drilling, Milling

ABSTRACT: The behavior of the tool-clamping interface was investigated in high-performance conditions in drilling and in end-milling operations. The main interest was in characterizing the behavior of typical clamping unit, evaluating the influence of their properties on the process performance parameters. Five clamping systems were characterized experimentally. Two of them were tested in drilling, and two were tested in different milling operations. The relations between the clamping system properties, the cutting conditions and process performance parameter were defined. This may contribute to a better evaluation of the tool-clamping role in high performance cutting, and to an improved clamping design.

## 1. INTRODUCTION

High performance machining may be defined as a process in which the influence of dynamic features of the system components is no longer negligible and must be taken into account. This situation may exist due to one or more of the following factors: High spindle speed, large feed rates, and relatively flexible cutting tools. The dynamic behavior of machine and cutting tool elements was investigated extensively during the years by [1] [ 2 ] [ 3 ] and others, mostly in the context of stability-limits research. In this work, the behavior of the tool-clamping interface was investigated, in high performance conditions, in drilling and in end-milling operations, under proper cutting conditions (no chatter). The main interest was in characterizing the behavior of different clamping units, evaluating the influence of their properties on the process performance parameters.

Some research has been done in the past by Rivin et al [4] concerning the tool holder behavior, and by the authors [ 5] concerning the clamping unit behavior, indicating that the compliance of tool-clamping interface unit should not be ignored.

Published in: E. Kuljanic (Ed.) *Advanced Manufacturing Systems and Technology*,
CISM Courses and Lectures No. 372, Springer Verlag, Wien New York, 1996.

In the following, a procedure for evaluating the clamping unit features is described, and the transfer function of five clamping units is presented. Results of **drilling** experiments are shown, revealing the behavior of the drill-clamping unit during the process, especially in the penetration phase. The relations between the clamping unit features, the drilling conditions and the hole-location accuracy will be demonstrated.

In **milling**, the role of the dynamic features of the clamping-end mill unit is demonstrated applying two clamping units, under different milling geometries. The relations between the (dynamic) cutting force excitation, clamping dynamic features and end-mill response are demonstrated. All the above is directed to reach a better understanding of the role of the interface unit in high performance machining, leading to the improvement of cutting process planning as well as to a rational development of the clamping units.

## 2. CLAMPING UNIT CHARACTERIZATION

### 2.1 Problem Description
Investigating the features of the tool-clamping interface, several common industrial units were tested, as shoown in Fig. 1.
(1) A standard spring collet
(2) A standard "Weldon" chuck, having one clamping screw.
(3) A modified "Weldon" chuck having two clamping screws.
(4) A hydraulic chuck
(5) A thermal shrinkage chuck, having 0.1 mm pressure fit plugged in at 600°C tested as some "reference", closest to the ideal clamping).

Since the tool end response is the multiplication (in frequency domain) or convolution (in time domain) of the **exciting force** and the **structure transfer function**, the clamping unit should be described in terms of **transfer function.**

Characterizing the clamping unit may be done either theoretically or by means of Experimenal Modal Analysis (EMA), and it involves the following problems:
(a) In a theoretical model by means of F.E.A., a detailed description of some units (spring collet, hydraulic chuck) is very complicated. It includes non-linear contact points, frictional and viscous damping, complicated geometries etc.
(b) In both theoretical and experimental models (acquired by experimental modal analysis), the question of generalization arises, i.e., how to define parameters of the clamping unit, predicting its behavior while clamping different tools. Facing these problems, it was decided that at this phase an impulse test, of a standard cylindrical rod, will serve as a fast simple means for clamping comparison as to static and dynamic stiffness, equivalent damping, etc.

An approximate linear model, **including equivalent stiffness and damping** parameters will be developed in a second phase (not introduced in this paper).

### 2.2 Experimental transfer function
Test procedure
Five clamping systems were tested as described above. The clamping interface to the spindle fixing was identical in all clamping units, designed to be as rigid as possible. A modal impulse test was carried out, while clamping a carbide rod having 12.7 mm diameter and 90 mm overhang. The transfer functions are shown in Fig. 2.

The static compliance, the dynamic compliance (governed also by the system damping), and the first natural frequency are clearly defined [ 5 ].

The first mode shapes acquired by the EMA are shown in Fig. 3. The contribution of the clamping unit compliance to the rod-deflection may be noticed, especially as we compare the real mode shapes to the theoretical, ideally-clamped rod.

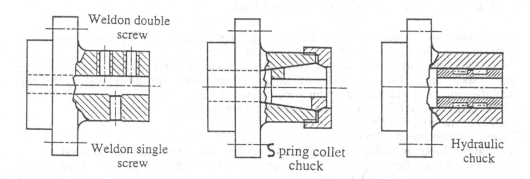

Fig. 1    Various clamping units

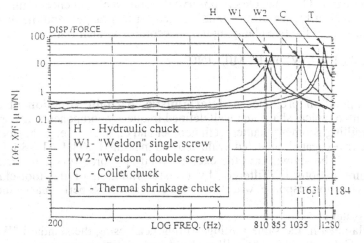

Fig. 2    Transfer functions of various clamping units.

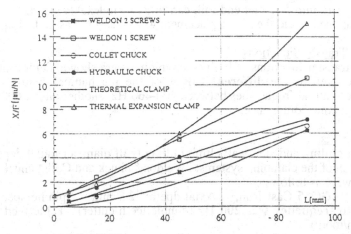

Fig. 3.    First mode shape of a rod in various clamping units

## 3. CLAMPING UNIT BEHAVIOR IN DRILLING

The behavior of various clamping units in high performance drilling was described in
details in [ 5 ]. Here, a shortened summary will be brought.
The drilling was carried out using a horizontal milling/drilling head, yileding 8000 rpm
spindle speed and 15kw . The drilling axial and lateral (radial) forces, as well as the drill
end deflection in x y (radial) directions were measured. Fig. 6.
A solid-carbide twist drill of 12.7 mm diameter and 90 mm long was used.
Workpiece material was GG25 cast iron.
The main results are as follows:
- The **lateral force** in **drilling**, especially in the penetration phase is of **random nature**
(as a result of manufacturing errors, unproper cutting in the chisel area etc.). Therefore, it
may be treated in **statistical terms** only.  In Fig. 4 the cutting forces and the drill end
deflection are shown during the penetration phase, leading finally to the hole location error.
In Fig. 5, average radial location errors in two different clamping units  are shown in
various cutting conditions.  One may clearly observe the dominant influence of the feed per
revolution, the marginal influence of the spindle speed (in this range), and the difference
between the performance of the two clamping units (according to their transfer functions).

## 4. CLAMPING UNIT BEHAVIOR IN MILLING

### 4.1 General:
The end milling operation produces **a deterministic, periodic function** of cutting
(exciting) forces, with a typical shape and well defined frequency components, according to
the spindle speed, milling geometry, cutter teeth number and cutting      edge geometry.
Milling forces were investigated, modeled evaluated and measured [ 6 ] [ 7 ] [ 8 ] [ 9 ].
In the following it will be shown how the exciting (cutting) force function, created by
certain milling combinations is "filtered" by transfer function of the tool-clamping
producing the tool-vibration response, which is in this case a process performance index.

### 4.2 Experimental System
An end-mill was clamped in the same milling drilling head using the standard "Weldon"
chuck and also the Collet chuck.  The milling head and dynamometer were oriented to
enable milling operations: Fig. 12.  The end mill overhang was 50 mm.
 The milling cutter: a 16 mm diameter end-mill carrying two indexable inserts.
The x y z force components, and the x y displacement were measured as in drilling.

### 4.3 Clamping Unit Transfer Functions
In Figs. 7 & 8, the transfer functions of both clamping units are shown, demonstrating the
**similar nature** as well as the **different values** of the mechanical features:   static
compliance, a pair of natural frequencies between 400-460 Hz, another one at 600-700 Hz,
all having typical dynamic compliance.

### 4.4 Milling experiments.
Experiments of slot milling and side-wall climb milling (half diameter radial depth), were
carried out using each of the clamping systems: "Weldon " chuck and Collet chuck.
Milling conditions were as follows:
Workpiece material: GG25 Cast Iron,    Axial dpeth; 2.5 mm,   Spindle speed: 6000
rpm(100 Hz), **Tooth frequency was 200 Hz** (double teeth cutter).  Table feed was 630
mm/min (0.053 mm/tooth)
Synchronours sampling was carried out at a sampling rate of 9000 Hz (90 points/rev.)

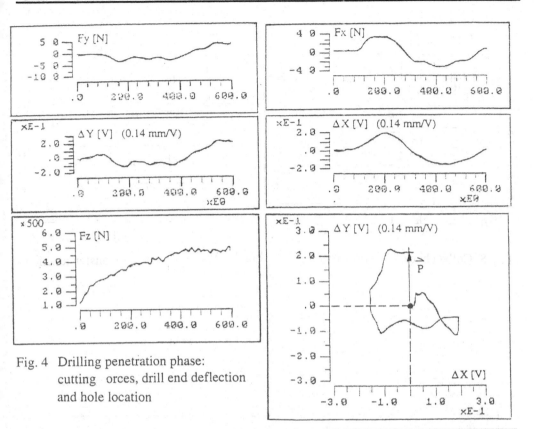

Fig. 4 Drilling penetration phase:
cutting orces, drill end deflection
and hole location

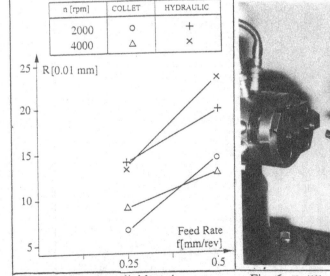

Fig. 5 Average radial location error
in collet and hydraulic chucks

Fig. 6 Drilling measurement system

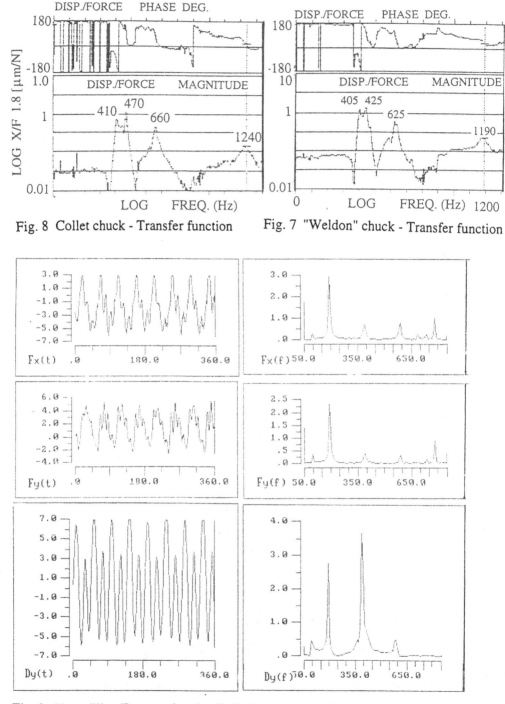

Fig. 8 Collet chuck - Transfer function

Fig. 7 "Weldon" chuck - Transfer function

Fig. 9 Slot milling Force and end-mill displacement results

Each experiment was repeated 4-5 times, and the results were averaged.
Fig. 11 presents a qualitative description of the milling forces as measured in [ 5 ].
Fig. 9 presents the force and deflection functions of a "Weldon" chuck in slot milling in time and frequency domain; y - feed direction, x - perpendicular direction.
In this example one may observe the following:
a. The typical force function includes the main component at tooth frequency
(200 Hz) and smaller but still significant higher-lobes (at 400, 600 Hz). There are also tool runout component (at 100 Hz), and some high frequency dynamometer components.
b. The large magnification of the 400, 600 Hz components of the exciting force, due to clalmping natural frequencies expresses in the deflection curves.
In Fig. 10 force and deflection functions are shown, while using the "Weldon" chuck in side-wall milling. In this example, the following may be observed:
a. The force function is different now: the "y" force (feed direction) sign is changing, therefore an additional 400 Hz (twice-tooth frequency) excitation occurs.
b. The clamping-unit transfer function amplifies this component both in x and y directions.
The **combination** of **milling geometry** (force function shape), **cutting conditions**, and **clamping unit features** leads in this case to **increased vibrations** in both directions.
In Table 1, the maximum values of force and vibrations are presented. They were averaged in 4-5 experiments each, over four spindle rotations, in time domain.
Force values in identical operations are the same in both chuck, which means that the influence of the force-deflection feedback mechanismn is negligible (no chatter).
The difference in deflection between the two chucks is evident, and matches their transfer functions.

| Milling Operation | No. of Tests | Measured Parameter | Weldon Average Value | Collet Average Value |
|---|---|---|---|---|
| | | Fx - perp. | 0.83 | 0.86 |
| | | Fy -feed | 0.68 | 0.68 |
| Slot Milling | 5 | Dx - Perp. | 0.89 | 0.62 |
| | | Dy - feed | 1.11 | 0.42 |
| | | fx - Perp. | 0.84 | 0.80 |
| Side    Wall | 5 | Fy - feed | 0.64 | 0.65 |
| Milling | | Dx - perp. | 0.94 | 0.56 |
| | | Dy - feed | 1.03 | 0.40 |

Table 1: Averaged values of max force and deflection

## 5. CONCLUSIONS

The role and influence of the tool-clamping interface properties was investigated in drilling and milling operations, in high performance conditions.
The combined effect of the exciting force, (as produced by the specific process), and the clamping unit transfer function was demonstrated. This effect should be taken into account in any high performance process planning, tool design and performance evlauation.

## 6. REFERENCES

1. Tobias, S.A., 1965. Machine Tool Vibration. Blackie and Sons Ltd.

2. Weck, M., 1994. "Werkzeugmaschinen-Fertigungssysteme". VDI Verlag.

3. Gygax, P.E., 1980. "Cutting Dyanamics and Process Structure Interactions Applied to Milling", Wear Journal, Vol. 62, pp. 161-185.

4. Rivin, E., Agapiou, J., 1995. "Toolholder/Spindle Interfaces for CNC Machine Tools", CIRP Annals, Vol. 44/1 pp. 383-387.

5. E. Lenz, J. Rotberg, R.C. Petrof, D.S. Stauffer, K.D. Metzen, 1995. "Hole Location Accuracy in High Speed Drilling. Influence of Chucks and Collets". The 27th CIRP Seminar on Design Control and Analysis of Mfg. Systems Proceedings. pp. 382-391.

6. Armarego, E.S., Deshpande, N.P., 1991. "Computerized End Milling Force Predictions with Cutting Models". CIRP Annals, Vol. 40/1, pp. 25-29.

7. Smith, S., Tlusty, J., 1991, "An Overview of Modeling and Simulation of the Milling Process", ASME Journal of Engineering for Industry, Vol. 113, pp. 169-175.

8. J. Rotberg, S. Shoval and A. Ber, 1995, "Fast Evaluation of Cutting Forces in Milling", Accepted for publication in the Intl. J. of Advanced Manufacturing Technology.

9. Schulz, H. , Herget, T. 1994. "Simulation and Measurement of Transient Cutting Force Signal in High Speed Milling", Production Eng., R&D in Germany, Vol. I/2, pp.19-22.

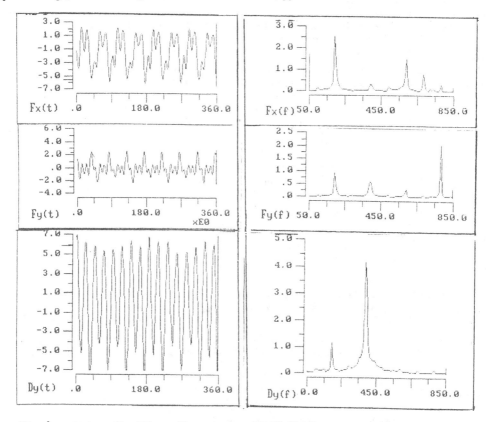

Fig. 10   Side wall milling - Force and end-mill displacement results

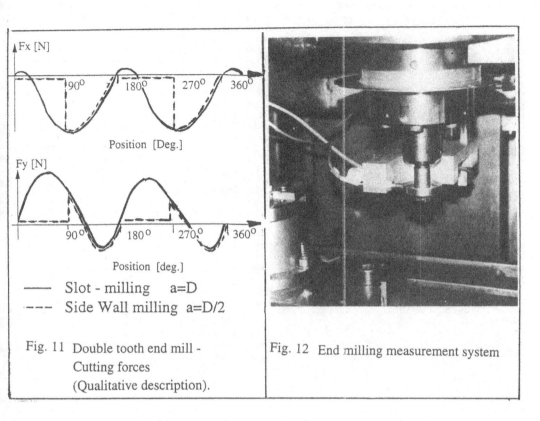

Fig. 11  Double tooth end mill -
Cutting forces
(Qualitative description).

Fig. 12  End milling measurement system

# A METHODOLOGY TO IMPROVE FMS SATURATION

M. Nicolich, P. Persi, R. Pesenti and W. Ukovich
University of Udine, Udine, Italy

KEY WORDS: Production Planning, Flexible Manufacturing Systems, Tooling, Mathematical Programming.

ABSTRACT: The problem of improving the saturation of an FMS cell is considered. Limited tool buffer capacity turns out to be a relevant constraint. A hierarchic approach is formulated, contemplating at the higher level the determination of "clusters", i.e., sets of parts that can be concurrently processed. At lower levels, clusters are sequenced, linked, and scheduled. While for the latter subproblems solution methods taken from the literature are used, two original mixed integer programming problems are formulated to determine clusters. The proposed methods are then discussed on the basis of computational experience carried out on real instances.

# 1   INTRODUCTION

The present paper discusses some production programming and scheduling problems studied for a flexible cell at the Trieste (Italy) Diesel Engine Division of Fincantieri, a major Italian state company in the shipbuilding sector.

The Diesel Engine Division produces and assembles diesel and gas engines for ships and electric generation plants. Its normal production is about 50 engines per year. Typical costs for a single product range from about 3 to 6 million dollars for fast and slow

Published in: E. Kuljanic (Ed.) *Advanced Manufacturing Systems and Technology*,
CISM Courses and Lectures No. 372, Springer Verlag, Wien New York, 1996.

engines, respectively. Orders arrive about one year before due date; then supplying
row parts requires six monts, production and assembly three months each.

The FMS cell under study is composed of four machines: two machining centers, a
vertical lathe and a washer. Each machine has an input and an output buffer for
parts. Also, each machine, except the washer, has a tool buffer with 144 positions
each. Tools are loaded and unloaded automatically. Pallets carrying parts or tools are
moved automalically too.

# 2    IMPROVING THE SATURATION LEVEL OF THE FMS CELL

The FMS cell is a crucial resource for the production system, as it can process parts
as much as three times faster than ordinary job shops, due to its ability to comply
with changes both in production mix and in lot size. However, the cost per hour of its
machines is sensibly higher (about 1.5 times) than the cost of a machine of an ordinary
job shop. As a consequence, the FMS cell requires to be saturated as much as possible.
The cell started its production in 1991. It operates on three shifts, eigth hours each,
from 6 a.m. of Monday to 2 p.m. of Saturday; the night shift is not supervised by
operators, thus giving about a 50% yield.

For each week from 1992 to 1995, the saturation level of the cell, defined as

$$\text{saturation level} = \frac{\text{produced parts, measured in processing hours}}{\text{hours available for the three operating machines}},$$

has been recorded. Although the nominal saturation level indicated by the supplier of
the cell (Ex–cell–o, Germany) is 70%, the recorded value is about 50%.

In this section the original approach devised to improve the saturation level of the FMS
cell is discussed.

## 2.1    PROBLEM FORMULATION

The problem of improving the saturation level of the FMS cell has the following elements:

**Input data:** the monthly production plan, which specifies parts to be produced, assigns operations to machines and determines operation times;

**Output decisions:**

- parts to be concurrently produced;
- operation scheduling;

**Objective:** maximizing the saturation level of the cell;

**Constraints:** complying with the operational rules of the cell, in particular with its limited resources, such as buffer capacities.

Concerning constraints, it turned out that a critical resource is the tool buffer capacity: since each part normally requires several dozens of tools on each machine, the possibility of concurrently working different part types is severely affected, thus reducing in practice the flexibility of the cell. This is the central issue of the problem considered.

## 2.2   A HIERARCHICAL APPROACH

A hierarchic approach has been devised for the just formulated problem. It decomposes the whole problem in four simpler subproblems, that, when solved in turn, give the solution of the whole problem. The subproblems are arranged in a hierarchic order (see for instance [1]), according to their detail level, from the most to least aggregate. The solution of each level subproblem provides the data for the subproblem of the following level. The subproblems are:

**clustering:** in this phase the set of all parts required by the production plan is partitioned into subsets ("clusters") of parts that may be concurrently processed;

**cluster sequencing:** in this phase the most appropriate sequence of the clusters which are determined in the previous level is sought;

**cluster linking:** in this phase, according to the sequence decided in the previous level, the transition from each cluster to the following one is determined;

**scheduling parts within each cluster:** in this phase each part is scheduled within the cluster to which it was assigned in the first level.

The merit of the proposed hierarchic approach is to allow to tackle with a complex problem through the sequence of simpler subproblems; on the other hand, this entails some degree of suboptimality, as the solution provided at higher levels affects the performance obtainable at lower levels, and, as a consequence, for the whole problem. However, this is the price to be paid in order to deal with the whole problem, which is so complex to be unaffordable by any global approach.

The clustering and scheduling subproblems have been considered in detail, and will be discussed here. The other two subproblems seemed to have less relevance, as far as their impact on the global solution is concerned.

## 3   CLUSTERING MODELS

In this section the two models are presented, which have been formulated for the clustering problem. As it has been pointed out, concurrent processing of different part types is limited by the capacity of the tool buffers of the operating machines. Two different models have been devised for this problem. They basically have the same decision variables and constraints. They only differ for the objective function: for a given production plan, the former minimizes the global lead time, while the latter seeks to produce a good machine balance.

The problem data for either model are as follows:

- the set $P$ of part types to be produced in the considered horizon;

- the set $M$ of machines;

- the set $U$ of available tools;

- the set $C = \{1, 2, \ldots, |C|\}$ of required clusters.

- the aggregate production plan, in terms of the number $N(p)$ of parts that must be produced for each part type $p \in P$;

- the processing times, in terms of the time $T(p, m)$ required to process a part of type $p \in P$ on machine $m \in M$;

- the tools needed for each part and each machine, in terms of a matrix $a(p, u, m)$, defined as

$$
a(p, u, m) = \begin{cases} 1 & \text{if tool } u \in U \text{ is required to process a part of type } p \in P \\ & \text{on machine } m \in M \\ 0 & \text{otherwise} \end{cases}
$$

- the number of positions $k(m)$ in the tool buffer size for each machine $m \in M$;

- the tool size, in terms of the buffer positions $b(u)$ taken by each tool $u \in U$.

## 3.1   MINIMIZING THE GLOBAL LEAD TIME

This model has the following (nonnegative integer) decision variables:

- the number $x(p, c)$ of parts of type $p \in P$ assigned to the cluster $c \in C$;

- the tool loading matrix $z(c, u, m)$, defined as

$$
z(c, u, m) = \begin{cases} 1 & \text{if tool } u \in U \text{ is loaded in the buffer of machine } m \in M \\ & \text{for cluster } c \in C \\ 0 & \text{otherwise} \end{cases}
$$

Furthermore, the following auxiliary (nonnegative continuous) variables are defined:

- the load (in terms of processing time) $\tau(m, c)$ of each cluster $c \in C$ on each machine $m \in M$;

- the load $\tau(c)$ of the most loaded machine for each cluster $c \in C$.

Then the problem of determining the $|C|$ clusters minimizing the global lead time is formulated as a mixed integer programing problem:

$$\min \sum_{c \in C} \tau(c) \tag{1a}$$

$$\sum_{c \in C} x(p, c) \geq N(p) \qquad \forall p \in P \tag{1b}$$

$$\sum_{p \in P} a(p, u, m)x(p, c) \leq Kz(c, u, m) \qquad \forall u \in U, m \in M, c \in C \tag{1c}$$

$$\sum_{u \in U} z(c, u, m)b(u) \leq k(m) \qquad \forall m \in M, c \in C \tag{1d}$$

$$\tau(m, c) = \sum_{p \in P} x(p, c)T(p, m) \qquad \forall m \in M, c \in C \tag{1e}$$

$$\tau(c) \geq \tau(m, c) \qquad \forall m \in M \tag{1f}$$

In practice, problem (1) takes a given number of clusters and assigns each part of the production plan to a cluster; furthermore, the tools to be loaded on each machine buffer are determined, complying with their capacities. Among the several possible assignments, the one minimizing the global lead time is sought. The lead time is given by processing times only, thus disregarding transportation, fixing and washing times, as they are at least of one order of magnitude smaller. Thus the lead time of a cluster is the total processing time of the most loaded machine. Note that at this aggregation level idle times due to the processing sequence are not considered.

In particular, (1e) expresses the total processing time for each cluster and each machine as the sum of the processing times of all parts assigned to that cluster on that machine. Then (1f) takes the lead time of the most loaded machine for each cluster. The global lead time is then minimized by (1a). Furthermore, (1b) guarantees that at least as much parts are produces as required by the production plan. Also, (1c) guarantees that all necessary tools are loaded on each machine ($K$ is a suitably big number), and (1d) expresses the tool buffer capacity.

The model (1) allows to find the minimum number of clusters by solving it for increasing values of $|C|$: the first feasible solution gives the minimum number of clusters.

Although model (1) correctly formulates the problem of minimizing the global lead-time, it is rather heavy for the computational resources it requires. In particular, too many integer variables $z(c, u, m)$ may result, even for practical problems of limited size. To overcome this drawback, another version of problem (1) has been devised, by introducing new decision variables $y(p, c)$ expressing the type–cluster assignment:

$$y(p, c) = \begin{cases} 1 & \text{if part of type } p \text{ is assigned to cluster } c \\ 0 & \text{otherwise} \end{cases}$$

Then a new constraint is introduced:

$$x(p,c) \leq N(p)y(p,c) \qquad \forall p \in P, c \in C, \tag{2}$$

stating that if $x(p,c) > 0$, then part type $p$ must be assigned to cluster $c$. Furthermore, the constraint (1c) is modified as

$$a(p,u,m)y(p,c) \leq z(c,u,m) \qquad \forall p \in P, m \in M, u \in U, c \in C. \tag{3}$$

Note that variables $z$ need not to be integer anymore, since the l.h.s. of (3) can assume only 0 or 1 values. Furthermore, there is no need for the big $K$ constant.

Since the number of integer variables is now much smaller than in (1), the latter model turns out to be much more convenient that the former as far as computation times are concerned.

## 3.2   BALANCING WORKLOADS

The above model can be easily modified in order to balance machine workloads. It suffices to introduce another constraint

$$\tau(m,c) \geq \alpha\tau(c) \qquad \forall m \in M, c \in C, \tag{4}$$

expressing that, for each cluster, the workload for each machine cannot be larger than the load of the most loaded machine, multiplied by the "balance factor" $\alpha$. Considering $\alpha$ as a new variable to be maximized, a two objectives nonlinear programming problem results.

This problem may be conveniently tackled ranging the objectives in a lexicographic way, with the balance factor at the higher level; then $\alpha$ can be seen as a parameter, and several problems with the additional constraint (4) can be solved for increasing values of $\alpha$, until no feasible solution is found. Then the last found solution minimizes the global leadtime subject to the maximum balance factor.

The programming problems of the two above models have been solved using Cplex on a DEC Alpha 3000 workstation, for several instances taken from practical cases. Cpu times and workloads experienced on the basis of a large number of trials, performed on real cases, show that the proposed approach is both efficient and effective.

## 4   CLUSTER SEQUENCING AND LINKING

The subproblem of the second level in the hierarchy consists in determining the most appropriate sequence in which clusters, determined at the first level, have to be processed. In order to determine this sequence, some factors are considered, that were neglected at the upper, more aggregated, level. In particular, tool loading times appear to be the appropriate performance index for cluster sequencing.

More specifically, for any pair of clusters, the number of non common tools (i.e., tools that are used only by one of them) is a sensible way to assess the burden of tool changing when they are processed one after the other. Minimizing the total number of tool changes leads to a Travelling Salesman Problem [4], that can be conveniently solved by approximate methods [3]. However, due to the fact that the number of clusters is in practice always very low for our application, no particular importance has been attributed to this problem.

Once clusters have been sequenced, they could be "linked", in the sense of processing in a concurrent way the ending transient of the first one with the beginning transient of the following one. However, also in this case this subproblem appeared to be quite irrelevant, due both to the fact that transients are relatively short, and clusters are quite different, thus leaving few margins to the possibility of concurrent operations.

# 5   DETAILED SCHEDULING WITHIN CLUSTERS

In this section the problem is faced of finding the most appropriate scheduling of the parts assigned to each cluster, as they have been determined at the upper level.

For each part type, the production plan specifies the operations that have to be performed on each machine, and their sequence. Due to te reduced number of parts involved, it is natural to model the sequencing problem as a job shop problem [2]. Note that also in this case, being at a lower level of our hierarchic approach, more detailed elements are considered as in upper levels: in particular, processing sequence on machines, that was not contemplated to determine clusters, is now taken into account. For this subproblem, local dispatching rules have been considered [5]. In particular, six of them turned out to be quite interesting:

1. Erliest Due Date (EDD);

2. Minimum Lateness (ML);

3. Random;

4. Starvation Avoidance (SA);

5. First In First Out (FIFO);

6. Shortest Remaining Processing Time & Starvation Avoidance (SRPT&SA).

Typical results that have been obtained on real instances show that SA maximizes the saturation index, while SRPT minimizes WIP. As a natural consequence, SRPT&SA provides the most equilibrate results.

# 6   CONCLUSIONS

Motivated by the problem of improving the saturation level of a flexible cell of Divisione Motori Diesel, Trieste, a general hierarchic approach has been proposed to the production planning and scheduling problem when tools are a scarce resource.

The subproblem corresponding to the upper level of the proposed approach has been modelled as a mixed integer programming problem, and solved using a general purpose mathematical programming package. The other subproblems have also been modelled and standard solution methods have been proposed for them.

Several computational experiments have been performed on real instances, showing the effectiveness of the proposed approach.

# ACKNOWLEDGEMENTS

The authors are grateful to the managers of Divisione Motori Diesel of Fincantieri for their support to this research.

# REFERENCES

1. Bitran, G.R., and Tirupati, D.: Hierarchical Production Planning, in: Graves, S.C., Rinnooy Kan, A.H.G., and Zipkin, P.H. (eds.): *Handbooks in Operations Research and Management Science, Vol. 4: Logistics of Production and Inventory*, North–Holland, 1993.

2. Blazevicz, J., Ecker, K., Schmidt, G., and Węglarz, J.: *Scheduling in Computer and Manufacturing Systems*, Springer, 1993.

3. Jünger, M., Reinelt, G., and Rinaldi, G.: The Traveling Salesman Problem, in: Ball, M.O., Magnanti, T.L., Monma, C.L., and Nemhauser, G.L. (eds.): *Handbooks in Operations Research and Management Science, Vol. 7: Network Models*, North–Holland, 1995.

4. Lawler, E.L., Lenstra, J.K., Rinnooy Kan, A.H.G., and Shmoys, D.B.: *The Traveling Salesman Problem*, Wiley, 1985.

5. Panwalkar, A., and Iskander, W.: A survey of scheduling rules, *Operations Research* 25, pp.45–46, 1977.

# AUTONOMOUS DECENTRALIZED SYSTEM
# AND JIT PRODUCTION

**T. Odanaka**

**Hokkaido Information University, Hokkaido, Japan**

**T. Shohdohji and S. Kitakubo**

**Nippon Institute of Technology, Japan**

**KEY WORDS:** Autonomous Decentralized Systems, Multi-Echelon Inventory, JIT Production

**ABSTRACT:** One of the clues to the autonomous decentralized system is biological system that exists in nature. A multi-echelon production inventory systems are worked out as one of the example. At first, we shall define simply the autonomous decentralized system. Secondary, we discuss the multi-echelon inventory production system that is the theoretical bases of Just-In-Time productions. We shall state the theorem on the theoretical foundation and the implication of this theorem. Thirdly, we shall consider the problem that two installations arranged in series, with a set up cost in transportation between them.

## 1. INTRODUCTION

With the progress of frontier science in recent years, large scale complicated systems have appeared on the stage and demands for flexibility, diversity, reliability and so forth are put forward. For these purposes conventional concentrated control systems which bring all information to one control center are no longer able to meet the demands, because exponential scale increase of control center will be inevitable in this massive concentrated control system, so that once a possible accident happened in a control and communication structure, the whole control system would be broken down through stoppage and aberration of it. It is under these circumstances that a new concept of system which is able to work out these problems with the name of ads attracts attention recently[1].

Published in: E. Kuljanic (Ed.) *Advanced Manufacturing Systems and Technology,*
CISM Courses and Lectures No. 372, Springer Verlag, Wien New York, 1996.

In a new concept of this system there is no control center that integrates the whole system but each element of it which constitutes the system is dispersed but can carry out the duty of it through the cooperation of each element in a autonomic way, discharging what is called for, therefore, once something was wrong in a part of the system and environments of it changed, cooperative adjustment may be realized in a autonomic fashion and may be able to fulfill its whole duty, so that interest in it is growing among all quarters. However, this approach is on the way to unification of system theory and composition principle to realize this artificial ads has not been disclosed as yet.

One of the clues to this approach is biological system that exists in nature and it will be necessary to incorporate this autonomous decentralized concept to system making of each field. Although some approaches are being considered in the formulation of ads, a production system is worked out as one of the examples.

## 2. AUTONOMOUS DECENTRALIZED SYSTEMS

In the formulation of this autonomous decentralized systems the following two concepts are above all important.
    (i) intelligent level of an individual.
    (ii) relationship between fields that constitutes order and cooperation among individuals.

Intelligence that an individual has means individual autonomic behavior and intelligence that makes it possible to establish order, keeping cooperation with other individuals. It is essential for an individual to have some intelligence in a behavior to keep order as a whole, observing his own behavior and that of others, and act according to the situations.

Let us next consider about relationship of cooperation between the total system and individuals. In order to do a thing on system, each element or individual that is a member of it has to be integrated and behave in a reciprocal fashion. Therefore, it will be important to formulate mathematical equations representing relationship between order of the system as a whole and cooperation between the individuals.

## 3. JIT PRODUCTION

As mentioned before, in ads it is required for each unit to equip itself with autonomy and for the whole factory with cooperation and harmony between the units, simultaneously. How to formulate a system with such adjustment processes is extremely important but difficult. Pulling system in JIT production (Just-In-Time production) formula does not generate flow of entire production automatically but it has a scheduling program by MRP (material requirements planning), that is, simultaneous programmed instruction from the program center to all processes, all dealers and parts manufacturers. Under the pulling system each process is nothing but the so-called discentralized one. On the other hand the program center provides the final assembling process only with production program information based on orders of selling shops, so that the program center does not have to send simultaneously program information concerned to the processes preceding the final one.

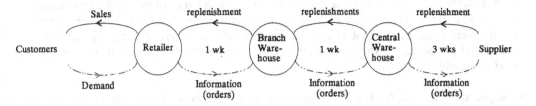

**Figure 1.    Information and Stock Flow in a Base Stock System**

In the upper stream processes above the final process only through mutual exchange of information by Kanban production program is carried out independently[2], [3]. In other words each process acts autonomously but total harmony of the whole factory can be maintained automatically (see Figure 1).

On the other hand the headquaters considers prearranged program designed for the whole processes, all dealers and all parts manufacturers. These programmed values are leveled down to obtain programmed production value by operation days, which are given as daily programmed value of an operative production month. These values are not ordered but estimated. Actual orders are issued by Kanban.

It should be noted that this kind of system can be realized by both scheduling by MRP and operation program based on pulling system by Kanban formula. Autonomy in this approach is constrained by minute adjustment within 10 per cent up and down the scheduled value of the program center. These narrow ranges are the values of autonomy.

## 4. MULTI-ECHELON INVENTORY PRODUCTION SYSTEM,
### − A THEORETICAL BASIC OF JIT MANUFACTURING −

Production Planning and control system arise when departments confine themselves to considering the material-flow only in their own functional areas (purchasing, production and marketing) and ignore the effects of their decisions upon the enterprise as a whole. Based on the traditional functions of purchasing, production and marketing, the logistical system is generally divided into three areas of the competence:
  a) the material flow in purchasing and the therewith associated information flow,
  b) the flow of material and goods through production and the therewith associated flow of information,
  c) the flow of goods from the firm to its customers and the therewith associated information flow.

The isolated pursuit of functional objections enhances the probability that solutions arrived at will be merely suboptimal ones. That can mean high inventories and long throughout times. Considering this weakness of traditional organization structure, we see the necessity for viewing the chain of material and goods-flow in a holistic manner. This is the fundamental concern of system. The role of system is to plan and control the total flow of materials and goods from the acquisition of the raw materials up to the delivery of the finished products to the ultimate users, and the related counter-flows of information that both record and control the movement of materials. System can therefore be seen as the art of managing the flow of material and goods-getting the products where they are needed, at the time they are wanted, and at reasonable cost.

An important feature of the traditional art of managing the flow of materials in enterprise is the division of the responsibility into several functional subsystem. A major disadvantage of this method is that sometimes optimization efforts remain confined to the individual subsystem, efforts upon the enterprise as a whole are not taken into consideration. The consequences are frequently high inventories and long throughout times.

System can help us to avoid these efforts, it involves the consideration of the total flow of material from the receipt of raw materials though manufacturing and processing stage up to the delivery of the finished products. There are two difficult concepts for managing the flow of material and goods in production, which are the concept of material stock optimization and the concept of material flow optimization. With the concept of material stock optimizations one can, though optimization of inventory, attempt to guarantee quick deliveries, circumvent the unpleasant consequence of equipment break down insure constant utilization of production unit even when several variations of demand exists, etc.

The traditional systems of production planning and control are based on this concept. Recently the concept of material flow optimization has attained importance. Stock keeping is now seen as the "root of all evil," when inventories are available, the management has little incentive to prove the operating system, e.g. reduce set up times, to increase process and product quality reliability, etc[4].

Therefore the aim is first to eliminate waste(e.g. high set up times) in the operating system and afterwards to install new planning concept, e.g. JIT. Although a stockless production is the ultimate goals, trade-offs must be taken into consideration. Finally, production logistics has to carefully evaluate the consequences of these trade-offs, thereby increasing the competitive advantage of an enterprise. In the following section we will discuss the fundamental concept of production planning and control systems.

## 4.1 Concept of Production Planning and Control Systems

As we have seen, production system span all activities which are concerned with planning and controlling the material and goods flows from the raw material inventory to production and between all the production units up until the final-product-inventory. We can distinguish two different concepts for the realization of these tasks, which are the concept of material stock optimization and the concept of material flow optimization. The traditional concept of material stock optimization tries to attain a certain customer service level by investing inventories. The benefit of high inventories is that goods can be quickly delivered to the customers, in the case the right products are at hand. For a given service level the stock should be as minimal as possible. That is the reason for the name: material stock optimization.

The draw backs of the concept are:

1) High material stock and product stock piles demand a great deal of turn over capital and negatively affect the liquidity of the enterprise.
2) Inventories increase the risk of having unsellable products, since forecasts sometimes prove incorrect.
3) High inventories prevent the weakness of production processes (high set up times, unsynchronized production processes, imperfect flexibility in production) from becoming apparent.

Therefore waste is not eliminated. Inventory can therefore be judged from two viewpoints:
1) In general the defenders of the concept of material stock optimization regard inventory in a positive manner.
2) Inventory makes it possible to guarantee a smooth production and quick deliveries and to avoid the negative consequences of breakdowns, they enable a constant utilization of production units, etc.

According to the opponents of this concepts, stock piles are the "root of all evil", because they prevent the elimination of long set up times, unsynchronized production processes, high reject rate etc. These opponents prefer the optimization to determine the modes of action. These involve a continuous process of improvements. Examples are
1) Minimizing of set up times, which enable smaller lot sizes to become economical.
2) New kinds and layout of machines which are able to produce parts or even products completely.
3) The synchronization of capacities,
4) The optimization of variants,
5) Quality assurance, etc.

A production-synchronous-procurement is also an example of this kind of rationalization. This reduces the throughout time dramatically, eliminates breakdowns and lowers stock piles. In summary this mode of action should make a JIT production possible; at any given time only the materials and products which are immediately needed at certain production stages or by customers are produced.

Now we have to pose the questions:
Under which conditions is a JIT production economical? Can inventory be justified under certain circumstances?

In the following sections, we will try to answer these questions by taking a theoretical model as a basis.

### 4.2 Theoretical Foundation
In parts and assembly industries, planning and controlling the flow of materials and goods is an especially complex task. The products consist of many parts and processing takes part in several production stages. The jobs must share capacities. An important function of management is therefore the condition and control of complex activities, including the machine sequencing problem. The well-known objectives-nominal throughout times, low inventories, high capacity utilization, ultimately associated with the goal of minimizing the

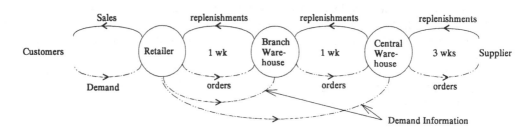

**Figure 2.   A Multi-Echelon System with Single Stage Information Flow**

sum of the differences of revenues minus expenses for all periods.

### 4.3 Dynamic Programming Formulation

We formulate our system to classify the system characteristics. We consider here the system simply multi-echelon inventory problem with a single item[5]. Following variables are used:

$n$ : period,

$x_r$ : finished goods inventory at the $r$th system in the $n$th period $(r = 1, 2, \cdots, m)$,

$z \doteq y - x_r$ : production (or transportation) order quantity at the $r$th system in the $n$th period,

$\xi$ : demand at final stage in the $n$th period.

Consider the case of $m$ processes arranged in series, so that process 1 receives stock from process 2, 2 from 3, and so on (see Figure 2). Stock matters the system through highest process, material and Kanban being removed at each step to satisfy random demands. If the following assumptions are made, then the optimal policies for the material and Kanban may be calculated easily.

1.  $c(z)$ is the cost of shipping units from a single process to the next one. This is proportional to the quantity shipped $z$, that is $c(z) = c \cdot z$.
2.  Excess demand is backlogged.
3.  The expected holding and shortage cost to be charged to each process during a single periods are functions of the stock at that process plus all stock in transit and on hand at lower process.

There will be $m$ processes, with process $r$ receiving stock from process $r + 1$. The shipment time from process $r - 1$ to process $r$ will be assumed to be a one period. Demand for the item occurs only at the lower process 1 with a density $\phi(\xi)$. The expected holding and shortage cost function will be $L(x_r)$ respectively with $x_r$, representing stock at $r$th system is described by the pairs $(x_1, x_2, \cdots, x_m)$, overall minimum cost function $C_n(x_1, x_2, \cdots, x_m)$ that satisfy

$$C_n(x_1, x_2, \cdots, x_m) = C_n(x_1) + q_n(x_2) + \cdots + q_n(x_{m-1}) + C_n(x_m), \tag{1}$$

and production order quantity $z_n(x_r)$ at the $r$th system in the $n$th period is for $n = 1, 2, \cdots$ under the some assumption,

$$z_n(x_r) = \begin{cases} \bar{x}_n - x_r, & \text{for } x_r \leq \bar{x}_n, \\ 0, & \text{for the otherwise,} \end{cases} \tag{2}$$

where for $r = 1, 2, \cdots, m-1$ in branch warehouse,

$$q_n(x_r) = \min_{y \geq x_r} \left\{ c_r(y - x_r) + L_n(x_r) + \Lambda_n(x_r) + \alpha \int_0^\infty q_{n-1}(y - \xi)\phi(\xi)d\xi \right\},$$

$$\text{where } \Lambda_n(x_r) = \begin{cases} c_r(x_r - \bar{x}_n) + \alpha \int_0^\infty \left\{ q_{n-1}(x_r - \xi) - q_{n-1}(\bar{x}_n - \xi) \right\}\phi(\xi)d\xi, & (x_r \geq \bar{x}_n), \\ 0, & (x_r < \bar{x}_n). \end{cases} \tag{3}$$

The critical value $\bar{x}_n$ is computed as the unique solution of the equation:

$$c_r(1-\alpha) + \alpha \int_0^\infty \left\{ L_n'(\bar{x}_n - \xi) + \alpha \int_0^\infty L_{n-1}'(\bar{x}_n - \xi_1 - \xi)\phi(\xi_1)d\xi_1 \right\}\phi(\xi)d\xi = 0. \tag{4}$$

Also, in the retailer,

$$C_n(x_1) = \min_{y \geq x_1} \left\{ c_1(y - x_1) + L_n(x_1) + \alpha \int_0^\infty C_{n-1}(y - \xi)\phi(\xi)d\xi \right\}. \tag{5}$$

The critical value $\bar{x}_n$ is computed as the unique solution of the equation:

$$c_1(1-\alpha) + \alpha \int_0^\infty L_1'(\bar{x}_n - \xi)\phi(\xi)d\xi = 0, \tag{6}$$

and in central warehouse,

$$C_n(x_m) = \min_{z \geq 0} \left\{ c_1(z) + L_m(x_m) + \Lambda_n(x_m) + \alpha \int_0^\infty C_{n-1}(x_m + z - \xi)\phi(\xi)d\xi \right\}. \tag{7}$$

The critical value $\bar{x}_m$ is computed as their unique solution of the equation:

$$c_1(1-\alpha) + c_r(1-\alpha) + \alpha \int_0^\infty \left\{ L_m'(\bar{x}_m - \xi) + \alpha \int_0^\infty L_{m-1}'(\bar{x}_m - \xi_1 - \xi)\phi(\xi_1)d\xi_1 \right\}\phi(\xi)d\xi = 0. \tag{8}$$

The physical meaning of the critical value $\bar{x}_m$ is the critical number of Kanban. The first assumption we shall consider involves the stocking of only one item. We shall assume that orders are made at each of a finite number of equally spaced items and immediately fulfilled. After the order has been made and filled, a demand is made. This demand is satisfied as far as possible, with excess demand loading to a penalty cost. The principal consequence of discussion is the following important theorem:

**Theorem** If $h(t)$ and $p(x)$ are convex increasing and $c(z) = c \cdot z$, and installation numbers is $m$, then the optimal policy is the form described in equation (2) where $x$ is determined satisfying equations (4), (6), and (8). The functions $h(t)$ and $p(x)$ are the expected holding cost function and the expected shortage cost function, respectively.

Please refer to the reference [6], because we have no space to explain with some numerical results.

Our problem now is to attempt to in corporate a set up cost associated with the transportation of items from installation 2 to installation 1. It is therefore appropriate to ask, still on the intuitive level, for the part played by the assumption of no set up cost in the previous policy. First of all, the lack of a set up cost was responsible for the simple description of optimal policies at the lower level in terms of a sequence of single critical numbers $\bar{x}_3, \bar{x}_4, \cdots$ . If a set up cost in transportation were included in the problem, the optimal policy would no longer be of this simple form.

Then the optimal policies are of the (S, s) type, with a pair of numbers, $S_n$ and $s_n$, relevant for each period.

## 5. CONCLUSION

The statements show that the efficiency of a JIT production depends on the selection of the dimension of capacities. In cases where significant seasonal variations, long lead times or huge set up times exist, stock-keeping can be more advantageous than a JIT production.

## REFERENCES

1. Tanaka, R., Shin, S., and Sebe, N.: "Controllability for Autonomous Decentralized Control," *Seiken*(The University of Tokyo)/*IEEE Symposium on Emerging Technologies and Factory Automation*, Tokyo, November 6-10, 1994, 265-272

2. Monden, Y.: *Toyota Production System*, Industrial Engineering and Management Press, 1983

3. Odanaka, T.: "On the Theoretical Foundation of Kanban in Toyota Production System," *Memories of Tokyo Metropolitan Institute of Technology*, No. 4, (1991), 79-91

4. Odanaka, T.: "On Approximation of Multi-Echelon Inventory Theory," *Memories of Tokyo Metropolitan Institute of Technology*, No. 3, (1989), 43-46

5. Odanaka, T.: *Optimal Inventory Processes*, Katakura Libri, Inc., 1986

6. Odanaka, T., Yamaguchi, Keiichi., and Masui, Tadayuki.: "Multi-Echelon Inventory Production System Solution," *Computers and Industrial Engineering*, 27 (1994) 1-4, 201-204

# SELECTION OF MACHINE TOOL CONCEPT FOR MACHINING OF EXTRUDED ALUMINIUM PROFILES

**E. Røsby**

**Norwegian University of Science and Technology, Trondheim, Norway**

**K. Tønnessen**

**SINTEF, Production Engineering, Trondheim, Norway**

KEY WORDS: Aluminium extrusions, Manufacturing cells, Machining, Machine tools

ABSTRACT: Aluminium extrusions offer flexibility in product design and are extensively utilised in structural applications and machine design. The actual aluminium alloys are easily machinable and machining of extruded aluminium profiles is developing as an important production method. It is therefore necessary to know the process factors to meet quality requirements and to remain competetive. Experience from a project in a company shows that organizing the machining department into cells increased efficiency and simplified planning and control. This paper describes important factors in the machining of extruded profiles and focuses on decision criterions for the choice of new machining systems. One of the factors is the clamping of the workpiece, either as full length profile or as cut to final length. The complexity of the profiles and the batch sizes are also important factors.

## 1. INTRODUCTION

Aluminium alloys are effectively utilized in the form of extruded profiles, as hot extrusion can produce very complex hollow and solid sections. A range of alloys and heat treatments are available to cover a broad spectrum of strength requirements. The application of profiles in the manufacturing of aluminium components includes machining operations as cutting, drilling, milling and tapping. Punching is not extensively used for hollow sections, but for open sections this is a process often employed to produce various shapes of holes and slots. Punching necessitates costly equipment, but renders possible high production rate.

The commonly used alloys do not display machinability problems in the conventional sence, since tool wear and cutting forces are low. The dominating problems are related to chip formation, as aluminium tends to produce long, continous curls, [1]. Heat treatable alloys, however, contain high percentages of alloying elements, and are more easily

Published in: E. Kuljanic (Ed.) *Advanced Manufacturing Systems and Technology*, CISM Courses and Lectures No. 372, Springer Verlag, Wien New York, 1996.

machinable. Aluminium profiles are more machinable in heat treated tempers than in softer tempers.

Typical products are aluminium profiles used in buildings and transportation, as door and window frames and components for cars, trains, airplanes and boats. The aluminium part production must be economically even at small batches, since there is an increasing demand of complexity of the products and towards one-of-a-kind production.

The aim of the paper is to present the central criterions used when small and medium sized enterprises are faced with the problem of choosing between different concepts for machining extruded profiles. A systematic approach is presented and advice is given, based on experiences made during a project run in a company.

The basic problem when planning the production of aluminium parts is to choose between the 'high technology model' and the 'small factory model'. The 'small factory model' in this scence utilizes simpler, often manually operated machines organized in manufacturing cells. The 'high technology model' involves highly automated machines, and there are two different concepts which have evolved the last years; the bench type and the flowline. The bench type machine tool has one spindel and the profile is fixed on the bench during the machining. The flow line has three to four spindles and the profile is fed progressively into the work space of the spindles.

## 2. SPECIAL FEATURES OF THE PROCESS

The process of machining extruded profiles has some significant features which separate it from other machining operations. Parts requiring surface treatment must be anodized before machining, because anodizing of full length profiles is more economic than that of singel parts. The advantage of anodizing before machining is also simpler handling of fewer parts. Anodizing requires electric contact points, leaving marks on the profile. The marks must be cut off, thus making it impossible to anodize parts that are already machined. Anodizing the parts after machining will of course reduce the risk of getting scratches in the surface, but it will give marks from the points at which the part is fixed to the electric conductors.

Machining of extruded profiles differs from machining of solid or hollow bars, which are usually machined in turning centers. Often the total surface of the parts produced from bars are machined, thus leaving a surface integrity decided by the machining process itself. This is not valid for the machining of separate holes and slots in profiles, as the main visible surface is produced by the extruding process or the anodizing. The milling tool is entered perpendicular to the surface, leaving a slot suitable for fixing it to other parts or for mounting of other functions. The angle of the spindle can also be varied relative to the surface of the profile.

What separates these machines from conventional CNC machines and milling centres? As aluminium requires low machining forces, there is no need for powerful spindels. Since the tolerances are not too narrow, there is not a need of extremely rigid constructions. The number of available tools is limited to a handful. The control system is simpler, since the machined features are limited to holes, slots and cutoffs. Despite the small diameter of the tool, high cutting speed is obtained through high spindle speed. Since the profiles are usually machined in a hard temper, burrs are minimized when the tool is sharp.

## 3. CHOICE OF PRODUCTION CONCEPT

Faced with lack of capacity or failing profitability, the machining department is forced to reorganize using the available equipment or to provide new machines. The following procedure is recommended:

1. Study of products and processes. All the products to be machined must be examined to reveal information on the following points: number of parts and variants per batch and year, throughput times, technical requirements like tolerances and surface specifications.
   Output: requirements to be set to the machining concept.
2. Evaluate alternative solutions. Several 'high technology models' and 'small factory models' must be regarded. Costs and savings must be calculated for each alternative.
   Output: chosen machining concept.
3. Elaborate the solution. The chosen concept must be planned in detail.
   Output: investment plans, layout, operator tasks and surrounding organization.

There are developed several methods for the design of manufacturing cells using group technology, [2,3]. In the next chapter, different organization models and machining concepts are dicussed. The alternatives are the ones to be evaluated in point 2 above.

## 4. WORK SHOP ORGANIZATION AND SPECIAL MACHINE TOOLS

### 4.1 Manual machines and functional layout

Traditionally, the production processes are carried out using manually operated machines. After the design phase, a technical drawing of the product is the basis for the planning and manufacture of the product. In the planning phase the sequence of the machining operations is optimized to utilize the available resources in the best way possible. During the planning of the operations, it is necessary to get as many products as possible out of each profile. The waste material must be reduced to a minimum to ensure optimal economical production. Usually, the length of the profiles are 6 meter. Often the first operation is cutting of the full length profiles to ease the handling in the workshop. The subsequent operations are milling, drilling, tapping and countersinking, and finally packing. As the numerous machines are arranged by function, this may be called functional layout. This is characterized by inefficient manual handling and intermediate storing of the semifininshed parts between each station. Logistics is of course a challenge when the number of variants is high and the batch sizes are small. A high degree of manual handling is monotonous and less effective than automated handling.

The machine tools may also be automated. For the cutting of the profiles, special saws with programmable feeding are frequent. In the case of simple cross sections of the profiles, the technique of cutting several profiles at time is efficient. When boring multiple holes, special machines using several spindles may be used. The spindles are then manually positioned at the start of each batch.

## 4.2 Manual machines and manufacturing cells

Reduced handling and increased productivity may be achieved by organizing the production department in manufacturing cells. The principle of cell production is very relevant when machining extruded aluminium profiles. The production department is then organized into almost independent cells responsible for groups of parts depending on the types of machines in the cell. Benefits regarding cell production are simple material flow, simple planning and control and high motivation among the workers. Special features for cell production are:

- Each cell is dedicated to produce one type of products, analysed by group technology
- Order of priority is 'first in - first out'
- The product is machined to a finished state before leaving the cell
- There is a limited number of machine tools inside each cell to keep it surveyable and easy to control
- The organizing of the cells is supposed to be simpler, requiring more responsibility at cell level
- Splitting the production of a batch is not allowed. That increases necessary control and is an unwanted effect

Organizing the production department as manufacturing cells will increase the number of necessary machine tools and often offer excess production capacity. However, simple machine tools are not too expensive and the cost of capital is less than labour cost. The problem of intermediate stock is reduced, but not eliminated. Each operation will still need handling and waiting. The structure is, on the other hand, more flexible. For the case of profiles already surface treated, manual handling can be gentle to avoid surface scratches. Visual inspection can be done at site. The choice of organization affects the need of new machines. If the workshop is organized in cells, there is less need for investments in highly automated machines. Manual mounting of the workpiece into the machines makes it possible to machine products that are already bent to its final form.

The following two chapters focus on some technical aspects of the two different concepts in the 'high technology model'. The latest generation of these two types of machine tools are highly automated CNC machines in order to finish the machining of the parts once the profile has been mounted into the machine. Because aluminium alloys used for extrusions of this type have good machinability, high speed spindels are used in both concepts.

### 4.3 Bench type with one spindle

One of the main types of advanced machine tool concepts is the bench type with one spindle on a gantry and fixed workpiece. This concept of machining holds the workpiece fixed in one position until the machining of all the parts out of one profile is finished. There is only one spindle, which can move in six axis and approach the aluminium profile from up to six sides, by pulling the last finished workpiece forward and apart from the rest of the profile, see Figure 1.

Advantages with this system is high machining precision and good surface quality of the profile. The front side of the aluminium profile is the free surface and the other sides are clamping surfaces. By fixing the workpiece this way, the risk of introducing scratches on visual surfaces are reduced, as the profile never moves relative the supporting surfaces. A drawback is long transfer times because the spindle support must move to all positions.

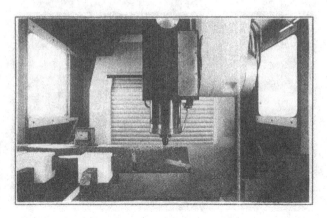

*Figure 1. One spindle gantry machining (Handtmann)*

Mounting and clamping is critical to avoid scratches in the anodized layer. Manual mounting ensures careful handling and control of the clamping. It is possible to use palettes to fix complicated sections.

### 4.4 Flowline concept with moving workpiece

Multi spindle machining with moving workpiece is the other concept in the 'high technology model'. Normally, the profiles are automaticly loaded into the gripper from the loading magazine. Up to four spindles machine four sides of the profile simultaneously. Each spindle can move in three directions. It is not possible to machine the profiles from the two sides at the ends of the workpiece. A gripper holds and feeds the profile to position it near the milling tools. When the milling process is completed, the part is cut off in the integrated saw and stored on the unloading magazine. Cutting of the profiles can be done at arbitrary angles. The magazines are manually loaded and unloaded, see Figure 2.

Chips between the fixtures and the profile can reduce the positioning accuracy. The risk of getting scratches in the anodized surfaces is also higher because chips can attach to the surface of the profile. During feeding and clamping this can cause damages and rifts even though the clamping system is made of soft material and the profile is fed on rubber cylinders. This may be partly avoided by reducing the amount of cutting fluide and machine as dry as possible. It is only possible to machine straight profiles. Products that are already cut and bent can not be clamped in this machine. The concept can be integrated with a CAD/CAM system. The CNC codes for the machining can be generated from the CAD drawing of the product.

*Figure 2. Flowline concept with moving workpiece and multiple spindles (Extech)*

## 5. CLAMPING AND SURFACE QUALITY

Clamping and fixturing of the workpiece is a central problem when machining extruded profiles. Because of the low elastic modulus of aluminium, there is a risk of deforming thin wall profiles in the clamping equipment. The excerted forces on the profiles must be low, and they must be distributed over an area and not act in one point, causing local plastic deformations. Machining forces are low, reducing the need of high clamping forces.

Due to the nature of the extrusion process and the heat treatment, the shape of the extruded profile vary. In spite of stretching after extrusion, the profile is not completely straight. There is also thickness variation over the cross section, resulting in wall thickness variations along the length of the profile. This affects of course the tolerances of the machined product.

It is difficult to set specific requirements to an anodized or lackered surface after machining. The requirements should be easy to verify in order to speed up control. Often the demands are qualitative, like 'no visible scratches allowed', but these criteria are diffuse and not uniquely defined. The production of machined profiles are thus on the 'safe side' to avoid scratches and surface damages. Only visible surfaces need to be free of marks, but handling must be gentle and this slows down the production rate.

Machining of materials with low Young's modulus, like aluminium alloys, normally gives large burrs. It is difficult to define and quantify the size of machining burrs without thorough inspection. At the same time it is hard to predict the resourses required to deburr the parts. In practice, this is no problem, because the burrs are small. Nevertheless, it is important to know parameters like burr root thickness and burr length. The knowledge of how different cutting parameters influence the burr formation is important.

Modern machining concepts, like those described here, use a milling tool for most machining operations, exept cutting and tapping. Small diameter holes are drilled with a flute drill, but larger holes are milled with a circular motion after the milling tool has penetrated the profile wall. In this way the burrs formed as the tool enters the material are removed as the tool enlarges the hole to the specified diameter. The large burrs formed by a drilling tool as it enters and penetrates the workpiece are thus avoided. The two different machining concepts bench type and flowline both use this technique and they are equal when it comes to problems regarding burr formation.

## 6. CONCLUSIONS AND COMMENTS

Even though there are automated machines available on the market, the concept of machining profiles using simpler machines organized in manufacturing cells still is an alternative when investments have to be kept at a minimum. It is not necessary to introduce new machines without evaluating other possibilities, using the present equipment. Manual handling operations can be combined with visual quality control of the surfaces, and traditional production concepts are on the safe side when it comes to scrap because of surface damages.

If the conclusion is to invest in the 'high technology model', there are basically two different solutions. In fact there are many, since the suppliers are numerous. The dilemma is to choose a concept which does not violate the surface requirements. The bench type machine gives the possibility to clamp profiles that are already cut and slightly bent. To be on the 'safe side' when it comes to scratches on anodized surfaces, the bench type should be preferred, since the profile does not move relative the clamping equipment.

If the products are always machined starting with a full length profile, the flowline concept is effective. Since there is some sceptisism to use this system for high quality surfaces, a test using the actual products should be conducted.


## ACKNOWLEDGEMENT

Thanks are due to the Research Council of Norway for financial support.

## REFERENCES

1.  Metals Handbook, American Society for Metals, 8th Edition, 1967, vol. 3 Machining
2.  Kamrani, A. K., Parsaei, H. R.: A methodology for the design of manufacturing systems using group technology, Production Planning and Control, 1994, vol. 5, no. 5, 450-464
3.  Onyeagoro, E. A.: Group technology cell design: a case study, Production Planning and Control, 1995, vol. 6 , no. 4, 365-373

# MACHINE LOCATION PROBLEMS
## IN FLEXIBLE MANUFACTURING SYSTEMS

**M. Braglia**

**University of Brescia, Brescia, Italy**

KEY WORDS: Machine Layout, Flexible Manufacturing System, Heuristic, Optimization

ABSTRACT: In this paper we deal with an algorithm for the facility placement problem relevant to one of the most commonly used layout schemes in flexible manufacturing systems: the single-row layout problem. The objective function that we want to minimize is the backtracking. As a result we obtain a heuristic easy to implement which generates good quality solutions in acceptable response times. It turns out that, in particular cases, it outperforms an alternative algorithm recently developed to the same end.

## 1. INTRODUCTION

Manufacturing facilities which make use of new technologies and philosophies such as group technology (GT) or flexible manufacturing systems (FMSs), in designing activities require more attention than in the past. The development of an optimal machine layout constitues an important step in designing manufacturing facilities due to the impact of the layout on material handling cost/time, machine flexibility, and productivity of the workstations. According to the GT philosophy, the FMS facility layout problem involves three steps: (i) part family and machine cell formation , (ii) detailed layout within the cell, and (iii) the arrangements of the cells. The first problem is relevant to the clustering of machines, based on the part families, in manufacturing cells (or departments). The second

Published in: E. Kuljanic (Ed.) *Advanced Manufacturing Systems and Technology*, CISM Courses and Lectures No. 372, Springer Verlag, Wien New York, 1996.

one concerns the optimal placement of the facilities within each cell. Finally, the third one is similar to the block layout problem.

Our aim is to present a heuristic expressly developed to treat the second step in the particular case of the linear single-row machine layout, one of the most implemented schemes in FMSs (Figure 1). In this configuration, machines are arranged along a straight line where a material handling device (MHD) moves the items from one machine to another. The success of this configuration is due to the major efficiency of the handling device flow path [1]. Of course, also a circular machine layout (see Figure 2), where equipments are placed on a circumference served by a robot positioned in the centre of the circle, can be treated with our heuristics. To this end, it will be sufficient to rectify the path of the handling robot.

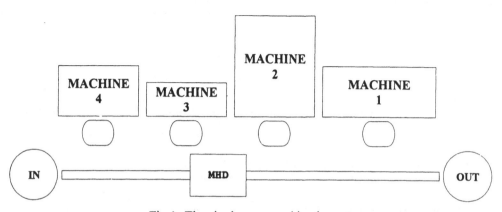

Fig. 1: The single row machine layout

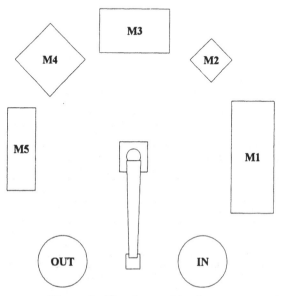

Figure 2: Circular machine layout

The single row layout problem (SRLP) is NP-complete and, therefore, it can be addressed with heuristics. In this paper we propose a new heuristic for minimising the *backtracking* movement in an FMS. (We recall that backtracking means movement of some parts from a machine to another one that precedes it in the sequence of placed machines). To this end, we use data accumulated in an **F** flow matrix (also referred to as "travel chart", "cross chart", or "from-to chart") and we consider the distances between two generic machines equal to one. Therefore, the fundamental quantity to be determined is the *frequency* $f_{ij}$ of parts displacements between machine $i$ (MI) and machine $j$ (MJ) in the cell.

Rigorously, the mathematical formulation of the problem is the following [2]. Given a flow matrix F, whose generic element $f_{ij}$ is the number of total parts moves from machine $i$ to machine $j$, the model requires each of M unique machines to be assigned to one of M locations along a linear track in such a way that the total backtracking distance is minimized. Assume, as mentionned, that machine locations and facilities are equally-spaced. Then, the general form of the model may be stated as follows:

minimize $\qquad c(x) = \sum_{i=1}^{M}\sum_{k=1}^{M}\sum_{j=1}^{M}\sum_{h=1}^{M} d_{ikjh} x_{ik} x_{jh}$

subject to $\qquad \sum_{i=1}^{M} x_{ik} = 1, \qquad k = 1, 2, ..., M$

$\qquad\qquad\qquad \sum_{k=1}^{M} x_{ik} = 1, \qquad i = 1, 2, ..., M$

$\qquad\qquad\qquad x_{ik} = 0 \quad \text{or} \quad 1, \qquad \text{for all } i, k.$

The decision variable $x_{ik}$ equals 1 if machine $i$ is assigned to location $k$, and 0 otherwise. The distance parameter $d_{ikjh}$ is defined by

$$d_{ikjh} = f_{ij}\delta_{kh}$$

where

$$\delta_{kh} = \begin{cases} k - h & \text{if } h < k, \ i, j = 1, 2, ..., M \\ 0 & \text{otherwise.} \end{cases}$$

The constraints ensure that each machine is assigned to one location and that each location has one machine assigned to it.

Backtracking adversely impacts the movement cost and productivity of a manufacturing facility as it causes the movement in the flow line to resemble the movement in a job shop layout. Moreover, different material handling devices from those in use may be required, and queues may appear. The study of the facility layout is important as increased machine flexibility and product diversification create additional complexity in scheduling and material handling. Several heuristics have been developped to this end (e.g., [1], [4], [6], [7]). An excellent review on this subject can also be found in [3].

To evaluate the performances, our algorithm is compared with a procedure developed by Kouvelis, Chiang and Yu in 1992 [8]. The results are obtained using several sets of problems with flow matrices characterized by different sizes and different flow densities. It turns out that our technique outdoes the heuristic of Kouvelis, Chiang and Yu, when using low density flow matrices, or when the system is characterized by high densities and few machines. For this reason, our heuristic can also be considered a complementary technique to that of Kouvelis, Chiang and Yu.

## 2. THE KOUVELIS, CHIANG AND YU HEURISTIC

As described in [3], Kouvelis, Chiang and Yu suggest two different techniques for minimizing the backtracking in a SRLP. In particular, we consider the first heuristic, here simply called KC_1. According to this heuristic, the flow matrix is given another form that reflects the difference between the entries above and below the diagonals of the matrix. The machine having the largest row sum in the trasformed matrix is selected and placed next in the layout. The trasformed matrix is updated after each placement.

## 3. A NEW HEURISTIC

The steps of the heuristic proposed here are the following:

Step 1) in correspondence of a given *random* initial solution, X*, represent the backtracking cost in an $n \times n$ matrix;

Step 2) for each machine calculate the sum of the elements of the corresponding row in this matrix;

Step 3) order the $n$ machines by decreasing sums;

Step 4) take the first two machines and schedule them in order to minimize the partial backtracking value as if there were only these two machines;

Step 5) for k = 3 to $n$ do:

Step 6) insert the $k$-th machine at that place, among the $k$ possible ones, which minimizes the partial backtracking value;

Step 7) if the final solution X gives better result than X*, then X* = X and goto Step 1, else stop.

## 4. EXPERIMENTAL RESULTS

Figure 3 reports our results in terms of *relative percentage deviations* (RPDs) defined as

$$RPD = \frac{F_{KC\_1} - F_{NEW}}{F_{KC\_1}} \cdot 100$$

where $F_{KC\_1}$ and $F_{NEW}$ are the backtracking values provided by KC_1 heuristic and present heuristic, respectively. All the values are obtained as average over 100 repetitions and are relevant to 6x5=30 classes of problems, caractherized by different problem sizes (i.e., number of machines) and flow densities (i.e., the percentage of zero terms in the flow matrix). The maximum number of machines we have considered is 30, which is quite sufficient for this kind of problem. In fact, the number of machines in a cell or a line on the facility floor is usually small (e.g., [3]). The non-zero frequencies are randomly generated from a uniform distribution U[1,20]. For each element $f_{ij}$ a random number between 0 and 1 is generated. If this random number is greter than the flow density value, we sample the non-zero frequencies, otherwise $f_{ij} = 0$. The distance between two machines is always equal to 1 and clearance is included in this distance.

As one can see, for low densities, the average improvements of our method are not negligible. Our heuristic outperforms the KC_1 algorithm also for medium/high densities (i.e., 0.75) and problems with few machines (i.e., 5, 10 and 15). The KC_1 heuristic is better for very high densities (i.e., 0.99) while the gap is reduced when we consider problems with higher size.

It is worth noting that the results of this new heuristic are obtained with interesting computation times. In fact, the response times required are less than one second also for problems with 30 machines (values obtained using a DIGITAL DEC 3000/600 Alpha computer).

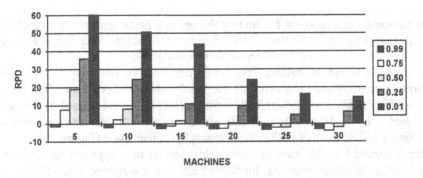

Fig.3: Average RPD values

To better evaluate our results, we conclude with an economical observation on the considerable importance of the layout problem. As reported in [10], Tompkins and Reed estimate that from 20% to 50% of the total manufacturing operating cost is due to the material handling system. As a consequence, even a small improvement of the performances of the heuristic must be considered as very interesting.

The results of Figure 3 are confirmed by the values of Table 1 which report the number of instances in which our heuristic is found to give the best solution for each class of 100 problems. As one can see, there are regions where our heuristic dominates the KC_1 heuristic. We recall that one speaks of *dominance*, when a heuristic produces better solutions with higher probability. (Concerning the concepts of dominance, the reader may refer to [9]).

|        | 5       | 10       | 15       | 20      | 25      | 30      |
|--------|---------|----------|----------|---------|---------|---------|
| **0.99** | 45 (8)  | 40 (0)   | 34 (0)   | 23 (0)  | 13 (0)  | 14 (0)  |
| **0.75** | 64 (12) | 60 (0)   | 47 (0)   | 29 (0)  | 30 (0)  | 16 (0)  |
| **0.50** | 65 (19) | 71 (1)   | 57 (0)   | 52 (0)  | 36 (0)  | 36 (0)  |
| **0.25** | 48 (41) | 73 (1)   | 72 (0)   | 77 (0)  | 65 (0)  | 78 (0)  |
| **0.10** | 16 (84) | 68 (18)  | 77 (2)   | 77 (0)  | 72 (0)  | 70 (0)  |

Tab.1: Number of instances in which the new heuristic is found to give the best solution (enclosed in brackets the number of instances in which the heuristic is found to give the same solution than KC_1 procedure)

## 5. CONCLUSIONS AND REMARKS

We have presented an algorithm for finding the optimal placement of machines in a single row layout, that is the configuration which minimises the backtracking distance travelled by a material handling device in a flexible manufacturing system.

Our algorithm is found to behave very well, with interesting computation times, and outperform another (recent) algorithm developed for the same problem when considering low flow matrix densities or medium/high densities (i.e., 0.75) and classes of problems with few machines (i.e., 5, 10 and 15). In this sense, we can consider the two heuristics as complementary of one another in the (Size,Density) problem space (Figure 4).

Future work should be addressed to the development of new appropriate versions of the heuristic able to minimise other cost functions such as the weighted sum of bypassing and backtracking, or the total movement distance (time) of the material handling device.

## ACKNOWLEDGEMENTS

This work has been supported by MURST and CNR. The calculations have been made on a DIGITAL DEC 3000/600 Alpha computer of the Department of Physics, University of Parma.

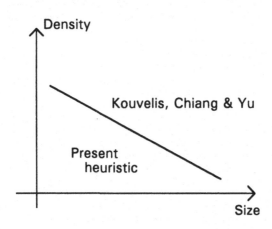

Fig.4: Classes of problems where the two heuristics dominate

## REFERENCES

1. Heragu, S.S., and Kusiak, A.: Machine layout problem in flexible manufacturing systems, Operations Research, 36 (1988), 258-268.
2. Sarker, B.R., Wilhelm, W.E., and Hogg, G.L.: Backtracking and its amoebic properties in one-dimensional machine location problems, Journal of the Operational Research Society, 45 (1994), 1024-1039.
3. Hassan, M.M.D.: Machine layout problem in modern manufacturing facilities, International Journal of Production Research, 32 (1994), 2559-2584.
4. Hollier, R.H.: The layout of multi-product lines, International Journal of Production Research, 2 (1963), 47-57.
5. Groover, M.P.: Automation, Production Systems, and Computer Integrated Manufacturing, Prentice-Hall International Inc.,1987.
6. Sarker, B.R., Han, M., Hogg, G.L., and Wilhelm, W.E.: Backtracking of jobs and machine location problems, in Progress in Material Handling and Logistics, edited by White, J.A., and Pence, I.W., Springer-Verlag, Berlin, 1991.
7. Kouvelis, P., and Chiang, W.: Optimal and heuristic procedures from row layout problems in automated manufacturing systems, Working paper, The Fuqua School of Business, Duke University, 1992.
8. Kouvelis, P., and Chiang, W.: A simulated annealing procedure for single row layout problems in flexible manufacturing systems, International Journal of Production Research, 30 (1992), 717-732.
9. Lin, S.: Heuristic programming as an aid to network design, Networks, 5 (1975), 33-43.
10. Afentakis, P., Millen, R.A., and Solomon, M.M.: Dynamic layout strategies for flexible manufacturing systems, International Journal of Production Research, 28 (1990), 311-323.

# WORK-IN-PROCESS EVALUATION IN JOB-SHOP AND FLEXIBLE MANUFACTURING SYSTEMS: MODELLING AND EMPIRICAL TESTING

A. De Toni and A. Meneghetti
University of Udine, Udine, Italy

KEY WORDS: work-in-process, lead time, job-shop, flexible manufacturing system

ABSTRACT   The authors investigate how work-in-process can be reduced when passing from a job-shop to a flexible manufacturing system. As explained in the paper, the problem can be brought back to analyzing how lead time can be shortened. Lead time evaluation formulas for a job-shop and a flexible manufacturing system are therefore proposed. In the case of job-shop systems the formula is empirically tested by comparing estimated values with data collected in a manufacturing firm. Thus, by comparing the relations proposed, the factors concurring to reduce work-in-process when passing from a job-shop to a flexible manufacturing system are identified: the different ways of processing in the two systems and the possibility of reducing lot size.

## 1. INTRODUCTION

The authors are particularly interested in investigating how work-in-process can be affected by a change in the production system, i.e. when passing from a job-shop environment to a flexible manufacturing one. In the following sections, evaluation formulas for work-in-process in the two systems analyzed are proposed and finally compared in order to find the factors concurring to reduce its amount.

## 2. WORK-IN-PROCESS EVALUATION

According to Little's law [1], the following parameters are significant for estimating the value of work-in-process *(WIP)* in a system with a stationary process:

Published in: E. Kuljanic (Ed.) *Advanced Manufacturing Systems and Technology*, CISM Courses and Lectures No. 372, Springer Verlag, Wien New York, 1996.

$a_i$    = average lead time of the i-th item lot produced in the system [days];
$m_i$   = average value of raw materials per lot for the i-th item [$/lot];
$l_i$    = average value of direct costs per lot for the i-th item [$/lot];
$\Delta T_i$ = average time between the arrivals of two consecutive item i lots to the system [days/lot].

In fact, if the value is supposed to be added to item lots linearly with time as shown in fig. 1, then [2]:

$$WIP_i = \frac{a_i}{\Delta T_i}\left(m_i + \frac{l_i}{2}\right) \quad [\$] \tag{1}$$

The value of $WIP_i$ is therefore represented by the area of the trapezium in fig. 1. The evaluation formula (1) shows how $WIP_i$ [$] is the product of the average number $(Q_i)$ of lots in the system provided by Little's law $(Q_i = a_i/\Delta T_i)$ and the average value of each lot. On average 50% of the final value of each lot is already present and 50% remains to be added. The most difficult parameter to determine in equation (1) is the average lead time. Hence the reason why lead time evaluation formulas for job-shop and flexible manufacturing systems are proposed in the following sections.

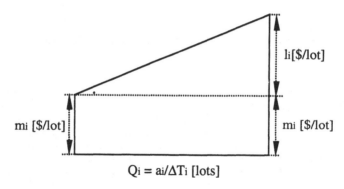

$Q_i = a_i/\Delta T_i$ [lots]

Fig. 1 - WIP when value is added to item lots linearly with time

## 3. LEAD TIME EVALUATION IN A JOB-SHOP SYSTEM

It would be useful to express lead time as a function of some known parameters relative to process cycles and job-shop characteristics. The authors propose the evaluation formula explained in section 3.1 and provide empirical test results in section 3.2.

### 3.1   MODELLING

When considering the i-th item belonging to the j-th product family manufactured in the system:

$$a_i \cong \frac{n_i}{X_j} \tag{2}$$

where:    $a_i$   =   lead time of the i-th item lot [days];

$n_i$ = number of cycle operations for the i-th item [operations/lot];

$X_j$ = average number of operations completed per day on each lot of the j-th family [operations/(lot*day)].

Parameter $n_i$ can be easily deduced from each item routing, while particular attention must be paid to the number of operations completed per day on each lot of the j-th family($X_j$).

The first to mention a similar parameter was Corke [3], who talked about the possible utility of the average number of operations per lot and day in lead time evaluation for job-shop systems. He considered this parameter as a characteristic of the whole job-shop, without distinguishing product families, and ascribing it values which ranged between 0.2 and 0.4 operations/(lot*day). Also on the grounds of empirical evidence (as shown in the following section) it appears that when a single value is ascribed to parameter $X$, no significant errors are made, provided that the items are characterized by a high degree of homogeneity in their manufacturing cycles. When, on the other hand, different families are processed in the job-shop system, $X$ must be related both to the manufacturing system management and the product characteristics.

In fact from relation (2) it follows that:

$$X \cong \frac{n_i}{a_i} = \frac{n_i}{\sum_{h=1}^{n_i} t_h} = \frac{1}{\bar{t}_i} \tag{3}$$

where: $t_h$ = duration of the h-th operation of item i cycle, including run time, set-up time, moving time and queue time [days];

$\bar{t}_i$ = average duration of a cycle operation for the i-th item [days].

Parameter X can therefore be related to the average duration of a cycle operation. However, each machine in the system differs for processing times, work loads and therefore queue length. Thus, lead time depends on which machines are used to process the i-th item. Since items belonging to the same family have similar manufacturing cycles, the error introduced by considering the average duration of an operation for a whole family is less than that given by calculating the average duration of an operation for all the products the system produces. Hence the reason why the authors relate X to item families and not to the whole system as Corke proposes.

From relation (2) it follows that for a given item family, lead times in a job-shop system approximately differ only for the number of cycle operations, i.e. the number of displacements in the system. Lot size is not involved in (2); this is due to the fact that most of the time an item spends in a job-shop system is wasted waiting in queue to be processed, thus independently from its size. Moreover items belonging to the same family are usually manufactured in lots of similar entity. Omitting the variability of time with lot size does not therefore lead to a considerable mistake in the approximation.

## 3.2    EMPIRICAL TESTING

The authors tested relation (2) in a local firm, which produces centrifugal ventilators and where component parts manufacturing is organized as a job-shop system.

Data on several families of items were collected over a three month period. To obtain the value of $X_j$ for each j-th family every day the number of operations performed for the family and the number of its lots present in the system were measured. Then X was calculated as the mean of the values obtained each day by dividing the number of operations performed by the number of lots in the system.

The authors furthermore measured the average lead time for each item produced in the observed period and deduced the number of their cycle operations from their routings.

The results of the four most representative families are shown in the tables below, which also provide lead time evaluation when a single value of $X$ ($X=0.29$) is calculated for the whole system.

| Part number | number of operations | Actual Lead Time [days] | Estimated Lead Time [days] $X_i=0.34$ | Error | Estimated Lead Time [days] $X=0.29$ | Error |
|---|---|---|---|---|---|---|
| 10100110 | 3 | 9.9 | 8.8 | 12.5% | 10.3 | 3.9% |
| 11000110 | 3 | 9.85 | 8.8 | 11.9% | 10.3 | 4.4% |
| 12800110 | 3 | 6.6 | 8.8 | 25.0% | 10.3 | 35.9% |
| 17130110 | 3 | 10.1 | 8.8 | 14.8% | 10.3 | 1.9% |
| 20100110 | 2 | 5.2 | 5.9 | 11.9% | 6.9 | 24.6% |
| 22200110 | 4 | 10.3 | 11.8 | 12.7% | 13.8 | 25.4% |
| 23200110 | 4 | 9.9 | 11.8 | 16.1% | 13.8 | 28.3% |
| 25000110 | 2 | 6.1 | 5.9 | 3.4% | 6.9 | 11.6% |
| 26300110 | 2 | 5.75 | 5.9 | 2.5% | 6.9 | 16.7% |
| 29000110 | 2 | 5.3 | 5.9 | 10.2% | 6.9 | 23.2% |
| average | 2.8 | 7.9 | 8.24 | 12.1% | 9.64 | 17.59% |

Table 1 - Actual and estimated lead times for side panel family

| Part number | number of operations | Actual Lead Time [days] | Estimated Lead Time [days] $X_i=0.29$ | Error | Estimated Lead Time [days] $X=0.29$ | Error |
|---|---|---|---|---|---|---|
| 11600130 | 1 | 3.2 | 3.4 | 5.9% | 3.4 | 5.9% |
| 11800130 | 1 | 3.3 | 3.4 | 2.9% | 3.4 | 2.9% |
| 12000130 | 1 | 3.1 | 3.4 | 8.8% | 3.4 | 8.8% |
| 12200130 | 1 | 3.8 | 3.4 | 11.8% | 3.4 | 11.8% |
| 12800130 | 1 | 3.8 | 3.4 | 11.8% | 3.4 | 11.8% |
| 13100130 | 1 | 3.2 | 3.4 | 5.9% | 3.4 | 5.9% |
| 14000130 | 1 | 3.4 | 3.4 | 0.0% | 3.4 | 0.0% |
| 14500130 | 1 | 3.45 | 3.4 | 1.5% | 3.4 | 1.5% |
| 15000130 | 1 | 3.6 | 3.4 | 5.9% | 3.4 | 5.9% |
| 16300130 | 1 | 2.5 | 3.4 | 26.5% | 3.4 | 26.5% |
| average | 1 | 3.33 | 3.4% | 8.1% | 3.4% | 8.1% |

Table 2 - Actual and estimated lead time for back panel family

| Part number | number of operations | Actual Lead Time [days] | Estimated Lead Time [days] $X_j=0.21$ | Error | Estimated Lead Time [days] $X=0.29$ | Error |
|---|---|---|---|---|---|---|
| 21000200 | 4 | 18.8 | 19.0 | 1.1% | 13.8 | 36.2% |
| 23190200 | 3 | 12.1 | 14.3 | 15.4% | 10.3 | 17.5% |
| 24000200 | 6 | 21.6 | 28.6 | 24.5% | 20.7 | 4.3% |
| 24090200 | 3 | 12.2 | 14.3 | 14.7% | 10.3 | 18.4% |
| 25090200 | 3 | 12.1 | 14.3 | 15.4% | 10.3 | 17.5% |
| 26300200 | 5 | 23.4 | 23.8 | 1.7% | 17.2 | 36.0% |
| 27100200 | 4 | 20.9 | 19.0 | 10.0% | 13.8 | 51.4% |
| 28000200 | 4 | 18.4 | 19.0 | 3.2% | 13.8 | 33.3% |
| 29000200 | 4 | 18.9 | 19.0 | 0.5% | 13.8 | 37.0% |
| 29092200 | 2 | 11.8 | 9.5 | 24.2% | 6.9 | 71.0% |
| average | 3.8 | 17.02 | 18.08 | 11.07% | 13.09 | 32.26% |

Table 3 - Actual and estimated lead times for nozzle family

| Part number | number of operations | Actual Lead Time [days] | Estimated Lead Time [days] $X_j=0.82$ | Error | Estimated Lead Time [days] $X=0.29$ | Error |
|---|---|---|---|---|---|---|
| 11000260 | 6 | 6.1 | 7.3 | 16.4% | 20.7 | 70.5% |
| 12000260 | 4 | 5.6 | 4.9 | 14.3% | 13.8 | 59.4% |
| 12200260 | 4 | 5.6 | 4.9 | 14.3% | 13.8 | 59.4% |
| 13100260 | 6 | 6.0 | 4.9 | 22.4% | 20.7 | 71.0% |
| 18000260 | 6 | 6.4 | 7.3 | 12.3% | 20.7 | 69.1% |
| 19000260 | 6 | 6.0 | 7.3 | 17.8% | 20.7 | 71.0% |
| 42091260 | 3 | 3.65 | 3.7 | 1.4% | 10.3 | 64.6% |
| 43600260 | 3 | 3.4 | 3.7 | 8.1% | 10.3 | 67.0% |
| 55094260 | 4 | 4.0 | 4.9 | 18.4% | 13.8 | 71.0% |
| 57194260 | 4 | 4.2 | ·4.9 | 14.3% | 13.8 | 69.6% |
| average | 4.6 | 5.09 | 5.38 | 13.97% | 15.86 | 67.26% |

Table 4 - Actual and estimated lead times for fan family

## 4. LEAD TIME EVALUATION IN A FLEXIBLE MANUFACTURING SYSTEM

For a flexible manufacturing system formed by working centers, the "scheduling factor" *(SF)* can be defined as the average number of hours per day of work capacity that can be dedicated to each production lot [4].

The high investments required for fixtures and tools to provide the system with the maximum degree of flexibility, i.e. the possibility of simultaneously processing items of the same type, lead to limit their availability. Therefore different items must be simultaneously processed in order to saturate the system. Thus, the scheduling factor is calculated by dividing the work capacity of the system with the number of different item lots that have to simultaneously be present.

Like $X$, also $SF$ must be related to each product family processed by the system, since a different number of fixture and tools can be provided to each family.

Therefore lead time for a flexible manufacturing system can be evaluated in the following way:

$$a_i \cong \frac{r_i}{SF_j} \qquad (4)$$

where:  $a_i$  = lead time for the i-th item lot [days];
  $r_i$  = total average run time for the i-th item, i.e. the total hours of work needed by the system to process a lot of a given size [hours/lot];
  $SF_j$  = the scheduling factor for the j-th family [hours/(lot*day)].

A formal comparison between lead time in a job-shop (JS) and in a flexible manufacturing system (FMS) can thus be made.

There appears to be a correspondence between the average number $(X_j)$ of operations per lot performed daily for a given family in a job-shop and the "scheduling factor" $(SF_j)$ for a FMS.

Since in a job-shop environment most of the time is spent waiting for the items to be processed by each machine of the routing, lead time is substantially determined by the number of operations to be performed, i.e. the number of displacements from one centre to another, and can be considered lot size independent (see equation 2). On the other hand, in a flexible manufacturing system inter-operation waiting times are not considerable and lead time appears to be strictly related to lot size (see equation 4). Hence, for a given family lead time is a fixed parameter when a job-shop system is involved, but is considered as variable one when a flexible manufacturing system is analyzed.

## 5. PASSING FROM A JOB-SHOP TO A FLEXIBLE MANUFACTURING SYSTEM

By analyzing relationship (1), (2) and (4) it is possible to identify which factors concur in reducing work-in-process when passing from a job-shop to a flexible manufacturing system with identical capacity.

Considering item data $(m, l, \Delta T)$ unchanged and omitting, for semplicity, the subscript related to the i-th item being analyzed, it can be written:

$$\frac{WIP_{JS}}{WIP_{FMS}} = \frac{\frac{a_{JS}}{\Delta T}\left(m + \frac{l}{2}\right)}{\frac{a_{FMS}}{\Delta T}\left(m + \frac{l}{2}\right)} = \frac{a_{JS}}{a_{FMS}} = \frac{n/X_j}{r/SF_j} \qquad (5)$$

Hence, the reduction of WIP is associated with shorter lead time one has when passing from one system to the other.

A first reduction can be ascribed to the different ways of processing in the two systems: a job-shop has fixed lead times, while a flexible manufacturing system has variable ones, with no queue time between operations (see equation 5). A numeric example is given in table 5, which shows how it is possible to shorten the lead time from the 28 days in the job-shop to the 9 days in the flexible manufacturing system (see columns 2 and 3).

A further reduction can be obtained by reducing lot size, due to the dependent nature of lead time in a flexible manufacturing system. If $k$ is the "size-reducing factor", then lead time can be shortened with the same factor until a single unit lot is reached (see columns 5 and 6 of table 5).

The following relationship, in fact, exists:

$$r_{FMS} = \frac{r_{JS}}{k} \qquad (6)$$

| | Job-Shop | FMS no lot-size reduction (k=1) | FMS with lot-size reduction (k=9) | FMS with lot-size reduction (k=99) |
|---|---|---|---|---|
| **Independent variables** | | | | |
| $S_i$ = lot size [units] | 99 | 99 | 99/9=11 | 99/99=1 |
| $n_i$ [operations/lot] | 7 | 1 | 1 | 1 |
| $c_i$ = run unit time [hours] | $0,\overline{63}$ | $0,\overline{63}$ | $0,\overline{63}$ | $0,\overline{63}$ |
| $X_j \left[\dfrac{\text{operations}}{\text{lot} \cdot \text{day}}\right] \; SF_j \left[\dfrac{\text{hours}}{\text{lot} \cdot \text{day}}\right]$ | 0,25 | 7 | 7 | 7 |
| $m_i$ [$/lot] | 54 | 54 | 54/9 | 54/99 |
| $l_i$ [$/lot] | 36 | 36 | 36/9 | 36/99 |
| $\Delta T_i$ [days/lot] | 1 | 1 | 1/9 | 1/99 |
| **Dependent variables** | | | | |
| $r_i = S_i \cdot c_i$ [hours/lot] | 63 | 63 | 63/9 | $63/99=0,\overline{63}$ |
| capacity $= \dfrac{r_i}{\Delta T_i}$ [hours / day] | 63 | 63 | 63 | 63 |
| $a_i = \dfrac{n_i}{X_j}$ [days] $\quad a_i = \dfrac{r_i}{SF_j}$ [days] | 28 | 9 | 9/9=1 | 9/99=0,091 |
| $Q_i = \dfrac{a_i}{\Delta T_i}$ [lots] | 28 | 9 | 9 | 9 |
| $WIP_i = a_i \left[\dfrac{1}{\Delta T_i}\left(m_i + \dfrac{l_i}{2}\right)\right]$ [$] | 28·72 | 9·72 | 9/9·72 | 9/99·72 |

Table 5 - Formal analogy between Job-Shop and FMS

The coefficient of WIP reduction can, therefore, be calculated:

$$\frac{a_{FMS}}{a_{JS}} = \frac{\dfrac{r_{JS}/k}{SF_j}}{\dfrac{n}{X_j}} = \frac{1}{k \cdot SF_j} \cdot \left(\frac{r_{JS} \cdot X_j}{n}\right) \tag{7}$$

The terms in brackets represent characteristics of the job-shop system being abandoned, while $k$ and $SF$ describe the flexible manufacturing system to which one passes and represent the variables granting a WIP reduction. The scheduling factor, in fact, takes into account the different nature of a flexible manufacturing system as compared to a job-shop one, which is instead characterized by the average number of operations per lot and day $(X)$. The size-reducing factor describes the possibility of reducing lot-size thanks to the lower set-up times in FMS.

From equation (7) it can be deduced that minimum lead time, and consequently minimum level of work-in-process, is related to the maximum values of the size-reducing factor $(k)$ and the scheduling factor $(SF)$.

The maximum value which can in theory be given to $k$ is equal to lot size [in table 5 $k = 99$]; the actual value assumed by this parameter does not however depend on FMS characteristics but rather on the requirements of the upstream and downstream stages.

On the other hand the value of the scheduling factor is related to the amount of investments on fixtures and tools. If the flexible manufacturing system is provided with the maximum degree of flexibility, than $SF$ is equal to the work capacity of the system.

When trying to limit the investments needed for the FMS, the value of $SF$ that is associated with the same lead time of the job-shop being abandoned must be considered as a lower bound.

It is, therefore, possible to determine the minimum acceptable value for $SF$ as follows:

$$a_{FMS} = a_{JS} \Rightarrow \frac{n}{X_j} = \frac{r/k}{SF_j} \Rightarrow \left(SF_j\right)_{min} = \frac{r}{k} \cdot \frac{X_j}{n} \tag{8}$$

An investment on fixtures and tools leading to a scheduling factor less than $SF_{min}$ provides the flexible manufacturing system with a lead time which is worse than that of the job-shop.

## REFERENCES

1. Little, J.D.C.: A proof for the queueing formula: $L = \lambda W$, Operations Research, 1961, 383-387.

2. De Toni, A.: Metodologie di stima del work-in-process in sistemi produttivi job-shop, Proceeding of the "Workshop sull'innovazione industriale", edited by Pagliarani G. and Gottardi G., CEDAM, Padua, 1986.

3. Corke, D.K.: Production Control is Management, Arnold Ltd., London, 1969, 205-231.

4. De Toni, A.: Problemi di gestione e controllo degli FMS, PhD thesis, University of Padua, 1987.

# ON THE APPLICATION OF SIMULATION AND OFF-LINE PROGRAMMING TECHNIQUES IN MANUFACTURING

**R. Baldazo, M.L. Alvarez, A. Burgos and S. Plaza**

Escuela Tecnica Superior de Ingenieros Industriales y de
Ingenieros de Telecomunicaicones, Bilbao, Spain

**KEY WORDS:** Simulation, Programming, Robot. Off-Line

**ABSTRACT:** The integration of Simulation and Off-line Programming techniques in highly automated manufacturing systems , like in the Automobile industry, is justified for the multiple benefits that it reports.

The Simulation improves the design of the manufacturing system, facilitates the study of design modifications and guarantees realibility and flexibility of the system.

On the other hand the Off-line Programming avoids the expensive production stops for traditional programming, keeps the programmers far from a dangerous and sometimes toxic ambient of the plant and reduces the final preparation of the system. In the Automobile sector this is a key factor for reducing cost and increasing the productivity.

Whenever an off-line programming procedure is to be introduced in a manufacturing system, a lot of problems can arise due to the fact that most of the times the design of the system is already done and the off-line techniques is not taken into account.

This paper reflects some of the problems encountered during the feasability study of an implementation of off-line programming techniques for a welding line in the Automobile industry. Most of them are related to the modelization of the geometrical elements and the different formats and providers of the necessary information. This claims the necessity of a good information management in the design of a plant, and the stablisment of standard data base for all of the suppliers.

## 1. INTRODUCTION

At present, the application of Simulation and Off-Line programming of robotized processes to the industrial environment is very scarce. The satisfactory experience that has been obtained with Simulation and Off-Line programming in the field of machine-tools (CAD/CAM) makes it possible, based upon similarity, the application of this method to robotized processes, allowing for a greater synchronism between the programming and production stages, and providing important quality improvements for the products, as well as important reductions in the production time (which, in turn, involve a cost reduction).

Having robots available as a production means, is becoming more and more common in the industrial environment, both as a basic manufacturing tool (welding robot, painting robot, assembly robots, etc.) and as an auxiliary element in production (machine load and unload, line feeding, etc.). The automobile industry is an excellent example of the constant increase of the number of robots in its assembly lines.

**Applying Simulation and Off-Line programming to Robotics** is one of the fields in which this technology proves to be more beneficial, both when designing and programming robotized processes.

**The design and modifications on robotized installations** have been carried taking prior experiences as a point of departure, and with the help of a very specialized staff. However, once the cell has been applied, it is very likely that certain parts will require some alterations and modifications. By means of a Simulation, the validation of the installation designs, and of the modifications, will be possible, even before their physical application. On the other hand, by means of this Simulation, we will also be able to verify the behavior of the cell movable elements, their scope, and the hypothetical collisions with the rest of the elements that may happen.

Traditional programming requires the use of robots as program generating tools. This involves stopping the assembly line, and the presence of the operator in charge of this line. Off-Line programming, on the other hand, allows for the creation and clearing of programs **outside the assembly line**. Programs are performed on the simulated plant, and trajectories can be defined accurately, and optimized as well. This technique is even more commendable when dealing with environments which are dangerous for the operator's physical well being, or at areas to which conventional programming finds it difficult to have access (such as welding or painting cells).

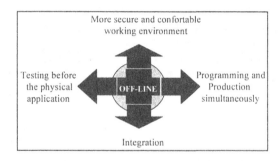

In the automobile industry, stopping production involves big economic losses. **Design changes** are common, and daily production must be quite high in order for installations to be as profitable as possible, and in order to obtain a competitive price for the vehicle. Simulation and Off-Line programming have an impact on all these areas, taking production closer to the final objective, that is, obtaining higher benefits.

The advantages that this kind of technology offers are summarized below:

- Time improvement:
    - Installation designs and modifications are validated before their physical application.
    - Stopping at the assembly line during the robot programming is avoided.
    - Robot programs are optimized, improving thus the cycle times.
- Quality improvement in:
    - Designs.
    - Job positions.
    - Programs.
    - Information organization.
    - Final product.
- Cost improvement, which, in turn, involves a benefit increase.

However, as we will show later, this is not as easy as it looks, and introducing this technology involves an important change in mentality, not only as far as robot programming (which requires a higher training) is concerned, but also as far as the whole production environment (which will have to get adjusted to the new production philosophy) is concerned.

## 2. STAGES IN SIMULATION AND OFF-LINE PROGRAMMING.

When applying Simulation and Off-Line programming techniques, assuming that it is necessary to compile first all kind of information (both geometrical and distributional information, and information about the process being performed at the working cell), a number of steps should be followed:

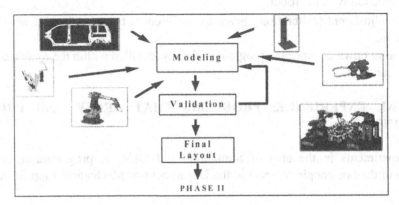

- Graphic representation of the cell components, which will be designed within the system, by means of importing the geometrical data from other CAD systems, or by using elements which are available in the system libraries.
- Plant layout definition.
- Validation of this design, checking the robots accessibility, tool kits, tweezers, grips, etc.

Once the final layout has been obtained, a second stage would start, comprising:

- The definition of the trajectories to be followed by all movable elements.
- Simulation.
- Optimization, detection of collisions among the elements, mistakes in the design of tool-kits, grips, tweezers, etc., positioning mistakes, and cycle time analysis.

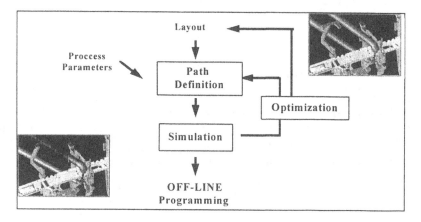

Once the trajectories are defined, and once the correct functioning of the system is checked upon, the robot programs will be obtained, and thus the Off-Line Programming stage would start:

- Post-processing, translation of the program from the language which is characteristic of the Simulation system, into the language which characterizes the robot controller.
- Communication with the robot.
- Program adjustment (calibration), given the differences between the real world and the Simulation.

From this point onwards, once the program is already installed within the robot's controller, production can start.

## 3. ACTUAL EXPERIENCE: PROBLEMS THAT AROSE AND PROPOSED SOLUTIONS

Several experiments in the area of Simulation and Off-Line programming have been carried out in the last couple of years at the Department of Mechanical Engineering of the

Higher Technical School of Industrial and Telecommunication Engineers at Bilbao. These experiences have been developed in the environment of automobile industries. In particular, we have worked on intermittent welding processes, concerning the rough assembly of car bodies.

The goal of this paper is describing the results of these applications, making both the problems that have appeared and the solutions that have been proposed in order to facilitate the Simulation process, so that a greater implementation of this technology is favored. The problems could be classified into several groups, according to the steps that are necessary in order to cover a Simulation project.

### 3.1. Information Management

When developing a Simulation and programming project, an important point that should be considered is the great amount of information that is being handled. It will be necessary to compile all kinds of information, such as geometrical information (plans, CAD files, neuter formats, etc.), the information concerning the mechanical, kinetic and dynamic behavior of mechanisms, trajectory information, process information, etc. Of course, all these pieces of information are distributed among the different departments in the company, and even outside the company itself (suppliers); moreover, a large number of people is in charge of its handling. Thus, a problem arises when trying to compile information.

The solution that has been proposed is setting up a way of working such that all those areas or departments of the company, which may be able to provide information when the Simulation is being carried out, should have available the mechanisms that are necessary to fulfill that task. Thus, new elements will be included in the common way of working of the company's staff, in order to register the information which the Simulation team needs, so that Simulation can be facilitated, and so that no additional cost is added due to the compilation.

Throughout time, and singularly in the automobile area, some design modifications appear, which need to be incorporated to all those places where reference to them is made. The different company departments, design and production, and even suppliers, may carry out small design modifications. Thus, it is essential for our Simulation and Off-Line Programming system to have updated data available. In order to solve this problem, a correct quality management system is necessary, together with an adequate configuration management, which will ensure that different double information doesn't exist, and that all elements have a clear reference. Thus, improving the communication system within the company, and as far as suppliers are concerned, is crucial.

### 3.2. Modeling
### Integration

When considering the **initial modeling problems**, one of the most important ones is the lack of standardization as far as information is concerned; standardization would allow us to reutilize the cell geometric data among the different systems used for the design, in a quick and reliable way.

Very often, data are available in different manners; they may appear in different CAD systems (2D or 3D), or in paper. When information is available in a CAD system, the problems that we find are the typical ones, concerning the information transfer among different CAD systems (which involves the need of translators). When information appears represented in planes, the design will have to be carried out, which involves an increase in both time and cost.

Given the fact that Simulation and Off-Line systems have a module for component design (a CAD module) available, the solution to this problem of information transfer will be provided by the development of these programs, which will progressively incorporate translators, in order to facilitate the direct transfer of information, coming from other systems.

**Information duplicity**

On the other hand, we found that the time devoted to modeling was longer than intended, due to the problem mentioned above, in the preceding section, and also due to the repetition of parts of geometry which were already present in other components. This was difficult to control, given the high number of components and component elements which make up a cell.

Creating a component data base, integrated within the system, has been posited as a solution to get rid of this lack of co-ordination which affects geometric data. The goal is to facilitate the cell design, taking advantage of prior designs, with similar sub components, by means of a simple codification.

This data base will contain those elements which may have a repeated role in the modeling of a robotized system (grips, tool-kits, engines, etc.), accumulating the effort made for prior developments, in order to minimize component modeling costs.

Just as robot libraries exist, the future tendency would be for suppliers of tool-kits, grips, tweezers, etc., to provide this kind of libraries.

**3.3. Simulation**

Among the problems which are **Simulation specific**, the presence of elements which do not appear in the Simulation (such as electrical or refrigeration conductions both in the robot and in the welding tweezers) should be considered. These elements just mentioned cannot be modeled, since they are not rigid elements, which are interposed between the robot access points, and the welding tweezers, so that trajectory modification is necessary. These systems don't take into consideration the loading capacity of robots during the Simulation either; that is, if a given robot is not able to support the weight of some welding tweezers, this will not be appreciated in the Simulation. Thus, a parallel study should be carried out in order to ensure it.

**3.4. Programming**

Both robots and Simulation systems have their own system available. All this language variety makes the application of numerous specific translators necessary. Depending on the kind of robot controller which is used in the system, a bi-directional relation should be set up between the programs that are obtained in the neuter language of the Simulation and Off-Line Programming system, and the specific languages that robots have. The level of this kind of developments does not have the necessary quality when translations are performed directly from the neuter Simulation language of the system to that of the robot.

### 3.5. Testing in the plant
**Measurement**
One of the problems that are specific to Off-Line Programming is measurement; this is a task which aims at balancing automatically the positioning mistakes which are due to inaccuracies of the robot, or to variations when placing the objects, by adjusting the simulated program to the real system. It is necessary to ensure the exact positions when placing robots, mainly repeatability, resolution, and accuracy.

**Communication**
Another problem is robot-computer communication. Due to the fact that the development platforms for both of them are different, communication problems may arise when trying to carry out program control or transfer operations.

### 4. CONCLUSIONS

Within the automobile field, an area in which robotized processes are widely implemented, and where stopping production may cause great economic losses, the application of this technology will provide important benefits, allowing for a greater synchronism between the programming and production stages; in this way, the adjustment to the market changes will be very fast. However, we should also be aware of the fact that the implementation of this technology in companies requires a change in the company's way of working, in order to get adjusted to the needs of these systems, and in order to obtain the real benefits that they offer.
As far as the future of these systems is concerned, it is related to the development of the graphic representation systems upon which they are bases, and to the improvement of areas such as data standardization and calibration.

### 5. REFERENCES
1. Baldazo R., Alvarez M. L., Burgos A.; Simulación y Programación Off-Line de Líneas Robotizadas; XII Congresso Brasileiro e II Congresso Iberoamericano de Engenharia Mecânica 1995.
2. Sorrenti P., May J.P.; Simulación y Programación de Sistemas Robotizados de Soldadura; AI/RR n 66, p. 60-66. 1992.
3. Dr. John Owens; Microcomputer-Based Inductrial Robot Simulator and Off-Line Programming System; Robotics Today , Vol 8, n° 2, 1995.

4. Rivas, J.; Una Herramienta que puede Ahorrar Tiempo y Dinero; Especial CIM, n°243, pp. 75-78. 1994.

5. Saïd M. Megahed; Principles of robot Modelling and Simulation; Ed. John Wiley & Sons, 1993.

6. Readman, Mark C.; Flexible Joint Robots; Ed. Prentice Hall, 1991.

7. ABB Industria, S.A.; Simulación Gráfica de Células Robotizadas; Revista Española de Electrónica, pp. 38-40. 1991.

8. Williams; Manufacturing Systems; Ed. Halsted Press, Jonh Wiley & Sons, 1988.

# SELECTION CRITERIA FOR GEOREFERENCING DATABASES IN INDUSTRIAL PLANT MONITORING APPLICATIONS

C. Pascolo

CO.R.EL. Italiana, Udine, Italy

P. Pascolo

University of Udine, Udine, Italy

KEY WORDS:     Plants Management, Simulation, Data Base Management Systems, Multimedia Integration.

ABSTRACT: This report attempts to identify numerical models suitable for describing the technological networks underpinning industrial plants and for realizing production line monitoring systems.

Two previous reports submitted to the proceedings of the 1993 AMST Conference highlighted the shortcomings of models available on the market and at the same time proposed an innovative method based on the GEER model.

This study presents the results obtained from applications of the GEER model over two years and compares them with other systems currently available on the market to assess the model's efficiency. Graphic representations of sites and plants are usually based on a layer approach. This involves artificial linking of databases to the graphic layer in order to manage alphanumeric data that are at least as important as the graphic data themselves.

Technical managers have pointed out that in these systems the graphic data, which are of course necessary to the operator, provide a visual representation only and need to be supported by an alphanumeric component. Moreover, the representation-layer technique has caused considerable difficulties both in creating multi-user systems and in safeguarding congruency with the data and graphic layers.

Many different types of data are processed by technical management, including site representations, alphanumeric data, scale drawings and drafts as well as information useful for the maintenance, upgrading and links with internal networks.

During the course of our research work, it became necessary to define a specific model. Other models using representation layers linked to relational databases did not meet the needs that had emerged. A new model, called GEER, was therefore developed along with a compatible user-interface language.

Published in: E. Kuljanic (Ed.) *Advanced Manufacturing Systems and Technology*, CISM Courses and Lectures No. 372, Springer Verlag, Wien New York, 1996.

These ideas led us toward the definition of a system that would be able to describe and manage CIM plants and would perform appropriate actions when necessary.

One relevant aspect of the management of industrial plants is related to the ability to properly describe both the layout of the plant and the interactions that take place between its components. Management activities require a uniform approach to this information which is denied by traditional database systems.

Information systems with the ability adequately to define, manage, store and retrieve structured, georeferenced multimedia information will provide technical managers with the capability they require.

Such systems will be able to manage CIM plants and to perform appropriate actions in response to sensorial stimuli.

## 1. INTRODUCTION

Application programs currently available can organize the descriptive graphic database in a number of ways: layer-oriented, object-oriented or structured object-oriented.

This distinction is more than merely methodological. As will be shown, the procedures used to organize graphic databases influence the implementation phase to such a degree that in many cases it becomes impossible to achieve the management objective identified.

The foregoing assertion will be borne out by the following examination of the differences in layer-oriented, object-oriented and structured object-oriented data structures.

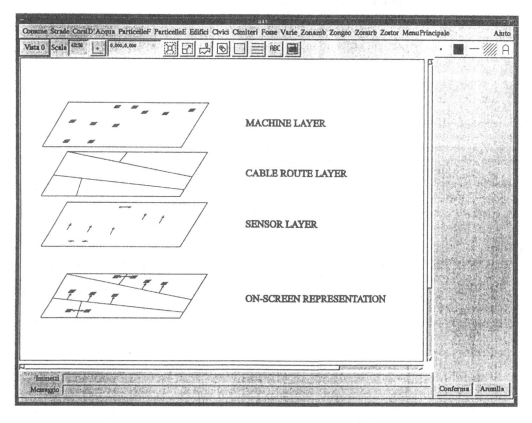

Figure 1. Graphic illustration of a layer-oriented system.

## 2. COMPARISON OF MODELS

The most significant examples for model comparison are to be found in information-technology solutions adopted to implement descriptive information systems for industrial plants and their technological and installation network system.

In layer-oriented systems, each layer contains an undifferentiated set of bitmap-coded graphic data (lines, points, etc.) representing all items belonging to a single class (machine layer, electric cable layer, etc. see fig. 1).

Where the technical office wishes to interact with a single object, it will have to retrieve all objects belonging to the same class, or in other words the entire layer, with consequent penalties of various kinds. This is because in layer-oriented systems, each layer contains an undifferentiated set of bitmap-coded graphic data representing all the items that belong to the same class. In contrast in object-oriented systems, each item is distinct within the database and may therefore be treated independently.

The substantial difference between these two approaches influences the response speed of their various application programs (see table 1). The excessive time complication of layer-oriented search algorithms is a consequence of their having to perform readings on bulk memory whereas the more sophisticated search algorithms for management structured object-oriented management perform readings mainly on random-access memory, which has access times that are five orders of magnitude faster.

In addition, it is also possible with object-orientation to set up functional links of belonging, existence and location between objects that are functionally dependent on each other (see fig. 2).

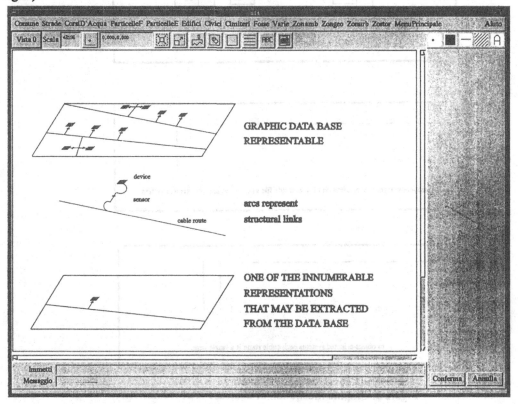

Figure 2. Graphic illustration of an object-oriented system.

The next figure highlights the differences between layer-oriented (fig. 3b) and object-oriented (fig. 3c) graphic data acquisition and memorization procedures.
Figure 3b shows that it is necessary to manipulate the entire file in order to represent and/or update one item. In contrast in figure 3c only the second record is retrieved (see also table 1 - response times).
In layer-oriented systems, matters are complicated each time the graphic part interacts with management data.

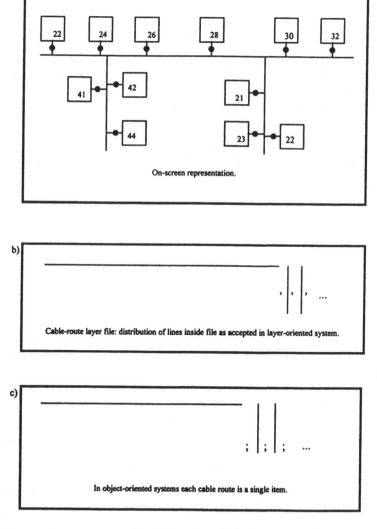

Figure 3. Graphic illustration of database in coordinates.

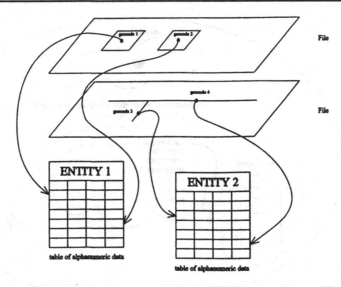

Figure 4. In layer-oriented systems, the geocode is the link between the graphic datum and alphanumeric datum.

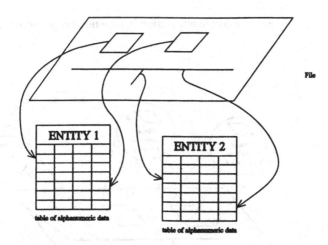

Figure 5. In object-oriented systems, the graphic datum and the alphanumeric datum are memorized in different files.

As may be seen from fig. 4, the links between alphanumeric data (machines and cables) and the graphic files are the *geocodes*. Geocodes have no structural links with the graphic part because individual graphic items are not physically distinct from each other (see fig. 3b). The above figures also show that it is not possible to create a multi-use facility. Two or more workstations cannot operate simultaneously on items belonging to the same class because these will not be distinguished within the database (see fig. 3b).

It should also be noted that this model poses severe problems regarding database reliability (consistency and congruence). It is notoriously difficult to maintain consistency over separate sections of databases where each section refers to the same items memorized, however, on different archive structures (see fig. 4).

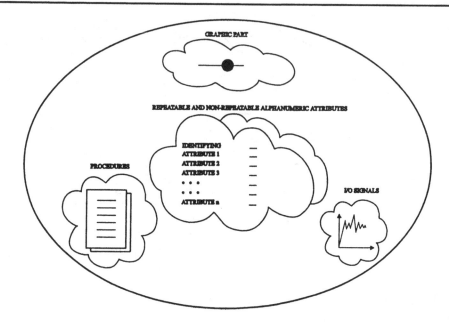

Figure 6. Object identification in structured object-oriented models.

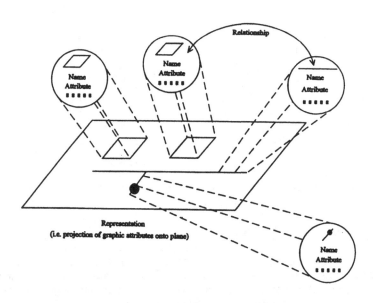

Figure 7. In structured object-oriented systems, data of all types regarding one object

are memorized in a single variable-length logical record.

There is a significant improvement in overall performance on adoption of a conventional object-oriented approach. Since data may be referenced directly to the graphic representation of the object, the drawbacks associated with the need for geocodes disappear. However the separation of the graphic part from the alphanumeric part in conventional object-oriented systems (see fig. 5) means that problems relating to database consistency and multi-user access, which is complicated in some cases and practically impossible in others, remain unsolved.

There was therefore a need to define a better structured record in the form of a variable-length logical alphanumeric record that could treat all attributes in a homogeneous manner.

The natural approach to object-orientation is through a description of the object itself (see fig. 6). This will have a graphic representation part, a variable-length descriptive part including repeatable attributes and must be able to contain the history of the object and its relationships with other objects. The object may also be provided with images (diagrams) and other data held to be necessary for planning and control.

An information system that defines an object in the way described in fig. 6 will create a single homogeneous raster-graphic-alphanumeric database (see fig. 7). It will also be possible to set up an effective multi-user facility, problems of database consistency will be resolved and moreover required management and monitoring performance levels will be achieved (see table 1). The theoretical support for the realization of such a (structured object-oriented) system is a graphic relational DBMS such as Graphically Extended Entity Relationship (GEER) [2,3].

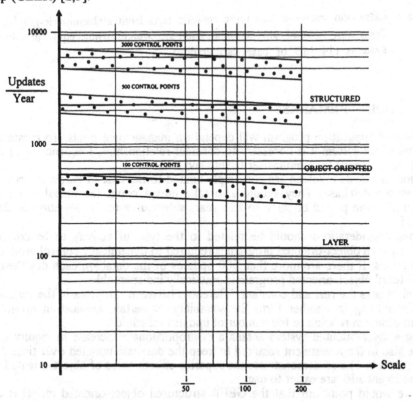

Figure 8. Limits of reliability in relation to the number of variations required to maintain data updated (logarithmic representation).

|                                                                | LAYERS   | OBJECTS (NO INDEX) | STRUCTURED OBJECTS (WITH INDEX) |
|----------------------------------------------------------------|----------|--------------------|----------------------------------|
| NETWORK ON 1/100 WINDOW (MACHINES, CABLE ROUTS, SENSORS)       | 3' 11"   | 23"                | 15"                              |
| RECORDING UPDATING GRAPHIC PART OF NETWORK OBJECT              | 54"      | 8"                 | 6"                               |
| MULTI-USER FACILITY                                            | NO       | NOT ALWAYS         | YES                              |
| INTEGRITY OF ENTIRE DATABASE (ALPHANUMERIC-GRAPHIC)           | CRITICAL | CRITICAL           | YES                              |

Table 1. Extraction and on-screen representation tests from alphanumeric-graphic database. Tests were carried out at equal batabase weight using equivalent hardware systems (14 MB of base cartography).


## 3. CONCLUDING REMARKS

The choice of application program will depend on management goals. To create a control and management system it is necessary to consider relationships, frequency of updates, the need (or lack of need) for a multi-user facility and so on.
In addition, the costs involved are not simply limited to the cost of hardware, software and setting up the database. They should also take into account the actual commitment of personnel over the period of three to five years subsequent to the adoption of the system concerned.
The above considerations should be related to the type of activity to be controlled. In general terms, an object-oriented program is required to dynamically control production and machine cycles. If there are more than 500 updates of the database each day then a multi-user structured object-oriented program is practically indispensable.
Table 1 illustrates the real and constant differences between run times in the various system types, highlighting in another form the suitability of certain application programs with respect to others in relation to the number of updates scheduled.
Adopting a layer-oriented system entails a disproportionate increase in inquiry times and therefore also in the investment required to keep the database updated over time. Run-time analysis (see tab. 1) also demonstrates the superior effectiveness of object-oriented systems, which, we might add, are easier to use.
Finally, we would point out that the GEER structured object-oriented model is currently being evaluated for experimental use in the construction of manufacturing territory systems and for the coordinate mapping of multiple-location factories.

REFERENCES

1. Pascolo C.: *Costruzione di un sistema informativo per la gestione del territorio* ("Construction of an information system for territory management"), from the proceedings of the "Seconda Conferenza Nazionale Informatica" CISPEL, Santa Flavia (Palermo), 1990

2. Pascolo C., Pascolo P.: The graphic extended entity relationship data model and its applications in plants design and management, from the proceedings of "The third international conference on advanced manufacturing system and technology" CISM Udine, 1993

3. Pascolo C., Pascolo P., Casco G., Nalato N.: Plant management applications of CORAD/GEER, from the proceedings of "The third international conference on advanced manufacturing systems and technology" CISM, Udine, 1993

4. Pascolo C., Pascolo P.: Seeking quality in G.I.S., from the proceedings of "The Fifth European Conference on Geographical Information Systems EGIS, Paris, 1994

# OCTREE MODELLING
## IN AUTOMATED ASSEMBLY PLANNING

**F. Failli and G. Dini**
**University of Pisa, Pisa, Italy**

KEY WORDS: Assembly Sequence Detection, Assembly Planning, Solid Modelling, Octree.

ABSTRACT: In this paper a not-conventional approach to solid modelling, specifically oriented to the problems of automated assembly planning, is presented and discussed. This technique is based on octree modelling. Using the octree modelling tool, the usual capabilities of an assembly process planning system can be improved. The problems related to the effective use of octree modelling in assembly planning are analysed. A simple but meaningful example of application of this method is also shown.

## 1. INTRODUCTION

In the field of automatic assembly, many efforts to optimise the planning phase have been performed. Many software systems have been developed and some of them are able to detect the available assembly sequence of a product [1]. The main input for this kind of systems is a representative model of the product. Two kinds of model are usually adopted: an absolute or a relational model.

In the absolute model, all the morphological information about the product are present, but the information about the reciprocal positions of its elements are not immediately

Published in: E. Kuljanic (Ed.) *Advanced Manufacturing Systems and Technology*,
CISM Courses and Lectures No. 372, Springer Verlag, Wien New York, 1996.

available. The most diffuse absolute model is the CAD model of the product, directly obtained during the design stage.

In the relational model only positional information are available. Typical positional information are the contact or alignment between two elements along an axis of the coordinate system. The relational models are often simpler than the absolute ones: they have to usually store a smaller number of information. However, using these models, the product is not completely defined and only simple analyses are possible. The absolute models are more complex, but using this approach a more accurate simulation of the disassembly process can be performed (for instance, the disassembly trajectory can be studied).

This paper proposes an absolute modelling method based on octree encoding [2,3,4,5]. Using this method, together with a simple relational model, the usual performance of an automated assembly planning system can be improved.

## 2. OCTREE MODELLING: BASIC CONCEPTS

The octree modelling is based on a volume digitising process. The simplest method to digitise a volume is to divide the space in a grid of blocks (voxels): blocks are equal and each block has a specific code. In this domain, the object is represented by the codes of all blocks included in the volume of the object. This method is really simple, but has a great disadvantage: the number of blocks strongly increases with the volume of the object and with the resolution of the model too. In this situation the grid of blocks is dense as the accuracy of desired details in the model increases, obtaining a time-consuming procedure that requests a large memory to store the model.

To reduce this problem the octree representation method can be used.

An octree model is generated with an automatic recursive process: the process starts considering a cubic work space including the object to be modelled. This work space is subdivided in eight equal blocks (octants). The intersection between each octant and the object is analysed. Three conditions are possible:

1. no intersection exists between the octant and the object (white octant);
2. an intersection exists between the octant and the object, but the octant is not completely included in the object (grey octant);
3. the octant is completely included in the object (black octant).

In case 1 the octant is discarded. In case 2 the octant is again subdivided in eight sub-octants and the process continue. In case 3 the octant is added to the octree model without further subdivision.

The memory and computing time saving is obtained as a direct consequence of inclusion in the model of octants of different size: a black octant included in the object does not generate smaller blocks. A further generation of blocks occurs only when a grey octant is detected (i.e. a higher resolution is requested). In this way the total number of blocks in the model of the object can be drastically reduced. The process stops when the minimum dimension, previously stated for the blocks, is reached. At the end of the process the set of black octants represents the model of the object.

If N successive octant subdivisions (levels) have been required to model an object, the model is stored in N binary files, where each block is represented using the octal code to minimize the memory requirements; the greater blocks are stored in the lower level files, the smaller blocks are stored in the higher level files.

In Fig.1 an example of steps for an octree model creation is shown.

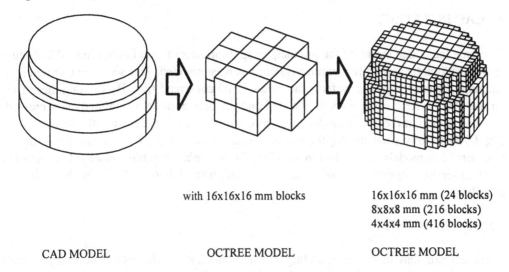

|                        | with 16x16x16 mm blocks | 16x16x16 mm (24 blocks) |
|                        |                         | 8x8x8 mm (216 blocks)   |
|                        |                         | 4x4x4 mm (416 blocks)   |
| CAD MODEL              | OCTREE MODEL            | OCTREE MODEL            |

Fig.1 - Example of octree model creation

It has to be emphasized that, in example of Fig.1, if only 4x4x4 mm blocks are used, the total number of blocks is: 24*32+216*8+416=2912 instead of 24+216+416=656.

## 3. OCTREE IN ASSEMBLY PLANNING: DESCRIPTION OF THE METHOD

The goal of an assembly planning system is the detection of the best assembly sequences of a product. A common procedure adopted for this purpose is to simulate the disassembly of the product by translations of its elements along collisions free directions. The system detects these directions, performing checks on interference and connections between the elements. The assembly sequences are obtained by upsetting the disassembly sequences (decomposition method). Although the basic idea is often the same for various systems, the specific algorithm for sequence detection can be very different.

At the Institute of Mechanical Technology of the University of Pisa an automated assembly planner named FLAPS (Flexible Assembly Planning System) was developed [6,7,8]. In the first version of this system a relational model of the product was used.

In a previous work [9] the improving of the performance of this system by using octree was proposed for the first time. In particular, the capability to detect non-rectilinear disassembly trajectories was emphasized.

Although the relational model is not sufficient to perform each possible analysis of the product, it is partially kept. In particular, the information concerning contact and connection relationships among the elements are conveniently use in order to make faster the sequence detection process.

In the next paragraphs further features of octree encoding oriented to solid modelling for assembly planning are discussed and some meaningful examples are proposed.

## 3.1 CAD INTERFACE

To create the octree model of the product, an appropriate software linking the CAD system and the automatic assembly planning system has been realized. This software (written in C language, as the rest of the system) can create CAD command files. By running these files on the CAD system each block is detected and the octree model is automatically created. For the creation of the octree model the steps reported in section 2 are strictly followed.

The CAD is only used during the octree model construction phase. Thus, using octree representative models stored in binary files, the assembly planning system can perform solid modelling operation also running on computers where a CAD package is not installed.

## 3.2 MODEL ACCURACY

A critical problem in octree encoding is the evaluation of the model accuracy, which depends on the minimum dimension of the blocks used to represent the object. In the octree modelling two contrasting requirements are present:

- for an accurate modelling, the minimum dimension of the blocks should be as little as possible; it means that a high number of blocks in the model is requested;
- for a fast running of the software, the number of the blocks in the model should be as small as possible.

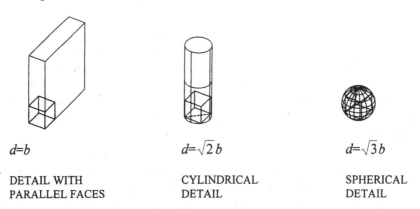

$d=b$                              $d=\sqrt{2}\,b$                          $d=\sqrt{3}\,b$

DETAIL WITH                  CYLINDRICAL                 SPHERICAL
PARALLEL FACES             DETAIL                          DETAIL

Fig.2 - Different kinds of details with their critical dimension ($d$) related to block edge length ($b$).

Therefore a minimum block dimension, which allows to consider each detail in the product using the minimum number of blocks, has to be detected.

Three typical situations are reported in Fig.2: details with parallel faces (blocks), cylindrical details, spherical details. To correctly model the details, the following features have to be considered: in the first case the edge length, in the second case the diagonal of a face, in the third case the diagonal of the block. To assure a correct modelling of all the situations, the following relation must be fulfilled:

$$b_{min} \leq 0.5 \frac{d_{min}}{\sqrt{3}} \qquad (1)$$

Where $b_{min}$ is the minimum block dimension and $d_{min}$ is the minimum detail dimension. Furthermore, $b_{min}$ must be an acceptable value to create an octree grid, i.e. must be equal to work_space_dimension/$2^n$, being $n$ an integer number. $d_{min}$ has to be multiplied by 0.5 because the particular situation reported in Fig.3 can occur. To show the situation clearly 2D representation is used, but the relation (1) is valid in 3D. If the object is positioned in the octree grid as in Fig.3.a, the current resolution is sufficient and the minimum dimension of the block is equal to the minimum dimension of the detail . But, if the object is positioned as in Fig.3.b, to take the detail into consideration, at least higher level of blocks is required.

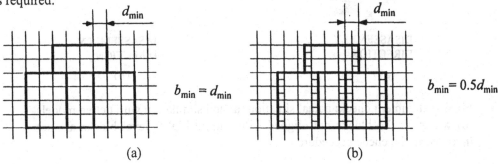

(a)                                                                 (b)

Fig.3 - Example of evaluation of minimum block dimension ($b_{min}$) in an octree model

## 3.3 DETECTION OF INTERFERENCES BETWEEN COMPONENTS

The fundamental phase in assembly planning is the assembly sequence detection. The algorithm used in FLAPS, basically works as follows:

- an attempt to disassembly an element of the octree model of the product is performed: an element is not disassemblable if any other element interferes with it along the current disassembly direction.
- if the element is not disassemblable an attempt with a new element is performed; if the element is disassemblable it is removed from the model and a new attempt with a new element is again performed.

It results that a fundamental operation that the system has to perform, is the detection of interferences between elements moving along a disassembly direction.

Because an interference during the disassembly of an element occurs when it is aligned with any other element along the current disassembly direction, the interference detection can be performed by an alignment check (Fig.4.a).

The octree model allows this operation because the position of each block in the model is available from its code. Therefore it is simple to control if any alignment between components exists by checking the alignments of the blocks in octree models (Fig.4.b).

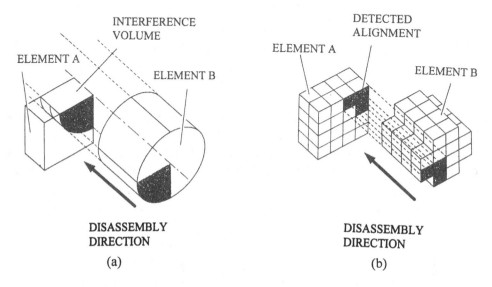

Fig.4 - Alignment check: actual situation (a) and simulation with octree models (b) where all the blocks not aligned with current higher level blocks are not involved in the check procedure.

The problem is the number of alignment checks to perform. If the model of the current element contains N blocks and the model of the rest of the product contains M blocks, for one interference control, the number of alignment checks is N*M. An octree model often contains more than 1,000 blocks, and during a disassembly sequence detection session hundreds of interference control have to be performed. So, the total number of alignment checks can be excessive for an effective use of the system.

To avoid this problem a particular algorithm has been developed: the blocks stored in the lower level file are used to fill a 3-D matrix. The matrix reproduce the spatial grid of octree model for the current higher level. If an element of the matrix is included in a lower level block, that element is filled with 1. On the contrary, it is filled with 0. Hence, each control between files is very fast because for each higher level block only a short series of checks between element in the matrix have to be perform. Thus, the whole interference control between elements A and B is subdivided in n*m controls, where n and m are the number of files (i.e. levels) forming the octree models of A and B, respectively. The value $n*m$ is always much lower than N*M: with 9 levels (i.e. 9 files) a workspace of 512x512x512 mm can be modelled with an accuracy of 1 mm.

Because the octree model is totally included in the object volume, the octree model is smaller than the modelled object; in this situation, little volume interferences can be undetectable using octree models. To avoid this problem an enlarged octree model of the element to disassembly can be used (this enlargement can be also used to reduce the number of detected sequences: if, during its disassembly, an element moves close to another one, by enlargement an interference can be generated; in this way the critical sequences can be discarded by an appropriate enlargement). The enlargement factor is set by the user considering, for instance, the tolerance of critical elements or the trajectory error of the robot.

## 4. EXAMPLE OF APPLICATION

The chosen benchmark is a manually operated stop valve actually produced in industry. The high production volumes make it suitable for automatic assembly. A section of the CAD model is reported in Fig.5.

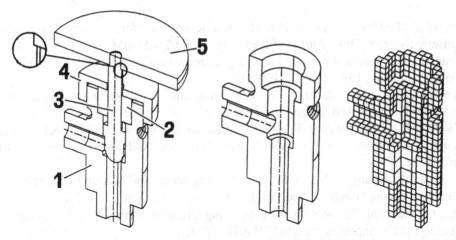

Fig.5 - Section of the stop valve chosen as benchmark with an example of octree model for the element n.1

To model this benchmark a 512x512x512 mm workspace has been used. The detail to consider is highlighted in Fig.5. If $d_{min} = 2$ mm, applying (1), a minimum block dimension of 0.5 mm results. In Fig.5 is also reported an example of octree modelling for the element n.1.

Using octree model the system can recognize that only a monodirectional disassembly is possible (this analysis was already present in the previous release of FLAPS but it was performed by different tools). Then the disassembly sequences are detected. For this benchmark only a sequence is feasible and no subassembly is available.

The detected assembly sequence is: 1, 3, 2, 4, 5. Each element have to be assembled with a movement along the negative versus of Y axis.

## 5. CONCLUSIONS

The feasibility of an assembly planning using octree modelling is demonstrated. In particular the following considerations can be made:

- the correct use of the system is not dependent from the available CAD package (only the octree model creation is CAD-dependent);
- choosing a high accuracy for octree models, time-consuming procedures can result;
- any analysis of the product depending on the shape of its component is possible; in particular the interference control operation is really fast to perform in octree domain.

ACKNOWLEDGEMENTS: this work has been supported by the Italian Ministry of University and Scientific and Technological Research (MURST).

## REFERENCES

1. Santochi, M., Dini, G., Failli, F.,: STC A Cooperative Work on Assembly-Planning Software Systems, 1995, Annals of the CIRP, Vol.44/2: 651-658.
2. Meagher, D.,: Geometric modeling using octree encoding, Computer Graphics And Image Processing, 1982, Vol. 19: 129-147
3. Gargantini, I.,: Linear Octtrees for Fast Processing of Three-Dimensional Objects, Computer Graphics And Image Processing, 1982, Vol. 20: 365-374
4. Walsh, T. R.,: Efficient Axis-Translation of Binary Digital Pictures by Blocks in Linear Quadtree Representation, Computer Vision, Graphics, And Image Processing, 1988, Vol. 41: 282-292
5. Chen, H. H., Huang T. S.,: A Survey of Construction and Manipulation of Octrees, Computer Vision, Graphics, And Image Processing, 1988, Vol. 43: 409-431
6. Dini, G., Santochi, M.,: Automated sequencing and subassembly detection in assembly planning, 1992, Annals of the CIRP, Vol.41/1: 1-4.
7. Santochi, M., Dini, G.,: Computer-aided planning of assembly operations: the selection of assembly sequences, 1992, Robotics and CIM, Vol. 9, No. 6: 439-446.
8. Dini, G., Failli, F., Santochi, M.,: Impianto automatico di montaggio Sistema integrato per progettazione e pianificazione, 1995, Automazione e Strumentazione Anno XLIII N°4: 115-123.
9. Dini, G., Failli, F., Santochi, M.,: Product modelling in assembly process planning: use of octree representation in assembly sequence detection, 1995, Proc. of 11th International Conference on Computer-Aided Production Engineering, London, 20-21 Sept. 1995, 185-190.

# TECHNOLOGY SUBSYSTEM IN THE INFORMATION SYSTEM OF INDUSTRY

N. Majdandzic, S. Sebastijanovic, R. Lujic and G. Simunovic
Mechanical Engineering Faculty, Slavonski Brod, Croatia

KEY WORDS: Technology, Information system, Integration information system

ABSTRACT: The paper shows the structure of the Integration information system for mehanical, electro, wood and food processing industry, and the place of the subsystem for product, technology and materials definition - DEPTO in the global Information system of industry.
Subsystem DEPTO organizes data for different kinds of tehnology (manual work, machining, welding, assemblly), different automation levels of equipment (NC, CNC, production centers EPS) and diferent kinds of production (individual, small-scale, large-scale).

## 1. INTRODUCTION

In the last 10 years the Faculty of mechanical engineering in Slavonski Brod, has developed the information system for different production enterprises - ASIP.
Today, the aim of Information system has been changing: from separate programes called information islands, to package programmes of universal application (Computer Aided Design - CAD, Computer Integrated Office - CIO, Computer Aided Manufacturing - CAM ....) and Management Information System - MIS to Computer Integrated Manufacturing - CIM [1].
AISP has been developed for manufacturing firms in metal processing, electrical, timber, construction industries, erection companies, processing industry, meat industry and bread production.

Published in: E. Kuljanic (Ed.) *Advanced Manufacturing Systems and Technology*, CISM Courses and Lectures No. 372, Springer Verlag, Wien New York, 1996.

Several types of IS, depending on manufacturing or servicing type of programme of the manufacturing unit, might be defined [2]:
-    IS for manufacturing units with single, small - scale and large - scale production in metalprocessing, electro and wood industry,
-    IS for erection enterprises in construction and assembly of plants and objects,
-    IS for enterprises in processing industry and power plants (oil refineries, cement factories, sugar factories, thermal power plants, thermal heating plants, hydropowerplants as well as factories making bricks, glass and chemical products, etc.),
-    IS for groups of firms and factories in agriculture and food processing industry,
-    IS for large state corporations.

Fig. 1 represents the structure of the AISP for production systems in metal procesing, electrical, timber and construction industries [2].

AISP of this group of enterprises contains the following subsystems:
- Common data base BAZAP,
- Sales and calculation PROKA,
- Product, technology and materials definition DEPTO,
- Purchasing and material stocks NAZAL,
- Production planning and assembly PLAPE,
- Production monitoring PRAPE,
- Quality assurance OSKVE,
- Maintainance of capacities ODKAP,
- Financing FINIS,
- Accounting and bookeeping RINIS.

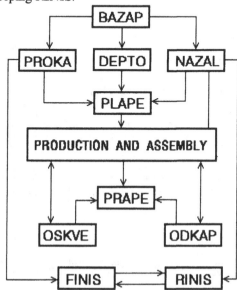

Fig. 1.  Structure of AISP for  production system in  metal procesing

It presupposes some or all of the mentioned types of production capacities in manufacturing process of the enterprise:
- universal machines and manual working places,
- numerical control machines (NC, CNC, DNC and processing centres),
- flexible technological systems.

It is to be expected that the structure of production systems in developing countries will be retained for many years to come and that some technologies of processing (manual work of fitters, preparation for assembly, assembly of complex plants, overhaul, maintenance, manufacture of non-standardized tools, some of heat treatment operations and anti corrosion protection etc.) will also remain at the low level of automation in industries of developed countries.

Therefore it is neccessary to introduce information technologies in the first stage of development of the enterprises, and later on, in the second development stage find out the possibilities to develop the system of computer integrated manufacturing [1].

On the way to CIM concepts we must make some intergration (Fig.2).

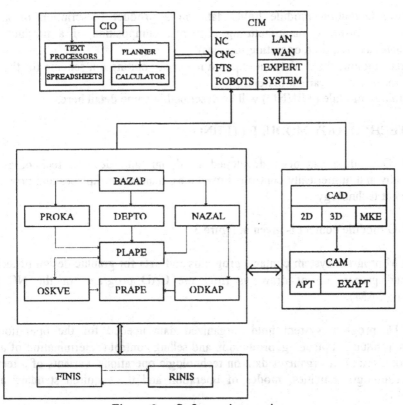

Figure 2    Software integration

The first level of integration is the connecting of the business (FINIS, RINIS) and production subsystems. (PROKA, DEPTO, NAZAL, PLAPE, OSKVE, PRAPE, ODKAP).

The second level of integration is the connecting of AISP and CAD/CAM system.

The third level of integration is connecting AISP, CAD/CAM, CIO (Computer Integrated Office), expert system and production capacities (NC, CNC machines, robots and flexibile technology systems).

## 2. CONTENTS OF THE DEPTO SUBSYSTEM

The DEPTO subsystem organizes data and programs needed to define the structure of products, operations of technology and standards of materials.
The DEPTO subsystem contains three modules:
- the Product Definition module,
- the Technology module,
- the Materials module.

The Product Definition module holds data on a product (description of a product, illustration of a product, manufacturing elements, composition of a product, required auxiliary materials, required operating supplies).
The Materials module holds data on required materials, cutting out scheme for the common starting material and on variant material.
The Technology module (TEHNO) will be described in some detail here.

## 2.1 THE TECHNOLOGY MODULE (TEHNO)

The TEHNO module has been developed to design and develop technology for the conventionally and numerically controlled machines, manual workplaces and protection and heat treatment technology.

The TEHNO module content is given in figure 3.

The GRAFM program system contains programs and data for graphic design of technologic sketches and preparation of drawings from the CAD program for the NC program development(CAM).

The KONTE program system holds organized data needed for the operation of the subsystems: production planning, production and selling control (determination of a product manufacturing price). It organizes data on technologic operations, variants of a technologic process, technologic activities, modes of operation and the required standard and non-standard tools.

The NUSTE program system contains programs and data needed to develop programs for operation of the numerically controlled machines.

The OPTRA program system contains programs and data for optimization of modes of operation of machines in accordance with the performance possibilities of the machines and tools.

The OPTPA program system automatically selects the optimal technologic process.

The IZDET program system makes it possible to generate and print technologic documentation while the program system POSTP contains programs for postprocessing of the programs for the operation of NC machines.

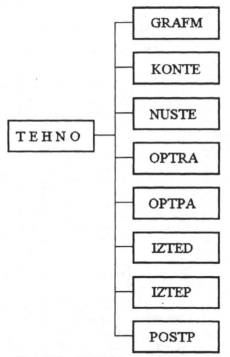

Fig. 3 The TEHNO module content

## 2.2 INTEGRATION OF THE APPLICATIONS PACKAGE PROGRAMS INTO THE AISP OF A COMPANY

Integration of the applications package programs (CAD,CAM,ES) into AISP is exemplified by the integration of the DD/NC programs for the purpose of optimizing the layout of manufacturing elements on common starting material and of automatically generating the program for CNC machines. It is shown in figure 4.

In the Definition of product, technology and materials module, structure of the product is defined.

After that by the OPKROJ program in an interactive mode of operation, the layout of parts on common starting material, defined by its dimensions, is determined. The layout procedure also integrates the gas cutting elements so that after determining the layout the technologic parameters are also obtained and sent back to the DEPTO subsystem to be used as the data for determining the operation and time of cutting and for defining the list of materials through the cutting out scheme.

The Selling and calculation subsystem - PROKA takes standards from DEPTO, price from the accounting subsystem RINIS and then calculates the cost price. Based on the monthly plan it supplies the elements necessary for making an order in the Purchasing and selling subsystem - NAZAL.

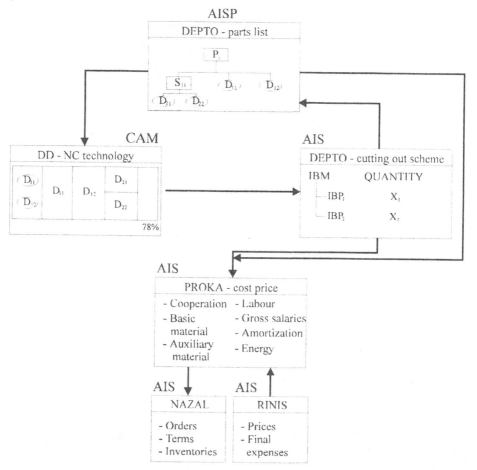

Fig.4 Integration of the DD/NC package into AIS

## 3. CONCLUSION

With increased power and responsibilities, managers also show more interest in introducing information technology into their firms. The concepts of development and similarity of subsystems in various types of manufacturing companies are defined. The necessary integrations of applications package programs into the AIS of a company are given as well as one of the possible strategies of the program integration into the CIM system.

## REFERENCES

1. Scheer, A.W.: CIM - Der Computergesteuerte Industriebetieb, Springer - Verlag, 1987.
2. Majdandzic, N.: Kompjuterizacija poduzeca, Sveuciliste u Osijeku, Strojarski fakultet, Slavonski Brod, 1994.

# INTEGRATION CAD/CAM/CAE SYSTEM FOR PRODUCTION ALL TERRAIN VEHICLE MANUFACTURED WITH COMPOSITE MATERIALS

**G. Vrtanoski, Lj. Dudeski and V. Dukovski**
University "Sv. Kiril i Metodij", Skopje, Macedonia

KEY WORDS: Composite materials, Integrated CAD/CAM/CAE system, vehicle

ABSTRACT: This paper deals with the problems concerning the computer integrated design, structural analysis and manufacturing. The developed system is presented by the concrete example of all terrain vehicle. The new design of all terrain vehicle body was manufactured with composite materials and is installed to the existing metal chassis and driving group.

## 1. INTRODUCTION

The rapidly growing up of science and computer technology have a dramatic impact on the development of production automation technologies. Nearly all modern production systems are implemented today using computer systems. This tendency also has a very strong requirement from the properties of constructive materials. The optimization of classical materials and their characteristics have a limit, so the new solutions were taken developing quality new materials. Composite materials are unique materials that could be content compromise requirement for a good reliability, light weight, high statical and dynamical characteristics.

The use of composite materials helps the development and production of new body in a short period and without heavy investment. Due to the heavy involvement of the manual work the overall design and manufacturing process of the vehicle is characterized with the

Published in: E. Kuljanic (Ed.) *Advanced Manufacturing Systems and Technology*, CISM Courses and Lectures No. 372, Springer Verlag, Wien New York, 1996.

great deal of lead time and man hours. Design and production of different types of vehicles in small series are rapidly growing busyness for small and medium size companies.

Analyzing those parameters and use the PC and PC based computer software was developed a new integrated CAD/CAM/CAE system. In each phase of this system was shown the development and production of all terrain vehicle. The new body was manufactured by composite materials and installed on an existing metal chassis and driving group, with appropriate changes caused by the specific demands for a new vehicle design.

## 2. INTEGRATED CAD/CAM/CAE SYSTEM

Integrated CAD/ÇAM/CAE system deals to involve the use of the digital computer to accomplish certain functions in design and production. The combination of CAD, CAM and CAE in the term CAD/CAM/CAE is symbolic of efforts to integrate the design, structural analysis and manufacturing functions in a firm into a continuum of activities, rather than to treat them as three separate and disparate activities, as they have been considered in the past. The manufacturing process of composite materials structure is very specific process. But however, the phases of overall design and manufacturing process for products made by classical materials are very similar as a product made by composite materials. The structure of integrated CAD/CAM/CAE system for vehicle manufactured by the composite material is presented in a Figure 1.

The overall design and manufacturing process consists of several phases [1,2,6]:
- Conceptual design of the new vehicle based on market demands, existing standards and recommendations;
- Geometry modeling of the new body design and existing chassis;
- Engineering analysis of the chassis and body structure (FEA);
- Evaluation and computer graphics presentation of the new design and its animation for better marketing of a new product;
- Development of the appropriate design documentation;
- Manufacturing of the new vehicle.

## 3. COMPOSITE MATERIALS

In the area of research and development a new advance design material the most special place was taken from the composite materials. The literature [3,4] which treat this problematic, composite material was defined as an artificial materials system composed of a mixture or combination of two or more macroconstituents differing in form and or material composition and that are essentially insoluble in each other.

Composite materials are two phases' materials and on the microstructure level they consist two materials: the reinforced fibers and matrix resin. The macrostructure level of the composite material a developed with layers made by reinforced fibers and bounded with the matrix resin. Engineering methods of calculation was developed on the macrostructure model and according on this, the layer was taken as a basic type of element in the finite element method [4,7,8].

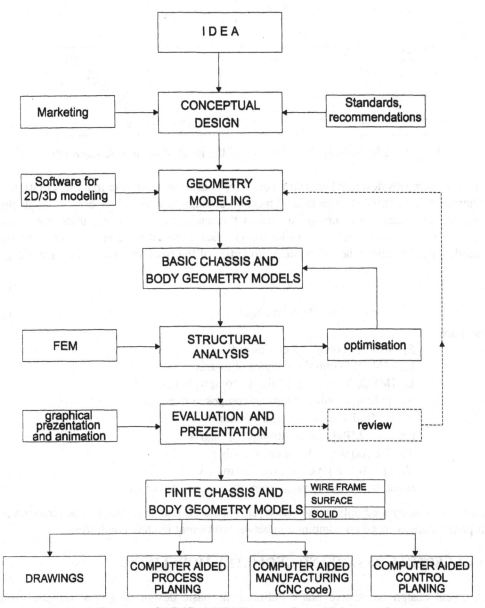

Figure 1. Integrate CAD/CAM/CAE system for vehicle manufactured
with composite materials

The layer of composite material made with the unidirectional disconnected fibers a orthotropic material (isotropic perpendicular in the fiber plane). This material is characterized with two principal axes, one long to the fiber (longitudinal) and second perpendicular to the fiber (transversal) (Figure 2.).

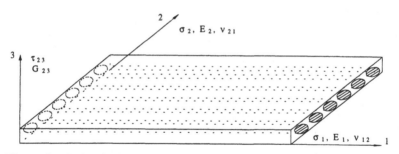

Figure 2. Mechanical characteristics in the layer of composite materials

In the literature was described several theories for analyzing composite materials [3,4]. One of them is rule of mixture. The rule of mixture belongs to classical theories of analyzing composite materials. This method describes the analytical way to calculate the elastic modulus (Young's modules) and tensile strength. Before to calculate and analyze with this method, it must to know the mechanical properties of the fiber and matrix as their volume ratio.

$$E = \alpha E_f V_f + E_m V_m \tag{1}$$

$$\sigma_e = \beta \sigma_{fu} V_f + (\sigma_m)_{efu} V_m \tag{2}$$

Where are:

E$_f$ - [MPa], Young's modules of fiber
E$_m$ - [MPa], Young's modules of matrix
E - [MPa], Young's modules of composite material
$\sigma_e$ - [MPa], tensile strength of composite material
$\sigma_{fu}$ - [MPa], tensile strength of fiber
$(\sigma_m)_{efu}$ - [MPa], tensile strength of matrix
V$_f$ - [ x 100 %], volume ratio of fiber
V$_m$ - [ x 100 %], volume ratio of matrix
$\alpha$ and $\beta$ - factors depend from the arrangement of fibers.

This theory was applied only in the one layer, not for all real structure. In real structure all components are exposed on combination between different loading conditions.

## 4. CONCEPTUAL DESIGN AND GEOMETRY MODELING

Conceptual design is creative phase defining the basic design of a new vehicle. More different variations of the vehicle could be developed on the ground of the market investigation and competitive products, as well as of the existing standards and recommendation for the developed type of the vehicle. In this case the AutoCAD software has been used for defining the different conceptual variations [2,10].
The basic geometry modeling is a phase in which the shape and dimensions of major vehicle elements are geometrically defined. Having defined them, they are subject to further structural analysis. In the particular case suitable geometry models of the chassis

and body are developed using AutoCAD drafting and modeling software. Such approach enables the following phases of the development process, after being verified by the structural analysis, avoiding the need of their multiple drafting during the different phases of the new vehicle development.

## 5. STRUCTURAL ANALYSIS

The design process of the certain structural entity of the vehicle includes its stiffness and strength analysis. The finite element method is the most widely used method in the world practice. The main objective of the analysis is to develop suitable model of construction that will be able to define the stress and strain state of the overall system as realistic as possible. Those defined states refer to possible loads of the vehicle which could occur at certain regimes during its exploitation.

The main characteristic of the presented composite body is the use of polyester composite structure of different thickness reinforced at certain part with ribs and close profiles.

In the essence of the composite body manufacturing is a very simple technology of processing a polyester resin, reinforced with glass fiber or other type fibers. The mechanical properties of the obtained composite structure are directly related to the used components. They depend on: mechanical properties of reinforced fibers and matrix resin, the percentage of the reinforced fibers in the structure, degree of the adhesive ability, the section and orientation of the fibers in the matrix. The table 1 are presented minimal mechanical properties of reinforced polyester composite materials [3,4].

Table 1. Minimal mechanical properties of polyester composite materials

| thickness mm | tensile strength MPa | flexible strength Mpa | Young's modulus MPa | density kg/cm$^3$ | volume ratio of fibers |
|---|---|---|---|---|---|
| 5 | 45 | 112 | 4900 | 1.8 | 25 % |
| 6.5 | 84 | 133 | 6300 | 1.8 | 28 % |
| 8 | 95 | 140 | 7000 | 1.8 | 30 % |
| 9 | 105 | 154 | 8400 | 1.8 | 34 % |

Increased resistibility and absorption of the impact energy could be achieved by the usage of matrix resin, fillers and reinforced fibers. The composite body feature has good mechanical properties in relation with weight, good antirust and insulation properties, possibility to manufacture big parts (decreasing the duration and the process of assembly) and decreasing of noise.

For the purpose of this analysis ALGOR commercial software package has been used because it meets the necessary requirements and has references for such type of analysis [9]. The ALGOR software package has a module for exchange DXF files with AutoCAD. The basic geometry of the vehicle model including chassis and body is transferred from AutoCAD thought the use of DXF. For the purpose to get more realistic picture of the stress and strain state the chassis and composite body are interconnected with joint elements of specified characteristics. The chassis and composite body independent operation is provided by those joint elements.

Three types of finite elements have been applied for the body structure modeling with the ALGOR software [9]:

                    - plate/shall, triangular and quadrangular finite elements of thin shell type.
                      This element is for complete composite body modeling.
                    - beam space element. This element is for reinforcement of composite body
                      as well as chassis modeling
                    - bar space element. This element is for modeling suspension and some
                      parts of the steering mechanism.

The basic models of the composite materials and chassis model are shown in Figure 2 and 3.

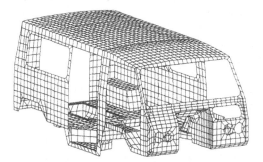
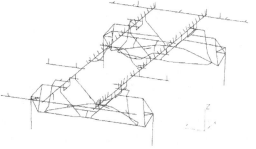

Figure 3. FEA model of composite body          Figure 4. FEA model of chassis

During the loads modeling, the arrangement of the useful loads and loads caused by vehicle equipment have been chosen very carefully. The most appropriate design solution based on strength properties of whole vehicle as well as for certain characteristic sections has been determined by several successive iterative analyses.

## 6. EVALUATE AND FINITE GEOMETRY MODEL

After each iterative analysis, the basic geometry model undergoes certain specified corrections which lead to the finite geometry model. As a result a 3D wire frame model is transferred from ALGOR FEA to AutoCAD via DXF. Surface and solid models in AutoCAD are formed with the use of 3D Modeling and AME (Advanced Modeling Extension). Those models are used in the further phases for computer integrated design and production system [10,11].

The graphical visualization and animation of vehicle have great influence on the evaluation of the overall design. The investors and other technical persons would be show interest in the aesthetics and value of new design. They have so easier understand design presented in three dimensional models. This techniques help to have short time need for manufacturing prototype physical model, also save investments need to involve and start the production process, much faster made appropriate changes on the geometry model from the market pressers, and so simply and easy on the computer graphical display explain and solve the non understanding parts and details from vehicles. Using 3D Studio software program some deficiency in the design are eliminated and the new vehicle for marketing purpose is presented in Figure 5 and 6 [11].

Figure 5.  3D design of composite body          Figure 6.  3D design of chassis

The solid model combined with wire frame model and drafting capabilities of AutoCAD is used for elaborating the final technical documentation. Detailed description of the vehicle in a form of 2D technical drawing is shown by such technical documentation.

## 7. MANUFACTURING PROCESS AND DEVELOPMENT CNC PROGRAMME

The surface model has been also used to generate DXF file, which on the other hand helps to produce CNC machine program. Using AutoCAD and their programming language AutoLISP are developed a new program PCNC. The program PCNC generate file which is consists tool path and instructions for the programming three axes CNC cutter machine. Using direct numeric control and program TNC the generate file is translated to the computer unit of the CNC machine. The CNC machine could be used for the model production, required for body shaping as well as for some complex body parts which in general if manufactured by means of manual methods need much more time. In the Figure 7. is shown generate tool path for a 3D complex mold using AutoCAD and PCNC. The details from the generate file also for the same mold is presented in the Figure 8 [6].

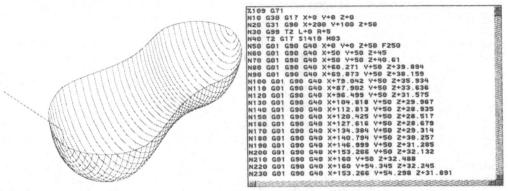

Figure 7. Generate path tool for complex mold    Figure 8. Generate file from the PCNC program

## 8. CONCLUSION

The presented integrate CAD/CAM/CAE system for development, design and production of the vehicle with composite materials provide possibilities for reduction of the vehicle development time, labor required and improvement of quality. Taking into consideration widely used PC and PC based AutoCAD software and well-known ALGOR FEA software, the computer based development and production of vehicles with composite materials body becomes reality for the small enterprises as well.

## REFERENCE

1. Spur G.; Krause F.: **CAD - Technik**, Carl Hauser Verlag, 1984
2. M.P.Groover, E.W.Zimmers: **CAD/CAM: Computer - Aided Design and Manufacturing**, Prentice Hall,1984
3. R.L.Calcote: **The Analysis of Laminated Composite Structure**, Van Nostrand-Reinhold,1969
4. K.G.Sabodh, V.Svalbonas & G.A.Gurtman: **Analysis of Structural Composite Materials**, Marcel Dekker, New York,1973
5. J.Pawlowski: **Vehicle Body Engineering**, Business Book Limited, London,1969
6. V.Dukovski, Lj.Dudeski, G.Vrtanoski: **Computer Based Development and Production System for Vehicles with Composite Materials Body**, Ninth World Congress IFToMM, Milan,1995
7. Rao S.S.: **The Finite Element Method in Engineering**, Pergamon Press S.E., 1988
8. S.G.Advani, J.W.Gillespie: **Computer Aided Design in Composite Material Technology III**, Elsevier, London, 1992
9. ALGOR FEA - System: **Processor and Stress Decoder Reference Manual**, Pittsburgh, 1991
10. AutoCAD release 12 **User's Guide**, Autodesk Inc.,1993
11. S.Elliot,P.Miller, & G.Pyros:**Inside 3D Studio release 3**, NRP Indiana,1994

# MICROFABRICATION AT THE ELETTRA SYNCHROTRON RADIATION FACILITY

**F. De Bona**
University of Udine, Udine, Italy

**M. Matteucci**
C.N.R. Sezione di Trieste, Trieste, Italy

**J. Mohr and F.J. Pantenburg**
Institut fuer Mikrostrukturtechnik Forschungszentrum Karlsruhe, Karlsruhe, Germany

**S. Zelenika**
Sincrotrone Trieste, Trieste, Italy

KEY WORDS: MEMs, Microfabrication, LIGA, Deep X-Lithography

ABSTRACT

The first micro-structures produced at Elettra (Trieste, Italy) with deep x-ray lithography are presented in this work. The most important details concerning LIGA technique and in particular deep x-ray lithography using synchrotron radiation are described. The scanning electron microscopy images of test micro-structures made of polymethylmethacrylate characterized by a height of 200 µm and lateral dimensions in between 10 µm and 200 µm are presented. In all the considered cases good accuracy with high aspect ratios and nanometric roughness is observed.

## 1. INTRODUCTION

Nowadays micromechanics is probably one of the most promising and rapidly growing fields among the new emerging technologies. In fact, the possibility of reducing the size of mechanical structures to the micro-domain opens a wide variety of possible applications. In the biological and medical fields microelectromechanical systems (MEMs) have been developed to obtain catheters for intraluminal diagnostics and surgery [1], eye surgical tools, blood flow sensors inside arteries [2] and microgrippers for cell manipulation [3]. MEMs could find several applications also in aerospace industry, for the production of accelerometers and gyroscopes [4] or for delta-wing control [5]. At present MEMs are already used in automotive industry for airbag accelerometers and pressure sensors [6]; in fact, apart from the miniaturization, micromechanics generally makes it possible to achieve two other important assets: closer integration with microelectronics and mass production, thus improving reliability and reducing costs. Other promising fields of application of

Published in: E. Kuljanic (Ed.) *Advanced Manufacturing Systems and Technology*, CISM Courses and Lectures No. 372, Springer Verlag, Wien New York, 1996.

MEMs are robotics [3], molecular engineering [7], fiber [8] and integrated [9] optics, fluid technology [10] and microconnector arrays [11].
The success of micromechanics has been made possible by the development of a huge variety of microfabrication processes (see Tab.I). Most of these techniques have been derived from "traditional" microelectronics technology. In fact the first mass fabrication of MEMs started with the process of wet-chemical etching, successively replaced by anisotropic dry-etching processes by means of low-pressure plasma or ion beams. At present microelectronics derived technologies in the MEMs area are essentially of two types: surface processes and bulk processes.

## Tab. I Unit and Integrated Microfabrication Processes

### Integrated processes

| | |
|---|---|
| **-Bulk micromachining** | photolithography, wet-chemical anisotropic etching |
| **-Surface micromachining** | photolithography, evaporation, sputtering and CVD, selective wet and dry plasma etching, sacrificial layer |
| **-LIGA** | x-ray lithography, electroforming, moulding |

### Unit processes

| | |
|---|---|
| **-Photolithography** | UV, x-rays, e-beam, etc. |
| **-Micro stereolithography (IH)** | |
| **-Beam machining processes** | laser ablation, focused ion beam (FIB), fast atom beam (FAB), etc. |
| **-Etching techniques** | wet chemical etching, dry physical etching, reactive ion etching (RIE), etc. |
| **-Deposition techniques** | diffusion of dopants, implantation, epitaxy, chemical vapour deposition, (CVD), physical vapour deposition (PVD), electrodeposition, Langmuir-Blodgett (LB) |
| **-Bonding techniques** | fusion, anodic, adhesive, etc. |
| **-Micro electro-discharge (EDM)** | |
| **-Mechanical micromachining** | turning and grinding |

Surface processes (see Tab. I) are those that make it possible to obtain a silicon microstructure by means of depositions of thin films, as an additive technique, and selective etching of the thin films, as a subtractive technique. The thin film system usually consists of a structural layer at the top of a sacrificial layer; in this way the etching of the sacrificial layer allows tridimensional surface structures to be made such as microbeams, microsprings and lateral mobile microelements [6].
Bulk microfabrication is based on photolithographic etching techniques. The most popular materials for bulk micromachining are silicon, glass and quartz. Even if wet chemical etching is still the dominating bulk machining technique, dry etching techniques are rapidly growing. The main drawback of bulk micromachining, compared to surface micromachining, is that it requires double sided processing of the wafer to make tridimensional structures [6].
As microelectronics generally deals with planar silicon structures, the main limitations of microelectronics derived technologies are related to the fact that traditional mechanical materials such as metals can not be used and truly tridimensional structures can not easily be obtained. Recently, to overcome these drawbacks several new "non-traditional"

processes of microfabrication of components or even complete microsystems have been developed. Among them the most interesting from the mechanical point of view are microstereolithography (IH), electro-discharge machining (EDM) and LIGA.

In the IH process a thin film of photopolimerizing polymer is first formed and then it is exposed to an ultraviolet beam, which scans the surface to cure it, forming thin layer structures. The process is repeated to build up layers of cured resin to form an arbitrary tridimensional geometry. The polymeric structures fabricated with this process can be used directly or as a cast for a metal moulding process.

EDM is based on the erosion of the material to be machined by means of a controlled electric discharge between an electrode and the material. Electroconductive materials (metals) or semiconductors (silicon) have to be used [12].

LIGA (from the German Lithographie = lithography, Galvanoformung = electroforming, Abformung = moulding) is a combination of lithography, electroforming, and moulding and permits the fabrication of micro-structures from metals, polymers and ceramics.

Since IH and EDM are not presently suitable for mass production, the LIGA process still maintains the advantages of mass scale production typical of the microelectronics technology; for these reasons nowadays LIGA is considered one of the most promising microfabrication techniques [12], [8]. In the following the basic characteristics of the LIGA process will be described and a detailed description of the lithographic step using synchrotron radiation will be given. The scanning electron microscopy (SEM) images of test micro-structures produced at Elettra will then be presented.

## 2. LIGA TECHNOLOGY

LIGA technology has been developed in Germany at the Karlsruhe Nuclear Research Center (now Forschungszentrum Karlsruhe) with the goal of manufacturing extremely small nozzles for the separation of uranium isotopes [13]. The LIGA technology is a manufacturing process by which a polymeric material (resist), which changes its dissolution rate in a liquid solvent (developer) under high-energy irradiation, is exposed through an X-ray mask to highly intensive parallel X-rays, i.e. to synchrotron radiation (Fig. 1). The mask consists of an absorbing pattern of a high X-ray absorption material (i.e. large atomic number - usually gold) applied onto a thin membrane of a low atomic number material (usually beryllium, titanium or diamond) which is thus largely transparent to X-rays; generally a cross-linked polymethylmetacrylate (PMMA), which is polymerized as a thick layer directly on an electrically conductive substrate, is used as the resist material. The irradiation of PMMA results in chain scissions of the polymer and consequently the irradiated regions can be dissolved in a proper developer and a high aspect ratio relief structure of PMMA is obtained. In the second step of the process the resulting structure is used as a template in an electroforming procedure where metal is deposited onto the electrically conductive substrate in the spaces between the resist structures. The obtained metallic structure can be either the final microstructure product or it can be used as a mould insert for multiple reproduction of polymer templates by reaction injection moulding. Finally, metallic micro-structures are manufactured in a secondary electroforming process.

The plastic structure obtained in the moulding process can be also used as a "lost form" for the fabrication of ceramic micro-structures; after filling the plastic form with ceramic slurry, the ceramic structure is formed by the usual drying and firing process [8].

The LIGA process makes it possible not only to fabricate micro-structures whose cross sectional shape is constant with height, but also conical, pyramid-shaped or even more complicated geometries, using processes such as multiple irradiation with different angles of incidence [14].

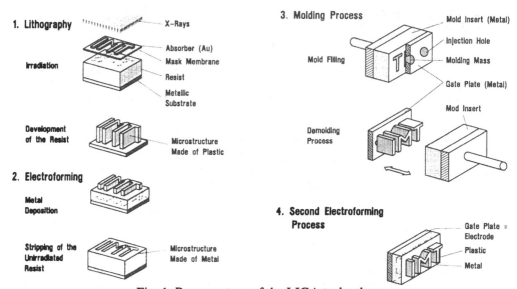

Fig. 1: Process steps of the LIGA technology

Using PMMA as resist material, aspect ratios (defined as the ratio of the height of the resist structure to its smallest lateral dimension) of up to 150 or even more can be obtained with precision better than ±100 nm for heights up to 500 μm [13], [8]. The structure height could be even several millimetres depending on the synchrotron spectrum. Structure walls' smoothnesses of down to 30-50 nm have also been achieved.

## 3. DEEP X-RAY LITHOGRAPHY WITH SYNCHROTRON RADIATION

In the LIGA technology deep X-ray lithography constitutes the most important fabrication step. In comparison to traditional lithographic processes used for the manufacturing of integrated circuits, in the case of LIGA the thickness of the resist can be chosen to be larger up to a factor of 1000, since synchrotron radiation of short wavelength is used for the pattern transfer. In fact, compared to conventional X-ray sources, synchrotron radiation is characterized by high intensity and good parallelism thus permitting high penetration depth with extremely good accuracy and aspect ratios.

Synchrotron light is the radiation produced by highly relativistic electrons forced to follow curved trajectories by means of magnetic fields. The synchrotron light beam from bending magnets is characterized by a very small vertical emission angle and a wide horizontal emission angle; therefore usually 100x100 mm substrates can be irradiated putting the lithography station 10-20 meters from the source and scanning the mask and the substrate periodically in the vertical direction. Apart from the X-ray scanner, the lithographic apparatus consists of an evacuated metal tube connected to the storage ring and a filter system in front of the exposition chamber. The possibility of inserting additional filters permits to modify the beam spectrum according to the experimental requirements. Vacuum windows made of beryllium are also used to separate the storage ring, where the pressure is maintained around $5 \cdot 10^{-8}$ Pa, and the scanner, that is generally filled with helium at a small overpressure in order to provide convective cooling of the mask and the substrate, which are heated by the absorption of the intense synchrotron flux. Additional valves and stoppers are then used for safety reasons.

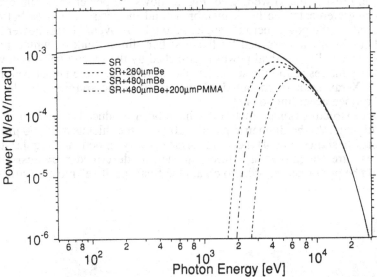

Fig. 2: Power spectrum of the synchrotron radiation (SR) from a bending magnet of Elettra upstream and downstream from the vacuum windows, the mask membrane and the resist

Synchrotron radiation has a broad radiation spectrum, covering a range from the visible light to the hard x-rays; generally this spectral distribution has to be modified, as it is not optimized for the irradiation process. Using PMMA as a resist material, the minimum and the maximum dose required to have a good development and no irradiation below the mask absorbing pattern are well known [8]; therefore, for a defined resist height and filtering configuration, the actual doses at the surface and in the depth of the resist can be evaluated and compared with the required values. If the synchrotron radiation power spectrum of the source is known, using Bear's law, the power spectrum after the filtering elements (Be windows, mask membrane, mask absorber, etc.) can be evaluated (Fig.2); then, integrating in energy and multiplying by the absorption coefficient of the resist material, the dose of irradiation and the irradiation time at different resist depths can be easily obtained. If the ratio between the minimum and the maximum dose is too high, extra filters must be added, even if this produces an increase of the irradiation time. However the suppression of the longer wavelengths has always to be performed; in fact such radiation is useless for irradiation purposes, as it has a short penetration depth and it is absorbed by the mask membrane thus producing an undue heating of the mask, with its consequent expansion and therefore a reduction of the accuracy of the fabrication process.

Recently it has been observed that the high energy components of the radiation spectrum have also to be reduced as they could produce secondary electrons at the substrate interface thus inducing undesired irradiation processes [15]. In that case it is necessary to remove the high energy photons by using grazing incidence mirrors [16].

## 4. FIRST MICRO-STRUCTURES OBTAINED AT ELETTRA

Irradiations were performed at the Elettra synchrotron radiation source, operating at an energy of 2 GeV with an electron beam current of 200 mA. The X-ray beam used in the experiment had a vertical aperture of 1 mrad and an horizontal aperture of 7 mrad. The power spectrum of the beam is represented in Fig.2; two beryllium windows (thickness: 200 μm and 80 μm respectively) were used. Fig. 2 shows the power spectrum before and after these filtering elements, the mask membrane and the resist. The area below these curves corresponds to the power per unit angle. Altough the overall beam power is 57 W. a great part of it (25 W) is absorbed by the first beryllium window, which has to be therefore water cooled. The residual power is absorbed by the second beryllium window, by the mask and by the resist and the substrate. The irradiations were performed using the Jenoptik's DEX X-ray scanner mounted 12 meters from the source. The developing process was performed according to [17].

Several test micro-structures made of PMMA have been produced. Fig. 3 shows a gear wheel with involute teeth; the diametric pitch is 200 μm, the thickness is 100 μm. As the close-up view clearly shows (Fig. 4) the tooth surface is very smooth and regular. It can be noticed that these are "negative" structures, in fact, as described previously, electro-deposition should be performed in order to obtain the final "positive" metallic structure.

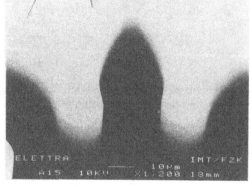

Fig. 3: Overall view of microgears                Fig. 4: Close-up view of a gear tooth

Deep X-ray lithography permits to obtain very high aspect ratios. In Fig. 5, 6 and 7 this concept is stressed. Fig. 5 shows a cross with smallest bare width of 10 μm, Fig. 6 shows a single stand-in wall of the same width and Fig. 7 shows column structures with 20 μm in diameter and a height of 200 μm.

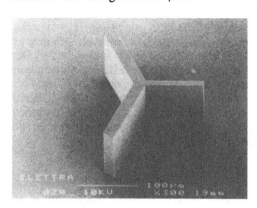

Fig. 5: Test structure A                          Fig. 6: Test structure B

All these structures were obtained using a beryllium mask membrane (thickness: 200μm) patterned with gold and a PMMA resist with a thickness of 200 μm.
Fig.8 shows a grating structure used for a IR-spectrometer developed at the Forschungszentrum Karlsruhe. In this case a beryllium membrane with a thickness of 500 μm was used. The great precision of the fabrication process is here enhanced: the grating steps have in fact an height of 2 μm and a thickness of 100 μm (aspect ratio: 50); wall roughness lower the 10 nm can be also observed.

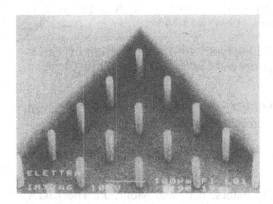

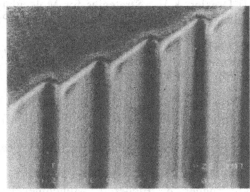

Fig. 7: Array of micropins                    Fig. 8: Grating structure

Several other test structures (microconnectors, micro turbines, etc.) have been produced. In all these cases good accuracy has always been obtained, thus confirming that Elettra is very well suited for performing this technique. Future work will consist of experiments dedicated to a increased optimization of the synchrotron spectrum, particularly concerning the reduction of the high energy components, while in parallel the first micromechanical products are under detailed design.

## ACKNOWLEDGEMENTS

The authors wish to thank the Jenoptik company for providing the X-ray scanner.
This work was partially supported by E.C. grant ERBCHRX-CT-930394/130.

## REFERENCES

[1] Lim, G., Minami, K., Sugihara, M.: Active Catheter with Multilink Structure Based on Silicon Micromachining, Proc. IEEE Micro Electro Mechanical Systems Conf., Amsterdam (NL), (1995), 116-121

[2] Rapoport. S.D., Reed, M.L. and Weiss, L.E.: Fabrication and Testing of a Microdyamic Rotor for Blood Flow Measurements, J. Micromech. Microeng., 1 (1991), 60-65

[3] Dario, P., Valleggi, R., Carrozza, M.C., Montesi, M.C., and Cocco, M.: Microactuators for Microrobots: a Critical Survey, J. Micromech. Microeng., 2 (1992), 141-157

[4] Brown, A.S.: MEMs: Macro Growth for Micro Systems, Aerospace America, October (1994), 32-37

[5] Liu, C., Tsao, T. and Tai, Y.C., Leu, T.S., Ho, C.M., Tang, W.L., Miu, D.: Out-of-Plane Permalloy Magnetic Actuators for Delta-Wing Control, Proc. IEEE Micro Electro Mechanical Systems Conf., Amsterdam (NL), (1995), 7-12

[6] Ohickers, P., Hannebor, A., Nese, M.: Batch Processing for Micromachined Devices, J. Micromech. Microeng., 5 (1995), 47-56

[7] Drexler, K.E.,: Strategies for Molecular System Engineering, in: Nanotechnology, edited by Crandall, B.C. and Lewis, J., MIT Press, (1994), 115-143

[8] Ehrfeld, W. and Lehr, H.: Deep X-Ray Lithography for the Production of Three-dimensional Microstructures from Metals, Polymers and Ceramics, Radiat. Phys. Chem., 3 (1994), 349-365

[9] Uenishi, Y., Tsugai, M., Mehregany, M.: Micro-Opto-Mechanical Devices Fabricated by Anisotropic Etching of (110) Silicon, J. Micromech. Microeng., 5 (1995), 305-312

[10] Shoji, S., Esashi, M.: Microflow Devices and Systems, J. Micromech. Microeng., 4 (1994), 157-171

[11] Rogner, A., Eicher, J., Munchmeyer, D., Peters, R.P, and Mohr, J.: The LIGA Technique - What are the New Opportunities, J. Micromech. Microeng., 2 (1992), 133-140

[12] Dario, P., Carrozza, M.C., Croce, N., Montesi M.C., and Cocco, M.: Non-Traditional Technologies for Microfabrication, J. Micromech. Microeng., 5 (1995), 64-71

[13] Becker, E.W., Ehrfeld, W., Munchmeyer, D., Betz, H., Heuberger, A., Pongratz, S., Glashauser, W., Michel, H.J. and von Siemens, R.: Production of Separation-Nozzle Systems for Uranium Enrichment by a Combination of X-Ray Lithography and Galvanoplastics, Naturwissenschaften, 69 (1982), 520-523

[14] Mohr, J., Bacher, W., Bley, P., Strohrmann, M. and Wallrabe, U.: The LIGA-Process - A Tool for the Fabrication of Microstructures Used in Mechatronics, Proc. 1er Congres Franco-Japonais de Mecatronique, Besancon, France (1992)

[15] Pantenburg, F. J, Mohr, J.: Influence of Secondary Effects on the Structure Quality in Deep X-ray Lithography, Nucl. Instr. and Meth, B97 (1995), 551- 556

[16] Pantenburg, F.J., El-Kholi, A., Mohr, J., Schulz, J., Oertel, H.K., Chlebek, J., Huber, H.-L.: Adhesion Problems in Deep-Etch X-Ray Lithography Caused by Fluorescence Radiation from the Plating Base, Microelectr. Eng., 23 (1994), 223 - 226

[17] Mohr, J., Ehrfeld, W., Munchmeyer, D. and Stutz, A.: Resist Technology for Deep-Etch Synchrotron Radiation Lithography, Makromol. Chem., Macromol. Symp., 24 (1989), 231-240

# THERMAL RESPONSE ANALYSIS
# OF LASER CUTTING AUSTENITIC STAINLESS STEEL

**J. Grum and D. Zuljan**

**University of Ljubljana, Ljubljana, Slovenia**

KEYWORDS: Laser Cutting, Austenitic Stainless Steel, Thermal Phenomena

ABSTRACT: Efficient control of laser cutting processes is closely related to knowledge of heat effects in the cutting front and its surroundings. Similarly to other machining processes using high energy densities, in laser cutting processes it is very important to monitor the heating phenomena in the workpiece material due to heat input. In laser cutting processes with oxygen as an auxiliary gas, cutting energy is a combination of laser beam energy and the energy of the exothermic reactions occurring in the cutting front. The presence of oxygen in the process increases cutting efficiency, but it also causes additional physical processes in the cutting front which render a more detailed analysis of the cutting phenomena difficult. The aim of the article is to analyze the emission of infrared rays from the cutting front with a photo diode, statistically analyze the temperature signals and optimize the laser cutting process based on a critical cutting speed.

## 1. INTRODUCTION

The use of laser as a cutting tool has been expanding in the last few decades mostly because of its numerous technological advantages. One of the main advantages of laser sources is high-intensity of the laser beam which can be adapted to very different machining conditions and materials machined. The laser cutting process belongs to non-conventional machining techniques such as electro-thermal processes.

Theoretical investigations of temperature in the vicinity of the cutting front were carried out for gas welding by Rosenthal [1] and for laser cutting by Rykalin [2], Schuöcker [3] and

Published in: E. Kuljanic (Ed.) *Advanced Manufacturing Systems and Technology*, CISM Courses and Lectures No. 372, Springer Verlag, Wien New York, 1996.

Arata [4]. Rykalin's analysis was limited to a circular laser source, such as Gaussian source, and to the determination of the temperature on the surface of the cutting front. Olson [5] very carefully analysed the cutting front for which he plotted isothermal lines and then determined the thickness of the molten and recrystallized layers of the workpiece material. One of his important findings is that in case of a high temperature gradient, a thin layer of the molten and recrystallized material and a small thickness of the heat affected zone are obtained, which assures a good and uniform quality of the cut.

Rajendran [6] used thermoelements for temperature measurement in the vicinity of the cutting front and then analysed the so-called temperature cycles. He found that the cut quality is strongly related to the temperature gradient in heating and cooling.

Chryssolouris [7] gave a survey of various ways of sensing individual physical phenomena in the workpiece material during laser cutting processes. He studied various possibilities of temperature measurement and acoustic emission perception and presented various methods of on-line monitoring of the temperature in the cutting front and acoustic emission in the workpiece material with a survey methods for controlling laser cutting processes.

Nuss et al.[8] studied the deviations in the size of round roundels in laser cutting different steels with a $CO_2$ laser in pulsating and/or continuous operation. The deviation was gathered with regard to the precision of NC-table control and direction of light polarisation. Toenshoff. Samrau [9] and Bedrin [10] investigated the quality of the cut by measuring the roughness at varying laser source power and varying workpiece speeds. They also studied the quality of the cut while changing the optical system focus position with respect to the workpiece surface. Thomassen and Olsen [11] studied the effects produced on the quality of the cut by changing the nozzle shape and oxygen gas pressure.

On the basis of research investigations which consisted of temperature measurements with thermocouples, we established the temperature cycles from which we defined the temperature gradient and designed the temperature fields with isotherms. From these data it was possible to assess the quality of the laser cut. In addition to temperature measurement with a thermocouple in the vicinity of the cutting front, an infrared pyrometer was used to measure the temperature in the cutting zone itself [12].

In addition to the temperature signals, analysis was also made of the surface. The data in the histograms of the surface of the cut show that the surface signal variance is increasing with increasing roughness. From the normalised auto correlation function we can note that the surface signal variance is related to roughness [13,14].

## 2. EXPERIMENTAL PROCEDURE AND RESULTS

This article will present, some of the results obtained in indirect temperature measurement by means of infrared radiation density emitted from the cutting. The aim of the temperature measurement is to achieve monitoring the thermal phenomena in the cutting  front in the laser cutting process with coaxial supply of oxygen as the auxiliary gas, which provides, besides laser beam energy, additional exothermal heat which additionally affects the temperature changes in the cutting front.

For this purpose, the following activities were carried out:
- assembling of a sensor for measurement of infrared radiation from the cutting front;
- elaboration of a system of analysis and assessment of measurements of temperature signals from the cutting front;
- macro and microanalyses of the cut edge based on measurement of geometrical characteristics of the cut;
- optimisation of the laser cutting process on the basis of mean values of temperature signals at various cutting speeds

Experimental testing was carried out on a laser machining system ISKRA-LMP 600 with a laser power of up to 600 W and with a positioning table speed from 2 to 50 mm/s. In laser cutting, oxygen was supplied as auxiliary gas, having the role of contributing additional exothermal heat to the process and blowing the molten metal and the oxides away from the cutting front.
The investigations were carried out by applying a commonly used austenitic stainless steel alloyed with chromium and nickel 18/10 designated A276-82A according to ASTM standard. The other steel grade studied was a low-carbon structural steel A620 according to ASTM standard.
In order to study the processes in the cutting front and investigate the quality of cut, certain parameters were selected as process constants and other parameters as process variables. In Table 1, we can find constant process parameters, e.g. laser power, characteristics of the optical system, and characteristics of the shape and arrangement of the nozzle designed for coaxial supply of oxygen as auxiliary gas, oxygen pressure, and workpiece size.

Table 1: Constant machining conditions of the laser machining system

Table 2: Laser cutting process variables

| MACHINING CONDITIONS | |
|---|---|
| Laser power | PL = 450 W |
| Focal distance of the lens | zf = 63.5 mm |
| Focal point/workpiece distance | g = 0 mm |
| Nozzle/workpiece distance | s = 2.0 mm |
| Nozzle diameter | d = 1.0 mm |
| Oxygen pressure | $P_{O2}$ = 4.5 bar |
| Test pieces/ plates | 100 x 100 mm |

| Material thickness D[mm] | Cutting speed v[mm/s] |
|---|---|
| 0.6 | 35 40 45 50 |
| 0.8 | 30 35 40 45 |
| 1.0 | 25 30 35 40 |
| 1.5 | 20 25 30 35 |

Fig. 1 shows the measuring set-up for measurement of IR radiation from the cutting front, including components for capturing, storage, analysis, and assessment of the temperature signals. A temperature signal is proportional to energy flow density of infrared radiation, spreading from the cutting front, which is detected by a sensor for IR radiation. The sensor consists of a photo diode CONTRONIK-BPX 65 with an electromagnetic radiation sensibility range of from 0.4 to 1.0 μm, an amplifier, and a transformer. During the laser cutting process, such a sensor is directed towards the cutting front so that it intercepts the

radiation from the cutting front and its surroundings and transforms it into a temperature signal TS expressed in millivolts /mV/. The value of the temperature signal in cutting a given kind of material and a given thickness depends on:
- density of IR radiation of the overheated material, of the molten pool, and the plasma in the cutting front due to energy input by the laser beam;
- additional density of IR radiation from the cutting front due to exothermal reactions.

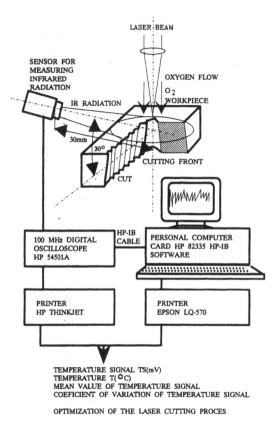

TEMPERATURE SIGNAL TS(mV)
TEMPERATURE T($^{o}$C)
MEAN VALUE OF TEMPERATURE SIGNAL
COEFICIENT OF VARIATION OF TEMPERATURE SIGNAL

OPTIMIZATION OF THE LASER CUTTING PROCES

Fig. 1: Measuring set up for temperature signal measurement by means of IR radiation

The measuring assembly, particularly the sensor for density measurement of IR radiation, shows a satisfactory response function in the frequency range of up to 200 kHz, which corresponds to our measuring requirements regarding the output temperature signal from the photo diode. The output signal obtained from the photo diode is called the temperature signal and is expressed in mV. The measuring range for the temperature signal is limited by sampling duration T=0.2 s, which gives us 2000 data on the temperature signal, considering that the interval duration selected for signal digitization is t = 0.1 ms. With regard to the temperature signal value, the maximum horizontal value of the voltage measured on the oscilloscope U=390 mV is selected, which is then sorted out into 250 classes in increments of temperature signal sensing $\Delta U$=1.5626 mV. The frequency measuring range of the temperature signal was selected with reference to our previous studies and the study carried out by Schuöcker /3/ who analyzed the variation of thickness of the molten layer.

The thickness variation of the molten layer along the cutting front depends on the temperature signal variation, i.e. its temperature. The expected frequencies of variation in the molten layer thickness depend on the cutting speed and vary between 100 and 800 Hz /3/. The temperature signal from the photo diode is finally led into a 100 MHz digital oscilloscope where it is digitized and stored on a floppy disk for subsequent statistical processing.

Fig. 2 shows bar charts of the mean values of temperature signals TS in mV (black)and magnitude of standard deviations of temperature signals STD-TS in mV (white) after cutting austenitic stainless steel having various thicknesses.

A comparison of the data provided by the temperature signal or temperature in the cutting front in cutting the investigated materials with various cutting speeds shows that:

– The mean values of temperature signals in cutting low-carbon structural steel range from 155 mV up to 175 mV, with a standard deviation of about 14 mV.

– The mean values of temperature signals in cutting austenitic stainless steel of various thicknesses range from 145 mV up to 185 mV, with a standard deviation of about 40-45 mV. The sole exception is the temperature signal in cutting thicknesses of 1.0 mm and 1.5 mm with the highest speeds where the mean value of the temperature signal decreases even to 110 mV and 75 mV respectively.

– The low mean values of temperature signals measured in cutting austenitic stainless steel, having thicknesses of 1.0 mm and 1.5 mm with the highest cutting speed can be considered as a warning of the troubles in cutting caused by a too low energy input.

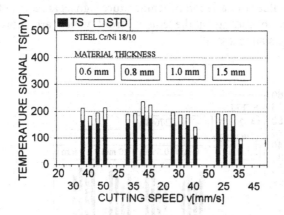

Fig. 2: Bar charts of the temperature signal measured for various steels, various material thicknesses, and various cutting speeds

A consequence of the low energy input is a low quality of the cut. At a specific laser source power, critical cutting speeds for individual kinds of materials and different thicknesses can be defined. The critical cutting speed is the highest cutting speed which, with regard to the value of the temperature signal, still ensures a good quality of the cut. The cutting speeds lower than the critical one produce a too high energy input in the cutting front which results in an increased cut width and in formation of a deeper heat affected zone.

Fig. 3 shows the coefficient of variation of the temperature signal from the cutting front in cutting low-carbon steel with a thickness of 2 mm   and austenitic stainless steel having various thicknesses, i.e. 0.6, 0.8, 1.0, and 1.5 mm. The coefficient of variation of the temperature signal CV provides more information than standard deviation of the temperature signal, therefore, it is recommended for the description of signal variation  by numerous authors. Coefficient of variation CV is a quotient obtained by dividing the standard deviation STD by the mean value of the temperature signal sample:

$$CV = \frac{STD}{TS} \cdot 100 \quad [\%$$

Based on the calculated coefficients of variation of temperature signals, the following observations about the laser cutting process can be made:
- In laser cutting a 2mm thick low-carbon steel at the selected cutting speeds, the calculated coefficients of variation of the temperature signal range from 8 to 9 %.
- In laser cutting austenitic stainless steel of various thicknesses at different cutting speeds, the calculated coefficients of variation are much higher and range from 23 to 32 %.
- The coefficient of variation of the temperature signal is relatively high in both steel grades analyzed; therefore, it can be stated that in both cases there is a very important variation of the temperature signal in the vicinity of the mean value.
- Exceptionally high values of variation of temperature signals are a result of a number of exothermal reactions occurring in the cutting front which are more important and stronger in austenitic stainless steel.

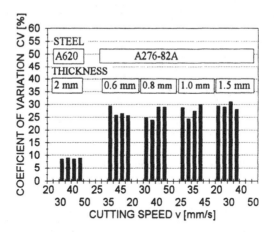

Fig. 3: Coefficient of variation of temperature signals  for low-carbon steel and austenitic stainless steel at various cutting speeds

Fig. 4 shows the magnitude of the mean values of temperature signals as a function of the cutting speed in cutting various thicknesses of austenitic stainless steel. Quick variations of temperature signals with material thicknesses of 1.5 mm and 1.0 mm can easily be observed so that critical cutting speeds can be determined for both cases uniformly. If a material thickness is 1.5 mm, the critical cutting speed is 30 mm/s, and if the material thickness is 1.0 mm, the speed is 35 mm/s. It is more difficult, however, to determine critical cutting speeds for smaller material thicknesses since the mean values of the temperature signal at higher cutting speeds are even higher than those at lower cutting speeds, which is the result of reduction in cut width. Thus a critical cutting speed of 45 mm/s could be chosen for a material thickness of 0.8 mm; for the smallest material thickness, however, additional tests should be carried out at higher cutting speeds in order to determine its critical value.

The procedure for determining the critical cutting speed is as follows:
- A specified cutting speed is selected, the temperature is measured by means of IR radiation density in the cutting front, and its mean value is determined.
- The procedure is repeated by continuous or stepwise changing of cutting speeds.
- On a rapid decrease of the mean value of the temperature signal, the critical cutting speed is achieved which, in accordance with our criteria, represents the optimum cutting speed.

The proposed procedure allows the determination of the critical cutting speed during the cutting process itself, which can be utilized for controlling the process. The procedure is both practicable and simple. By changing laser source power and/or optical and kinematic conditions it is possible to determine the optimum laser cutting conditions. The same optimization procedure of the laser cutting process may be used also for other related materials.

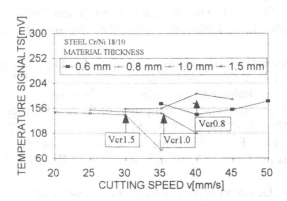

Fig. 4: Determination of critical cutting speed in laser cutting of various thicknesses of austenitic stainless steel

## 3. CONCLUSIONS

The analysis of heat effects in laser cutting the studied steels is extremely important since it explains very complex physical phenomena in the cutting front and provides a way for determining optimum cutting conditions. In laser cutting, the amount of energy input into the cutting front varies due to oscillations in laser source power, changes in the heat released in exothermal reactions, and heat losses. The results of the present study show that thermal changes in the cutting front may successfully be monitored by the temperature signal measurement via IR radiation density. The described critical cutting speeds are the optimal cutting speeds for the analysed lasers cutting process and austenitic stainless steel.

## REFERENCES

1. Rosenthal D.: Mathematical Theory of Distribution during Welding and Cutting, Welding Journal, vol. 20, 1941, 220 - 225

2. Rykalin N., Uglov A. , Kokora A.: Laser Machining and Welding, Mir Publishers, Moscow, 1978

3. Schuöcker D.: The Physical Mechanism and Theory of Laser Cutting, The Industrial Laser Annual Handbook, PennWell Books, Tulsa, Oklahoma, 1987, 65 - 79

4. Arata Y., H. Maruo, Miyamoto I., Takeuchi S.: New Laser-Gas-Cutting Technique for Stainless Steel, IIW Doc.IV-82, 1982

5. Olsen F.: Cutting Front Formation in Laser Cutting, Annals of the CIRP, vol. 37, no. 2, 1988, 15 - 18

6. Rajendran H.: The Thermal Response of the Material during a Laser Cutting Process, Proceeding of the 6th International Conference on Application of Laser and Electro - Optics, ICALEO'87, IFS Publishing, California, 1987, 129 - 134

7. Chryssolouris G.: Sensors in Laser Machining, Keynote Papers, Annals of the CIRP, vol. 43, no. 2, 1994, 513 - 519

8. Nuss R., Bierman S., Geiger M.: Precise Cutting of Sheet Metal with CO2 Laser, Laser Treatment of Materials, Ed.:B.L.Mordike, Deutsche Gesselschaft für Metallkunde, Oberursel, 1987, 279 -288

9. Tönshoff H. K., Semrau H.: Effect of Laser Cutting on the Physical and Technological Properties of the Surface of Cut, Laser Treatment of Materials, Ed.: B.L. Mordike, Deutsche Gesselschaft für Metallkunde, Oberursel, 1987, 299 -308

10. Bedrin C., Yuan S. F., Querry M.: Investigation of Surface Microgeometry in Laser Cutting, Annals of CIRP, vol. 37, No. 1, 1988, 157 -160

11. Thomssen F. B., Olsen F. O.: Experimental Studies in Nozzle Designe for Laser Cutting, Proc. of the 1st Int. Conf. on Laser in Manufacturing, Brighton, 1983, Ed.: Kimm M. F., 169 -180

12. Zuljan D., Grum J.: Detection of Heat Responses in Laser Cutting, EURO MAT'94, 15th Conference on Materials Testing in Metallurgy, 11th Congress on Materials Testing, Coference Proceedings, vol. IV., Ed.: B. Vorsatz, E. Szöke, Balatonszéplak, Hungary, 1994, 1077 - 1082

13. Grum J., Zuljan D.: Thermal Response Analysis of Cutting Metal Materials with Laser, MAT-TEC'93, Improvement of Materials, Technology Transfer Series Ed.:A.Niku-Lari, Subject Ed.:T. Ericsson, G. Pluvinage, L. Castex, Paris, France, 1993, 217 - 225

14. Grum J., Zuljan D.: Thermal Response Analysis of Laser Cutting Austenitic Stainless Steel, MAT-TEC'96, Improvement of Materials, Technology Transfer Series Ed.:A.Niku-Lari, Subject Editor: Jian Lu, Paris, France, 1996, 301 - 311.

# OPTIMIZATION OF LASER BEAM CUTTING PARAMETERS

**R. Cebalo**
University of Zagreb, Zagreb, Croatia

**A. Stoic**
University of Osijek, Osijek, Croatia

KEY WORDS : Laser beam cutting parameters, cut quality, cutting costs

ABSTRACT: It is possible to achieve reducing of laser beam cutting costs by correct determination cutting parameters. Production costs have a lot of sources and maximal reducing is achievable by doing multifactor analysis of influence factors. In this paper, influence of cutting speed and cutting gas pressure shall be analyzed. Each of these parameters strongly influence the obtained cut quality. Surface roughness, as indicator of cutting ability, is usually functional value which is dependent on cutting parameters. For its optimization it is necessary to determine optimal cutting parameters. Following reducing of production costs is possible by optimization of material utilization and laser cutting head path. Reducing of cutting time (number of machine hours) is obtainable by selection of good cutting order of cutout, and by reducing a number of free movement.

## 1. INTRODUCTION

Base approach of modern production is to rationally manage with material and time. Development and application of CNC tool machine demand definition of optimal parameters, and regarding that definition, interdependence of parameters in the form of mathematical model is necessary.

Conventional determination of technological parameters on the base of technologist experience and machine producer suggestions is not recommended. That is the reason for application of optimization methods. First step in optimization process is definition of mathematical model which is very complex because of effect a lot of influence factors on cutting process. More accurate mathematical model include linear and nonlinear parts, define by function of second order. In practice, these model are sufficient for description of technological processes, if possibility of description exists at all [1].

Mathematical models are given by goal function and are base of optimization process (determination of optimal parameters) in which we are determining the extreme value of goal function. Analysis of influence factors on cut edge quality are more represented in the literature but mathematical model given by goal function are rather infrequently.

Published in: E. Kuljanic (Ed.) *Advanced Manufacturing Systems and Technology*,
CISM Courses and Lectures No. 372, Springer Verlag, Wien New York, 1996.

## 2. OPTIMIZATION OF CUTTING PARAMETERS

Cut quality , as indicator of chosen cutting regime, is expressed as cutting width, high of slag, roughness etc. Goal function for optimization of cutting parameters in this case  is surface roughness. There are more optimization processes and in this paper derivation of function is used.

### 2.1. DETERMINATION OF INFLUENCE FACTORS

From all influencing factors, influence of three factors on surface roughness Ra [μm] are examined :
- cutting speed  v, m/min
- measuring location  h, mm
- cutting gas pressure (O₂)  p, bar.

Cutting material is steel QStE380N, thickness 3 mm. Cutting length is 150 mm. Roughness measuring device is Perthometer (T. Hobson, model Surtronic 3).

Optimization model is shown in fig. 1.

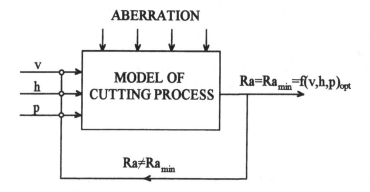

Fig 1. Scheme of input/output parameters

### 2.2. TEST PLAN

A large number of cut ability testing use method for test planing. According to previous testing, mathematical model for roughness determination is given by:

$$R_a = C \sum_{i=1}^{k} f_i^{p_i} \tag{1}$$

where are :    C, $p_i$ - cut ability factors
$f_i$ - parameter value
k - number of parameters

Equation (1) is  model  of cut ability function for description of phenomena in the tested area. Using test plan and data processing it is possible to determine function dependence of measured value (surface roughness) on input parameters and to check the accurance of seted model.

‘or varying chosen parameters (independence variable) v,h,p on five level it is necessary to do $N=2^3+6+6=20$ tests.

According to central composition test plan it is possible to get regression equation define with polynom of second order :

$$Y = \sum_{i=0}^{k} b_i \cdot x_i + \sum_{1 \le i < j}^{k} b_{ij} \cdot x_i \cdot x_j + \sum_{i=1}^{k} b_{ii} \cdot x_i^2 \qquad (2)$$

what can be expressed as follows :

$$Y = b_o + b_1 x_1 + b_2 x_2 + b_3 x_3 + b_{12} x_1 x_2 + b_{13} x_1 x_3 + b_{23} x_2 x_3 + b_{11} x_1^2 + b_{22} x_2^2 + b_{33} x_3^2 \qquad (3)$$

where are : $b_i$ , $b_{ij}$ - regression coefficients
$x_i$ - coded value of parameters.

Coding of parameters is carried on :

$$X_i = \frac{p_i - p_{im}}{\dfrac{p_{imax} - p_{imin}}{2}} \qquad (4)$$

where : $x_1$ is coded value of cutting speed (1,5m/min $< v <$3m/min), $x_2$ is coded value of cutting gas pressure (0,8bar $< p <$2 bar) and $x_3$ is coded value of measuring location (0,8mm$< h <$2,2mm).

## 2.3. REGRESSION ANALYSIS RESULTS

Mathematical processing of the roughness measuring results is carried out by using multiple regression analysis with software package Statgraphics ver. 4.2. Calculated value of coefficient of regression and review of coefficient significance are given in table I.
Multiple regression factor R=0,966214 shows good interdependence of theoretical and measured results. According that, functional dependence (5) sufficiently accurate describe surface roughness in the tested area.

Table I Regression analysis results for model with statistically significant coefficients of regression

| Variables | \multicolumn{9}{c}{Multiple regression factor R=0,966214 Factor of determination $R^2$=0,933557} |
|---|---|---|---|---|---|---|---|---|---|
| | C | $X_1$ | $X_2$ | $X_3$ | $X_1X_2$ | $X_2X_3$ | $X_1^2$ | $X_2^2$ | $X_3^2$ |
| Regression coefficients | 4,1664 | -0,9514 | 0,3755 | 1,6896 | -0,6312 | -0,2643 | 0,4930 | 0,9350 | 0,2488 |
| Standard errors | 0,2918 | 0,1936 | 0,1936 | 0,1936 | 0,2529 | 0,2529 | 0,1884 | 0,1884 | 0,1884 |
| t-value | 14,2776 | -4,9142 | 1,9398 | 8,7273 | -2,4909 | -1,0451 | 2,6165 | 4,9621 | 1,3204 |
| significance level | 0,0000 | 0,0003 | 0,0671 | 0,0000 | 0,0243 | 0,2972 | 0,0192 | 0,0003 | 0,1942 |

Regression equation is :

$$Ra = 4,166441 - 0,95141X_1 + 0,37555X_2 + 1,689623X_3 - 0,630125X_1X_2 -$$
$$-0,26438X_2X_3 + 0,49306X_1^2 + 0,93507X_2^2 + 0,2488X_3^2 \qquad (5)$$

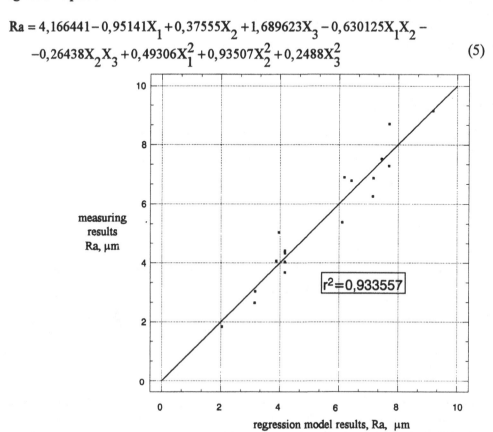

Fig. 2 Comparison of theoretical (calculated from regression model) and surface roughness
measured results

Increase of surface roughness in dependence of measuring location is shown in fig.3.

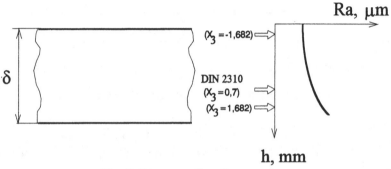

Fig. 3. Increase of surface roughness

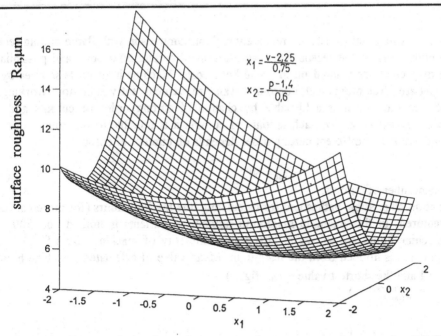

Fig. 3 Surface roughness dependence on cutting parameters (measured at 2/3 δ)

Goal function $F_c=Ra=f(x_1,x_2)$ in allowable area of cutting parameters variation $-1,682 < x_i < 1,682$, reach minimum $F_c=Ra_{min}$ for coded value $X_1=X_{10}$ and $X_2=X_{20}$ (or physical value of cutting parameters $v=v_{opt}$ and $p=p_{opt}$).

Derivating a mathematical model according to:

$$\frac{\partial Ra}{\partial x_i} = 0 \text{ ili } \frac{\partial F_c}{\partial x_i} = 0 \qquad (6)$$

or:

$$\frac{\partial Ra}{\partial x_1} = 0, \frac{\partial Ra}{\partial x_2} = 0 \text{ i } x_3 = 0,7 \qquad (7)$$

we obtain a equation system :

$$-0,9514-0,630125X_2 + 0,98612X_1 = 0 \qquad (8)$$
$$0,18785-0,630125X_1 + 1,87014X_2 = 0 \qquad (9)$$

From the equation system (8) and (9) follows the optimum values of parameters and surface roughness :

- coded parameters $X_{10}=1,1477$ and $X_{20}=0,28626$
- phisical value $v_{opt} = 3,11$ m/min and $p_{opt} = 1,57$ bar
- surface roughness $Ra = Ra_{min} = 5,0089$ μm.

# 3. OPTIMIZATION OF MATERIAL UTILIZATION AND LASER CUTTING HEAD PATH

Basically cutting costs consists of material costs and cutting costs (machine hour costs and manual hour costs). Material costs were previous more treated dealing with increasing of material utilization. Laser cutting hour costs are relatively high in comparison with flame cutting technology or another similar. That is the reason for quantification of cutting hour costs and possibility of its reducing. Cutting costs are vaulted on machine hour costs. Machine hour costs sources by cutting head movement are : free movement, contour working movement, entry in contour working.

Reducing of free movement is achievable by change of cutting order of cutouts and by free movement number reducing. For each layouts of cutouts it is possible to determine more cutting order. Possible number of different cutting order depends on number of cutouts :

$$n_{comb} = N! \tag{5}$$

where N is the number of cutouts.

For determination all possible pathes and for calculating the machine hours (for more cutouts) a lot of time is required. In this example maximum number of variants is limited on 500 random. Calculated machine hours are shown in histogram. Similarity of machine hours frequency curve with Gaussian curve is noticed from the histogram. Mean value of calculated machine hours of all 500 variants is $\bar{t}$ and the shortest value is $t_{sh}$. (fig.4 )

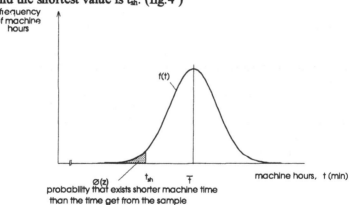

Fig.4. Frequency of calculated machine hours

Probability that exists shorter machine time than the time get from the sample (darker area on fig. 4) can be provide form :

$$z = \frac{|t_{sh} - \bar{t}|}{\sigma} \tag{10}$$

and reading after statistical tables follows $\emptyset(z)$.

Reducing of free movement number is achievable by partial cutting of cutout. Number of cutted between two free movement is grater or equal one. For reducing free movement number on that statement it is necessary to lead the cutout in contact.

Reducing of contour working movement is achievable by determination of two jointly segment which belong to different cutouts (cutting two segments with one cut).

According to /2 / it is possible to calculate laser cutting (machine) costs per hour or per unit length of cutout. This way for costs calculation is applicable for cutting head path simulation.

## .1. RESULTS OF OPTIMIZATIONS

n the test example the free movement machine hours costs were 178,3 DM/h and the working machine hour costs were 197,3 DM/h. For test cutting three specimens with unregular shape were used. After calculation of machine hour costs and material costs for different layouts of specimens, t is possible to quantificate the total cutting costs and its portions as shown in fig. 5.

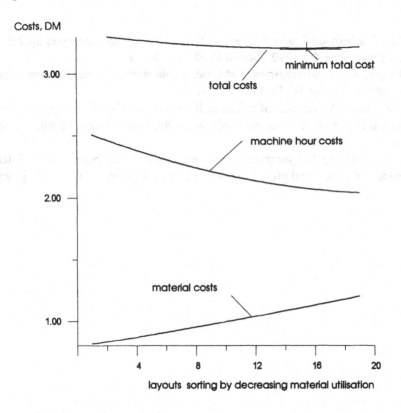

Fig. 5. Optimum cutting costs and its portion of different costs

According to fig. 5 it is seen that total costs are different for different layouts. By decreasing of machine hour costs it is seen that material costs increase. Optimum total cost is combination of costs between maximum material utilization and minimum cutting head path (machine hour costs).

## 4. CONCLUSIONS

Using test plan method and data processing, according to multiple regression analysis, it is possible to determine the functional dependence between surface roughness and cutting parameters ( cutting speed and cutting gas pressure). By derivating a goal function it is possible to find the cutting parameters which results with minimum surface roughness value. After the simulation of disposing and calculation of cutting head path for cutouts it is possible to determine minimum cutting costs. Optimal costs are the cost portion combination between cost for maximum material utilization and costs for minimum cutting head path.

## LITERATURE

[1]    Sakic, N. : Mathematical modelling and optimising of welding processes applying computing devices,  Zavarivanje 4/5 1990, Zagreb 1990, p. 205-211

[2]    Niederberger, K :Economic Aspects of Laser Applications, Obrada materijala laserom, Zbornik radova, Opatija 1-3.06.1995., 233-237.

[3]    Cebalo, R.; Stoic, A.: Analisis of influence factors of laser beam cutting on cut edge quality, International Conference, Laser material processing, Proceedings,  Opatija 1-3.06.1995., p. 71-77.

[4]    Cebalo, R.; Stoic, A.: NC programing and optimizing of laser beam cutting, International Conference, Laser material processing, Proceedings,  Opatija 1-3.06.1995., p. 95-102.

# LOW-VOLTAGE ELECTRODISCHARGE MACHINING - MECHANISM OF PULSE GENERATION

I.A. Chemyr, G.N. Mescheriakov and O.V. Shapochka
State Academy of Refrigeration, Odessa, Ukraine

KEY WORDS: Pulses generation. Erosion polarity. Dies machining.

ABSTRACT: This paper deals with the development of new electrodischarge processes which can be named LOW-VOLTAGE EDM and are distinguished by high metal removal rate and simplicity of equipment design. Conditions are found when stationary low voltage electrical arc can be converted into discrete discharges thus providing pulses generation. The role of metal "bridges" is shown as the factor providing polarity of erosion control resulting compensation of electrode wear.

New practical technologies are proposed which can compete with the existing EDM, as well as with the mechanical machining.

## 1. INTRODUCTION

The process which was investigated can be named LOW-VOLTAGE EDM (LV-EDM) and is characterised by peculiar mechanism of the pulses forming and control of the transient nature of their development [1,2].

The following points are significant in this paper:

The stationary electrical arc fed by the direct current circuit can be converted into spontaneously interrupted pulses which are generated under specific conditions of relative movement of electrodes or cross section flow of technological fluid involving the arc into migration slipping along the electrodes surface. Pulses generation depends on a number of inlet data, such as velocity of arc slipping, open voltage, gap size and is of the random nature being the mixture of the arc discharges and short circuits. Phenomenon which is responsible for the current circuit interruption always occur at the anode solid metal - molten metal boundary and is determined by the specific current density across the discharge channel. The generated pulses

Published in: E. Kuljanic (Ed.) *Advanced Manufacturing Systems and Technology*, CISM Courses and Lectures No. 372, Springer Verlag, Wien New York, 1996.

posses the ability to be self-adjusted to the changeable inlet data.

Practical production technologies based on the LV-EDM application are described for consideration of production organizations.

## 2. INVESTIGATION OF THE LOW-VOLTAGE EDM

### 2.1. THE NATURE OF PULSES GENERATION

An oscillogramm shown in Fig.1 has been obtained when cylindrical rolls were rotating without slipping, each having independent drive with the fixed distance between them (fixed gap). Surfaces of the rolls of convexed profile provided single channel discharge process. Direct voltage has been applied. The rotation of specimen rolls helped to recieve the unfolded picture of the process thus providing observation of the changes in the pulses evolution which was resulted by the open voltage (Ui) and the gap size (Sn) variation. The arc discharges, short circuits and intervals are being observed on the oscillogramm.

**Fig.1** Oscillogramm showing the nature of pulses generation

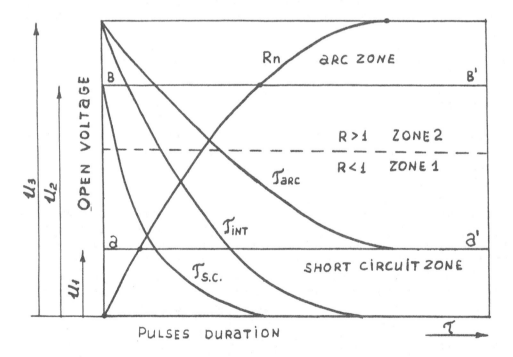

**Fig 2.** Diagramm of pulses generation.

Short circuits have place when the gap is overlaped by the molten metal which is moving in the electrical field from anode to cathode.

The bridge of molten metal exists a very short time and than is replaced by the intervals or the arc discharges. Discontinued current circuit can be restored again by the bridge or the arc discharge which occurs with a certain ignition delay.

Bridge, as a rule, ceases the arc burning. Bridge or discharge in turn, interrupt the interval. Pulsation is distinguished by the duration of each element and ratio of their numbers.

Fig.2 shows these qualitative relatioinships derivated from the oscillogramm on Fig.1.

Duration of pulses elements depends on the open voltage of current circuit (Ui) and the gap size (Sn). When Ui reaches some critical value the pulsation completely disappears.

With the reduction of Ui bellow 6 V the arc can not appear and the process is characterised only by the bridges existance.

To access technological pecularities of LV-EDM, the ratio Rn = Narc/Ns.c. must be analysed. Here Narc and Ns.c. are the arc discharges and the short circuits numbers existing per the time unit. Ratio Rn is subdued to the random situation in the gap, but its average meaning depends on the Ui and Sn (frontal gap).

The curve Rn characterises the change of these relations. With the change of Ui the Rn is subjected to the regeneration. In the zone (1) which is related to the low Ui the bridge pulses are prevailing and Rn < 1, whereas in zone (2) Rn > 1. Bridge and arc pulses are different in contribution to the metal removal and electrode wear therefore the ratio Rn can exist which provides reasonable removal rate and controlled wear.

Control of the outlet data can be provided also by the variation of Sn with the fixed U.

## 2.2. PHYSICAL EVOLUTION OF THE GENERATED PULSES

To clarify the nature of circuit interruption the following investigation was completed. The experimental device was constructed which provided installation of two metal electrodes one beind solid iron [Fe], the second of molten lead [Pb]. The bull-end size of electrode and the temperature of the molten metal were fixed. This model was ment to imitate the interface boundary between solid and the molten metal of electrodes.

It was found that circuit can be spontaneously interrupted when the current density in the boundary zone will reach the certain critical value. Thereafter the current pulsations inevitably occur to be similar to that which were observed above [Fig. 1].

In fact, during the pulse heating the boiling at the anode boundary zone was observed. The gas bubles being charged are rushing to the opposite electrode, thus involving the molten metal into motion. Therefore the force of electrical nature exists which can enter into interruction with mechanical links stipulated by the boundary interphase absorbtion.

Reciprocal action of two electrical fields can infringe their equilibrium and therefore destruction of the boundary links will occur. As it was found the separation of molten and solid metal can be observed when current density reaches the certain value. Number of experiments have shown that it can be greatly increased whenever interphase mechanical links are reinforced.

## 2.3. INVESTIGATION OF THE PULSE HEATING DEVELOPMENT

Two thin cylindrical specimens [Fe-Fe] were mated having width equal to that of anticipated size of molten spot [0.2...0.3 mm]. It could provide the melting throughout the specimens width. The loose contact was kept between them.

The low voltage was applied and control of the current was provided. Development of heating was registered by high speed camera. It was found that during approximatly 10 sec one side metal melting could be observed which was related to the anode. The photo is shown in Fig.3. Meanwhile cathode was remaining "cold". Assymmetry of the pulse melting even when specimens were identical [Fe-Fe] can be explained by the very powerful Thomson effect. This phenomenon is characterised by the physical interruction of the heat fields and electron stream and depends on their mutual directions. The great gradient of temperature is of paramount importance for achiving the high assymmetry of heating. Therefore the significant electrical convection of heat energy can be seen on the picture, thus stipulating high polarity of heating.

$(-)$

$(+)$

**Fig 3.** Assymmetry of electrodes heating.

In case Cu[-] - Fe[+] were mated the polarity of pulse heating was still higher. Significant is the fact that the highest degree of heating polarity always was observed when those cathodes were used which were recognised as the best in EDM practice. The most effective were graphite compositions.

The observed phenomenon of one side heating is developing during each short circuit [Fig. 1] which appears in course of spontaneous pulses generation.

## 2.4. LV - PROCESS WITH THE MOVING DISCHARGES

It is important to know how the relations which are shown above [Fig. 2] can be changed in case discharges are involved into the migration. It is known that the diametre of the freely burning arc can be found from the equation:

$$D = \frac{I \cdot H}{A \cdot V_n} \tag{1}$$

where A - coefficient which depends on the environment existing around the arc channel. From (1) a relation can be derived:

$$D = \frac{\sqrt{I}}{V_n} \tag{2}$$

which shows that arc diameter and, therefore, current density can be controlled when velosity [Vn] is changeable.

Discharges along with the metal bridges can be involved into migration by the magnetic field or by the powerful liquid stream across the gap [1], or by electrodes relative motion.

Concentration of energy in the discharge channel involves the changes of pulses duration, growing pulses frequancy and process of pulses generation becomes more stable. Any controlled factor which provides energy concentration can change the ratio:

$$R_n = \frac{Narc}{Ns.c.} \tag{3}$$

and therefore helps to optimize removal rate and electrode wear.

Fig.4 shows the traces of discharges migration.

**Fig.4** Traces of discharges migration

## 3. MACHINE TOOLS DESIGN AND SELECTED TECHNOLOGIES

The application of LV-EDM is shown bellow for machining of machine parts, which are characterised by the wide industrial application and at the same time are distinguished by the high labour consumption. The purpose was to develop technologies and equipment, which could compete with the conventional mechanical and EDM machining and to obtain higher quality of articles in the service.

Three examples are submitted, each being demonstration of scope of LV-EDM application.

### 3.1. MACHINE TOOLS AND TECHNOLOGY FOR THE NEEDLES TOOTH SHAPING

The shaping electrode is located on the rotating table having the variation of speed, which is provided by the drive. On the flat surface of the electrode the concentric grooves are produced by the profiled cutter, which are similar to the tooth to be shaped. The cutter ought to be periodically shifted across the electrode by steps, which are equal to the tooth pitch of the needles. The next support is carrying the tooth comb, which is pressed to the electrode tooth surface and therefore can by the microcutting to resume periodically profile of the tooth grooves. The needle workpiece is fixed by the magnetic forces. In the design of the working supports the self adjustment of the gap size is foreseen so that the application of the feedback loop is not needed.

Free vibration of the electrod is yielded which can be controlled by the variation of the mass and the rigidity of vibrating elements. Discribed technology can be used for the shaping or reshaping of the small diametres of taps, files and hack-saw plates and also to complete heat threatment of cutting edges.

Developed technology is distinguished by a number of important advantages:

– Rotary principle of machine tool structure which provides continuous processing;
– Provision of perfect tooth form of cavities along with the facilities to variate the front cutting angle;
– Simultaneous machining of several jobs and agregation of different kind of technological processings.

### 3.2. RESHAPING THE PROFILES OF RAILWAY LOCOMOTIVE TYRES.

It is well known that carrying profile of locomotive wheels are subjected to wear and local deformation. At the same time, their surfaces in course of service can acquire very high hardness, which create the obstacles for the mechanical cutting. It can be shown that the LV process is justified to solve the problem. Machining is implemented without rolling out the locomotive wheel sets. There fore the machine tool must be located in the pit and the locomotive is weighed up to liberate the wheels for the rotation. Two saddles are available which are carrying the tool-electrodes. They have plate form with the thickness about 20 mm. Their bull ends are shaped in accordance with the wheels profile. Thus, machining is proceeding along the overall wheels profile. The saddles are equiped with the device which provides adjustment of two electrodes in a working position for the preliminary and finishing machining. Machining is very similar to the grinding, therefore only cross feed is needed which is exercised by the moving saddles with the feed-back loop drive. Tool-electrode is located in the slides thus it can "swimm" in the cross direction and easy compeled to vibrate. Vibration is used as the means to provide the changes of the gap size thus securing the favourable variation of the ratio (3):

Cross magnetic field as the means to involve discharges into the migration can secure stability of pulses generation. Machining process by its efficiency can compete with mechanical cutting.

### 3.3. PORTABLE DEVICE FOR MACHINING DIE CAVITIES AND PRESSFORMS

This is an example of LV-Technology, when electrodes are not alloted with the relative motion. The direct current is applied and the feed is provided by the following-up device. Electrode and the job are located in the hermetic camera consisting of two sections which are separated by the swimming piston. The piston is fixed to the spindel, which can slip down providing feed of the working electrode. Two sections of the camera can provide counterbalance of two forces which are acting in the sections in the opposite directions. Balance of the pressures facilitates the high sensibility of the following - up system. Technological liquid [TL] must be pumped into both sections and has to penetrate into the gap. Velosity of the TC flow across the gap is the moving factor which can be used to control all machining outlet data,such as the removal rate, surface finish, electrode wear. Moving factor [MF] can be used for any section of open voltage U1, U2, U3. In case U>U3, stationary arc burning appears which however can be supressed and pulses generation restored by the increasing of the pressure in the gap. Removal rate is 3 - 5 times higher compared with the mechanical milling of dies cavities. Machining set shown in fig. 5. is portable, its weight does not exceed 20 kg while the working area is about 250.250 mm.

**Fig. 5.** Portable LV-EDM device for cavities machining.

## 4.CONCLUSIONS

– The low-voltage process of the pulses generation is existing, which is determined by the peculiar physical phenomenon occuring at the solid-molten metal interphase boundary.

– Spontaneous interruction of the current circuit appears when specific current density across the discharge channel aquires the certain critical value.

– Existance of the self-excited current pulsation is stipulated by the applied open voltage, which also determine the pulses duration, their physical peculiarities and frequancy.

– Spectrum of the pulses generation with the different ability of the wear compensation and efficiency of the removal rate can be created by the programmed and cycled changes of the open voltage or the gap size.

– Multy-channel nature of the process can provide realisation of the high power and anables to incorporate into the gap the current of thousand Ampers.

– The scope of LV-process for the industrial application is illustrated by the described technologies.

## 5.REFERENCES

1. G. Mescheriakov, N. Mescheriakov, V. Nosulenko. Physical and Technological Control of Arc Dimentional Machining. CIRP Annals, vol 57, 1, 1988.

2. G. Mescheriakov. Electro-physical Process in the Electro-Pulse Metal Cutting. CIRP Annals, vol XVIII, 1970.

# NUMERICAL MODEL FOR THE DETERMINATION
# OF MACHINING PARAMETERS
# IN LASER ASSISTED MACHINING

**F. Buscaglia,  A. Motta and M. Poli**

**C.N.R. - ITIA, Milan, Italy**

KEY WORDS: Laser Machining, Machining of Advanced Materials

ABSTRACT: LAM is a technology proposed in the early '80s for increasing the cutting speed during the machining of traditional materials; the subsequent applications corcerned the machining of hard metals as steel alloys, nickel based alloys and ceramics. In ITIA experimental investigations on nickel based alloys and ceramics were performed, and a numerical model was realized and validated at the same time. In this paper the mentioned model is presented, together with the conclusions about the ways to set all the parameters involved in the turning.

## 1. INTRODUCTION

With LAM (acronym of Laser Assisted Machining) is intended a turning operation by which the material is heated, at least until a temperature of 500 K (as stated in [1]), in order to lower the mechanical characteristics of the material and to make the machining easier. Up to now all the studies performed (for ex. [1], [2], [3] and [4]), and also experimentations done at the CNR-ITIA of Milano, agree about the possibility of decreasing cutting force up to 30%.

The mentioned studies concentrate only on an analysis of experimental kind, evidencing the lack of an analitic tool for the optimization of the whole parameters involved. This work proposes a numerical model, realized by finite volume method, to investigate temperature

Published in: E. Kuljanic (Ed.) *Advanced Manufacturing Systems and Technology*,
CISM Courses and Lectures No. 372, Springer Verlag, Wien New York, 1996.

distribution inside a sample machined with laser assistance. The developed model is a useful tool for identifying almost all working parameters (cutting speed, depth of cut, feed, laser specific power, laser spot dimensions etc); the target during the development is to create an easy to use model, that allows an agile adaptation of the grid to the specific geometry and finally that allows to obtain the distribution with reasonably small calculation times (about 30 min on a DEC 3000 AXP model 400 workstation).

## 2. EXPERIMENTAL CONDITIONS

The experimental conditions we referred to, during model definition, follow. A laser assisted machining operation of a cilyndrical sample of Inconel 718, 74.16 mm in diameter is considered. The physical characteristics of the nickel based alloy used are: specific heat equal to 501.3 J/(kgK), density equal to 8220 kg/m$^3$ and thermal conductivity depending on temperature in accordance with the equation

$$k=11.068+1.5964E\text{-}02\ T$$

obtained by interpolation of the experimental behaviour.

The laser supplies a power of 384 W (absorbed by the material) and is focused on an elliptical spot with axes 1.7 mm and 2.0 mm long. A carbide tungsten tool is used, for which the builder furnished a constant thermal conductivity of 50 W/(mK); the rake angle is equal to 6°. The feed is equal to 0.25 mm/rev and the tool to spot distance is 5 mm.

This study considers 3 values of depth of cut (1, 2 and 3 mm) with speed values of 15, 25 35 and 45 m/min.

## 3. THE MODEL

The numerical model is realized with the software PHOENICS 2.1 by CHAM Ltd., and considers only a part of the whole cylinder: this portion has to be great enough in order to ensure the correctness of the solution, but not too much, in order to avoid a very long calculation time. The domain obtained with the above rules for a depth of cut equal to 2 mm is 2 mm long in axial direction, 2.5 mm long in radial direction and embraces an angle of 1 rad (see fig. 1); the mesh is made by 32x40x70 cells along the just mentioned directions. Three heat sources are considered: the laser beam, one placed in the chip formation zone and one due to the tool-chip friction placed at their interface. The laser is represented as a square source (side equal to 1.6 mm) with release of a costant power on the whole surface. The deformation zone, or primary zone, is characterized by a shear angle $\phi$ of 31.7° (see for ex. [5]) and is 0.183 mm in width (see [6]); the integral value of the released heat in the time unity is given by the product $F_s x V_s$ of the cutting force component along the shear plane times the velocity component along the same direction.

The friction, or secondary, zone dimensions are identified by depth of cut and length of contact, set to 0.5 mm. Into the secondary zone the energy dissipation takes place both at the interface between tool and chip because of friction, and into a volume of finite depth due to deformation.

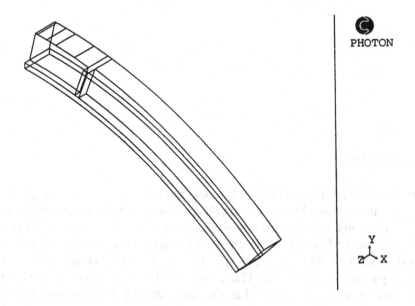

fig. 1

As the depth of the secondary zone is very small (about 0.02 mm), the released heat is neglected, so that only a source on the surface is considered; the power supplied is given by the product FxV, where F and V are respectively the cutting force and velocity component along tool face. In the model the heat generated in the secondary zone is introduced as a constant source on the whole surface considered, while for the primary zone the force is related to temperature, following the formula (given for ex. in [7])

$$F = k_s A$$

where A is the surface of contact between tool and uncutted chip and $k_s$ is the shear flow stress, considered a function of temperature as shown in the following graphic in fig. 2 (experimental and interpolated curves).

In our model the tool is not directly represented, but its presence is considered by the subdivision of the heat generated into the secondary zone into two parts, proportionally to the material and tool thermal conductivities; this implies that we suppose that the tool dimensions are great enough, with respect to the chip, so that heat fluxes are not obstructed.

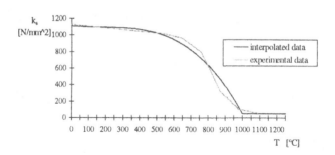

fig. 2

## 4. THE VALIDATION

The validation of the model was realized by comparison between the calculated distribution of temperature and the one obtained by experiment. The experiments were performed with a cylinder of titanium machined so that it was identical in shape to the cylinder worked during LAM sessions. The surface created by the tool is heated by a laser beam (with the same characteristics used for LAM) and the temperature reached in the centre of the spot is read by an optical pyrometer, available at ITIA. The model was changed accordingly with the new conditions; it is to be noted that the modifications involve only some costant parameters used for the description of the material. The pyrometer reads the temperature inside of a square spot ( 0.2 mm of size) and can do up to 5000 acquisition in 1 second. The measure of the temperature is based on the hot surface emission of radiation: the radiation emitted is focused on a diffraction grid and two images are created on two diodes, corresponding to two different wavelength (950 and 650 nm). These ones give as output an electrical signal linearly proportional to the incident radiation (i.e. to the temperature). The first diode is utilized to acquire a range of low temperature (1000-2000 °C), the second for an high range. In order to be sure to measure a temperature higher than the threshold of 1000°C, the pyrometer acquisition spot coincides with the center of the laser spot.

The comparison between the temperature read by pyrometer and that calculated shows the very good precision of the model (maximum error within the range of ±5% of the actual value).

The validation of a model realized for Inconel 718 by experiments on titanium is correct, as already remembered, because the only differences between the two models are some constant parameters, that do not modify the distribution of temperature (only the local values).

## 5. ANALYSIS OF CALCULATION RESULTS

The obtained results are shown in the following pictures as graphs of temperature on cutting edge.

Figures 3, 4 and 5 show the distribution of temperature for depth of cut equal to 1 mm, 2 mm and 3 mm respectively and for the velocity values indicated in each picture. These graphs show that:
-   the depth of cut must not be too high with respect to spot dimensions, because laser effects are negligible in the external zones of the tool created shoulder if the above condition is not satisfied;
-   the velocity increase lowers the maximum value of the temperature and lets the obtaining of distribution of growing uniformity on the cutting edge;
-   the maximum temperature is reached in the middle of the cutting edge, apart for depth of 1 mm; for this value an effect of heat accumulation takes place, and so the maximum shifts toward the upper edge of the tool created surface.

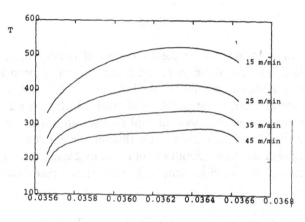

fig. 3

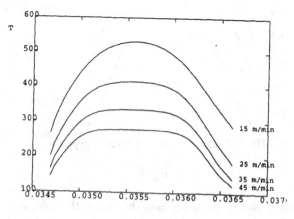

fig. 4

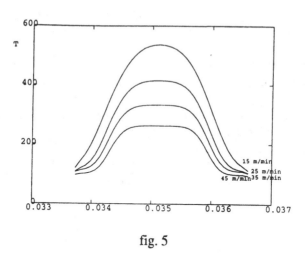

fig. 5

These distributions are obtained under the hypothesis that laser beam hits the shoulder exactly in the middle. The condition just stated can be not rigorously respected while machining, because of the vibration either of the lathe or of the final focusing lense. For this reason two more situations are investigated, by shifting a little upward, or downward, the laser spot. If beam dimensions and those of tool created surface are very different, these shifts do not change significantly the temperature distribution, apart a transfer of the curves. If the dimensions are near the same (depth of cut equal to 2 mm for conditions here tested) the spot shift determines great modifications in the temperature distribution (see fig. 6)

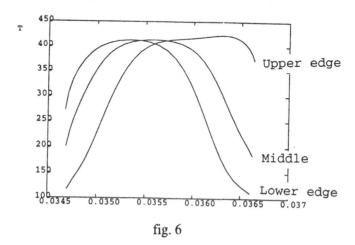

fig. 6

The results were analized also to study the heat flux inside the chip, in order to identify the machining conditions for minimizing tool wear (that exponentially depends on temperature,

as stated in [8]). Because of this we realized some simulations which consider only the laser as heat source. Fig. 7 shows the distribution of temperature on the cutting edge and on a parallel line placed immediately into the chip (at the least distance allowed by the mesh).
The curves in the graph show that an increase of speed decreases the heat flux directed from chip to tool. This condition is positive because allows the tool to machine in the same conditions it would experience when working a material of low mechanical characteristics without laser assistance.

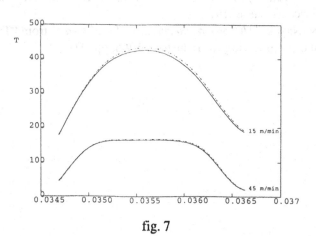

fig. 7

## 6. CONCLUSIONS

From what said in the previous paragraph, it is possible to conclude that:
- it is necessary to machine with high laser power (at the limit of melting in the hottest point of the laser spot), so to lower as much as possible material resistance;
- it is necessary to machine at the maximum speed (at least in the range considered) so to obtain distribution as uniform as possible, and also to reduce laser heat flux from chip to tool;
- the laser spot must be focused in the middle of the tool created shoulder and their dimensions have to be almost the same;
- the feed and spot-tool distance have to be properly choosen, in order to let the heat supplied by the laser to spread enough inside the material (in this sense the two parameters mentioned are function of cutting speed, too).

## REFERENCES

1. I.Y. Smurov, L.V. Okorokov "Laser assisted machining"
2. M. Albert "Laser on a lathe" from "Modern machine shop", May 1983, pp. 50-58

3. S. Copley, M. Bass, B. Jau, R. Wallace *"Shaping materials with lasers"* from "Laser materials processing", North-Holland Publishing Company, 1983

4. F. Cantore, S.L. Gobbi, M. Modena, G. Savant Aira *"L'assistente laser"* from "Rivista di meccanica", 1993, pp. 92-99

5. M.E. Merchant *"Mechanics of the metal cutting process. II. Plasticity conditions in orthogonal cutting"* from "Journal of Applied Physics", Vol. 16, 1945, pp. 318-324

6. P.L.B. Oxley *"Mechanics of machining - An analytical approach to assess machinability"* John Wiley and Sons, 1989

7. G.F. Micheletti *"Tecnologia meccanica - Vol.I - Il taglio dei metalli"* Unione Tipografico-Editrice Torinese, 1975

8. N.H. Cook, P.N. Nayak *"The termal mechanics of tool wear"* from "Transaction of the ASME - Journal of Engineering for Industry", 1966, pp. 93-100

# DEEP SMALL HOLE DRILLING WITH EDM

## M. Znidarsic and M. Junkar
### Faculty of Mechanical Engineering, Ljubljana, Slovenia

KEY WORDS: small hole drilling, inductive learning, decision tree

## ABSTRACT

This paper discusses a technological application of deep small hole drilling with EDM where the ratio between depth and width is about 10. Many different authors suggested new approaches when producing small and deep holes (d≤1mm) which is a rather complicated technological task. Additional special generators, extra vibration of the electrode, additional relative motion and many other solutions have been proposed. In this research we studied the influence of high pressure of the dielectric in the gap to improve flushing conditions and to perform better technological results. Process evolution is studied by identification of pulse series by means of high frequency digital oscilloscope connected to the computer. We developed a special chamber-device in which dielectric pressure is augmented locally. In this way the drilling of small holes is enabled on a classical ED sinking machine. Process behaviour is studied by means of voltage pulse parameters and the technological characteristic of machined hole. Voltage and current pulses play predominant role in the optimisation of EDM process. The results will be applied as a data base for the automatic control of the gap distance and flow of the dielectric that are the most common actions of the operator. The experiments are evaluated by the expert in order to develop a Technological Knowledge Data Base (TKDB), as well as decision support system for adaptive control of the process.

## 1. INTRODUCTION

Small hole drilling is a rather difficult process in most technological applications [1,2,3,4]. The producing of deep small holes (d<1mm, h/d>10) is often associated with unfavourable metal and other particles, tool cooling and especially the stiffness the tool in conventional machining processes [5]. Some of the nonconventional machining processes like laser beam machining (LBM) or electron beam machining (EBM) enable producing of deep and small

Published in: E. Kuljanic (Ed.) *Advanced Manufacturing Systems and Technology*, CISM Courses and Lectures No. 372, Springer Verlag, Wien New York, 1996.

holes. The problem of costs and insufficient accuracy and costs still remains. Electro discharge machining process (EDM) is convenient for machining electro conductive materials. It is made up to produce irregular shapes in the workpiece from the beginning of EDM technology. Present technological inconveniences in small hole drilling with EDM are:

- working accuracy
- electrode material
- to clamp and to position the electrode
- to attain acceptable dielectric pressure in deep small holes
- the selection of dielectric
- the working condition selection

## 2. SMALL HOLE DRILLING WITH EDM

The radical technological problem in producing deep small holes with EDM is to attain efficient working conditions in the working gap by means of flushing out the produced particles [2,3]. For this purpose we developed special chamber-device (Figure 1.) to enable EDM process locally without submerging the workpiece into the dielectric and to achieve higher dielectric pressure in the gap.

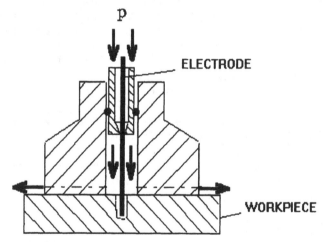

**Figure 1**: Dielectric flow in chamber-device

The technological database for existing EDM sinking machine is based on standard experiments (diameter of the electrode d=20mm). The optimal working conditions for small hole drilling (d>1mm) differ from standard conditions essentially. We have chosen 5 different working conditions to define proper working parameters with tested Ingersoll 80P sinking machine (tab.1.).

| regime | sign. | tp/μs/ | ti/μs/ | ui/V/ | τ/ / | ie/A/ |
|--------|-------|--------|--------|-------|------|-------|
| 1 | M1,4,4,10 | 50 | 48 | 180 | 0.96 | 4.5 |
| 2 | M1,4,4,7 | 50 | 34 | 180 | 0.68 | 4.5 |
| 3 | M1,6,4,7 | 50 | 34 | 180 | 0.68 | 6.5 |
| 4 | M1,8,4,6 | 50 | 30 | 180 | 0.60 | 13 |
| 5 | L1,9,4,10 | 510 | 450 | 180 | 0.88 | 19 |

Tab.1.:Working conditions

Electrodes for tests were made from electrolytic copper. We used the electrodes with diameter d=20mm for standard experiments and with diameter d=0.65mm in the case of deep small hole drilling tests. Workpieces were made from tool steel OCR12 hardened and annealed to 60-62 HRc. High speed digital oscilloscope was capturing the pulse series (Figure 2). Each serie included 320 pulses in progression and was transferred to the computer.

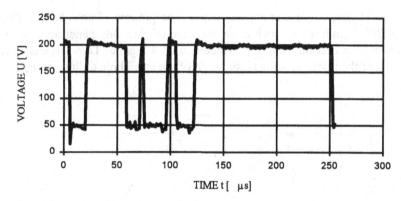

Figure 2: The sample of the pulse serie in progression

The chosen machining parameters enabled the comparison between standard process and small hole drilling with EDM (Figure 3.). The derived conclusions are:

- the material removal rate $V_W$ is higher when using standard electrode (d=20mm) in all working regimes. The difference is still increasing with higher maximum discharge current $\hat{i}_e$=4.5A (1.regime) to $\hat{i}_e$=19A(5.regime);

- the electrode wear $V_e$ is lower when testing small hole drilling with EDM at 2.,3. and 4. regime. This results probably derived from the fact that the 2nd, 3rd and 4th regimes are rather fine regimes.

- the relative electrode wear $V_W/V_e$ increases especially at 5th regime. The 5th regime's working conditions are too rough ($t_i$=450μs and $\hat{i}_e$=19A). The cumulated heat increases melting of the small electrode (d=0.65mm). The appropriate small hole drilling working conditions are attained at shorter pulses ($t_i$=50μs).

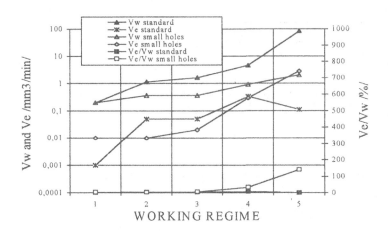

Figure 3.: The removal rate Vw, wear Ve and relative wear Ve/Vw versus working regimes

The removal rate is not eligible for machining estimation. The process of small hole drilling is estimated better with $V_p$ [mm/min] which is the measure of electrode penetration speed (Figure 4.). We achieved the best machining parameters with 4.regime. These results were applied in the next step of the research - the recognition of the small hole drilling process.

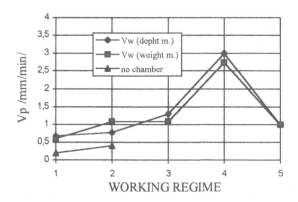

Figure 4.: The electrode penetration speed $V_p$ versus working regime

## 3. DISTINGUISHING THE SMALL HOLE DRILLING PROCESS

The goal of this research is to gain as many as possible information about deep small hole drilling process [6,7,8,9]. We wanted to distinguish between the standard process and small hole drilling process. We have defined the samples of each process. Each sample consisted of 8 pulses in progression. Each of the tested working conditions (regimes 1 to 5) was recorded with at least 40 samples. The appointed portions of characteristical pulses in each sample were classified by means of statistical analysis. The characteristical EDM pulses are: open voltage pulses A, effective discharge B, arcs C and short circuits D. The evaluated results enable us to define the discrepancy between standard EDM process and deep small hole drilling process with EDM (Figure 5.).

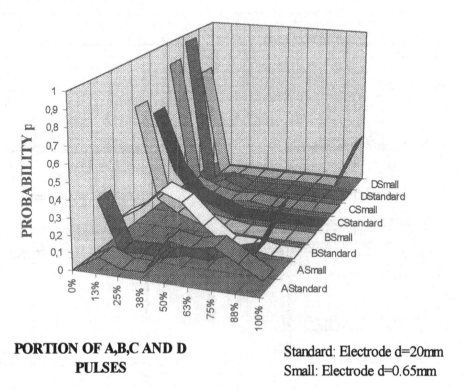

**PORTION OF A,B,C AND D PULSES**

Standard: Electrode d=20mm
Small: Electrode d=0.65mm

Figure 5: The probability of characteristical pulse portions in comparison with standard EDM process and deep small hole drilling process.

At this point the research was focused on the study of voltage pulses in progression only. The process classification was based on the computed pulse area which is as a significant process attribute. Namely the small hole drilling process is completely different comparing with standard process. So we described both processes with the examples (8 pulses in progression) consisted on probabilities of certain pulse area (-1.6 to 18 mV·s). The examples representing the tested working regimes were ranged into three classes: INEFFECTIVE, LESS EFFECTIVE and EFFECTIVE process. We evaluated the most effective working conditions (regime 4) in the case of small hole drilling. The decision of the probability of defined pulse area as a process attribute derived from the pulse analysis. For example: pulse with area of 17 mV·s is typical open voltage, pulse with area of 8 mV·s is effective discharge with long $t_d$ (60% of $t_e$), pulse with area of 1 mV·s is an arc and pulse with area of 0 mV·s is short circuit. The samples of probability trends enable the process ponderation (Figure 6.). It is obvious that small hole drilling process is completely different compared with standard ED sinking process. The portion of effective pulses is about a half of the portion attained in the case of standard EDM sinking. The process optimization parameters for small hole drilling process need to be prescribed separately.

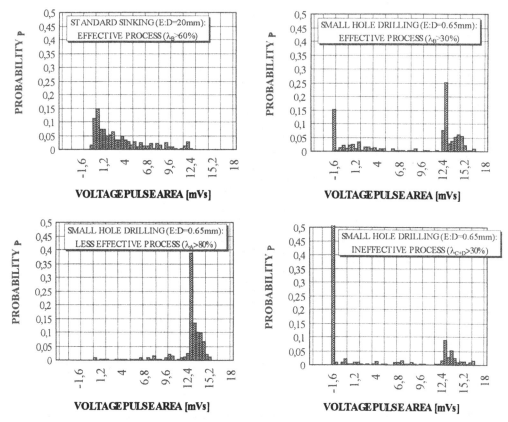

Figure 6.: Probability trends versus voltage pulse area

## 4. CONCLUSIONS

The results of deep small hole drilling tests were conducted to complement the technological database on classical EDM sinking machine. The conclusions are based on the following findings:

- the classical EDM device is appropriate for deep small hole drilling with certain limits like high accuracy of machined hole;

- the special chamber-device was developed to enable EDM process on workpieces which are not under the dielectric level and to improve flushing conditions in the working gap;

- the small hole drilling process demands different working parameters in comparison with standard EDM process. The portion of arcs and short circuits is increased, due to bad flushing conditions especially when producing deep small holes (h/d>10);

- at certain conditions it is possible to work out a hole with diameter d=0.65mm and depth h=6.5mm in the time t<3.5min. This results enable comparison with other competitive machining processes directly;

- the process recognition enables to define the optimization parameters. The accomplished experiments contributed to the development of a Technological knowledge data base (TKDB) and the decision system for the process control.

## 5. LITERATURE

1. Masuzawa, T.; Kuo, C.; Fujino, M.: Drilling of Deep Microholes using Additional Capacity, Bull. Japan Soc. of Prec. Engg., Vol.24, No.4, Dec. 1990,275-276

2. Masuzawa, T.; Tsukamoto, J.; Fujino, M.: Drilling of Deep Microholes by EDM, Annals of the CIRP, Vol. 38/1/1989,195-198

3. Toller, D. F.: Multi-Small Hole Drilling by EDM, ISEM 7, 1983,146-155

4. Žnidaršič, M.; Junkar, M: Crater to Pulse Classification for EDM with the Relative Electrode to the Workpiece Motion, EC'94, Poland 1994

5. Roethel F., Junkar M., Žnidaršič M.: The Influence of Dielectric Fluids on EDM Process Control, Proceeding of the $3^{rd}$ Int. Machinery Monitoring & Diagnostic Conference, Dec. 9-12, Las Vegas, USA, 1991, 20-24

6. Junkar, M.; Sluga A.: Competitive Aspects in the Selection of Manufacturing Processes, Proceedings of the $10^{th}$ Int. Conference on Applied Informatics, 10-12 February, Innsbruck, Austria, 1992, 229-230

7. Junkar, M.; Filipič, B.: Grinding Process Control Through Monitoring and Machine Learning, $3^{rd}$ Int. Conference "Factory 2000", University of York, UK, 27 - 29 July, 1992, 77-80.

8. Junkar, M.; Komel, I.: Modeling of the Surface Texture generated by Electrical Discharge Machining, Proceedings of the 12th IASTED International Conference on Modeling, Identification and Control, Insbruck, Austria, Acta Press 141 - 142

9. Junkar, M.; Filipič, B.; Žnidaršič, M.: An AI Approach to the Selection of Dielectric in Electrical Discharge Machining, presented at the 3.rd Int. Conf. on Advanced Manufacturing Systems and Technology, AMST 93, Udine, 1993. 11-16.

# INFLUENCE OF CURRENT DENSITY ON SURFACE FINISH AND PRODUCTION RATE IN HOT MACHINING OF AUSTENITIC MANGANESE STEEL

**S. Trajkovski**

University "Sv. Kiril i Metodij", Skopje, Macedonia

KEY WORDS: Hot Machining, Tool-Life, Production Rate, Surface Finish

ABSTRACT: The paper deals with the influence of current density on surface finish, machinability, and production rate in hot machining of austenitic manganese steel with electric contact heating. The experimental test shows the existence of optimal current density for which the maximal tool ife and minimal roughness can be achieved. From the experimental test can also be seen that in hot machining the influence of feed rate on surface finish is much smaller than in case of conventional machining.

## 1. INTRODUCTION

In hot machining, heat is applied at the workpiece material in order to reduce the shear strength in the vicinity of the shear zone. Briefly, the technique of electric contact heating consists of passing a relatively large current (AC or DC) between the cutting tool and workpiece, heat being generated on the workpiece material by the Joule-effect.

Austenitic manganese steel is characterized by the high resistance to abrasive wear and considerable work - hardening effect which causes a very poor machinability.

Applying additional heat in the cutting zone it is attended primarily to reduce the work hardening effect and the shear strength of machined material in order to obtain a better machinability.

Published in: E. Kuljanic (Ed.) *Advanced Manufacturing Systems and Technology*, CISM Courses and Lectures No. 372, Springer Verlag, Wien New York, 1996.

## 2. EXPERIMENTAL CONDITIONS

### 2.1 Workpiece-material

Casting specimens of manganese steel (heat treated), with following chemical composition: 1.2% C, 11.7% Mn, 0.66% Cr, 0.96% Si, and the following mechanical characteristics: 580 Mpa tensile strength, 372 Mpa yield strength, HB 200-220, 31% relative elongation and 24% reduction of area, were used.

### 2.2 Tool-material

SINTAL - throwaway cemented carbide tips: HV-08 (ISO-K10), HV-20 (ISO-K20), SV08 (ISO-P10), SV20 (ISO-P20), and coated tips with TiC+TiN coating: TNC-H$^{PLUS}$ , TNC-S$^{PLUS}$, all with tips geometry SNMA 120408 and TNC-H$^{PLUS}$ SPUN 120308 (on the base of ISO-K25 and ISO-P20) were used.

When the carbide tips are clamped on the tool holder the principle cutting angles were as follows:
- for the tool tips SNMA: rake angle $\gamma=6°$, inclination angle $\lambda=-6°$,
- for tool tips SPUN: $\gamma=6°$, $\lambda=-6°$, and
- for both tool tips clearance angle $\alpha=6°$, principle cutting angle $\kappa=75°$ and the auxiliary cutting angle $\kappa_1=15°$.

### 2.3 Experimental arrangement

Figure 1 shows the experimental arrangement for electric contact heating. A relatively large D.C. passes through the chip tool interface during a turning operation, by connecting a welding type transformer-rectifier across the tool and workpiece. Amperage was measured by calibrated shunt connected with a mV meter (600A/60mV).

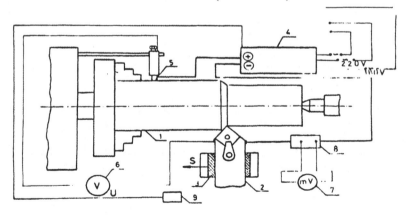

*Fig. 1 Circuit diagram for D.C. electric contact heating. 1- workpiece, 2- cutting tool, 3- tool isolation, 4- D.C. rectifier, 5- graphite-copper brush, 6- V meter, 7- mV meter, 8- calibrated shunt, 9- switching device.*

## 3. TOOL-LIFE

The tool life tests were performed in turning operations using a new short time method developed by author [4]. The obtained results show that the tool tips of grade P (SV-08 and SV-20) are more wear resistant specially on the rake face, but they are more unreliable. After short period of machining (3-5 min) failure of cutting edge appears. Fig. 2 shows the wear process developed on the rake face and the flank face of the cutting tool.

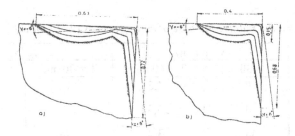

*Fig. 2 Tool wear development in machining austenitic manganese steel with electric contact heating. a) for HV-20 SNMA; b) for TNC-H$^{PLUS}$ SNMA*

Results obtained in tool life test are expressed as a ratio of tool life in hot machining ($T_h$) to the tool life in conventional machining ($T_c$). Figure 3 shows the change of the relative tool life ($T_h/T_c$) with the current density in machining with tool tips TNC-H$^{PLUS}$ SNMA. From fig.3 can be observed existence of two optimal current densities (one for 80 A, and the other for 50 A) in which the maximal relative tool life can be obtained.

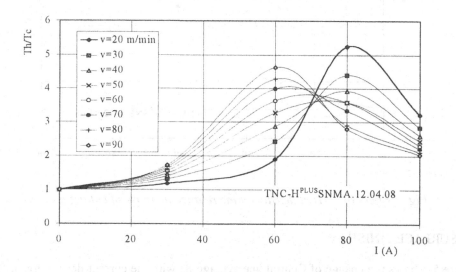

*Fig. 3 Effect of heating current density on relative tool life for different cutting speeds*

With the increase of the cutting speed above 60 m/min (for the given conditions) the optimal current density is removing to the smaller values.

In the table 1 are given the calculated cutting speeds from the Taylor's equations, obtained from the tool life tests, with tool tips types TNC-H$^{PLUS}$ SNMA for different current densities I (A), and the $v_h$ / $v_c$ - ratio, where $v_c$ - cutting speed in conventional machining and $v_h$ in hot machining. Figure 4 presents the change of $v_h$ / $v_c$ with the current density for different tool life.

*Table1 Cutting speeds (v) and the relative increase of cutting speeds ($v_h/v_c$) for different tool life*

| I (A) | Taylor's equ. | v (m/min) | | $v_h$ / $v_c$ | |
|---|---|---|---|---|---|
| | | T = 30 min | T = 60 min | T = 30 min | T = 60 min |
| 0 | $T^{0.33} v = 94$ | 30,6 | 24,3 | 1 | 1 |
| 30 | $T^{0.36} v = 115$ | 33,8 | 26,3 | 1,1 | 1,08 |
| 60 | $T^{0.41} v = 178$ | 44,2 | 33,2 | 1,44 | 1,37 |
| 80 | $T^{0.29} v = 126$ | 47 | 38,4 | 1,54 | 1,58 |
| 100 | $T^{0.3} v = 116$ | 41.8 | 34 | 1,37 | 1,4 |

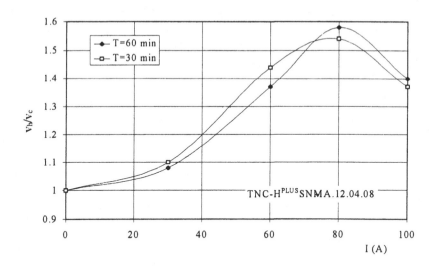

*Fig. 4 Effect of current density on the relative increase of cutting speeds.*

## 4. SURFACE FINISH

Figure 5a shows the change of Central line average $R_a$ with the current density for different tool tips and figure 4b the change of Ra with the current density in machining with TNC-H$^{PLUS}$ SNMA tool-tips for different cutting speeds.

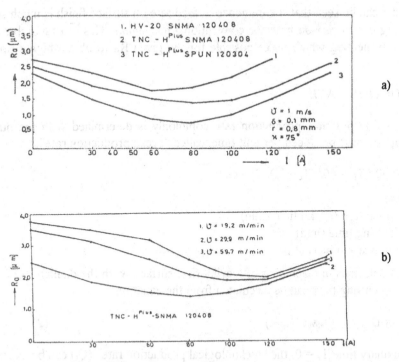

a)

b)

*Fig. 5 Effect of current density on surface finish*

From figure 5 can be observed the existence of optimal current density in which the smallest value of Ra can be achieved.

Figure 6 shows the change of Ra with the feed rate in conventional and hot machining.

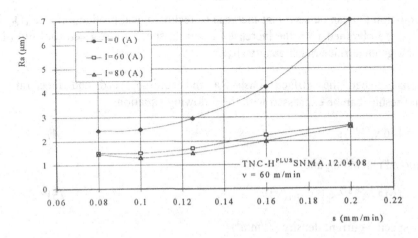

*Fig. 6 Effect of feed rate on surface finish for different current densities.*

From figure 6 can be seen that the influence of feed rate on surface finish is much smaller in hot machining in comparison with the conventional machining. This is a favorable effect of electric current heating which make possible for the given Ra to obtain higher production rate.

## 5. PRODUCTION RATE

In the production practice, production rate commonly is determined with the number of pieces (or operations) produced per unit time, called "cyclic production rate".

$$Q_c = 60 / T_0 = 60 / (T_c + T_n) \quad \text{(piece/h)} \tag{1}$$

where:
$T_0$ - operation (cycle) time (min)
$T_c$ - cutting time (min)
$T_n$ - auxiliary time (min)

For turning operations in machining of a cylindrical surface, with the diameter D (mm) and length L (mm) cutting time can be calculated from the equation:

$$T_c = \pi D L / 1000 \, v \, s \quad \text{(min)} \tag{2}$$

assuming auxiliary time $T_n = 0$, the "technological production rate" $(Q_t)$ can be expressed:

$$Q_c = Q_t = 60 \, 1000 \, v \, s / \pi D L = C \, s \, v \tag{3}$$

where :
v - cutting speed (m/min)
s - feed rate (mm/min)
C - constant depending of the size of machined surface.

From the equation (3) can be seen that increase of technological production rate $(Q_t)$ in hot machining can be obtained with the increase of cutting speed for a given tool life (T) and feed rate for a given surface finish quality $(R_a)$.

The relation between the surface finish Ra and cutting speed and feed rate using experimental results can be expressed with the following equations:

$$Ra = 166 \, v_c^{-0.386} \, s_c \tag{4}$$

for conventional machining and:

$$Ra = 1013 \, v_h^{-0.677} \, s_h^{0.657} \, q_{EO}^{0.345} \tag{5}$$

where $q_{EO}$ - specific current density (A/mm$^2$)

Solving equations (5) and (4) for $\underline{s}$ can be obtained:

$$S_h = (R_a/1013)^{1.522} \; v_h^{1.03} \; q_{EO}^{0.325} \tag{6}$$

and

$$S_c = (R_a/166)^{0.873} \; v_c^{0.337} \tag{7}$$

According to the equation (3) and the equations (6) and (7) the relative production rate can be written:

$$\frac{Q_{th}}{Q_{tc}} = \frac{v_h}{v_c} \frac{S_h}{S_c} = 0.0023 \; Ra \; \frac{v_h^{2.03} \; q_{EO}^{0.52}}{v_c^{1.337}} \tag{8}$$

For T = 30 min, $R_a$ = 3 μm and $v_c$ = 30.6 m/min (table 1)

$$Q_{th}/Q_{tc} = 0.0000485 \; q_{EO}^{0.525} \; v_h^{2.03} \tag{9}$$

and for T = 60 min, $R_a$ = 3 μm and $v_c$ = 24.3 m/min,

$$Q_{th}/Q_{tc} = 0.000066 \; q_{EO}^{0.525} \; v_h^{2.03} \tag{10}$$

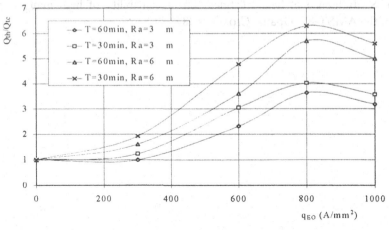

*Fig. 7 Effect of current density on the relative production rate.*

## 6. CONCLUSIONS

It is evident that machinability of materials which are difficult for machining such as austenitic manganese steels, can be improved using electric contact heating. There exists an optimal current density for every tool material and cutting conditions, after which the tool life fails. The temperature rise in the tool-chip interface due to electric current has favorable effects on surface finish and the production rate.

Favorable effects were obtained using TiC-TiN coated cemented carbides with WC-Co base which are normally used for machining of cast iron.

As higher is the cutting speed, the lower additional heat is needed to obtain the optimal cutting temperature for maximal relative tool life, relative cutting speed and relative production rate in hot machining.

REFERENCES

1. Trajkovski, S.: Investigation of machinability of high manganese steels at elevated temperatures, Research Project supported by the State department of Science, Skopje, Macedonia, 1983

2. Trajkovski, S.: Machinability of austenitic manganese steel at elevated temperatures using electric current heating, Proceedings of the 7th ICPR, Windsdor, Canada, 434-440

3. Trajkovski, S.: Distribution of heat sources in the cutting zone in hot machining by electric current heating, Proceedings of the 9th ICPR, Cincinnati, Ohio, USA, Vol. II, 1987, 1619-1623

4. Trajkovski, S.: Flank wear intensity method for determination of tool life equation, Proc of the 9th ICPR, Cincinnati, Ohio, USA, Vol. II, 1987, 1770-1775

5. Trajkovski, S.: Influence of heat treatment on machinability of high manganese steel, Proceedings of the AMST-87, Opatia, Croatia, 1987, 79-83

# EXPERIMENTAL STUDY OF EFFICIENCY AND QUALITY IN ABRASIVE WATER JET CUTTING OF GLASS

E. Capello
C.N.R., Milan, Italy

M. Monno and Q. Semeraro
Polytechnic of Milan, Milan, Italy

KEY WORDS: AWJ Machining, Glass, Cutting, Surfaces, Quality.

ABSTRACT: Abrasive Water Jet (AWJ) cutting of glass is considered as one of the most promising industrial applications of this non-traditional manufacturing process. The overall problem addressed in this study is the dependence of quality and efficiency of the AWJ glass cutting on the process variables. In fact, in the evaluation of the AWJs suitability to the glass industry, as well as of other traditional or non-traditional cutting processes, the competition is mainly played in terms of efficiency of the machining and of the quality of the generated kerf. The paper describes and discusses the results of a three stage experimental study on AWJ glass machining. The first part of the paper deals with the problem of the characterization of the quality of the generated kerf. The second part aims to the study of the efficiency of the cut, and an empirical model which can be used to predict the maximum cutting speed is presented. The third part of the paper presents a study on the effect of the process variables on the quality of the kerf. At last, a regression analysis has been performed in order to identify statistical models that relate the process parameters to the quality of the resulting kerf.

## 1. INTRODUCTION

The suitability of Abrasive Water Jet (AWJ) machining to the glass industry is well known and several applications can be found in many sectors. In fact, AWJ is mainly used in glass machining when traditional manufacturing processes fail, that is, for

Published in: E. Kuljanic (Ed.) *Advanced Manufacturing Systems and Technology*, CISM Courses and Lectures No. 372, Springer Verlag, Wien New York, 1996.

example, in cutting complex plane profiles (with curvature radii < 50 mm) or in multi-layered glass cutting. Due to the intrinsic flexibility of the AWJ systems, several sectors of the glass industry might be involved in the use of an AWJ system for different operations, such as cutting, piercing or drilling.

Only few studies concerning the efficiency of cutting brittle materials with AWJs have been presented in the literature and the aspects concerning the "quality" of the generated kerf have been partially investigated up to now. The paper presents an experimental study on the influence of the cutting parameters on efficiency and quality of the cut. This study has been divided into three parts: the first part aims to the definition of a set of macro and micro geometric parameters that can be used to characterise, or "measure", the quality aspects of the generated kerf. The identified parameters are: a quality score which is obtained using a quantitative classification of the cracks and flaws caused by AWJ cutting, the kerf taper and parameters related to the topography and morphology of the side surfaces of the kerf (waviness and roughness).

The second part of the paper aims to the identification and validation of a statistical prediction model for the "threshold feed rate", that is the maximum cutting head feed rate by which the jet completely cuts the workpiece (and does not generate a blind groove). The threshold feed rate is one of the most interesting process variables from the industrial and operative point of view, because productivity is a critical factor when comparing AWJ cutting to other cutting processes.

Finally, the third part is an analysis and a characterisation of the quality of the generated kerf, in terms of taper and surface finish. In this study the influence of the process variables has been investigated. A family of statistical models has been identified and validated, which has shown the possibility to establish a direct relationship between process variables and quality of the cutting results.

## 2. QUALITY OF THE CUT

In order to perform a statistical analysis of the influence of the process variables on the quality of the cut, a set of indexes has been identified to quantitatively "measure" the quality of the resulting kerf.

Damage and not-through-passing cuts: As a first approach, a class model has been defined (see figure (1)). An acceptable cut (that is through passing and without severe damage) can be classified at least in class 2; it should be noticed that traditional manufacturing processes reach class number 0 or 1 maximum.

Taper: The taper of the kerf has been defined as (see figure (2)):

$$\gamma = \frac{a - c}{t} \qquad (1)$$

Roughness and waviness: The surfaces generated by AWJ cutting are characterised by the presence of striations in the lower part of the kerf, while the upper part is dominated only by roughness (see figure (3)). Therefore, in order to qualify the AWJ surfaces it is necessary to evaluate both roughness and waviness parameters at least at two different kerf depths. The investigated parameters are $R_a$, $R_z$, $R_t$, $R_s$, $R_{sk}$, $W_t$, $S_m$.

## 3. ANALYSIS OF THE EFFICIENCY OF THE CUT

### 3.1 Experimental Analysis

The Threshold Feed Rate (TFR) is the maximum traverse rate of the cutting head by which the abrasive jet passes through the workpiece and the generated kerf completely separates the workpiece into two parts. Then, the cut can be classified in class number 2 of figure (1). The operative procedure used has led to the assessment of the TFR with a tolerance of ±25mm/min. As a first investigation to the problem of the TFR evaluation, an experimental plan has been executed, varying 5 factors (see table (1)). The analysis of the experimental data has pointed out that all the investigated process variables significantly influence the TFR, except the abrasive mass flow rate. Moreover, it has been found that the second order interactions between factors are significative. The result of this experimental plan is reported in table (1). From this table it can be noticed that:

1. Given the glass thickness, the TFR increases as the grain size decreases.
2. The glass thickness is the most relevant factor: in fact increasing the thickness of the glass, the TFR decreases drastically.
3. The water pressure deeply influences the TFR.
4. The focuser diameter has a relevant influence on the TFR.
5. The abrasive mass flow rate mildly influences the TFR.
6. There is a small positive interaction between pressure and abrasive mass flow rate.

An interesting result is that the TFR increases with the mesh number, that is as the particle size decreases. In order to investigate this particular result, the width of the kerf at the entrance side of the jet and the taper of the kerf have been measured. It has been found that both these entities remain almost constant for all cuts performed at the TFR. The mass of eroded material in a single cut can be expressed as (see figure (2)):

$$W = \rho_m \frac{a+c}{2} s \cdot t \tag{2}$$

The same quantity can be expressed in terms of erosion process as the product of the abrasive mass flow rate $m_a$, the erosion ratio $g_m/g_a$ (the ratio between the mass per unit of time of the material $g_m$ eroded by the abrasive mass $g_a$ delivered to the workpiece in the time unit) and the time $\tau$ (duration of the cut):

$$W = \frac{g_m}{g_a} m_a \tau \tag{3}$$

Dividing both sides by the length of the cut $s$, one obtains:

$$\frac{W}{s} = \frac{g_m}{g_a} m_a \frac{\tau}{s} = \frac{g_m}{g_a} \frac{m_a}{u} \tag{4}$$

For a cut performed at the threshold feed rate TFR:

$$\frac{W_{TFR}}{s} = \frac{g_m}{g_a} \frac{m_a}{u_{TFR}} \tag{5}$$

where $u_{TFR}$ is the TFR. Substituting equation (5) in equation (2), and considering that the kerf taper has been evaluated using formula (1), one obtains:

$$u_{TFR} = \frac{g_m}{g_a} \frac{m_a}{\rho_m} \left( at - \frac{\gamma_{TFR} t^2}{2} \right)^{-1} \tag{6}$$

As stated before, the kerf taper $\gamma_{TFR}$ and the width $a$ of the upper side of kerfs obtained at the TFR are almost constant and, therefore, the TFR increases if the erosion ratio $g_m/g_a$ increases, that is if the mass of material eroded by the unit of abrasive mass increases. The influence of the grain size and of the process parameters on the erosion ratio is not known, since many physical aspects of the erosion phenomena, due to their complexity, are still unknown. In order to identify a predictive model of the TFR it is therefore necessary to use a statistical model based on experimental data.

### 3.2 Predictive Model of the TFR
The previous analysis has been used to select the process variables that significantly influence the TFR and should be included in a predictive model. It has been found that the effect of the abrasive mass flow rate on the TFR is very small and can be neglected. Moreover, since the second order interactions between the process variables are significative, a linear double logarithm model has been analysed and validated:

$$\ln u_{TFR} = a_0 + a_1 \ln P + a_2 \ln t + a_3 \ln G + a_4 \ln d_f \tag{7}$$

where $a_i$ ($i=0..4$) are the regression parameters. The linear regression has led to the following exponential equation, which is the empirical representation of equation (7) and can be used to evaluate the influence of the process variables on the erosion ratio:

$$u_{TFR} = 0.917 \cdot P^{1.67} \cdot t^{-1.25} \cdot G^{-0.39} \cdot d_f^{-1.62} \tag{8}$$

The model has been validated and the regression coefficient is $r^2=0.986$; the residuals are not correlated (confidence level 98.7%) and normally distributed (confidence level 88.3%). The pure error test, which expresses the coherence between the model and the data, has been executed and verified.

## 4. ANALYSIS OF THE QUALITY OF THE KERF

### 4.1 DOE and ANOVA
Aiming to the investigation of the quality of AWJ kerfs, a new experimental plan has been designed (see table (2)). The experimental cuts have been performed in a random sequence, in order to reduce the effect of any possible systematic error. Each cut has been replicated twice. The quality of the kerf has been measured in terms of taper and surface finish. In particular, the surface parameters used to characterize the quality of the kerf surfaces are the ones previously identified in the first part of the work. These surface parameters of the kerf have been evaluated using a profile recording instrument Perthen S6P (sampling length 17.5 mm, cut off length 2.5 mm). The results of the ANOVA are reported in table (3). These results clearly show that:

1. The grain size is the most relevant process variable for all the quality factors.
2. The kerf taper is deeply influenced by the feed rate and the grain size.
3. $R_a$, $R_z$, $R_q$, $R_t$, (that will be referred to as the $R_x$ family) are strongly influenced by the four variables and by some interactions, while the values of $R_{sk}$ are extremely widespread but none of the process variables has a significative influence on them.
4. $W_t$, $S_m$, are influenced neither by the mass flow rate nor by the interactions, but the variance explained by the remaining factors is a small part of the global variance. This implies that a predictive model is not very significative and of small practical interest.

Based on these results, a second experimental plan has been designed and executed, using the scheme reported in table (4). The roughness parameters family $R_x$ has been measured at different kerf depths $h$. The collected data have been used to perform a linear regression between process variables and roughness parameters. The aim of this regression analysis is to find a common predictive model with a simple mathematical structure for all the roughness parameters, that is a model for the $R_x$ family.

4.2 Predictive model of the surface finish

The search for a significative model common to the $R_x$ family has lead to the definition of the following equation:

$$\ln R_x = a_{0x} + \ln G \cdot (a_{1x} \ln G + a_{2x} \ln P + a_{3x} \ln m_a + a_{4x} \ln u + a_{5x} \ln h) \tag{9}$$

for the Barton Garnet abrasive and

$$\ln R_x = a_{0x} + a_{1x} \ln P + a_{2x} \ln m_a + a_{3x} \ln u + a_{4x} \ln h \tag{10}$$

for the Olivina sand, where, as explained, $R_x$ expresses the family of roughness parameters, and $a_{ix}$ are the regression coefficients for each component of the family. As can be seen, the presence of the grain size $G$ in the model (equation (9)), valid for the Garnet abrasive, has led to a complicated structure of the model. In fact, the grain size deeply influences the effect of the other process parameters (second and higher order interaction). On the contrary, the model for Olivina sand (equation (10)) is simply the exponential model, since this abrasive is commercially available only in mesh # 70. Tests on the hypothesis of applicability of the linear regression have been executed and verified. The values of the regression coefficients are reported in table (5), together with the estimated parameters. In table (6) are summarised the trend between the relative parameter and the predicted value. As expected, the signs remain the same for Olivina sand and Garnet. As an example, figures (4, 5) show the predicted vs. actual values of $R_a$, and the residuals vs. actual values graphs for the garnet model. The other parameters of the $R_x$ family show similar graphs.

5. CONCLUSIONS

The overall problem addressed in the paper is the influence of the process variables on the efficiency and quality of the AWJ cutting of glass. To this end an experimental study has been conducted in order to investigate the relationship between six process variables and the threshold feed rate (TFR) and the quality of the kerf (taper and surface morphology). From the analysis of the experimental data it can be observed that:

- The efficiency of the AWJ machining of glass (related to the TFR) strongly depends on the glass thickness, on the water pressure and on the abrasive grain size.
- The quality of the machining strongly depends on the abrasive grain size and on the feed rate.
- The abrasive mass flow rate mildly effects the efficiency and the quality of the cut.

From these results it can be stated that the critical parameters that must be carefully selected are the feed rate and the grain size. In particular, the grain size has a deep influence on the efficiency and the quality of the cut. A decrease in the grain size leads to an increase of efficiency of the machining and of the quality of the results (surface roughness and kerf taper). Moreover, in order to predict the efficiency and the quality of the AWJ machining, a set of experimental models has been identified and validated. The proposed models can be used to predict the results of the machining and to evaluate the suitability of AWJs to the glass industry.

## ACKNOWLEDGEMENTS

This work was carried out with the fundings of the italian M.U.R.S.T. (Ministry of University and Scientific and Technological Research) and CNR (National Research Council of Italy). The authors are grateful to Dott.sa G. Boselli for her help in reviewing the final manuscript.

## REFERENCES

1. S. Bahadur and R. Badruddin "Erodent Particle Characterization and the Effect of Particle Size and Shape on Erosion", *Wear*, vol. 138, 1990, pp. 189-207.
2. S. Yanagiuchi and H. Yamagata, "Cutting and Drilling of Glass by Abrasive Jet", *8th International Symposium on Waterjet Cutting*, Durham, UK, 1986.
3. A.J. Sparks and I.M. Hutchings, "Effects of Erodent Recycling in Solid Particle Erosion Testing", *Wear*, vol. 162-164, 1993, pp. 139 - 147.
4. D.J. Whitehouse, *Handbook of Surface Metrology*, Institute of Physics Publishing, N.Y., 1994.
5. M. Hashish, "Characteristics of Surfaces Machined with Abrasive Water Jet", *Transactions of the ASME*, vol 113, July 1991.
6. E. Capello, M. Monno, Q. Semeraro, "On the Characterisation of the Surfaces Obtained by Abrasive Water Jet Machining", *13th International Symposium on Jet Cutting Technology*, BHRA, Cranfield, UK, 1994

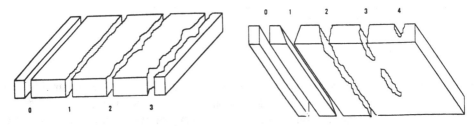

Figure 1 - Quality class model of the generated kerf.

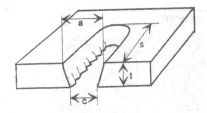

Figure 2 - Geometrical characteristics of the kerf.

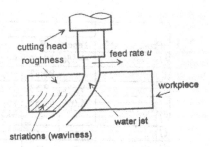

Figure 3 - Generation of AWJ surfaces.

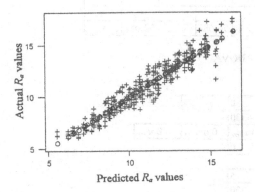

Figure 4 - Predicted values of $R_a$ vs. actual values.

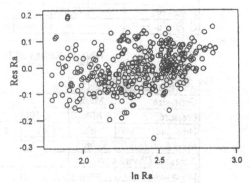

Figure 5 - Residuals of $R_a$ vs. $\ln(R_a)$.

| $d_f$ | $t$ | Abr. | Mesh | $P$ 200 $m_a$ 5 | $P$ 200 $m_a$ 10 | $P$ 300 $m_a$ 5 | $P$ 300 $m_a$ 10 |
|---|---|---|---|---|---|---|---|
| | | Oliv. | 70 | 1.1 | 1.15 | 2 | 2.15 |
| | | | 50 | 1 | 1.05 | 1.8 | 1.9 |
| | 5 | Garn. | 80 | 1.3 | 1.3 | 2.35 | 2.6 |
| 1.2 | | | 120 | 1.45 | 1.5 | 2.9 | 3.2 |
| | | Oliv. | 70 | 0.25 | 0.275 | 0.45 | 0.5 |
| | | | 50 | 0.25 | 0.275 | 0.45 | 0.5 |
| | 15 | Garn. | 80 | 0.35 | 0.35 | 0.65 | 0.75 |
| | | | 120 | 0.35 | 0.35 | 0.65 | 0.75 |
| | | Oliv. | 70 | 0.3 | 0.5 | 0.6 | 0.7 |
| 2 | 5 | | 50 | 0.4 | 0.6 | 1.1 | 1.2 |
| | | Garn. | 80 | 0.5 | 0.6 | 1.1 | 1.2 |
| | | | 120 | 0.6 | 0.6 | 1.1 | 1.2 |

Table 1 - Experimental results of the TFR [m/min]

| Grain size | Mesh | 50 | 120 |
|---|---|---|---|
| Pressure | MPa | 200 | 300 |
| Feed rate | m/s | 0.2 $u_{TFR}$ | 0.8 $u_{TFR}$ |
| Mass flow rate | kg/min | 0,3 | 0,6 |

Table 2 - Experimental plan for the analysis
of the quality of the kerf

| Proc. variable | Quality parameter | | | | | | | |
|---|---|---|---|---|---|---|---|---|
| | Taper | $R_a$ | $R_q$ | $R_t$ | $R_s$ | $R_{sk}$ | $W_t$ | $S_m$ |
| $G$ | 99,5% | 99,5% | 99,5% | 99,5% | 99,5% | = | 99,5% | 99,5% |
| $P$ | = | 99,5% | 99,5% | 99,5% | 99,5% | = | 99,5% | 99,0% |
| $u$ | 99,5% | 99,5% | 99,5% | 99,5% | 99,5% | = | 99,5% | 99,5% |
| $m_a$ | 75,0% | 99,5% | 99,5% | 99,5% | 99,5% | = | = | 97,5% |
| $G - P$ | = | = | 75,0% | 75,0% | 95,0% | = | = | 75,0% |
| $G - u$ | = | = | = | = | = | 90,0% | 99,5% | = |
| $G - m_a$ | = | = | = | 99,0% | = | = | 75,0% | 95,0% |
| $P - u$ | = | 99,0% | = | 75,0% | 95,0% | = | 97,5% | = |
| $P - m_a$ | = | 99,5% | 75,0% | 75,0% | = | = | = | = |
| $u - m_a$ | 75,0% | 97,5% | 75,0% | = | 95,0% | = | = | = |
| $G - P - u$ | = | 75,0% | 75,0% | 75,0% | 75,0% | 75,0% | 95,0% | = |
| $G - P - m_a$ | = | 95,0% | 95,0% | = | = | = | 90,0% | = |
| $G - u - m_a$ | 75,0% | 99,5% | 99,5% | = | 99,0% | = | = | = |
| $P - u - m_a$ | = | 75,0% | = | = | = | = | = | = |
| $G - P - u - m_a$ | = | = | = | 90,0% | 75,0% | = | 95,0% | = |

Table 3 - Results of the ANOVA

| Abrasive: **Bartons garnet** | | | | | |
|---|---|---|---|---|---|
| Grain size | Mesh | 50 | 80 | 120 | |
| Pressure | MPa | 200 | 300 | | |
| Feed rate | m/s | $0.2\,u_{TFR}$ | $0.4\,u_{TFR}$ | $0.6\,u_{TFR}$ | $0.8\,u_{TFR}$ |
| Mass flow rate | kg/min | 0,3 | 0,6 | | |
| Abrasive: **Olivina sand** | | | | | |
| Glass thickness (t | mm | 5 | 15 | | |
| Pressure | MPa | 200 | 300 | | |
| Feed rate | m/s | $0.2\,u_{TFR}$ | $0.4\,u_{TFR}$ | $0.6\,u_{TFR}$ | $0.8\,u_{TFR}$ |
| Mass flow rate | kg/min | 0,3 | 0,6 | | |
| Roughness measurement position | | | | | |
| Distance (h) | mm | $1/3\,t$ | $2/3\,t$ | | |

Table 4 - Experimental plan designed for the regression analysis

| | Garnet | | | | Olivina sand | | | |
|---|---|---|---|---|---|---|---|---|
| | $R_a$ | $R_q$ | $R_t$ | $R_s$ | $R_a$ | $R_q$ | $R_t$ | $R_s$ |
| $a_0$ | 2,29 | 2,59 | 4,54 | 4,10 | 0,20 | 0,57 | 2,22 | 2,15 |
| $a_1$ | -0,42 | -0,39 | -0,35 | -0,33 | 0,36 | 0,37 | 0,34 | 0,29 |
| $a_2$ | 0,03 | -0,12 | 0,04 | 0,03 | 0,39 | 0,40 | 0,38 | 0,32 |
| $a_3$ | 0,08 | -0,13 | 0,08 | 0,07 | -0,11 | -0,15 | -0,09 | -0,05 |
| $a_4$ | -0,12 | 0,05 | -0,10 | -0,09 | -0,22 | -0,24 | -0,22 | -0,21 |
| $a_5$ | -0,12 | 0,08 | -0,10 | -0,09 | | | | |
| $r^2$ | 0,92 | 0,89 | 0,82 | 0,84 | 0,91 | 0,90 | 0,82 | 0,84 |

Table 5 - Estimated parameters and regression coefficients

| | Garnet | Olivina sand |
|---|---|---|
| | $R_x$ | $R_x$ |
| $G$ | + | |
| $u$ | + | + |
| $P$ | - | - |
| $m_a$ | - | - |
| $h$ | + | + |

Table 6 - Trend of the effects of the process variables on the Rx family

# TECHNOLOGICAL ASPECTS OF LASER CUTTING
# OF SHEET METALS

**M.R. Radovanovic and D.B. Lazarevic**
**University of Nis, Nis, Yugoslavia**

KEY WORDS: Laser Cutting, Process Parameters, Cut Quality

ABSTRACT: Technological problems faced in the field of laser machines' application to contour sheet cutting lie in insufficient knowledge of the laser technique in addition to the absence of both sufficiently reliable practical data and knowledge about the parameters affecting the forming process itself. One consequence of this is the fact that laser machines are not as much used as they should be regarding the possibilities they offer. The knowledge of the laser cutting process and its dependence on various factors will provide for a rise in forming quality as well as in its liability to manufacturing

## 1. INTRODUCTION

The increased market competition as well as the need to provide for a desired level of product quality as one of the highly demanding requirements of modern manufacturing is directly related to the introduction of new or high technologies. In this sense, the laser technology has managed to impose itself very quickly as indispensable in almost all manufacturing industries. It has found a wide application in metal working industry for material forming, measurement and quality control. In material forming laser machines are mostly used for contour cutting of sheet metal, drilling, surface hardening and welding. The laser cutting is one of the largest applications in metal working industry. It is based on the precise sheets cutting by means of a focused beam of laser rays. The laser beam is a new universal cutting tool able to cut almost all known materials. With respect to various other

Published in: E. Kuljanic (Ed.) *Advanced Manufacturing Systems and Technology*, CISM Courses and Lectures No. 372, Springer Verlag, Wien New York, 1996.

procedures (such as gas cutting, plasma cutting, sawing and punching), its advantages are numerous, namely, a narrow cut, minimal area subjected to heat, a proper cut profile, smooth and flat edges, minimal deformation of a workpiece, the possibility of applying high velocities, intricate profile manufacture and fast adaptation to changes in manufacturing programs. That is why comprehensive research of both theoretical basis and experimental aspects of the laser cutting is being carried out.

## 2. LASER CUTTING

Laser cutting is based on applying a highly concentrated light energy obtained by laser radiation that is used for metal forming by melting or evaporation. Laser cutting processes make heat action fully effective (namely, heating, melting, evaporation), that is those that are produced by the laser beam affecting a workpiece surface.

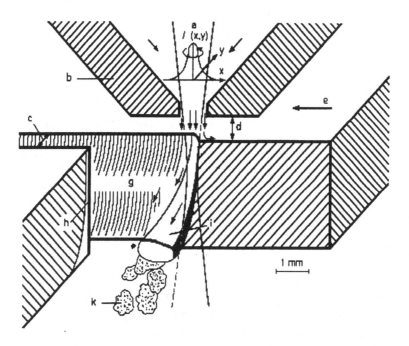

(a) focused laser beam; (b) nozzle con; (c) cut width; (d) distance of the nozzle con from the workpiece surface; (e) cutting velocity; (f) focused laser beam at the focus; (g) molten material flow; (h) heat zone; (i) cutting front; (k) molten material particles

Fig. 1: Schematic Drawing of Laser Cutting Process

The laser beam' effect upon a workpiece material can be divided into several characteristic phases:

-  absorption of the laser radiation in the workpiece surface layer and transformation of the light energy into the heat one,
-  heating of the workpiece surface layer at the place subjected to the laser beam,
-  melting and evaporation of the workpiece material,
-  removal of the break-up products, and,
-  workpiece cooling after the completion of the laser beam' effect.

A desired cut is obtained by moving the laser beam along a given contour. Since our desire is to remove the evaporated and molten material from the affected zone as soon as possible, the laser cutting is performed with a coaxial current of the process gas. The gas blowing increases the cutting velocity for as much as 40%. Fig. 1 gives the schematic drawing of the laser cutting process.

By combining the laser as the light radiation source and the machine providing motion, in addition to the applied numerically controlled system, it is possible to provide for a continual sheet cutting along the predetermined contour.

A very important indicator of the laser cutting is balance of power in the cut zone and the used heat conditioned by it. The balance of power with the laser cutting is given by the expression:

$$P_L = P_R + P_O + P_P - P_S \tag{1}$$

where: $P_L$- laser radiation power; $P_R$- power used for inducing the molten state; $P_O$- power led away by the molten material and process gas; $P_P$- power lost due to its passing through the workpiece, $P_S$- power obtained by the exothermic reaction.

Part of the laser beam power is lost due to its passing through the workpiece, by the molten material and process gas. Still, its greatest part is absorbed and used for inducing melting and evaporation of the material at the cut point. The absorbed energy quantity is mostly depending on thermal and physical properties of the workpiece material; the choice of material also depends upon them. Absorption is essential only at the first moment of the interaction between the laser beam and the workpiece material. Later on, heat diffusion is of crucial influence.

In the laser cutting operation, in addition to the heat obtained by focusing the laser beam, the process gas is also used for removing the molten material from the cutting zone, to protect the lenses from evaporation and to aid the burning process. The useful power can be increased in the case that the process gas is oxygen due to the exothermic reaction. The energy balance of the exothermic reaction is given in the following equations:

$$Fe + 1/2 O_2 \Rightarrow Feo - 3,43 kJ/g \tag{2}$$

$$3FeO + 1/2 O_2 \Rightarrow Fe_3 O_4 - 1,29 kJ/g \tag{3}$$

Thus this energy presents a greater part of the overall energy used for melting the workpiece material. In the cutting process, during the focused laser beam' movement with respect to the workpiece, a part of the power $P_p$ is used in the cut direction since it preheats the cut place. Due to the rapid heating and cooling in the cut zone the molten surface layer hardens thus affecting the cut quality. As a function of the power density there is a break-up starting-point at which the process of the material removal begins. What is often meant by breaking-up is the achievement of melting point on the workpiece surface. At powers greater than this one the material is removed by evaporation.

For laser cutting, the most commonly applied machines are $CO_2$ lasers (90% of all sheet cutting lasers for continual and impulse working regimes) due to their high productivity in the modern manufacturing. With respect to their effectiveness the $CO_2$ lasers are among the top ones in the laser techniques. Their efficiency ratio reaches 20% and their wave length radiation is 10,6μm absorbing a great number of materials. They are used for cutting all sorts of metal (carbon steels, stainless steel, alloyed steels, aluminium, copper, brass, titanium, etc.) and non-metal (plastics, rubber, leather, textile, wood, cardboard, paper, asbestos, ceramics, graphite, etc.).

## 3. WORKING QUALITY

Working quality obtained by laser cutting is determined by the shape and dimension precision as well as by cut quality. The workpiece shape and dimensions' accuracy are determined by the characteristics of the coordinate working table as well as by the control unit quality as in the case of NC or CNC laser sheet cutting machine. The cut quality refers to the cut geometry, the cut surface quality and physical and chemical characteristics of the material in the surface cut layer.

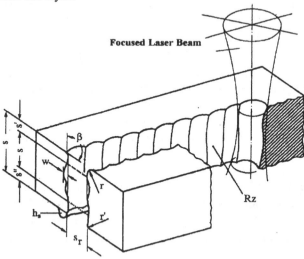

Fig. 2: Schematic Drawing of the Laser Cut

The cut geometry comprises the following: cut width, cut sides' inclination and rounding out of the cut edges. The surface quality includes the accessed roughness, waviness and deviation of the shape - surface error. The physical and chemical properties of the material in the surface cut layer refer to the surface layer formed in the laser cutting process due to the heat effect of the laser beam upon the workpiece material. What is observed is the material's microstructure as well as its hardness, delayed strains, oxide layer thickness and slag's deposits. Fig. 2 gives a schematic drawing of the laser cut.

The cut width is an essential characteristic of the laser cutting process giving it advantage over other sheet cutting procedures. The cut width of metals is small; it ranges $0,1 \div 0,3$ mm with steel sheets' cutting. The cut width increases along with the sheet thickness. The cut sides' inclination also determines the cutting quality. The cutting of material by means of the focused laser beam is characterized by narrowing of the cut. Its size depends on many factors, primarily on the focal distance of the focusing lenses as well as on defocalization, in addition to the properties of the workpiece material and the laser beam's polarization. In order to determine quantitatively the cut sides' inclination the cut sides' inclination tolerance (u) and the cut sides' inclination angle ($\beta$) are used. The cut edges at the laser beam entrance side are rounded out due to the Gauss distribution of radiation intensity over the laser beam cross-section. The edges' rounding-out is very small; the cut edge rounding radius ranges from $0,5$ mm to $0,2$ mm with steel sheets cutting; the round increases along with a rise in sheet thickness.

The laser cut surface reveals a specific form of unevenness. As either semicircular grooves or proper grooving they are the consequence of the focused laser beam shape, the cutting velocity and formation process, as well as of the removal and hardening of the molten material at the cut place. Observation of the cut surface can reveal two zones: the upper one in the area of the laser beam entrance side and the lower one, in the area of the laser beam exit side. The former is a finely worked surface with proper grooves whose mutual distance is $0,1 \div 0,2$ mm while the latter has a rougher surface characterized by the deposits of both molten metal and slag. That is why it is determined to measure roughness of the cut surface at the distance of one third of sheet thickness from the upper cut edge. There is a difference between the cut surface roughness in the direction of the laser beam from that in the direction perpendicular to the laser beam axis, that is in the cutting direction. The former is of no crucial importance in considering the problem of the cut surface roughness due to the fact that the laser is applied to thin sheet cutting. The latter is a more obvious phenomenon that can be observed and analyzed. The parameters that are most often used for accessing the surface roughness are: ten point height of irregularities ($R_z$) and mean arithmetic profile deviation ($R_a$).

The laser cutting is a high-temperature process causing a noticeable yet small heat damage of the material surrounding the cut zone, that is an insignificant change of the basic properties of the workpiece material. The shape of the changes upon the materials induced by the laser radiation can be of various forms. The changes may involve the crystal structure as micro and macro cracks of the material on its surface or as zones molten together or evaporated. A great number of metals are characterized by two or more crystal structures stable at various temperatures. High temperatures cause polymorphous modifications to change into one another along with the change of properties of the basic material. Since the laser cutting is actually the thermal way of cutting then the structure of the material changes in the cut zone. Changes of hardness in the surface cut layer are due to the fact that the workpiece material is heated to high temperatures exceeding the critical transformation points with the onrush of the laser beam. After the passing-through of the laser beam the process of self-cooling occurs causing a rapid cooling of the heated surface layer.

In most cases the laser thin sheet cutting is successful in removing material from the cut zone with no slag produced. In sheets of greater thickness and some kinds of materials deposits of the molten metal slag appear along the exit cut edge. The distribution of this slag primarily depends on the molten material viscosity that could not be removed from the cut zone by process gas.

The experimental research of the work quality with the laser cutting [2] the expressions are obtained to determine indicators of the cut quality of the laser cutting shown in Table 1.

Table 1

| LASER CUTTING | | |
|---|---|---|
| Cut Width | $s_r$ (mm) | $s_r = 0{,}321 \cdot \dfrac{P_L^{0.406} \cdot s^{0.259}}{v^{0.394}}$ |
| Cut Edge Inclination | $\beta$ (°) | $\beta = 0{,}226 \cdot \dfrac{v^{1.138} \cdot s^{0.805}}{P_L^{0.540}}$ |
| Cut Edge Rounding Radius | $r$ (mm) | $r = 0{,}039 \cdot \dfrac{v^{0.414} \cdot s^{0.610}}{P_L^{0.189}}$ |
| Roughness | $R_z$ (µm) | $R_z = 12{,}528 \cdot \dfrac{s^{0.542}}{P_L^{0.528} \cdot v^{0.322}}$ |
|  | $R_a$ (µm) | $R_a = 2{,}018 \cdot \dfrac{s^{0.670}}{P_L^{0.451} \cdot v^{0.330}}$ |
| Slag Height | $h_s$ (mm) | $h_s = 0{,}005 \cdot \dfrac{v^{1.687} \cdot s^{1\,530}}{P_L^{1.311}}$ |
| where: $P_L$(kW)-laser power, s(mm)-sheet thickness, v(m/min)-cutting velocity | | |

The experiments have been performed on the $CO_2$ laser sheet cutting machine with the CNC control, namely "Technological Laser Systems TLS-1A" manufactured by the company "Optical Technologies," Plovdiv, Bulgaria. The machine consists of the technological $CO_2$ laser HEBAR-1A, the portal coordinate work table TLU-1000, the work head and control unit CNC ZIT 500M. The technical characteristics of the $CO_2$ laser are: radiation wave length 10,6μm, zone of the continual power regulation 0,2÷1,3kW, continual work regime, beam divergence less than 4mrad, beam diameter 22mm, mode $TEM_{00}$, circular polarization. The optimal laser power is 0,8kW  The focusing system lens is of 28mm in diameter and of focal distance 125mm. The nozzle con opening is 1,6mm. The material used for examination is low carbon steel Ust 13/Werkst.No 1.0333.5 (DIN). The work process is carried out by the oxygen process of 98% in purity.

Fig. 3 shows the cutting velocity change along with the sheet thickness change for various dross heights at the laser power of 800W and the process gas pressure of 80 kPa. Fig. 4 shows the cutting velocity change along with the laser power change for a variety of sheet thickness at the process gas pressure of 80kPa and the dross height of 0,2mm  The cutting velocity rapidly decreases when sheets of greater thickness are cut. While cutting a sheet of concrete thickness the cutting velocity can increase at the expense of increasing the allowed dross height. The cutting velocity can also increase along with the laser power. However, it has to be remembered that the laser almost always works with an optimal radiation power so that this way of increasing the cutting velocity is not desirable.

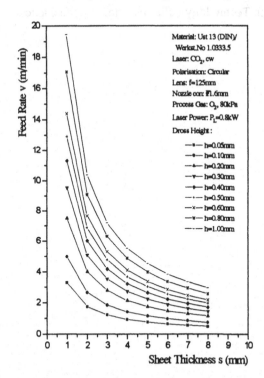 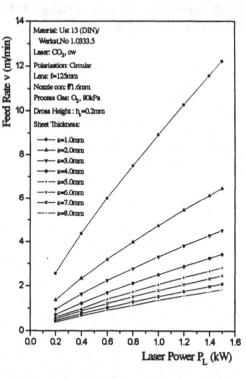

Fig.3. Cutting Velocity Change Due to Sheet Thickness Change for Various Dross Heights at Laser Power of 800W and Process Gas Pressure of 80kPa

Fig.4. Cutting Velocity Change Due to Laser Power Change for Various Sheet Thickness at Process Gas Pressure of 80kPa and Dross Heights of 0,2mm

## 4. CONCLUSION

Technological problems related to the application of laser machines to continual sheet cutting are in insufficient knowledge of the laser technique application as well as due to absence of sufficiently reliable practical data and knowledge about the parameters influencing the work process itself. Therefore, in order to contribute to practical data, this paper gives results of the experimental research referring to the determination of: quality indicators of the cuts obtained by laser cutting, cutting velocity changes due to the ones of sheet thickness for various dross heights at particular laser power as well as changes of cutting velocity induced by those of laser power for various degrees of sheet thickness and at particular dross height.

## REFERENCES

1. Lazarevic,D., Radovanovic, M.: Nonconventional Methods; Metal Forming by Removal, Mechanical Engineering Faculty, Nis, 1994
2. Radovanovic,M.: Automatic Design of Laser Technology , Ph. dissertation, Mechanical Engineering Faculty, Nis, 1996

# MICROFABRICATED SILICON BIOCAPSULE
# FOR IMMUNOISOLATION OF PANCREATIC ISLETS

**M. Ferrari**

University of California, Berkeley and San Francisco, CA, U.S.A.

**W.H. Chu**

University of California, Berkeley, CA, U.S.A.

**T.A. Desai**

University of California, Berkeley and San Francisco, CA, U.S.A.

**J. Tu**

University of California, Berkeley, CA, U.S.A.

KEY WORDS: BioMEMS, Biohybrid Organs, Microfabrication, Biocompatibility, Islets of Langerhans, Silicon

ABSTRACT: Immune rejection rapidly destroys cellular transplants, unless the immune system is pharmocologically depressed. This is typically achieved through administration of cyclosporine, which in itself causes a level of collateral damage that is harmful to cell transplant-based therapeutics of Type I Diabetes. Encapsulation of pancreatic islets of Langerhans with an artificial membrane has been proposed as a means of cell immunoprotection after transplantation. Currently, organic biocapsule materials are used in this context, and have presented significant problems that are related to capsules biodegradation and limited biocompatibility. With this background, we have introduced silicon-based biocapsules. Previous studies have shown the viability and functionality of various cell lines within our biocapsule microfabricated environment. Here, a biocapsule fabrication protocol is disclosed, and novel results are presented that specifically address the insulin-secreting capability of islet cell clusters within silicon-based biocapsules for the therapy of diabetes.

## 1. INTRODUCTION

The inadequacy of conventional insulin-therapy for the treatment of the chronic disease of Type I Diabetes has stimulated research on alternative therapeutic methods. The most physiological alternative to insulin injections is the transplantation of the whole pancreas or portions thereof, namely the pancreatic islet of Langerhans. About one to two percent of the mass of the pancreas is composed of islets of Langerhans which secrete the hormones insulin, glucagon, and somatostatin into the portal vein. The beta cells of the islets secrete insulin in response to increasing blood glucose concentrations. In diabetes, insulin secretion is either impaired or destroyed entirely. Ideally, transplantation of pancreatic islet

Published in: E. Kuljanic (Ed.) *Advanced Manufacturing Systems and Technology*, CISM Courses and Lectures No. 372, Springer Verlag, Wien New York, 1996.

cells (allografts or xenografts) could restore normoglycemia. However, long-term reversal of diabetes with cell transplantation still remains an elusive goal. An underlying problem is the shortage of donor human islets. Although the problem may be solved by using animal donors (xenografts), this presents additional difficulties due to graft rejection and auto-immune destruction [1]. A possible solution to this problem has been to protect the islets by an artificial porous capsule which would allow the diffusion of both glucose and insulin, while preventing the passage of larger entities such as antibodies, lymphocytes, and specific transplant rejection molecules.

The requirements for such a microcapsule are numerous. In addition to well-controlled pore size, the capsule must exhibit mechanical stability, nondegradability, and biocompatibility. All encapsulation methods to date have used organic semipermeable membranes [2-7]. These membranes have invariably exhibited deficiencies in their insufficient resistance to organic solvents, inadequate mechanical strength, and uncontrolled pore size distributions -- all of which lead to short-term destruction of cell grafts[2-11].

Our research has focused on the use of silicon-based microcapsules for the immunoisolation of pancreatic islets. This biological containment capsule is achieved by applying fabrication techniques originally developed for the technology of Micro Electro Mechanical Systems (MEMS) and integrated circuits. By virtue of their biochemical inertness and relative mechanical strength, silicon and its oxides offer an alternative to the more conventional organic biocapsules. Of further advantage is the application of standard microfabrication techniques to provide the silicon capsule with extremely well-defined pore sizes. In particular, we have developed several variants of microfabricated diffusion barriers, containing pores with uniform dimensions as small as 20 nm [12]. Pores of this size seem suitable for application in xenotransplantation, as the typical dimension of human immunoglobins and major histocompatibility complex antigens is several hundred Angstroms, while insulin, glucose, oxygen and carbon dioxide molecules have dimensions less than 35 Angstroms.

Although such a silicon-based microcapsule seems promising, many challenges aside from fabrication must be faced in the development of the capsule. Most importantly, interactions between the capsule and the cells contained in the capsule must be characterized. It is necessary for the encapsulated cells to remain stable and viable in order to perform their intended metabolic function. Therefore, capsules must be fabricated with pore sizes that are large enough for the timely diffusion of necessary cell nutrients and oxygen. Furthermore, the capsule must not elicit excessive inflammatory or chronic reactions from the host when in contact with the biological environment. Fibroblastic encapsulation must not reduce pore patency below a limit of proper functionality.

In previous works, we have reported on preliminary investigations of silicon capsule biocompatibility. Direct cell contact tests and long-term bulk material implants have indicated a sufficient degree long-term biocompatibility [13]. Previous studies on silicon biocompatibility have been in agreement [14-17]. In this communication, we discuss the fabrication protocols for cell-containing microfabricated environments, and report on the viability and functionality of pancreatic islets of Langerhans within our silicon-based biocapsules.

## 2. CELL CULTURE WAFER FABRICATION

Cell culture wafers have been fabricated in order to evaluate the response pancreatic islets to biocapsules-simulating environments. Culture wafer microfabrication involves standard bulk processes in order to obtain the several mm-wide square pockets within a (100) wafer. In figure 1, a preferred embodiment is shown wherein the pores of the diffusion membrane are photolithographically defined.

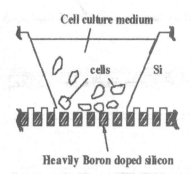

**Figure 1: A schematic diagram of a cross-sectional area of an unit of the cell culture wafer (not to scale).**

Figure 2: A micrograph of the Design I cell culture wafer. The etched holes on the filter membrane are 3 $\mu$m in diameter.

Photolithography is not amenable to the fabrication of pores with dimensions smaller than 1 micrometer. To reach a desired pore size in the tens of nanometers range, we have developed a strategy based on the use of a sacrificial oxide layer, sandwiched between two structural poly layers, for the definition of the pore pathways[12]. This strategy defines a multitude of viable embodiments for filters, culture wafers, and biocapsules. A possible fabrication protocol exemplifying this concept is summarized next.

A 0.38$\mu$m of silicon dioxide is grown on a 4 inch (100) silicon wafer and photolithographically defined to produce the hexagonal trench pattern defining the location of the reinforcement ridges (Fig. 3(a)). These trenches will be filled with the deposited polysilicon. The length, width and depth of the hexagonal trenches are 100$\mu$m, 4$\mu$m and 6$\mu$m, respectively. Patterned photo resist and the silicon dioxide are used as the etch mask for this plasma etching step. The wafer is then cleaned and stripped off the remaining silicon dioxide on its surface using buffered oxide etch. The wafer is wet oxidized again in pure oxygen environment at 1000°C for 1 hour. This produces a 0.38$\mu$m of silicon dioxide on the silicon wafer (Fig. 3(b)). This silicon dioxide layer is later used as an etch-stop for the final back-side silicon anisotropic etch.

A 3$\mu$m thick undoped polysilicon is LPCV-deposited at 605°C and 300m torr on the oxidized silicon wafers. Silane (SiH4) gas is used in this undoped silicon deposition. The wafer is cleaned and annealed at 1000°C in an $N_2$ environment for 1 hour, and heavily boron doped at 1125°C for 6 hours (Fig. 3(c)).A 30 minute 950°C low temperature

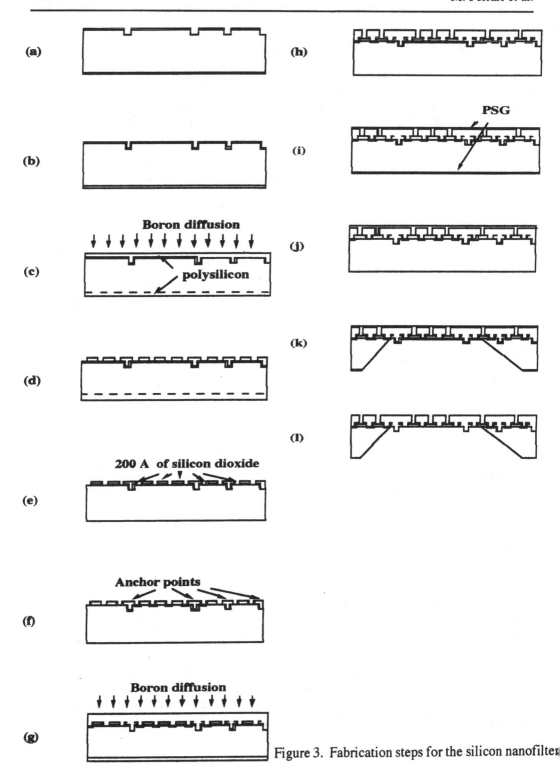

Figure 3. Fabrication steps for the silicon nanofilter

oxidation eases the removal of the borosilicate glass, generated during p$^+$ diffusion, on the polysilicon surface. Rectangular polysilicon holes (20x4 $\mu$m$^2$) (Fig. 2) are photolithographically defined on the deposited p$^+$ polysilicon (Fig. 3(d)).

Low temperature oxidation is used to produce a silicon dioxide layer of uniform thickness which will be removed in the final step. The thickness of this layer determines the maximum sized particles that can pass through the filter. Because of the well controlled oxidization environment, the variation of pore channel oxide layer is less than 10% over a 4 inch wafer. By changing the oxidation time, temperature, and gas composition, pores ranging from several tens of nanometers to several micrometers can be fabricated easily.

Several rectangular (12x10 $\mu$m$^2$) anchor points are defined on top of this thin oxide layer (Fig. 3(f)). A second polysilicon layer (6 $\mu$m) is then deposited and heavily doped usedthe previously described recipe (Fig. 3(g)). The second layer is anchored to the first deposited p$^+$ polysilicon layer through the defined anchor points. Similarly, a 30 minute 950°C low temperature oxidation is done and the borosilicate glass is removed from the silicon wafer. Square polysilicon (3x3 $\mu$m$^2$) holes are patterned photolithographically and plamsa etched into the second p$^+$ polysilicon (Fig. 3(h)).

Finally, the wafer is cleaned and wet oxidized for 1 hour at 1000°C to get 0.38$\mu$m thick silicon dioxide layer. Three micrometers of phosphorous silicate glass (PSG) is deposited on the front side of silicon wafer to provide further protection against the final silicon anisotropic etching (Fig. 3(i)). Etch windows are opened on the backside of the silicon wafer (Fig. 3(j)). Since the filter membrane structure is connected on the front side, it is not necessary to use a double-side aligner to define the position of the backside etch holes. The wafer is etched in ethylene-diamine-pyrocatechol (EDP) at 100°C for about 10 hours. The anisotropic silicon etching will stop automatically at the silicon and silicon dioxide interface (Fig. 3(k)). Buffered silicon dioxide etch is then used to dissolve silicon dioxide and to generate the pores for the nanofilters (Fig. 3(l)). Final rinse in deionized water removes the residual acid from the filter. The biocapsules and culture wafers employed in this study utilize pore paths obtained via the use of a sacrificial layer, as illustrated above, but differ in their final structural embodiment.

## 3. CELL VIABILITY AND FUNCTIONALITY RESULTS

The viability and functionality of rat islets of Langerhans within the porous pockets of the silicon culture wafer were monitored in vitro. Figure 4 is a schematic diagram showing the microfabricated cell culture wafer with several cells inside the pocket. The bioseparation membranes used in the experiments described below are similar to the ones presented in section 2. All silicon culture wafers were autoclaved for twenty minutes prior to use.

Figure 4. A schematic of the culture wafer suspended in medium with tube supports

*Viability of Islets of Langerhans in Culture Wafer Pockets*
Pancreatic Islets of Langerhans were isolated from five day old male Sprague-Dawley rats
by the collagenase method of Hellerstrom et al.[18] and cultured for two weeks. Islets
were hand picked under a microscope and two to three islets were then transferred into each
pocket of the silicon culture wafer and in standard six-well plates as controls. The
morphological state of the pancreatic islets were examined daily. Overall, pancreatic islets
in the silicon pockets and the control dishes appeared to have similar morphology. The
islets were all intact and round in appearance. Islets in pockets showed no dissolution or
dissociation. Islets in both the culture wafer pockets and standard culture dishes exhibited
slight necrosis in their central portions after 72 hours.

*Cell Functionality/ Membrane Permeability to Glucose and Insulin*
The ability of pancreatic islets of Langerhans to secrete insulin in response to glucose
diffusion through capsule pores was studied in both microfabricated half-capsule and full-
capsule environments. Islet cells were cultured in the pockets of silicon culture wafers,
with pore sizes of 3 $\mu$m and 78 nm, respectively, as well as in standard culture dishes as
controls. Approximately five to ten islets were placed inside each pocket. At time zero,
16.6 mol glucose/l was added to the RPMI medium underneath the wafer pockets.
Glucose supplemented medium was allowed to diffuse to the islets to stimulate insulin
production. After one hour, aliquots of medium from underneath the pores of the culture
wafer were collected and analyzed for insulin concentration using a radioimmunoassay kit
(Coat-a-Count®). Insulin concentrations in the medium were compared between the
unencapsulated islets and the islets in both 3 $\mu$m and 78 nm pore-sized half culture wafers.
Experiments were repeated four times for each group. The insulin concentration for both
the 3 $\mu$m and 78 nm half culture wafers was comparable to the unencapsulated islets
(Figure 5), suggesting that glucose is able to sufficiently pass through the pores of the
wafer pockets to stimulate islets for insulin production. Additionally, it appears that the
environment of the silicon pockets does not impede islet functionality and insulin secretion,
as compared to unencapsulated islets. Figure 6 compares the secretion of insulin for
control and half-encapsulated islets at 24 and 72 hours.

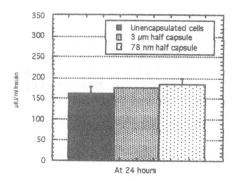

Figure 5. Insulin concentration after a
one hour glucose stimulation

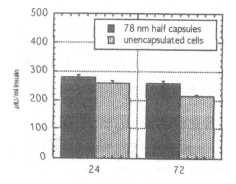

Figure 6. Insulin Secretion at 24 and
72 hours

Preliminary studies were also done on fully encapsulated islets. Full capsules were obtained by joining two half-capsule units containing cell-prefilled pockets. Performance was again monitored by measuring insulin concentration after glucose stimulation of the capsules in cell culture medium. As seen in Figures 7 and 8, fully encapsulated islets remain viable and functional as long as free-floating control islets. In addition, islets in the half capsules exhibited superior viability, in terms of both survival time and insulin production. This latter result is attributed to direct oxygenation of the islets from the free surface.

## 4. CONCLUSION

In this study, a microfabricated biocapsule for the immunological isolation of cell transplants is introduced and characterized. The biocapsule is determined to be sufficiently biocompatible and nondegradable for the intended purposes, and to provide sufficient diffusion of nutrients, glucose, and insulin for islet cell longevity. Preliminary results indicate that microfabricated porous silicon environments, providing partial and full containment of islets of Langerhans, maintain viability and functionality of the islets.

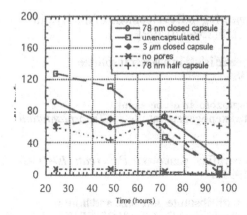

Figure 7. Insulin Release over time for encapsulated and free islets

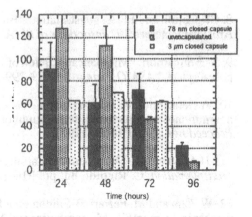

Figure 8. Insulin release from closed capsules over 96 hours

## 5. ACKNOWLEDGEMENTS

Financial sponsorship for this research program was provided by MicroFab BioSystems, which is hereby gratefully acknowledged. Special thanks to the Whittier Institute for Islet Research in San Diego, CA., for the isolation of islets. Our gratitude to other researchers of the Berkeley Biomedical Microdevice Center, for their encouragement, help, suggestions, and support: Derek Hansford, Tony Huen, Lawrence Kulinsky, Debbie Sakaguchi, Jason Sakamoto, and Miquin Zhang.

## REFERENCES

1. KJ Lafferty, "Islet cell transplantation as a therapy of Type I Diabetes Mellitus," *Diab. Nutr. Metab.* 2:323-332 (1989).

2. P Soon-Shiong et al. "Successful reversal of spontaneous diabetes in dogs by intraperitoneal microencapsulated islets" *Transplantation* 54:769-774, n. 5, 1992.

3. P Lacy et al. "Maintenance of normoglycemia in diabetic mice by subcutaneous xenograft of encapsulated islets," *Science*, 254:1728-1784, 1992.

4. "Living Cure" by P.E. Ross, *Scientific American*, 18-20, June 1993.

5. CJ Weber et al. "Xenografts of microencapsulated rat, canine, porcine, and human islets," *Pancreatic Islet Cell Transplantation*, C. Ricordi, ed. pp. 177-189. 1991.

6. R Calafiore et al. "Immunoisolation of porcine islets of Langerhans with alginate/polyaminoacid microcapsules," *Horm. Metab. Research* 25:209-214, 1990.

7. M Goosen et al. "Optimization of microencapsulation parameters: semipermeable microcapsules as an artificial pancreas," *Biotechnology Bioengineering* 27:146-150, 1985.

8. CK Colton and ES Avgoustiniatos. "Bioengineering in Development of the Hybrid Artificial Pancreas," *Transactions of the ASME*, 113:152-170,1991.

9. ME Sugamori. "Microencapsulation of pancreatic islets in a water insoluble polyacrylate," *ASAIO Trans.* 35:179-799, 1989.

10. A Gerasimidi-Vazeo et al. "Reversal of Streptozotocin diabetes in nonimmunosuppressed mice with immunoisolated xenogeneic rat islets," *Transplantation Proceedings*, 24:667-668, 1992.

11. JJ Alman et al. "Macroencapsulation as a bioartifical pancreas," *Pancreatic Islet Cell Transplantation*, C. Ricordi, ed. pp. 216-222, 1991.

12. W. Chu and M. Ferrari, ``Silicon nanofilter with absolute pore size and high mechanical strength'', Microrobotics and Micromechanical Systems 1995, *SPIE Proceedings* Vol. 2593.

13. M. Ferrari, WH Chu, T Desai, D Hansford, T Huen, G Mazzoni, M Zhang. "Silicon nanotechnology for biofiltration and immunoisolated cell xenografts," MRS Proceedings, Fall 1995 (to appear).

14. R Normann et al. "Micromachined silicon based electrode arrays for electrical stimulation of the cervical cortex" MEMS '91, Nara, Japan, 1991, pp. 247-252.

15. B Lassen et al. "Some model surfaces made by RF plasma aimed for the study of biocompatibility," *Clinical Materials* 11:99-103, 1992.

16. YS Lee et al. in *Mat. Res. Soc. Symp. Proc.* Vol. 110, pp. 17-22, 1989.

17. J. Balint et al. in *Mat. Res. Soc. Symp. Proc.* Vol. 110, pp. 761-765, 1989.

18. Hellerstrom C, Lewis NJ, Borg H, Johnson R, Freunkel N. "Method for large scale isolation of pancreatic islets by tissue culture of fetal rat pancreas," *Diabetes*, Vol. 28, pp. 766-769, 1979.

# MATHEMATICAL MODEL FOR ESTABLISHING THE INFLUENCE OF AN EXTERNAL MAGNETIC FIELD ON CHARACTERISTICS PARAMETERS AT ECM

C. Opran and M. Lungu

Polytechnical University of Bucharest, Bucharest, Romania

KEY WORDS: ECM, Magnetic Field, Characteristics Parameters

ABSTRACT: This paper deals with the electro-chemical machining in a medium with imposed magnetic field. It contains the analitical model of magnetic field distribution from working gap, the influence of magnetic field on electric and hydrodynamic parameters,as well as the influence on the productivity and the machining accuracy.

## 1.INTRODUCTION

At present electro-chemical machining has not a great area of utilization because complex interrelations between its characteristic parameters are partially known or even unknown. As a result electro-chemical processes that lead to anodic dissolution of the material piece are very hard to control and conduct, and machining accuracy is relativelly low. [3],[4],[5].
With a view to improving these deficiencies the authors have done researches regarding electro-chemical machining with externally imposed magnetic field (ECM-MF). For this, it has been taken into account the following:
a) The possibility of an electric and hydrodinamic parameter control and favorable conditions for raising productivity and accuracy of ECM by carrying out the electro-chemical processes in the presence of an exterior magnetic field deliberately generated.
b) The fact that ECM unfolds in the presence of an interior magnetic field generated as a result of specific ECM conduction phenomena such as electronic tool and piece conduction, ionic conduction in electrolyte, and electric field intensity variation.

Published in: E. Kuljanic (Ed.) *Advanced Manufacturing Systems and Technology*, CISM Courses and Lectures No. 372, Springer Verlag, Wien New York, 1996.

## 2. STUDY OF MAGNETIC FIELD DISTRIBUTION IN THE GAP

For an evaluation of ECM-MF effects first we have to establish the distribution of the exterior magnetic field in the working gap, taking into consideration the following conditions and assumptions:
a) The electrolyte solution is a homogeneous linear and isotropic medium being, from an electro-magnetic point of view, a magnetic liquid. [1]
b) The permanent process of material removal is considered a succession of elementary processes with small periods, when the tool does not move.
c) The movement effects of anodic particles in relation with electrolyte solution are neglected. These particles move at the same speed as the solution.[1]
d) The development system of electrical and magnetic processes is quasi-stationary because the variation frequency in time of characteristic parameters is smaller than $10^{12}$Hz.
The following research considered the magnetic field generated by electric coils situated close to the gap. There were studied different ways of setting a coil or a system of coils in gap vicinity, choosing the solution in Fig.1a and Fig.2a.
The magnetic field is generated by a coil outside the working gap, with different positions against the frontal gap. Coil 3, is moving with the tool at the speed $v_f$, and the lines of force may be directed towards processing piece 2. (Fig.1b) or towards tool 1 (Fig.2b), depending on the aims of machining.
The position of the coil relative to the gap and to the form of the surfaces that limit it determines the uniform or non-uniform nature of the field. In this way, if the coil contains the whole gap in it and the distance a>(5-10)mm, (Fig.1a), the magnetic field is uniform in gaps limitated by plane surfaces (Fig.1b,c,d) and non-uniform if the surfaces are not plane. If the gap is exterior to the coil, the magnetic field in the gap is non-uniform (Fig.2b,c,d).
In order to construct the spectrums for the force lines in the gap, it has been noticed that these lines pass through mediums with different magnetic behaviour. The tool has a diamagnetic behaviour if it is made of copper and alloys, or a ferromagnetic one if it is made of steel.
The electrolyte solution is a magnetic liquid and has paramagnetic behaviour. If the magnetic permeability of the tool of the electrolyte and of the piece are $\mu_T$, $\mu_L$, $\mu_W$, (H/m) respectively, then for tools from diamagnetic materials: $\mu_T < \mu_L < \mu_W$. When the lines of force are passing from a medium into another the spectrum of lines as well as the magnetic induction will change. This paper presents the spectrum of field lines in frontal (Fig.1b,2b), normal (Fig.1c,2c) and lateral gaps (Fig.1d,2d).
Knowing the theoretic value of the intensity of the uniform magnetic field inside the coil, H(A/m), we can calculate the effective value of the magnetic field intensity $H_L$ in every point of the gap and every moment according to relation (1).

$$\overline{H}_L(x,y,z,t) = K_\mu \, K_\alpha \, \overline{H} \tag{1}$$

and the respective magnetic induction $B_L$(T) acording to relation (2)

$$\overline{B}_L(x,y,z,t) = \mu_L \, \overline{H}_L(x,y,z,t) \tag{2}$$

where $K_\mu$ is an coefficient that takes into account the magnetic behaviour of the mediums

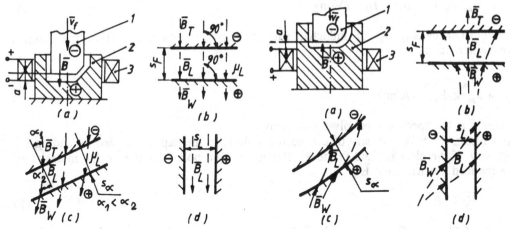

Fig.1 Spectrum of uniform magnetic field lines directed towards pieoe at ECM-MF

Fig.2 Spectrum of non-uniform magnetic field lines directed towards tool at ECM-MF

crossed by lines of force when they enter the gap, and $K_\alpha$ is a non-uniformity coefficient of the magnetic field determinated by the form of the active surface of the tool and by the value of distance a, (Fig.1a,2a) according to relation (3)

$$K_\alpha = K_\alpha(x,y,z,t) \tag{3}$$

The coefficient $K_\alpha$ has a constant value determinated for gaps limited by plane surfaces and has the value $K_\alpha = 1$ for the frontal gap.

## 3. DETERMINATION OF MAGNETIC FIELD INFLUENCE ON ELECTRICAL PARAMETERS

The magnetic flux of the field generated by the coil is variable because the coil moves at the same time with the tool, and the anodic surface continually modifies its form and dimensions. As a result, induction phenomena are produced in the gap. These cause the electrically-induced field, with $E_{im}$ intensity.

The local form of magnetic induction law applied for continuity domains, in the gap, for quasi-stationary regime of the magnetic field, allows the determination of the induced electric field intensity $E_{im}(V/m)$ depending on gap magnetic field induction $B_L(T)$ and the flowing speed of the electrolyte $v_L(m/s)$ according to relation (4).

$$\text{rot } \overline{E}_{im} = \text{rot } (\overline{v}_L \times \overline{B}_L) \tag{4}$$

Vectorial fields $\overline{E}_{im}$ and $\overline{v}_L \times \overline{B}_L$ are either identical or different through a scale field gradient since the vectorial fields have the same vector. If this scale field is considered constant, the gradient becomes null and relation (4) becomes:

$$\overline{E}_{im} = \overline{v}_L \times \overline{B}_L \tag{5}$$

or $E_{im} = v_L \, B_L \, \sin(\overline{v}_L, \overline{B}_L)$ (V/m) $\qquad\qquad$ (6)

The current density $\overline{J}_{im}$ of inducted current, may be determined from the electric conduction law:

$\overline{J}_{im} = \kappa \cdot \overline{E}_{im}$ $\qquad\qquad$ (7)
or $J_{im} = \kappa \cdot E_{im}$ (A/m²) $\qquad\qquad$ (8)

where $\kappa$ is the specific electrolyte conductivity.
As a result, the total intensity of the electric field $\overline{E}_t$ in the gap is obtained by adding the intensity of the electric field produced through anodic polarization $\overline{E}$ to the intensity of the induced electric field $\overline{E}_{im}$:

$\overline{E}_t = \overline{E} + \overline{E}_{im}$ $\qquad\qquad$ (9)

On ECM-MF, the flowing speed of the electrolyte $\overline{v}_L$ is obtained as the sum of two components: component $\overline{v}_p$ determined by the pressure forces of circulation system of the electrolyte and component $\overline{v}_m$, due to the electromagnetical forces created by the externally imposed magnetic field with the help of a special coil:

$\overline{v}_L = \overline{v}_p + \overline{v}_m$ $\qquad\qquad$ (10)

The speed $\overline{v}_p$ may be determined either experimentally or analitically on the viscous liquids mechanical law. It has perpendicular direction on the local normal at the anodic surface. As the electrolyte is a magnetic liquid [1], $\overline{v}_m$ component may be determinated by integrating the movement equation of anodic particles with a medium mass $m_a$ and an electric charge $Q_a$. [1]

$m_a (d\overline{v}_m / dt) = Q_a \, (\overline{E} + \overline{v}_m \times \overline{B}_L)$ $\qquad\qquad$ (11)
where: $B_L = B_x \cdot \overline{i} + B_y \cdot \overline{j} + B_z \cdot \overline{k}$ $\qquad\qquad$ (12)

The solution of vectorial equation (11) using magnetic field laws is given by (13),(14),(15)

$v_{mx} = (Q_a / m_a) \, B_x \, (\overline{B}_L \cdot \overline{E}) \, t / B_L^2$ (m/s) $\qquad\qquad$ (13)
$v_{my} = (Q_a / m_a) \, B_y \, (\overline{B}_L \cdot \overline{E}) \, t / B_L^2$ (m/s) $\qquad\qquad$ (14)
$v_{mz} = (Q_a / m_a) \, B_z \, (\overline{B}_L \cdot \overline{E}) \, t / B_L^2$ (m/s) $\qquad\qquad$ (15)

The relations (13),(14),(15), point out to the fact that only when $(\overline{B}_L \cdot \overline{E}) > 0$ (the angle between $\overline{B}_L$ and $\overline{E}$ is smaller than $90^0$) the externally imposed magnetic field produces the acceleration of the electrolyte flow.
Otherwise the flow slows down, with turbionary phenomena with negative effects on ECM productivity and accuracy. That is why the following research will consider only the practical applications where magnetic field lines are directed towards the tool. (Fig.2) At ECM-MF with uniform magnetic field in frontal gap limited by plane surfaces (Fig.3a) we emphasise the following:

$\overline{E} = -\overline{E}_F \cdot \overline{k}$ $\qquad\qquad$ (16)

$$\overline{B}_L = -\mu_L \cdot K_\mu \cdot H \cdot \overline{k} \qquad (17)$$

$$\overline{v}_p = v_p \cdot \overline{1} \qquad (18)$$

$$\overline{v}_m = -(Q_a/m_a) \, E_F \, t \cdot \overline{k} \qquad (19)$$

Making the substitution of relations (16)-(18) in relation (5) it results:

$$\overline{E}_{im} = -v_p \cdot \mu_L \cdot K_\mu \cdot H \cdot \overline{j} = -F_1 \cdot H \cdot \overline{j} \qquad (20)$$

where: $F_1 = v_p \cdot \mu_L \cdot K_\mu > 0$ $\qquad (21)$

Using (9) and (20) it results:

$$\overline{E}_t = -F_1 \cdot H \cdot \overline{j} - E_F \overline{k} \qquad (22)$$

where $\overline{E}_F$ is the polarisation electric field intensity in the frontal gap. It may be noticed that $\overline{E}_{im} \perp \overline{E}_F$. Thus at ECM-MF in uniform magnetic field growths of electric field intensity $\overline{E}_t$ ($\overline{E}_t > \overline{E}_F$) are produced, without the modification of the normal component $E_{imz}$ which determines the intensity of the electro-chemical processes.

For machining in non-uniform magnetic field in the same frontal gap (Fig.3b), it is considered that the coil is cylindrical and has symmetry ax Oz.

The field generated by it has axial symmetry. E and $v_p$ may be calculated with relations (16) respectively (18). Using notation $\beta$ and in Fig.3b, it results:

$$\overline{B}_L = \mu_L \cdot K_\mu \cdot H \, (\overline{i} \, \sin\beta \, \cos\varphi - \overline{j} \, \sin\beta \, \sin\varphi - \overline{k} \, \cos\beta) \qquad (23)$$

from relations (16),(23),(13),(14),(15) we obtain:

$$v_{mx} = (Q_a/m_a) \cdot t \cdot E_F \, \sin\beta \cdot \cos\beta \cdot \cos\varphi \; (m/s) \qquad (24)$$

$$v_{my} = -(Q_a/m_a) \cdot t \cdot E_F \, \sin\beta \cdot \cos\beta \cdot \sin\varphi \; (m/s) \qquad (25)$$

$$v_{my} = -(Q_a/m_a) \cdot t \cdot E_F \cdot \cos^2\beta \qquad (m/s) \qquad (26)$$

In this case the greatest influence on electro-chemical processes development comes from Oz component, $E_{imz}$ of $\overline{E}_{im}$ vector. This is due to the fact that $E_{imz}$ has the same direction as $\overline{E}_F$ vector. From relations (23)-(26) and (5) it results:

$$E_{imz} = -v_p \cdot \mu_L \cdot K_\mu \cdot H \cdot \sin\varphi = -F_2 \cdot H \; (V/m) \qquad (27)$$

where: $F_2 = v_p \cdot \mu_L \cdot K_\mu \cdot \sin\varphi > 0$ $\qquad (28)$

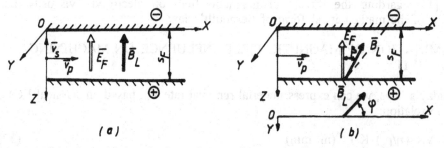

$(a)$ $(b)$

Fig.3 Model of machining in frontal gap with magnetic field: a) uniform, b) non-uniform, at ECM-MF

It may be noticed that at ECM-MF with non-uniform magnetic field, growths of normal component of electric field intensity in the gap, proportional with H are produced. From (7) and (27) we can formulate similar observation about current density $J_{imz}$ in the conditions above as in relation (29):

$$J_{imz} = -\kappa \cdot F_2 \cdot H \quad (A/m^2) \tag{29}$$

The same relations and conclusions were obtained for normal and lateral gap zones. It results that for the general case of any surfaces machining in non-uniform magnetic field, E vector of anodic polarisation electric field intensity is oriented after the local normal at anodic surface and its sense is towards the tool. Thus the greatest influence on electric parameters and also on electro-chemical processes development, is that of the $E_{im}$ vector projection of the induced magnetic field on the local normal, whose versor is n. By using (5), (23),(27) it results the following form of the normal component $E_{im,n}$ of vector $E_{im}$:

$$E_{im,n} = \overline{E}_{im} \cdot \overline{n} = F_n \cdot H \quad (V/m) \tag{30}$$

where: $F_n = F_n(\mu_L, K_\mu, K_\infty, Q_a/m_a, \beta, t, v_p) > 0$; $\overline{n}$ has the sense towards the tool. The density of the induced current $J_{im,n}$ looks similary:

$$J_{im,n} = \kappa \cdot F_n \cdot H \quad (A/m^2) \tag{31}$$

Coresponding to relations (30) and (31) it results that at ECM-MF in non-uniform magnetic field,growths of normal component of electric field intensity and current density in gaps proportional with H are produced.

## 4. DETERMINATION OF MAGNETIC FIELD INFLUENCE ON HYDRODINAMIC PARAMETERS

From relations (13),(14),(15) it may be noticed that at ECM-MF the magnetic field lines lead the electrolyte flowing in gap, and that is so because $v_{mx}$, $v_{my}$, $v_{mz}$ components are direct proportional with the $B_x$, $B_y$, $B_z$ components. As a result, the flow of anodic particles, and also of the electrolyte, becomes uniformly accelerated in the magnetic field specially created, and the acceleration is proportional with the $B_L$ vector components. The results of our research presented herewith confirm and develop the ones formerly presented [1] regarding the effects of magnetic field on electrolyte viscosity and modification of the mathematical form of Bernoulli's law.

## 5. DETERMINATION OF MAGNETIC FIELD INFLUENCE ON MACHINING PRODUCTIVITY

Using Faraday's laws, we can express material removal rate $V_W$, based on classical ECM, according to relation:

$$V_W = V/t = (\eta/\rho_p) \cdot K \cdot I \quad (m^3/min) \tag{32}$$

where: $\eta$ is current efficiency (%); $\rho_p$-mass density of piece material ($Kg/m^3$);

K-electro-chemical equivalent of piece material (Kg/C); I-working current (A); t-machining time (min); V-volum of removed material (m³).

On ECM-MF, within the same electrical parameters regulated at machining equipement, $V_{WM}$ corresponds $V_W$, that can be evaluated taking into account the fact that in the same conditions the working current has the modified value $I_M$, according to relation:

$$I_M = (J + J_{im,n})/A_{Wtot} \quad (A) \tag{33}$$

Using relation (31) and (33) we will obtain:

$$I_M = I + \kappa \cdot F_n \cdot H/A_{Wtot} \quad (A) \tag{34}$$

where $A_{Wtot}$ is the area of current flow (m²)

In this conditions $V_{WM}$ on ECM-MF coresponds to (35)

$$V_{WM} = (\eta/\rho_p) \ K \cdot (I + \kappa \cdot F_n \cdot H/A_{Wtot}) \quad (m^3/min) \tag{35}$$

or taking into account relation (32) and noting

$$C_{WM} = (\eta/\rho_p) \cdot K \cdot \kappa \cdot (F_n/A_{Wtot}) \tag{36}$$

where $C_{WM}$ is a positive parameter that depends on externaly imposed magnetic field characteristics, and using relations (32)-(36) it results the final form of material removal rate $V_{WM}$ on ECM-MF:

$$V_{WM} = V_W + C_{WM} \cdot H \quad (m^3/min) \tag{37}$$

From relation (37) it results that material removal rate on ECM-MF is bigger the on classical ECM, depending on externaly imposed magnetic field parameters and grows directly proportional with H.

## 6. DETERMINATION OF MAGNETIC FIELD INFLUENCE ON MACHINING ACCURACY

Present experimental researches have confirmed that at complex surface machining the smaller the gap the higher the machining accuracy [3], [4], [5], and the more reduced surface flaws are [5].

On classical ECM, the frontal gap may be analytically determined based on Ohm's law for ionic conductors, as in (38)

$$s_F = U_E \cdot \kappa/J \quad (mm) \tag{38}$$

where: $U_E$ (V) is the gap voltage.

The present researches show that on ECM-MF, the frontal gap will be $s_{FM}$, according to relation (39), (40)

$$s_{FM} = U_E \cdot \kappa/(J + J_{im,n}) = U_E \cdot \kappa/(J + \kappa \cdot F_n \cdot H) \quad (mm) \tag{39}$$

It results that $s_{FM} < s_F$, because $\kappa \cdot F_n \cdot H > 0$, and this demonstrattes that on ECM-MF, favourable conditions for machining accuracy are created.

## 7. CONCLUSIONS

The following main conslusions are obvious:

1. The most important effects on electrical parameters are obtained on non-uniform magnetic field machining, when the lines of the field are orientated towards the tool (Fig.2a).

2. Electric field intensity and current density raise proportionally with the exterior magnetic intesity, a fact demonstrated by relation (27)...(31).

3. The flowing of the electrolyte becomes uniformly accelerated in the magnetic field, being conducted by the lines of the field, a fact demonstrated by relations (10)...(15).

4. Material removal rate raises proportionally with exterior magnetic field intensity, a fact demonstrated by relations (35)...(37).

5. Machining accuracy raise a fact yielded implicitely by relations (39).

## 8. REFERENCES

[1] Enache,St., Opran,C., 1989, The Mathematical Model of the ECM with Magnetic Field, Annals of the CIRP, 38/1:207-210.

[2] Mc.Geough,J.A., 1974, Principles of Electrochemical Machining, Chapman and Hall, London.

[3] Rajurkar,K.P., Schnoeker,C.L., 1988, Some Aspects of ECM Performance and Control,Annals of the CIRP, 37/1:183-186.

[4] Rajurkar,K.P., Wei,B., Kozak, J., 1995, Modelling and Monitoring Interelectrode Gap in Pulse Electrochemical Machining, Annals of the CIRP, 44/1:177-180.

[5] Snoeys,R., Staelens,F., Dekeyser,W., 1986, Current Trends in Non-conventional Material Removal Processes, Annals of the CIRP, 35/2:467-480.

# COMPUTER AIDED CLIMBING ROBOT APPLICATION FOR LIFE-CYCLE ENGINEERING IN SHIPBUILDING

**R. Mühlhäusser**

**Institute for Production Systems and Design Technology
Fraunhofer-Gesellschaft, Germany**

**H. Müller and G. Seliger**

**Technical University of Berlin, Germany**

KEY WORDS: Life-cycle Engineering, Mobile Climbing Robot, Shipbuilding

ABSTRACT: The competitive nature of the shipbuilding-industry has been intensified in recent years. To increase the productivity and quality of shipbuilding the modernisation of the manufacturing processes is an important requirement for companies. Quality and safety within shipbuilding and operation can be significantly improved using the high potential of information technology for life-cycle oriented production and operation. Approaches towards life-cycle modelling and design are mainly focusing on the product design and development. An extension of life-cycle modelling results in life-cycle engineering which covers manufacturing, maintenance and inspection over the whole product-life-cycle. Especially in ship-building and operation a high standard of quality and safety is required, whilst barriers of place, time and ship's one-of-a-kind nature have to be overcome. Therefore, recognised gaps within the computer integration and production tools in ship-industry have to be filled. The mobile climbing robot system, described in this paper, is designed according to these requirements.

## 1. INTRODUCTION

The ship building market has become increasingly competitive, with Korean shipbuilders increasing market share, and the US funding it's moves from naval to marine production, factors which are threatening European shipyards. To survive and prosper in the

Published in: E. Kuljanic (Ed.) *Advanced Manufacturing Systems and Technology,*
CISM Courses and Lectures No. 372, Springer Verlag, Wien New York, 1996.

continuously growing global markets, the European ship industry needs a modernisation in shipbuilding and shipping as it has been stated in one of the Task Forces set up by the European Commission. The development of new technologies should improve competitiveness, safety and environmental protection whilst protecting employment.

Quality and safety within ship-building and operation can be significantly improved using the high potential of information technology for life-cycle oriented production and operation. Approaches towards life-cycle modelling and design are mainly focusing on the product design and development. An extension of life-cycle modelling and design results in life-cycle engineering which covers manufacturing, maintenance and inspection over the whole product-life-cycle. Especially in ship-building and operation a high standard of quality and safety is required, whilst barriers of place, time and ship's one-of-a-kind nature have to be overcome. A higher level of computer integration has to be achieved to perform life-cycle engineering by an effective and efficient exploitation of life-cycle modelling and design and therefore, recognised gaps within the computer integration and production tools in ship-industry have to be filled. Realising a higher level of computer integration, the development aims at

- using time variable product models for life-cycle engineering,

- adapting product models by reverse engineering within ship-industry and

- providing real world data by the use of innovative robot tools suited for ship production and maintenance.

Time-variable product models have to represent ships at every stage of their life-cycle. These product models can be used for the different engineering tasks over the product's life. Thereby, the adaptation of product models based on measurement and data acquisition guarantees an increased model accuracy.

Within the ship industry the assembly, disassembly, inspection or maintenance tasks such as gas cutting, welding, painting, annealing or ultrasonic testing and leak finding are mostly performed manually under difficult and demanding working conditions. Because of the dimensions of the products and their modules, conventional stationary robot types would have to be of huge proportions, requiring high investment. Consequently, no computerised tools suitable for integration are used in this area yet. Performing manufacturing, assembly or inspection tasks, the small climbing robot prototype of the IWF can be integrated as a tool for life-cycle engineering to realise measurements and data acquisition.

## 2. STATE-OF-THE-ART

Since ships can be characterised as one-of-a-kind products the concept on the application of computer integrated technologies is quite different from other industries. This was resulting in a tailored development and research on product models for ships which have been stated as the essential part of computer integrated technologies within shipbuilding [1,2]. Although the now existing STEP-based product model standard is widely used in various tailored computer aided applications within shipbuilding, there is still a gap concerning the integration of non-tailored applicable technologies.

Quality control and management has reached strategic importance in nearly all industries and has led to developments of various computer aided quality management and control tools (CAQ). An extension of quality oriented product and process development results in newer approaches on life-cycle modelling and design [3,4]. The objective of life-cycle modelling is to achieve high quality standard and environmental conscious products regarding the whole product-life-cycle. Especially in the case of ship-industry the high environmental and human risk has to be minimised by a high quality standard over the whole product-life-cycle. Although, Quality Control is implemented in ship production and inspection the application of QC as well as CAQ results in isolated solutions not part of the computer integrated infrastructure. The information received within inspection processes is not used in any way for the product model or the application of CAQ neither to improve the model accuracy nor to represent quality information of the product nor to use life-cycle modelling for the different engineering tasks towards product-life-cycle.

The main objective is to improve manufacturing quality by an increased model accuracy concerning the geometric properties towards the manufacturing progress reducing production time and costs. Although geometric properties of the ship parts are enormous important within shipbuilding especially within the final assembly in the dry dock or in the berth which is an extremely expensive bottleneck for most shipyards, the application of geometric measurement methods for work in progress QC is neither implemented nor integrated in the IT infrastructure.

Research and development on automated manufacture or inspection techniques applicable for large scale steel constructions such as ships have been carried out in recent years. Most of the approaches are focusing welding and material inspection and therefore are based on climbing robot developments, either attached by vacuum cups, magnetic wheels, electro-magnetic feet, or even using the propulsive force of a propeller to remain attached to a surface. Except the magnetic wheeled design none of these prototype developments are suitable to be used in shipbuilding environment. Either they are too sensible or they are not able to perform continuous movements on the surface. Concerning the information flow it can be stated that currently all approaches are island solutions not suitable to be integrated in the IT infrastructure. Neither the planning of the processes executed by the climbing robots nor feedback mechanisms to use inspection data for product models are possible.

Computer aided design and manufacturing (CAD/CAM) is widely introduced in European shipyards as well as an increasing number of robots are installed for shipbuilding. Computer Aided Robotic (CAR) combines the application of robots and CAD/CAM, allowing the off-line planning and simulation parallel to the running production. Although CAR has already been successfully applied in other industries such as automotive or aircraft industries, todays ship-industry cannot realise the advantages. The application of this technology becomes even more important within ship-industry since ships are one-of-a-kind-products. Indeed, there are a lot of industrial applied powerful CAR-tools including model libraries for the most common robots but none of them can be efficiently used for ship-industry. The major limitation are given by the missing interfaces from the CAD/CAM-systems for design and work-preparation in ship-industry to available CAR-systems. This has again led to some island solutions which can only be used for one special kinematic-type robot for defined tasks.

## 3. APPROACH

The outlined technical limitations within ship-industry are defining the scope of our approach on the computer integrated planning, execution and analysing of manufacturing, maintenance and inspection processes over the whole product-life-cycle. The main innovations within this approach are the following:

- The feedback coupling of product information to adapt the product model based on model matching mechanisms regarding material and geometric properties to increase model accuracy.

- Performing work-in-progress inspection instead of end-of-pipe inspection to increase manufacturing quality as well as to reduce production time and costs.

- The closed loop design, planning, manufacturing, inspection and feedback to model management, allowing improved product quality.

- Application of life-cycle modelling within ship-industry including the continuous analysing and updating of product models over the whole product-life-cycle from production towards inspection to prevent damage by increased information quantity and quality.

- The planning and simulation of manufacturing and automated inspection technologies parallel to the running ship-production to reduce production time.

- The continuous use of a common base tool over product-life-cycle represented by the climbing robot used in manufacturing, maintenance and inspection for the efficient planning and execution of the single processes.

The overall system design is shown in figure 1.

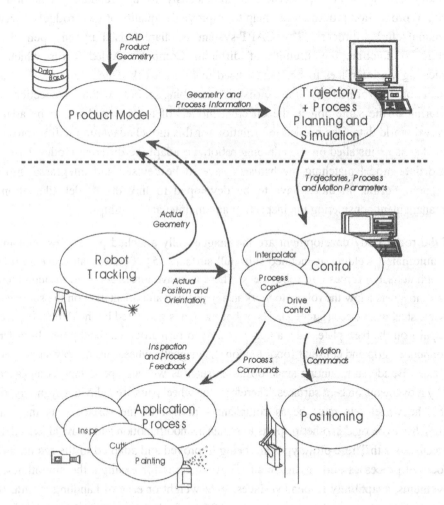

Figure 1: System design

The application of flexible mobile systems that work autonomously can help to reduce production time and cost as well as to improve human working conditions and product quality. The system does not focus one single process, but can be specifically programmed and equipped with different special tools. As ships can be characterised as an one-of-a-kind product that is tailored to the requirement of each customer the use of a Computer Aided Process and Trajectory planning system (CAPT) is absolutely necessary. The tasks to be performed by the climbing robot have to be planned off-line before the manufacturing date.

Process models are used to find the optimal operating of the robot combining the planning of trajectory and application processes. Simultaneously to the running production the simulation of production processes can help to improve the quality of the product as well as of the manufacturing process. The CAPT-system is characterised by an open system architecture to combine the strength of different Computer Aided Tools which are communicating via interfaces to be standardised in future work. Due to the various tasks which can be performed by the proposed climbing robot within inspection and manufacturing of the same object, the different models of the system have to be adapted based on real world data. The needed information for this model adaptation will be provided by different sensors installed on the climbing robot. To adapt the different models based on real world data model matching mechanisms have to be devised and integrated into the CAPT-system. Further methods have to be developed to link the model adaptation to quality management considering the inspection and maintenance of ships.

Base of the robot body development are the magnetically attached prototypes built up at IWF for automated welding at inclined ship hull surfaces [5]. These small wheeled robots provide high adhesive forces and robustness. These adhesive forces achieved through the use of rare earth magnets allow the robots to work in any position and orientation on the surface of a magnetisable steel workpiece (figure 2). The robot motion is generated by the different speed of the wheelpairs on the base plate. Like a tank it is able to turn around a fixed point. In order to ensure continuous grip and constant robot motion the magnetic wheels are flexibly suspended in different axes. Beside surmounting small obstacles such as welding spots, high compliance is required by movements on bent surfaces. Therefore, the wheel pairs are induvidually mounted on a rocker. The wheel sets must be as compliant - normal to the surface - as the whole construction has to be rigid in other respects to obtain a smooth motion and to avoid jerks during tracking a curved path.These prototypes are being improved and adapted to the requirements of the focused processes and environment. That concerns for example the smoothness of turn movements, adaptibility to bend surfaces, light-weight or ease of handling in industrial environment. The robot weight must not exceed 25 kg for workers protection rules. It will have a minimum load capacity of its own weight under unfavourable surface conditions.

Figure 2: Welding prototype MoRoMAG at a gas tank

The robot tracking unit enables the robot to follow an highly accurate trajectory on the ship hull surface by constant position measurements. The application process quality heavily depends on the tracking accuracy. The system will have to deal with dust, vibration, paint mist and splatters or other influences. Solutions by laser triangulation, laser polar tracking, 3D camera image interpretation and ultrasonics trilateration have been considered and tested.

The motion control is dealing with three-dimensional planned trajectories and three-dimensional tracked position data. Interpreting the tracking data, it determines the robot's path deviation as well as the geometric hull surface deviations. For further motion control, the trajectories and path deviations are reduced in dimension into a plane problem which is similar to other mobile robot projects. The control algorithms also include navigation problems, position calculations from tracking data, process quality data fusion and motor control.

## 4. CONCLUSION

In order to increase productivity and quality in shipbuilding industries we were following an approach towards the modernisation of manufacturing, assembly and inspection processes. Life-cycle Engineering in shipbuilding and operation requires new computer integrated tools in production and maintenance. The proposed robot is based on a prototype construction of the IWF using magnetic wheels. The system design includes the integration into the company IT which provides both, efficient robot programming and application as well as model adaptation for production monitoring and product and process improvement. The programming of the robot within a computer aided process and trajectory planning system can be done simultaneously to the running production. The quality can be verified within the CAPT-system by the simulation of the production processes before manufacturing. Future work will be conducted in several directions. The prototypical integration into a commercial available CAPT-system has to be extended regarding the requirements of the end-user. To adapt the different models based on real world data newer model matching mechanisms have to be devised and integrated into the CAPT-system. Methods have to be developed to link the model adaptation to quality management considering the inspection and maintenance of ships. Finally, the transfer of the system to other industries building large products like pressure vessels, off-shore platforms, tanks or power stations will be considered. There is a major importance for the production and maintenance of tanks containing dangerous goods as liquid gas or petrol and for the nuclear industries.

## REFERENCES

1. Kuo C., 1992, Recent Advances in Marine Design and Applications, *Computer Applications of Shipyard Operation and Ship Design - VII*, Elsevier Science Publishers B.V. (North-Holland), 1992, 13-24

2. Koyama T.,1992, The Role of Computer Integrated Manufacturing for Future Shipbuilding, *Computer Applications of Shipyard Operation and Ship Design - VII*, Elsevier Science Publishers B.V. (North-Holland), 1992, 3-12

3. Spur G., 1996, Life Cycle Modeling as a Management Challenge, Life Cycle Modelling for Innovative Products and Processes, Chapmann & Hall, London, 1996, 3-13

4. Krause F.-L. & Kind Chr., 1996, Potentials of information technology for life-cycle-oriented product and process development, Life Cycle Modelling for Innovative Products and Processes, Chapmann & Hall, London, 1996, 14-27

5. Seliger, G., Müller, H., 1994, Programming and navigation of a mobile welding robot, *Proc. of the 25th Int. Symp. on Industrial Robots-ISIR*, 25:749-753

# THE USE OF THREE-DIMENSIONAL FINITE ELEMENT ANALYSIS IN OPTIMISING PROCESSING SEQUENCE OF SYNERGIC MIG WELDING FOR ROBOTIC WELDING SYSTEM

**M. Dassisti, L.M. Galantucci and A. Caruso**

**Polytechnic of Bari, Bari, Italy**

KEY WORDS : synergic MIG welding robotic system; Finite Element analysis, industrial application

ABSTRACT: The paper presents a Finite Element Analysis performed to optimise an industrial welding process for the production of an header, used in boilers for conventional thermal power plants. The production level was in fact below the actual production capability of the robotic MIG welding system utilised, due to a number of set-up problems.

The approach followed was to built two complete parametric three-dimensional FE model; the one of a section of the header was made to determine the optimal welding condition. The complete model of the whole structure was instead intended for distortion analysis. By means of a validation of the models, the output of the simulation runs were found to be in good agreement with the experimental evidence. The promising results obtained in forecasting the main control parameters for this process, even with these simple thermal models, allows to be confident on the possibility to adopt the same FE tool as a decisional support for the frequent set-up phase, due to the company job-production type.

## 1. INTRODUCTION

The problem of design optimisation of conventional welding processes can be essentially stated as the selection of the operating parameters in such a way the quality requirements are satisfied, assuring also the satisfaction of one or more constraints (namely the minimisation of cost, or of the time spent ; etc.). Despite the apparent simplicity of the problem statement, it may require a significant effort to solve it. First of all, the operating parameters, even for the best experienced welding, are so many that it become an expensive effort to find the optimal operating conditions ; a more comprehensive treatment of the influencing parameters for conventional welding processes can be found in the huge amount of literature available (see, e.g., [1]). Depending on the specific application, some factors

Published in: E. Kuljanic (Ed.) *Advanced Manufacturing Systems and Technology,*
CISM Courses and Lectures No. 372, Springer Verlag, Wien New York, 1996.

might have a stronger influence than others, such as the optimisation problem can be stated as a multivariable multiobjective non-linear problem, which seldom can be treated in a mathematical form. Further, quite often the objectives conflict and/or compete with one another (for instance, as quality and cost). The cost of finding the optimal solution of this kind of problem (where it is possible!) is so high, with respect to the income of the operation, that sub-optimal conditions are in most of the cases accepted as good.

The cost can be thought as proportional to the number of factors (variables) considered. In this way, the easier empirical approach adopted in the industrial practice (the trial and error strategy) usually takes into account no more than one or two variables at time in the expensive process of browsing among the several sub-optimal solutions. This approach becomes even more effective if it is essentially tied to what is called the 'know-how', that is the amount of operative experience; this latter allows to reduce the order of the problem by taking several variables as fixed in a tight range. The optimisation process thus only becomes a long stepping process spread over several years. The customer's requirement acts as an explicit constraint to this process. The true problem behind this is the cost of the process, which sometimes may dramatically reduce the profit margins.

The other approaches adopted so far, essentially belong to two broad categories : the mathematical and the experimental approach.

As experimental approach we intend all the set of scientific methods applied to performing practical experiments: several techniques for design of experiments have been developed so far (see, e.g., [2]). The main limitation of these approaches is the number of experiments required, which sometimes are not sustainable for the practical industrial application. Furthermore, there is a limitation of the forecasting equations, which usually holds in the strict range of values where the experiments where carried out.

Among the mathematical approaches, the analytical one tries to solve the optimisation problem via appropriate models of the reality : the number of limiting assumption that has to be taken in order the mathematical problem to be tractable, make the solution quite far from the practical applicability. Another mathematical approach is the numerical one, which basically relies on modelling of the process via discrete set of differential equations : the most common one is the finite element method , even though the finite difference and recently the boundary element approach have been explored. Despite the approach basically requires some simplifying assumptions, these are not so far from reality as in the case of the analytical approach. This fact results to be a significant advantage, because it is possible that results are quite more useful for industrial application than in the other cases. Furthermore, it allows to analyse several different applications with minor changes in the structure of the simulation models, thus requiring small efforts.

In the paper, we adopt the F.E. approach to select an optimal operating welding condition for the process analysed, once the FE model has been validated. The basic hypothesis that is behind this approach is that the scale of the problem, at a macro level with respect to the weld zone size, does allow to compensate for the errors embedded in the use of a thermal model. This means that as far as real industrial problems have to be addressed this approach, which requires anyway a final limited experimental validation procedure, may contribute to the economy of the problem-solving process.

## 2. THE INDUSTRIAL PROCESS ANALYSED

The F.E. analysis carried out in the paper concerns a type of commercial header, produced by the TERMOSUD S.p.A. of Gioia del Colle (BA), an Italian company of the ANSALDO ENERGIA group. The header is used in boilers for conventional thermal power plants. The finished product consists essentially of a cylindrical pipe and a number of tube stumps welded on upper side of it, in a layout arrangement of two or three rows. They are used as a mounting facility for piping the heat-exchanger tubes. The synergic MIG welding process is performed by a couple of NC-controlled robot arms, acting simultaneously (see Figure 1).

The company is characterised by a job production; batches are usually made up of the headers and tubes belonging to the same boiler (about 170 headers of 4 different types for each boiler). Usually boilers vary quite frequently, depending on customer's requirement. This fact leads to frequent changes in the header characteristics, namely material, number and diameter of tube stumps, diameter of the header, pipe thickness and length .

Figure 1. The operation of MIG welding on the header (by permission of TERMOSUD)

The analysis was carried out to solve some problems arose during the set-up phase, which lowered the productivity level allowed by the robotic system. In particular, the occurrence of a permanent bending distortion (in the case presented its maximum value is about 10[mm] at the centre). This required an elastic positive-contact bracing on a curved surface (realised using the special jig visible in Figure 1) to mechanically correct the bending distortion. When the magnitude of the imposed bending was not correct, a flame straightening has to be performed [3]. Another set-up problem was the selection of the optimal welding parameters, since a partial penetration were required by design.

The object of the present analysis is to optimise the production process; thus no attention has been paid to the engineering choices.

## 3. THE FINITE ELEMENT ANALYSIS

Two different FE models have been built; a local three-dimensional model of a section of the header, to analyse in detail the welding process at a micro-level, and a complete 3-D model of the structure. In both cases no simplification was allowed given the asymmetry of the process performed. A uncoupled thermo-mechanical analysis has been carried out for both the models, based on the assumption that dimensional changes during welding are negligible and that mechanical work done during welding is insignificant compared to the thermal energy change.

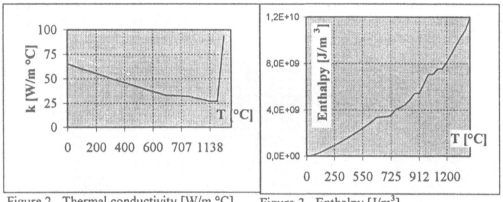

Figure 2.  Thermal conductivity [W/m °C]          Figure 3.  Enthalpy [J/m³]

Finite Element models have been build in parametrical form, such that a different diameter, length, thickness, number and diameter of tube stumps, can be modelled  by simply specifying a variable value. Material properties can also easily be changed, even though no data are available close to the melting temperature [4]. In Figures 2 and 3, the thermal properties adopted for the analysis are shown [5]. The convective heat transfer coefficient has been assumed constant to 15[W/m² °C].   Radiative heat exchanges has been neglected.

As concern the mechanical analysis, the deformation process is assumed to be strain rate independent ; a thermo-elastic-linear work-hardening constitutive model is used in the mechanical analysis for both the models: see Figure 4 for the stress-strain curve  and Table I (material constant values are linearly interpolated when values are not given for a temperature).   The Poisson ratio has been taken constant equal to 0.3, as well as the density, taken to be 77000[N/m³].

The results presented in the paper refer to an header consisting of  a cylinder of ASTM 106 (B grade) steel,   2565[mm] length, external diameter of 168.3[mm] and thickness of 16[mm]. 84 tube stumps  made of ASTM 210 (A1 grade) steel are welded in three rows of 28 tubes each.

| Table I. Thermal expansion coefficient $\alpha (x\ 10^{-6})$ | |
| --- | --- |
| T [°C] | $\alpha$ [1/°C] |
| 20 | 2.76 |
| 100 | 7.96 |
| 300 | 11.56 |
| 420 | 12.5 |
| 682 | 13.99 |
| 736 | 14 |
| 774 | 11.25 |
| 900 | 14 |
| 1200 | 15 |

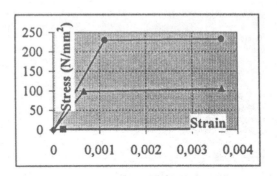

Figure 4. Stress-strain curves (T=20-600-1400[°C]).

## 3.1 THE LOCAL MODEL

A thermal F.E. model of a section of the header has been built, in order to study the effects of the local surrounding structure on the single-pass weld of the tube stump (see Figure 6). An equivalent heat-exchange coefficient has been assumed for the boundary surfaces to take into account the presence of the rest of the pipe.

The weld joint has a geometry continuously varying around the tube, and so for the volume to be filled (Fig.5-b), resulting from the intersection of a cylinder (the pipe) orthogonal to a truncated cone (the mill that is used to shape the weld groove according to Fig.5-a).

The selection of the optimal operating parameters to assure a constant depth of penetration is not a trivial problem, given the geometrical complexity of the surrounding structure.

In the synergic type of MIG welding, the independent variables selected for the F.E. analysis are welding speed and power, considering also the expertise of the company. The distribution of the heat input has been assumed according to the meshing refinement adopted (see Fig.5-a and b) ; the weld seam has been divided into 16 sectors.

The optimal welding speed has been determined in few runs by taking fixed the input power, such as a constant penetration of 2[mm] was reached. This is represented in Figure 7 in terms of 'delay' time per sector. The depth of penetration has been determined by tracing the isothermal at the fusion temperature of 1400[°C]. The resulting optimal processing time for each tube stump is 76[sec], with a power of 5[kW]. Experimental evidence confirmed the result, satisfying the weldment the quality requirements.

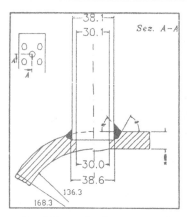

Figure 5-a. T branch-pipe connection
(measures [mm])

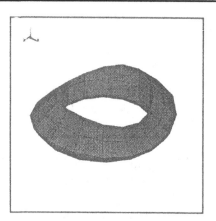

Figure 5 -b. Volume of the weld groove
(FE representation with 16 sectors).

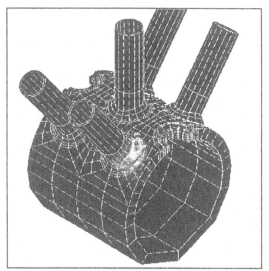

Figure 6. Model of a section :
temperature contours (2416 elements)

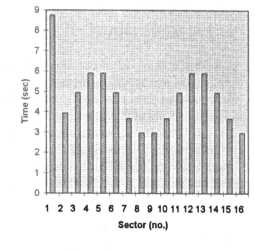

Figure 7. Optimal delay times per sector.

## 3.2 THE FE MODEL OF THE WHOLE STRUCTURE

As concern the analysis of the bending distortion, a complete three-dimensional model  of
the whole header was built,  for the asymmetry of the process. Given the size of the pipe, a
coarser mesh was adopted, to make the problem computationally treatable (see Figure 8).
The model has been built using iso-parametric brick  elements with one degree of freedom
(temperature, T) for the thermal analysis, and   three d.o.f. for the mechanical analysis
(displacements along the three axis Ux, Uy, Uz) [6]. The validation made  by comparing the
predicted bending distortion with the actual one, shows very good   results (see Figure 9)

despite the coarser mesh. A careful selection of the constraints was made, which gave a significant influence.

Several welding sequencing strategies (with two simultaneous welding at time to reduce bending) have been analysed without simulating the elastic bracing, in order to verify the possibility to reduce at the minimum the bending distortion. None of the several runs gave a significant reduction of the permanent distortion. It has been verified that, for the operating conditions adopted, even performing only one welding in the less favourable condition at centre of the pipes, a permanent bending distortion of about 0.2[mm] results. These facts confirm the bending shrinkage expected by theory [3], being the entire welding processes equivalent to a single-side longitudinal welding.

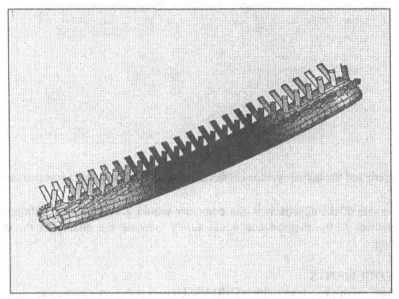

Figure 8. FE model of the whole header : pipe plus tube stumps (1846 elements)

## 4. DISCUSSION AND CONCLUSIONS

The F.E.A. approach adopted is starting to be a quite standard practice in design analysis in several fields of the welding technology. This practice however is still far from being used also for process optimisation scope, as an alternative to the traditional trial & error approach in the common industrial practice. In the paper we present such an application, of a F.E. thermal model which allowed to reduce significantly the set-up times of a robotic welding operation. This was possible once the model was validated by comparing in few cases the results with very experimental evidence. The parametrical feature of the models allows also simply extend the analysis of the production process for other similar products with no significant efforts.

Some final consideration about the potential economics allowed might give insight on the advantages of this 'hybrid' approach. From very simple considerations, it has been calculated that net potential production savings for the company, are of the order of magnitude of 6.500[US$] per year. This roughly estimation has been made based on the volume of production of about 7 boilers per year, considering also the average cost for software development and code licence fee, neglecting the cost for the hardware.

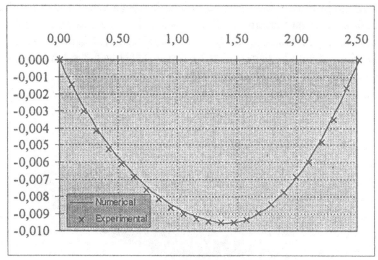

Figure 9. Predicted vs. actual values of deviation from the linear axis (measures in [m])

An extensive use of the approach at the company would give a higher confidence on the true potentialities of the method and could surely increase the quality of the design and manufacturing.

AKNOWLEDGEMENTS

The Authors wish to tank the TERMOSUD S.p.A, in particular ing. Ernesto CHIARANTONI, for the availability in supporting all experimental activities. The Authors are also grateful with Prof. Attilio Alto for his fruitful suggestion and encouragement.
This research is partially funded by the contribution of the Italian Ministry of Research.

REFERENCES
1. Lancaster, J.F.: Metallurgy of Welding, Chapman & Hall; Cambridge, 1993
2. Taguchi, G. : System of Experimental Design, 1-2, UNIPUB/Krons Int. Pub., U.S.A., 1987
3. Radaj, D. : Heat Effects of Welding, Springer-Verlag, Berlin, 1992
4. Alto, A. and Dassisti, M. : Role of some approximation assumptions in finite element modelling of arc welding processes, II Convegno AITEM, Padova, Italy, 18-20 settembre 1995
5. Toulokian, Y. S.; Ho, C.Y. :_Thermophysical Properties of Matter: Specific Heat-Metallic Elements and Alloys, New York, 1970
6. ANSYS, Analysis Inc. User Manual for Revision 5.0A, 1993

# MONITORING CRITICAL POINTS IN ROBOT OPERATIONS
## WITH AN ARTIFICIAL VISION SYSTEM

**M. Lanzetta and G. Tantussi**

**University of Pisa, Pisa, Italy**

KEY WORDS: Stereo Vision, 3D Trajectory Reconstruction, Real-Time Tracking

ABSTRACT: Stereo Vision for assisted robot operations implies the use of special purpose techniques to increase precision. A simple algorithm in the reconstruction of 3D trajectories by real-time tracking is described here, which has been extensively tested with promising results. If the form of the trajectory is known *a priori*, the interpolation of multiple real-time acquisition yields an increase of precision of about 25% of the initial error, depending on the uncertainty in locating the points within the image. The experimental tests which have been performed concern the case of a straight trajectory.

## 1. STATE OF THE ART AND POSSIBLE APPLICATIONS

The importance of Artificial Vision for industrial applications is increasing even for small and medium enterprises for the reduction of hardware costs. It allows performing a closed loop process control not interfering with the observed system. Unfortunately this technology is limited by the low resolution of sensors; for this reason even a small increase in precision becomes attractive.

In this article a general algorithm for this purpose is described, which has been extensively tested at different space resolutions and camera configurations.

The main advantages of this algorithm are:

Published in: E. Kuljanic (Ed.) *Advanced Manufacturing Systems and Technology*, CISM Courses and Lectures No. 372, Springer Verlag, Wien New York, 1996.

- easy to implement;
- fast;
- versatile.

The idea is to increase precision by multiple acquisitions, but not to interfere with the operations and in particular not to increase the acquisition time, the Observed Object (OO) is followed in *real-time* on its trajectory which precedes a critical point. This implies that a great computing power must be available, which is not a tight constraint when high accuracy is needed. The information to be exploited in the process concerns the kind of trajectory of the OO which is supposed to be known in parametric form. In common applications the presence of approximately straight lines is usual.

The Galilean relativity allows the application of this algorithm both to a moving camera (for instance in a robot hand-eye configuration [1 -2 ] or ego-docking [3 ] for autonomous robot or vehicle navigation [4 -5 ]) and to a fixed camera observing a moving object. Some examples of the latter are:
-  robot gripper obstacle avoidance [6 ] or objects localisation and catching;
-  in the case of autonomous navigation the so called echo-docking.

This algorithm can be applied both to trajectories in two and three dimensions.

Since in experimental tests it has been shown that it is resolution-independent, it can be applied in many different fields:
- mobile robot docking [7 ] for the compensation of the errors due to incorrect positioning in the execution of a robot program relative to the robot base;
- closed loop control of assembly operations [8 -9 ];
- tracking and control of an AGV position [10 ];
- robot calibration, [11 ]; in the case of hand-eye configuration, Stereopsis can be achieved even with just one *moving* camera, by the observation of a known pattern [2].

For all these cases it is necessary to know the OO trajectory in a parametric form.

## 2. SYSTEM CALIBRATION AND STEREOPSIS

To reconstruct the position of a point in 3D, more than one view is necessary or one view and at least one of the three coordinates.

In this article two cameras have been used with different configurations. The OO in the experiments was the sensor of a measuring machine. Since at any instant the OO coordinates are provided by a measuring machine, an *absolute* system has been defined coincident with its main axes. The transformation from three space coordinates $W = \begin{pmatrix} X & Y & Z \end{pmatrix}^T$ to four image coordinates $w_{1,2} = \begin{pmatrix} x & y \end{pmatrix}^T$ (a couple for each camera) is defined by the system calibration [12 ].

This can be achieved by calibrating each camera separately through the minimisation of the squared errors of known 3D points projected onto the plane of view [13 ]. It sould be emphasised that the law of projection of a point with the *pinhole* camera model is non-linear. Expressing the problem in *homogeneous* coordinates we get the following expression

$$\begin{pmatrix} x_{h_{i,c}} \\ y_{h_{i,c}} \\ z_{h_{i,c}} \\ k_{i,c} \end{pmatrix} = \begin{pmatrix} m_{11} & m_{12} & m_{13} & m_{14} \\ m_{21} & m_{22} & m_{23} & m_{24} \\ m_{31} & m_{32} & m_{33} & m_{34} \\ m_{41} & m_{42} & m_{43} & m_{44} \end{pmatrix} \cdot \begin{pmatrix} X_i \\ Y_i \\ Z_i \\ 1 \end{pmatrix} \tag{1}$$

where $h_i$ stands for homogeneous coordinate of the $i$-th point of the $c$-th camera, with

$$x_{i,c} = x_{h_{i,c}} / k_{i,c} \; ; \; y_{i,c} = y_{h_{i,c}} / k_{i,c}.$$

Eliminating $k_{i,c}$ from the first two lines in equation (1) yields

$$m_{11}X_i + m_{12}Y_i + m_{13}Z_i + m_{14} - m_{41}x_{i,c}X_i - m_{42}x_{i,c}Y_i - m_{43}x_{i,c}Z_i - m_{44}x_{i,c} = 0$$
$$m_{21}X_i + m_{22}Y_i + m_{23}Z_i + m_{24} - m_{41}y_{i,c}X_i - m_{42}y_{i,c}Y_i - m_{43}y_{i,c}Z_i - m_{44}y_{i,c} = 0$$

$$(2)$$

The unknowns in this expression are the $m_{j,k}$, the elements of the matrix of projection of a point from 3D space onto the camera plane of view. To find an exact solution of the system almost six control points are needed. To get the maximum performances by a real system it is suitable to use a high number of points. Experimental tests have shown that about 30-40 points were enough for the utilised system.

Once the system has been calibrated, given a couple of point projections, it is possible to estimate the 3D coordinates calculating the pseudo inverse of the projection matrix. This latter is constituted of the calibration parameters taken from a couple of equations like the (2) for each camera $c$ [14]. From equation (2) an expression in the form

$$A_{2c \times 3} \cdot W = b_{2c \times 1} \tag{3}$$

can be derived. Equation (3) can be solved by the inversion of matrix $A$ in the sense of least squared errors in order to find the unknown $W$ vector.

It sould be noticed that the explained calibration model does consider translation, rotation, scale and perspective projection of a point in 3D space from/to one or more view planes, but it does not consider other effects which may introduce even a high degree of uncertainty, such as optical aberrations, lack of a suitable lighting and focus.

The application of this analytical model does not depend on the relative position between the camera and the observed system.

To achieve more general results, no further mathematical correction has been applied in experimental tests beside the Stereo Vision algorithm. In order to reduce the effect of optical aberration, the described algorithm has been applied with the fragmentation of the working space in smaller volumes and performing a different calibration on each of them.

Different configurations have been tested. The configuration which provides the lowest error (e. g. the Euclidean distance between the measured and the real point), was achieved when the three measured coordinates had the *same* accuracy. The best condition is with the camera optical axes perpendicular to each other.

In the case of two cameras forming an angle of about 75°, perpendicular to the Z axis and symmetrical with respect to the XZ plane, the accuracy on X coordinates were about half of the Y coordinates and about one third of the Z ones.

## 3. THE ALGORITHM

For the application of this algorithm, the following items are required:
- compensation of errors which can be approximated by a function of higher order than the calibration model or the trajectory form;
- knowledge about the *real* trajectory (e. g. if it is really a straight line or a curve);

otherwise the result of interpolation is to improve some trajectory parts and to worsen others.

The algorithm can be summarised in the following steps:
1. Acquisition of the point coordinates in 2D
2. Interpolation for movement compensation
3. Recovery of the 3D point coordinates
4. Calculation of the interpolated trajectory
5. Correction of points

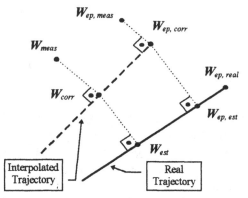

The chosen criterion to correct a measured point $W_{meas}$, is to project it perpendicularly onto the interpolated trajectory. The reason which inspired this idea is that the most accurate point belonging to the interpolated line is the closest one.

In order to treat fewer data, the real trajectory is described by its endpoints. Thus just two exact points are enough to test the algorithm on several measured points; this is very useful for practical applications.

For the algorithm estimation the following exact information is exploited:
– the known trajectory endpoints $W_{ep}$;
– for particular straight lines (parallel to the main axes), the two coordinates which remain constant during displacement.

For each measured point $W_{meas}$ , an estimation of the exact one $W_{est}$ , is obtained by projecting onto the known real trajectory the corresponding *corrected* point $W_{corr}$ .

For the trajectory endpoints $W_{ep,meas}$ whose corresponding real points $W_{ep,real}$ are known, in experimental tests it has been shown that the distance between $W_{ep,est}$ , the estimated one and the real endpoint $W_{ep,real}$ is lower than 2/10 of initial error.

Considering the application of the described algorithm to a straight trajectory, the least squared errors straight line in parametric form, for any straight line non parallel to the main axes, is given by

$$\begin{cases} X = t \\ Y = s \cdot t + q \\ Z = n \cdot t + p \end{cases} \tag{4}$$

with, for $i \geq 2$,

$$\begin{pmatrix} s & q \\ n & p \end{pmatrix} \cdot \begin{pmatrix} X_1 & \cdots & X_i \\ 1 & \cdots & 1 \end{pmatrix} = \begin{pmatrix} Y_1 & \cdots & Y_i \\ Z_1 & \cdots & Z_i \end{pmatrix}$$

Given a measured point $W_{meas} = \begin{pmatrix} X_{meas} & Y_{meas} & Z_{meas} \end{pmatrix}^T$, the corrected point $W_{corr}$ is found by substituting in equation (4)

$$t = \frac{X_{meas} + (Y_{meas} - q) \cdot s + (Z_{meas} - p) \cdot n}{1 + s^2 + n^2}$$

## 4. EXPERIMENTAL DATA

The process performances, viz. the maximum space resolution and the OO maximum speed are limited respectively by the sensors resolution and by the computing power of the Artificial Vision system. The used system is made by a high performance acquisition and processing card. A complete frame with a resolution of 756x512 is acquired every 40 ms. One of the cameras sends a synchronisation signal to the other one and to the frame grabber that switches between them at the synchronisation frequency. This implies that to get the position of a point at instant $j$ an interpolation of coordinates $j$-1 and $j$ of the other camera is necessary. For this reason, after the acquisition of $N$ couples of points, one gets $M=2N-2$ measured coordinates (the last one is static and the one before is taken during deceleration). The interpolation between point $j$-1 and $j$ depends on the trajectory form, which is supposed to be known.

The A/D conversion and grabbing produce a negligible delay but does not affect the frequency of acquisitions of 25 non-interlaced frames/s.

The system is able to locate within a grayscale image of the indicated width and height, a pre-defined model in about 210 ms. This time can be reduced to about 1/10 by limiting the search area. In this specific application, the area reduction can be performed considering that in the time between two couples of acquisition, the OO moves just a few pixels from the present position along the pre-defined trajectory.

The system is able to locate within the image all the matches over a minimum score without increasing the search time. The option of following more than one point can be exploited

- to increase precision;
- to describe the trajectory of complex objects the central point of which is not visible or unknown;
- to retrieve the inclination of an object;
- to track multiple patterns for monocular robot hand-eye configurations [15 -16 ].

In the next step the found coordinates of the OO in the image are used to calculate the corresponding 3D point performing the product of the pseudo inverse of matrix $A$ and $b$ in equation (3).

The Vision system has been programmed in order to follow the OO after it has entered the field of view at the beginning of its known trajectory and interrupts the acquisition when the OO stops. At this time the measured 3D trajectory has already been reconstructed and the corrected one is calculated. The parametric information on the corrected trajectory can now be used either to correct the end point or the whole sequence of points. The interpolated trajectory computation and the measured points correction are performed in less than 80 ms on a 120 MHz Pentium based PC.

In real-time tracking, line-jitter phenomenon can occur, viz. different lines in a frame are acquired with the OO at different positions. Line-jitter effect is to limit the OO speed relative to a camera because the OO model matching within the resulting image is affected by a greater error. Both the movement compensation and the correction through the described algorithm reduce the error of points belonging to a trajectory in space described at speeds up to about 40 mm/s. Below this value, the improvement coming from the application of this algorithm does not depend on the OO speed.

Different trajectories have been followed at different speeds in the range between 1 and 40 mm/s. For the same trajectory this implies acquiring an inversely proportional number of points in the following range: $150 \geq M \geq 10$.

The estimation of the algorithm has been performed on the *known* final point of a sample of 26 straight trajectories inside a working space of about $300^3$ mm$^3$.

1. An average reduction of 25% of the initial error the mean value of which is 1.13 mm, with a standard deviation of 0.47, has been achieved. This data consider the presence of less than 4% of negative values. The error was due to a bad approximation of the measured trajectory by a straight line; the other endpoint accuracy was improved.

2. For over 80% of these trajectories the error between the estimated and the real coordinates $W_{ep,est}$ and $W_{ep,real}$ (see figure) was lower than 20% of the initial error. As a consequence we can state that $W_{est}$ represents a good estimation of the unknown real point corresponding to a measured one.

An extension of 1. to any measured point of the whole sequence is that by projecting it onto the interpolated trajectory we achieve a remarkable correction in most cases. In all tests, this has been shown by a reduction of standard deviation of about 50% on the examined sample.

A consequence of 2. is that projecting a *corrected* point onto the real trajectory, we can estimate its corresponding exact point coordinates.

Finally we can state that given a parametric description of a trajectory, we can estimate the errors on a sequence of measured points (the corresponding real point of which are unknown), their mean value, standard deviation, etc. by computing the distance between $W_{meas}$ and $W_{est}$.

To get a higher precision, a higher space resolution (e. g. a higher camera resolution or a smaller workspace) is required and more sophisticated techniques of optical aberration compensation and a sub-pixel analysis are needed.

## 5. EXTENSIONS

In order to achieve a computation time reduction, this algorithm could be applied directly finding the interpolated trajectory as it will appear after projection on the camera sensors, e. g. if the trajectory is a straight line, finding the interpolated line inside the image, if the trajectory is a circular arc, finding the ellipse arc, etc. This sort of analysis would probably involve a different numerical approach to Stereo Vision, considering visual maps [6], epipolar lines and other primitives [14], and geometric relations between the *absolute* and the *camera* reference systems in order to optimise the overall process.

Dealing with a parametric description of primitives in space instead of points could allow providing a correction to the robot directly in this form.

Multiple Stereo algorithms [17 ] are available to optimise the search in image; a direct use of features extracted from the image instead of operating on the coordinates could represent a shortcut.

In this article the least squared errors approach has been used; the experimental tests have been performed on a straight trajectory. A different kind of interpolation can be applied according to a different trajectory and error distribution (e. g. with low weights for less accurate points). Furthermore the benefits coming from the use of Kalman filter which is suitable for time dependent problems can be investigated.

## 6. CONCLUSIONS

A simple algorithm to increase accuracy in the case of an object moving on a trajectory the parametric description of which is known, has been described. The algorithm has been extensively tested on straight trajectories with different camera configurations. The interpolation of points both for movement compensation and for the trajectory calculation allow an increase of accuracy which depends on the initial error distribution.

It has been shown that a simple perpendicular projection onto the interpolated trajectory gives a suitable correction to most of the points of the whole sequence.

In order not to worsen some parts of the measured trajectory by the application of this algorithm, the following condition must be satisfied: absence of higher order discrepancies between

- the real trajectory and its mathematical parametric description;
- the real 3D points coordinates and the stereo reconstruction model.

Since the increased accuracy remains about the same on wide ranges of number of measured points, it has been shown that it is OO speed-independent.

If the OO inclination in the trajectory after the observed one is known, for instance in the case of the coupling of two parts, the angle between the interpolated trajectory and the exact one represents the correction to apply before the coupling.

The increase of accuracy can be exploited by increasing the field of view (thus compensating the reduction of spatial resolution) to monitor several critical points and trajectories with just one couple of cameras.

Once the interpolated trajectory is calculated, the *absolute* OO position can be reconstructed even *after* it exits one of the camera fields of view or if the localisation reliability of a camera has significantly decreased in that view area.

The method to test the described algorithm can also be used to test the performances of a general Artificial Vision system by employing just a few exact data (e. g. the trajectory endpoints).

## REFERENCES

1 . Tsai, R.Y.; Lenz, R.K.: A New Technique for Fully Autonomous and Efficient 3D Robotics Hand/Eye Calibration, IEEE Journal of Robotics and Automation, 3 (June 1989) 3, 345-358

2 . Ji, Z.; Leu, M.C.; Lilienthal, P.F.: Vision based tool calibration and accuracy improvements for assembly robots, Precision Engineering, 14 (July 1992) 3, 168-175

3 . Victor, J.S.; Sandini, G.: Docking Behaviours via Active Perception, Proceedings of the 3rd International Symposium on Intelligent Robotic Systems '95, Pisa, Italy, July 10-14 1995, 303-314

4 . Matthies, L.; Shafer, S.A.: Error Modeling in Stereo Navigation, IEEE Journal of Robotics and Automation, RA-3 (June 1987) 3, 239-248

5 ., Kanatani, K.; Watanabe, K.: Reconstruction of 3-D Road Geometry from Images for Autonomous Land Vehicles, IEEE Transactions on Robotics and Automation, 6 (February 1990) 1, 127-132

6 . Bohrer, S.; Lütgendorf, A.; Mempel, M.: Using Inverse Perspective Mapping as a Basis for two Concurrent Obstacle Avoidance Schemes, Artificial Neural Networks, Elsevier Science Publishers, 1991, 1233-1236

7 . Mandel, K.; Duffie, N.A.: On-Line Compensation of Mobile Robot Docking Errors, IEEE Journal of Robotics and Automation, RA-3 (December 1987) 6, 591-598

8 . Nakano, K.; Kanno, S.; Watanabe, Y.: Recognition of Assembly Parts Using Geometric Models, Bulletin of Japan Society of Precision Engineering, 24 (December 1990) 4, 279-284

9 . Driels, M.R.; Collins, E.A.: Assembly of Non-Standard Electrical Components Using Stereoscopic Image Processing Techniques, Annals of the CIRP, 34 (1985) 1, 1-4

10 . Petriu, E. M.; McMath, W.S.; Yeung, S.K.; Trif, N.; Biesman, T.: Two-Dimensional Position Recovery for a Free-Ranging Automated Guided Vehicle, IEEE Transactions on on Instrumentation and Measurement, 42 (June 1993) 3, 701-706

11 . Veitschegger, W.K.; Wu, C.-H.: Robot Calibration and Compensation, IEEE Journal of Robotics and Automation, 4 (December 1988) 6, 643-656

12 . Tsai, R.Y.: A Versatile Camera Calibration Technique for High Accuracy 3D Machine Vision Metrology Using Off-the-Shelf TV Cameras and Lenses, IEEE Journal of Robotics and Automation, RA-3 (August 1987) 4, 323-344

13 . Fu, K.-S.; Gonzales, C.S.; Lee, G.: Robotics, Mc Graw-Hill, 1989

14 . Ayache, N.: Artificial Vision for Mobile Robots, The MIT Press, Cambridge, Massach., London, Engl., 1991

15 . Fukui, I.: TV Image Processing to Determine the Position of a Robot Vehicle, Pattern Recognition, Pergamon Press Ltd., 14 (1981) 1-6, 101-109

16 . Zhuang, H.; Roth, Z.S.; Xu, X.; Wang, K.: Camera Calibration Issues in Robot Calibration with Eye-on-Hand Configuration, Robotics & Computer-Integrated Manufacturing, 10 (1993) 6, 401-412

17 . Kim, Y. C.; Aggarwal, J.K.: Positioning Three-Dimensional Objects Using Stereo Images, IEEE Journal of Robotics and Automation, RA-3 (August 1987) 4, 361-373

# CAM FOR ROBOTIZED FILAMENT WINDING

**L. Carrino, M. Landolfi and G. Moroni**
University of Cassino, Cassino, Italy

**G. Di Vita**
Centro Italiano Ricerche Aerospaziali, Capua, Italy

KEY WORDS: CAD/CAM, Filament Winding, Fiber Placement, Robotics, Composite Materials.

ABSTRACT: Filament winding is a particular process to produce composite workpieces. Nowadays, the deposition of the composite filament is done using special CNC machine able to generate axisymmetric or quasi-axisymmetric parts. Non-axisymmetric parts are, instead, made by hand. However, the interest in manufacturing of 3D complex shaped components by filament deposition (e.g. parts of aerospace industry) is bringing to the development of a new deposition technology. In these new systems a robot winds a composite filament on a complex support through complex trajectories, in a way similar to the human hand in an hand made object. At the University of Cassino, with the collaboration of CIRA, a robotized filament winding process for non-axisymmetric parts is studied. Parallel to it, a software prototype to define and simulate the process is under development. In this paper the preliminary results in the development of this CAD/CAM prototype are presented.

## 1. INTRODUCTION

Filament winding is a particular process to produce composite workpieces. It consists of the deposition of a composite material filament on a mandrel with adequate shape. Nowadays, it is used to produce axisymmetric parts on lathe-like numerically controlled machines: the mandrel rotates on its axis and the pay-out eye is moved with 2 to 5 degrees of freedom. From the coordination of the mandrel rotation and the pay-out eye movement it is possible

Published in: E. Kuljanic (Ed.) *Advanced Manufacturing Systems and Technology*, CISM Courses and Lectures No. 372, Springer Verlag, Wien New York, 1996.

to deposit a composite filament along a path which guarantees the requirements of strength and stiffness for the part.

This automatic process is characterized by a high repeatability and quality, and it has been successfully used to produce fiber-reinforced parts such as rocket-motor pressure vessels, launch tubes, storage tanks, pipes and other aeronautic components. Its main disadvantages are the high cost, justified only for high production volumes, and the limited obtainable shapes.

To reduce costs, to increase flexibility and to use this technology also for non-axisymmetric part, some example of the use of a robot to move a feeding head were presented in [2,3,5]. In all this studies, however, the considered parts are characterized by having the possibility to be wound following a convex geodetic path [4,5]. In particular, the whole surface of the mandrel has to be covered as in the tube-like components: the main examples are the filament winding of T-shaped and elbow tubes. The issues associated with the concave winding are investigated only at CCM [7] where a special feeding head and a thermoplastic winding process are considered.

The increase of flexibility and the reduction of cost may be reached also by developing adequate CAM software modules able to help the user in defining the part program for the filament winding machines or robots. Among the few example of such systems [1,2,4,6,8], the more relevant results were reached by the work done at CIRA [1,9] and at the University of Leuven [2].

In the first case, the Arianna software is able to generate collision free trajectories for the pay-out eye of a traditional 2 to 5 axes filament winding machine. The filament paths may or may not be geodesic paths, they are calculated solving a set of differential equations and the feasibility of these paths are guaranteed by the control of the slippage tendency ratio. The considered shapes are axisymmetric shapes, but the possibility to have deviation from the axisymmetric shape is considered. However those deviation do not lead to concave surface sections. The fiber bridging problem is well treated.

Cawar [2] is a software module to calculate the fiber path trajectory both in a traditional axisymmetric filament winding process and in a prototype of robotized tape winding cell for asymmetric components. In the first case semi-geodesic fiber paths are considered, while in the second only geodesic paths are possible. A heuristic collision avoidance method is implemented, but it do not cover all the possible collisions which may occur: a detailed detection is done in the final simulation stage.

At the University of Cassino, a robotized filament winding process for asymmetric complex parts is studied. In this research, a software prototype to define and simulate the process is under development. In section 2 the preliminary implementation of this CAD/CAM software is presented, in section 3 some examples are discussed.

## 2. DESIGN OF THE PROCESS FOR A FILAMENT-WOUND PART

The architecture of the CAD/CAM system under development at the University of Cassino in collaboration with CIRA is shown in figure 1. It respects the basic principles of Concurrent Engineering. An adequate interface enables the user to design a filament-wound part by defining not only its shape and geometry but also the main technical features like the desired fiber orientation in each part zone, the admitted tolerance of this orientation and the dimensional and geometric tolerances of the part. Starting from these data, an intelligent module should be able to generate alternative filament paths which represent different real filament-wound parts by approximating, if possible, the design requirements. If no feasible path is able to respect the requirements, the user will be asked to change some parameters or even the component shape.

Interacting with the system the designer should be able to perform a complete stress analysis on the composite workpiece with a given exact fiber orientation and, simultaneously, queries the system on the manufacturability of the part by running the robot trajectory generator module. This module will generate, if possible, the collision free trajectories of the robot to correctly wind the fiber on the form and will give an estimation of the process cycle time and cost. Final modules will perform a complete simulation of the process and will generate the part program for the robot (in this example the Val II language for a Puma 562 will be used).

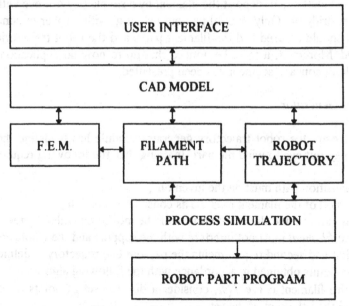

*Figure 1* - System Architecture

The actual implementation of the system is limited to a simple part representation and filament path generator, and to a more complete robot trajectory generator. They are discussed in the following.

*Part Representation and Filament Path Generator*

A part is defined through the shape of its support and by instantiating its dimensions. Few parametric part families are actually defined in the system, one for each element chosen to benchmark the complete prototype. To each part a particular composite filament (roving) may be associated. Each roving is characterized by the shape and dimensions of its cross-section, and by its fiber and resin materials from which the final structural behavior and the maximum friction force between the filament and the support will depend.

Actual part solid model may be generated as a composition of prismatic and cylindrical primitives. To each primitive a preference in fiber orientation is given. A simple algorithm is used to define a filament path respecting the requirements, if possible. The possible paths are geodesic paths: straight lines for planar surfaces, circular or helix arcs for cylindrical surfaces. The continuity of the filament tangent is guaranteed by a smooth approximation. Therefore, the deposition of the filament will be stable.

By an appropriate visualization of the results, the designer could evaluate the part and generate alternative solutions.

It must be pointed out that, in the actual implementation, a complete filament path is represented as a union of a set of curve segment (straight line, circle and helix). The resulting entire curve must not present any concave segment. In this situation, in fact, it is not possible to find methods to deposit the filament by a simple pay-out eye without causing the so called fiber bridging. Only, special pay-out eye (e.g. with a roller in contact with the support surface) should be used and the filament paths and the robot trajectories should be ad hoc generated. Moreover, it must be noticed that up to now no application considering the deposition along concave segment has been presented.

*Robot Trajectory Generator*

Given a filament path, the robot trajectory generator module has to define the path to be followed by the pay-out eye to wind the part. In doing this, the following requirements must be met:

1. the deposition path must be the given one;
2. the tension of the filament must be as constant as possible;
3. the pay-out eye and the robot arm should be moved on collision free trajectories;
4. the *free filament* must not interfere with the support and the whole environment.

To respect the first and second requirements the pay-out eye trajectory is defined through a discrete set of via points obtained in accordance with the following algorithm:

- Given the entire filament curve $\Gamma(s)$, consider a discrete set of points of $\Gamma$, say $P_i$ the vector defining the i-th point of the set;
- Compute $\tau_i$, the tangent unit vector to $\Gamma$ in $P_i$;
- Compute $E_i$, the position vector of the pay-out eye when the deposition point is $P_i$, as:

$$E_i = P_i + d\,\tau_i \qquad \text{where } d \text{ is a positive parameter.}$$

Therefore, each via point of the pay-out eye trajectory will be at a given distance from the deposition point on the tangent in that point to the filament path. To keep the filament tension constant the parameter $d$ must be kept constant or should be increased gradually.

To guarantee a collision free trajectory of the pay-out eye, each via point is kept out of a safety volume including the support. The volume is created with a positive offset of the support solid model characterized by a distance $r$, given as user input. If a via point previously computed is found to be inside the safety volume the distance $d$ is increased to let the point be on the volume boundary. This is not sufficient to have a collision free path, in fact nothing is known about what happens to the pay-out eye between two subsequent via points. If a linear interpolation between two via points is supposed, it is possible to verify if this line segment intersects the safety volume. If this happens a further set of intermediate via points is introduced on the surface of the safety volume to move the pay-out eye on a collision free path (see figure 2).

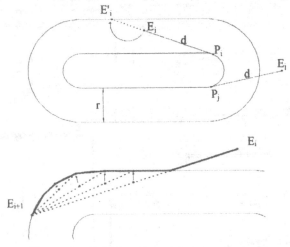

*Figure 2* - Collision free trajectroty generation.

Actually not implemented, a verification of the collision between the entire robot arm and the environment has to be done. Of course, starting from the pay-out eye via points the possible configurations of the robot arm have to be determined. Eventual collisions could be detected intersecting the safety volume with the volume representing the robot arm movement. This volume may be computed as the sweep of the robot arm along the trajectory given by the defined set of via points. If a collision is detected a change either in robot arm configuration or in the position of the via points should be defined.

Another important verification that has to be done is the detection of any possible collision of the *free filament* with the support and the whole environment. The *free filament* may be defined as the segment of the filament that is between the deposition point and the pay-out eye. This kind of verification is very important if a complex shaped workpiece has to be manufactured. In fact, even if a collision free path is possible for the robot, this is not sufficient to guarantee that, before the deposition, the filament is not colliding with the support itself or with other components in the environments (e.g. fixture devices). To verify if such a situation occurs a simple line/solid classification is implemented, where the line segment is the free filament and the solid is the generic obstacle (support, etc.).

Finally, given a feasible filament path and a feasible robot trajectory and known the imposed velocity of the pay-out eye (experimentally determined), it is straightforward to evaluate the needed cycle time and the associated cost of the wind process.

## 3. EXAMPLES

The prototype under development is implemented according to the object oriented paradigm using the Visual C++ compiler and the ACIS solid modeler. It runs on a Intel Pentium based personal computer. In the following some implemented examples are discussed.

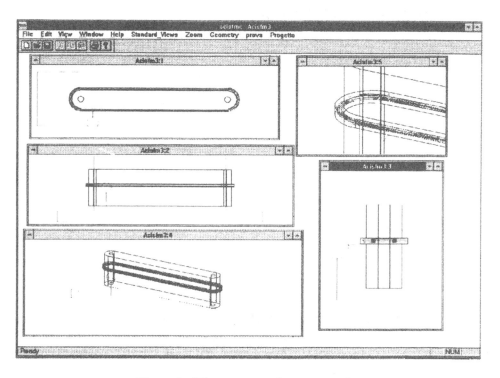

*Figure 3* - Filament wound part example.

In figure 3 a very simple object is considered. It is composed of two parallel prismatic elements connected by two semi-cylindrical elements. The section is rectangular and the dimensions are parametrized. In the figure the form of the support, the filament paths and a sphere representing the pay-out eye are shown. In figure 4 the associated pay-out eye trajectories are represented. It is possible to see the filament paths, the safety volume and the robot trajectories. The same figure shows a particular of the filament path: note that the paths follow geodesics - straight line parallel to the main direction for the two prismatic elements, a circular arc in one cylindrical element and an helix arc in the other cylindrical element. The filament path generator creates subsequent layers of fibers and considers the fibers parallel to each other in the some layer.

Figure 5 shows a more complex object, that can be considered as a folded version of the previous part.

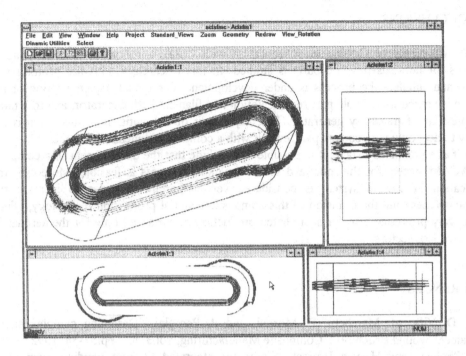

*Figure 4* - Filament path, safety volume and robot trajectory.

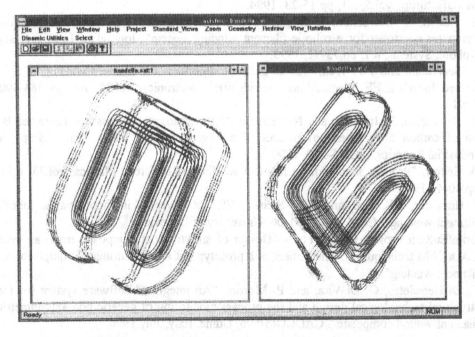

*Figure 5* - Filament path and robot trajectory for a more complex object.

## 4. CONCLUSIONS

At the University of Cassino in collaboration with CIRA, a robotized filament winding process for non-axisymmetric parts is studied. In this research, a software prototype to define and simulate the process is under development. The actual implementation of the system is limited to a simple part representation and filament path generator, and to a more complete robot trajectory generator. The major limit of this initial implementation is the ability to deal only with convex part, even if with a complex tridimensional shape.

Even if at a preliminary stage, it is possible to conclude that the development of a complete CAD/CAM system for the robotized filament winding process is feasible. Moreover, the application of such a system is considered very important in order to increase the competitiveness and the diffusion of this composite material production technology, which seems very promising also for small batch production of parts, not only for the aeronautic and aerospace sectors.

## REFERENCES

1. G. Di Vita, M. Marchetti, P. Moroni, and P. Perugini, "Designing complex shape filament-wound structures". Composite Manufacturing, vol.3, n.1, pp.53-58, 1992.
2. J. Sholliers and H. van Brussel, "Computer-integrated filament winding: computer-integrated disign, robotic filament winding and robotic quality control". Composite Manufacturing, vol.5, n.1, pp.15-23, 1994.
3. E. Castro, S. Seereeram, J. Singh, A.A. Desrochers, and J.T. Wen, "A real-time computer controller for a robotic filament winding system". Journal of Intelligent and Robotic Systems, n.1, pp.73-91, 1993.
4. S. Seereeram and J.T. Wen, "An all-geodetic algorithm for filament winding of a T-shaped form". IEEE Transactions on Industrial Electronics, vol.38, n.6, pp.484-490, 1991.
5. S. Tornincasa, R. Ippolito, and N. Bellomo, "New trends in robotics: the robotized lay up of carbon fiber tapes on surfaces with large curvature". Proc. 21st Symp. on Industrial Robotics, pp.215-220, 1990.
6. A. Seifert, "Process simulation in filament winding". Reinforced Plastics, vol.35, n.11, pp.40-42 1991.
7. J. Hummler, S.K. Lee, and K.V. Steiner, "Recent advances in thermoplastic robotic filament winding". CCM Report 91-06, University of Delaware.
8. Brite/EuRam Project BREU0114 - "Design of structures in composite materials with CAD/CAM techniques - Achievement of a prototype of a fully automated equipment for filament winding".
9. C. Di Benedetto, G. Di Vita, and P. Moroni, "An integrated software system for the structural-technological design and the process monitoring of generic non-axisymmetric filament wound composite". CADCOMP 96, Udine, Italy, July 1996.

# AN AUTOMATIC SAMPLE CHANGER
# FOR NUCLEAR ACTIVATION ANALYSIS

**F. Cosmi**

**University of Udine, Udine, Italy**

**V.F. Romano**

**Federal University of Rio de Janeiro**

**and**

**COPPE-PEM, Rio de Janeiro, Brazil**

**L.F. Bellido**

**SUFIN / Nuclear Engineering Institute, Rio de Janeiro, Brazil**

KEY WORDS: Robotics, Integrated Systems, Radiation Analysis

ABSTRACT: This paper presents an Automatic Sample Changer for nuclear activation analysis of standard samples. An innovative concept for the changer is described here, involving several design aspects of modular components such as Cartesian manipulator, detector mechanism, standard sample-support, auxiliary and data acquisition systems.

## 1. INTRODUCTION

Techniques used to analyze chemical compositions of materials are often based on destructive processes or require a lot of intermediary steps to prepare the samples for the tests.

A technique known as Analysis by Nuclear Activation (ANA) can be used in organic and inorganic materials. This process consists of:

1) transforming stable isotopes of a sample (target) into radioisotopes, through nuclear reactions with neutrons and/or accelerated particles (protons, deuterons, and so on), from a cyclotron or reactor.

2) measuring by means of a detector the induced radioactivity of the sample (gamma-ray). Measuring errors in the order of 1% can be obtained.

Minerals like gold, tungsten, uranium and molybdenum are sensitive to this technique. Sediments, graphite, biological materials and ceramics are also often analyzed.

Some of the advantages of this technique over the conventional ones are mentioned below:

• it is a non destructive method;

• it requires samples with very small masses (in the order of milligrams);

Published in: E. Kuljanic (Ed.) *Advanced Manufacturing Systems and Technology,*
CISM Courses and Lectures No. 372, Springer Verlag, Wien New York, 1996.

- more than 30 chemical elements can be determined simultaneously;
- it can be used both for liquid and solid samples;
- it can be used both for organic and inorganic materials.

Since the half lives of radiation originated in this processes are of short time duration, the frequency spectrum of the irradiated samples, obtained by the data acquisition devices, must be performed in a small period of time.

Information obtained by ANA technique regards the description of the elements present in the sample and their decay curve.

## 2. INTEGRATED AUTOMATIC MEASURING SYSTEM

The automatic measuring facilities, composed by a sample changer and its auxiliary systems, should be able to:

- optimize the period of time between counting;
- allow long terms programmed analysis;
- reduce human exposure to radiation;
- increase the feasibility of sample measuring.

The sequence describing ANA technique within an integrated approach can be characterized by the following steps:

(a) system parametrization:

. evaluation of system parameters and data setting;

. analysis programming and control;

. set of the diagnose routines.

(b) samples irradiation:

. implemented in cyclotron or reactor;

. samples installed in standard supports;

(c) positioning of the sample-supports into a sample recipe:

. standard supports for different types of samples;

. positions defined with respect to a local reference;

(d) motion of a sample-support from the sample recipe to the detector:

. performed by a Cartesian manipulator with gripper;

(e) positioning of the sample-support for detection:

. sample-support fixed in the detector mechanism (DM);

. high-precision displacements;

. position of the sample support in the detector is a function of gamma-ray intensity.

(f) sample measuring:

. hyper-pure germanium detectors;

. automatic procedure to perform a multi-canal data acquisition.

(g) sample-support reallocation in the sample recipe:

. DM in home position;

. Cartesian arm motion;

(h) start of a new cycle for the other samples.

In order to provide full automation capabilities, the integration of all subsystems (manipulator, detector mechanism, programming and data acquisition facilities) is required.

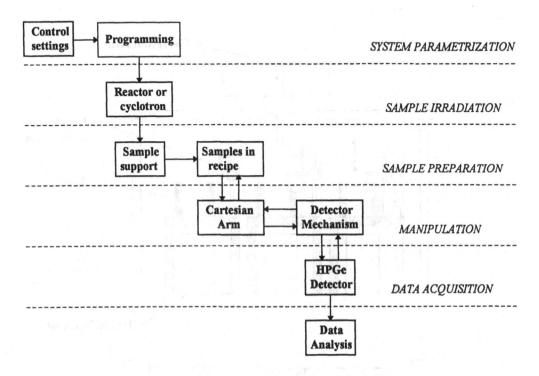

Figure 1: Scheme of the Integrated Measuring System

## 3. AUTOMATIC SAMPLE CHANGER CONFIGURATION

A modular design concept is used for the Automatic SAmple Changer (ASAC), which provides the possibility of system expansion by modules implementation, according to necessity and system capabilities. Shields are necessary to avoid site contamination. They are located in the detectors and around the sample recipe. A brief description of ASAC layout is presented in figure 2.

The sample support consists of a modular auxiliary device used to hold a sample for grasping and transportation tasks.

After a classification of different kinds of samples frequently used for ANA, a standardization was performed in order to optimize the dimensional variables. Samples were organized in four dimensional groups:

- liquid samples: plastic cylindrical container (10 mm diameter);
- solid samples with a section area of 1"x 1";
- solid samples with cylindrical section of 1" diameter;
- solid samples of small dimensions: aluminium paper.

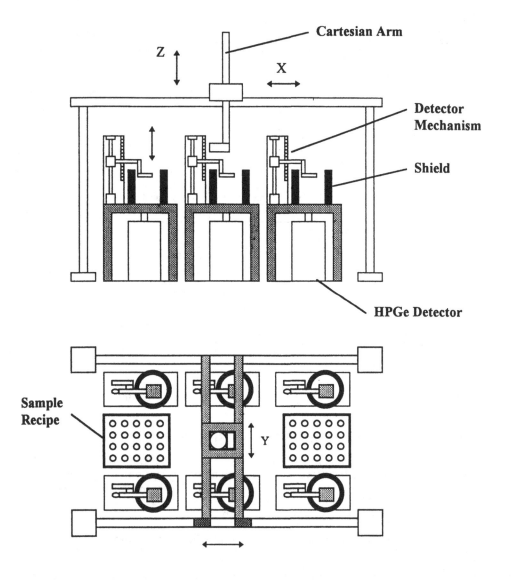

Figure 2: ASAC layout.

The next step was the development of an interchangeable sample support consisting of *upper*, *intermediary* and *lower* part. The sample-support, that is in direct contact with the sample, has been designed in order to:

- allow gripper fixation to the Cartesian manipulator (intermediary part);
- connect the sample recipe to the detector mechanism (lower part);
- in case of *radioactive leak*, only the upper part must be eliminated.

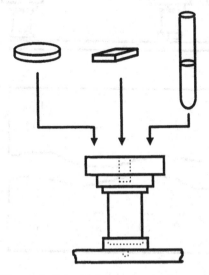

Figure 3: Standard samples and sample-support.

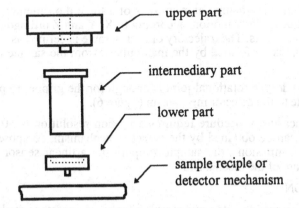

Figure 4: Interchangeable parts of sample-support.

Sample-supports are placed in a sample recipe in a n x m matrix form (figure 5). Full information about samples and their relative locations in the recipe are necessary to the control system, so that the Cartesian manipulator can connect a specific sample-support and move it to a pre-defined detector.

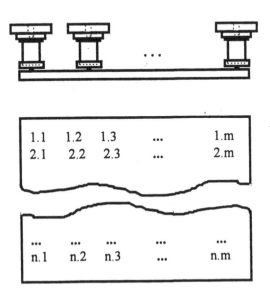

Figure 5: Sample recipe with n x m sample-supports.

The Cartesian manipulator consists of a three d.o.f manipulator and a one d.o.r. gripper. The manipulator motion, according to X, Y and Z orthogonal directions, with dc servo motors actuators. The trajectory of each sample is defined as a function of time, and a cyclic routine is performed by the manipulator from the sample recipe to the detector mechanism.

A ± 90 degree rotational joint is enough for the gripper to provide the necessary approach angle to the detector mechanism (figure 6).

The measure procedure requires a motion resolution of 0.1 to 0.2 mm. These requirements can be obtained by the detector mechanism, composed by a step motor, a ball screw transmission , the sample support and a linear sensor. Figure 7 shows the general structure of the mechanism.

CONCLUSION

This project presents an automatic sample changer to operate in nuclear environment. The modular components provide easy maintenance and assembling, reducing costs for system implementation. The analysis by nuclear activation technique can be applied in a wide range of materials. Our research is mainly interested in minerals and organic materials. A prototype version of ASA with four detectors is currently being developed at IEN.

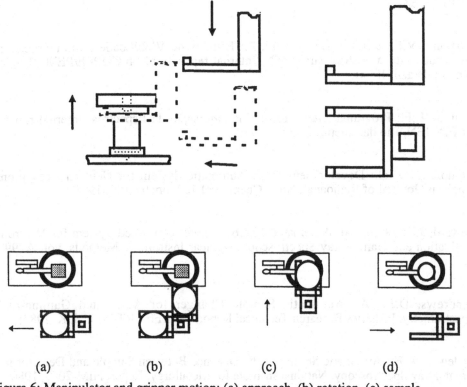

Figure 6: Manipulator and gripper motion: (a) approach, (b) rotation, (c) sample
deposition, (d) manipulator back to home position.

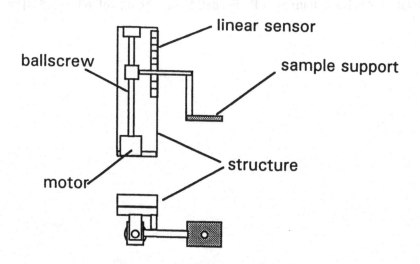

Figure 7: Detector mechanism scheme.

REFERENCES

1. Romano, V.F., Bellido, L.F, Cosmi, F.: Estudos de Viabilidade de um Trocador Automático de Amostras para o IEN, Internal report LabRob-COPPE/PEM, Rio de Janeiro, Brazil, 1994;

2. Bellido, L.F.: Programa RAE - Rotina de Aquisição de Espectros, Internal report SUFIN-IEN, Rio de Janeiro, Brazil, 1995;

3. Munoz, L., et al.: Development of an Automatic System for Neutron Activation Analysis, Journal of Radioanal. Nucl. Chem., vol 167, pp. 97-100, 1993;

4. Edward, J.B., et al.: An Automated Microcomputer-controlled System for Neutron Activation and Gamma-ray Spectroscopy, Nuclear Instrum. & Methods, vol. A299, 1990

5. Andrews, D.J.,: An Automatic Sample Changer for Automated Gamma-ray Spectrometry, Fisheries Research Technical Report - MAFF-FRTR-77", UK, 1984;

6. Andeweg, A.H.: Automatic Setting of the Distance Between Sample and Detector in Gamma-ray Spectroscopy, National Institute for Metallurgy, Report NIM-2069, 1980;

7. Cross, J.B., et. al.: An Automated Gamma-ray Counting System for Fast Neutron Activation Analysis, Journal of Radioanal. Nucl. Chem., vol. 63, pp.155-158, USA, 1981;

# ON THE SIMPLIFICATION OF THE MATHEMATICAL MODEL
## OF A DELAY ELEMENT

**A. Beghi and A. Lepschy**
**University of Padua, Padua, Italy**

**U. Viaro**
**University of Udine, Udine, Italy**

KEY WORDS: Model Reduction, Feedback, Time Lag, Step and Frequency Response.

ABSTRACT: Transportation lags are often present in manufacturing processes. The design of the related control system is easier if reference is made to a proper rational approximation of the transcendental transfer function of the delay elements. To this purpose, both an analytic method and an empiric procedure are suggested. The first approximates the step response of a unity-feedback system including the delay element in the direct path, whereas the second is based on the direct inspection of the Bode phase plot of the delayor. The results compare favorably with those obtained using the standard Padé technique.

## 1. INTRODUCTION

In the analysis and synthesis of control systems, the designer has often to deal with delay phenomena, by which an input acting for $t \geq t_0$ affects the output only for $t \geq t_0 + T$, where $T$ is the so-called dead time or time lag. In terms of Laplace transforms, the considered time-shift operation is represented by the factor $e^{-Ts}$, which is therefore the transfer function of the delay element (delayor).

As is known, many manufacturing systems exhibit transportation lags; this is typically

Published in: E. Kuljanic (Ed.) *Advanced Manufacturing Systems and Technology*,
CISM Courses and Lectures No. 372, Springer Verlag, Wien New York, 1996.

the case when a piece is successively operated at different working stations located
along a conveyor: an operation executed at a station affects the operation performed
at the next one after a time interval that depends on the distance between the stations
and the speed of the conveyor.

The problem of approximating the trascendental transfer function $e^{-Ts}$ by means of
a rational function has a long history (see, e.g., [1]). In the field of dynamic system
simulation, much work has been done since the Fifties (see, e.g., [2],[3],[4]), when the
main simulation tool was the electronic analog computer, particularly suited for implementing circuits characterized by rational transferences. Today, simulations are usually
performed with the aid of digital computers which are not subject to such limitations.
Nevertheless, the considered approximation problem is still of interest [5],[6],[7], with
particular regard to the synthesis of control systems using standard techniques that
are applicable only to finite-dimensional systems, i.e., systems characterized by rational
transfer functions.

Various methods have been suggested to derive a rational approximant of $e^{-Ts}$. The
most popular is probably the one based on the Padé procedure [7]. It consists in determining a rational function $G_P(s)$ with numerator of degree $m$ and denominator of
degree $n$, whose first $m + n + 1$ MacLaurin series expansion coefficients match the
corresponding coefficients of

$$e^{-Ts} = \sum_{i=0}^{\infty} \frac{(-T)^i}{i!} s^i . \tag{1}$$

By this method, it is possible to obtain both proper ($m \leq n$) and non-proper ($m >
n$) models with different characteristics. In particular, when $m = n$ the magnitude
$|G_P(j\omega)|$ of the frequency response of the Padé approximant is identically equal to 1,
like $|e^{-Tj\omega}|$, whereas for $m = 0$ (all-pole approximant) its step response is equal to zero
together with its first $n - 1$ derivatives at $t = 0$, like the step response of the ideal
delayor. On the other hand, the Padé procedure does not ensure the stability of the
approximant of a stable original system and, in fact, for certain values of $m$ and $n$,
$G_P(s)$ may have a non-Hurwitz denominator.

The other approximation methods have either an empiric or an analytic basis and
can be classified according to the specific response considered (typically, the frequency
response $e^{-Tj\omega}$ or the step response $\delta_{-1}(t - T)$ - cf., e.g., [8]).

In this paper, two simple but effective procedures for approximating the transfer function of the delayor are presented. The first refers to the step response of a unity-feedback
system with the delayor in the forward path and leads to rational approximants that
match well the ideal frequency response (magnitude equal to 1 at all frequencies and
phase interpolating $-T\omega$ over a suitable frequency interval). The second is an empiric
procedure based on the inspection of the delayor's phase plot. The results obtained are
finally compared to those obtainable using the Padé technique.

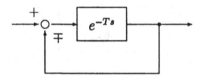

Figure 1: Delay element inserted into a unity-feedback loop.

## 2. APPROXIMATION VIA FEEDBACK

It is easily seen that the step response $w_{-1}(t)$ of a unity-feedback system with a delay element in the direct path (Fig. 1) can be decomposed for $t > 0$ into a step and a square wave in the case of negative feedback (Fig. 2(a)) or into a step, a ramp and a saw-tooth wave in the case of positive feedback (Fig. 2(b)), whereas a periodic term is not present in the step response of the isolated delayor. It is therefore natural to approximate the feedback system of Fig. 1 by means of a system with (proper) rational transfer function $W(s)$ whose step response retains the non-periodic components of the original response together with a suitable number of the first harmonics of the periodic component. The (proper) rational approximant of $e^{-Ts}$ will then be obtained as the transfer function $G(s)$ of the direct path of a unity-feedback system whose closed-loop transfer function is $W(s)$, i.e.,

$$G(s) = \frac{W(s)}{1 - W(s)} \tag{2a}$$

in the case of negative feedback, or

$$G(s) = \frac{W(s)}{1 + W(s)} \tag{2b}$$

in the case of positive feedback.

By expanding into Fourier series the periodic component of $w_{-1}(t)$ in Fig. 2(a) (negative feedback), we get

$$w_{-1}(t) = \frac{1}{2}\delta_{-1}(t) - \frac{2}{\pi}\sum_{i=1}^{\infty}\frac{1}{2i-1}\sin\left[(2i-1)\frac{\pi t}{T}\right], \qquad t > 0. \tag{3}$$

By differentiating (3), transforming term by term, and retaining the first $k$ terms in the summation, we obtain the following rational approximant of even order $2k$ for the closed-loop transfer function:

$$W_{2k}(s) = \frac{1}{2} - \frac{2}{T}\sum_{i=1}^{k}\frac{s}{s^2 + [(2i-1)\pi/T]^2} \tag{4}$$

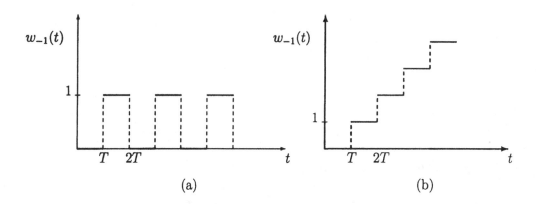

Figure 2: Step response $w_{-1}(t)$ for (a) negative feedback and (b) positive feedback.

which may be rewritten as

$$W_{2k}(s) = \frac{1}{2} - \frac{N_{2k-1}(s)}{D_{2k}(s)} \tag{5}$$

where

$$N_{2k-1}(s) = \frac{2}{T} \sum_{i=1}^{k} s \prod_{\substack{j=1 \\ j \neq i}}^{k} \left\{ s^2 + \left[ \frac{(2j-1)\pi}{T} \right]^2 \right\} \tag{6}$$

and

$$D_{2k}(s) = \prod_{i=1}^{k} \left\{ s^2 + \left[ \frac{(2i-1)\pi}{T} \right]^2 \right\}. \tag{7}$$

Using (2a) we thus arrive at the following even-order approximant of $e^{-Ts}$:

$$G_{2k}(s) = \frac{\frac{1}{2} - \frac{N_{2k-1}(s)}{D_{2k}(s)}}{1 - \left[ \frac{1}{2} - \frac{N_{2k-1}(s)}{D_{2k}(s)} \right]} = \frac{\frac{1}{2} - \frac{N_{2k-1}(s)}{D_{2k}(s)}}{\frac{1}{2} + \frac{N_{2k-1}(s)}{D_{2k}(s)}} = \frac{D_{2k}(s) - 2N_{2k-1}(s)}{D_{2k}(s) + 2N_{2k-1}(s)} \tag{8}$$

which is a stable Blaschke product since: (i) its numerator and denominator have the same even part $D_{2k}(s)$ and opposite odd parts $\pm N_{2k-1}(s)$, and (ii) the denominator of the third fraction in (8) is Hurwitz because $N_{2k-1}(s)/D_{2k}(s)$ appearing in the second fraction is an odd positive-real function (cf. (4)-(7)) that may assume the value $-1/2$ only for Re $[s] < 0$.

Similarly, by expanding into Fourier series the periodic component of the step response in Fig. 2(b) (positive feedback), we get

$$w_{-1}(t) = \frac{1}{T} \delta_{-2}(t) - \frac{1}{2} \delta_{-1}(t) + \sum_{i=1}^{\infty} \frac{1}{i\pi} \sin \left[ \frac{2i\pi t}{T} \right], \qquad t > 0. \tag{9}$$

Again, after differentiation, Laplace transformation, and truncation, we obtain the following rational approximant of odd order $2k+1$ for the closed-loop transfer function:

$$W_{2k+1}(s) = -\frac{1}{2} + \frac{1}{Ts} + \frac{N_{2k-1}(s)}{D_{2k}(s)} \qquad (10)$$

where

$$N_{2k-1}(s) = \frac{2}{T}\sum_{i=1}^{k-1} s \prod_{\substack{j=1 \\ j\neq i}}^{k}\left\{s^2 + \left[\frac{2\pi j}{T}\right]^2\right\}, \qquad D_{2k}(s) = \prod_{i=1}^{k}\left\{s^2 + \left[\frac{2\pi i}{T}\right]^2\right\} \qquad (11)$$

from which, according to (2b), the following odd-order approximant of $e^{-Ts}$ is derived:

$$G_{2k+1}(s) = \frac{-\frac{1}{2} + \frac{1}{Ts} + \dfrac{N_{2k-1}(s)}{D_{2k}(s)}}{1 - \frac{1}{2} + \frac{1}{Ts} + \dfrac{N_{2k-1}(s)}{D_{2k}(s)}} = \frac{2(D_{2k}(s) + TsN_{2k-1}(s)) - TsD_{2k}}{2(D_{2k}(s) + TsN_{2k-1}(s)) + TsD_{2k}}. \qquad (12)$$

In this case too, $G_{2k+1}(s)$ turns out to be a stable Blaschke product.

## 3. A HEURISTIC PROCEDURE

As seen in the previous section, both the suggested approximant and the Padé one are Blaschke products whose poles and zeros are in complex conjugate pairs, except for one pole (and the corresponding zero) in the odd-order models. This leads to a perfect reproduction of the magnitude of the delayor's frequency response and to a good fit of its phase within a given frequency band.

The standard "asymptotic" approximation of the Bode phase diagram of these Blaschke products is formed by a sequence of segments of different slopes connecting points whose phase is $2\ell\pi$, $\ell = 0, 1, \dots, k$ (even-order models) or $0$ and $(2\ell + 1)\pi$, $\ell = 0, 1, \dots, k$ (odd-order models). This observation suggests to obtain another Blaschke product approximating $e^{-Tj\omega}$ as follows:

(i) the Bode phase diagram of $e^{-Tj\omega}$ is subdivided into stripes of breadth $2\pi$, except for the first stripe of breadth $\pi$ in the case of odd-order approximants (Fig. 3);

(ii) the course of $-T\omega$ (as a function of $\log \omega$) in each stripe is approximated by a segment of suitable position and slope, so that the entire diagram can be considered to be the sum of a number of subdiagrams formed by two horizontal straight half-lines connected by such slanted segments;

(iii) each subdiagram, except for the first in the case of odd-order approximants, is then regarded as the standard approximation of the phase plot of a factor of the type:

$$G_i(s) = \frac{1 - 2\frac{\xi_i}{\omega_{ni}}s + \frac{1}{\omega_{ni}^2}s^2}{1 + 2\frac{\xi_i}{\omega_{ni}}s + \frac{1}{\omega_{ni}^2}s^2} \qquad (13)$$

where $\omega_{ni}$ is the abscissa of the central point of the slanted segment and $\xi_i$ is chosen according to its slope; the first subdiagram of the odd-order approximants corresponds to a factor of the form

$$G_1(s) = \frac{1 - T_1 s}{1 + T_1 s} \tag{14}$$

where $T_1$ is related to the position of the first segment;

(iv) the overall approximant $G_H(s)$ is finally obtained as

$$G_H(s) = \prod_i G_i(s). \tag{15}$$

The above heuristic procedure is illustrated in Fig. 3 with reference to $T = 1$ and to a 4th-order approximant. Its transfer function turns out to be characterized by the following 4 parameters: $\omega_{n1} = 3.2, \xi_1 = 0.6, \omega_{n2} = 8.5, \xi_2 = 0.2$. The deviation of its phase from $-T\omega$ is compared in Fig. 4 to the phase deviations for the 4th-order Blaschke products obtained according to the analytic procedure of Section 2 and to the Padé procedure.

## 4. CONCLUSIONS

The problem of approximating $e^{-Ts}$ by a rational function has often been considered in the literature [1]-[7]. Two new solutions have been presented in this paper; they are characterized by an easy implementation and a good accuracy.

The first method (Section 2) starts from the approximation of the step response of a unity-feedback system with the delay element in the direct path. Both in the case of negative and in that of positive feedback, this response contains a periodic component that can be approximated by truncating its Fourier series expansion. The corresponding approximation of $e^{-Ts}$ is finally obtained according to relation (2a) or (2b), and turns out to be a stable Blaschke product formed by pairs of complex conjugate poles (and zeros) except for one pole (and a zero) in the case of odd-order approximants.

An approximant of the same form has also been obtained in a heuristic way (Section 3) by fitting a sequence of segments with suitable slopes to the Bode plot of the delayor's phase. A term of type (13) or (14) is then associated with each segment, thus arriving at the desired Blaschke product (15).

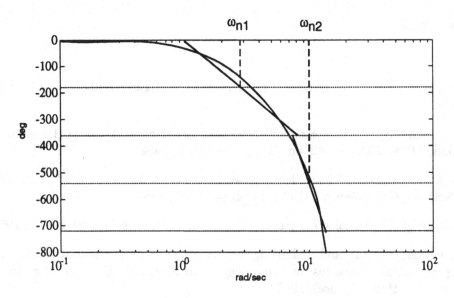

Figure 3: Heuristic procedure for constructing a 4th-order approximant.

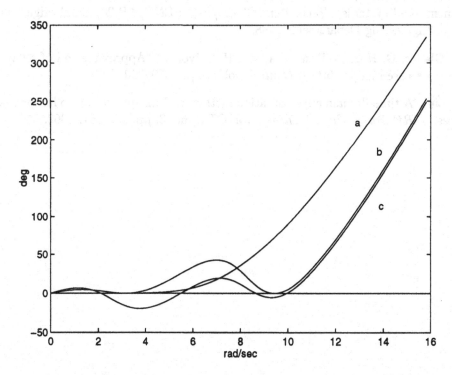

Figure 4: Phase deviations from $-T\omega$ for (a) the Padé model, (b) the model of Section 2, and (c) the heuristic model of Section 3.

REFERENCES

1. O. Perron, *Die Lehre von den Kettenbrüchen.* Stuttgart: Teubner, 1913. 3rd ed. 1957.

2. L. Storch, "Synthesis of constant-time-delay ladder networks using Bessel polynomials," *Proc. IRE*, vol. 42, no. 11, pp. 1666-1675, 1954.

3. W.J. Cunningham, "Time delay network for an analog computer," *IRE Trans. Electronic Computers*, vol. EC-3, no. 4, pp. 16-18, 1954.

4. C.H. Single, "An analog for process lag," *Control Engineering*, vol. 3, Oct. 1956.

5. K. Glover, J. Lam, and J. Partington, "Balanced realizations and Hankel-norm approximation of systems involving delays," in *Proc. 25th Conf. on Decision and Control*, (Athens, Greece), 1986.

6. K. Glover, J. Lam, and J. Partington, "Rational approximation of a class of infinite-dimensional systems," Tech. Rep. CUED/F-INFENG/TR.20, Cambridge University, Engineering Department, 1988.

7. C. Glader, G. Hognas, P. Makila, and H. Toivonen, "Approximation of delay systems - a case study," *Int. J. Control*, vol. 53, pp. 369–390, 1991.

8. B. Liu, "A time-domain approximation method and its application to lumped delay lines,", *IRE Trans. Circuit Theory*, vol. CT-9, no. 3, pp. 256-261, 1962.

# MODEL REDUCTION BASED ON SELECTED MEASURES
# OF THE OUTPUT EQUATION ERROR

## W. Krajewski
### Polish Academy of Sciences, Warsaw, Poland

## A. Lepschy
### University of Padua, Padua, Italy

## U. Viaro
### University of Udine, Udine, Italy

KEY WORDS: Linear Continuous-Time Multivariable Systems, Reduced-Order Models, Matrix Fraction Descriptions, Equation Error.

ABSTRACT: This paper is concerned with the problem of constructing reduced-order models of a stable continuous-time multivariable system by minimizing the $L_2$ norm of suitably weighted equation errors. To this purpose, the approximants are represented by either left or right matrix fraction descriptions. According to the adopted weighting and approximant description, the resulting model matches different sets of both first- and second-order information indices of the original system, which allows a great flexibility in the choice of the order and characteristics of the approximant.

## 1. INTRODUCTION

Automatic control plays a fundamental role in modern manufacturing systems. To synthesize the related control laws, it is often mandatory to avail ourselves of a sufficiently simple model of the plant to be controlled. For this reason, the problem of model reduction has become a very important topic in feedback systems theory and has attracted considerable attention over the past twenty years [1].

Among the most popular approaches, we may mention those based on the retention of

Published in: E. Kuljanic (Ed.) *Advanced Manufacturing Systems and Technology*, CISM Courses and Lectures No. 372, Springer Verlag, Wien New York, 1996.

some meaningful parameters of the original system [2] and those based on the minimization of an approximation error norm [3].

In the recent literature, remarkable interest has been devoted to the construction of reduced-order models that preserve a set of both first- and second-order information indices of a given original system [4].

The first-order information is usually provided by Markov parameters and/or time moments. In the case of continuous-time systems, the former correspond to coefficients of the asymptotic series expansion of the transfer function, whereas the latter correspond to coefficients of its McLaurin series expansion.

The second-order information is usually provided by entries of the impulse-response Gramian or of the Gram matrix [5]. The "essential" second-order information is supplied by the energies of the impulse response and its derivatives or the transient parts of its integrals, which correspond to the diagonal entries of the mentioned matrices. In fact, all the other entries can be obtained from these and from a corresponding set of first-order indices.

The last methods are computationally simple and lead to stable reduced models of a stable original system. The present authors have been concerned with such techniques [6], [7] using a direct approach according to which the parameters of an input-output description of the reduced model are computed from the first- and second-order information indices to be retained.

The same results can be obtained by minimizing the $L_2$ norm of a weighted equation error, instead of the norm of the output error (which would be more demanding from the computational point of view). This approach is adopted in the following with reference to multi-input multi-output (MIMO) continuous-time reduced models represented by either a left or a right matrix fraction description (LMFD or RMFD, respectively). It is shown that RMFD's are more convenient when the number of inputs is smaller than the number of outputs. Moreover, a greater flexibility of the reduction procedure is achieved by considering suitably weighted equation errors.

## 2. NOTATION AND PROBLEM STATEMENT

The problem considered in this paper is that of determining a reduced model of a given linear, time-invariant, asymptotically stable, strictly proper original system of order $n$ (dimension of a minimal realization) with $m_i$ inputs and $m_o$ outputs. Its (transformed) output vector $\hat{y}(s)$ is related to its (transformed) input vector $\hat{u}(s)$ by the transfer-function matrix $\hat{W}(s)$, i.e.,

$$\hat{y}(s) = \hat{W}(s)\hat{u}(s) \tag{1}$$

where $\hat{W}(s)$ has MacMillan degree $n$.

The corresponding time-domain functions will be denoted by $u(t), y(t)$ and $W(t)$; the last is thus the $m_o \times m_i$ impulse-response matrix. A minimal realization of (1) will be denoted by

$$\dot{x}(t) = Fx(t) + Gu(t), \tag{2}$$

$$y(t) = Hx(t) \tag{3}$$

where $x(t)$ is the state vector of dimension $n$.

The approximant (with the same numbers of inputs and outputs) will be represented by an LFMD as

$$\hat{y}_a(s) = A^{-1}(s)B(s)\hat{u}(s) \tag{4}$$

or by an RFMD as

$$\hat{y}_{\tilde{a}}(s) = \tilde{B}(s)\tilde{A}^{-1}(s)\hat{u}(s) \tag{5}$$

where $A(s), B(s), \tilde{A}(s)$ and $\tilde{B}(s)$ are the following matrix polynomials:

$$A(s) = A_q s^q + A_{q-1}s^{q-1} + \ldots + A_1 s + A_0, \tag{6}$$

$$B(s) = B_{q-1}s^{q-1} + B_{q-2}s^{q-2} + \ldots + B_1 s + B_0, \tag{7}$$

$$\tilde{A}(s) = \tilde{A}_q s^q + \tilde{A}_{q-1}s^{q-1} + \ldots + \tilde{A}_1 s + \tilde{A}_0, \tag{8}$$

$$\tilde{B}(s) = \tilde{B}_{q-1}s^{q-1} + \tilde{B}_{q-2}s^{q-2} + \ldots + \tilde{B}_1 s + \tilde{B}_0, \tag{9}$$

with $A_i \in R^{m_o \times m_o}$, $\tilde{A}_i \in R^{m_i \times m_i}$ and $B_i, \tilde{B}_i \in R^{m_o \times m_i}$.

If the above MFD's are irreducible, then the order of any minimal realization of the reduced model is equal to the degree $m_o q$ of $det\,A(s)$ or $m_i q$ of $det\,\tilde{A}(s)$, respectively [8]. In order to ensure true simplification, the degree $q$ must therefore satisfy the inequality $q < \frac{n}{m_o}$ or the inequality $q < \frac{n}{m_i}$, respectively. As a consequence, the order of the reduced model is a multiple of $m_o$ in the case of LMFD's and a multiple of $m_i$ in the case of RMFD's. These considerations are in favour of LFMD's when $m_o < m_i$ and of RFMD's when $m_i < m_o$.

Under the adopted assumptions, $\hat{W}(s)$ admits both the asymptotic series expansion:

$$\hat{W}(s) = \sum_{i=1}^{\infty} C_{-i}s^{-i} \tag{10}$$

whose coefficients are the so-called Markov parameters, and the MacLaurin series expansion:

$$\hat{W}(s) = \sum_{i=0}^{\infty} C_i s^i. \tag{11}$$

As is known, the Markov parameters $C_{-i}$ are equal to the coefficients of $\frac{t^{i-1}}{(i-1)!}$ in the MacLaurin series expansion of $W(t)$, i.e.,

$$C_{-i} = W^{(i-1)}(0) \tag{12}$$

where the exponent between brackets denotes the order of differentiation.

Instead, the coefficients $C_i$ with nonnegative subscripts are related to the system time moments $M_i$ as follows

$$C_i = \frac{(-1)^i}{i!} M_i , \quad M_i = \int_0^\infty t^i W(t) dt. \tag{13}$$

The reduced model will be derived by minimizing the (squared) $L_2$ norm of the weighted equation error:

$$E_k(s) = \frac{1}{s^k}[A(s)\hat{W}(s) - B(s)] \tag{14}$$

or

$$\tilde{E}_k(s) = \frac{1}{s^k}[\hat{W}(s)\tilde{A}(s) - \tilde{B}(s)]. \tag{15}$$

Obviously, for $k = 0$ the standard equation error is obtained. As is known, the expression "equation error" is due to the fact that the reduced transfer matrix $\hat{W}_a(s)$ satisfies the equations $A(s)X(s) - B(s) = 0$ and $X(s)\tilde{A}(s) - \tilde{B}(s) = 0$ whereas $\hat{W}(s)$ does not, so that $E_0(s)$ and $\tilde{E}_0(s)$ are the deviations from zero of the respective left-hand sides of these equations when $X(s) = \hat{W}(s)$.

More precisely, the indices to be minimized are:

$$J_k = \frac{1}{2\pi} \int_{-\infty}^{\infty} tr E_k(j\omega) E_k^*(j\omega) d\omega \tag{16}$$

or

$$\tilde{J}_k = \frac{1}{2\pi} \int_{-\infty}^{\infty} tr \tilde{E}_k^*(j\omega) \tilde{E}_k(j\omega) d\omega. \tag{17}$$

## 3. MAIN RESULTS

Let us form recursively the sequences of functions:

$$W_{-l}(t) = \int_\infty^t W_{-(l-1)}(\tau) d\tau, \quad l > 0 \tag{18}$$

$$W_l(t) = \frac{dW_{l-1}(t)}{dt}, \quad l > 0 \tag{19}$$

starting from

$$W_0(t) = W(t). \tag{20}$$

The entries of (18) correspond to the transient part of the system responses to the canonical inputs (integrals of the impulse), whereas the entries of (19) are the responses to the derivatives of the impulse.

From the above functions, the following $m_o \times m_o$ matrices of second-order indices are obtained:

$$P_{ij} = \int_0^\infty W_i(t)W_j^T(t)dt = P_{ji}^T. \tag{21}$$

With these building blocks we can form the matrix:

$$P_{[-k:q-k]} = \begin{bmatrix} P_{-k,-k} & \cdots & P_{-k,q-k} \\ \vdots & & \vdots \\ P_{q-k,-k} & \cdots & P_{q-k,q-k} \end{bmatrix} = P_{[-k:q-k]}^T \in R^{m_o(q+1) \times m_o(q+1)} \tag{22}$$

which plays for MIMO systems a role similar to that played by the $q$-th order impulse-response Gramian for single-input single-output (SISO) systems.

Matrix (22) can conveniently be computed as

$$P_{[-k:q-k]} = \mathcal{O}_{[-k:q-k]} \mathcal{W}_c \mathcal{O}_{[-k:q-k]}^T \tag{23}$$

where

$$\mathcal{O}_{[-k:q-k]} = \begin{bmatrix} HF^{-k} \\ \vdots \\ H \\ \vdots \\ HF^{q-k} \end{bmatrix} \tag{24}$$

and $\mathcal{W}_c$ is the controllability Gramian associated with the original system (2)-(3), which is the solution of the Lyapunov equation:

$$FX + XF^T + GG^T = 0. \tag{25}$$

Let us now turn our attention to the minimization of (16).

In order for this index to be finite, error (14) must not exhibit poles at the origin or at infinity, which implies that matrix coefficients $B_i$'s in (7) must satisfy the conditions:

$$B_{k-j} = \sum_{i=0}^{k-j} A_i C_{k-i-j}, \quad j = 1, 2, \ldots, k \tag{26}$$

$$B_{k+j} = \sum_{i=k+1+j}^{q} A_i C_{j-i+k}, \quad j = 0, 1, \ldots, q - k - 1. \tag{27}$$

In this case, by denoting the row block vector of matrix coefficients $A_i$'s in (6) by

$$\underline{A} = [A_0, A_1, \ldots, A_q] \in R^{m_o \times m_o(q+1)} \tag{28}$$

index (16) can be written in the form:

$$tr \underline{A} P_{[-k:q-k]} \underline{A}^T. \tag{29}$$

If (22) is positive definite, quadratic form (29) admits a unique minimum corresponding to the solution of the set of linear equations:

$$\sum_{i=0}^{q} A_i P_{i-k,j} = 0, \quad j = -k, -k+1, \ldots, q-k \tag{30}$$

obtained by setting to zero the derivative of (29) with respect to (28).

Therefore, a (unique) reduced model of form (4) can be derived, e.g., by: (i) choosing $A_q = I$, (ii) solving (30) in terms of the remaining $A_i$'s after discarding the matrix equation for $j = q - k$, and (iii) determining coefficients $B_i$'s according to (26)-(27).

This approximant can be realized by the triple $\{F_a, G_a, H_a\}$, where $F_a$ is the $m_o q \times m_o q$ block companion matrix whose last block row is $[A_0, A_1, \ldots, A_{q-1}]$, $G_a$ is the $m_o q \times m_i$ column block vector whose blocks are (from top) $-C_{k-1}, \ldots, -C_0, C_{-1}, \ldots, C_{-q+k}$, and $H_a$ is the $m_o \times m_o q$ row block vector whose $m_o \times m_o$ blocks are zero matrices except for the $(k+1)$-th block which is an identity matrix.

By assuming that (4) is irreducible, the pair $\{F_a, G_a\}$ is controllable. In this case, it can be proved that the matrix obtained from (22) by deleting its last block row and column is the unique positive definite matrix satisfying the Lyapunov equation $F_a X + X F_a^T + G_a G_a^T = 0$, which implies that the reduced model is asymptotically stable and matches the relevant second-order indices (21) of the original system (besides the corresponding first-order indices $C_i$'s, as shown by (26)-(27)).

A similar procedure can be adopted with reference to the derivation of an RMFD like (5) and to the minimization of index (17).
To this purpose, let us define the following $m_i \times m_i$ matrices of second-order information parameters:

$$\tilde{P}_{ij} = \int_0^\infty W_i^T(t) W_j(t) dt = \tilde{P}_{ji}^T \tag{31}$$

and form the block matrix:

$$\tilde{P}_{[-k:q-k]} = \begin{bmatrix} \tilde{P}_{-k,-k} & \cdots & \tilde{P}_{-k,q-k} \\ \vdots & & \vdots \\ \tilde{P}_{q-k,-k} & \cdots & \tilde{P}_{q-k,q-k} \end{bmatrix} = \tilde{P}_{[-k:q-k]}^T \in R^{m_i(q+1) \times m_i(q+1)} \tag{32}$$

which can be computed as

$$\tilde{P}_{[-k:q-k]} = C_{[-k:q-k]}^T \mathcal{W}_o C_{[-k:q-k]} \tag{33}$$

where

$$\mathcal{C}_{[-k:n-k]} = \begin{bmatrix} F^{-k}G \dots & G & \dots & F^{q-k}G \end{bmatrix} \tag{34}$$

and $\mathcal{W}_o$ is the observability Gramian associated with the original system. Matrix $\mathcal{W}_o$ is the solution of the Lyapunov equation:

$$F^T X + X F + H^T H = 0. \tag{35}$$

In order for index (17) to be finite, it is necessary that:

$$\tilde{B}_{k-j} = \sum_{i=0}^{k-j} C_{k-i-j} \tilde{A}_i, \quad j = 1, 2, \dots, k \tag{36}$$

$$\tilde{B}_{k+j} = \sum_{i=k+1+j}^{q} C_{j-i+k} \tilde{A}_i, \quad j = 0, 1, \dots, q - k - 1. \tag{37}$$

In this case, by denoting the block vector of matrix coefficients $\tilde{A}_i$'s in (8) by

$$\underline{\tilde{A}}^T = [\tilde{A}_0^T, \tilde{A}_1^T, \dots, \tilde{A}_q^T] \in R^{m_i \times m_i(q+1)} \tag{38}$$

index (17) can be written in the form:

$$tr \underline{\tilde{A}} \tilde{P}_{[-k:q-k]} \underline{\tilde{A}}^T. \tag{39}$$

If (32) is positive definite, the unique minimum of (39) is attained for those values of coefficients $\tilde{A}_i$'s satisfying the set of linear equations:

$$\sum_{i=0}^{q} \tilde{P}_{i,j-k} \tilde{A}_j = 0, \quad j = -k, -k+1, \dots, q - k. \tag{40}$$

A reduced model of form (5) is thus obtained by normalizing all coefficients $\tilde{A}_i$'s to one of them, e.g., setting $\tilde{A}_q = I$, by solving (40) for the remaining $\tilde{A}_i$'s after discarding the last matrix equation, and by computing coefficients $\tilde{B}_i$'s according to (36)-(37).

In this case, the reduced model can be realized by the triple $\{\tilde{F}_a, \tilde{G}_a, \tilde{H}_a\}$, where $\tilde{F}_a$ is the $m_i q \times m_i q$ block companion matrix whose last block column consists of (from top) $-\tilde{A}_0, -\tilde{A}_1, \dots, -\tilde{A}_{q-1}$, $\tilde{G}_a^T = [0 \dots I \dots 0]$ with $I$ in $(k+1)$-th position, and $\tilde{H}_a$ is the $m_i \times m_i q$ row block vector: $[-C_{k-1}, \dots, -C_0, C_{-1}, \dots, C_{-q+k}]$.

Again, if the pair $\{\tilde{F}_a, \tilde{H}_a\}$ is observable, the approximant is asymptotically stable and matches the relevant first- and second-order information indices of the original system.

## 4. CONCLUSIONS

Some model reduction techniques considered with interest in the recent literature can be regarded as particular cases of a general procedure, based on the minimization of the weighted equation error norms (16) and (17), from which new effective variants can be derived. The adopted approach refers to matrix fraction descriptions of MIMO continuous-time systems and is characterized by remarkable flexibility, strictly related to the choice of $k$ in (14) and (15).

# References

1. Fortuna L., Nunnari G., and Gallo A.: Model Order Reduction Techniques with Applications in Electrical Engineering, Springer-Verlag, London, 1992.

2. Bultheel A. and Van Barel M.: Padé techniques for model reduction in linear system theory: a survey, J. Comp. Appl. Math., vol. 14, 1986, 401-438.

3. Glover K.: All optimal Hankel-norm approximations of linear multivariable systems and their $L^\infty$−error bounds. Int. J. Control, vol. 39, no. 6, 1984, 1115-1193.

4. de Villemagne C. and Skelton R.E.: Model reductions using a projection formulation, Int. J. Control, vol. 46, no. 6, 1987, 2141-2169.

5. Sreeram V. and Agathoklis P.: On the computation of the Gram matrix in the time domain and its application, IEEE Transactions on Automatic Control, vol. AC-38, no. 9, 1995, 1516-1520.

6. Krajewski W., Lepschy A., and Viaro U.: Reduction of linear continuous-time multivariable systems by matching first- and second-order information, IEEE Transactions on Automatic Control, vol. 39, no. 10, 1994, 2126-2129.

7. Krajewski W., Lepschy A., and Viaro U.: Model reduction by matching Markov parameters, time moments, and impulse-response energies. IEEE Transactions on Automatic Control, vol. 40, no. 5, 1995, 949-953.

8. Kailath T.: Linear Systems, Prentice-Hall, Englewood Cliffs, NJ, 1980.

## 4. CONCLUSIONS

Some model reduction techniques considered with interest in the recent literature can be regarded as particular cases of a general procedure, based on the minimization of the weighted equation error norms (16) and (17), from which new effective variants can be derived. The adopted approach refers to matrix fraction descriptions of MIMO continuous-time systems and is characterized by remarkable flexibility, strictly related to the choice of $k$ in (14) and (15).

# References

1. Fortuna L., Nunnari G., and Gallo A.: Model Order Reduction Techniques with Applications in Electrical Engineering, Springer-Verlag, London, 1992.

2. Bultheel A. and Van Barel M.: Padé techniques for model reduction in linear system theory: a survey, J. Comp. Appl. Math., vol. 14, 1986, 401-438.

3. Glover K.: All optimal Hankel-norm approximations of linear multivariable systems and their $L^\infty$−error bounds. Int. J. Control, vol. 39, no. 6, 1984, 1115-1193.

4. de Villemagne C. and Skelton R.E.: Model reductions using a projection formulation, Int. J. Control, vol. 46, no. 6, 1987, 2141-2169.

5. Sreeram V. and Agathoklis P.: On the computation of the Gram matrix in the time domain and its application, IEEE Transactions on Automatic Control, vol. AC-38, no. 9, 1995, 1516-1520.

6. Krajewski W., Lepschy A., and Viaro U.: Reduction of linear continuous-time multivariable systems by matching first- and second-order information, IEEE Transactions on Automatic Control, vol. 39, no. 10, 1994, 2126-2129.

7. Krajewski W., Lepschy A., and Viaro U.: Model reduction by matching Markov parameters, time moments, and impulse-response energies. IEEE Transactions on Automatic Control, vol. 40, no. 5, 1995, 949-953.

8. Kailath T.: Linear Systems, Prentice-Hall, Englewood Cliffs, NJ, 1980.

# CONTROL OF CONSTRAINED DYNAMIC PRODUCTION NETWORKS

**F. Blanchini and F. Rinaldi**
University of Udine, Udine, Italy

**W. Ukovichk**
University of Trieste, Trieste, Italy

KEY WORDS: Inventory system, Production systems, Feedback control, Uncertain inputs

ABSTRACT: In this paper we consider multi–inventory production systems with control and state constraints affected by unknown demand or supply flows. Unlike most contributions in the literature concerning this class of systems, we cope with uncertainties in an "unknown–but–bounded" fashion, in the sense that each unknown quantity may take any value in an assigned range. For these situations we perform a worst-case analysis. We show that a "smallest worst–case inventory level" exists and it is associated to a steady state control strategy. Then we consider the problem of driving the inventory levels to the smallest worst-case ones. For this problem, we give necessary and sufficient conditions and we show that convergence occurs in a finite number of steps for which an upper bound is found.

## 1. INTRODUCTION AND PROBLEM STATEMENT

Many relevant problems concerning production, transportation and distribution of goods can be addressed by network models (or extended network models) in which nodes represent storage capabilities and arcs (or generalized arcs) represent production or transportation activities whose main goal is to satisfy the demand. If the demand is known on a given time horizon, then the dynamic flow problem can be handled via the well known time-expanded network method (see [1] and [5]). Unfortunately, the demand is often unknown, and this fact has led to use stochastic models to handle

Published in: E. Kuljanic (Ed.) *Advanced Manufacturing Systems and Technology*,
CISM Courses and Lectures No. 372, Springer Verlag, Wien New York, 1996.

problems of this kind. For a review of these models the reader is referred, among the others, to [6], [2] and the references therein. In this paper, uncertainties are modeled in a different way, as it has been done in [3]. Production and demand are assumed to have a known range of allowed values, but no knowledge is given on which allowed value will actually be taken. These unknown–but–bounded specifications for uncertainties are quite realistic in several situations. In general, upper and lower bounds for production yields and demands can be inferred from historical data or from decision makers' experience much more easily and with much more confidence than empirical probability distributions for the same quantities. In other cases, these bounds are explicitly stipulated in supply contracts. In this paper we report, in an abridged form, the result presented in [4] to which the reader is referred for proofs, details and a complete list of reference.

The discrete–time dynamic model that describes our class of systems has the form

$$x(t+1) = x(t) + Bu(t) + Ed(t), \tag{1}$$

where $B$ and $E$ are assigned matrices, $x(t)$ is the system state whose components represent the storage levels in the system warehouses, $u(t)$ is the control, representing controlled resource flows between warehouses, and $d(t)$ is an unknown external signal representing the demand, or more in general non-controllable flows. We assume that the following constraints are assigned. Both storage levels and control components must be nonnegative and upper bounded:

$$x(t) \in X = \{x \in \Re^n : 0 \le x \le x^+\}, \tag{2}$$

$$u(t) \in U = \{u \in \Re^q : 0 \le u \le u^+\}; \tag{3}$$

the external uncontrolled inputs are unknown but each included between known bounds

$$d(t) \in D = \{d \in \Re^m : d^- \le d \le d^+\}, \tag{4}$$

where $x^+$, $u^+$, $d^-$ and $d^+$ are assigned vectors.

The state variables of the system represent the amount of certain resources in their warehouses, each one represented by a node. These resources are raw materials, intermediate and finished products, as well as any other resource used in the production processes. Production processes are represented as "flow units". Formally, a flow unit is an activity which, in unit time, takes amounts $\mu_i$, $i = 1, \ldots, k$, of resources from $k$ source warehouses $(s_1, \ldots, s_k)$ and generates $\nu_j$, $j = 1, \ldots, h$, amounts of products in destination warehouses. Such a flow unit can be associated to an hyper-arc (a subset of nodes each associated to a certain coefficient, see Fig. 1).

Both source and destination nodes may be non homogeneous: for instance, in the source we can consider materials, capitals, auxiliary goods, tools, work force, while in the destination nodes we can have both final products or productions remains.

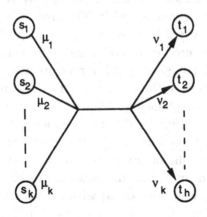

Figure 1: Flow unit

Furthermore, the warehouses can include also abstract objects such as orders. For instance, a production unit can take one purchase order from the order queue of a certain final object and some raw materials from their storage locations, and produce the required object and some production remainders. The model may also consider flow units which enter into (or exit from) the system from (to) the external environment which may represent purchases, supplies or demands.

The flow units may be of two kinds: some of them may be governed by the system manager, and some may depend from external factors. For instance demands and orders typically depend from external factors, while production and transportation operations may be controlled. For this reason, the inputs and outputs of our system are split in two vectors: $u$ the control and $d$ the external demand. The former can be controlled, the latter is exogenously determined. So a controlled flow unit is represented by a column of the matrix $B$ whose coefficients are the numbers $-\mu_i$ and $\nu_j$ described above. Similarly, an uncontrolled flow unit is represented by a column of the matrix $E$. It is natural to assume that both controlled and uncontrolled flows unit have a lower and an upper bound. There is only an interpretation difference between the controlled flow unit constraints, which are system limits, and the bounds on the uncontrolled flow units, which represent the uncertainty margin specifications.

To enlighten our model, we introduce now a simple example that will be reconsidered later. Consider the system represented in Fig. 2, having three warehouses represented by the three nodes 1, 2 and 3 denoted by $A$, $B$ and $AB$, respectively. Nodes 1 and 2 contain two goods A and B. The arcs $u_1$, and $u_2$ with the constraints $0 \leq u_1 \leq \bar{u}_1^+$ (the reason why a bar has been put on $\bar{u}_1^+$ will be explained soon) and $0 \leq u_2 \leq u_2^+$ represent the productions of A and B, respectively, in a unit time. The arc $u_3$, with the constraint $0 \leq u_3 \leq u_3^+$ represents a flow units which takes some amount of A and B to produce the same amount of AB in node 3. The presence of the arc $u_4$, $0 \leq u_4 \leq u_4^+$ is explained as follows. The resources A and B are provided separately by two production lines. The line producing A has a capacity of $u_1^+$, and the line producing B has a capacity of $u_2^+$. Moreover, there is an additional flexible capacity $u_4^+$ which can be split in any proportion between the two production lines. This situation is modelled by adding this flexible capacity to the line $u_1$, to obtain $0 \leq u_1 \leq u_1^+ + u_4^+ \doteq \bar{u}_1^+$, while the arc $u_4$ with capacity $u_4^+$ (the same added to $u_1$) represents a re–distribution. If the arc $u_4$ works at full force, this means that the flexible capacity is employed all to produce B, while if it works at 0 force, this flexible capacity is all used to produce A. The arcs $d_1$, $d_2$ and $d_3$ represent demands of A, B and AB. Again we have re–distribution arcs $d_4$ and $d_5$ which represent certain amounts of demand which can unpredictably require A or AB, and B or AB, respectively. These demands must be provided of proper limits $d_i^- \leq d_i \leq d_i^+$, derived by the available sale projections. The matrices $B$ and $E$ for this system are clearly

$$B = \begin{bmatrix} 1 & 0 & -1 & -1 \\ 0 & 1 & -1 & 1 \\ 0 & 0 & 1 & 0 \end{bmatrix}$$

and

$$E = \begin{bmatrix} -1 & 0 & 0 & -1 & 0 \\ 0 & -1 & 0 & 0 & -1 \\ 0 & 0 & -1 & 1 & 1 \end{bmatrix}$$

In this setting, a game between two players $\mathcal{P}$ and $\mathcal{Q}$ is considered. At each time, player $\mathcal{P}$ decides the flow $u(t)$ and player $\mathcal{Q}$ decides the flow $d(t)$. The information about $X$, $U$ and $D$ is known to each player. The game is the following. For a certain initial distribution $x(0)$ of the commodity within the nodes at time $t = 0$, player $\mathcal{P}$ first chooses a flow distribution $u(0)$ according to (3), and then player $\mathcal{Q}$ chooses a flow distribution $d(0)$ according to (4). These moves produce a new distribution $x(1)$ of the commodity according to (1). Then the two players choose new flows $u(1)$ and $d(1)$ in their feasible ranges in order to produce $x(2)$, and so on. The basic problem considered in this paper is that of finding a feasible initial condition set and a feasible feedback control for player $\mathcal{P}$, that is a function of the form $u(t) = \Phi(x(t), t)$, such that the constraints (2)-(4) are always fulfilled.

**Definition 1** *Given the constraints* (2), (3) *and* (4),*we say that* $\Phi : X \times \mathcal{N} \to U$ *is an admissible control strategy, and that* $X_0 \subseteq X$ *is an admissible initial condition set,*

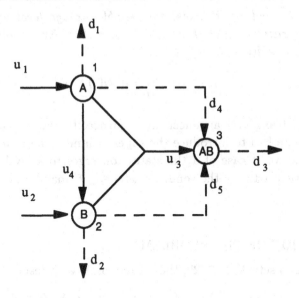

Figure 2: The system structure

*if for all $x(0) \in X_0$ and for all $d(t) \in D$, $t \geq 0$, the sequences $x(t)$ and $u(t)$ produced by (1) when $u(t) = \Phi(x(t), t)$ are always feasible, in the sense that $u(t) \in U$ and $x(t) \in X$.*

It is clear that the permanence of the state $x(t)$ and the control $u(t)$ in their allowable ranges is a a basic condition for the system. Our next problem is to choose, among all the admissible controls, the one which minimizes the storage amount, i.e. the work–in–progress (WIP). To this aim, we introduce the following definition.

**Definition 2** *The vector $\chi \leq x^+$ is a feasible storage level (FSL) if there exists an admissible control $\Phi$ such that*

$$x(t) \leq \chi \quad \text{for all } t \geq 0 \tag{5}$$

*for some initial condition $x(0) \in X_0$ and for all $d(t)$ as in (4).*

Clearly, the existence of an admissible strategy is equivalent to the existence of the FSL $\chi = x^+$. However, as is well known, one of the main goals in production control is keeping the stored resources at the lowest level. Thus a cost of the form

$$J = c^T \chi$$

is introduced for the FSL where $c \in R^n$ is an assigned vector with positive components. We assume now that the largest admissible initial condition set $X_0$ is given and we consider the problem of finding the optimal FSL, and driving the state below the optimal FSL.

**Problem 1** *Find (if it exists) the feasible storage level $\chi^{opt}$ of minimum cost and find (if it exists) a feedback control $\Phi$ such that, for all admissible initial conditions $x(0) \in X_0$, and $d(t) \in D$ (cf. 4)*

$$\limsup_{t \to \infty} x_i(t) \le \chi_i^{opt} \tag{6}$$

Clearly, being $d(t)$ an uncertain sequence, the state evolution $x(t)$ may have no limit, thus we had to introduce the superior limit. The concept above roughly means that, in the worst case of $d$ the states converges to a level which is below $\chi^{opt}$. We will show that actually the condition $x(t) \le \chi^{opt}$ may be assured in a finite number of steps.

## 2: SOLUTION OF THE PROBLEM

Given two sets $X, S \subset \Re^n$, the *erosion* of $X$ with respect to $S$ is defined as

$$X_S = \{ x \in \Re^n : x + s \in X \; \forall s \in S \}. \tag{7}$$

As a first step, we provide conditions for the existence of an admissible strategy.

**Theorem 1** *[3] [4] There exist an admissible control and an admissible initial condition set if and only if the following two conditions are satisfied:*

$$X_{ED} \ne \emptyset, \tag{8}$$

$$ED \subseteq -BU. \tag{9}$$

*Moreover, the largest initial condition set (i.e. the set of all the initial conditions) for which the game is favorable to player $\mathcal{P}$ is given by*

$$X_0 = (X_{ED} - BU) \cap X, \tag{10}$$

*and any function $\Phi(x)$ such that*

$$\Phi(x) \in U \tag{11}$$

*and*

$$x + B\Phi(x) \in X_{ED} \qquad \text{for all } x \in X_0, \; t \ge 0, \tag{12}$$

*is an admissible control.*

Now we are able to provide an expression for the optimal storage level $\chi^{opt}$. Then we give necessary and sufficient conditions for the existence of a control $\Phi$ that drives the state to the optimal level and we show that convergence occurs in a finite number of steps. For convenience, for $a, b \in R^n$, we denote the following parameterized hyper-box $X$

$$X(a, b) \doteq \{x \in R^n : \; a \le x \le b\}. \tag{13}$$

Furthermore we define the vectors $\delta^-$ and $\delta^+$ in $R^n$ whose components are

$$\delta_i^- \doteq \min_{d\in D} E_i d, \quad \text{and} \quad \delta_i^+ \doteq \max_{d\in D} E_i d. \tag{14}$$

where $E_i$ denotes the $i$th row of $E$. It turns out that

$$X(a,b)_{ED} = \{x: \quad a - \delta^- \le x \le b - \delta^+\}.$$

With these notations we have the following [4].

**Theorem 2** *For any vector cost $c > 0$, the minimum cost FSL is given by*

$$\chi^{opt} = \delta^+ - \delta^-. \tag{15}$$

The Theorem above points out the following property. The set $X(0, \chi^{opt})$ is independent on $c > 0$. Furthermore its erosion is reduced to a singleton

$$X(0, \chi^{opt})_{ED} = \{\bar{x}\} \quad \text{where} \quad \bar{x} \doteq -\delta^-. \tag{16}$$

The vector $\bar{x}$ characterizes what we call the *central strategy*. In other words, if $\chi = \chi^{opt}$, then the control is obtained by pointing to the *central state* $\bar{x}$, in the sense that $u(x)$ is selected in such a way

$$u = u(x): \quad x + Bu = \bar{x}. \tag{17}$$

Selecting the control in this way is a *both necessary and sufficient condition* to keep (in the worst case sense) the state below the minimal level, say in $X(0, \chi^{opt})$. Let us now consider the problem of the convergence of the storage amounts to the minimal levels. In order to exclude trivial cases, in the following we assume

**Assumption 1** $X^{opt} \doteq X(0, \chi^{opt}) \neq X$.

If $X^{opt} = X$ then the problem of the convergence does not arise and the only feasible strategy is the central one. Under Assumption 1, there exist admissible initial states which are not in $X(0, \chi^{opt})$. This also implies that there exits $x^0 \in X_0$ such that $x_0 \notin X^{opt}$ (see [4]). Define the vector $\theta$ as

$$\theta \doteq x^+ - \delta^+ + \delta^- \ge 0. \tag{18}$$

Under Assumption 1, the vector $\theta \ge 0$ has at least one positive component. The next theorem gives a necessary and sufficient condition for the existence of a control law which drives the states inside $X^{opt} = X(0, \chi^{opt})$ from all admissible initial conditions.

**Theorem 3** *[4] There exists an admissible control strategy such that condition (6) holds for every initial state $x(0) \in X_0$ and every sequence $d(t)$ as in (4), if and only if the conditions of Theorem 1 hold, and there exists $\epsilon > 0$ such that*

$$ED + \epsilon Z \subset -BU, \quad \text{where} \quad Z \doteq X(0, \theta) = \{x: \quad 0 \le x \le x^+ - \delta^+ + \delta^-\}. \tag{19}$$

To introduce a control law which assures convergence to the least storage level we need to consider the set

$$\Omega(x) = \quad \text{the set of all the solutions of} \quad \begin{cases} \min \lambda : \\ \bar{x} \leq x + Bu \leq \bar{x} + \lambda\theta, \quad u \in U. \end{cases} \tag{20}$$

Then any function $\Phi : R^n \rightarrow R^q$ such that

$$\Phi(x) \in \Omega(x). \tag{21}$$

assures the required convergence. Moreover, denoting by $\lceil \mu \rceil$ the smallest integer greater or equal $\mu \in R^n$, if a control law of this form is applied, then for all initial conditions $x(0) \in X_0$ at most

$$T = 1 + \lceil \frac{1}{\epsilon} \rceil,$$

steps are needed to drive $x(t)$ in $X^{opt}$ and keep it inside for $t \geq T$. The control implementation requires solving on-line the linear programming (LP) problem (20).

## 3: EXAMPLE

Let us consider the example in Section 1, with the following data

$$x^+ = [\; 130 \quad 120 \quad 150 \;]^T,$$

$$u^+ = [\; 170 \quad 50 \quad 100 \quad 70 \;]^T,$$

$$d^- = [\; 15 \quad 20 \quad 60 \quad 0 \quad 0 \;]^T,$$

$$d^+ = [\; 25 \quad 30 \quad 80 \quad 20 \quad 10 \;]^T,$$

The vectors $\delta^+$ and $\delta^-$ are immediately computed to be

$$\delta^+ = [\; -15 \quad -20 \quad -30 \;]^T, \quad \delta^- = [\; -45 \quad -40 \quad -80 \;]^T,$$

so that $\chi^{opt}$ for this system is readily computed to be

$$\chi^{opt} = \delta^+ - \delta^- = [\; 30 \quad 20 \quad 50 \;]^T.$$

The corresponding vector $\theta$ is

$$\theta = x^+ - \chi^{opt} = [\; 100 \quad 100 \quad 100 \;]^T.$$

The largest $\epsilon$ such that $ED + \epsilon Z \subset BU$ with $Z = X(0, \theta)$ is $\epsilon = 0.3$. Now $\lceil \frac{1}{\epsilon} \rceil = 4$, thus at most 5 steps are necessary for convergence. For all $x(0) \in X_0$, the state evolution is bounded by the following inequalities if the strategy in (21) is applied

$$\begin{bmatrix} 0 \\ 0 \\ 0 \end{bmatrix} \leq \begin{bmatrix} x_1(t) \\ x_2(t) \\ x_3(t) \end{bmatrix} \leq \begin{bmatrix} 130 - t25 \\ 120 - t25 \\ 150 - t25 \end{bmatrix}, \quad t = 1, 2, 3, 4. \tag{22}$$

For $t \geq 5$, we have the "steady–state" solution which fulfills the conditions $0 \leq x(t) \leq \chi^{opt}$. The set $X_0$ for this system is defined by the set of inequalities

$$
\begin{array}{rcccccl}
0 & \leq & x_1 & & & \leq & 130 \\
0 & \leq & & x_2 & & \leq & 120 \\
0 & \leq & & & x_3 & \leq & 150 \\
-\infty & \leq & & x_2 & + \; x_3 & \leq & 260 \\
-12.50 & \leq & 0.5x_1 & 0.5x_2 & + \; x_3 & \leq & 262.5
\end{array}
\tag{23}
$$

Finally the control law is obtained by solving for each $x$ the problem in (20). Note that this problem is an LP problem in 4 variables and 7 two–side constraints. Problems of this kind are solved almost immediately, even by a personal computer.

## 4. CONCLUSIONS

We have considered the problem of driving the storage amount of a production-distribution system under the minimal feasible level in the presence of uncertain demand, whose bounds are known. A control strategy which assures convergence in the worst case sense has been proposed. The result described in this paper can be extended in several directions. For instance, the problem in which failures may occur in the system and the cases in which some resources admit integer values only can be considered. The reader is referred to [4] to have more details on these further developments.

Acknowledgment. Work supported by C.N.R. under grant 94.00543.CT11.

# References

[1] J. E. ARONSON, A survey of dynamic network flows, *Annals of Operations Research*, Vol. 20 (1989), pp. 1–66.

[2] D. P. BERTSEKAS, *Dynamic Programming and Optimal Control*, Athena Scientific, Belmont, Massachusetts, 1995.

[3] F. BLANCHINI, F. RINALDI AND W. UKOVICH, A Network Design Problem for a Distribution System with Uncertain Demands, *SIAM J. on Optimization, to appear*

[4] F. BLANCHINI, F. RINALDI, W. UKOVICH, Least storage control of multi–inventory systems with non-stochastic unknown inputs", *Submitted*.

[5] J. B. ORLIN, Minimum convex cost dynamic network flow, *Mathematics of Operations Research*, Vol. 9 (1984), pp. 190–207.

[6] E.L. PORTEUS: Stochastic Inventory Theory, in: D.P. HEYMAN, M.J. SOBEL, EDS. *Handbooks in Operations Research and Management Science, Vol. 2: Stochastic Models*, NORTH–HOLLAND, 1990.

# A COORDINATE MEASURING MACHINE APPROACH
# TO EVALUATE PROCESS CAPABILITIES
# OF RAPID PROTOTYPING TECHNIQUES

**R. Meneghello, L. De Chiffre and A. Sacilotto**
**University of Padua, Padua, Italy**

KEYWORDS: Process Capabilities, Coordinate Measuring Machine, Rapid Prototyping, Free-Form Surfaces.

ABSTRACTS: A CMM measuring approach is presented which is specifically aimed at evaluating process capabilities of Rapid Prototyping techniques and generally suitable for Free-Form Fabrication technologies. This approach is based on software and hardware facilities available on advanced CMMs that include continuous scanning and evaluation of normal deviations from the CAD model used as the reference object. In this work the method is applied to components manufactured using LOM and SLS techniques of Rapid Prototyping. The paper focuses on the procedure of evaluating the natural tolerance and the dimensional variability of the process, analysing systematic distortions and form errors by the use of a common geometrical parameter.

## 1. INTRODUCTION

Evaluation of the process capabilities of a new technology is generally based on the assessment of the geometrical accuracy achieved by the process. The geometric quality shall be duly referred to linear, form and position tolerances as well as to the surface roughness of industrial components.

With regard to the Rapid Prototyping (RP), real advantages of the technique consist of almost no limitations to the complexity of the part as well as of reduced limits to the least attainable thickness and to the form of surfaces [3].

In process capabilities studies on RP techniques, different approaches have been proposed including the conventional benchmark by the 3D System [1- 2], the user part presented by

Published in: E. Kuljanic (Ed.) *Advanced Manufacturing Systems and Technology*,
CISM Courses and Lectures No. 372, Springer Verlag, Wien New York, 1996.

Kruth [3], by Lart [4], the component proposed by Schmidt [5] and the model by Childs [6].

However, no analysis has been carried out to qualify the RP-prototype in terms of free-form surfaces and the geometric feature complexity of real components [5 - 6].

Difficulties in establishing the accuracy of free-form surfaces with coordinate measuring machines (CMM) can be summarised in:

a) no single benchmark can present all geometric features of "real parts" (thin walls, flat surfaces, holes, etc.);

b) a measuring strategy shall be defined in order to assure the repeatability of the measuring conditions if a number of real parts, representative for the current production, is selected.

In this paper the authors propose a method, a measuring strategy and a data analysis, as a base for the design of a comparative benchmark to evaluate the dimensional quality of a prototype without any limitations of form, dimension and complexity. The universality of the principle, based on a single parameter to express geometrical accuracy, and the availability of software and hardware facilities should promote process capability studies of any Free-Form technology.

## 2. METHOD

The measurement principle is based on comparing an actual point probed on the surface of the prototype with the corresponding nominal point belonging to the reference model surface. Such a measurement should be performed independently from position, extension and curvature of the surface in order to assure the minimum uncertainty of results and to compare results of different geometrical features.

The characteristics of the method have been identified as follows:

I) the physical reference model shall be substituted by a numerical model of the surface to be measured,

II) the direction of probing any surface feature must be normal to the nominal surface,

III) the evaluation of deviation is to be performed by calculating the orthogonal distance (normal deviation) between actual and nominal points.

Thus no substitute element has to be defined and fitted into the probed points.

Hardware and software facilities based on a Zeiss UMM550 coordinate measuring machine equipped with the HOLOS® software package have made such a method now available [7].

The CAD model is imported via VDA protocol in a graphic interactive user interface, HOLOS, from where CNC-controlled measurements can be automatically defined. Single-point measurements as well as continuous scanning are performed.

Single-point measurements are defined by grid points, $P^i_{CAD}$ (see fig.1), on the surface of the CAD model. Spatial coordinates and normal vectors are then established. Measuring paths are automatically defined to probe the points with respect to the normal vector directions.

After probing, the normal deviation ($\Delta$) between nominal $P_{CAD}$ and estimated $P_M$ (see figs. 2, 3) is evaluated as:

$$\Delta = \left[ \cos X \cdot \left( X_M - X_{CAD} \right) + \cos Y \cdot \left( Y_M - Y_{CAD} \right) + \cos Z \cdot \left( Z_M - Z_{CAD} \right) \right] \qquad (1)$$

where cos X, cos Y, cos Z are the direction cosines of the normal vector in $P_{CAD}$ ($X_{CAD}$, $Y_{CAD}$, $Z_{CAD}$).

A positive value of this parameter has to be interpreted as exceeding material over the nominal surface.

The probed point $P_T$ does not coincide with the target point $P_R$, corresponding to $P_{CAD}$ in the actual surface, or with the estimated point $P_M$ (see fig.3). The unknown error $\varepsilon_\Delta$ can be assumed as a function of: i) the curvature of the actual surface, ii) the angle between the normal vector in $P_R$ and the direction of probing, iii) the radius of the probe sphere and iv) the friction between the probe sphere and the workpiece surface.

The error $\varepsilon_\Delta$ is of great importance in free-form surface measurement as factors i), ii) and iv) vary continuously in any point probed, increasing the uncertainty of the result and thus reducing the possibility to compare results relevant to different prototypes as for shape and complexity [8].

The normal probing represents the condition to minimise and control $\varepsilon_\Delta$ as it allows to limit the variability of factors i), ii) and iv).

## 3. PROCEDURE

The measurement strategy is shown in figures 4, 5 and 6.

The first phase (see fig.4) is aimed at selecting surfaces on the prototype. Here a surface is established by one or more surface features as defined in the ISO 5459 [9].

Surfaces are classified in three categories in terms of curvature, spatial alignment, normal deviation and complexity degree: class A, class B and class C. Class A and B surfaces are used to align the CNC coordinate system to the reference CAD system ($\Sigma_{CAD}$) via a 3D best fit of probed points (see fig.5). Depending on the size of the prototype, the number of probings (single-point) is fixed within the range 500-1000 to limit the required measuring time to 3 hours; the probings percentage assigned to class A and B surfaces is approximately 50%.

As a results, form and position errors, generally superimposed, are minimised by independently setting rotational and/or translational degrees of freedom, during best fitting.

In the second phase all surfaces are measured. Single-point probing is generally preferred as to assure repeatability of results, when complex free-form surfaces are to be measured, as to limit processing time. Continuous scanning is applied when the waviness of the surface is investigated to assess its periodicity and amplitude (see fig.7).

A preliminary analysis of data is carried out by HOLOS via numerical and graphical representation (see figs.10, 11, 12). The complete processing of data is performed in a PC spread-sheet module.

## 4. NORMAL DEVIATIONS ANALYSIS

The analysis, based on normal deviations of probed points, is aimed at assessing the statistical accuracy capability of the process. Free curvature of surfaces would extend the knowledge of in-plane (2D) and 3D-regular form accuracy capability.[1-5]

Parameters Δ Mean, Δ Outer and Δ Inner based on normal deviation (Δ) are graphically described in figure 8 for a flat surface. When relating to free surfaces the parameters follow the definitions in [10].

Measurement results are summarised in table 1 and in figure 9.

| SURFACE | NAME | TYPE | Δ Mean | Δ Out | Δ Inn | DEV.st | Nr points |
|---------|------|------|--------|-------|-------|--------|-----------|
| S1 | S1-ALTO | B | -0.410 | -0.265 | -0.628 | 0.086 | 117 |
| S2 | S2-FLANGIA | A | 0.000 | 0.083 | -0.072 | 0.029 | 306 |
| S3 | S3-FLANGIA-inn | B | 0.096 | 0.177 | 0.027 | 0.035 | 100 |
| S5 | S5-BOCCOLA | A | -0.023 | 0.130 | -0.182 | 0.065 | 120 |
| S6 | S6-RETRO | C | -0.839 | -0.737 | -0.976 | 0.066 | 65 |
| S7 | S7-MENSOLA-bordo | C | 0.067 | 0.309 | -0.237 | 0.147 | 168 |
| S8 | S8-FONDO | C | -0.237 | -0.140 | -0.360 | 0.051 | 78 |
| S9 | S9-MENSOLA-int | C | 0.164 | 0.253 | 0.091 | 0.035 | 96 |
| S10 | S10-MENSOLA-est | C | -0.070 | 0.094 | -0.158 | 0.050 | 96 |
| S11 | S11-FRONTE | A | 0.063 | 0.190 | -0.185 | 0.129 | 27 |
| S12 | S12-INNESTO | C | -0.037 | 0.077 | -0.129 | 0.051 | 32 |
| General | PROTOTYPE | | -0.078 | 0.309 | -0.976 | 0.248 | 1205 |

Table 1. Normal deviations of surfaces and prototype [mm].

## 5. DIMENSIONAL ANALYSIS

The dimensional characterisation is aimed at evaluating process capabilities of RP techniques with respect to the dimensional accuracy.

A scale factor is proposed as dimensional accuracy index, defined as:

$$\text{SCALE FACTOR} = \frac{\text{Mean actual dimension}}{\text{Nominal dimension}}$$

This parameter is calculated for each of the three principal directions (X, Y and Z) measuring opposite both regular and free-form surfaces. The mean normal deviation (Δ) is projected on the axis directions and added to the nominal coordinate of the grid point centre or of other reference point. The operation is performed in both opposed surfaces and the actual dimension is then evaluated.

| Direction | Surf. | Nominal dim. | Mean Ac.dim | Mean error | Outer error | Inner error | Scale Fact. |
|-----------|-------|--------------|-------------|------------|-------------|-------------|-------------|
| X | S2, S3 | 35 | 35.096 | 0.096 | 0.261 | -0.045 | 1.003 |
| Y** | S11, S6 | 86.199 | 85.452 | -0.748 | -0.522 | -1.126 | 0.991 |
| Z** | S1, S8 | 65.542 | 64.897 | -0.645 | -0.404 | -0.984 | 0.990 |
| Sx | S9, S10 | 5 | 5.094 | 0.094 | 0.347 | -0.067 | 1.019 |

Table 2. Dimensional deviation and scale factors (**=nominal surfaces are not parallel). Sx points out the dimensional value of a thin wall thickness, normal to the X direction. [mm]

In this way, the scale factors describe the mean percentage variation of dimensions along the axis ensuring a repeatable evaluation of the process anisotropy (see fig.10). Table 2 outlines a summary of results related to scale factors.

## 6. FORM ANALYSIS

The third phase investigates the process capabilities in the reproduction of free and regular geometric forms. The results of this study are shown in table 3. Furthermore, a continuous scanning approach is used to evaluate systematic errors (a waviness effect in fig.7; a shrinkage effect in fig.10; a bending effect in fig.11), which are previously underlined by single-point measuring.

|  | Surface | Form error |
|---|---|---|
| S1 | S1-ALTO | 0.363 |
| S2 | S2-FLANGIA | 0.155 |
| S3 | S3-FLANGIA-inn | 0.150 |
| S5 | S5-BOCCOLA | 0.312 |
| S6 | S6-RETRO | 0.238 |
| S7 | S7-MENSOLA-bordo | 0.546 |
| S8 | S8-FONDO | 0.219 |
| S9 | S9-MENSOLA-int | 0.162 |
| S10 | S10-MENSOLA-est | 0.252 |
| S11 | S11-FRONTE | 0.375 |
| S12 | S12-INNESTO | 0.206 |

Table 3. Form error of the protoype surfaces [mm].

## 7. CONCLUSIONS

A principle for the assessment of RP geometric accuracy has been proposed and tested by using a CAD measuring interface, HOLOS, with a coordinate measuring machine.

The principle is based on: i) the CAD model used as the reference object, ii) the direction of probing normal to the surface, iii) the orthogonal distance (normal deviation $\Delta$) as the parameter to be investigated. This parameter is particularly suited for the measurement of free-form surfaces as it does not require any substitute element to be mathematically defined and fitted into probed points.

A measuring strategy has been proposed which may help the RP techniques researchers to perform a selection of surfaces of the prototype, an accurate alignment and a complete evaluation of results. By combining normal deviations and nominal points coordinates in the CAD model it is possible to determine the natural tolerance of the process as well as dimensional tolerances along axis direction and free-form surfaces errors as specified in the standards.

With regard to RP, the method should help researchers to extend the range of current investigations from an in-plane (2D) and 3D-regular form accuracy estimation to a complete 3D description of any industrial prototype.

## REFERENCES

1. Jacobs, P.F.: Fundamentals of StereoLitography, SME, 1992, 306-315.
2. Gargiulo, E.P.: Stereolithography Process Accuracy, Proceedings of the 1st European Conference on Rapid Prototyping, Nottingham, 1992, 187-201.
3. Kruth, J. P.: Material Incress Manufacturing by Rapid Prototyping Techniques, Annals CIRP 40/2, 1991, 603-614.
4. Lart, G.: Comparison of Rapid Prototyping System, Proc. of the 1st European Conference on Rapid Prototyping, Nottingham, 1992, 243-254.
5. Schmidt, L. D.: Rapid Prototyping Technology benchmarking results, RP Technology Clinic, Michigan, 1992.
6. Childs, T.H.C. et alt.: Linear and Geometric Accuracies from Layer Manufacturing, Annals CIRP 43/1, 1994, 163-166.

7. Breyer, K.-H. et alt.: Holometrische Koordinatenmesstechnik - Neue Ansatze zur umfassenden Messung beliebiger Formelemente, VDI Berichte 1006, 1992.

8. Frowis, R., Porta C. et alt.: Praktischer und theoretischer Nachweis der Richtigkeit von Freiformflachenmessungen auf Koordinatenmessgeraten, VDI Berichte 1258, 1996.

9. ISO 5459 - 1981: Technical drawings - Geometrical tolerancing. Datums and datum-systems for geometrical tolerances.

10. ISO 1101 - 1983: Technical drawings - Geometrical tolerancing.

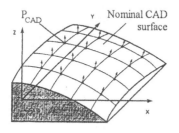

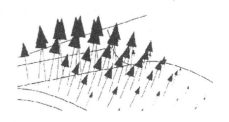

Figure 1. Representation of grids points with associated normal vectors.

Figure 2. Vectors representing the estimated normal deviations after measurements.

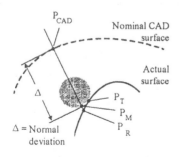

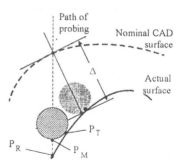

Figure 3. Representation of $P_{CAD}$, target ($P_R$), probed ($P_T$) and measured point ($P_M$) in normal path (left) and in workpiece related (right) probing.

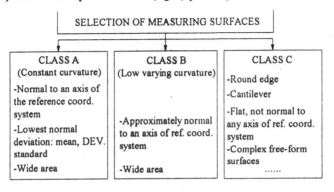

Figure 4. Surfaces classification.

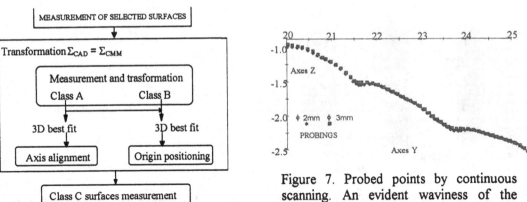

Figure 5. Measurement procedure.

Figure 7. Probed points by continuous scanning. An evident waviness of the surface is represented. Two different probe spheres are used: $\Phi$ 2mm and 3mm. [mm]

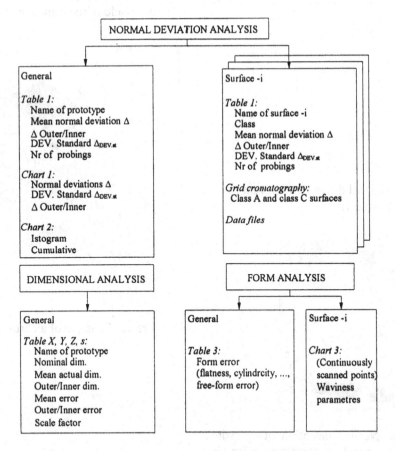

Figure 6. Analysis of data and classifications of geometrical errors

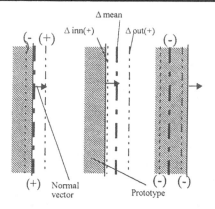

Figure 8. Description of positive and negative values for the normal deviations. Case of flat surface.

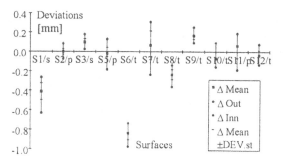

Figure 9. Normal deviations relevant to surfaces in table 1.

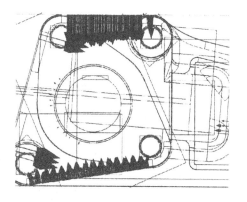

Figure 10. Normal deviation of opposite surfaces (Z axes): a shrinkage effect is evident.

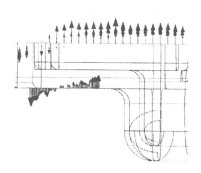

Figure 11. Evident bending of a cantilever -like surface. Single-point probings on the top-side and continuous scanning on the bottom-side of the cantilever.

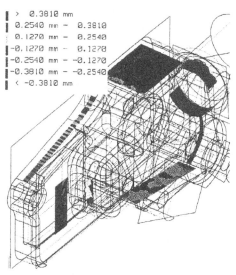

Figure 12. Example of a chromatography.

# AN INTEGRATED APPROACH TO ERROR DETECTION AND CORRECTION ON THE CURVES AND SURFACES OF MACHINED PARTS

**A. Sahay**

**School of Business and Technology, Salt Lake City, UT, U.S.A.**

KEY WORDS: Error Analysis, CAD/CAM, Curve/Surface Design, Point Correspondence, Transformation, Parametric Approximation.

ABSTRACT: This research has been able to integrate the design, manufacturing, and measurement processes, thus providing an integrated approach to error detection and correction problem. Based on the results of these studies the research suggests new methods to reduce such errors by creating a better basis for the machining specifications. The outcome of the research has provided new methods for improvements in the accuracy of machined parts. Mathematical models are developed to identify the errors. Since the curved boundaries do not have distinct feature points, the errors are determined through establishing point correspondence between the designed shape and manufactured shape. Both the designed shape and the manufactured shape has point representation. A transformation algorithm is used to transform the measured points over the design points. This determines the error. An optimal approximation curve is fitted through the measured and matched points. Through successive approximation a new set of points is obtained. This set can be used to revise the part program (created earlier to machine the part based on the design data) such that the successive parts can be machined within specification and with minimum of errors.

## 1.INTRODUCTION

Error analysis is broken down in deterministic errors and random errors. Using a somewhat simplified definition, deterministic errors are those to which a definite cause may be assigned,

Published in: E. Kuljanic (Ed.) *Advanced Manufacturing Systems and Technology*, CISM Courses and Lectures No. 372, Springer Verlag, Wien New York, 1996.

and, in most instances, reduced or limited by controlling the cause. In contrast the random errors have no assignable cause and therefore would seem to be beyond control.

The major objectives of most of the research in the area of error analysis is to (i) determine the machining errors on a given machine that cause significant errors in the parts, and (ii) to determine the accuracy improvements that can be expected by correcting or minimizing these machining errors.

Error correction technique are divided into two categories based on the error identification schemes used. These are precalibrated error compensation and active error compensation [1]. In the first case, the errors are determined or measured before or after the machining process. The known errors are then compensated for and corrected during subsequent operations.The other error correction technique is known as active error compensation. In this technique,the determination of errors and their compensation are done simultaneously, with machining. In-process gauging is used for the detection of errors. Mathematical modeling is then used to establish a relationship between errors and sources. The active error compensation approach has not been widely used because of its complexity.

This research uses mathematical models to detect the errors but does not use in-process gauging.All the current methods to improve the accuracy through the error compensation control one or two dominant deterministic parts of the error for a specific type of machine.

## 2. LITERATURE REVIEW

Researchers have applied a number of different methods and approaches to the error compensation problem . The basic principles and the approach of some of these studies are the same.The methods used for error correction include statistical method [2], methods based on random process analysis [3], forecasting techniques [1], computer-aided accuracy control [4], methods based on mathematical modeling[5]. In recent years, several studies have been done relating to the error correction problem in coordinate measurement machines. Some of these techniques, for example [6] have been used by other researchers to study and analyze the error problems in machine tools.

## 3. THE PROPOSED METHOD

The central concept proposed here is that it is possible and desirable to change the machining specifications so that the parts can be machined within specification and with minimum of errors. This is done by analyzing the differences between the design specifications and the corresponding actual measurements and creating a new set of machining specifications.

In this research, the total error introduced in machined parts are determined using mathematical models. The determination of error consists of several steps. It involves: (1) designing the parts involving curves and surfaces using appropriate curve and surface designing techniques, (2) developing the part programs to machine the designed parts,(3) measuring the part in the coordinate measurement machine and extracting the measured part dimensions, (4) relating design points to measured points by establishing the point correspondence between the design set and measured set of data and transforming the measured set over the design set, and finally,

(5) determining the error vectors whose components are the differences between the specified part dimensions and the dimensions obtained after machining.

The errors determined using the approach discussed above will consist of both systematic and random errors. Once the errors are determined, this research focuses on finding a better machining specification by fitting an optimal approximation curve which also minimizes the random errors to some extent. For this, a model is suggested and discussed. Some parts were machined and the data were used to test the working of the models and the computer programs.

## 4. MODELS AND ALGORITHMS

This section provides a discussion on the models and algorithms used to solve the error detection and correction problem. The models and algorithms include:
(1) Curve and surface design algorithms
(2) Part programs for machining the parts
(3) Error determination model in co-ordinate measurement machine (CMM)
(4) Algorithms to determine the point correspondence between the design points and measured points
(5) An algorithm to calculate the errors based on the transformation of the measured point set over the design point set
(6) A hypothesis test procedure to test the null hypothesis that the design points and the corresponding measured points are the same having no significant error against the alternate hypothesis, that the design points and corresponding measured points are different having significant errors, and
(7) An algorithm to revise the specification using the least squares parametric approximation.

The sequence of the above models and algorithms is shown in Figure 1. A brief explanation of the models is also presented.

## 4.1 PART DESIGN

For designing the part, the technique of Bezier curve and surface has been used because of its many advantages in design. Also, the computations involved are less complex compared to the B-spline curves and surfaces which have wide applications in design and manufacturing. Bezier curves and surfaces are represented using parametric form [7]. Computer programs are written to design two-dimensional curves, space curve, and surfaces. The program for the surface also calculates the offset points that are need to machine the parts.

## 4.2 MACHINING, MEASUREMENT AND MEASUREMENT ERROR

Using the design data, part programs are developed to machine the specified shape. In the procedure recommended here, parts are then measured using a coordinate measurement machine (CMM). The measurement of the part on the measurement machine (coordinate measurement machine) requires that the measurement machine be error free, that is, there are no inherent systematic errors present in the machine. The systematic errors in the measurement machine is determined through a model. A rigid body assumption for the components of coordinate measurement machine (CMM) is used. This assumption means that the errors of the

X,Y, and Z carriages depend on their individual position, and not on the position of other carriages. In any coordinate measuring machine, if the measuring probe is displaced from one position to the other there is translational and rotational error. Thus, there are six error terms per axis (three rotation and three translation). In addition to this there are three squareness errors relating to XY, YZ, and XZ plane. Altogether there are a total of 21 error terms that need to be measured and corrected. The position error due to the displacement of the tip of CMM is related to the above error terms.

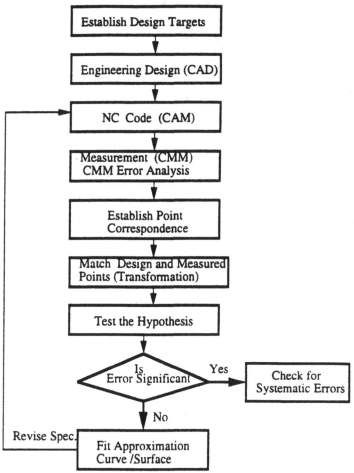

Figure 1    Models and Algorithms for Error Detection and Correction

## 4.3 POINT CORRESPONDENCE:MATCHING OF DESIGN POINT SET TO MEASURED POINT SET

When the part is measured in the CMM, the original part axis has different orientation on the measurement machine table. The point correspondence is also lost. When the part is set on the table of the measurement machine, it is not known which point on the designed part corresponds to the point on the machined part. In order to calculate the error vectors, the points on the machined part need to be related to the corresponding design points. This requires a match between the design points and measured points. Once  the match or point correspondence is determined, the transformation can be performed to map the measured points over the design space. The problem of  finding the point correspondence between the design and machined shape when both have curved boundaries has two steps: (1) Determine the feature points or the dominant points for the design and machined shape, and (2) Establish point correspondence (matching).

There are two widely used approaches to determine the feature points in curved objects. These are (1) polygon approximation method, and (2) the feature point detection through the curvature calculations. The first method is used here and is explained below.

Out of the many polygon approximation methods, the split method [ 8] is claimed to provide more accurate polygon representation. The goal here is to obtain a consistent polygon approximation to both the actual and machined curves (where both the design shape and the machined shape have point representation).

 A slight modification to this algorithm can produce very consistent polygons both for the original digitized boundary (design shape) and for the boundary points extracted from the machined object. The method by which this can be achieved is known as iterative end point fit. The end point fit proceeds as follows. Suppose, a set of $n$  points are given: { $P_1$, $P_2$,......,$P_n$}.
(a) Fit an initial line by connecting the end points of the set (the end points might be  the left and right  most points in the set, the top and bottom most points, or some other pair of distinguished points)
(b) Calculate the distances from each point to the line (say, AB)
(c) If all the distances are less than some preset threshold, the process is finished
(d) If not, find the point farthest from the line AB (say, C) and break the initial line into two lines AC and CB.
(e) The process is then repeated separately on the two new lines, with different thresholds.
(f) The final result is sequence of connected segments AC, CD, DB, etc.

## 4.4 TRANSFORMATION

Once the match between the design point set and the measured point set is obtained, transformation can be performed to map the design point set over the measured point set. This transformation can be done using the homogeneous coordinate transformation.The transformation of the measured data set over the design set is based on the principle of homogeneous coordinate transformation . If the design points and the corresponding measured points  are known then the transformation between these two sets of points can be represented by the following relationship:

$$[X\ Y\ Z\ 1] \begin{bmatrix} T_{11} & T_{12} & T_{13} & T_{14} \\ T_{21} & T_{22} & T_{23} & T_{24} \\ T_{31} & T_{32} & T_{33} & T_{34} \\ T_{41} & T_{42} & T_{43} & T_{44} \end{bmatrix} = [X'\ Y'\ Z'\ 1] = W'[L\ M\ N\ 1]$$

$$(1)$$

In the above equation, (X, Y, Z) are the design points and ( X' Y' Z') are the corresponding measured points. The matrix elements $T_{ij}$ are the elements of transformation matrix (the upper 3X3 matrix generates net rotation, the 1X3 matrix is for translation, and the 3X1 column matrix handles the projection effects) and W' is the homogeneous term. Multiplying out the terms in the above expression, three equations per point are obtained and there are 15 unknowns $T_{ij}$. The term $T_{44}$ is scale factor and is considered to be 1. Thus, at least five design points and five corresponding measured points are required to generate the required number of equations so that the terms in the [T] matrix can be determined. The system of equation generated in this procedure has the following form:

$$[A] [T] = [B] \tag{2}$$

Equation (23) shows that the transformation matrix can be set for any number of data points, and the 15 unknowns of [T] matrix can be determined from the above relationship.

$$[A]^T [A] [T] = [A]^T [B] \tag{3}$$

## 4.5 HYPOTHESIS TEST

A hypothesis test is conducted at this point where the null hypothesis is that the design points and the corresponding measured point are identical having no significant error. If the null hypothesis is rejected, there is indication that the design points and the corresponding measured points have significant error, and the systematic errors must be investigated further. The test procedure is explained below.

The model of the part is stored in the data base in form of points. The part is machined, measured and point correspondence is established (using the algorithms discussed earlier). The corresponding measured points are transformed over the design point set. The design points and the corresponding transformed measured points may not be equal because of the errors present in the measured set of points. The error is determined as the difference between the design points and the corresponding transformed measured points. The magnitude of the error should be determined so that the test statistic can be defined. The hypothesis test can be stated using appropriate symbols in the following way:

$H_0$ : the design points and the corresponding transformed measured points have no
     significant error
$H_1$: the design points and the corresponding transformed measured points have
     significant error

where, $H_0$ is the null hypothesis and the $H_1$ is the alternate hypothesis. To estimate the error, the differences between the design points and the corresponding measured points is determined. The average of the differences, D-bar, (the differences between the design points and the corresponding measured points) follows a t-distribution with (n-1) degrees of freedom when $s_D$ , the standard deviation of the differences of the sample, is known. In this case, the test statistics are given by

$$t_{n-1} = \frac{\text{D-bar}}{S_D / \sqrt{(n)}} \tag{4}$$

where, n= sample size, $t_{n-1}$= appropriate t value from the t-table for selected level of significance, and D-bar = $\Sigma D_i / n$ ($D_i$ are the differences between design and corresponding measured points) , and where $S_D^2$ is the variance of the differences. The average and the standard deviation of the differences are calculated from the data and equation (4) is used to test the hypothesis.

## 4.6 REVISING THE SPECIFICATION

Once it is determined through the hypothesis test that no significant systematic errors are present, the specifications are revised using the algorithm in this section. This algorithm fits an optimal approximation curve through the measured set of points. This algorithm is based on a least square parametric approximation. Through successive approximation, a revised or better set of points is obtained. This point set is used to revise the part program created earlier so that the parts can be machined within the prescribed tolerances and with minimum of errors.

Suppose n corresponding measured points $P_i$ ( $X_i$, $Y_i$ ) or $P_i$ ($X_i$, $Y_i$, $Z_i$ ), i= 0,1,...,n are obtained. An optimal approximation curve is to be fitted through these points. The approximation curve can be a Bezier curve of appropriate degree. In this research,the approximation curve is the same as the design curve. Suppose, a Bezier curve of degree n is chosen. This curve can be represented as

$$Y(u) = \sum_{i=0}^{n} b_i f_i \tag{5}$$

If the sum of absolute values of error vectors is minimized using the least squares method, the objective function can be written as;

$$E = \sum_{i=0}^{n} E_i^2 = \Sigma ( P_i - Y(u_i) )^2 \tag{6}$$

Taking the derivative of the above equation and setting it equal to zero, $\delta E / \delta b_i = 0$, we can solve for the unknowns.

If the error vectors in equation (6) are defined as the distance from the given point $P_i$ to a corresponding point on the approximation curve $Y(u_i)$, these error vectors are not perpendicular to the approximation curve. The requirement in the above case is to make the error vectors normal to the approximation curve. This can be done by changing the parameter values u. If the parameter values are changed by some rule, a change in the original value of point on the curve

$Y(u_i)$ will occur which will cause convergence of the error vectors to the normal of the approximation curve. A change in parameter values will change the approximation curve $Y(u)$, so each time a change in set of parameter values is initiated, a new approximation curve is calculated and the angle of the resulting error vectors is calculated. When all the error vectors are normal or approximately normal to the approximation curve, the process is stopped.Some methods of changing the parameters are suggested by Strauss and Henning, Hoschek, and Pratt [9,10].

All the algorithms discussed above will lead to the identification of errors and revision of the specification.

5.0 SUMMARY

This research has (1) suggested models and algorithms to reduce errors in machined parts by creating a better basis for machining specification, (2) provided methods and means to deal with random errors, (3) provided a method of relating design points to measured points through point correspondence and transformation (4) investigated the methods of detecting the feature points in the curved objects which is a major requirement to match the design points to the measured points having curved boundaries, (5) provided a method that can be applied to curves and surfaces which enabled the investigation of errors in all three axes of the tool, and (6) been able to integrate the design, manufacturing, and measurement processes, thus providing an integrated approach to error detection and correction problem.

REFERENCES

1.Eman, K. F., "A New Approach to Form Accuracy Control in Machining," International Journal of Production Research, Vol. 24, No. 4, 1986.
2.Nevelson, M. S., "Factors Affecting Machine Accuracy and Selection of a Control Algorithm," Production Engineering Research Association, 1973.
3. Raja, J., Whitehouse, D. J., "An Investigation into the Possibility of using Surface Profiles for Machine Tool Surveillance," International Journal of Production Research, Vol. 22, No. 3, pp. 453-466, 1984.
4.Dufour, P., Groppetti, R., "Computer Aided Accuracy Improvement in Large NC Machine Tools," Proceedings of the 21st International MTDR Conference, Swansea, U.K., 1980, pp. 611-618.
5. Kurtoglu, G. and Sohlenius, "The Accuracy Improvement of Machine Tools," Annals of the CIRP, Vol. 39/1, 1990.
6. Zhang, G., et al., "Error Compensation of Coordinate Measurement Machines," Annals of the CIRP, Vol. 34, 1985.
7.Mortensen, M. E., Geometric Modelling, John Willey and Sons, New York, 1985.
8.Duda, Richard O. and Hart, Peter E., Pattern Classification and Scene Analysis, John Willy and Sons, 1973.

9. Hoschek, J., " Spline Approximation of Offset Curves," Computer Aided Geometric Design 5, Elsevier Science Publishers B.V., North Holland, pp. 33-40,1988.
10. Pratt, M. J., " Smooth Parametric Surface Approximations to Discrete Data," Computer Aided Geometric Design 2, 1985.

# A NEW ON-LINE ROUGHNESS CONTROL
# IN FINISH TURNING OPERATION

## C. Borsellino, E. Lo Valvo, M. Piacentini and V.F. Ruisi
## University of Palermo, Palermo, Italy

KEYWORDS: Roughness control, Wear minor cutting edge, Image processing.

ABSTRACT: A new on-line control method for the wear state of sintered carbide tools for finish turning operations is proposed.
The suggested technique is based on the acquirement and processing of the minor cutting edge image where grooves appears because of the contact between cutting tool and the part of the workpiece that has been work-hardened during the cutting operation.
The changing in the number of grooves on the minor cutting edge is related with the tool damage with the aim to control the roughness of the workpiece.
The technique can be implemented on automated machining systems to reduce the costs of survey and control of the cutting tools with the intent to operate a good tool replacement policy.

## 1. INTRODUCTION

As it is well known, in a turning operation the tool life is limited by chipping or by wear both originated by the interaction between tool and workpiece or by tool and chip, evaluated as flank wear and crater wear [1]. These wear parameters are efficient to control both qualitatively and quantitatively the tool deterioration and then, subsequently, the product quality in rough-turning. In finish-turning operations, instead, it has been already shown the non-suitability of them for the sake of control the dimensional accuracy and the

Published in: E. Kuljanic (Ed.) *Advanced Manufacturing Systems and Technology*,
CISM Courses and Lectures No. 372, Springer Verlag, Wien New York, 1996.

micro geometry of the worked surface; these last are the main factors that determine the quality of the final product [2,3], in fact, the machine members can work correctly only if the dimensional accuracy of the different members is match by a particular limit of finishing of the surfaces that must work coupled. In a turning operation surface roughness can be considered, from the geometrical point of view, as the envelope of the succeeding relative positions between tool and workpiece and its value, as theoretical mean value, of the surface roughness depends on tool geometry, on the tool position with respect to the workpiece and on feed [4].

The value of the actual roughness is greater than the theoretical one [5] because of several factors, between them one of the most significant is the wear of the minor cutting edge [6].

In sintered carbides tools these kind of wear manifests itself, as it is known, in nose wear and groove wear. The first one appears along the contact tool/workpiece during the cutting, i.e. it interests the active cutting edge and part of the tip of the insert and it has only the effect to shift the tool profile parallel to itself causing in the meanwhile an increase of the diameter of the workpiece at increasing time of cutting. The second one starts just at the beginning of cutting and it concerns the generation and growth of grooves perpendicular to the secondary edge.

The generation of these grooves is due to the contact between the secondary edge and the zone of the worked material that has been work-hardened during the cutting; the work-hardening is increased by the high deformation and high strain rate and, as known, it causes a localised increase of hardness and shear resistance so that the tool undergoes a concentrate wear and on its surface appears a groove which depth grows at increasing of cutting time.

As wear goes on, the groove reaches a depth such that the crest on the surface of the workpiece, protrudes in such a way to interest in the subsequent revolution another point of the minor cutting edge; following this way on the minor cutting edge appears a series of equally spaced grooves, with a relative distance equal to the feed. The number of grooves and their depth influence heavily the actual roughness of the workpiece surface.

The survey of the geometrical characteristics of the workpiece or of the tool wear plays a very important role in the automated machining systems; in fact, the in-process [7] or on-line [8] measurement techniques employed until now, needs great times and costs for the measurement of the roughness for the employment of complex software or expensive instruments such as: profilometers, tool-room microscopes, SEM [9], etc.

In the present paper a technique for on-line survey and analysis of the grooves on the minor cutting edge in finish turning operations is proposed; it is based on the acquirement and processing of the minor cutting edge image, then the changing in the number of grooves has been related, for the different cutting parameters adopted, with the roughness of the worked material.

## 2. EXPERIMENTAL TESTS

Several tests of continuous cutting under dry conditions were performed on a lathe (model

SAG 12 CNC), working AISI 1040 steel specimens; the properties of the employed material are reported in Tab. 1.

| | |
|---|---|
| Chemical composition: | C = 0.43 %, Mn = 0.76 %, Si = 0.28%, S = 0.027 %, P = 0.016 % |
| Tensile strength: | R = 700 N/mm$^2$ |
| Hardness: | HBN$_{(2.5/187.5)}$ = 208 |

Tab. 1: Characteristics of AISI 1040 steel

Sintered carbide inserts type TPUN 160308 (carbide P10 and P30) were used, mounted on a commercial tool holder with the following geometry:
- rake angle $\gamma = 6°$
- clearance angle $\alpha = 5°$
- side cutting edge angle $\psi = 0°$
- inclination angle $\lambda = 0°$
The cutting condition employed for each insert are:
-depth of cut d = 0.5 mm
-feed $f_1 = 0.05$ mm/rev, $f_2 = 0.1$ mm/rev
-cutting speed $v_1 = 3.3$ m/s, $v_2 = 4.22$ m/s

After each cut six longitudinal profiles in different radial position were measured by means of a profilometer (Taylor - Hobson, Series Form Talysurf).

On the insert was measured the flank wear and it was observed with a scanning electron microscope (SEM) to follow and evaluate the growth of the grooves on the secondary edge with respect both to their number and their geometry. The insert was subsequently positioned on support of a optical bench that allows to take the minor cutting edge image using a television camera CCD (Charged Coupled Device).

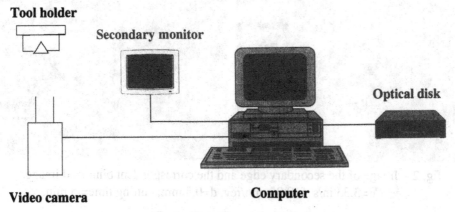

fig.1 - Experimental set-up

The system, shown in fig.1, is completed by an optic fiber light source and by a real time video digitizer board (512x480 pixel and 128 grey levels) installed on a personal computer. Each pixel show a zone of tool of about 1.0 μm along x-axis and 0.7 μm along y-axis. Such resolution is the best trade-off between the needing of an high resolution and the necessity to take the image of a zone of the insert that allows to see part of the insert tip and all the grooves that can be generated during the cutting operation.

## 3. GROOVES SURVEY METHODOLOGY

A methodology that employ the techniques of acquirement and processing of the images has been realised for the survey of the grooves in a precise and rapid way.

The part of the insert that is taken interest both the minor cutting edge and a part of the radius between the edges; the choice of illumination, relized with an optic fiber light source, is such to have the best contrast as possible.

At the end of every cut on the acquired image of the tool has been done a binarization, that consist in the ascribing, to the different grey tones of the image, the maximum or of the minimum value depending on the exceeding or not of a fixed threshold level conveniently chosen, obtaining the highlighting of the grooves generated during the cutting operation. Then the image has been enhanced with the median filter.

With an appropriate software, on the obtained image the grooves are counted and their number is stored and compared with the one of the previous cut.

For a better comprehension of the various steps of the proposed methodology in the following are reported the images of the secondary edge, the correspondent binarized images and the step-wise signal generated by the software for the counting of the grooves for an insert carbide P30; see figg. 2,...,6 .

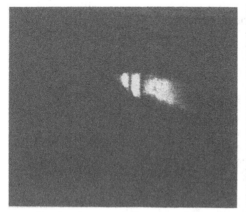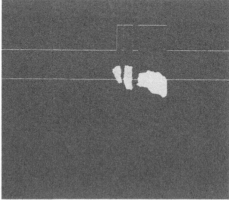

fig. 2 - Image of the secondary edge and the correspondent binarized image.
V=3.33 m/s, f=0.05 mm/rev, d=0.5 mm, cutting time= 5 min.

The number of grooves counted by the system is in perfect agreement with the one founded with the examination of the images obtained with a SEM, as it is shown in fig.7.

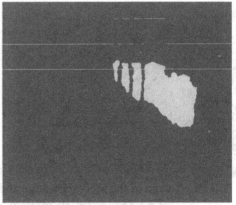

fig. 3 - Image of the secondary edge and the correspondent binarized image.
V=3.33 m/s, f=0.05 mm/rev, d=0.5 mm, cutting time= 22 min.

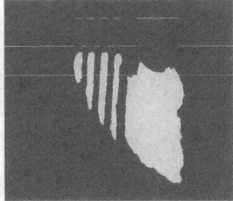

fig. 4 - Image of the secondary edge and the correspondent binarized image.
V=3.33 m/s, f=0.05 mm/rev, d=0.5 mm, cutting time= 35 min.

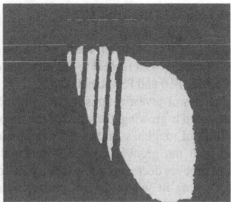

fig. 5 - Image of the secondary edge and the correspondent binarized image.
V=3.33 m/s, f=0.05 mm/rev, d=0.5 mm, cutting time= 45 min.

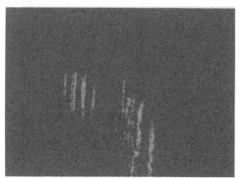
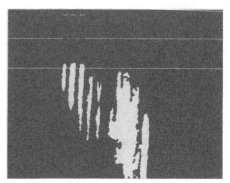

fig. 6 - Image of the secondary edge and the correspondent binarized image.
V=3.33 m/s, f=0.05 mm/rev, d=0.5 mm, cutting time= 80 min.

fig.7. - SEM observation of the insert reported in fig. 2.

## 4. ANALYSIS OF THE RESULTS

The trend of the Ra parameter, media of the six values acquired, is reported in fig.8 versus the time of cutting with respect to the different cutting parameters and for the two kind of insert (Carbide P10 and P30) employed.

For each series of investigations it can be noted that after the first minutes of the cutting where the trend is growing, the value of the Ra parameter is almost constant and lower than a fixed limit of roughness imposed by the project. In fig. 8. are reported also the graphs obtained with the grooves counting system described in the previous paragraph. The number of grooves decreases while the surface roughness of the workpiece increase.

During the time in which the Ra parameter is approximately constant, the number of grooves increases, but increases also their size until they begin to interfere each other leaving a zone of the minor cutting edge heavily weared with chipping, that cause a massive decay of the surface finishing conditions.

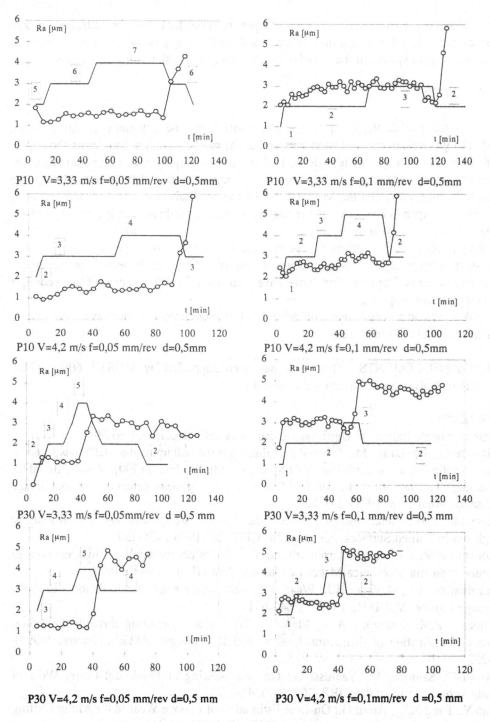

fig. 8 - Ra versus cutting time for the two kind of insert

This behaviour, encountered in each couple material-tool for the different cutting parameters employed, confirm that the proposed methodology is quite suitable to perform an on-line control system of the out of order of the tool in finish turning operations.

## 5. CONCLUSIONS

The proposed methodology for the on-line control, by the automatic counting of the number of grooves, of the tool wear state and of the worked material roughness show to be reliable for the different couple material-tool and for the employed cutting parameters. The technique allows to manage an appropriate tool replacement policy in very short times because the information on the wear state of the tool is obtained in a few seconds avoiding the long and expensive surveys of the roughness on the workpiece or the measure of the crater or of the flank wear of the inserts.

The time needed for the control is less than the one for the assembly - disassembly of a successive workpiece, so this technique of control can be implemented on automated machining systems. The repetitions done have confirmed the effectiveness of the technique of on-line control proposed.

The authors intend to verify the application field of such methodology also for the cutting with ceramic tools of other materials.

ACKNOWLEDGEMENTS   This work has been supported by MURST 60% (Italian Ministery of University and Scientific Research).

REFERENCES
1. Ente Nazionale Italiano di Unificazione, Rugosità delle Superfici, UNI 3963
2. Galante G., Piacentini M., Ruisi V.F. : Elaborazione dell'immagine dell'utensile per il controllo della finitura superficiale, X Congresso AIMETA, Pisa, (1990), Vol.II ,501-504.
3. Galante G., Piacentini M., Ruisi V.F.,: Surface roughness detection by tool image processing, Wear, 148 (1991), 211-220.
4. Lonardo P.M.: Relationships between the Process Roughness and the Kinematic Roughness in Turned Surfaces, Annals of the CIRP, 25 (1976), 455-459.
5. Lonardo P.M., Lo Nostro G.: Un criterio di finibilità dei materiali per le operazioni di tornitura, Industria Meccanica-Macchine Utensili, Anno IV n° 4, aprile 1977.
6. Pekelharing A.J., Hovinga H.J.: Wear at the end cutting edge of carbide tools in finish and rough turning, M.T.D.R., 1 (1967), 643-651.
7. Lonardo P.M., Bruzzone A.A., Melks J.: Tool wear monitoring through the neural network classification of diffraction images, Atti II Convegno AITEM, Padova (1995), 343-352.
8. Giusti F., Santochi M., Tantussi G.: On-Line Sensing of Flank and Crater Wear of Cutting Tools, Annals of the CIRP, 36 (1987), 41-44.
9. Yao Y., Fang X.D., Arndt G.: On-Line Estimation of Groove Wear in the Minor Cutting Edge for Finish Machining, Annals of the CIRP, 40 (1991), 41-44.

# APPLICATION OF THE CONTISURE SYSTEM FOR OPTIMIZING CONTOURING ACCURACY OF CNC MILLING MACHINE

**Z. Pandilov, V. Dukovski and Lj. Dudeski**
University "Sv. Kiril i Metodij", Skopje, Macedonia

KEY WORDS: Contouring accuracy, CNC machine tool, CONTISURE system

ABSTRACT: In recent years, a instrumentation circular profile tests has been specified to assess the contouring accuracy of CNC machine tools. Such an instrumentation type test is the CONTISURE system. In this paper, the influence of the position loop gain and mismatch of position loop gains for different machine axes are effectively studied. This work outlines a practical procedure for determining the position loop gain of the control system in order to minimize the resulting contouring errors.

## 1.INTRODUCTION

The contouring performance of CNC machine tool can be established by assessing its ability to move along a specified profile by the simultaneous movement of two or more axes.

When CNC machine tools are used for contouring applications, especially where high feed rates are used, significant dynamic errors can be introduced by the characteristics of the CNC controller and servo feed drive system. The assessment of such dynamic errors in CNC machines, has traditionally been undertaken by machining a standard circular test piece.

Published in: E. Kuljanic (Ed.) *Advanced Manufacturing Systems and Technology*, CISM Courses and Lectures No. 372, Springer Verlag, Wien New York, 1996.

Such a test piece is outlined in some of the national machine tool standards (British, American, Japanese), where the circular profile is produced by the simultaneous motion of two linear axes.

An alternative approach to the machining test, specified in recent British and US machine tool standards, is an emulation by instrumentation techniques of the circle test. Such an instrument type test is the CONTISURE system developed by Burdekin [1,2].

Although instrumentation techniques generally check the machine in no-load condition, they offer certain advantages over cutting conditions. In particular, tools and test specimens are not consumed and the time consumed in metrologising the test piece after machining is eliminated.

## 2. THE CONTISURE HARDWARE AND SOFTWARE SYSTEM

The CONTISURE system is shown schematically in fig.1.

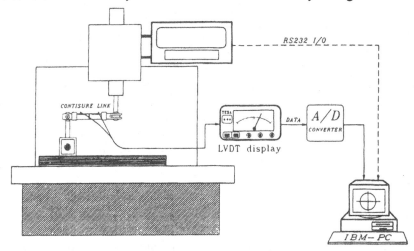

Fig.1 CONTISURE hardware set-up on CNC milling machine

The system comprises two high precision reference spheres, rigidly mounted at the spindle and table positions. A transducer link of carbon fibre construction and containing two precision transducers, is located kinematically between the two reference spheres. These two trancducers contact directly onto the two spheres, and the summation of their outputs represents the change in distance of the two reference spheres, as the machine performs a circular contouring operation. The absolute distance between the two spheres can be established by setting the transducer link against a calibrated setting block. This feature, which is unique to the CONTISURE system, ensures the complete traceability of data to be maintained.

The data acquisition and analysis software offer the user a complete flexibility. The number of sampled data points can be selected, up to a maximum of 12000 per 360 degrees scan. An analysis in the form of least squares best fit circles, can also be perform on data obtained for 360 degrees scans as well as for partial arcs. This feature eliminates the need for precise set-up of the sphere datum with respect to the programmed circle.

It is also essential that the start and end points of the circular contour should be selected, so that these do not coincide with the axis reversal points. The reason for this is that significant lost motion errors may occur at these points, and additional transients errors, resulting from the servo control system, may not be detected. In this respect the software is completely flexible and enables the start and end points to be freely selected. A start position of 22 degrees from the X axis was therefore used for all tests.

The approach to the start point of the circular profile should, if possible, be representative of that used under practical machining conditions. A tangential approach to the start and exit points on the profile is therefore assumed by the software.

## 3. THE INFLUENCE OF POSITION LOOP GAIN AND MISMATCH OF POSITION LOOP GAINS FOR DIFFERENT MACHINE AXES ON OPTIMIZING THE CONTOURING ACCURACY OF CNC MILLING MACHINE

One of the most important factors which influences the dynamical behavior of the feed drives for CNC machine tools is position loop gain or Kv factor. Tracking or following error depends on the magnitude of the Kv-factor. In multi-axis contouring the following errors along the different axes may cause form deviations of the machined contours. Generally, position loop gain Kv should be high for faster system response and higher accuracy, but the maximum allowable gains are limited due to undesirable oscillatory responses at high gains and low damping factor which produce significant transient errors and accuracy started to decrease again. Usually Kv factor is set up by the machine tool manufacturer.

But the question is, whether the set-up value of the Kv-factor is always optimal? Generally, contouring error of circular contour, according [3,4], could be analitically approximately calculated with following equations:

$$ec = \left| R \cdot \left[ 1 - \sqrt{\left( \frac{v}{60 \cdot R \cdot Kv} \right)^2} \right] \cdot 10^3 \right| \ \mu m \qquad (1)$$

or $\quad e_c = \dfrac{1}{2 \cdot R} \cdot \left( \dfrac{v}{60 \cdot K_v} \right)^2 \cdot 10^3 \quad \mu m$ $\qquad\qquad\qquad$ (2)

where: $e_c$-maximal contouring error from the nominal radius μm, R-radius of the circle mm, v-feedrate mm/min, $K_v$-position loop gain $s^{-1}$.

These equations do not take into consideration the influence of nonlinear phenomena, such as lost motion, stick motion and stick-slip, on the magnitude of the contouring errors [5]. That is a reason why real values of the contouring errors are significantly higher. Real contouring errors can be obtained only experimentally.

Contouring measurements with CONTISURE test equipment have been undertaken on a FGS32 CNC milling machine with HEIDEHANN 355 TNC controller, in order to illustrate a methodology which could generally be applied to any CNC machine. Only two sets of axes have been considered (X and Y). The same procedure can be repeated for other axes. A relatively short link of 150 mm, was used for all tests.

In the tests the feedrate was constant v=600 mm/min and the $K_v$ factor in the controller was changed in the range of $4 s^{-1}$ to $130 s^{-1}$. The tests were done in two directions clockwise (CW) and counterclockwise (anticlockwise) (CCW). The results of tests are given in Table 1.

Table 1

| $K_v$ $s^{-1}$ | 4 | 6 | 8 | 10 | 20 | 28.3 | 30 | 40 | 50 |
|---|---|---|---|---|---|---|---|---|---|
| $e_c$ μm (CW) | 46.8 | 39.5 | 38.3 | 32.2 | 20.7 | 19.6 | 16.5 | 14.7 | 13.6 |
| $e_c$ μm (CCW) | 50.7 | 45.5 | 37.5 | 36.2 | 25.2 | 22.2 | 20.3 | 17.8 | 13.3 |
| $K_v$ $s^{-1}$ | 60 | 70 | 80 | 90 | 100 | 110 | 120 | 130 | |
| $e_c$ μm (CW) | 12.1 | 11.1 | 10.8 | 10.5 | 10.2 | 10.3 | 10.4 | 10.5 | |
| $e_c$ μm (CCW) | 12.4 | 10.8 | 10.7 | 10.4 | 10.2 | 10.4 | 10.5 | 10.6 | |

From Table 1 it is obvious that optimal experimental value for $K_v$ factor is $100 s^{-1}$. $K_v$ factor, set up by the machine manufacturer, was $28.3 s^{-1}$.

We can see that increasing position loop gain $K_v$ in the range of 4 to $100 s^{-1}$ decreases maximal contour deviation from nominal radius. But also we can see that the values for $K_v$ in the range of 110 to $130 s^{-1}$ increase contouring error. This can be explained by the fact that transient errors become dominant. Further analyses have shown that with increasing the position loop gain from 28.3 to $100 s^{-1}$ the maximal contouring deviation was decreased from 19.6 (CW)/22.2 (CCW) μm to 10.2 (CW)/10.2 (CCW) μm. Figs. 2 and 3 show the results mentioned above.

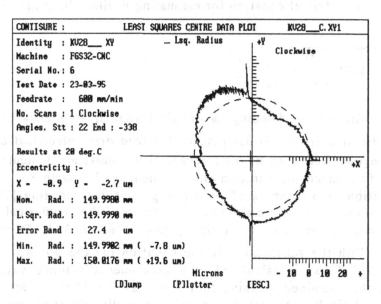

Fig.2 The result of a measured circular test (feedrate =600 mm/min, radius of the circle=150 mm, position loop gain=28.3 s$^{-1}$, clockwise direction)

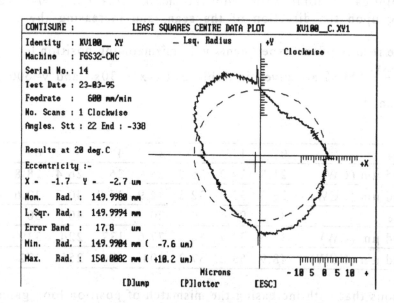

Fig.3 The result of a measured circular test (feedrate =600 mm/min, radius of the circle=150 mm, position loop gain=100 s$^{-1}$, clockwise direction)

In ref. [6] an analytical equation for estimating position loop gain is given:

$$Kv = \frac{1}{4 \cdot \zeta^2 \cdot \left( \dfrac{2D}{\omega} + \dfrac{2D_m}{\omega_m} + \dfrac{T}{2} \right)} \tag{3}$$

where $\xi$-position loop damping, $\omega$ -nominal angular frequency of the feed drive electrical parts $s^{-1}$, D-damping of the feed drive electrical parts, $\omega_m$-nominal angular frequency of the mechanical transmission elements $s^{-1}$, $D_m$ - damping of the mechanical transmission elements and T-sampling period s.

Position loop damping of $\xi=0.7$ is preferable according [7]. That is the value which gives minimal contouring errors. Other numerical values of the examined system are: $\omega =1000s^{-1}$, D=0.7, $\omega_m=663s^{-1}$, $D_m=0.17$, and T=0.006s. With the substitution in the equation (3) the position loop gain value Kv=103.85 $s^{-1}$ is calculated. The experimentally tuned value of Kv-factor on the examined machine tool axis was Kv=100$s^{-1}$. The difference between analytically calculated and experimentally obtained value of Kv-factor is around 4%, which is completely acceptable.

Another parameter which influences the contouring accuracy is the mismatch of position loop gains for different machine axes. This will result in an elliptical contour path with the major axes lying +/-45 degrees, depending upon the direction of the scan, and increasing the contouring errors.

.The results of the experiments with mismatching position loop gains $a = \dfrac{Kvx - Kvy}{Kvx} \cdot 100$ % are given in Table 2. (Kvx = $30s^{-1}$ and v=600 mm/min are constant.)

Table 2

| a % | 0 | 1 | 2 | 3 | 4 | 5 | 6 | 7 |
|---|---|---|---|---|---|---|---|---|
| error band μm (CW) | 21.8 | 24.3 | 25.6 | 25.9 | 26.2 | 26.4 | 28.8 | 29.6 |
| error band μm (CCW) | 25.5 | 32.4 | 32.8 | 34.9 | 35.9 | 40.7 | 42.6 | 44.9 |
| a % | 8 | 9 | 10 | 20 | 30 | 40 | 50 | |
| error band μm (CW) | 31.5 | 32.7 | 38.9 | 77.6 | 137 | 224 | 339 | |
| error band μm (CCW) | 48.9 | 53.2 | 57.3 | 104 | 170 | 255 | 372 | |

It is obvious that with increasing the mismatch of position loop gains of the axes, the error band rises up. The best case is when the position loop gains are identical (a=0). Figs. 4 and 5 show the results of circular test when the difference between position loop gains for X and Y axes is a=20%.

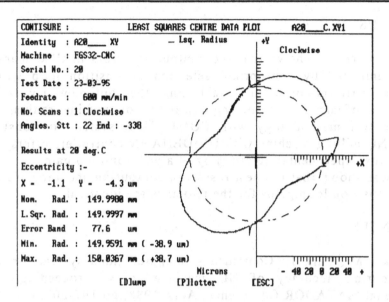

Fig. 4 The results of the measured circular tests with gains mismatched a=20% (clockwise direction, feedrate =600 mm/min, radius of the circle=150 mm, position loop gains Kvx=30 s$^{-1}$ and Kvy=24 s$^{-1}$)

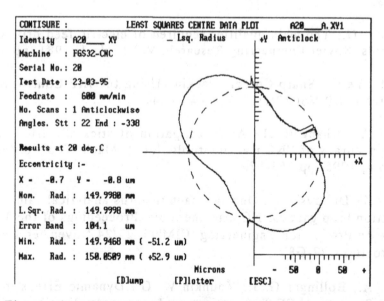

Fig. 5 The results of the measured circular tests with gains mismatched a=20% (anticlockwise direction, feedrate =600 mm/min, radius of the circle=150 mm, position loop gains Kvx=30 s$^{-1}$ and Kvy=24 s$^{-1}$)

## 4. CONCLUSION

The work has shown that the contouring errors in CNC machine tool can be minimized by appropriate selection of position loop gain in the controller. Criteria used in establishing the optimum Kv value was minimization of maximal contouring deviation from nominal radius.

The test methodology with CONTISURE system, demonstrated on FGS32 CNC milling machine with HEIDEHANN controller, offers a general approach for experimental determining of a position loop gain.

It was shown that the best results in contouring accuracy are provided when the position loop gains for the two axes are identical.

## REFERENCES

1. Burdekin M., Park J.: Contisure-a computer aided system for assessing the contouring accuracy of NC machine tools. Proceedings of 27th International MATADOR Conference, April 1988, pp.197-203.

2. Burdekin M., Jywe W.: Application of Contisure for the Verification of the Contouring Performance of Precision Machines. Proceedings of the 6th international Precision Engineering Seminar, Braunschweig, Germany 1991, pp.106-123.

3. Andreev I. G.: The main qualities required of electric feed drives for NC machine tools. Soviet Engineering Research, Vol.1, No.1, 1981, pp.59-62.

4. Weck M., Ye G.: Sharp Corner Tracking Using the IKF Control Strategy. Annals of the CIRP Vol.39/1/1990, pp.437-441.

5. Tarng S. Y., Chang E. H.: An investigation of stick-slip friction on the contouring accuracy of CNC machine tools. Int. J. Mach. Tools & Manufact. Vol.35, No.4, 1995, pp.565-576.

6. Pandilov Z., Dukovski V.: One approach towards analytical calculation of the position loop gain for NC machine tools. Proceedings of the 3rd Conference on Production Engineering, CIM'95, 23/24 November 1995, Zagreb, Croatia, G77-G83.

7. Poo -N. A., Bollinger G. J., Younkin W. G.: Dynamic Errors in Type 1 Contouring Systems. IEEE Transactions on Industry Applications, Vol.IA-8, No.4, July/August 1972, pp.477-484.

# FUNCTIONAL DIMENSIONING AND TOLERANCING:
## A STRONG CATALYST
## FOR CONCURRENT ENGINEERING IMPLEMENTATION

**S. Tornincasa, D. Romano and L. Settineri**
**Polytechnic of Turin, Turin, Italy**

KEY WORDS: GD&T, Concurrent Engineering, CMM inspection

ABSTRACT: Design drawings embody, in selected components and tolerance specifications, product function and costs. They have also a great integrating potential inside the company as they affect every function involved in product development. In this paper an industrial application is presented in the field of automotive industry; it demonstrates that the functional dimensioning analysis, using GD&T design tools, provides a better product design at maximum allowable tolerances and, enhancing the understanding of product function at any level in the company, is a basic vehicle of interfunctional integration and a key factor in Concurrent Engineering practice.

## 1. FUNCTIONAL DIMENSIONING PARADIGM

The design drawing has a great integrating potential inside the company as it affects every function involved in product development. If the design is not clear and understood by everyone, the organization will not work efficiently. The quality of the drawing affects the product functionality and cost; flawed drawings are likely to create excess waste, or even lead to the rejection of parts which would otherwise have been functionally acceptable.

Functional dimensioning is a philosophy of dimensioning and tolerancing based on the prioritization of part functions. Thinking primarily in terms of the function the designer has the clear objective to minimize costs without deteriorating the function itself. Furthermore, understanding part function in drawings enables a better and more timely communication

Published in: E. Kuljanic (Ed.) *Advanced Manufacturing Systems and Technology*,
CISM Courses and Lectures No. 372, Springer Verlag, Wien New York, 1996.

between design and other departments, being a key factor in Concurrent Engineering practice.

The design language based on a "plus and minus" system of dimensioning and tolerancing is insufficient to consistently convey design intent. For example, the combination of all locations of the axis for a round hole that will fit over a smaller size round pin is not a square zone, but a cylindrical one. Moreover, the larger the hole (as long as it is within its size tolerance) the larger the tolerance that it may benefit whilst fulfilling its function properly. This is the basic idea underlying the Maximum Material Condition (MMC) principle.

Geometric Dimensioning and Tolerancing (GD&T) is the new drafting language which translates the aforementioned conceptual framework through a system of internationally accepted symbols and notations.

If we consider how a system made of several parts functions, instead of considering each single part individually, this new dimensioning philosophy prompts us to identify the effects of each single part tolerance on the overall system; and thus the largest possible tolerance for the single part dimension can be selected.

Prioritizing the function implies that dimensions should define the part without specifying manufacturing processes (Process Independence Principle); however clearer evidence of the functions helps the process engineer in selecting the most economical methods to process the parts.

## 2. INSPECTING GEOMETRIC TOLERANCES

Geometric tolerance specifications do not only translate designer's intent, as they embody precise information for the definition of consistent assembly and inspection plans.

In this section the assessment of how conformance is driven by tolerance specification will be dealt with and a spectrum of problems arising when CMMs are used for inspecting parts will be discussed. Two main consequences on dimensional control of parts derive from the new tolerancing standard [6,7].

1. Functional dimensioning requires that tolerance independence be violated, meaning that tolerances may interact with themselves (MMC is a typical example); this leads to additional complexity in the part verification task: measurements sequences are to be carefully planned and measurement data properly elaborated.

2. The standard can handle unambiguously imperfect shape part verification. This is a critical guarantee of the interchangeability of parts, because it ensures that dimensional control done on the same part in different laboratories will deliver the same acceptance/rejection response.

It is possible to distinguish between a case where a reference system is necessary and where it is not. Form tolerances need no reference system since the tolerance zone depends on the feature itself and a scalar quantity only, consequently no datum appears in the control frame. Position and orientation tolerances are controlled by constructing a local reference system, named DRF (Datum Reference Frame), on the basis of the features of physical parts referenced as datum in the control frame.

Note that the order of datum calls is essential for establishing the right system; each datum constrains one or more degrees of freedom but the set of datum of an individual tolerance may or may not fully constrain the coordinate system for locating and orientating the tolerance zone.

One last point remains to be questioned: how can a mathematical representation of a datum be obtained from the actual feature? Here the genuine functional inspiration of geometric tolerancing is fully realized. The underlying dictum is: *when controlling a feature imagine its mating counterpart in a hypothetical assembly*. This means: if the feature is a hole think of the largest perfect-form shaft that can pass through it, if it is a shaft think of the smallest perfect-form hole that still fits the shaft, if it is a flat surface think of a perfect-form plane which makes stable contact on the surface. Thus the actual mating envelope of the feature is designated to establish datum, as if inspection was a simulated assembly (the order of datum calls reflects the order of an assembly sequence).

In the context presented above hard gauges appear to be the most natural inspection technique. Unfortunately, hard gauges are not always feasible (Least Material Condition principle is a well known example) and sometimes they have a poor cost justification.

In inspection practice CMMs are so widely used in today's industry that it is worthwhile examining the specific problems which occur when they have to control geometric tolerances. CMMs provide a sample of discrete measurements which allows for a description of the manufactured part geometry. The collected data can be elaborated by means of software routines to verify the conformance to geometric tolerances. This is the popular "soft gauges" technique. Sophisticated versions of on-line control are currently being investigated by directly loading CMM data back into a CAD environment where the perfect-form model of the part also exists. This might seem a brilliant solution to the dimensional control problem if the hidden issue is not questioned: to what extent are CMM measurements consistent with the new tolerancing principles? Or, equally, is the soft-control of geometric tolerances operated by CMM reliable enough? In this framework two primary problems cannot be evaded.

1. Relying on a finite set of data, the reconstructed part's geometry as well as its related soft gauge are only approximations of what they are intended to be. A poor set of measurements is likely to lead to grossly incorrect results; conversely large sample set sizes are impractical with the current CMM technology. Hence the problem of the sufficiency of the sample set has to be addressed in detail; a substantial statistical appraisal is foreseen to be applied in this subject [2,3]. It was noted elsewhere [1] that if models of the manufacturing processes involved were available, smaller sample set sizes would be feasible but the principle of process independence in tolerancing would be totally disregarded.

2. Even if a sufficient measurement sample is taken the subsequent treatment of data must be consistent with the aforementioned actual mating envelope principle. Most of the algorithms implemented in CAT (Computer Aided Testing) packages traditionally use a best-fit approach based on least-square approximations. This practice clearly defies the actual mating envelope principle and, ultimately the underlying functional paradigm. Computational engineers will thus be prompted to renew their software equipment.

## 3. AN INDUSTRIAL APPLICATION OF GEOMETRIC TOLERANCING

### 3.1 Redimensioning an old drawing

In Fig. 1 shows the fly wheel of a diesel engine, carrying the sensor charged with phasing the electronic injection. The functional need translates into a duplex angular tolerance: the first one between the crank pin n° 6 (see Fig. 2) and the dowel hole for the coupling of the crankshaft with the fly wheel, the second one between the dowel hole on the fly wheel and the circumferential hole housing the sensor. In Fig. 2 the old dimensioning of the crankshaft is depicted; the holes for the fastening of the fly wheel to the crankshaft are positioned by means of a location tolerance, while the dowel hole shows an angular tolerance of ±15' with a radial error of ±0.025 mm. The same dimensioning criterion is adopted for the fly wheel, the only difference being a radial error on the dowel hole of ±0.05 mm.

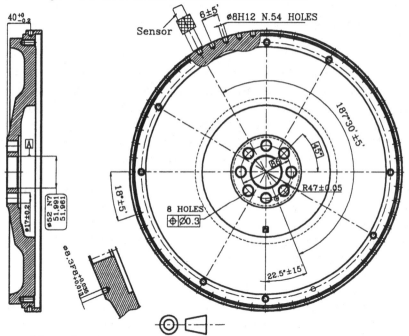

Fig. 1: fly wheel dimensioned using the old system

This dimensioning system can be criticized thus:

a) since the position of the dowel hole on the fly wheel with respect to the circumferential hole is functionally critical, the tolerance zone is small (0.135 mm$^2$) and unusually elongated, which results in increased production costs;

b) given that location and dimensional tolerance of the fastening holes on both parts are not related, the tolerance zone is not the maximum allowed by the function, so that good parts might be rejected;

c) the holes' control procedure is not reliable, due to different possible interpretations allowed for by the dimensioning system.

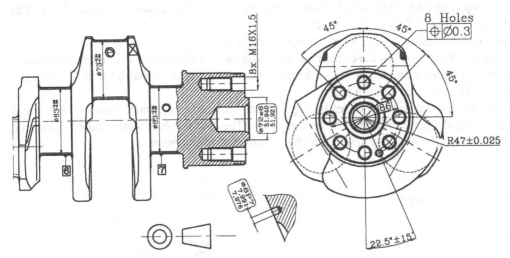

Fig. 2: crankshaft dimensioned using the old system

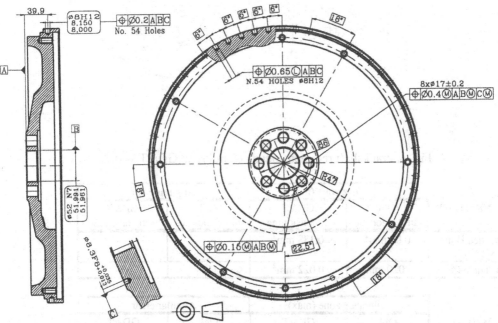

Fig. 3: fly wheel redimensioned according to GD&T system

Figs. 3 and 4 display the two parts according to the principles of functional dimensioning; the relevant data are listed in Tab. I. The following considerations hold:

a) the dowel hole on the fly wheel benefits from both a larger tolerance zone (see Tab. I) and a circular shape, which better fits into the spatial distribution of the manufacturing error;

b) by implementing MMC principle on the fastening holes the tolerance zone nearly doubles (0.125 instead of 0.070 mm2, Tab. I);

c) the control procedure can be now uniquely defined;

d) the functional need on the circumferential holes has been identified in a minimum distance to be kept between the edges of two adjacent holes; this has been translated in the adoption of the LMC principle on each hole.

It can be seen that the new dimensioning system produced two simultaneous advantages: a smaller angular error and larger tolerance zones. The first result is welcomed by designers, whilst the second is mostly appreciated by manufacturing engineers.

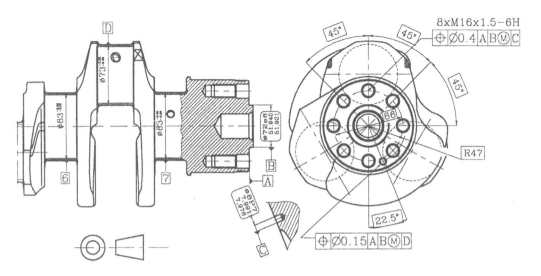

Fig. 4: crankshaft redimensioned according to GD&T system

| CRANKSHAFT | Tolerance zone (max) | | Angular error | |
|---|---|---|---|---|
| | Old dimensioning | GD&T dimensioning | Old dimensioning | GD&T dimensioning |
| Threaded Holes M16x1.5 | 0.07 mm$^2$ | 0.138 mm$^2$ | --- | --- |
| Dowel hole Ø8P7 | 0.020 mm$^2$ | 0.022 mm$^2$ | ±15' | ±6' |

| FLY WHEEL | Tolerance zone (max) | | Angular error | |
|---|---|---|---|---|
| | Old dimensioning | GD&T dimensioning | Old dimensioning | GD&T dimensioning |
| Clearance Holes Ø17±0.2 | 0.07 mm$^2$ | 0.570 mm$^2$ | --- | --- |
| Circumferential holes Ø8H12 | 0.135 mm$^2$ | 0.160 mm$^2$ | ±5' | ±6' |
| Total angular error: | | | ±20' | ±12' |

Tab. I: comparison of results obtained using the two dimensioning systems

## 3.2 Inspection procedure

In the following flow-chart the inspection strategy translating the GD&T principles is depicted.

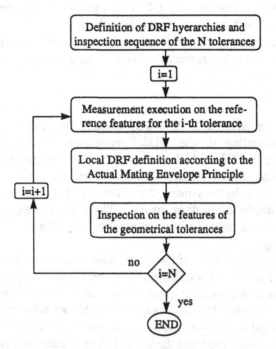

The inspection of the two parts (i.e. crankshaft and fly wheel) was operated on a CMM. For the sake of brevity, only the steps followed for the crankshaft inspection are listed below:

A) First DRF building:

- Contacting the end face of the crankshaft against a flat surface; definition of the planar datum **A** (plane $ZY$ in the DRF) through the measured co-ordinates of 6 points on the flat surface.
- Definition of the datum **B** corresponding to the axis of the smallest cylinder circumscribed to the centering element on the end face of the crankshaft by touching 50 points on the centering element.
- Locating the origin of the DRF on the intersection point of datum **B** with datum **A**.
- Definition of the datum **D** corresponding to the axis of the smallest cylinder circumscribed to the crank pin n°6 by touching 50 points on it.
- Rotation of the DRF around its $X$ axis so that its $Z$ axis intersects datum **D**.

B) Verification of dowel hole and second DRF building:

- Definition (in the previously built DRF) of the largest cylinder inscribed to the dowel hole (datum **C**) by touching 50 points inside the hole.
- Comparing the position of the previously defined cylinder with the virtual condition boundaries. The tolerance is respected if the cylinder contains these boundaries.

- Definition of the second DRF by rotating the first one around its X axis such that its Z axis intersects datum C.

C) Verification of the fastening holes:

- Definition (in the previously built DRF) of the largest cylinders inscribed to the fastening holes by touching 50 points inside each one.
- Comparing the position of the previously defined cylinders with the virtual condition boundaries. The tolerance is respected if the cylinders contain these boundaries.

## 4. CONCLUSIONS

The paper describes the advantages of redimensioning the design of a crankshaft and relevant fly wheel according to the GD&T principles. Furthermore, the inspection procedure, operated with a CMM, which conforms to GD&T is reported.

A noticeable enlargement of the tolerance zone as well as an improvement in the functional capabilities is demonstrated. In this context the possibility of sharing the derived advantages between design and manufacturing departments is likely to generate negotiations between them which result in a spontaneous integration process useful for an eventual application of Concurrent Engineering.

GD&T allows for a unique definition of the dimensional control procedure. However, when only discrete measurements are available, some restrictions apply. A great deal of theoretical and computational work is expected in order to adjust the use of CMMs to geometric tolerance verification. However, a qualitative higher enhancement will perhaps be triggered by technological advance rather than by conceptual efforts. Progress in optical measuring technology, which promises faster and denser measurements to be available, seems the most feasible option for near future.

## REFERENCES

[1] Voelcker, H., 1993, "A Current Perspective on Tolerance and Metrology", Manufacturing Review, Vol. 6, No. 4

[2] Nigam, S. D. and Turner, J. U., 1995, "Review of statistical approaches to tolerance analysis", Computer Aided Design, Vol. 27, No. 1

[3] Kurfess, T. R. and Banks, D. L., 1995, "Statistical verification of conformance to geometric tolerance", Computer Aided Design, Vol. 27, No. 5

[4] Chirone, E., Orlando, M. and Tornincasa, S., 1995, "Il disegno funzionale e le tolleranze geometriche", Proc. of IX ADM Congress, Caserta, Italy

[5] Yan, Z. And Menq, C.-H., 1994, "Evaluation of geometric tolerances using discrete measurement data", Jour. Of Design and Manufacturing, Vol. 4

[6] ASME (The American Society of Mechanical Engineers), "Dimensioning and Tolerancing" 1982, *ANSI Y14.5M-1982*, New York, USA

[7] ASME (The American Soc. of Mech. Eng.), "Mathematical Definition of Dimensioning and Tolerancing Principles", 1994, *ANSI Y14.5.1M-1994*, New York, USA

# METHODOLOGY AND SOFTWARE FOR PRESENTING THE MODEL OF MECHANICAL PART BY ASSEMBLING PRIMITIVES

V. Gecevska and V. Pavlovski
University "Sv. Kiril i Metodij", Skopje, Macedonia

KEY WORDS: Primitives, Modelling, DLL, Mechanical parts, Graph

## ABSTRACT

This paper presents the research of creating the methodology for modelling mechanical parts. The method is analysed and the software is made for presenting the model of mechanical part, where the integrated information about geometry, topology and technology are. The model of the mechanical part [1] is used as a base to create the CIM concept. The information, which is a part of the model is also the necessary entrance in each module of production system designing.

## 1. INTRODUCTION

In our research [3], we were used parametric method in mechanical elements projecting in aspect of the projection and manufacturing. The primitives are the a base in the research and primitives connecting are the base elements for the procedure development for modelling rotational mechanical parts with unrotational deviation of the shape. Primitives are the scope of geometrical parameters and are technologically oriented, connected with forms reaching all the information from the higher level. The information from the model is entrance in module of production system designing. This module is based on the procedure for automatic process planning and graph representation.

Published in: E. Kuljanic (Ed.) *Advanced Manufacturing Systems and Technology*, CISM Courses and Lectures No. 372, Springer Verlag, Wien New York, 1996.

## 2. MODELLING OF THE MECHANICAL PARTS BASED IN THE PRIMITIVES

The parametric projection is done by creating 40 different primitives. Each primitive is described with parameters and they are connected among themselfs with relationships: - until, -on, -in. The model is created by using primitives and it contains the information, about the geometry, the quality of the processing surface, the tolerances of the dimensions and the topological characteristics.
Primitives are described by the characteristic groups. We can analyse the characteristic groups with specific meaning, as certain sub groups of the geometrical elements from the mechanical parts. They can be grouped by the following aspects [2]:
            - the process of designing and  projection,
            - planing of the parts production and assembling.
Respecting the aspect of designing and projection of the part, the modelling is based on the fundamental geometric elements (straight line, circle, plane) or other separate complicated geometrical entities.
In designing of the parts, respecting the aspect of production, the modelling is based on the formation the geometrical entities with specific meaning which comes out from the moving kinematics of the cutting tool and manufactured workpiece.
The information about the mutual interconnection of the parts, during manufacturing of the product, represents the base for formation of the geometrical entities intended for planing the assembling.
In order to define the combination of process - action from which has a characteristic group of geometrical elements come out, then we use the fact that the geometrical elements projecting is a results of the actions during the process of consideration which is used in the actual process of production system. It is obvious that each combination of the process - action has the characteristic groups of geometrical elements.

## 3.   SOFTWARE FOR MODELLING AND REPRESENTATION

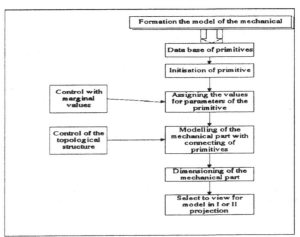

Figure 1:  Structure for software module for modelling of the mechanical part

In order of general division of mechanical product parts as rotational and non rotational, within the modeler for rotational parts whit unrotational shapes.

Software determination of the developing modeler [3] is done by the compiler Borland C++ v.4.0   (Fig. 1.). The best solution for icons choosing from the aspect of the screen's vectoring  is the use of matrix whit primitives icons.

The icons are grouped in four matrix, for four different groups of the primitives: basic, internal, external and external. All the icons together create the user menu (toolbar) in the interactive graphic editor (Fig.2).

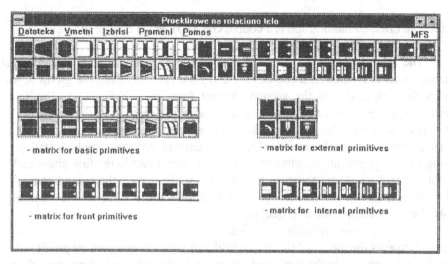

Figure  2:  Structural toolbar with matrix of the primitives icons

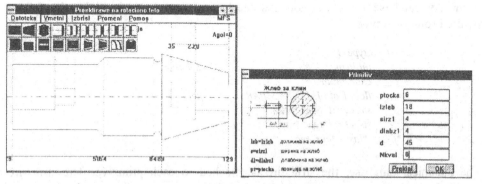

Figure 3.  Represenatation of the mechanical part model

## 4. METHODOLOGY FOR MODELLING OF MECHANICAL PARTS

It is necessary and basic for constructing one mechanical part to provide definition and connection of the separate primitives from which it is made into one part. As a result of that we have to make a methodology based on mathematical methods during the research.

The graphs law theory and linear algebra are implemented directly in the forming model of the mechanical part. Oriented multi-graphs, which are connected together typically, are used for describing the mechanical parts structure, with primitives.

By definition, oriented multi-graph is a couple G=(X,U) which consist of:
- definitive unempty group X,
- The relation U in X, where the elements in U are ribs of graph which are connecting directed couples in shape (x,y). x and y are elements from the group X, where x is always the first and y is the second element from the couple (x,y). The same two elemented subgroup from X may appear once or more as a member of the family U.

For the creation of a model of the product, it is necessary to enable correct storing of all data, concerning the whole model. This way enables different implementation in the following phases; graphical presentation of the mechanical part in the first phase and choice of suitable technological procedure for manufacturing the part, in the second phase.

The data storing for the mechanical parts, in this software, is done with dynamically data base. All the information's are arranged as:

*base ----- syllable ----- field*

The base represents the model of the mechanical part which is designed by primitives.

Syllable is each primitive in the base, and arranging of the syllable is done according to the function 'find'.

Field is a value of each parameter, which takes a part in primitive's defining.

Dynamically database for each modelled mechanical part is declared by the following structured rotational type:

```
struct rot type {
        int Prim;                         - syllable
        char *data;
        float Data [MAXNUMPAR+1];     -field
        struct rot_type *prior;
        struct rot_type *next;
} *first, *last, *cur_rec;
```

In the dinamical database, the structure is attached to each primitive. The structure is formed of geometric parameters which are defining the primitive and its tehnological characteristics such as dimension tolerance and surface roughness. If the picture of one mechanical part, includes several primitives, then that parameter could be initialised more then once with different values, which is provided from the dinamical database.

Two basic activities are creating data structures and algorithms which will work with these structures. There are several ways for structuring the data such as single and double connected list, binary tree, strings with priority and     other more complicated data structures.

Storing and sorting the data for the whole product model, in fact defining the dynamical database is made by creating two multi-graphs linked between each other as a double linked list (DLL).

In DLL data is stored in nodes, and the connection between them is provided by ribs (Fig.4.) in different directions. Null nodes are at the beginning and at the end, and between them data nodes are settled. Two pointers are setting out every node, the first one is NEXT and it is providing connection with the following node, and the other is PRIOR which is providing connection with the previous node. At the same time two pointers are arriving from two neighbour nodes, but exception are two NULL nodes, because from the beginning node only one NEXT pointer is setting out and only one NEXT pointer is arriving at the end node.

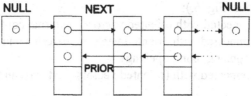

Figure 4:    Structure of double connected data list

In this research, this kind of DLL is created for the first group of basic primitives which are located along the axis of rotation and are implemented by adding them one by one. The primitives of this group are nodes of DLL. In fact, each node contain data for the parameters and attributes values for the certain primitive (Fig. 5.).

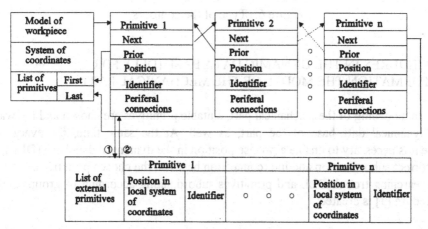

Figure 5:    Structure for sorting the data for primitives from the first group

The second group of primitives is systematised in one directed vector DATA [x1, x2, ...] which is connected with two pointers "to" and "from" to DLL for the first group of primitives. DLL is created during the research and is specific type because two pointers are included, the first is setting out from one DLL node, (node (1) - Fig. 5.) and the other one is arriving to DLL node (node (2) - Fig. 5.). These additional pointers are providing inserting data for selected primitives from the second group into certain place between sorted data in DLL.

This specific type of DLL structured database is formed also for the inside contour where the nodes of the inside cylinders are connected with the vector which is carrier of the structured data for possible inserted additional unrotating parts. These two basic DLL for the inside and outside contour are connected between them with NEXT and PRIOR pointer in node, which is data carrier for a primitive where the inside contour is perforated.

The model of the product in the base of the modeler can be presented in few ways depending of the contour (Fig. 6.):
    1. One-DLL connected with oriented vectors - external contour.
    2. Two-DLL connected with oriented vectors - external and internal contour
       (penetrated opening, blind opening).
    3. Two-DLL connected with (oriented vectors - external and two internal blind
       opening).

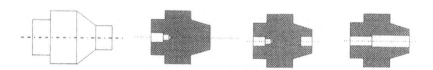

Figure 6: Type of the contours

## 5. PROCEDURES IN THE DYNAMIC DATA BASE (DLL) FOR
## FORMATION THE MODEL OF THE MECHANICAL PART

During the modelling of the mechanical part, suitable primitives are chosen and the way they fill the dynamical data base of the part, as well. At the same time, for every entered primitive it is necessary to enable a precise position in the dynamical data base (DLL), i.e. to enable correct subordination and interconnection between the chosen primitives.
For the searching trough DLL and primitives subordinating according to groups and type, the procedure [3] is created.

Basical algorithms in the procedure are:

-searching algorithm ( it finds entered syllable (primitive) and returns pointer to it),
-subordinate algorithm (examines the type of a entered syllable and returns information about the type througn parametar 'group').

*struct rot_type \*find (struct rot_type \*from, int way, int kakov)*
*struct rot_type \*from* - pointer for syllable in DLL, where is started the searching
*way* - mode for searching primitives in DLL
*kakov* - parametr for subordinate the primitives into type

Searching algorithm determinates the method at searching through DLL, and depending of the way (way - forward or backward) the pointer *info shows to every primitive in DLL, step by step in forward or in backward.

Subordinate algorithm through the parametar "group", enables subordination of the new entered primitive according of the type and its connection to svitable primitive in DLL.

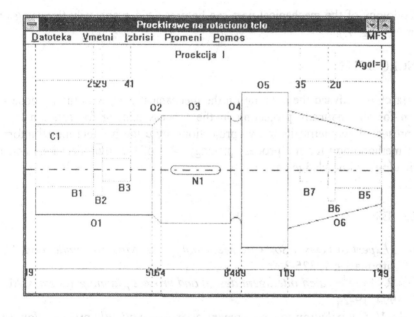

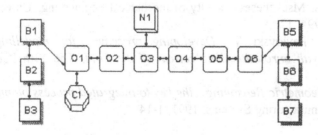

Figure 7: Graph of the model for mechanical part

The topological structure rules for creating the exact structure of the mechanical rotational part with unrotationaling retreat of the form, are the base for the procedure of primitive's connecting. The structured rules for formation the model of the mechanical part, are related on the way of the mutual interconnection of the former entered primitives [4].

The procedure for interconnection of primitives contains algorithm [3] which is based on few functions which contain rules for different groups and types of primitives.

```
     void Prva-proekcija ( void )
{
  osnovni I ();                      - interconecting the basic primitives
  nadvoresni I ();              - interconecting the external primitives
  leva_kontura I ();                 - interconecting the internal primitives
  celni I ();                        - interconecting the face primitives
  desna_kontura I ();                - interconecting the internal primitives
}
```

The methodology of the mechanical part can be presented graphically (by using graph), as shown on the (Fig. 7.).

## 6.  CONCLUSIONS

In the research is analysed the creating on the mechanical part, with the purpose - forming the system ·for automatically projecting in the process planing for rotational parts. The model is presented geometrically in two projections; by using product model information, it is made a methodology for the process planing, rules of the logic decisions and rules for using technological knowledge base.

REFERENCES:

1. Raeth P. *Expert Systems a Software Methodology for Modern Application*, IEEE, Los Alamitos-California,1991; 325-360.
2. Phillips R. *An integrated intelligent design and process planning system*, CAE Journal, December 1994, 19-24
3. Gecevska V. *Contribution to development of an automatically process planing systems for rotational parts*, Msc. theses, Faculty of mechanical engineering, University Sv Kiril i Metodij, Skopje, 1994; 120.
4. Gecevska V., Pavlovski V.; *Developing procedure for modelling rotationally mechanical parts with unrotationalyng retreat of the form*, The 13th ICPR, Jerusalem, Izrael, 1995
5. Chang T. C. *Geometric Reasoning - the key to integrated process planing*, 22th CIRP Int. Seminar on Manufacturing Systems, 1991, 1-14

# DESIGN OF HYDRAULIC PUMPS
# WITH PLASTIC MATERIAL GEARS

**A. De Filippi and S. Favro**
**Polytechnic of Turin, Turin, Italy**

**G. Crippa**
**NYLTECH Italia, Milan, Italy**

KEY WORDS: Power Gears, Hydraulic Pumps, Engineering Plastics.

ABSTRACT: Engineering plastics are already employed to some extent for small power drives. Problems arise, however, when greater stresses are applied. This paper examines the possibilities of their use instead of steel as gearwheels for hydraulic pumps, particularly since they would be less expensive and less noisy. Calculations showing that this is feasible with a glass-fibre reinforced polyamide are presented.

## 1. INTRODUCTION

By contrast with metals, polymers are endowed with viscoelasticity, and hence the deformation they undergo when subjected to a load is determined not only by its magnitude, but also by its rate and duration of application.
Plastic as opposed to steel gears have both *advantages*:
- less expensive fabrication
- good self-lubricating capacities
- lower mass for volume
- high damping capacity
- high resilience
- corrosion resistance

and *drawbacks*:
- low strength and modulus of elasticity values
- marked influence of temperature on these figures
- sensitivity to humidity.

Published in: E. Kuljanic (Ed.) *Advanced Manufacturing Systems and Technology*,
CISM Courses and Lectures No. 372, Springer Verlag, Wien New York, 1996.

## 2. FATIGUE AND WEAR RESISTANCE OF PLASTIC GEARWHEELS

An idea of the factors on which tooth strength depends can be drawn from the literature on the fatigue resistance of cogged wheels:
a) there does not appear to be a bending fatigue limit (strain below which the tooth has an infinite duration), one reason being that testing has not always been continued long enough to determine this point;
b) fatigue resistance is greatly influenced by the number of teeth: the more teeth there are (and hence the lower the modulus for the same pitch diameter), the higher the resistance;
c) fatigue resistance is influenced by the temperature, the degree of lubrication, and the process used to manufacture the cogged wheel.

Wear resistance is assessed in various ways: from the bending resistance of the tooth, or from wear itself, especially under dry conditions or during the starting transient.
Six conclusions were reached in a detailed study [1] of the behaviour of plastic cogged wheels:
a) torque values of 20 to 50 Nm can be transmitted by glass-reinforced plastic wheels for more than $10^7$ cycles;
b) plastic wheels usually break as the result of fatigue, though the effect of wear cannot be neglected. This depends on the materials in contact and the lubrication conditions;
c) lubrication procedure (grease or immersion in oil) does not seem to have a significant influence on fatigue resistance;
d) glass-fibre reinforcement may sometimes result in unacceptable wear;
e) wear is reduced by the addition of $MoS_2$.
f) even better performance is obtained when carbon fibres are employed.

## 3. METHODS FOR DETERMINING THE SIZE OF PLASTIC WHEELS

Standards for the design of plastic gearwheels have been laid down in several countries: VDI in Germany, AGMA in the USA, BSI in Britain. There are also methods devised by resin manufacturers (BASF, Du Pont, Celanese Plastics, Hoechst), and individual researchers (Niemannn, Chen). One or more of the following parameters are always considered:
- the bending stress of the tooth
- the contact pressure
- the tooth deformation.
Only the VDI standard and the Niemann and BASF methods, however, take all three parameters into account. Niemann's method [2] in particular appears to be one of the most complete and in good agreement with the experimental results.

The following polymer materials are most commonly used:
a) polyamide: excellent wear, impact and fatigue resistance, combined with a low friction coefficient; when reinforced with fibres, it offers acceptable dimensional stability, stiffness and strength up to 150 °C;
b) acetal resin: dimensional stability, stiffness, fatigue and abrasion resistance;

c) polycarbonate: exceptional impact resistance;
d) polyester resin: remarkable dimensional stability;
e) polyethylene: made in different density versions; low cost, good stiffness and hardness, but poor dimensional stability and reduced temperature resistance;
f) epoxy and phenol (thermosetting) resins: mostly employed with charges and reinforcing fibres.

## 4) GEARS FOR OIL PUMPS

The feasibility of using a polymer instead of steel for the gears of a specified oil pump was determined by calculating the geometrical proportions of the teeth in function of the flow rate, and calculating tooth strength according to Niemann. The values of the stresses on the flanks and at the root of the teeth were then compared with the experimental values of the Woehler curves.

### 4.1 *Pump geometry*
In a cogged wheel pump, the volumetric flow rate is proportional to the product of the number of teeth by the square of the modulus. The use of fewer teeth thus requires a higher modulus. The drawback of this reduction (which has been taken down to 10 teeth) is that there is interference during engagement that must be prevented by using corrected toothing. This, however, means that the head of the tooth must be thinner and hence less practical in terms of tightness and volumetric efficiency.
A detailed account of this question can be found in the literature. All that needs to be said here is that the Merritt and Enims methods for the correction of toothing give cogs with better wear resistance, whereas the Tuplin method results in teeth that are too pointed, though it has the advantage of creating a large compartment volume.
The pump specifications were:

| | | | |
|---|---|---|---|
| flow rate | 40 l/min | max pressure | 5 MPa |
| motor power | 4 kW | speed | 1500 rpm. |

The size of the gearing was determined by applying three basic criteria to give a total of 23 combinations of the geometrical factors (Table I):
a) small number of teeth (10-12) and high modulus (4.5 - 5.5 mm) with corrected toothing (Enims and Merritt) in keeping with the present manufacturing trend for steel teeth;
b) large number of teeth (30-40) and small modulus (2-3 mm) with normal tooth proportions;
c) in-between tooth and modulus values (17-20 and 3.5 mm respectively), with Enims and Merritt correction.

### 4.2 *Determination of tooth size*
The next step was to calculate according to Niemann the strength of a cogged wheel in PA6 polyamide with 35% glass fibres and supplemented with $MoS^2$. The maximum operating temperature was assumed to be 80 °C, which is normal in the tank of an ordinary hydraulic pump.

**TABLE I**

| CASE | No OF TEETH | MODULUS (mm) | FACE WIDTH (mm) |
|------|-------------|--------------|------------------|
| 1 (E) | 12 | 5 | 15.72 |
| 2 (M) | 12 | 5 | 15.72 |
| 3 (E) | 12 | 4.5 | 19.44 |
| 4 (M) | 12 | 4.5 | 19.44 |
| 5 (E) | 11 | 5 | 17.16 |
| 6 (M) | 11 | 5 | 17.16 |
| 7 (E) | 11 | 4.5 | 21.14 |
| 8 (M) | 11 | 4.5 | 21.14 |
| 9 (E) | 10 | 5.5 | 15.56 |
| 10 (M) | 10 | 5.5 | 15.56 |
| 11 (E) | 10 | 5 | 20 |
| 12 (M) | 10 | 5 | 20 |
| 13 (E) | 20 | 3 | 26.22 |
| 14 (M) | 20 | 3 | 26.22 |
| 15 (E) | 17 | 3.5 | 22.61 |
| 16 (M) | 17 | 3.5 | 22.61 |
| 17 (N) | 40 | 2 | 30 |
| 18 (N) | 35 | 2 | 34.42 |
| 19 (N) | 30 | 2 | 40.18 |
| 20 (N) | 30 | 2.5 | 25.71 |
| 21 (N) | 40 | 3 | 13.5 |
| 22 (N) | 35 | 3 | 15.3 |
| 23 (N) | 30 | 3 | 17.86 |

**(E): ENIMS   (M): MERRITT   (N): NORMAL.**

The following fatigue resistance values at this temperature were derived from the results in [1]:

$$C = 40 \text{ Nm} \quad N \approx 10^7 \text{ cycles}$$
$$C = 24 \text{ Nm} \quad N \approx 10^8 \text{ cycle}$$
$$C = 9 \text{ Nm} \quad N \approx 10^9 \text{ cycles.}$$

The last two values were extrapolated, since the tests were  run from $5 \times 10^5$ to $2 \times 10^7$ cycles. The operation is none the less regarded as acceptable, because the presence of a possible fatigue limit not reached in the tests would lead to a situation more favourable than that extrapolated.

The results show that the bending stress at the base of a tooth (fig. 1) is never a factor that limits attainment of its assumed maximum life. The stress induced by contact of the flanks (fig. 2), on the other hand, prevented it in 21 cases, with the exception of conditions 17 and 21.

This fact, however, is in conflict with the experimental results, which indicate that the true critical variable is bending of the teeth. There are two possible explanations of this contradiction:

- the limit value of the stress due to pressure on the flanks was actually greater than the calculated value;

- the question of wear may predominate in the case of wheels with few teeth and hence a high modulus, since they generate higher sliding speed and contact pressure values. The calculations, in fact, show that the problem raised by this pressure decreases in importance as the number of teeth increase.

Three conclusions, therefore, can be drawn from these results:

a) all the wheels examined can be used for up to 100 hours without problems;

b) all the wheels can be accepted for up to 1,000 hours, though the wear limits for those with few teeth are now close;

c) only wheels with a large number of teeth can be adopted for periods of 1,000 to 10,000 hours. Here, too, the limit values for contact pressure and stress at the base of the tooth are nearly reached.

## 5. ECONOMIC ASSESSMENTS

Costs were assessed for the two geometrical combinations ensuring attainment of the maximum service life, assuming an output of 35,000 pieces per year and amortisation over the course of 4 years. The use of a one or two-cavity injection mould was also considered. The first solution ensures the geometrical identity of the moulded wheels (within the tolerance limits of the manufacturing process), the second results in a lower per unit cost (Table II). In order to evaluate the plastic gear cost and to simulate the injection process, the "Plastics & Computer International" (Milan, Italy) software was used and expecially the MCO (Moulding and Cost Optimization) program. The manufacturing schedule envisages comoulding of the plastic cogged wheel on a steel drive shaft. The wheel is held on the shaft by a groove and knurling. The cost estimates for the two combinations and the two moulding solutions set out in Table II indicate that this hydrid polyamide/steel drive offers a 25-27% saving compared with the current cost of about 10,000 £ per piece for an all-steel equivalent.

## 6. CONCLUSIONS

Two conclusions can be drawn from the calculations carried out in this study:

- replacement of steel by a polyamide resin reinforcement with glass fibres is a feasible solution within the limits of the requirements imposed, and the tooth life is more than acceptable;

- this alternative is economically sound, since the higher cost of the equipment needed for the injection mould is fully offset by a shorter manufacturing schedule.

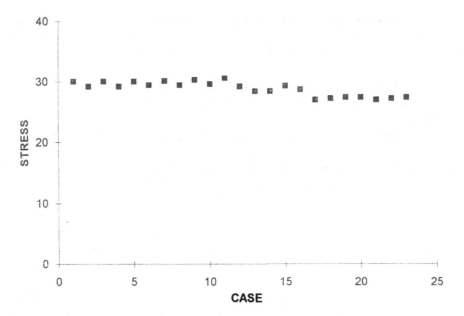

**Fig. 1. Bending stress as a function of gearwheel dimensions (limiting values: 31.6 Mpa for $10^9$ cycles).**

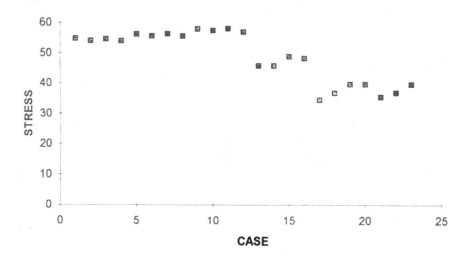

**Fig. 2. Stess due to flanks contact as a function of gearwheel dimensions (limiting value: 34.4 Mpa for $10^9$ cycles).**

**TABLE II**

| CONDITIONS | TOTAL COST (£/piece) |
|---|---|
| m = 2 mm; one-cavity mould | 7365 |
| m = 2 mm; two-cavity mould | 7275 |
| m = 3 mm; one-cavity mould | 7500 |
| m = 3 mm; two-cavity mould | 7370 |

## 6. REFERENCES

1. Crippa G., Davoli P.: Comparative Fatigue Resistance of Fiber Rinforced Nylon 6 Gears, ASME Sixth International Power Transmission and Gearing Conference, Phoenix, 1992

2. Niemann G., Winter H.: Elementi di macchine, Vol. II, Edizioni di Scienza e Tecnica, Milano, 1986

3. Favro S.: Progettazione e fabbricazione di particolari meccanici in plastica rinforzata con fibre, Graduation Thesys in Mechanical Engineering, tutor Augusto De Filippi

4. Crippa G., Davoli P.: Fatigue Resistance of Nylon 6 Gears, ASME Fifth International Power Transmission and Gearing Conference, Chicago, 1989

5. Du Pont: Design Handbook, 1992.

**The Graduation Thesis of dott. ing. S. Favro obtained an award from ASSOFLUID (the italian association of pneumatic and hydraulic devices builders).**

# CORRELATING GRINDED TOOL WEAR
# WITH CUTTING FORCES AND SURFACE ROUGHNESS

**R. Cebalo**

**Polytechnic of Turin, Turin, Italy**

**T. Udiljak**

**NYLTECH Italia, Milan, Italy**

KEY WORDS: Tool Wear, Artificial Tool Wear, Cutting Forces, Surface Roughness

ABSTRACT: These work presents the results of investigation conducted at Machine tool department of Faculty of Mechanical Engineering and Naval Architecture, Zagreb, Croatia. The aim of investigation was to examine the influence of grinded tool flank wear on cutting forces, surface roughness and coefficient of chip deformation (compression). The tests were done on lathe for ordinary and orthogonal cutting conditions, and by applying uncoated carbide inserts with, and without grinded flank wear. Comparison between influence of grinded and ordinary tool wear shows similarities when dealing with cutting forces, while the influence on surface roughness is difficult to compare. The tests were done on material Ck 15 (DIN 17006).

## 1. INTRODUCTION

The lack of clear and generally accepted physical explanation of tool wear process, continuously increasing number of CNC machine tools and FMS, same as significant growth in use of other types of unmanned manufacturing equipment, emphasis the significance of investigation in machining processes. Stochastic character of cutting process and random deviations of cutting conditions demands a large number of experiments and application of adequate statistical data processing. Higher level of automation, number of new materials for production equipment, new tool materials and tool geometry implies a need for fast, reliable, and economically feasible procedures for tool life testing and tool monitoring. Such procedures have their significance as methods for reaching predicted tool life, same as they have importance for validating or estimating mathematical models necessary for monitoring of cutting tool. This article deals with some aspects of possibility to replace ordinary wear with grinded wear in order to make testing procedures more economical.

Published in: E. Kuljanic (Ed.) *Advanced Manufacturing Systems and Technology*,
CISM Courses and Lectures No. 372, Springer Verlag, Wien New York, 1996.

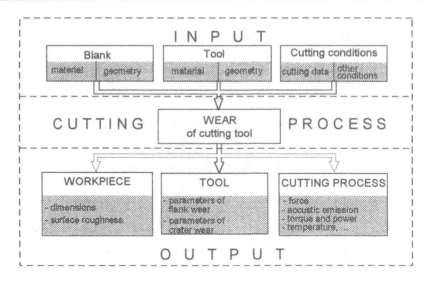

Figure 1.  Block sheme of tool wear process [3]

## 2. IMPORTANCE OF TOOL WEAR

Tool wear is result of interaction between tool, workpiece, and cutting conditions, figure 1. It is phenomena that must be considered in the phase of process planing and in the phase of process monitoring and process control [3].

For better and more reliable process planning it is important to have a cutting data as accurate as possible. Therefore it is necessary to investigate in cutting processes in order to reach more exact and physically based models, connecting input and output values of cutting process.

If not monitored, tool wear could cause process interrupts, damages or breakage of tool and damage of workpiece. Many researchers and industry experts have pointed out that most often cause of interrupts in cutting process on machine tools is tool. In order to avoid (prevent) the unwilling consequence or damages it is necessary to get timely information on tool condition. The importance of tool condition monitoring (TCM) is stressed in many investigations and articles [2, 4, 8]. One of the most comprehensive papers [1] presenting the results of investigation of TCM systems encompass 26 system manufacturers and 20 system users (a total of 1161 installed systems) of TCM in a range of industries. Those results and data on figure 2, which present utilization of TCM systems, are base for following statements:

- more than 80% of applied TCM systems are used for monitoring in turning and drilling processes;
- nearly 80% of TCM systems are used for monitoring of tool wear and tool breakage;
- all industrial applications of TCM systems are based on indirect methods for TCM;
  more than 80% of TCM failures have their cause in wrong selection, operator errors and interfacing;
- today's TCM systems are going in direction of applying multi sensor approach, and a new programming techniques for signal data processing.

It is one more proof that our ability for exact mathematical description of tool wear process is limited. At the same time it is one more impulse for further investigation of tool wear, and cutting process generally.

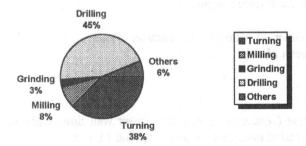

Figure 2. Distribution of monitoring systems by area of application [1]

## 3. EXPERIMENT

Cutting forces are very often used to indicate the condition of cutting tool. They are easy to measure, and are good parameters for evaluation of cutting process generally. Surface roughness is also very valuable source of information on cutting process, and it is, together with workpiece dimensions, also very often used. Trying to reduce the experiment expenses and experiment time, this article deals with possibility to replace ordinary wear with grinded tool flank wear in order to find the relations between tool wear and cutting forces, and between tool wear and surface roughness.

Two experiments were conducted, first one for measuring of cutting forces, and second one for measuring surface roughness. In both cases a central composite experiment with three independent variables on two levels, "$2^3$", was applied. Cutting speed $v_c$, feedrate $f$, and flank wear $VB$ were selected as independent variables, while all three component of cutting force and some parameters of surface roughness were measured as output variables.

### 3.1 Experimental conditions

**Workpiece:**
  material: Ck 15
  diameter:     D=40 mm (for cutting forces);   D=50 mm (for surface roughness)
**Cutting tool:**
  a) holder:     SCLCR 1616H09
    insert: CCMW 09T304 (SINTAL, Zagreb)
  b) holder:     STGCR 1616H11
    insert: TCMW 110204 (SINTAL, Zagreb)
  basic geometry elements of cutting edge:

$\gamma=0°$;    $\alpha=7°$;    $\lambda=0°$.
inserts grade:  SV25 (SINTAL-s mark corresponding to ISO P20-P30)

**Machine tool:**    Lathe type TES 3/2000 (for cutting forces)
                          Lathe type TES 1 (for surface roughness)

**Measuring instruments:**   Threecomponent analog force transducer "Fisher"
                                     Perthometer M4Pi

### 3.2 Mathematical model

For applied design of experiment (central composite experiment with three independent variables on two levels, "$2^3$"), mathematical model according to equation 1 is used.

$$R = C \cdot \prod_{i=1}^{n} f_i^{x_i}$$   (1)

The values for independent variables are presented in table 1. The minimal value of flank wear is 0 mm (new insert), what can not be used for calculations, so the value of *(1-VB)* is used. The experimental results for cutting forces are compared with calculation of forces according to Kienzle equation. Applied depth of cut was 2 mm for cutting forces, and 1.5 mm for surface roughness. Because previous experiments [11] proofs that grinded tool flank wear has little or no influence on passive force, orthogonal cutting were applied for measuring of cutting forces.

Table 1. The values for independent variables

|  | Cutting forces | | | Surface roughness | | |
|---|---|---|---|---|---|---|
|  | Cutting speed $v_c$ [m/min] | Feedrate $f$ [mm/r] | Flank wear $VB$ [mm] | Cutting speed $v_c$ [m/min] | Feedrate $f$ [mm/r] | Flank wear $VB$ [mm] |
| Maximal value | 106.8 | 0.3 | 0.5 | 165 | 0.3 | 0.5 |
| Minimal value | 51.5 | 0.16 | 0 | 83 | 0.16 | 0 |
| Medium value | 75.4 | 0.22 | 0.3 | 118 | 0.22 | 0.3 |

## 4. RESULTS

### 4.1 The influence of flank wear on cutting forces

After processing of experimental data, the parameter values, for applied mathematical model (2) are presented in table 2.

$$F_x = C \cdot v_c^{p1} \cdot f^{p2} \cdot (1 - VB)^{p3}$$   (2)

Table 2. Parameter values for mathematical model of cutting forces

| Cutting force $F_c$ | | | | Cutting force $F_f$ | | | |
|---|---|---|---|---|---|---|---|
| Tool holder SCLCR 1616H09, and insert CCMW 09T304 | | | | | | | |
| C | p1 | p2 | p3 | C | p1 | p2 | p3 |
| 3688.08 | 0 | 0.7432 | 0 | 695.82 | 0.133 | 0.4495 | -0.0803 |
| Tool holder STGCR 1616H11, and insert TCMW 110204 | | | | | | | |
| C | p1 | p2 | p3 | C | p1 | p2 | p3 |
| 2953.13 | 0 | 0.6696 | 0 | 981.62 | 0.0582 | 0.4663 | -0.0976 |

With data in table 2, the difference between feed force for VB=0 and feed force for VB=0.5 mm is 5.7% for insert CCMW, and 7% for insert TCMW. According to extended Kienzle equation, the influence of tool wear on cutting forces goes up to 30%, what have not been proved with this experiment.

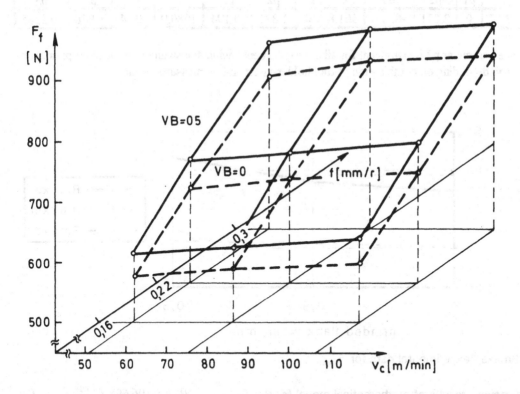

Figure 1.    Dependency of $F_f$ on $v_c$, $f$, and $VB$ for insert CCMW 09T304

During experiment, the coefficient of deformation $\theta$, was measured but no significant influence of grinded wear on coefficient of deformation was noticed. There was also no significant change in chip shape with change of grinded tool flank wear.

## 4.2 The influence of grinded flank wear on surface roughness

Another experiment were performed, with same tools, in order to examine the influence of artificial tool flank wear on parameters characterizing surface roughness. The same workpiece material were used, with small differences in cutting data, table 1. The results obtained for $R_{max}$ are compared with theoretical results according to equation 4.

Table 3. Parameter values for mathematical models of surface roughness parameters

| $R_a$ | | | | $R_z$ | | | | $R_{max}$ | | | |
|---|---|---|---|---|---|---|---|---|---|---|---|
| Tool holder  SCLCR 1616H09, and insert  CCMW 09T304 | | | | | | | | | | | |
| C | p1 | p2 | p3 | C | p1 | p2 | p3 | C | p1 | p2 | p3 |
| 42.62 | 0 | 1.482 | 0 | 107.4 | 0 | 1.025 | 0 | 106.7 | 0 | 0.973 | 0 |
| Tool holder  STGCR 1616H11, and insert  TCMW 110204 | | | | | | | | | | | |
| C | p1 | p2 | p3 | C | p1 | p2 | p3 | C | p1 | p2 | p3 |
| 129.1 | 0 | 2.071 | -0.226 | 501.8 | 0 | 1.871 | -0.194 | 1080.9 | -0.14 | 1.868 | -0.2285 |

The parameter p2 is significant for all the experiments, while the significance of other parameters vary, depending on combination of tool holder-insert and on measured value.

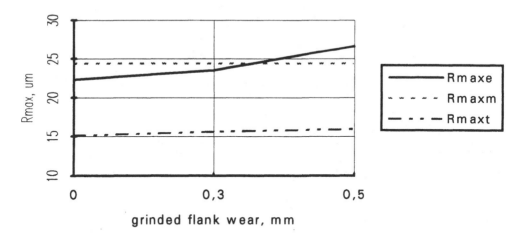

Rmaxe - experimental data for $R_{max}$

Rmaxm - results of mathematical model for $R_{max}$:     $$R_{max} = 106.69 \cdot f^{0.9726} \qquad (3)$$

Rmaxt - theoretical values for $R_{max}$:     $$R_{max} \approx \frac{f^2}{8 \cdot r_\varepsilon (1 - VB \cdot tg\alpha)} \qquad (4)$$

Figure 4.  Comparison of $R_{max}$ values

## 5. CONCLUSION

This article investigate some aspects of possibility to replace ordinary tool flank wear with grinded tool flank wear for orthogonal turning.

Experiments showed that for cutting processes with dominant flank wear, obtained results are mostly in good agreement with results obtained with ordinary weared cutting tools, although the influence of tool wear is less then for non orthogonal cutting. For conducted experiment grinded flank wear was of no influence for the coefficient of deformation or the chip forms.

Comparison between influence of grinded and ordinary tool wear shows similarities when dealing with cutting forces, while the influence on surface roughness is difficult to compare. The flow for $R_{max}$ calculated according to mathematical model and theoretical model is very similar, while the experimental results are different. The replacement of ordinary flank wear with grinded flank wear could be recommended for cutting processes when subject of investigation is the influence of flank wear on feed force $F_f$ (and partly cutting force $F_c$). For more reliable application, the experiments should be performed for variety of workpiece materials, tool geometry, and tool materials.

## REFERENCES

1. Byrne, G.(2), Dornfeld, D.(2), Inosaki, I.(1), Kettler, G., König, W.(1), Teti, R.(2): Tool Condition Monitoring (TCM) - The Status of Research and Industrial Application, Annals of The CIRP, 41(1995)1, 541-568
2. Kluft, W.: Tool monitoring systems for turning (in german), doctoral thesis, TH Aachen 1983.
3. Udiljak, T.: Tool wear - necessity to predict and to monitor, 6th DAAAM Symposium, Krakow, 1995, 335-336
4. Waschkies, E., Sklarczyk, C., Hepp, K.: Tool Wear Monitoring at Turning, Journal Of Engineering for Industry, Vol. 116, 1994, 521-524
5. Chen N.N.S., Pun W.K.: Stresses at the Cutting Tool Wear Land, Int. J. Mach. Tools Manufact. Vol. 28 No. 2 1988
6. Cebalo, R.: Identification of materials and automatically determination of cutting data for turning (in croatian), Suvremeni trendovi proizvodnoga strojarstva, Zagreb, 1992, 15-20
7. Kuljanic, E.: Effect of Stiffness on Tool Wear and New Tool Life Equation, Journal of Engineering for Industry, Transact. of the ASME 9, Ser. B, 1975, 939-944
8. Colgan, J., Chin, H., Danai, K., Hayashi S.R.: On-Line Tool Breakage Detection in Turning: A Multi-Sensor Method, Journal of Engineering for Industry, 116, 1994, 117-123
9. Ravindra H.V., Srinivasa Y.G., Krishnamurthy, R.: Modelling of tool wear based on cutting forces in turning, Wear, 169, 1993, 25-32
10. Damodarasamy, S., Raman, S.: An inexpensive system for classifying tool wear states using pattern recognition, Wear, 170, 1993, 149-160
11. Udiljak, T.: On possibilities for investigating the influence of tool wear on cutting forces, 4th DAAAM Symposium, Brno, 1993, 355-356

# TUBE BULGING WITH NEARLY "LIMITLESS" STRAIN

**J. Tirosh, A. Neuberger and A. Shirizly**

**Technion - Israel Institute of Technology, Haifa, Israel**

**KEY WORDS:** Tube Bulging, Tube Forming, Upper Bound, Normality Rule

**ABSTRACT:** Based on fundamental plasticity (the normality rule, the convexity of the yield function and material incompressibility) it becomes clear that there is a possibility to locate a distinctive loading path on the plastic yield function at which bulging of tube (by internal fluid pressure along with axial wall compression) can be performed without thinning of the tube wall. At this loading point the principal stress ratio remains $\beta = -1$ ( representing a pure shear) and consequently the original thickness is forced to remain unchanged throughout the bulging process. Testing this hypothesis is the aim of this paper. The experiments were done on Aluminum (Al 5052-0) tubes, using a specially-dedicated machine.

## 1. INTRODUCTION

The bulging of thin walled tubes can be considered as a typical example of sheet forming process. Its uniform deformation is limited by the eventual occurrence of plastic instability (like localized wall thinning, rupture, etc.). There are various ways by which such phenomena in sheet forming are analyzed, depending on the criteria used. For instance, Swift's 'diffuse necking' criterion [1] is based on the point at which the overall stretched sheet starts to unload. Hill's localized necking of rigid-plastic solid [2] is based on a bifurcation stress with a conjugate orientation of an extension-free line in the plane of the sheet (if exists). Marciniak and Kuczynski's [3,4], (shortly M-K), postulated the existence of a 'band of imperfection' across the sheet inside which strain rotation to plane

Published in: E. Kuljanic (Ed.) *Advanced Manufacturing Systems and Technology*, CISM Courses and Lectures No. 372, Springer Verlag, Wien New York, 1996.

strain condition leads eventually to plastic instability (recently reviewed comprehensively by Marciniak and Duncan [5]). Hutchinson and Neale's [6,7] proposed the 'long wavelength' treatment of the M-K-type imperfection-sensitivity analysis and also expanded the bifurcation analysis to include the incremental theory, beside the deformation theory.

## 2. DESCRIPTION

The schematic bulging process is shown in Fig.1.

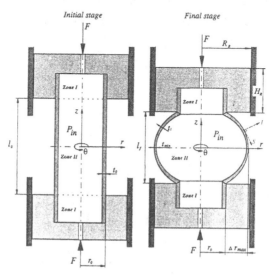

**Fig. 1:** The general scheme of bulging before and after the operation.

The Von-Mises yielding is formulated in terms of the two in-plane stress components (drawn in the above figure), which reads

$$\sigma_s^2 + \sigma_\theta^2 - \sigma_s \sigma_\theta = \sigma_0^2 \quad (\sigma_0 \text{ is the yielding in unidirectional tension}) \quad (1)$$

The point (d) on the associated ellipse represents the desired stress ratio where, by the normality rule, the two acting strains are equal in magnitude and opposite in sign, namely the stress ratio $\beta=-1$ leads to the strain ratio of $\rho=-1$ which is;

$$\varepsilon_s = -\varepsilon_\theta \tag{2}$$

In view of the material incompressibility , $\varepsilon_s + \varepsilon_\theta + \varepsilon_t = 0$, the third strain component $\varepsilon_t$ (the strain across the tube thickness) remains automatically zero. It means that a loading path which can be sustained at point (d) (given in Fig. 2) assures that no thinning will arise and the bulging can proceed, essentially without limits. Such a peculiar path is feasible if we can actually produce simultaneously positive hoop stress $\sigma_\theta$ (by internal fluid pressure) and meridional compressive stress $\sigma_s$ (by external compressive load) in a ratio of $\beta=-1$.

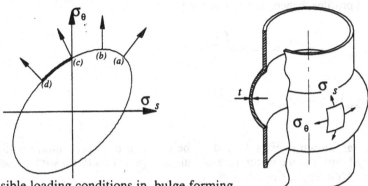

**Fig.2:** Possible loading conditions in bulge forming.

In order to generate the required stress ratio, a stress solution for the non-steady bulging process is needed. Since the geometry of the bulge is too complicated to have an **exact** solution we replace it by approximate solution. This is done next.

## 3. UPPER BOUND ANALYSIS

To maintain the desired stress ratio on the bulging process we need to synchronize the two independent load sources : the internal pressure, $p_{in}$, acting inside the tube and produces the circumferential stress $\sigma_\theta$, and the external **axial** pressure, $p_{ax}$, which furnishes, mainly, the meridional stress $\sigma_s$. Due to the large changes in the geometry during the process, the two sources are eventually interacting one with the other in generating the stresses. As mentioned before, an exact formulation of such evolutionary process seems intractable. Instead, an upper bound analysis is employed.

**The admissible velocity field**

Observations of actual tests indicate that the current amplitude of the bulged tube during the expansion process may be represented by a cosine-like profile. The difficulties in doing so lies in deciding about the appropriate geometrical boundary conditions of the profile. It certainly ranges between the two extremes described in Fig. 3: either with a smooth transition from the clamped edges (the 'continuous model') or with a sharp plastic hinge (the 'non-continuous model').

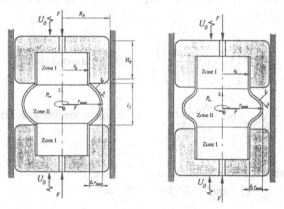

**Fig. 3:** The two admissible deformation modeling of the tube bulging process: The continuous and non continuous models.

The associated profiles, hence, are respectively

$$r(z, \Delta r_{max}) = r_0 + \frac{\Delta r_{max}}{2}\left[1 + \cos\left(\frac{2\pi z}{l_f}\right)\right]$$

(3)

$$r(z, \Delta r_{max}) = r_0 + \Delta r_{max}\cos\left(\frac{2\pi z}{l_f}\right)$$

(4)

It turned out, as shown in Figs 4a and 4b below, that both presentations (3) and (4) agree quite well with the measurements of bulged specimens (with some superiority to the plastic flow with hinges described by (4)

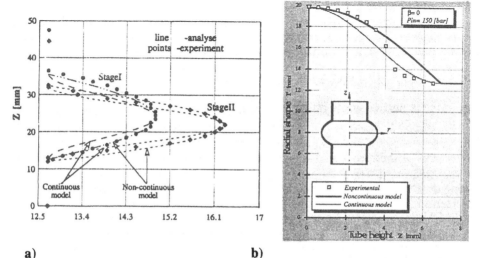

**a)**                         **b)**

**Fig. 4: a)** Comparison between the simulated configuration and the actual shape
**b)** Comparison between the geometrical changes in the two deformation models

Either of them can suit the purpose of devising an admissible velocity field. The outcome admissible velocity field is:

$$U_r^* = \begin{cases} \dfrac{d}{dt}\left\{r_0 + \dfrac{\Delta r_{max}}{2}\left[1 + \cos\left(\dfrac{2\pi z}{l_f}\right)\right]\right\} & @ \quad \text{continuous model} \\[3mm] \dfrac{d}{dt}\left\{r_0 + \Delta r_{max}\cos\left(\dfrac{2\pi z}{l_f}\right)\right\} & @ \quad \text{non continuous model} \end{cases}$$

$$U_\theta^* = 0$$

(5)

The boundary conditions for both models are:

$$U_z^* = -U_o \quad \text{at} \quad l_f/2, \quad \text{and} \quad U_z^* = 0 \quad \text{at} \quad z = 0.$$

The derived admissible strain rates are;

$\dot{\varepsilon}_\theta^* = \frac{1}{r}\frac{\partial U_\theta^*}{\partial \theta} + \frac{U_r^*}{r},$ ( but due to symmetry it is just $\dot{\varepsilon}_\theta^* = \frac{U_r^*}{r}$ ).From the strain ratio definition and the plastic incompressibility one gets the other strain rate components as

$$\dot{\varepsilon}_s = \rho\dot{\varepsilon}_\theta , \qquad \dot{\varepsilon}_t = -(\rho+1)\dot{\varepsilon}_\theta. \tag{6}$$

By using the above velocity field one can reach the upper bound expressions for the internal fluid pressure $p_{in}$ and the axial compressive pressure, $p_{ax}$, (the associated algebra is in Ref.8), as;

$$p_{in} = \frac{\left\{ \frac{2\pi l_0 t_0}{n\sqrt{3}}\left(r_0 - \frac{t_0}{2}\right)\sqrt{\rho^2 + \rho + 1}\int_v \sigma_0\left(\frac{2\varepsilon_\theta^*}{\sqrt{3}}\sqrt{\rho^2 + \rho + 1} + \varepsilon_0\right)^n \dot{\varepsilon}_\theta^* dv + \\ +4\pi\bar{\sigma}t_0\left(r_0 + \frac{t_0}{2}\right)\tan\left(\frac{\alpha}{2}\right) - \frac{U_0\beta\bar{\sigma}}{\sqrt{\beta^2 - \beta + 1}} \right\}}{U_0\frac{\pi}{4}\left(4r_1^2 - d_3^2\right) + \int_s\left(\bar{U}_z\sin a + \bar{U}_r\cos a\right)ds} \tag{7}$$

$$p_{ax} = \frac{R_r H_R m K_0}{r_0 t_0} - \frac{\beta\bar{\sigma}}{2\pi r_0 t_0\sqrt{\beta^2 - \beta + 1}}$$
$$+ \frac{\left\{ \frac{l_0 t_0}{n\sqrt{3}}\left(r_0 - \frac{t_0}{2}\right)\sqrt{\rho^2 + \rho + 1}\int_v \sigma_0\left(\frac{2\varepsilon_\theta^*}{\sqrt{3}}\sqrt{\rho^2 + \rho + 1} + \varepsilon_0\right)^n \dot{\varepsilon}_\theta^* dv + \\ +2\bar{\sigma}t_0\left(r_0 + \frac{t_0}{2}\right)\tan\left(\frac{\alpha}{2}\right) - \frac{U_0\beta\bar{\sigma}}{2\pi\sqrt{\beta^2 - \beta + 1}} \right\}}{r_0 t_0\left(U_0\frac{\pi}{4}\left(4r_1^2 - d_3^2\right) + \int_s\left(\bar{U}_z\sin a + \bar{U}_r\cos a\right)ds\right)} \tag{8}$$

The geometrical notation is given in Figs. 1 and 3. The material parameters are defined in the next paragraph. However the physical meaning of the above solutions (7,8) as to how the above pressures should vary simultaneously in order to follow certain stress ratios are given in Fig. 5

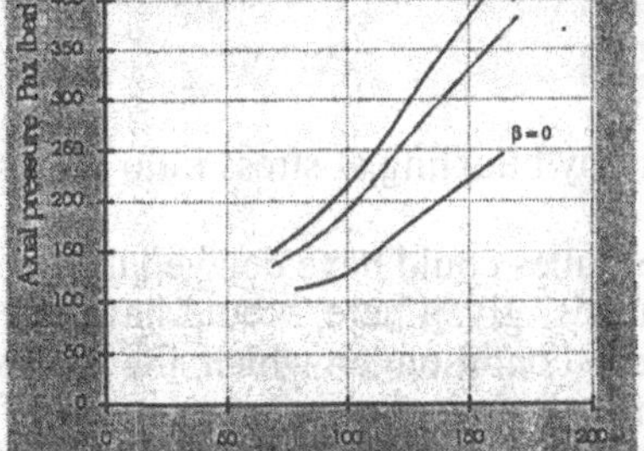

**Fig. 5:** The theoretical requirement for the magnitude of the two power sources of the bulging machine in order to provide a bulging operation with a constant principal stress ratio β employing the non continuous model (4) )

## 4. EXPERIMENTAL STUDY

The prime intention was to examine the technological benefit (if at all ) which may arise from working at a negative strain ratios of $\varepsilon_{min}/\varepsilon_{max}$ =-1 (or at least beyond $\varepsilon_{min}/\varepsilon_{max}$ =- 1/2 ). A series of experiments were conducted on a specially built machine capable of supplying both internal fluid pressure and external axial load, **independently**. The material of the tubes was aluminum (AL.5052-0), with a length of 50mm, external diameter of 1" (25.4 mm) and thickness of 0.035" (0.889 mm). The material properties were: Yield stress: 90 MPa, ultimate stress: 195 MPa, maximum elongation of 25 percent. By using the power low hardening, the constitutive equation becomes:

$$\bar{\sigma} = 340\bar{\varepsilon}^n \text{ [MPa], with } n=.223. \tag{9}$$

To facilitate the strain measurements, the tubes were imprinted with small circles and/or squares by photo chemical plating . The lubricant along the sliding interfaces was the commercial Molykote (a paste of oil with $M_0S_2$). Each bulging process followed a pre-computerized path for the internal pressure and the axial load, to provide a constant stress ratio. The stress ratios which were tested are: $\beta$ =-1, -1/2, 0.

## 5. DISCUSSION AND RESULTS

1) In order to achieve a stress path of pure shear $\beta$=-1 (which is also the strain ratio of $\rho$=-1), the bulging process necessitates a relatively high compressive load. This high load causes premature buckling of the tube, as shown in Fig. 6.

**Fig. 6**:An example of failure by buckling at stress ratio of $\beta$=-1/2.

It seems, though, that thicker tubes could have delayed the occurrence of early buckling, but the associated limit analysis, given here, would have then became less precise by deviating from the plane stress condition on which the analysis is based.
2) The aspiration for reaching 'infinite strain' was invoked from the normality rule at point (d) on Von -Mises locus shown in Fig. 2. Only at this point, due to incompressibility, the strain in the thickness direction should stay null. This peculiar condition has its equivalence in Hutchinson and Neale's bifurcation analysis [6,7] where the strain which minimizes the bifurcation stress was found to be

$$\varepsilon = \frac{n}{1+.\rho} \qquad \text{(n=the strain hardening exponent of (9))} \qquad (10)$$

Therefore, in point (d), where $\rho =-1$, the predicted strain according to (10) is indeed infinite.

3) The limit strain 'near' the stress path of $\rho =-1/2$ (which is $\beta =0$) was terminated by strain localization followed by rupture ( 'near' means that the unmeasurable frictional shear factor 'm' was assumed to be m=0.05 ). This is shown in Fig. 7 below:

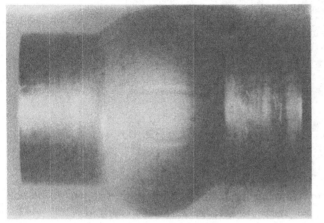

**Fig. 7:** Strain localization before ensuing of rupture at $\beta=0$ and a non-failed product.

The measured strain in sound products had reached the level of $\varepsilon_{\theta}\max =52\%$ . It is about 10 percent above the numerical value calculated by (10) based on the strain hardening exponent of the tested tubes (n=.22 for Al- 5052-0).

4) In spite of our experimental effort to reach 'infinite strain', it is seen (from the limited data collected till now in Fig. 8) that the recommended 'working zone' for bulging of thin tubes is the zone subtended in the vicinity of the path of $\rho = -1/2$ ($\beta=0$).

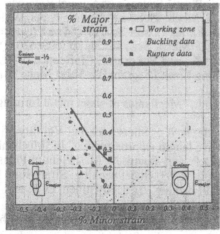

**Fig. 8:** The recommended 'working zone' (bright white) for bulging processes (with Al 5052-0. )

5) The same conclusion was reached by a similar work in Japan (as reported by Ken-ichi Manabe at al [11]) using aluminum tubes having somewhat different properties (Al 1050-H18, Al 1070- H18) and different sizes . The results of both works are given in Fig. 9

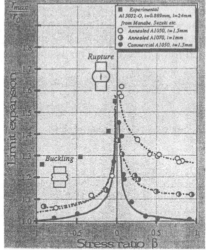

**Fig. 9:** The optimal path for maximizing the forming limit strain happens to be near β=0.

6) Apparently, there is no reason (beside the technical risk of premature buckling) why it is not possible, in principle, to overpass $\rho = -1/2$ and to get closer to the ratio of $\rho = -1$, at which 'infinite' strain is a theoretical outcome.

## REFERENCES

1. Swift, H.W., "Plastic Instability Under Plane Stress", J. Mech. Phys. Solids, vol. 1,pp. 1-18 , 1952.
2. Hill, R., "On Discontinuous Plastic States, with Special Reference to Localized Necking in Thin Sheets", J. Mech. Phys. solids, vol. 1, pp. 19-30 ,1952.
3. Marciniak, Z., and Kuczynski, K., "Limit Strains in The Processes of Strech-Forming Sheet Metal", Int. J. Mech. Sci., vol. 9, pp. 609-620, 1967.
4. Marciniak, Z., Kuczynski, K., and Pokora, T., "Influence of The Plastic Properties of a Material on The Forming Limit Diagram For Sheet Metal", Int. J. Mech. Sci., vol. 15, 789-805 ,1973.
5. Marciniak Z., and Duncan, J., "Mechanics of sheet metal forming", Edward Arnold Pub. 1992.
6. Hutchinson, J.W., and Neale, K.W., "Sheet necking: Validity of Plain Stress Assumption of the long Wave length Approximation" In "Mechanics of Sheet Metal Forming",(Ed.Koistinen, D.P. and Wang, N. M.), Plenum Press, pp.111-150,1978.
7. Hutchinson, J.W., and Neale, K.W.,"Sheet necking: Strain Rate Effects" In "Mechanics of Sheet Metal Forming", (Ed. Koistinen, D. P. and Wang, N. M.), Plenum Press, pp. 269-283, 1978.
8. Tirosh, J., Neuberger, A. and Shirizly, A. " On Tube Bulging by Internal Fluid Pressure with Additional Compressive Stress " to appear in Int.J. Mech. Sci., 1996.

# IMPROVING THE FORMABILITY OF STEEL
# BY CYCLIC HEAT TREATMENT

**B. Smoljan**
**University of Rijeka, Rijeka, Croatia**

KEY WORDS: Heat Cycling, Formability, Strain Hardening Parameters

ABSTRACT: The performance of cyclic heat treatment in steel formability improving has been investigated. The change of steel formability by cyclic heat treatment directly depends on structure fineness and thermal deformation hardening. Better mechanical properties are obtained by the appropriate cyclic heat treatment then by isothermal annealing.

## 1. INTRODUCTION

Making the structure finer and more uniform is the most effective way of improving the plastic properties of steel [1].

One of the most effective methods of the prior austenitic grain size refinement is heat cycling. Grain size refining of steel, by heat cycling is based on repeated alpha gamma phase transformations [2][3].

To obtain better effects of structure refining and greater cumulation of single cycle effects, a sample can be fast heated without austenite homogenization withholding at maximal temperature. This implies extra effects. At fast specimen heating, beyond the phase transformation, phase and temperature micro-deformation and thermal-diffusional

Published in: E. Kuljanic (Ed.) *Advanced Manufacturing Systems and Technology*,
CISM Courses and Lectures No. 372, Springer Verlag, Wien New York, 1996.

processes appear [1][2].

The grain refinement can be intensified by phase and temperature micro-deformation processes. Dislocation density is additionally increased by micro-deformation processes and sub-boundary structure can originate. The number of potentials centers of second phase nucleation increases by creating a sub-boundary in austenitic grain. In this way the creating of sub-boundary contributes to the refining of austenite. Additional austenite refining can be originated by recrystallization process, i.e., by heating of micro-deformed structure [2].

Phase transformation, phase and thermal micro-deformation, which are developed during the heat cycling are interactive processes, they affect each other, so that they by separate and by interactive manner can change mechanical properties of steel. However, in all these processes, the dislocation density increases, and strength coefficient can increase and formability of steel can be worsened.

Results of cyclic heat treatment directly depend on cycle number. To achieve same effects of cyclic heat treatment, more cycles are needed at a lesser rate of temperature changing. At cyclic heat treatment of not-alloyed steel, and if applied rate temperature change enables diffusinal $\alpha \rightleftharpoons \gamma$ transformations maximum ten cycles are needed.

The heating condition in heat cycling is determined essentially by heating rate and maximum cycle temperature. The rate of heating depends on the mode of heating, specimen dimension, and on the medium for heating.

## 2. HEAT TREATMENTS FOR IMPROVING THE STEEL FORMABILITY

Steel with good formability has small strength properties and high plasticity and toughness. Good formability can be obtained by the isothermal or cyclic heat treatment.

Heat treatments were done on the specimen with a circular cross section and with a specimen diameter of 16 mm and length of 60 mm. The investigations were done on steel Ck 35 (DIN), with elemental composition in wt % of 0,35 % C, 0,18 %Si, 0,36 %Mn, 0,011 %P, 0,008 %S.

The investigated heat treatments are shown in Fig. 1. The cyclic heat treatment was done by special apparatus [4]. Heat treatments and their parameters are shown in table 1.

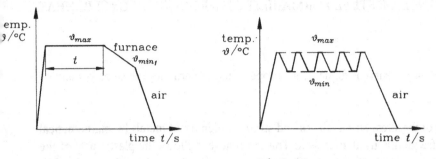

a) Izotermal annealing          b) Cyclic annealing

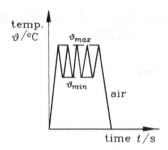

c) Cyclic heat treatment

Fig. 1. Heat treatments

Table 1. Heat treatment parameters

| N° | Heat treatment | Maximum temperature $\vartheta_{max}/°C$ | Time at $\vartheta_{max}$, $t/min$ | Minimum temperature $\vartheta_{min}/°C$ | Heating rate $v_h/°Cmin^{-1}$ | Cooling rate $v_c/°Cmin^{-1}$ | Cycles number $n$ |
|---|---|---|---|---|---|---|---|
| 1 | Isothermal annealing | 830 | 60 | 600 | 5 | 2 | 1 |
| 2 | Cyclic annealing | 780 | 5 | 680 | 5 | 3 | 5 |
| 3 | Cyclic heat treatment | 800 | - | 650 | 300 | 80 | 5 |

## 3. PERFORMANCE OF STEEL FORMABILITY IMPROVING BY CYCLIC HEAT TREATMENT

The true strength-strain curve is fundamental property that determines the steel formability [1].

The description of the stress-strain curves and strain-hardening of metals by mathematical expressions is frequently used approach. This is because it allows the plastic part of the curve to be treated by certain parameters that can be applied to the study of formability [5].

The most important and widely used application is the evaluation of stretch-formability by the n value, which is the exponent of the equation [5][6]:

$$\sigma = K\varepsilon^n$$

where:

$\sigma$ - true stress, $Nm^{-2}$
$\varepsilon$ - true strain
$K$- strength coefficient, $Nm^{-2}$
$n$- strain hardening exponent

The increase of steel formability released by cyclic heat treatments was estimated by comparing with the increase of steel formability released by isothermal annealing. The steel formability has been estimated on the base of true strength-strain curve. Stress-strain curves are determined by compression test. The true stress-strain curve has been determined by using the specimen with circular cross section (Fig. 2). Friction forces were reduced by filling-up holes with paraffin.

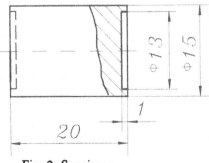

Fig. 2. Specimen

Results of strength coefficient and strain hardening exponent released by different heat treatment of investigated steel are given in Table 2.

Table 2. Strength coefficient and strain hardening exponent

| N ° | Heat treatment | Strength coefficient $K / Nm^{-2}$ | Strain hardening exponent, $n$ | Regression coefficient, $r$ |
|------|----------------|-----------------------|-----------------|----------------|
| 1 | Isothermal annealing (IA) | 915 | 0.18 | 0,962 |
| 2 | Cyclic annealing (CA) | 844 | 0,225 | 0,966 |
| 3 | Cyclic heat treatment (CHT) | 806 | 0,19 | 0,953 |

Stress-strain curves are shown in figure Fig. 3.

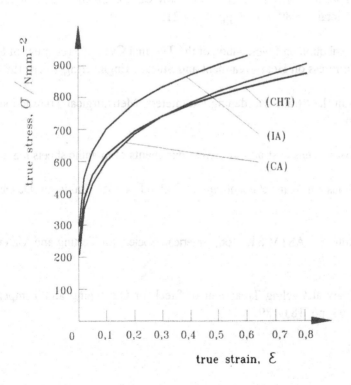

Fig. 3. Stress-strain curves

## 4. CONCLUSION

The increase of steel plasticity is directly in function of structure fineness. By application the cyclic heat treatment plasticity and strain-hardening exponent are increasing and strength coefficient is decreasing. The strain-hardening exponent increasing is lesser than decreasing the stress coefficient. Better formability properties of steel were achieved by the application the cyclic heat treatment than by conventional procedure of steel softening.

## REFERENCES

[1] Anashkin, A. et all: Heat Cycling of Carbon Steel Wire, Metal Science and Heat Treatment of Metals, 1987, vol. 2, pp. 10-14.

[2] Fedyukin, V.: Cyclic Heat Treatment of Steels and Cast Irons, Leningrad State Unv., Leningrad, 1984, pp. 51, (in Russian).

[3] Konopleva, E.: Thermal Cycling Treatment of Law-Carbon Steel, Metal Science and Heat Treatment of Metals, 1989, vol. 8, pp. 617-621.

[4] Smoljan, B.: Contribution on Investigation of the Thermal Cycling Treatment of Steel, 9th International Congress on Heat Treatment and Surface Engineering, Nice, 1994.

[5] Kleemola, H.: On the Strain- Hardening Parameters, Metallurgical Transactions, 5, 1973, p. 1863-1866.

[6] ..., ASTM E646-84, Tensile strain-hardening exponents of metallic sheets materials.

[7] Rose A. et al: Atlas zur Wärmebehadlung der Stähle I, Verlag Stahlesen, Düseldorf, 1958, pp. 128-134.

[8] Barsom J. and Rolfe S.:, ASTM STP 466, American Society for Testing and Materials, 1970, pp. 281.

[9] Smoljan B.: Thermal Cycling Treatment of Steel for Quenching and Tempering, Amst'93, Udine, 1993, pp. 183-189.

# MECHANICAL ANALYSIS OF A SOLID CIRCULAR PLATE SIMPLY SUPPORTED ALONG TWO ANTIPODAL EDGE ARCS AND LOADED BY A CENTRAL FORCE

**A. Strozzi**

**University of Modena, Modena, Italy**

KEY WORDS: Circular Plates, Deflections, Non Symmetrical Supports, Integral Equations

ABSTRACT: A mechanical analysis is presented for a solid circular plate, deflected by a transverse central force, and simply supported along two antipodal periphery arcs, the remaining part of the boundary being free. By exploiting a Green function expressed in analytical form, the original problem is formulated in terms of a Fredholm integral equation of the first kind, where the kernel is particularly complex. This initial formulation is then simplified, and two descriptions of this problem in terms of integral equations are achieved. In the first description, this plate problem is reformulated as an integral equation encountered in wing theory. Then, the same problem is expressed as a Fredholm integral equation of the second kind. Preliminary approximate analytical results and experimental measurements are also reported.

## 1. INTRODUCTION

The problems of thin circular plates deflected by transverse loads may be classified into four groups. The first set comprises circular plates axisymmetrically loaded and supported, a problem for which several closed form solutions are available, which are applicable to a variety of load distributions and types of constraining, [1]. The second class includes circular plates axisymmetrically constrained, but non symmetrically loaded. Particularly studied is the situation of concentrated loads, for which a solution technique of general applicability, based upon a series expansion of the load, is available, [2-4]. The third group encompasses circular plates loaded and constrained non symmetrically, a problem examined

Published in: E. Kuljanic (Ed.) *Advanced Manufacturing Systems and Technology*, CISM Courses and Lectures No. 372, Springer Verlag, Wien New York, 1996.

sporadically and only for particular situations, [1]. Finally, the fourth category includes circular plates loaded axisymmetrically, but non symmetrically constrained. Only a few papers dedicated to this problem are traceable in the technical literature, and often based on numerical treatments more than on analytical approaches, therefore furnishing results which are valid only for particular geometries. A solid circular plate supported along edge arcs has been studied in [5] with a numerical approach, and in [6] with a series expansion technique. In [7] an approximate series solution is obtained for a plate sustained by angularly equispaced supports. Finally, in [8] similar problems are treated from a mathematical viewpoint .

In this paper a situation falling into the fourth group is examined, namely a solid circular plate deflected by a transverse central load, and bilaterally simply supported along two antipodal edge arcs, Figure 1 . The Green function for this problem is connected to a circular plate deflected by a central transverse load and by two antipodal, equal forces which equilibrate the central load. The corresponding analytical expressions of the boundary deflections have been obtained in [9]. By exploiting this analytical Green function, the original plate problem is formulated in terms of a Fredholm integral equation of the first kind, where the kernel is particularly complex and therefore not suitable for an analytical solution. By analytical treatment, this original formulation is then simplified, and an integral equation with Hilbert kernel, similar to an equation encountered in wing theory is obtained. Starting from this result, two additional integral equations are derived, the first equation coinciding with that of wing theory, and the second equation being a Fredholm integral equation of the second kind. Although no closed form solutions could be obtained so far for the problem under scrutiny, the relative simplicity of the integral equations here derived justifies efforts in this direction.

Preliminary approximate analytical results and experimental measurements extracted from [10] are also reported.

## 2. THE FORMULATION IN TERMS OF INTEGRAL EQUATION

The edge deflection, $w(\theta)$ , of a solid circular plate loaded by a central transverse force, $P$ , and by two antipodal boundary loads $P/2$ equilibrating the central load is (Dragoni and Strozzi 1995) :

$$w(\theta)\frac{48\,\pi\,D\,(3+v)\left(1-v^2\right)}{P\,r_o^2} = 48\,(1+v)\left[\,ln(2\,sin\theta) - cos\theta\,ln\!\left(tan\frac{\theta}{2}\right)\right] + $$

$$12\,(1+v)^2\left[\pi|\theta - sin\theta| - \theta^2\right] - 2\,\pi^2\,(1+v)^2 + 3\,(1-v)^3 \tag{1}$$

where $P$ is the intensity of the central load, $D$ is the flexural rigidity of the plate, $v$ the Poisson's ratio, $r_o$ the outer radius, and $\theta$ is the angular coordinate, whose origin is defined by the point of application of one of the edge loads. The edge deflection, $w(\theta)$ , is expressed with respect to the plate centre.

Formula (1) constitutes the Green function for the problem of a plate loaded by a central force and simply supported along two antipodal edge arcs of angular semiwidth $\alpha$.

(The maximum value for $\alpha$ is $\pi/2$ , for which the plate becomes simply supported along its whole edge.) In fact, the distribution of the reaction force along the edge supports can be interpreted as an infinite series of infinitesimal edge forces. The reaction force distribution is correct when the edge deflection remains constant along the edge support, which is assumed as rigid. It is therefore possible to express this problem by superposing the various effects of the infinitesimal loads, and by imposing the constancy of the plate edge deflections along the supports. The unknown function is the reaction force distribution. The superposition of the effects leads to an integral equation.

In the following, the reaction force distributed along the edge supports is denoted with $F(\theta)$ . As already mentioned, the Green function describes a plate loaded by two antipodal forces and by an equilibrating central load. The infinitesimal force corresponding to the angular extent $d\theta$ is therefore $F(\theta)\, r_0\, d\theta$ , which in the Green function model is associated to an antipodal force of the same intensity and to an equilibrating central load of double intensity, $2\, F(\theta)\, r_0\, d\theta$ . A unity edge load is therefore connected to a central force of intensity 2 . In other words, the resultant force of each support equals $P/2$ . As a result, in the integral description of the title problem, number 48 in the left hand-side of equation (1) must be replaced by 24 . The integral equation is :

$$w(\theta)\frac{24\,\pi\,D\,(3+v)\left(1-v^2\right)}{r_o^3} = \int_{-\alpha}^{+\alpha}\left\{48\,(1+v)\left[\ln(2\,sin(\theta-\omega))-cos(\theta-\omega)\,\ln\left(tan\frac{(\theta-\omega)}{2}\right)\right]-\right.$$

$$12\,(1+v)^2\,(\theta-\omega)^2-2\,\pi^2\,(1+v)^2+3\,(1-v)^3\right\}F(\omega)\,d\omega+$$

$$12\,(1+v)^2\,\pi\int_{-\alpha}^{\theta}(\theta-\omega-sin(\theta-\omega))\,F(\omega)\,d\omega-12\,(1+v)^2\,\pi\int_{\theta}^{\alpha}(\theta-\omega-sin(\theta-\omega))\,F(\omega)\,d\omega$$

(2)

## 3. THE SIMPLIFICATION OF THE INTEGRAL EQUATION

In order to simplify the integral equation (2) , it is useful to compute its derivatives with respect to $\theta$ . The first derivative of the left hand-side, which represents the normalised deflection of the supported edge with respect to the plate centre, is :

$$\frac{d\,w(\theta)}{d\,\theta}\frac{24\,\pi\,D\,(3+v)\left(1-v^2\right)}{r_o^3} = \int_{-\alpha}^{+\alpha}\left\{48\,(1+v)\left[sin(\theta-\omega)\,\ln\left(tan\frac{(\theta-\omega)}{2}\right)\right]-\right.$$

$$24\,(1+v)^2\,(\theta-\omega)\right\}F(\omega)\,d\omega+$$

(3)

$$12\,(1+v)^2\,\pi\int_{-\alpha}^{\theta}(1-cos(\theta-\omega))\,F(\omega)\,d\omega-12\,(1+v)^2\,\pi\int_{\theta}^{\alpha}(1-cos(\theta-\omega))\,F(\omega)\,d\omega$$

The second derivative of the normalised deflection of the supported edge with respect to $\theta$ is :

$$\frac{d^2 w\,(\theta)}{d\,\theta^2}\,\frac{24\,\pi\,D\,(3+v)\left(1-v^2\right)}{r_o^3} = \int_{-\alpha}^{+\alpha}\left\{48\,(1+v)\left[1+\cos(\theta-\omega)\ln\left(\tan\left(\frac{\theta-\omega}{2}\right)\right)\right]-\right.$$

$$\left.24\,(1+v)^2\right\}F(\omega)\,d\omega +$$

(4)

$$12\,(1+v)^2\pi\int_{-\alpha}^{\theta}\sin(\theta-\omega)\,F(\omega)\,d\omega - 12\,(1+v)^2\pi\int_{\theta}^{\alpha}\sin(\theta-\omega)\,F(\omega)\,d\omega$$

The third derivative of the normalised deflection of the supported edge with respect to θ is :

$$\frac{d^3 w\,(\theta)}{d\,\theta^3}\,\frac{24\,\pi\,D\,(3+v)\left(1-v^2\right)}{r_o^3} = \int_{-\alpha}^{+\alpha}\left\{48\,(1+v)\left[-\sin(\theta-\omega)\ln\left(\tan\left(\frac{\theta-\omega}{2^\cdot}\right)\right)+\frac{1}{\tan(\theta-\omega)}\right]\right\}F(\omega)\,d\omega +$$

(5)

$$12\,(1+v)^2\pi\int_{-\alpha}^{\theta}\cos(\theta-\omega)\,F(\omega)\,d\omega - 12\,(1+v)^2\pi\int_{\theta}^{\alpha}\cos(\theta-\omega)\,F(\omega)\,d\omega$$

The sum of the first and third derivatives gives :

$$\left(\frac{d\,\delta\,(\theta)}{d\,\theta}+\frac{d^3\,\delta\,(\theta)}{d\,\theta^3}\right)\frac{24\,\pi\,D\,(3+v)\left(1-v^2\right)}{r_o^3} =$$

$$\int_{-\alpha}^{+\alpha}\left\{48\,(1+v)\frac{1}{\tan(\theta-\omega)}-24(1+v)^2\,(\theta-\omega)\right\}F(\omega)\,d\omega + \quad (6)$$

$$12\,(1+v)^2\pi\int_{-\alpha}^{\theta}F(\omega)\,d\omega - 12\,(1+v)^2\pi\int_{\theta}^{\alpha}F(\omega)\,d\omega$$

Since on a physical basis function $F(\theta)$ is symmetric with respect to θ , some further simplifications are possible. In particular :

$$\int_{-\alpha}^{\theta}F(\omega)\,d\omega - \int_{\theta}^{\alpha}F(\omega)\,d\omega = \int_{-\theta}^{\theta}F(\omega)\,d\omega = 2\int_{0}^{\theta}F(\omega)\,d\omega \qquad (7)$$

$$\int_{-\alpha}^{\alpha}(\theta-\omega)\,F(\omega)\,d\omega = \theta\int_{-\alpha}^{\alpha}F(\omega)\,d\omega - \int_{-\alpha}^{\alpha}\omega\,F(\omega)\,d\omega = \theta\int_{-\alpha}^{\alpha}F(\omega)\,d\omega \quad (8)$$

The resultant of each boundary reaction force $F(\theta)$ equals half the central load, $P$ , supposed known :

$$\int_{-\alpha}^{\alpha}F(\omega)\,d\omega = \frac{P}{2} \qquad (9)$$

In addition, since the support is assumed rigid with respect to the plate flexibility, the left hand-side of equation (6) vanishes. After simplifying the various constants, the simplest form of the integral equation is :

$$\int_{-\alpha}^{+\alpha} \frac{1}{tan(\theta - \omega)} F(\omega)\, d\omega = \frac{(1+v)}{2} \left\{ \theta \frac{P}{2} - \pi \int_0^\theta F(\omega)\, d\omega \right\} \tag{10}$$

which is an integral equation with Hilbert kernel.

## 4. THE REDUCTION OF THE INTEGRAL EQUATION (10) TO ONE OF WING THEORY

By putting :

$$\int_0^\theta F(\omega)\, d\omega = G(\theta) \quad ; \quad \int_0^\omega F(\tau)\, d\tau = G(\omega) \quad ; \quad \frac{d\,G(\omega)}{d\,\omega} = F(\omega) \tag{11}$$

the integral equation (10) becomes :

$$\int_{-\alpha}^{+\alpha} \frac{1}{tan(\theta - \omega)} \frac{d\,G(\omega)}{d\,\omega}\, d\omega = \frac{(1+v)}{2} \left\{ \theta \frac{P}{2} - \pi\, G(\theta) \right\} \tag{12}$$

By exploiting the symmetry of $F(\omega)$, one obtains :

$$\int_{-\alpha}^{+\alpha} \frac{1}{tan(\theta - \omega)} \frac{d\,G(\omega)}{d\,\omega}\, d\omega = \int_0^{+\alpha} \left( \frac{1}{tan(\theta - \omega)} + \frac{1}{tan(\theta + \omega)} \right) \frac{d\,G(\omega)}{d\,\omega}\, d\omega \tag{13}$$

Since :

$$\frac{1}{tan(\theta - \omega)} + \frac{1}{tan(\theta + \omega)} = \frac{2\,sin(2\theta)}{cos(2\omega) - cos(2\theta)} \tag{14}$$

the integral equation becomes :

$$\int_0^{+\alpha} \frac{sin(2\theta)}{cos(2\omega) - cos(2\theta)} \frac{d\,G(\omega)}{d\,\omega}\, d\omega = \frac{(1+v)}{4} \left\{ \theta \frac{P}{2} - \pi\, G(\theta) \right\} \tag{15}$$

By putting :

$$cos(2\omega)=t \quad ; \quad cos(2\theta)=x \quad ; \quad dt=-2\,sin(2\omega)\,d\,\omega \quad ; \quad \frac{d\,G}{d\,\omega}=\frac{d\,G}{d\,t}\frac{d\,t}{d\,\omega}=-2\frac{d\,G}{d\,t}\,sin(2\omega) \quad (16)$$

one obtains :

$$\int_{1}^{cos\,2\alpha} \frac{1}{t-x}\frac{d\,G(t)}{d\,t}\,dt = \frac{(1+v)}{8}\left\{\pi\,G(x) - \frac{a\,cos\,x}{2}\frac{P}{2}\right\} \tag{17}$$

To obtain a standard form for the integral equation (17), it is necessary that its integration limits are -1 and + 1 . To this aim, the new integration variables are employed :

$$t=\frac{1-cos(2\alpha)}{2}\gamma+\frac{1+cos(2\alpha)}{2} \quad ; \quad x=\frac{1-cos(2\alpha)}{2}\beta+\frac{1+cos(2\alpha)}{2} \tag{18}$$

The form of the integral equation adherent to the wing theory is :

$$\frac{(1+v)(cos(2\,\alpha)-1)}{32}G(\beta)-\frac{1}{2\,\pi}\int_{-1}^{1}\frac{1}{\gamma-\beta}\frac{d\,G(\gamma)}{d\,\gamma}\,d\gamma=-P\frac{(1+v)(1-cos(2\,\alpha))}{128\,\pi}a\,cos\left(\frac{1-cos\,(2\alpha)}{2}\beta+\frac{1+cos\,(2\alpha)}{2}\right) \tag{19}$$

## 5. THE REDUCTION OF THE INTEGRAL EQUATION (10) TO A FREDHOLM EQUATION OF THE SECOND KIND

For the Fredholm integral equation of the first kind :

$$\int_{-\alpha}^{+\alpha} \frac{1}{tan(\theta-\omega)}\,L(\omega)\,d\omega = R(\theta) \tag{20}$$

the analytical solution is known :

$$L(\theta)=\frac{k\,\alpha\,cos\,\theta}{\pi\sqrt{sin^2\alpha-sin^2\theta}}+\frac{2}{\pi\sqrt{sin^2\alpha-sin^2\theta}}\int_{-\alpha}^{\alpha}\frac{R(\omega)\sqrt{sin^2\alpha-sin^2\omega}}{sin(\theta-\omega)}\,d\omega \tag{21}$$

where $k$ is a generic constant. By treating the right hand-side in equation (10) as known, one obtains :

$$\frac{d\,G(\theta)}{d\,\theta}=\frac{k\,\alpha\,cos\,\theta}{\pi\sqrt{sin^2\alpha-sin^2\theta}}+\frac{(1+v)\,P}{2\,\pi\sqrt{sin^2\alpha-sin^2\theta}}\int_{-\alpha}^{\alpha}\frac{\omega\sqrt{sin^2\alpha-sin^2\omega}}{sin(\theta-\omega)}\,d\omega -$$

$$\frac{(1+v)}{\sqrt{sin^2\alpha-sin^2\theta}}\int_{-\alpha}^{\alpha}\frac{G(\omega)\sqrt{sin^2\alpha-sin^2\omega}}{sin(\theta-\omega)}\,d\omega \tag{22}$$

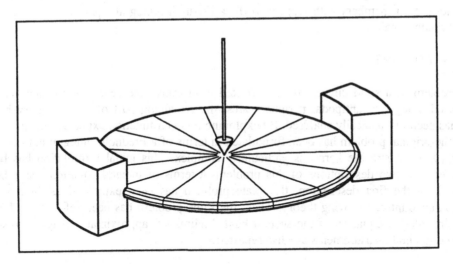

Fig.1. Circular plate deflected by a central transverse load and simply supported along two antipodal edge arcs.

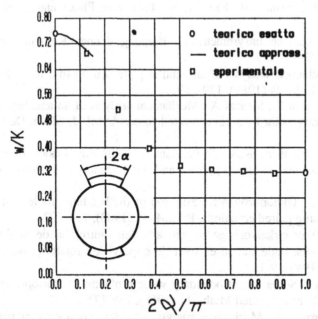

Fig.2. Preliminary analytical results and experimental measurements for the plate central normalized deflection $w(2\pi D)/(Pr_o^2)$ with respect to the arc supports, versus the normalized angular semiwidth of the supports, $2\alpha/\pi$.

where the first integral from the left could not be computed analytically so far. By integrating both members with respect to $\theta$ , a Fredholm integral equation of the second kind is finally obtained.

## 6. CONCLUSIONS

The problem of a solid circular plate deflected by a transverse central force, and simply supported along two antipodal periphery arcs, the remaining part of the boundary being free, has been mechanically analysed. By exploiting a Green function expressed in analytical form, the original problem has been formulated in terms of a Fredholm integral equation of the first kind, where the kernel is particularly complex. This initial formulation has been simplified, and two descriptions of this problem in terms of integral equations have been achieved. In the first description, this plate problem has been expressed as an integral equation encountered in wing theory. Then, the same problem has been reformulated as a Fredholm integral equation of the second kind. Preliminary approximate analytical results and experimental measurements are also reported.

## REFERENCES

1.Timoshenko, S.P., Woinowski-Krieger, S.: Theory of Plates and Shells. McGraw-Hill, London, 1959

2. Reißner H.: Über die unsymmetrische Biegung dünner Kreisringplatten. Ingenieur-Archiv, 1 (1929), 72-83

3. Strozzi A.: Mechanical analysis of an annular plate subject to a transverse concentrated load. J. Strain Analysis, 24 (1989), 139-149

4. Ciavatti V., Dragoni E., Strozzi A.: Mechanical analysis of an annular plate transversely loaded at an arbitrary point by a concentrated force. ASME J. Mech. Design, 114 (1992), 335-342

5. Conway H.D., Farnham K.A.: Deflections of uniformly loaded circular plates with combinations of clamped, simply supported and free boundary conditions. Int. J. Mech. Sci., 9 (1967), 661-671

6. Samodurov A.A., Tikhomirov A.S.: Solution of the bending problem of a circular plate with a free edge using paired equations. P.M.M, 46 (1983), 794-797

7. De Beer C.: Over cirkelvormige platen, aan den omtrek in de hoekpunten van een regelmatigen veelhoek ondersteund en over de oppervlakte rotatorischsymmetrisch belast. De Ingenieur, 59 (1947), 9-11

8. Gladwell G.M.L.: Some mixed boundary value problems in isotropic thin plate theory. Quart. Journ. Mech. and Applied Math., 11 (1958), 159-171

9. Dragoni E., Strozzi A.: Mechanical analysis of a thin solid circular plate deflected by transverse periphery forces and by a central load. Proc. Instns Mech Engrs, 209 (1995), 77-86

10. Strozzi A., Dragoni E., Ciavatti V.: Analisi flessionale di piastre circolari semplicemente appoggiate lungo tratti del contorno. Congresso in onore del prof. E. Funaioli, Bologna, 1996

# MACHINABILITY EVALUATION OF LEADED AND UNLEADED BRASSES: A CONTRIBUTION TOWARDS THE SPECIFICATION OF ECOLOGICAL ALLOYS

**R.M.D. Mesquita**

**Instituto Nacional de Engenharia e Tecnologia Industrial, Lisboa, Portugal**

**P.A.S. Lourenço**

**Instituto Superior Técnico, Lisboa, Portugal**

KEY WORDS: Machinability, Copper Alloys, Unleaded Brass, Chip Forms, Cutting Forces

ABSTRACT: Trends on materials development are reflecting the need for new specifications of materials composition considering the environmentally related behaviour of both, the manufacturing processes and material/product performance, disposal and recycling during the life cycle of the product. In particular, cast cooper - zinc - lead alloys, used in components for potable water systems and containing a significant level of lead are natural candidates to a reengineering process. The addition of lead is considered to be a key factor in order to improve the machinability of the alloys used in components machined in large batches. It was showed that contamination of potable water with lead occurs in these systems. The contamination of potable water with lead has' a deleterious effect on the nervous system, mainly during the early stages of human development. Consequently, new and more stringent standards, governmental and community directives (EEC directive 80/778), together with the pressure of consumers, are pressing the development of alloys with a reduced lead contents or even lead-free copper alloys. Reducing the contents of lead decreases the machinability of brasses. This paper presents the results of a research project aiming the establishment of the minimum lead content of a copper alloy that can be used for watermeter body manufacturing, together with the machinability reference data for ecological copper alloys development.

Published in: E. Kuljanic (Ed.) *Advanced Manufacturing Systems and Technology,*
CISM Courses and Lectures No. 372, Springer Verlag, Wien New York, 1996.

## 1. INTRODUCTION

Copper alloys and in particular brasses (high zinc-copper alloys) are selected for the manufacture of water valves and fittings, due to the combination of mechanical strength and corrosion resistance. Their excellent castability and low cost make it the dominant material for watermeter manufacturing. Free-cutting copper alloys are known to present an excellent machinability mainly due to the lead-contents (up to 3%). The lead is almost completely insoluble in the solidified alloys. It appears either as interdendritic islands or as discrete globules of pure lead surrounded by the corrosion-resistant lead-free copper alloy matrix [1]. It is generally recognised that there are three mechanisms by which lead addition can promote the free-cutting properties: brass embrittlement, lubrication at the tool/chip interface and cutting edge temperature reduction [2]. The interdendritic islands or the discrete globules of pure lead account for the introduction of heterogeneity's in a ductile matrix, providing conditions for shear instability and shear rupture (ductile fracture) as a result of the nucleation of voids at second phase particles within the shear band instability [3]. The same author (P. K. Wright), shows that the lead globules are deposited on the rake face of the tool. The main functions of the lead in the secondary shear zone is to provide regions of low strength providing internal lubrication at the tool/chip interface (reduced forces). Consequently a reduction in the cutting edge temperature is expected.

Together with the improvement of machinability, lead additions impart the castability of brasses, reducing the porosity levels. However, it can increase the incidence of short running [4].

Subsurface lead is protected from corrosion, and it is only the lead on the castings surfaces that can contaminate the water supply. Surface lead is found mainly on machined surfaces as a smeared layer, but also on the internal surfaces of hollow castings (valve and watermeter bodies) where metal in contact with hot cores solidifies last. Surface lead removal is carried out by a leaching process and it causes potable water contamination. U.S. Environmental Protection Agency (EPA) together with state regulatory bodies and European Union Directives (80/778) are setting limits on the amount of lead any plumbing product may contribute to the water. Some regulations place a maximum allowable lead content of 15 parts per billion (ppb) on US potable waters. Dresher et al. [5] showed that the leachate lead content produced after a 14-day exposure time of a copper-base alloy with a 1 to 1.5% lead content in water (pH8) can be as high as 100 ppb. When the lead content is reduced to 0.5%, the leachate lead content decreases to a 25 ppb level.

Pressures to eliminate lead-containing materials from potable water systems, will rise and eventually, only lead-free materials or lead-free like will be allowed in the near future. The development and economic use of these new alloys (ecobrass), can be achieved through the reengineering of the conventional copper-zinc alloys together with the reengineering of the part manufacturing process. In this context, an industrial research project is under development at INETI - Instituto Nacional de Engenharia e Tecnologia Industrial and IST - Instituto Superior Técnico, on behalf of B. Janz (watermeter manufacturer). The final scope of the project is the development of a die casting 60/40 unleaded brass with improved machinability and reduced dezincification. Preliminary work tried to establish the minimum lead content of the alloy that could be used for watermeter body manufacturing, together with the machinability reference data for further alloy composition refining. The second phase of the research project, in what machining behaviour is concerned, will establish the effect of bismuth on the machinability and is targeted towards further reductions of lead contents.

## 2. EXPERIMENTAL WORK

Three die cast alloys were produced with 1.997%, 0.877% and residual (0.059%) lead contents, in the form of cylindrical rods. A wrought 60/40 brass bar with 2.668% lead content was purchased and used for comparison purposes. All four materials were characterised by chemical analysis, physical and mechanical testing. The results of this characterisation is presented in table 1 for all the materials used in the machinability evaluation process.

| | Material M1 | Material M2 | Material M3 | Material M4 |
|---|---|---|---|---|
| Manufacturing route | Wrought | Cast | Cast | Cast |
| Nominal Composition | | | | |
| % Cu (Copper) | 59.18 | 60.08 | 62.11 | 61.63 |
| % Zn (Zinc) | 37.48 | 37.24 | 36.20 | 37.69 |
| % Pb (Lead) | 2.668 | 1.997 | 0.877 | 0.059 |
| Microstructure | | | | |
| % alfa phase | 72.3 | 72.5 | 74.6 | 78 |
| % beta phase | 27.7 | 27.5 | 25.4 | 22 |
| microhardness alfa HV | 89 | 95 | 87 | 115 |
| microhardness beta HV | 146 | 145 | 157 | 171 |
| Tensile properties | | | | |
| Tensile strength MPa | 397-409 | 359-370 | 353-360 | 351-357 |
| Yield strength MPa | 159-164 | 139-142 | 105-113 | 100-111 |
| Elongation in 30mm % | 27-30 | 29-35 | 41-44 | 57-65 |
| Brinell Hardness HB | 91.25 | 85.85 | 77.49 | 79.2 |
| Compression test | | | | |
| Strength coef. K | 872 | 836.26 | 835.12 | 792.96 |
| Strain hardening exp. n | 0.394 | 0.385 | 0.449 | 0.434 |

Table 1 - Chemical, physical and mechanical properties of leaded and unleaded copper alloys

Chemical composition was determined by optical emission spectroscopy using a Baird Spectromet 750. The volumetric fractions of $\alpha$ and $\beta'$ phases of the alloys were determined by Image Analysis. Phases microhardness were determined with a Reicherter MeF equipment using a 25g load. Tensile properties were determined using B6x30 specimens, according to DIN 50125 and a 50kN electromechanical Instron TT-CM-L machine. For every material, 5 specimens were tested. Specimens of the cast materials were removed from the outer diameter of the billets, in order to avoid the defects that usually arise in the central sections of the cast bars. The mechanical properties obtained are within the values that one can find in the literature, both for the wrought material C35600 (ASTM) - material M1 and for the cast material C85800 (ASTM), which is comparable to our M2 material. As expected, increasing the copper contents, increase the volume of the α phase, decreasing the strength of the alloy. However, ductility increases with the copper content together with the reduction of the lead contents. Consequently, from these mechanical properties one can expected significant differences on the machining behaviour of the copper alloys.

The machinability analysis consisted primarily in the evaluation of the effect of lead contents and, secondarily, of the machining parameters, on the machinability ratings, considering three machinability criteria - cutting forces, chip forms and surface roughness.

Machining tests were carried out according to the procedures established by ISO3685 (1993), when applicable, on a Gildmeister CTX 400 CNC turning centre, with 22 kW power and 5.000 rpm maximum spindle speed. Cutting forces were measured with a Kistler 9121 piezoelectric dynamometer. The measuring chain included also the charge amplifiers (Kistler 5011), a data acquisition board Metrabyte DAS1601. Data acquisition and processing was carried out using the Keitley EASYEST LX software. Chip forms and chip classification were evaluated using the method proposed by Zhang [6]. The level of chip breaking is assessed by an index - CPDI (chip packing density index) which is defined as a ratio between the chip packing density to chip material density. Surface roughness was measured with a Phertometer S4P, a drive PGK and a RFHTB-250 pickup. Table 2 presents the ranges of cutting conditions used during the machining experiments together with the characterisation of the geometry of the uncoated carbide cutting tool.

| Cutting Parameters | Material M1 | Material M2 | Material M3 | Material M4 |
|---|---|---|---|---|
| Bar Diameter [mm] | 100 | 48 | 62 | 62 |
| Cutting Speed [m/min] | 150, 300, 400 500, 600 | 150, 300, 400 500, 600 | 150, 300, 600 | 150, 300, 600 |
| Feed a [mm/rev] | 0.05, 0.1, 0.15 0.20, 0.25, 0.30 | 0.05, 0.1, 0.15 0.20, 0.25, 0.30 | 0.05, 0.1, 0.15 0.20, 0.25, 0.3 | 0.05, 0.1, 0.15 0.20, 0.25 |
| Depth of Cut [mm] | 2.5 | 2.5 | 2.5 | 2.5 |
| Cutting Tool | ISO Code | $\alpha$ [°] | $\omega$ [°] | $\chi$ [°] | $\sigma$ [°] |
| Toolholder | SDJCL 2020 K11 | 0 | 0 | -3 | 7 |
| Insert | DCGX 11T304-Al H10 | 20 (19.68) | 25 (19.65) | -3 | 7 |

Table 2 - Cutting parameters and cutting tools

The experimental results were processed by statistical techniques using analysis of variance.

## 3. RESULTS AND DISCUSSION

A statistically significant sample of chips, produced during the cutting tests, representing a wide range of cutting conditions (according to table 2), were collected and analysed. The analysis included the measurement of chip thickness together with the determination of the CPDI value. Fig.1 presents the effect of feed and workpiece material on the CPDI. It should be noticed that considering the definition of the index, elemental and short chips give rise to a higher packing index (CPDI) improving chip disposal. This characteristic is particularly important in the case of the manufacture of cast valve bodies where automated transfer lines are used, together with special shaped tooling and light feeds. A critical value was established taking into account the existing industrial expertise.

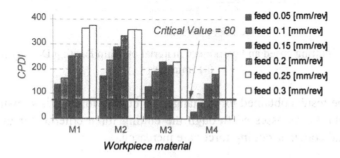

Fig. 1 - Effect of feed and workpiece material on Chip Packing Density Index (cutting speed 150 m/min)

From the figure, it is clear that for every tested material, increasing the feedrate increases the chip packing index, as expected, although the increasing shear angle, as determined from chip thickness measurements. Lead content has also a strong influence on chip form and dimension, characteristics that are assessed through the CPDI value. The unleaded material M4 has a poor machinability rating according to this criteria, being completely unsuitable to be used at low feeds, as required by the finishing process of the meter bodies. There is a correlation between the CPDI value of the tested materials and their ductility as measured through the elongation values presented in table 1. When one considers the chip form and dimension criteria, together with the empirically established critical value and the results obtained with material M3 (0.877% Pb) it can be concluded that an additional reduction in lead content can be considered in the design of the ecological alloy.

Increasing cutting speed decreases the CPDI value. Fig.2 presents this effect for a constant feedrate. This behaviour can be explained by the observed increase on shear angle with cutting speed. According to the theory of metal cutting when the shear angle increases, shear strain decreases. Consequently, chip forms can shift from discontinuous and broken types to continuous and snarled types. However, all the tested materials, produced acceptable chip forms at 0.15 mm/rev feedrate, when cutting speed was increased by a factor of 4. When we compare the effects of both, the feedrate and the cutting speed we can conclude that cutting speed is not a significant constraint on the design of the ecological alloy. The unleaded alloy can withstand further increases of cutting speed giving rise to acceptable values of the CPDI.

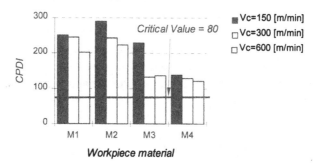

Fig. 2 - Effect of cutting speed and workpiece material on Chip Packing Density Index (feed - 0.15mm/rev)

Fig. 3 presents the results obtained for the tangential cutting force. It is shown the effect of lead on machinability, as assessed through the cutting force criteria: for every feedrate, decreasing the lead contents, cutting forces are increased.

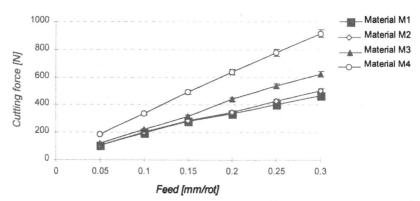

Fig. 3 - Effect of alloy composition and feedrate on tangential cutting force ( cutting speed: 150 m/min)

Mechanical and physical properties of the copper alloys presented in table 1, clearly show that the tensile and yield strength, together with Brinnel hardness decreased with the lead content. The percentage of the harder beta phase decreased also. However, cutting forces increased. This fact can be explained by the analysis of fig. 4. The figure presents the materials flow stress, as determined from the results of the compression test (refer also to data in table 1). The flow curve was expressed by the simple power curve relation and the flow stress was calculated considering the strain determined from chip thickness measurements at the particular cutting conditions. Plane strain conditions and the Von Mises yielding criterion was assumed.

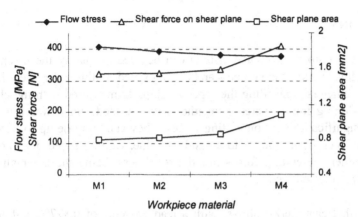

Fig. 4 - Effect of lead content on flow stress, shear force and shear plane area (feed:0.15 mm/rev, cutting speed:150m/min)

The experimental results show that material flow stress decreases with the lead content (from material M1 to M4). However, strain hardening increases (in particular for materials M3 and M4), inducing a departure from the single shear plane model to the more wider shear plane area. Chip thickness measurements showed that the average shear plane angle decreased and the shear plane area increased. Consequently, although the reduction on shear stress with lead contents, the required shear forces on the shear plane increased. Together with the effect of strain hardening on the orientation of the shear plane, the reduction of lead contents contributes to the increase of frictional forces on the rake face and, according to the Ernst & Merchant theory, to the decrease of shear angle and the increase of the cutting forces.

No significant effect of lead contents on surface roughness of machined parts was found. Roughness is determined by the feedrate, tool nose radius and the dynamic behaviour of the system. Consequently, surface roughness criteria was not used to qualify the machinability of the alloys.

The steep increase of cutting forces for the M4 alloy, together with the values of the CPDI, particularly at low feeds, does not allow its use for the manufacture of watermeter bodies. Alloy M3, although with a lower machinability, can be used as an alternative to the higher lead contents alloys (M1 and M2), provided that a 20% increase on cutting forces and power, at high feeds, is allowable. However, considering the constrains of the machining operations required for finishing the watermeter bodies, being the maximum depth of cut *(doc)* determined by the cast process and the feedrate limited by the forces allowable in the thin cast walls, the increase of machining power can be even smaller. The analysis of figure 1 and 3 indicates that the lead content can be further reduced.

## 4. CONCLUSIONS

Cutting forces and chip form related criteria can be used to qualify the machinability of copper-based alloys. Feed, lead contents and cutting speed, in this order, were identified to be the major parameters controlling the types of chips as measured with the chip packing density index. Cutting forces increase together with the feed, but it was found that lead contents has a significant effect on cutting forces. They can increase up to 82% when the lead contents reduces from 2.668% to a residual value. Cutting speed does not influence significantly neither the cutting forces nor the roughness, being the latter almost totally determined by the selected feed.

It is concluded that cast Cu/Zn alloys, with a lead contents of 0.877%, although with a lower machinability, can be used as an alternative to the alloy currently in use (1.997% lead contents), in order to reach a more selective and wider market. It is considered to be possible the development of Cu-Zn alloys having an adequate machinability and lead contents less than 0.877%, if state-of-the-art tooling and suitable control of cutting conditions is used.

## REFERENCES

1.  Dresher W.H., Peters D.T.: Lead-free and reduced-lead copper-base cast plumbing alloys - Part 1, Metall, 46 (1992) 11, 1142-1146.

2.  Stoddart C., Lea C., Derch W., Green P., Pettit H.: Relationship between lead content of Cu-40Zn, machinability and swarf composition determined by Auger electron spectroscopy, Metals Technology, 6 (1979) 5, 176-184.

3.  Wolfenden A., Wright P.K.: On the role of lead in free-machining brass, in "The Machinability of Engineering Materials", ASM, 1983

4.  Staley M., Davies D.: The machinability and structure of standard and low-lead copper-base alloys plumbing fittings, Proceedings Copper 90, Refining, Fabrication, Markets, The Institute of Metals, Sweden, 1990, 441-447

5.  Dresher W.H., Peters D.T.: Lead-free and reduced-lead copper-base cast plumbing alloys - Part 2, Metall, 47 (1993) 1, 26-33

6.  Zhang X. D., Lee L.C., Seah K.: Knowledge base for chip management system, J. Materials Processing Techn., 48 (1995) 1, 215-221

# SQUARE RING COMPRESSION:
# NUMERICAL SIMULATION AND EXPERIMENTAL TESTS

**L. Filice**

**University of Calabria, Cosenza, Italy**

**L. Frantini, F. Micari and V.F. Ruisi**

**University of Palermo, Palermo, Italy**

KEY WORDS: Expert Systems, Upsetting, Square rings, Buckling

ABSTRACT: In the upsetting process of square rings buckling problems often arise, depending on the geometry of the ring and on the frictional conditions at the punch-workpiece interface, yielding to the practical unacceptability of the forged component. Consequently in order to make up a set of rules to be introduced in an expert system able to assist the designer of cold forming sequences, a knowledge base has been built up, by means of a set of numerical simulations, the results of which have been validated by several experimental tests.

## 1. INTRODUCTION

Cold forming processes are always more used in the mechanical industries since they permit to obtain near-net shape forged parts characterised by shape and dimensions very close to the final desired ones, thus requiring little or no subsequent machining operations. The major problem to be faced by the cold forming designers of processes is represented by the choice of the most suitable operations sequence and, for each operation, by the selection of the operating parameters. Generally a cold forming sequence comprehends upsetting, forward and backward extrusion processes; moreover nowadays the geometry of the forged parts is more and more complicated since not only axisymmetrical components are forged, but also parts characterised by a complex fully threedimensional shape.

Published in: E. Kuljanic (Ed.) *Advanced Manufacturing Systems and Technology*, CISM Courses and Lectures No. 372, Springer Verlag, Wien New York, 1996.

In the past the choice of the operations sequence and of the operating parameters has been carried out just following the skill and the experience of the process designer and subsequently testing the validity of the selected sequence by means of a time consuming and expensive set of experimental tests. More recently the availability of powerful computer aided process planning techniques, based on expert systems or on neural networks, as well as the capabilities offered by the modern numerical codes, able to analyse in detail the process mechanics have suggested a different and more suitable approach to the design of a forming sequence. Such an approach is based on the preliminary application of a knowledge based expert system founded of a large number of simple technological rules, aimed to determine the set of technologically feasible forming sequences among all the possible ones, and on the subsequent finite element analysis of this reduced set of feasible sequences. The numerical simulations supply all these informations concerning the required loads and powers, the stresses on the dies, the distributions of the plastic strains and consequently the final mechanical properties of the forged part, which will represent the elements on which the final choice of the best forming sequence has to be carried out.

The above considerations show the central role of the knowledge based expert system, the capability of which to define the set of admissible sequences strongly depends on the number and on the goodness of the technological rules which represent the acquired knowledge.

In particular the production of fully threedimensional forged components frequently requires an upsetting process on a square hollow part of the component, obtained, in a previous step of the forming sequence, by means of a backward extrusion operation. Consequently a set of rules for the upsetting process of square rings has to be searched, since, depending on the geometry of the ring and on the frictional conditions at the punch-workpiece interface, the process could be carried out successfully or buckling problems could occur yielding to the practical unacceptability of the forged component.

In the literature only few studies on this subject can be found: Aku et al [1] carried out a series of experiments on the compression of prismatic blocks of plasticine and in particular on the case of square ring compression, observing different deformation modes depending on the aspect ratio of the workpiece and on the frictional conditions. Andrews et al [2] performed an experimental investigation of the axial collapse modes in the compression of aluminum cylindrical tubes. Park and Oh [3] tested a threedimensional commercial finite element code the upsetting process of an AA1100 square ring with a ratio of the height to the wall thickness equal to four. The numerical simulation suggested the insurgence of a buckling phenomenon in the ring walls. Finally Tadano and Ishikawa [4] carried out a numerical simulation of the upsetting process of thick cylinders.

In this paper a systematic study of the square ring compression process is carried out, taking into account two different materials, namely the AA6062 Aluminum alloy and a commercially pure Copper; a shape coefficient able to describe the geometry of the square ring is proposed and a set of numerical simulations has allowed to determine a "safe" and an "unsafe" zone in the shape coefficient vs. friction factor plane. Furthermore the validity of the numerical simulations has been confirmed by several experimental tests, carried out with different geometries and two lubricating conditions, which have highlighted a significant fitting of the numerical and the experimental results.

## 2. THE NUMERICAL SIMULATIONS

The complexity of the analysed problem has required the use of advanced and powerful numerical techniques: reliable algorithms, able to take into account the material nonlinearity, as well as variable contact and frictional conditions, are in fact required.

In particular in order to best analyse the proposed forming process two different models have been taken into account, namely an implicit and an explicit one. Following the former, a rigid plastic formulation has been employed which takes to a nonlinear system of equations due to the nonlinear material behaviour during plastic deformation. As a consequence the solution of such a system requires an iterative procedure in order to reach the convergence; in this case, the CPU time becomes very large and in particular it increases at increasing the number of nodes with an exponent between two and three.

On the other hand in the explicit approach a final set of linear equations is obtained, starting from the dynamic equilibrium equation of the assembled structure of finite elements and employing a proper time integration scheme. Moreover, if a lumped mass matrix is used, independent equations are obtained which can be solved one by one [5] [6]. In this way a great CPU time reduction is reached, since no iterative procedure is employed at each step; in particular the computational weight assumes a linear trend with the number of nodes. Anyway the integration scheme is conditionally stable and requires a time increment lower than a threshold which depends on the mesh density (i. e. the dimensions of the single element) and on the modelled material (i.e. the Young modulus and the density. As a consequence a very large number of steps would be necessary to follow the deformation path, losing the computational avantages upon the single time increment. The speed of the simulation is then increased for istance by artificially increasing the die velocity [6].

Both of the models have been employed in order to simulate the square rings upsetting with the aim to investigate the process mechanics; actually some considerations should be done as regards the application of the models. First of all, as expected, the explicit approach has resulted faster than the implicit rigid plastic formulation with a ratio 7 : 1; this is much important since complex 3-D simulations have been developed with high CPU times.

On the other hand the implicit approach has shown a better overlapping between the numerical results and experimental verifications since the metal flow has been followed much more faithfully. In this way the occurrence of shape defects has been highlihted as reported in the subsequent sections.

As far as the process mechanics is regarded, it should be observed that if the friction factor were equal to zero the ring would deform with each element moving radially outwards at a rate proportional to its distance from the centre; for low m values an outwards flow takes place at a lower rate, while for m larger than a critical threshold it is energetically favourable for only part of the ring to flow outwards and for the remainder to flow inwards towards the centre. As a consequence, depending on the the frictional conditions, the contact surface between the workpiece and the punch may have a region in which no relative motion occurs. For this reason the frictional tangential stresses are calculated as [7]:

$$\vec{\tau} = -mk\left\{\left(\frac{2}{\pi}\right)\tan^{-1}\frac{\left|v_{pw}\right|}{\alpha}\right\}\frac{\vec{v}_{pw}}{\left|v_{pw}\right|}$$

where $\left|v_{pw}\right|$ is the sliding velocity at the punch-workpiece interface, m is the friction factor, k is the local shear stress of the material and $\alpha$ is a small positive number.

## 3. DISCUSSION OF THE RESULTS AND CONCLUSIONS

In the square ring compression the process mechanics strongly depends on the geometry of the workpiece and on the frictional conditions at the punch-workpiece interface: these factors, in fact, affect the possibility to obtain a sound forged part without shape defects, or a defective component to be discarded due to the occurrence of an axial buckling mode. In their experimental study on the axial collapse modes in compression of aluminum cylindrical tubes, Andrews et al [2] had used a t/D vs. L/D diagram, being t the wall thickness, D the diameter and finally L the length of the tube. In this paper a new shape coefficient (SC) is used, defined as:

$$SC=A/\rho H$$

where A is the area of the anular section, $\rho$ is the inertia radius and H is the height of the specimen. In a previous paper the effectiveness of the proposed shape coefficient for the prediction of the insurgence of collapse modes in the upsetting process of cylindrical rings has been shown [8]. The coefficient, which decreases at increasing the slenderness of the specimen, does not depend on scale factors and can be easily applied and calculated for rings characterised by a geometry different with respect to the simple cylindrical case, i.e. with a square or an elliptical section.

Two different materials have been taken into account in the research, namely the AA6062 Aluminum Alloy and a commercially pure Copper: for these materials the following flow stress laws, obtained by means of compressive tests, have been assumed:

$$\bar{\sigma} = 592 \cdot \bar{\varepsilon}^{0.161} \qquad [N/mm^2] \qquad \text{(Aluminium Alloy)}$$

$$\bar{\sigma} = 350 \cdot \bar{\varepsilon}^{0.1} \qquad [N/mm^2] \qquad \text{(Copper)}$$

A set of numerical simulations at varying the shape coefficient and for four assumed values of the friction factor has been carried out. Table 1 reports the investigated geometries; the meaning of the symbols evident from the figure next to the table.

The simulations have been carried out up to a total punch stroke equal to 50% of the initial height for the Copper specimens, while they have been stopped at a stroke equal to 30% of the initial height for the Aluminum ones; the experimental tests, in fact, have indicated that after this stroke ductile fractures could occur on the outer lateral surface of the ring, depending on the frictional conditions at the tool-specimen interface.

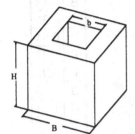

| b[mm] | B[mm] | H[mm] | A[mm$^2$] | ρ[mm] | SC |
|-------|-------|-------|-----------|-------|------|
| 12 | 20 | 25 | 256 | 6,73 | 1,52 |
| 10 | 20 | 25 | 300 | 6,45 | 1,86 |
| 12 | 24 | 25 | 432 | 7,75 | 2,23 |
| 10 | 24 | 25 | 476 | 7,51 | 2,54 |
| 12 | 28 | 25 | 640 | 8,79 | 2,91 |
| 10 | 28 | 25 | 684 | 8,58 | 3,18 |

Tab. 1 - Investigated rings geometries

The analysis of the deformed rings after the same punch stroke shows that, depending on the shape coefficient and on the frictional conditions, three main different deformation modes occur, as shown in fig.1, where the section obtained cutting the square ring with one of the vertical symmetry planes is reported. The first mode (A) corresponds to "safe" conditions: both the inner and the outer lateral surfaces remain almost plane and parallel after the deformation process. On the contrary the second (B) and the third (C) mode correspond to an "unsafe" process: more in particular, in the former case an internal bending phenomenon occurs, while in the latter one the presence of an internal folding can be easily distinguished.

It could be presumed that the internal bending would evolve in the internal folding in a further stage of the upsetting process, i.e. after a further punch stroke; nevertheless, taking into account the same punch stroke, the numerical simulations have highlighted the presence of the two different deformation modes described above.

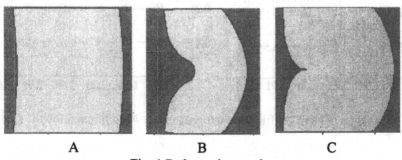

A                    B                    C

Fig. 1 Deformation modes

Furthermore it is necessary to outline that the corresponding accumulated plastic strain distributions show significant differences between the two configurations: fig.2 reports the effective plastic strain distributions calculated after a punch stroke equal to 50% of the initial height for two Copper specimens characterised by SC=1.52 and 2.23 respectively. The same friction factor f = 0.4 has been used for both the simulations.

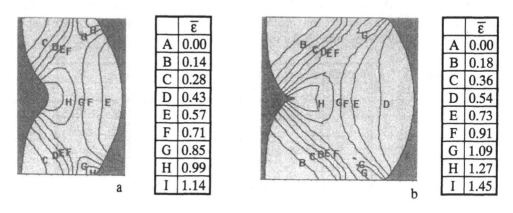

Fig. 2 - Equivalent plastic strain distributions: a) SC=1.52; b) SC=2.23

Fig.2 shows that for SC=2.23, a more dishomogeneous strain distribution is calculated, with large almost rigid zones at the contact with the punches and zones characterised by very high values of the accumulated deformation on the equatorial plane, where the folding phenomenon occurs.

On the other hand the slenderer specimen (SC=1.52) shows a more limited range between the highest and the lowest value of the accumulated plastic strain.

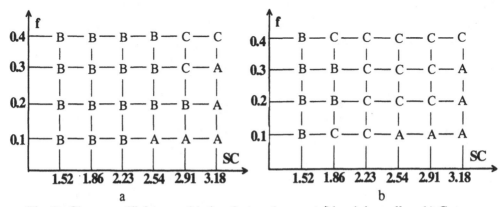

Fig. 3 - Shape coefficient vs. friction factor planes: a) Aluminium alloy; b) Copper

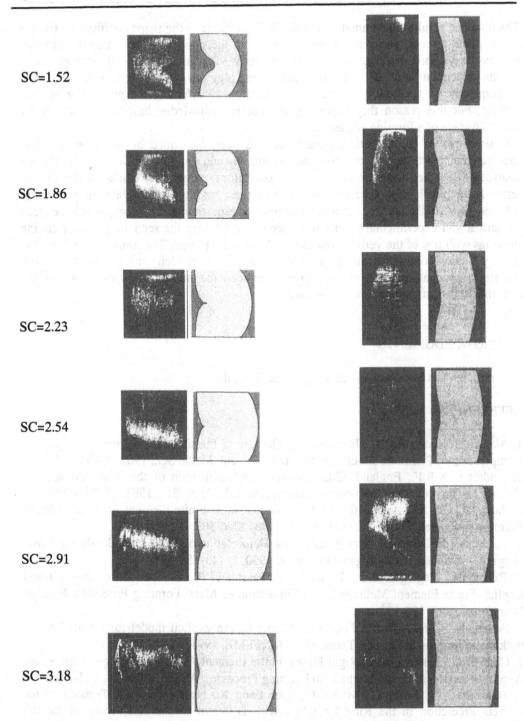

SC=1.52

SC=1.86

SC=2.23

SC=2.54

SC=2.91

SC=3.18

Fig. 4 - Comparison of the numerical and experimental results

The obtained results are summarized in the fig.3a,b where in the shape coefficient - friction factor plane three different areas corresponding to the previously described deformation modes can be easily distinguished. The figures show the combined effect that the geometry of the specimen and the frictional conditions play on the process mechanics and consequently on the quality of the forged component, affecting the possibility of flow defects. For this reason they represent an effective knowledge base to be used in the planning stage of a forming sequence.

The goodness of the numerical simulations has been confirmed by means of a set of experimental tests, carried out both on Aluminum and Copper speciemens for the six analysed SC values and with two frictional conditions, namely dry contact and a film of teflon at the punch-workpiece interface. The ring test has suggested a value of m equal to 0.1 and 0.4 for the above frictional conditions respectively. The comparison between numerical and experimental results is shown in fig.4 where the sections obtained cutting the rings with one of the vertical symmetry planes are reported. The results are referred to the dry friction condition (f=0.4) both for the Copper (on the left) and the Aluminum (on the right) specimens. A good overlapping between the numerical predictions and the experimental verifications can be observed.

ACKNOWLEDGMENTS

This research has been carried out using MURST funds.

REFERENCES

1. Aku, S.Y., Slater, R.A.C., Johnson, W., The Use of Plasticine to Simulate the Dynamic Compression of Prismatic Blocks of Hot Metal, Int. Jnl. Mech. Sci., 1967, 9, 495-525.
2. Andrews, K.R.F., England, G.L., Ghani, E., Classification of the Axial Collapse of Cylindrical Tubes under Quasi-Static Loading, Int. Jnl. Mech. Sci., 1983, 25, 687-696.
3. Park, J.J., Oh, S.I., Application of Three-Dimensional Finite Element Analysis to Metal Forming Processes, Trans. of NAMRI/SME, 1987, 296-303.
4. Tadano S., Ishikawa H., Non- Steady State Deformation of Thick Cylinder during Upset Forging, Advanced Technology of Plasticity, 1990, 1, 149-154.
5. Rebelo N., Nagtegaal J.C., Taylor L.M., Passmann R., Comparison of Implicit and Explicit Finite Element Methods in the Simulation of Metal Forming Processes, Proc. of Numiform, 1992, 123-132.
6. Alberti N., Cannizzaro L., Fratini L., Micari F., An explicit model for the analysis of bulk metal forming processes, Trans. of NAMRI/SME, 1994, 11-16.
7. Chen C. C., Kobayashi S., Rigid Plastic Finite Element Analysis of Ring Compression, Applications of Numerical Methods to Forming Processes, ASME, 1978, 28, 163-174.
8. Barcellona A., Filice L., Micari F., Riccobono R., Neural Network Techniques for Defects Prediction in the Ring Upsetting with Different Geometries, accepted for the publication on the proceedings of the 12th International Conference on CAD/CAM and Factories of the Future.

# NEW ADVANCED CERAMICS FOR CUTTING STEEL

**M. Burelli and S. Maschio**
**University of Udine, Udine, Italy**

**E. Lucchini**
**University of Trieste, Trieste, Italy**

KEYWORDS: Alumina, Zirconia, Mechanical Properties, Layered Composites

ABSTRACT

Layered ceramics are being devoloped as a new class of composite materials with enhanced strength, toughness and hardness, due to residual stresses built up during their production. In particular, by this method it is possible to improve the strength as well as the toughness without reducing hardness. This fact has been documented in this work which reports the results obtained on material consisting of two outers layers of alumina alternated by an inner layer of alumina containing different amounts of $CeO_2$-partially stabilized zirconia.

INTRODUCTION

In recent times, the use of monolithic ceramic oxides has become more frequent into the technologic field because such materials have good mechanical properties also at high temperature. In particular, alumina has, among them, the highest hardness, but this property is associated with a very low toughness so that pure alumina inserts suffer strong damages during manipulation before the cutting process, or on cooling after cutting when the thermal stresses cause strong degradation of the ceramic.

As for as the technologic field is concerned, the problem has been faced by the introduction of small amounts of zirconia, titanium carbides and nitrides or silicon carbides whiskers in alumina in order to improve the thermal conductivity and the toughness of the matrix.

Another approach to the problem consists on the realization of monolithic ceramic composites formed by alternate layers of alumina and alumina-zirconia.

Zirconia induces, during the manufacture of the composite, residual stresses due to the different thermal expansion coefficients of the layers, with compression stress in alumina and tension in alumina-zirconia layers.

The induced compression state reflects on the increase of strength, toughness and hardness of the composite preseving the insert integrity during manipulation before the cutting process and on cooling after cutting.

Published in: E. Kuljanic (Ed.) *Advanced Manufacturing Systems and Technology*, CISM Courses and Lectures No. 372, Springer Verlag, Wien New York, 1996.

In this work monolitic layered ceramics formed by two external layers of pure alumina alternate by a inner layer of alumina containing Ce-PSZ has been produced. The increase of the mechanical properties due to the residual stresses as a function of the inner layer composition has been evaluated and discussed.

ESPERIMENTAL PROCEDURE

Ceria partially stabilized zirconia (12Ce-PSZ Tosoh) and $\alpha$-Al2O3 (AKP-15 Sumitomo) were used as starting powders. The pure powders or their bleds were first flo-deflocculated and then milled as reported elsewhere(1,2). In this work a high quantity of binder (4%wt) and a longer milling time (2hrs) were used. After milling powders were dried and sieved. The high amount of binder is necessary to assure a good plasticity of the green samples and to build up a sufficiently strong interface between the layers during the pressing.
In this step of the production, the required amount of alumina necessary to realize the first layer, was introduced in a WC mould whose surfaces were polished down to 6$\mu$m diamond paste; at this point a very soft load was applied, in order to form the first interface, but not to press the powders; than the second layer of zirconia-added-alumina and the third of pure alumina were formed repeating the same procedure above described; finally the layered composite was uniaxially pressed at 120 MPa. A particular care must be kept in the extraction of the sample wich was then isostatically pressed at 200MPa.
The layers thickness was mantained constant, but it was changed the composition of the inner layer, adding different amonts os zirconia to alumina. In this comunication the thickness of the inner layers was 1.5 mm while that of the outers was 2.7 mm.
All the samples were fired in air for 1 hr at 1550°C with a heating rate of 10°C/min and cooled in the muffle.
Elastic modulus was measured by the resonance method with a self-made equippement. Flexural strength was determined by the four points procedure with a crosshead speed of 0.2mm/min on the average of 5 determinations. Toughness was evaluated by the ISB technique and hardness by a Vickers indenter applying a load of 200N.
Thermal expansion coefficient was measured with an alumina dilatometer up to the temperature of 1400°C with a heating rate of 10°C/min.

RESULTS AND DISCUSSION

In table I are reported flexural strength, toughness, hardness, elastic modulus and thermal expansion coefficient values measured on mobolithic materials which are important to evaluate and compare the mechanical properties of the layered composites.

| $ZrO_2$ %vol | $\sigma$ (MPa) | KIc (MPam$^{1/2}$) | Hv (GPa) | E (GPa) | $\alpha$20–1400 (°C$^{-1}$) |
|---|---|---|---|---|---|
| 0 | 238 | 3.31 | 16.5 | 326 | 6.1353 10$^{-6}$ |
| 20 | 240 | 5.01 | 12.6 | 273 | 6.6505 10$^{-6}$ |
| 40 | 380 | 6.11 | 11.8 | 252 | 7.3228 10$^{-6}$ |
| 60 | 500 | 7.88 | 11.6 | 230 | 8.0510 10$^{-6}$ |
| 80 | 500 | 10.24 | 11.0 | 220 | 9.1031 10$^{-6}$ |
| 100 | 509 | 9.85 | 10.5 | 186 | 10.2890 10$^{-6}$ |

Table I. Mechanical properties, and thermal expansion coefficients of the monolithic materials

It is worth to point out that strength and toughness increase with the amount of zirconia, but hardness is, at the same time, reduced, because hardness of zirconia is lower than that of alumina.

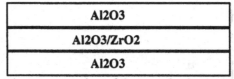

| Al2O3 |
| Al2O3/ZrO2 |
| Al2O3 |

In fig. 1 Schematic representation of the samples is reported

A schematic representation of the samples is reported ig fig. 1. Recently, for a similar configuration, Virkar et al.(3) proposed the following equations to determine the residual stresses in the three layers:

$$\sigma_1 = \frac{-E_1 E_2 d_2 \Delta\varepsilon_0}{(1-v)(2E_1 d_1 + E_2 d_2)} \quad (1)$$

$$\sigma_2 = \frac{2E_1 E_2 d_1 \Delta\varepsilon_0}{(1-v)(2E_1 d_1 + E_2 d_2)} \quad (2)$$

Where the symbol 1 refers to the outer and 2 to the inner layer, E is the elastic modulus, d is the thickness, $v$ is the poisson ratio and $\Delta\varepsilon_0$ must be calculated using the following equation

$$\Delta\varepsilon_0 = (\alpha_1 - \alpha_2)\Delta T \quad (3)$$

$\alpha_1$ and $\alpha_2$ are the average thermal expansion coefficients of the three layers which can be expressed as:

$$\alpha = \frac{1}{\Delta T}\int_{T_0}^{T} \alpha(T)dT \quad (4)$$

In fig. 2 is reported the trend of $\Delta\varepsilon_0$ as a function of the inner layer composition while fig. 3 shows the correspondent residual stress in the outer pure alumina and ineer alumina-zirconia layers.

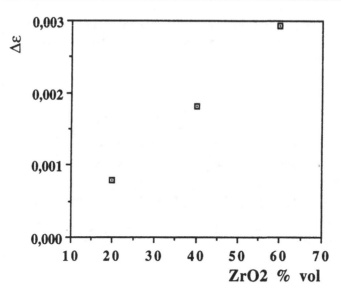

Fig. 2. $\Delta\varepsilon_0$ as a function of the inner layer composition being constant the thickness of the three layers.

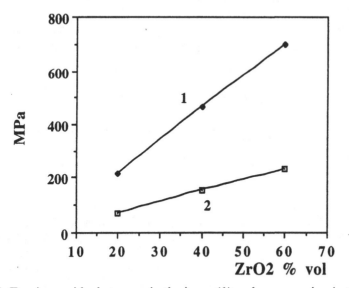

Fig. 3. Tension residual stresses in the inner (1) and compression in the outer (2) layers as a function of the inner layer composition, being constant their thickness.

Although these values can be considered known , nevertheless it is not possible to derive from these values the mechanical properties of the whole composite because some surface and edge phenomena, not yet well understood, compromise the mathematical model that have been used to to develope such equations. Therefore, actually, it is not possible to predict strength, toughness and hardness of the layered materials and it is necessary to verify their values by the traditional methods.

In literature some works reports strength and toughness of layered ceramics having similar geometry and sometimes an increase of about 300% or more with respect to the monolithic materials have been measured (4,5).

In our case we have not been able to reach so high values, but significant enhancements have been noted as it can be seen in fig. 4 where the strength as a function of the inner layer composition is reported (the strength of pure alumina is used as a comparision value). Composites containing more than 40% vol $ZrO_2$ showed diffuse cracks in the inner layer, because the residual stress exceeds the rupture strength of this material which breaks on cooling after sintering therefore, in this work, they were not tested.

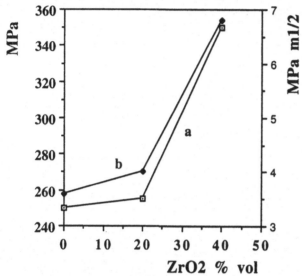

Fig. 4 Strength (a) and toughness (b) of the composite as function of the inner layer composition, being constant the layers thickness.

Fig. 4 also shows the toughness as a function of the inner layer composition and, as above, the value of pure monolithic alumina is referred.

It is possible to see that in samples having the inner layer containing 20% vol $ZrO_2$ the improvement of the mechanical properties with respect to pure alumina is limited, but it is not negligible with increasint the amount of zirconia.

Comparing values reported in fig. 2, 3 with those reported in fig. 4 it is possible to assume that high mechanical properties improvements in three layered ceramics can be obtained if $\Delta\varepsilon_0$ is higher than $1.5\ 10^{-3}$ and $\sigma_1$ is higher than 400 MPa respectively because under such conditions the residual stresses develope a significant improvement of strength and toughness of the composite. The not linear trend of strength and toughness with the inner layer composition will be furtherly studied.

As a final remark it must be noted that this class of composites associate to better strength and toughness a high harness (in our case 17GPa) which is one of the most important parameters that must be considered when cutting steel.

AKNOWLEDMENTS

The italian CNR is greatfully aknowledged for the financial support.

REFERENCES

1. M. Burelli, S. Maschio and S. Meriani, submitted to the J. Mat. Sci., (1996)
2. M. Burelli, S. Maschio and E. Lucchini, submitted to the J. Mat. Sci. Lett., (1996)
3. A. V. Virkar, J. L. Huang and R. A. Cutler, J. Am. Ceram. Soc., 70(3)164-70(1987)
4. D. B. Marshall, J. J. Ratto andF. F. Lange, J. am. Ceram. Soc., 74(12)2979-87(1991)
5. R.Lakshminarayanan, D. K. Shetty and R. A. Cutler, J. Am. Ceram. Soc.,79(1)79-87(1996)

# NEW GENERATION ARC CATHODIC PVD COATINGS "PLATIT" FOR CUTTING TOOLS, PUNCHS AND DIES

**D. Franchi**

**T.T. Ferioli & Gianotti, Division Genta-PLATIT, Caselette, Italy**

**F. Rabezzana**

**Metec Tecnologie S.n.c., Turin, Italy**

**H. Curtins**

**PLATIT AG, Grenchen, Switzerland**

**R. Menegon**

**STARK S.p.a., Trivignano Udinese, Italy**

KEYWORDS : Cutting tools, punchs, dies, Machining tests, PVD coatings

ABSTRACT: Today, thin film hard coatings are an indispensable element in the production of high quality tools, dies and mechanical components for various fields of industry. In the metalworking industry, they are applied as standard practice to a broad range of cutting tools and dies. Available hard-coatings options have increased dramatically over the last few years. In the production of hard coatings for the tool industry, three technologies are primarily used : High Temperature CVD, PVD and Plasma Enhanced CVD.
In the past decade , PVD (Physical Vapor Deposition) Methods have gained greatly importance, and in particular the more important industrial PVD Methods: electron-beam Evaporation, magnetron sputtering and Arc Evaporation.
Cathodic Arc technology has, in the past, suffered from a number of severe problems when implemented in industrial production environments, despite its indisputable basic Physical advantages for the Deposition of functional hard coatings.
Within the PLATIT concept a new type of Arc source has been developed, capable of overcoming most of these limitations. The PLATIT system has proved to satisfy the requirements of high quality combined with high productivity for cutting tools as well as the criteria of high density and low droplet number for mould injection applications.
The aims of the paper is to present some data related to the characterisation of innovative hard PVD "PLATIT" coatings for cutting tools, punchs and dies and to present the results of machining tests and forming tests performed with different tools and dies coated with these innovative PVD layers in comparison with std PVD coatings (TiN, TiCN).

Published in: E. Kuljanic (Ed.) *Advanced Manufacturing Systems and Technology*, CISM Courses and Lectures No. 372, Springer Verlag, Wien New York, 1996.

## 1. INTRODUCTION

Today, thin film hard coatings are an indispensable element in the production of high quality tools, dies and components for various fields of industry. In the metalworking industry, they are applied as standard practice to a broad range of tools. Available hard-coatings options have increased dramatically over the last few years.

Moreover the needs for greater productivity or to machine difficult to-cut workpiece materials call for the development of new, higher performance machine tools and cutting tools capable of higher tool life or cutting conditions, and in recent years considerable developments have taken place in the area of HSS and WC cutting tools coated by innovative hard thin films.

In the production of hard thin coatings for the tool industry, three technologies are primarily used : High Temperature CVD, PVD and Plasma Enhanced CVD.

In the past decade , PVD (Physical Vapor Deposition) Methods have gained greatly importance, and in particular the more important industrial PVD Methods: Electron-beam Evaporation, Magnetron Sputtering and Arc Evaporation.

Cathodic Arc technology has, in the past, suffered from a number of severe problems when implemented in industrial production environments, despite its indisputable basic Physical advantages for the Deposition of functional hard coatings.

Within the PLATIT concept a new type of Arc source has been developed, capable of overcoming most of these limitations. The PLATIT system has proved to satisfy the requirements of high quality combined with high productivity for cutting tools as well as the criteria of high density and low droplet number for the more critical applications.

In this paper the performance of different HSS and WC cutting tools, punchs and dies coated with innovative mono and multi-layers PLATIT Arc cathodic PVD films are presented.

The paper is organised as follows. In the section 2 the coating characteristics are described. The machining test conditions and the performance of std and innovative coated cutting tools are discussed in section 3 and finally the performance for coated punching and forming tools are discussed in section 4.

## 2. PVD COATINGS CHARACTERISTICS

The innovative coatings used for this study are obtained by the PLATIT cathodic arc PVD process.

The PLATIT coating system has been designed from the point of view of the user working under industrial conditions: easy operation, high coating quality combined with high productivity, complete automatic control and negligible maintenance work and standby (unloading- loading) times. The door to door cycle times are reduced by fast IR. heating up and fast cooling down system. The Deposition rates have been optimised for different applications (3-6 microns/h) while retaining a high standard of film quality The computer controlled system permits the operation either manually or fully automatic mode through a touch-screen interface.

One of the outstanding characteristics of the PLATIT coating system is the possibility to mix up a variety of components with different dimensions. In fact, in most coating systems one is limited to mixing small and large components because of the risk of over-heating and over-etching. In this system it is absolutely possible to coat together tools and parts with diameters ranging from diameter 10 mm up to 100 mm of higher, if similar coating thickness are allowed. This feature represents a considerable advantage for operating an industrial coating system economically.

The innovative **PLATIT** PVD method work with a rectangular large area source of dimension 150mm x 800 mm, with the goal to obtain the following advantages and characteristics [1-2]:

A- Poisoning of targets : is a problem strongly related to the size of the target and for increasing target area the problem becomes more and more difficult to control, this in particular when reactive gases are used. In the PLATIT method, this problem is solved by way of a special Arc source with a specially developed magnetic-field control system, so that the poisoning cannot get started in the first place, the Deposition rate remains constant within +/- 5% for the duration of a given coating process, and finally the Deposition rate likewise falls within +/- 5% throughout the service life of the target (200-300 batches).

B-Homogenous erosion: one of the clear advantages of long rectangular sources is that merely one source is able to deliver a large homogenous plasma (along one dimension) and a good thickness uniformity over a large distance (height). An important condition for the rectangular source is that a continuous Arc trace along the surface of the target is provided: a continuous movement of the Arc on the target is quite a necessary condition. Changes in direction of the Arc spot are in general inducing a higher emission rate of droplets. It is important to obtain a horizontal magnetic field distribution for guiding the Arc spot on the target having different strength: from the practical point of view an uniform erosion and good distribution is highly convenient because it assures the user a long lifetime and high yield for his target.

C-Magnetic field configuration : the PLATIT method use a Magnetic Arc Confinement (MAC) system and the task of the MAC is the generation of an adequate magnetic field configuration which can be varied and adapted for the individual situation (target material, Arc current, process parameters). The desired magnetic field configuration is established and controlled by a combination of permanent magnets and coils. Coils currents and monitoring of the source impedance are provided through power supplies and microprocessor. An optimal control leads to a situation for which a zone as large of possible on the target surface with a constant horizontal magnetic field strength is created.

The MAC control parameters allow, by means of changing the magnetic field strength and shape, to adjust the source impedance, i.e. to adjust the voltage for a given condition of Arc current and process parameters. **Fig.1** shows schematically one configuration implemented in a PLATIT cathode.

The Ti-based hard coatings obtained by this type of Cathodic Arc source show good adherence and can be controlled with respect to the hardness-tenacity behaviour and the droplet emission characteristics.

A particularity of the PLATIT process is the low internal stress level of the coatings produced. This makes it possible to deposit TiN or TiCN PVD films with thickness over the typical 3-4 microns without any risk of them peeling off or cracking. The thickest TiN coatings deposited so far were at 15-20 microns on HSS and WC substrates, and in fact there is basically no limitation, with the PVD Arc technology, to the Deposition time provided that the target is thick enough.

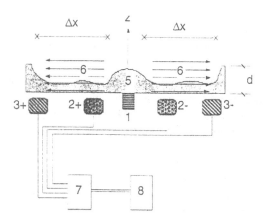

**Fig.1** Basic layout of magnetic Arc spot guiding system implemented in cathode of type PLATIT: (1) Permanent magnet, (2) and (3) coils, (5) target, (6) magnetic field lines generated by (1-3), (7) coil supplies, (8) microprocessor control.

The low internal stress levels of the coatings are achieved mainly by reducing the interfacial stress substrate-coating by adequate transition layers and optimising Deposition parameters and Arc source control during Deposition. In particular, no inert gases such as Ar are used: the use of such gases was found to increase considerably the internal stress level and therefore limit the thickness of the coatings.

**Table I** shows the characteristics of Ti-based coatings produced with the PLATIT technology.

| Coatings properties | Coatings material | | |
| --- | --- | --- | --- |
| | TiN | TiCN | $Ti_2N$ |
| Optimal Thickness (microns) | 1-20 | 1-8 | 1-5 |
| Hardness (HV 0,01) | 2.200-2.400 | 3.000-4.000 | 2.400-2.700 |
| Critical Load on HSS (N) | 60-80 | 50-70 | 50-70 |
| Friction Coefficient against 100C6 | 0,67 | 0,57 | - |
| Oxidation resistance (T °C, 1 hour in air) | 450-500 | 450-500 | 450-500 |

## 3. MACHINING TEST CONDITIONS AND PERFORMANCE RESULTS

The objective of this study was to define the real benefits of HSS cutting tools coated with innovative PLATIT PVD coatings, compared with the ones uncoated or conventionally coated. The real behaviour of innovative coated tools was evaluated by different laboratory and job-shop machinability tests organised for type of working.

During the production tests the machining cycle parameters, and the machining cutting conditions were optimised for each type of coated tool tested.

### 3.1 Milling tests

Job-shop dry rough and finish milling tests were performed with industrial production machine tools. The different cutting parameters are shown, with the results of machining tests in **Fig. 2**. Complex-shaped industrial workpieces of normalised carbon steel UNI Ck45 were used for these machining tests.

Commercial 10 mm diameter of integral HSS (AISI M42) mills has been used in production rough and finish milling tests, aiming at comparing the performance of std TiN and TiCN "market leader" PVD coatings, with the performance obtained by use of PLATIT TiN and TiCN coatings.

Evolution of mean flank wear VB (measured at regular time intervals) vs. milling length was observed during the milling operations. The different values are reported in **Fig.2** in terms of tool life (milling length) observed until reaching a flank wear VB = 0.20-0.25.

Results given in these Figures point out that the influence of innovative PLATIT coatings is very positive in terms of increase of tool life. The percentage variation of tool life obtained by use of these innovative coatings compared with std "market leader" PVD coatings is about of +60 -+350% in function of different cutting parameters used in the machining tests.

### 3.2 Disk saw tests.

The Disk saw tests were performed on a special laboratory machine tool equipped to control the different cutting parameters. The geometry of the disk saw cutters are shown in **Fig.3**. Workpieces of normalised alloyed steel (UNI 39NiCrMo3) were used in form of 35 mm diameter bar.

An evaluation and a comparison between standard HSS (AISI M2) disk saws "black oxide" surface treated, standard disk saws coated with a "market leader" TiN-PVD process, and disk saws coated with innovative TiCN and Ti2N "PLATIT" coatings has been performed with the cutting trials (see **Fig. 4**).

Evolution of tool life Vs tool wear, machine tool power absorbed and machined material surface roughness was observed during the cut operations according to the conditions in **Fig.4**. The different values are reported in **Fig.4** in terms of final tool life (n° of cuts) for the different solutions tested. Results show the increase of tool life in the case of the innovative PVD coatings.

The percentage variation of tool life obtained with the TiCN coating is about +330% versus the black oxide treated saws, and +21% versus the TiN coating "market leader". The increase obtained with the Ti2N is about +340% versus the black oxide treatment, and + 31% versus the TiN coating.

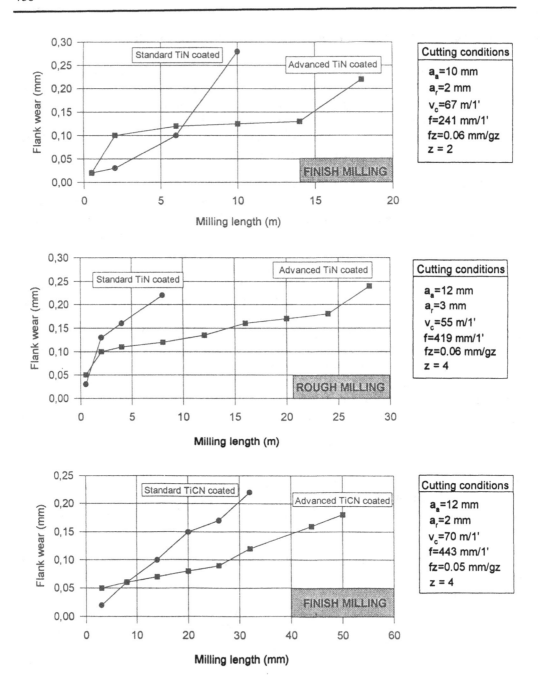

**Fig.2** : Industrial finish and rough milling tests ( HSS mill diameter = 10 mm, work material UNI CK45): comparison between std TiN and TiCN "market leader" coatings, with TiN and TiCN PLATIT coatings

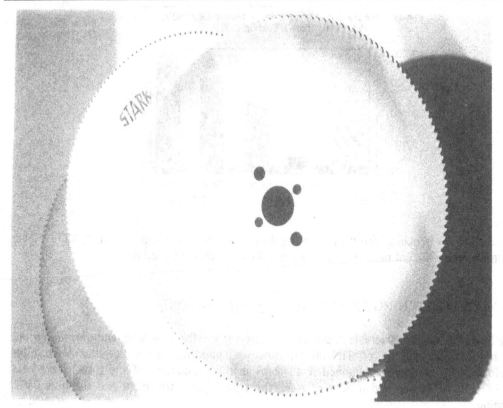

**Fig.3** Saw cutters geometry

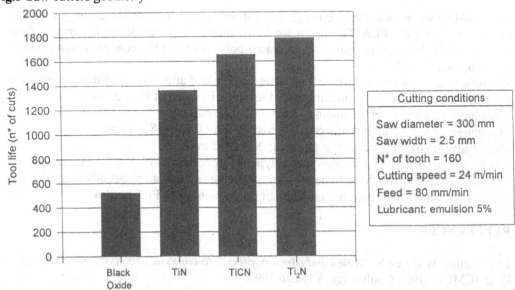

Cutting conditions

Saw diameter = 300 mm
Saw width = 2.5 mm
N° of tooth = 160
Cutting speed = 24 m/min
Feed = 80 mm/min
Lubricant: emulsion 5%

**Fig.4** Disk saw laboratory tests (saw material : AISI M2, work material : UNI 39NiCrMo3): comparison between std surface treatments and TiN PVD coatings, with TiCN and Ti$_2$N PLATIT coatings.

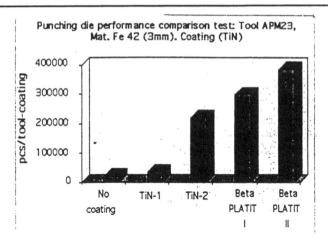

**Fig.5** APM 23 punching and deep-drawing dies coated with 8 microns Beta PLATIT (TiN) coatings with different parameters(Beta I and II), compared with std dies.

## 4. COATED PUNCHING AND FORMING TOOLS TESTING

**Fig.5** shows the considerable increase of performance for punching and forming tools coated with 8 microns of TiN. It is, however, important to notice that such high performance can only be obtained if all steps in the production of the tool have to be optimised: the choice and quality of steel, the type of heat treatment and the best PVD coating.

For mould injection tools the low roughness and the compactness of the coatings is of key importance. With the PLATIT system we have adjusted the Deposition parameters (low Arc current, high partial pressure and no macro-poisoning) and the coating design so as to obtain optimum conditions.
The following roughness before coating, after coating and after weak polishing is obtained for a 3 microns thick coating on punchs for the mould injection of PET bottles:
-Roughness of punchs before coating : Ra = 0,01-0,02 microns;
-Roughness of punchs after 3 microns TiN coating: Ra= 0.12-0.18 microns;
-Roughness of coated punchs after polishing: Ra= 0,02 microns.
It is important to state that for this special type of application of PET mould injection even very small coating errors have to be avoided, given the fact that an about 50x magnification of the error will result during the last phase of production of the PET bottles.

## REFERENCES

1.H. Curtins, W Bloesch, *"A new industrial approach to cathodic arc coating technology"*, Proc. ICMCTF 1995 Conference, S.Diego 1995.

2.D.Franchi, F.Rabezzana, H.Curtins *"New generation PVD PLATIT coatings for cutting tools, punchs and dies"*, Fourth Euro-Ceramics Conference, Riccione, 1995

# FINE BLANKING OF STEEL AND NONFERROUS PLATES

**J. Caloska and J. Lazarev**

**Faculty of Mechanical Engineering, Skopje, Macedonia**

KEY WORDS: Fine blanking, Cementite, Soft heating, Cutting Surface

ABSTRACT: The quality of cutting surface is measured depending upon the percentage of carbon and material structure. Examples from the following three groups of materials have been examined:
-carbon steel plates with a low percentage of carbon
-unalloyed carbon steel plates (0,19%C)
-alloyed carbon steel plates (0,48%C).
-nonferrous plates ( Al 99,5; EB1-Cu; CuZn 37 ).

## 1. INTRODUCTION

Fine blanking of steel plates is the working process that results with high quality of the cutting surface. In this paper are treated three materials with corresponding thermal treatment.Influence of the material structure before and after the thermal treatment to the quality of cutting surface has been examined.

The convenience of treatment by fine blanking is given by the relation

Published in: E. Kuljanic (Ed.) *Advanced Manufacturing Systems and Technology*, CISM Courses and Lectures No. 372, Springer Verlag, Wien New York, 1996.

$$K=Sg/S,$$

where:

  Sg - thickness of the high quality cutting surface, and
  S  - thickness of the material (Fig.1)

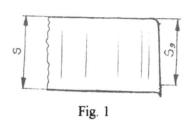

Fig. 1


## 2. THERMAL TREATMENT INFLUENCE ON THE FORMABILITY OF THE STEEL PLATES

The first group of materials with a low percentage of carbon (0,066% C), was represented by U St13 (DIN). This mechanical characteristics are convenient for the fine blanking treatment. The material structure). is given on Fig.2 (x500).

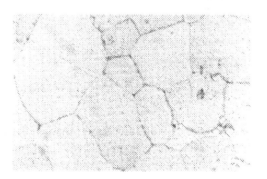

Fig.2

The size of the granule is 6-8 by ASTM and separated carbide is C-2 by ARMCO. The figure shows ferrite structure with a low level of separated perlite. To the boundary of the granules we can see separated tertiary cementite in a chain form.. Tertiary cementite is separated by the ageing process of material or by the termical treatment (glowing) of material.
Tertiary cementite has no influence on the cutting surface quality, because its granules are very small. That is why the thermal treatment for this material is not necessary.

Appearance of the cutting surface is given on Fig.3.

Fig.3

For this particular case, K=0.91, and the appropriate quality of the cutting surface is N8. The value of the relation K, shows that the quality of cutting surface depends of favorable mechanical characteristics as well as structure of the material. The presence of tertiary cementite into the carbon steel plates with low percentage of carbon, decreases the quality of cutting surface. Therefore, it has to be avoid, especially for more responsible parts.

The second group of the materials are unalloyed carbon steel plates, represented by R St 42-2 (DIN).

The third group of alloyed carbon steel plates is represented by 50 Mn 7 (DIN).

Their structures contain ferrite and perlite with laminated cementite ( see Fig.4).

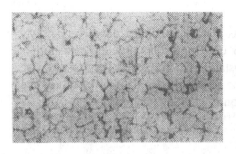
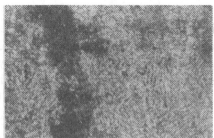

a) R St 42-2                    b) 50 Mn 7

Fig.4

Structure like this is not suitable for fine blanking treatment. The cutting edge of the tool is breaking the laminated cementite, which is extremely hard (750 HV). That is why the cutting surface have high edges ( Fig.5).

a) R St 42-2                                          b) 50 Mn 7

Fig.5

Our research shows that for the such as   material , there is necessary to conduct thermal treatment with soft heating (glowing) for transformation of laminated cementite into globular cementite. Thermal treatment with soft heating has been done periodically (Fig.6).

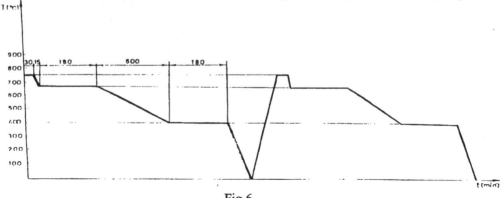

Fig.6

With this kind of  thermal treatment, steel plates are cooling down very slowly from the temperature higher than A1 point (A1=723 C). During this process, cementite is separated into globules on the existing austenite crystals. This process is very fast, so we avoid the appearance of the laminated cementite. The results of this thermal treatment are given on Fig.7, where 100% coagulation have been done.

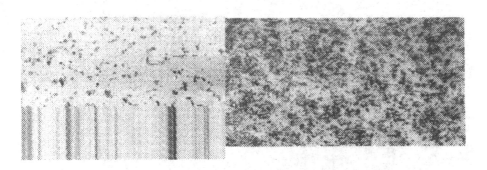

a) R St 42-2  b) 50 Mn 7

Fig.7

The cutting surface after the thermal treatment is given on the Fig.8.

a) R St 42-2  b) 50 Mn 7

Fig.8

## 3. FINE BLANKING OF NONFERROUS PLATES

The fourth group of materials was represented by Al 99,5 ( 99,5% Al ), EB1-Cu ( 99,9% Cu ) and CuZn37 ( 63,4% Cu; 36,4% Zn ). These materials have been used in the electrical engineering ( like bases, contact etc. ).

Structures of nonferrous plates have not been examined. Therefore the materials have been cut by fine blanking only, with prescribed regime.

The cutting surfaces are given in the photographs, (Fig. 9 ):

a) with conventional cutting, and

b) with fine blanking .

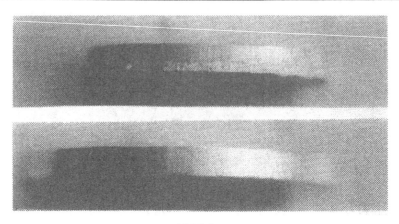

Al 99,5

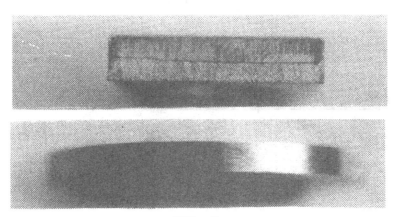

EB1 - Cu

CuZn 37

Fig. 9

The materials Al 99,5 and EB1-Cu have 100% smooth cutting surfaces, and the quality of the cutting surfaces is N8. The results are favorable for good formability with fine blanking.

From the Fig.9 can be noticed a small breaking layer on the spaceman. F - $\Delta$l diagram of this material shows that this material has required plastic characteristics ( $\varepsilon k = 28\%$ ), with higher hardness ( 141 HV ). That significantly influences the quality of cutting surface.

Beside that, the material contains 37% Zn, which is the highest boundary for good formability with the fine blanking.

Alloys with higher percentage of Zn, are harder and brittle , while the material with 50% Zn has strain, $\varepsilon k = 0$. The reason for this phenomenon is the new phase appearance, Cu5Zn8 , which has higher brittle.

According to the results of measurements, the relation $K = Sg/S = 89\%$.


4. CONCLUSION

The results of this research show that the material with higher percentage of carbon with laminated cementite structure, needs thermal treatment (soft heating) for transformation of laminated cementite into globular cementite. This treatment has given the quality of cutting surface N8 and relation K for material R St 42-2 is K=0,98, and for material 50 Mn 7, K=1.

Nonferrous materials can be successfully treated with the fine blanking, as well.


REFERENCES

1. V. Strezov, J. Lazarev:    Technological problems of fine blanking process, Faculty of Mechanical Engineering, Skopje, 1990
2. V. Strezov, J. Lazarev:    Technological problems of fine blanking and influence of geometrical and dynamically factories to the effects of fine blanking, Faculty of Mechanical Engineering, Skopje, 1989
3. J. Caloska: Examination of the formability to the materials with the fine blanking process, Skopje, 1993

# REMELTING SURFACE HARDENING OF NODULAR IRON

## J. Grum and R. Sturm
### University of Ljubljana, Ljubljana, Slovenia

**KEY WORDS:** Surface Remelting, Temperature Profiles, Microstructure, Microhardness

**ABSTRACT:** For the case of laser remelting surface hardening, a mathematical model was used to determine the temperature profiles in nodular iron and the depth of the modified layer. The temperature distributions and temperature gradients are presented for the studied depths. The mathematically obtained results are critically assessed and compared with the results in the experiments. The microstructural changes which can be predicted from the temperature profiles on heating and cooling are confirmed by microhardness measurements. The newly formed hard, fine-structured surface significantly increases corrosion and wear resistance of the material.

## 1. INTRODUCTION

In heat treatment with a laser beam, we have to achieve rapid heating and cooling rates of the surface layer. These are achieved already by rapid heat transfer into the remaining part of the cold material. By a correct choice of power density or energy input it is possible to achieve rapid local heating on to the temperature of austenite phase or even the temperatures of workpiece surface melting, which after cooling enables the formation of a modified layer of desired depth. The technology of laser remelting surface hardening provides a possibility of creating a hardened surface layer of greater depth, which makes the products more suitable for higher loads and raises their wear resistance. The depth of the remelted and hardened zone depends on the parameters of the laser beam which is defocussed and has in our case a Gaussian energy distribution. It also depends on workpiece

Published in: E. Kuljanic (Ed.) *Advanced Manufacturing Systems and Technology*,
CISM Courses and Lectures No. 372, Springer Verlag, Wien New York, 1996.

travelling speed, and its material properties defined by heat conductivity, specific density, specific heat, austenitization temperature or melting temperature. Our aim is to develop a mathematical model for practical purposes, describing the temperature evolution in the material by which it will be possible to predict the depth and width of the remelted and/or hardened trace. The calculated limiting temperatures, such as the temperatures of melting and austenitization, can be experimentally verified by measuring the size of particular traces, by microhardness measurements and microstructure analysis.

## 2. DETERMINATION OF TEMPERATURE DISTRIBUTION

Equations describing heat transfer in the laser remelting surface hardening process can refer to one, two or three dimensions. Ashby [1] presented a simple equation for the determination of temperature evolution $T(z,t)$ which describes the thermal conditions in the workpiece material around the laser beam axis. The equation considers the laser beam power P of the Gaussian source, its radius $r_b$ on the workpiece surface, workpiece or laser beam travelling speed $v_b$, absorptivity of the material A, distance from the workpiece surface z, time t, ambient temperatur $T_0$ and the thermal diffusivity of the material $\alpha$ and thermal conductivity of the material K:

$$T(z,t) = T_0 + \frac{A \cdot P}{2 \cdot \pi \cdot K \cdot v_b \cdot \sqrt{t \cdot (t + t_o)}} \cdot \left[ e^{-\left(\frac{(z+z_0)^2}{4 \cdot \alpha \cdot t}\right)} \right] \qquad [°C] \qquad (1)$$

Variable $t_0$ represents the time necessary for heat to diffuse over a distance equal to the laser beam radius on the workpiece surface [1]:

$$t_0 = \frac{r_b^2}{4 \cdot \alpha} \qquad [s] \qquad (2)$$

The variable $z_0$ measures the distance [1] over which heat can diffuse during the laser beam interaction time $t_i = \frac{2 \cdot r_b}{v_b}$ :

$$z_0 = \sqrt{\frac{\pi \cdot \alpha \cdot r_b}{2 \cdot e \cdot C \cdot v_b}} \qquad [m] \qquad (3)$$

where C is a constant, in our case defined as $C = 0.5$.

Since Ashby's equation could not be used for the determination of the characteristic temperature profile in the surface layer of the material, the equation was supplemented to some extent in order to reach an agreement between the calculated values and experimental results. We introduced certain physical facts about thermal conduction through the material reported also by some other authors [2,3,4]. Thus we obtained a relatively simple physico-mathematical model describing the temperature evolution $T(z,t)$ in the material depending on time and position, where we distinguished between the heating cycle and the cooling cycle.

1) The heating cycle conditions in the material can be described by the equation:

$$T(z,t) = T_0 + \frac{A \cdot P}{2 \cdot \pi \cdot K \cdot v_b \cdot \sqrt{t \cdot (t_i + t_0)}} \cdot \left[ e^{-\left(\frac{(z+z_0)^2}{4\alpha t}\right)} + e^{-\left(\frac{(z-z_0)^2}{4\alpha t}\right)} \right] \cdot \mathrm{erfc}\left(\frac{z+z_0}{\sqrt{4 \cdot \alpha \cdot t}}\right) \quad (4)$$

for $\quad 0 < t < t_i$

2) The cooling cycle conditions in the material can be expressed by the equation:

$$T(z,t) = T_0 + \frac{A \cdot P}{2 \cdot \pi \cdot K \cdot v_b \cdot \sqrt{t \cdot (t_i + t_0)}} \cdot \left[ e^{-\left(\frac{(z+z_0)^2}{4\alpha t}\right)} + e^{-\left(\frac{(z-z_0)^2}{4\alpha t}\right)} - e^{-\left(\frac{(z-z_0)^2}{4\alpha (t-t_i)}\right)} \right] \cdot \mathrm{erfc}\left(\frac{z+z_0}{\sqrt{4 \cdot \alpha \cdot t}}\right) \quad (5)$$

for $\quad t > t_i$

By equations (4) and (5) we can, in sufficient detail, describe the temperature evolution during the heating or cooling cycle of the material surface layer. Once, by means of limiting temperatures, the case depths or widths of the characteristic layers have been calculated, we can calculate the heating rate and especially the cooling rate in the material, according to the equation:

$$\dot{T} = \frac{dT}{dt} = \frac{T_1 - T_0}{t_1 - t_0} \qquad [°C/s] \qquad (6)$$

## 3. EXPERIMENTAL RESULTS

### 3.1. Temperature distribution

The physico-mathematical model of temperature evolution in laser remelting surface hardening was experimentally verified on a single trace on a nodular iron 400-12 according to the ISO standard with ferrite-pearlite microstructure. In the experiments we used a laser beam power of $P = 450$ W, with the beam radius on the workpiece surface $r_b = 0.633$ mm so that the achieved power density was 35700 W/cm$^2$. The laser light absorption coefficient was chosen according to recommendations in the literature as a constant value and was $A = 80\%$ [5]. The workpiece travelling speeds were chosen in a wide range from 2 to 42 mm/s in increments of 2 mm/s. The melting point temperature $T_m = 1190$ °C and austenitization temperature $T_a = 810$ °C were defined from phase diagrams while the densities of the iron were calculated considering the data on the proportion and density of the components of the matrix (ferrite: $\rho_f = 7860$ kg/m$^3$, pearlite: $\rho_p = 7846$ kg/m$^3$, graphite: $\rho_g = 2550$ kg/m$^3$ [6]). The data on specific heat $C_p = 507$ J/kgK, thermal conductivity $K = 35$ W/mK and thermal diffusivity $\alpha = 9.56 \cdot 10^{-6}$ m$^2$/s were taken from professional literature considering the carbon and silicon content in the iron, and the type of the basic microstructure of the iron [6].

In Figure 1.a temperature profiles on the surface of nodular iron 400-12 at different workpiece travelling speeds are presented. We can note a decreasing trend of maximum temperatures if workpiece travelling speeds are increasing. If the workpiece travelling speeds increase or if interaction times are shorter, the maximal temperatures on the workpiece surface are reached earlier, and after the beam interaction time, there follows a phase of rapid cooling of the heated material. Figure 1.b shows the temperature gradients or heating and cooling rates in the given conditions of heating. We can see that if the workpiece travelling speeds are increasing, the maximum surface temperatures achieved are, to be precise, lower, still the temperature gradients on heating and cooling are higher, which is followed by shorter completion times of martensite transformation.

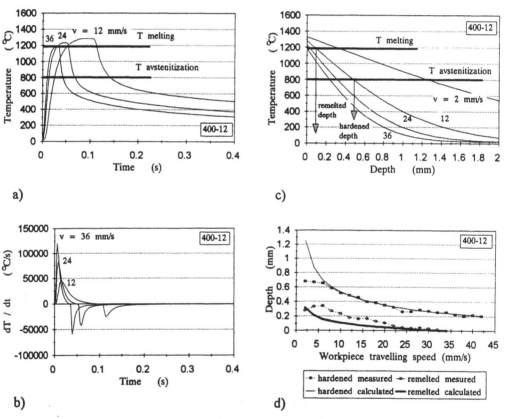

Fig. 1. a) Temperature profiles on the surface for various workpiece travelling speeds.
   b) Surface heating and cooling rates at different workpiece travelling speeds.
   c) Maximum temperature drop as a function of depth.
   d) Comparison of experimentally measured remelted and hardened zone depths with those calculated with the physico-mathematical model.

Another significant piece of information on the achieved effects of heat treatment is the depths of the remelted and hardened zones. This is defined on the basis of limiting temperatures on heating and under the assumption that sufficiently high cooling rates necessary for the formation of martensite structure are achieved. Figure 1.c shows the maximum temperature drop after the laser beam interaction time in the nodular iron at different workpiece travelling speeds. Knowing the melting and austenitization temperatures of nodular iron, we can successfully predict the depth of the remelted and hardened zone, see Fig. 1.c. Considering the fact that on the basis of limiting temperatures it is possible to define the depths of particular zones and that these can be confirmed by microstructure analysis, we can verify the success of the proposed physico-mathematical model for the prediction of remelting surface hardening. From Figure 1.d we can see that there is a good correlation between the calculated depth of the modified layer and the results of experimental measurements for the workpiece travelling speeds $v_b \geq 6$ mm/s. At smaller workpiece travelling speeds deviations appear between the calculations according to the physico-mathematical model and the experimental results due to higher heat losses, namely due to stronger radiation of heat into the environment and stronger influence of protective gas on workpiece surface cooling.

### 3.2. Microstructure

Once the optimal heat treatment conditions had been defined, the rest of the laser heat treatment was done so that a 30% overlapping of the remelted zone width was ensured (Fig. 2). The heat treatment was performed on roller specimens with a diameter of 37 mm and a width of 10 mm.

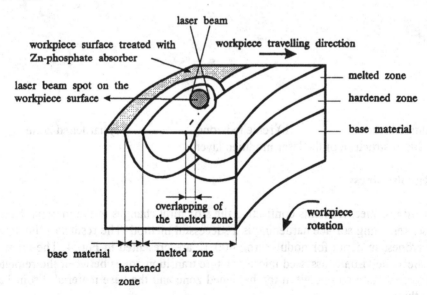

Fig. 2. Effects of the laser remelting surface hardening.

From Figure 3 we can see that in the modified trace of the laser machined nodular iron it is possible to distinguish two main zones:

1) Remelted zone consisting of austenite dendrites, ledeburite, individual coarse martensite needles and undissolved graphite nodules.

2) In the hardened zone only solid-state transformations take place. The hardened zone consists of martensite, residual austenite, ferrite and carbon in the form of nodules surrounded by martensitic shells. Some pre-conditions for the formation of the martensitic shells are the existence of a ferrite structure around the graphite nodules, suitably high heating rate beyond the austenitization temperature, and sufficiently high cooling rate. Since the whole process is running very fast, it is likely that only a smaller part of the austenite structure will become carbon-rich, getting the carbon from the graphite nodule through the diffusion process. Because of very rapid cooling rates, the carbon-rich austenite shell crystalizes into the martensite structure.

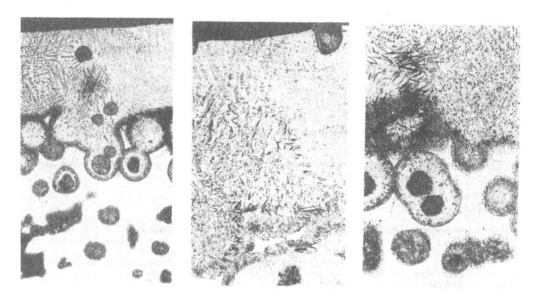

a) modified layer             b) remelted zone             c) hardened zone

Fig. 3. Microstructure of the laser modified layer.

### 3.3. Microhardness

Microhardness measurements confirming the structural changes in the material have shown that laser remelting surface hardening is a successful method. The results of the variation of microhardness in depth for nodular iron 400-12 are presented in Fig. 4. The arrows in the figure show the visually assessed microstructure transition zones between the remelted zone and hardened zone and between the hardened zone and the base material. From Fig. 4 we can see that:

1. The highest hardness (1000 $HV_{100}$) is achived in the surface layer to a depth of 100 μm in a single laser trace. Then it falls slightly and varies very uniformly with the depth of the modified layer (800 - 900 $HV_{100}$). On the bottom boundary of the hardened zone in the depth around 550 μm, the microhardness falls to a value around 250 $HV_{100}$, which represents the hardness of the base material.
2. The microhardness profile is much different in a process with a 30% overlapping of the remelted zone width. On the surface the microhardness is lowered and amounts in the entire depth of the remelted zone to about 800 $HV_{100}$. Another lowering of the microhardness values happens in the hardened zone down to 550 - 650 $HV_{100}$. The lowering of the microhardness values is effected by microstructure annealing caused by the repeated heating of the already modified trace. A characteristic in this case is a notable continuous transition of microhardness between the remelted and the hardened zone as well between the hardened zone and the matrix.
3. From the data on the microhardness variation it is possible to make a successful assessment of the depth of the remelted and hardened zones when the process is performed with a 30% overlapping of the remelted zone width.

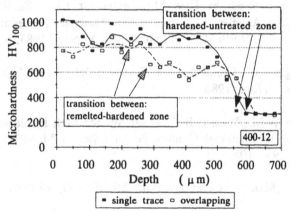

Fig. 4. Results of microhardness measurements in the modified layer for a single trace and for overlapped traces, $v_b = 12$ mm/s.

## 4. CONCLUSIONS

The mathematical model can be successfully used for the description of temperature conditions occurring in material after laser heat treatment. Deviations occurring between the physico-mathematical model and the measured depths of the remelted and hardened zones at lower workpiece travelling speeds can be attributed to an increased heat radiation into the environment and cooling effects of the protective gas which is supplied axi-symmetrically to the laser beam onto the workpiece surface. At lower workpiece travelling speeds and due to constant flow rate the protective gas has a greater influence on cooling or it enables higher transfer of heat from the workpiece surface into the environment. The experimental results have confirmed that even with a low laser source power it is possible to achieve a sufficient thickness of the modified layer if hardening is done by remelting the surface layer. The increased hardness of the remelted zone and hardened zone (up to 1000 $HV_{100}$) largely increases the wear resistance of the surface. On the basis of the results, it can be concluded that laser remelting surface hardening can be regarded as a highly successful method for increasing the hardness and wear resistance of nodular iron.

## REFERENCES

1. Ashby M.F., Easterling K.E.: The Transformation Hardening of Steel Surfaces by Laser Beams - I., Hypo-Eutectoid Steels; Acta Metall. Vol. 32, No. 11, 1984, p. 1935 - 1948

2. Gregson V.G.: Laser Heat Treatment; Laser Materials Processing, ed. by Bass M.; Center for Laser Studies, University of Southern California, Los Angeles, California, USA; Materials Processing - Theory and Practices, Vol. 3, Chapter 4, North-Holland Publishing Company, 1983, p. 209 - 231

3. Breinan E.M., Kear B.H.: Rapid Solidification Laser Processing at High Power Density; Laser Materials Processing, ed. by Bass M.; Center for Laser Studies, University of Southern California, Los Angeles, California, USA; Materials Processing - Theory and Practices, Vol. 3, Chapter 5, North-Holland Publishing Company, 1983, p. 236 - 295

4. Carlslaw H.S., Jaeger J,C.: Conduction of Heat in Solids; Second Edition 1959, New York, Clarendon Press, Oxford, 1986

5. Hawkes I.C., Steen W.M., West D.R.F.: Laser Surface Melt Hardening of S.G. Irons; Proceedings of the 1st International Conference on Laser in Manufacturing, November 1983, Brighton, UK, p. 97 - 108

6. Smithelss C.J. ed.: Metals Reference Book; 5th Edition, Butterworths, London & Boston, 1976

7. Grum J., Šturm R.: Laser Surface Melt-Hardening of Gray and Nodular Irons; Conference "Laser Material Processing", Opatija, Croatia, 1995, p. 165 - 172

8. Grum J., Šturm R.: Laser Heat Treatment of Gray and Nodular Irons, Journal of Mechanical Engineering, Ljubljana, Vol. 41, No. 11-12, 1995, p. 371 - 380

9. Grum J., Šturm R.: Laser Surface Melt-Hardening of Gray and Nodular iron; Conference MAT-TEC 96, Pariz, France, 1996, p. 185 - 194

10. Grum J., Šturm R.: Mathematical Prediction of Depth of Laser Surface Melt-Hardening of Gray and Nodular Irons, to be Published in Applied Surface Science, Proceedings of E-MRS Spring Meeting: Strasbourg, France, June 1996

# MACHINING WITH LINEAR CUTTING EDGE

**P. Monka**

**TU Kosice with seat in Presov, Presov, Slovakia**

KEY WORDS: Cutting Tool, Linear Cutting Edge, Surface Roughness, Geometrical Characteristics

ABSTRACT: The paper contains the results which have been achieved by cutting tool with linear cutting edge not paralel with the axis of the workpiece. The gained results of the measurements show that the investigated cutting tool enables to secure the same values of surface profile charakteristics of bearing steel 14 109.3 according to STN 414109 and corrosion resisting steel 17 241 according to STN 417241 as a classical cutting tool at finishing with the significant increase of the feed per revolution. It directly influences on the length of the technological operation time which is several times shortened.

NOTATION:

| | | |
|---|---|---|
| S | Tool major cutting edge (it's linear cutting edge not paralel with axis of workpiece) | |
| S' | Tool minor cutting edge | |
| $R_a$ | Arithmetical average deviation from a mean line | [μm] |
| $\lambda_s$ | Tool major cutting edge inclination | [°] |
| $\lambda_s'$ | Tool minor cutting edge inclination | [°] |
| $\alpha_o$ | Tool orthogonal clearance of major cutting edge | [°] |
| $\alpha_o'$ | Tool orthogonal clearance of minor cutting edge | [°] |
| $\gamma_o$ | Tool orthogonal rake of major cutting edge | [°] |
| $\gamma_o'$ | Tool orthogonal rake of minor cutting edge | [°] |

Published in: E. Kuljanic (Ed.) *Advanced Manufacturing Systems and Technology*, CISM Courses and Lectures No. 372, Springer Verlag, Wien New York, 1996.

| $r_\varepsilon$ | Corner radius | [mm] |
|---|---|---|
| $r_n$ | Rounded major cutting edge radius | [mm] |
| $r_n'$ | Rounded minor cutting edge radius | [mm] |
| $v_c$ | Cutting speed | [m.min$^{-1}$] |
| f | Feed per revolution | [mm] |
| STN | Slovak Technical Norm | |

## 1.INTRODUCTION

Despite of opinion of some experts, that the manufacturing of the workpieces by the machining is not perspective and that it's necessary to substitute it by chipless methods, is this technology irreplaceable by other technological methods in planty of cases.

The demanded accuracy and the quality of machined surface is not possible to achieve at the manufacture of the workpieces by machanical working or casting process, because the skin is adversely affected. That's why the machining for giving precision to the sizes is inevittable.

There are evident the productivity increase in machining technology during last years owing to automation of production , multitool machining and intensification of cutting conditions by means of new tool materials. The effectiveness of production will be increased by development of work organization in future time to the purpose of lowering of energy and material costs and unloading of living environment. [1]

The advancing of machining productivity, at the cutting speed fixed from the look of optimal durability, is possible by using of faster feeds. This is hindered by limiting conditions at the classical tools, which are given by required surface roughness.

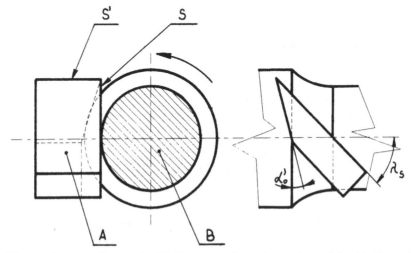

Fig. 1: The working cutting tool with linear cutting edge not paralel with the axis of the workpiece.    Designation:    A - cutting tip working by linear cutting edge
B - machined workpiece

The cutting tool with linear cutting edge not paralel with the axis of the workpiece (Fig.1) enables to achieve better values of surface roughness at multiple magnification of the feed.[2]

The most important problems, which are necessary to solve from the standpoint of machining technology with regard to tool equipment, are [1]:

a/ the right selection of the tool types and the number of tools.

The requirements for the tools are much major in automatized production as in convential production. The tools along with various tool holders must to enable good position of cutting edges with regard to the workpiece. Their basic signes are [3]:
- it's possible to set up of them outside of working place.
- they are quickly-clamping.
- the structural solution warrants their automatic manipulation not only in technological working place, but in system of intermediate manipulation,too.
- the tools are usable in various technological working place.
- tool units are unambiguously identified in the whole manufacturing system.

The tool used in automatized manufacturing systems must be constructed so, that the tool unit was staticaly and dynamicaly rigid for the all combinations of clamping and lengthening parts and they must to quarantee repeated true clamping of tool unit. [3]
It's typical for the present time, that the machining of some kind of the workpieces is realized by means of several tens, quite several hundreds of tools. The number of tools is possible to lower by standardization and unification of the working surfaces, what enables better manipulation with the tools and the reduction of the claims for informative system.

b/ automatical following of the cutting tool state during of the machining.

The most of present machines included in automatized systems are not provided by automatical identification of tool chipping and tool wear. That's why the additional meassuring operation is needed, what increases the dependence of the system on the man. Therefore future trend must to be directed to adaptive systems of the check of the workpiece sizes at the machining and to the equipments for compensation of tool wear.

c/ the optimalization of cutting conditions

The special problem in the machining technology is to specify of technological regimes what means to solve the technic-ekonomical optimalizing task. The calculation of cutting regime includes the appropriation of tool angle parameters, cutting speed, feed, depth of cut at the keeping of required accuracy, quality of machined surface and reliable operation of technological system at minimum costs for the machining. The use of the tool with linear cutting edge enables to increase machining productivity at the keepinng of good surface roughness quality.
The linear cutting edge paralel with workpiece axis is inclined in tool cutting edge inclination $\lambda_s$, what reduces the vibration evoked at the machining with this tool.

## 2.CHOICE OF PARAMETERS

The parameters of tool angles determine the shape of cutting part of tool and its position with regard to a workpiece. It's necessary to choose the true geometric parameters, because they very affect productivity and the quality of machined surface.

The individual parameters must to be chosen (at the planning of geometry of tool cutting part) so to be guaranteed:

a/ the good strength of cutting edge - it's mainly important at the material with lower bending strength, at the machining of high-strength materials and at intermittent cutting. During of the experiments with linear cutting edge was this conditions observed so, that the cuting edge S' (fig. 1) (which doesn't correspond strength conditions respect to the little cutting angle) doesn't take part in the machining.

b/ the maximum cutting edge life - it's necessary to choose the geometric parameters so, that the life is maximum at the keeping of all needed charakteristics. Therefore in practical tests the tool angle parameters were chosen according to [4].

c/ at once the minimum expenditure of energy and the suitable ratio of dimensions of cutting force components.

d/ the stability of cutting proces - mainly at the tools, which the rigidity respect to their construction is deficient in some of direction. This was in experiments achieved
- by the choise of the angle $\lambda_s$ according to results published in [5] (where the vibration is minimum)
- by the designing of rigid tool holder
- by the choise of more rigid kind of the clamping from two suggested variants.

e/ the accuracy of workpiece dimensions and the quality of machined surface. On the basis of preliminary results published in [2] is this kind of the tool very suitable for the achievement of high quality characteristics of machined surfaces.

## 3.PREPARATION OF EXPERIMENTS

The designing of suitable cutting tip was the first move before realization of experiments. The cutting tip for the tool with linear cutting edge was made from the cutting tip - type SNMN 12 0415 FR.
The tool cutting edge inclination was suggested on the basis of results published in [5] so to be kept the conditions a/, b/ and d/ of previous section. The used tool angle was defined by these following angles:

$$\alpha_0 = 10° \, , \, \lambda_s = \gamma_0' = 45° \, , \, \gamma_0 = \lambda_s' = 0° \, , \, \alpha_0' = 10°$$

The cutting tips for tool with linear cutting edge were spark erossion worked and next their skins were ground by diamond grinding wheel. They were lapping by diamond lapping compound with the grit size M3 and concetration S. It was done the measuring of the surface roughness on the tool face and the tool flanks after working.

The average values of surface roughness were:
- on the face $R_a = 0,21$ μm
- on the major flank $R_a = 0,21$ μm
- on the minor flank $R_a = 0,20$ μm

The measuring were done
- on the flanks - in the direction perpendicular on the primary cutting motion
- on the face - in two mutually perpendicular directions - in direction of the feed and in direction of the infeed.

The average values of radiuses on these cutting tips were:
$r_\varepsilon = 0,18$ μm
$r_n = 0,21$ μm
$r_n' = 0,08$ μm

The designing and the manufacturing of tool holder, which enables the good clamping of the cutting tip and the rigidity during of the machining (according to paragraphs d/ and e/ of previous section) were the next steps before realization of experiments. This tool-holder was milled from the tool shank with section 25x25 mm. The first suggested variant clamped the cutting tip by means of taper-head screw, but this fixing didn't assure good rigidity of the machining. Therefore it was used the second suggested variant of tool holder with the clamping by two-arms clamp.

The methodology of experiments was chosen so to be continued in the development of achieved results, which were published in [1].

4.CONDITIONS OF MACHINING AND MEASURING

The chemical composition of workpiece materials are shown in Table 1. [6, 7]
The practical authentication of the relation of $R_a$ on the cutting speed and on the feed per revolution was done under the conditions shown in Table 2.

5.RESULTS OF EXPERIMENTAL VERIFICATION

There was done following shapes of chips at the machining:
a/ a long helix conical chip at the machining by means of tool with linear cutting edge in the whole extent of cutting conditions
b/ a long continuous, helix annular chip at the machining by cutting tip WNMG 080416-NG in the whole extent of cutting conditions

The processing of measured values was done by means of software STATGRAFIC.

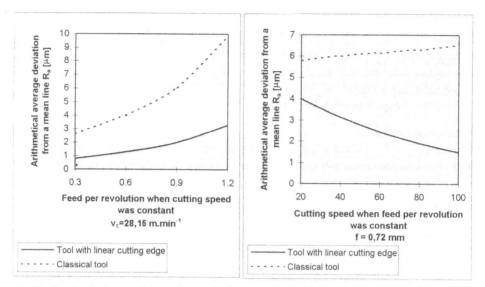

Fig. 2. Grafical relations of the arithmetical average deviation from a mean line $R_a$ on the feed per revolution and on the cutting speed for bearing steel 14 109.3.

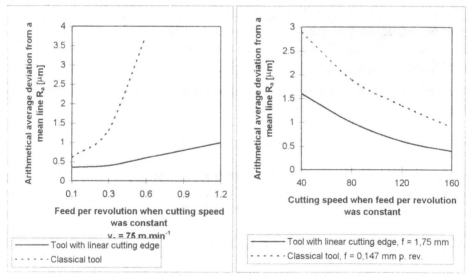

Fig. 3. Grafical relations of the arithmetical average deviation from a mean line $R_a$ on the feed per revolution and on the cutting speed for corrosion resisting steel 17 241.

Tab.1: Chemical composition of workpiece material [%]:

|         | C           | Mn          | Si          | P          | S          | Cr          | Ni           |
|---------|-------------|-------------|-------------|------------|------------|-------------|--------------|
| 14 109.3 | 0,90 - 1,10 | 0,30 - 0,50 | 0,15 - 0,35 | max. 0,027 | max. 0,030 | 1,30 - 1,65 | max. 0,30    |
| 17 241  | max. 0,12   | max. 2,00   | max. 1,00   | max. 0,045 | max. 0,030 | 17,0 - 20,0 | 8,00 - 11,00 |

Tab.2: The conditions of machining and measuring.

| | Classical tool | | Tool with linear cutting edge not paralel with the axis of workpiece | |
|---|---|---|---|---|
| **Cutting material:** | The cemented carbide composition coated by TiN-TiC-TiN similar S45 according to STN (similar as material P30-P50 according to ISO or similar as material C5 according to ANSI) | | The uncoated cemented carbide composition S30 according to STN (similar as material P30 according to ISO or similar as material C6 according to ANSI). | |
| **Metal cutting machine:** | The engine lathes SV 18 RD and SUI 40 | | | |
| **The cooling fluid:** | Emulsin H | | | |
| **The adjusting of the tool corner:** | in the workpiece axis | | 5 mm over the workpiece axis | |
| **Depth of cut :** | 0,5 mm | | | |
| **Constant feed per revolution:** | steel 14 109.3 $f = 0{,}72$ mm p. rev. | steel 17 241 $f = 0{,}147$ mm p. rev. | steel 14 109.3 $f = 0{,}72$ mm p. rev. | steel 17 241 $f = 1{,}75$ mm p. rev. |
| | when the relation of the $R_a$ on cutting speed was found out | | | |
| **Constant cutting speed:** | steel 14 109.3 $v = 28{,}15$m.min$^{-1}$ | steel 17 241 $v = 75$ m.min$^{-1}$ | steel 14 109.3 $v = 28{,}15$m.min$^{-1}$ | steel 17 241 $v = 75$ m.min$^{-1}$ |
| | when the relation of the $R_a$ on the feed per revolution was found out | | | |
| **Surface roughness measuring equipment:** | The measurements of surface roughness $R_a$ on the cutting tips and on the machined surfaces were done by means of profile meter HOMMEL TESTER T 1000 in parallel direction with the axis of workpiece. The total lenght of measuring section $L_t$ was 4,8 mm. | | | |

The relationship, betwen experimental obtained dependences for the tool with linear cutting edge and for the series productioned cutting tip WNMG 080416-NG (corner radius $r_{\varepsilon}$ = 1,6mm) with a cover IC635, follows from the grafical relations of the arithmetical average deviation from a mean line on the feed per revolution or on the cutting speed - Fig. 2 and Fig. 3.

It is evident from these grafical relations that the tool with linear cutting edge achieves the values of the arithmetical average deviation from a mean line $R_a$ some-times lower as a tool with corner radius 1,6 mm at the same conditions. The tool with linear cutting edge unite roughing and finishing cut, what enables to reduce the number of manufacturing operations and the direct manufacture time.

## REFERENCES

1. Vasilko, K. ; Bokučava, G. : Technológia automatizovanej strojárskej výroby. Alfa Bratislava, 1991

2. Monka, P. : Výsledky experimentálneho overovania novej geometrie sústružníckeho noža. Zborník konferencie "Počítačová podpora v technológii obrábania - PPTO 95", KSMaT SjF TU Košice, 1995

3. Pyramída. Robotizácia, automatizácia a pružné výrobné systémy. Obzor Bratislava, No. 3, 18, 1988

4. Vasilko, K.: A cutting tool with a new geometry of the cutting part and its applications. Transactions of the Technical University of Košice,No. 4, 1992

5. Vasilko, K.: Nové geometrické vzťahy medzi rezným nástrojom a obrobkom a ich vplyv na mikrogeometriu obrobeného povrchu. Preprinty vedeckých prác, No. 1/92, SjF TU Košice, 1992

6. STN 414109

7. STN 417241

# MODEL-BASED QUALITY CONTROL LOOP AT THE EXAMPLE OF TURNING-PROCESSES

O. Sawodny and G. Goch

University of Ulm, Ulm, Germany

KEY WORDS: Quality Control, Cutting Manufacturing, Turning

ABSTRACT: For a quality control in cutting manufacturing processes, a method is presented for the implementation of closed quality-control loops at turning process. Therefore, a model has been developed, which discribes the most important internal influences on the manufacturing process based on the known rules of the turning technologies. This model connects the geometrical deviations of the workpiece with the internal parameters of the turning operations and the parameter-settings of the related NC-program. After a workpiece is manufactured, it is measured and its deviations are evaluated. The internal process parameters are estimated quantitatively by extracting their individual values from the "geometrical response" to be measured on the workpiece. The estimated internal parameters lead to the optimal parameter settings of the NC-program, which guarantee the claimed tolerances. At the example of manufacturing shafts the effect of the control is shown. As a major advantage, the control needs no additional sensors and actors for the operating machines. The method is based on the traditional post-process-measurement-system. For the implemetation of the strategy, common machines and measurement devices can be used. SPC-methods can easily be integrated.

## 1. INTRODUCTION

The important skill to realize global quality management systems in industrial production leads to the demand of a quantifiable control of the quality characteristics in addition to the established methods. Therefore, the requirements for an effective strategy to control the quality in a closed loop structure are formulated and the special conditions in a common cutting manufacturing system are discussed. At the example of manufacturing shafts the general procedure to realize a model-based quality control is shown. At first, a detailed analy-

Published in: E. Kuljanic (Ed.) *Advanced Manufacturing Systems and Technology*,
CISM Courses and Lectures No. 372, Springer Verlag, Wien New York, 1996.

sis of the manufacturing process (turning operations) is necessary in order to evaluate the relevant process errors, aiming in a relationship between process errors and geometrical deviations. In a further step, the geometrical deviations and the investigated process-behaviour lead to a process model based on the well known fundamentals of manufacturing technologies. This model connects the geometrical deviations with internal process parameters. A control strategy is developed, which influences the geometry of the (subsequently produced) workpieces by extracting internal process parameters out of the the measured geometrical deviations of the (already produced) workpieces. Using the estimated model-parameters, the actual process inputs, which guarantee the claimed tolerances can be determined. At last, the results of experimental tests are presented to demonstrate the effect of the closed-loop quality control.

## 2. QUALITY CONTROL STRUCTURE IN CUTTING MANUFACTURING SYSTEMS

Up to now, the most common method to observe a given quality characteristic in cutting manufacturing systems is the SPC-method. But in fact, the SPC based on a statistical test related to the describing parameters of the assumed probability distribution, e.g. the expected mean value or the standard deviation, which depend on the investigated quality features. But the reason for the deviation of a quality characteristic cannot be determined [1]. In order to complete the quality control system, a quantitative conclusion backwards from the measured deviations to the optimal values of the process parameters is necessary. Therefore, the trend analysis of the SPC-results is not sufficient, as the reaction on process disturbances appears too late.

A possible alternative with quite reasonable results is in practice the analysis by a human expert. But, the results of this strategy depend on the subjectivity of the expert leading to a lower efficiency, reproducability and difficulties in documentation.

Further, the main objectives consists of a theoretical design of the system structure for the turning process, using the cutting process as the "plant", the common measuring devices (in separate air-conditioned measuring rooms) as the "measuring unit", and a model-based control-strategy (fig.1). The inputs of the control-strategy are the measurement results and the claimed tolerances. As the output, the controller generates the process inputs optimized in terms of the set tolerances and related to the corresponding NC-program. The control structure evaluates the update of the process inputs following the step of the produced workpieces or samples.

Figure 1:
Control structure
for the model-ba-
sed quality con-
trol.

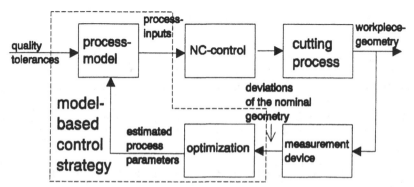

The control structure meets the two main purposes of a quality control. Due to the process model it is possible to generate starting values aiming at the given tolerances assuming ideal process conditions. The second purpose is to keep the quality characteristics of a currently produced object (and therefore the corresponding running production process) within defined tolerances by updating the process inputs, identical to the output of the controller.

The model itself describes the manufacturing process statically depending on the geometry of the workpiece. The control strategy is based on the estimation of process parameters by solving a nonlinear optimization problem. The dynamical effect of the control results in i) the calculation of the process inputs using the estimated process parameters of the last production step and ii) the application of these corrected process inputs for the next production step. In this sense, The method deals with an adaptive control structure based on the parameter scheduling idea [2]. To integrate the quality control in established manufacturing systems some additional aspects have to be considered. At first, the method must be independent of the used machine. The deviations of the geometry should be determined by the common measuring-equipment, i.e. no additional sensors should be necessary. At last, the common NC-controlled cutting machines are used. The quality control unit is directly connected to the NC-control of the machine without any additional actors.

## 3. PROCESS ANALYSIS AND MODELLING OF THE TURNING PROCESS

As an example for the development of a model-based control strategy the longitudinal outside turning process of solid shafts was chosen, controlling dimensional quality characteristics. To construct the process model, the main errors of turning operations and their effect on the process itself were considered (table 1).

| Process error | effect on the process |
|---|---|
| control errors<br>spindle errors<br>elasticity of the machine | deviation of nominal and real path → **offset deviation** |
| chattering | waviness, form of the workpiece surface (not considered) |
| eccentricity of clamping | changing depth of cut, chattering (not considered) |
| wear of tool | increasing cutting forces → **flexion deviation**, wear of cutting edge → **offset deviation** |
| temperature | microstructural change (not considered), continous material expansion → **offset deviation** |
| cutting forces | flexion of the workpiece → **flexion deviation** |
| tool material<br>workpiece material<br>cooling<br>changing geometry of the cutting edge | changing cutting forces → **flexion deviation** |

Table 1: Process errors and effects on the turning process.

In the next step these effects were investigated about their influences on the geometrical deviations, which where related to the quality attributes size, measures and form. Furtheron chatter errors were not considered.

The result was, that the major systematic process errors influencing the attributes form and geometry can be summarized in an offset deviation and an deviation due to the flexion of the clamped workpiece caused by the cutting-forces, whereas these deviations cover several different process errors.

Wear of the tool, deviation of workpiece and tool materials, change of the cutting-tool geometry and different cooling conditions: they all result in changed cutting-forces. Thus, these errors can all be summarized in the flexion deviation. The constant offset deviation covers the errors due to the wear of cutting edge, path errors of the NC-contouring control system and wrong settings of the tool reference points. The next step will be the modelling of these deviations by the equations of manufacturing engineering technologies.

For modelling the flexion, the passive cutting-force $F_p$ is the relevant cutting force, which points nearly in the normal direction of the workpiece surface (fig.2). To describe the machining operation of turning, the cutting-force law of Kienzle [3] and Meyer [4] is used, adding some necesssary modifications. An extension of the cutting-force law leads to the following equations.

$$F_p = \frac{1}{2-x} k_{a1.1} R f^{1-x} \left(\sqrt{\frac{a}{R}\left(2 - \frac{a}{R}\right)}\right)^{2-x} \quad ; \quad a \le R(1 - \cos\kappa) \tag{1}$$

$$F_p = k_{a1.1} f^{1-x}\left(\frac{1}{2-x} R(\sin\kappa)^{2-x} + \frac{\cos\kappa}{(\sin\kappa)^x}(a - R(1 - \cos\kappa))\right); \quad a > R(1 - \cos\kappa)$$

$k_{a1.1}$ is the cutting-force coefficient, R the radius of the cutting edge, $\kappa$ is the cutting edge angle, f the forward feed rate,a the depth of cut, 1-x the logarithmic slope. Experiments showed a good agreement between eq. (1) and measured data.

Figure 2:
Flexion of the workpiece in
the clamping.

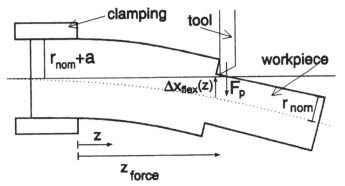

The flexion can be described with the following differential equation depending beside the cutting force on the workpiece geometry and the clamping.

$$\frac{d^2 \Delta x_{flex}}{dz^2} = \frac{F_p(z_{force} - z)}{EJ(z)} \tag{2}$$

$\Delta x_{flex}$ is the flexional deviation of the workpiece-axis, $z_{force}$ the z-coordinate of the effective interaction point of the cutting force, E the elastic modulus and J(z) the geometrical moment of inertia, which depends on the workpiece-radius r(z). The analytical solution of eq. (2) is only possible for very simple geometries like shafts with constant diameter. For shafts with cone or spherical contours eq. (2) is solved numerically. In experiments, the flexion model was validated. The modelling of the flexion connects the flexional deviation with the passive cutting-force and, furthermore, the passive cutting-force with the process parameters depth of cut a and the forward feed f. The offset error is directly compensated by changing start- and end-point of the path control. Based on these relationships between errors, deviations, and process parameters, a control strategy will be implemented.

## 4. DESIGN OF A CONTROL FOR MEASURES AND FORM OF SHAFTS

In this paper, only outside machining operations with a one-sided clamping are considered. The workpiece geometry is defined in sections of basic geometry elements (cylinders, cones and spherical surfaces). Each section can be tolerated differently.

The shafts are tolerated with the geometric characteristics in the different sections of the workpiece, i.e. the size feature "diameter" and the form feature of the "cylindricity" (in case of cones and spherical surfaces including the profile form deviation).

In a first step, the workpiece is turned roughly. The control only considers the last plane cut.

The expected geometry of the workpiece will be overlapped by a flexion and an offset deviation (see section 4.). The measured radius $r_{meas}$ at the z-coordinate $z_k$ will be

$$r_{meas}(z_k) = r_{nom}(z_k) + \Delta x_{flex}(z_k) + \Delta x_{offset} \qquad (3)$$

$r_{nom}$ is the nominal value of the radius at the coordinate $z_k$, $\Delta x_{flex}$ the flexion deviation and $\Delta x_{offset}$ the offset deviation. Since a passive cutting-force does always exist, an allowed flexion deviation is introduced relating to the set cylindricity (or profile form deviation). In addition, the allowed flexion deviation is compensated by cutting along the inverse (interpolated) line of the flexion curve (fig. 3). The resulting contour is showed in fig. 3, which is defined in the following as the ideal geometry of the workpiece section.

Figure 3: Example for the compensation of the flexion with an interpolated line and resulting ideal geometry in a cylindrical workpiece section.

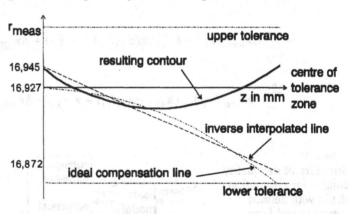

Based on the allowed flexion deviation and depending on the workpiece geometry, the numerical solution of the flexion curve eq.(2) yields sectionswise an allowed passive cutting-force $F_{pall}$. With eq.(1) the starting values for the process parameters $a_0$ and $f_0$ are determined by a Newton gradient method using the secondary condition of equal differential coefficients in the operating point. The cutting-force coefficients are set to their theoretical values.

Using the starting values, the first workpiece is produced. After its manufacturing it is measured on a conventional CMM (Coordinate-Measuring-Machine). Results are the actual radii of the measured circles at several z-coordinates. After the inverse of the flexion line is compensated and the nominal value of the radius is subtracted, a residual deviation $\Delta x_{meas}$ from the expected value is left. Regarding the error model, this deviation consists of a flexion and an offset contribution. The residual flexion error $\Delta x_{flex}$ itself depends on the passive cutting-force $F_p$, which again depends on the process parameters a and f.

This leads to an optimization statement that is used to estimate the coefficients of the cutting-force law $k_{a1.1}$ and 1-x and the offset error $\Delta x_{offset}$. They are calculated by finding the minimum of the objective function

$$Q = \sum_{k=1}^{n} (\Delta x_{meas}(z_k) - (\Delta x_{flex}(z_k) + \Delta x_{offset}))^2 \ . \tag{4}$$

$\Delta x_{flex}$ itself depends on the optimization parameters $k_{a1.1}$ and 1-x. n is the number of measured circles in the investigated workpiece section. The equation (4) is linear in terms of the parameters $\Delta x_{offset}$ and $k_{a1.1}$ and nonlinear to 1-x.

For the solution of this nonlinear optimization problem a sequential-quadratic-programming algorithm (SQP) is used [5]. It is a numerical, overlinearly converging method, which approximates the objective function locally by a quadratic function [6]. To decrease the computing time the problem is solved in two steps, where the nonlinear problem depends on the direct solution of the linear subproblem. In addition, the objective function (4) is overparameterized. That means, $k_{a1.1}$ and 1-x are not estimable in the same step. Additional measurements of other production steps with changed process parameters a and f are necessary to estimate both cutting-force parameters. For i production steps eq. (4) is then

$$Q = \sum_{k=1}^{n} (\Delta x_{meas}(z_k) - (k_{a1.1}g(a_1,f_1,1-x,z_k) + \Delta x_{off1}))^2 + \quad 1.\,meas.$$

$$+ \sum_{k=1}^{n} (\Delta x_{meas}(z_k) - (k_{a1.1}g(a_2,f_2,1-x,z_k) + \Delta x_{off2}))^2 + \quad 2.\,meas.$$

$$...+ \sum_{k=1}^{n} (\Delta x_{meas}(z_k) - (k_{a1.1}g(a_i,f_i,1-x,z_k) + \Delta x_{offi}))^2 \qquad i.\,meas. \tag{5}$$

Figure 4:
Structure of the quality control for production of shafts with defined measures and form.

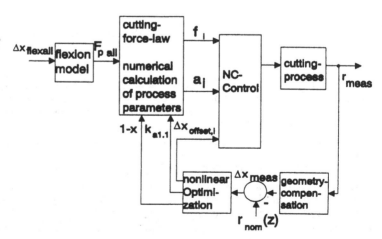

g is a nonlinear function depending on the logarithmic slope 1-x, the calculated (allowed) flexion curve, the allowed passive cutting-force, and the actual process parameters $a_i$ and $f_i$ in the production step i.

With the estimated cutting-force coefficients and the actual process parameters, the differences between the allowed and actual passive force can be evaluated. Thus, the new process parameters for the next production step can be determined. They are calculated like the starting values using the cutting-force law. But now, the estimated force coefficients are inserted. The process parameters are returned to the NC-control for the next production step. The estimated offset deviation is compensated directly. Therefore, the start- and end-point of the cutting track is changed. Figure 4 explaines the structure of the quality control.

## 5. EXPERIMENTAL RESULTS

Fig. 5 shows an exemplary workpiece of a shaft with two different diameters and an intermediate cone section. The following figures 6 to 8 demonstrate the effect of the model-based control. Fig. 6 to 8 are restricted to the most difficult cone section of the workpiece. The tolerance leads to an allowed flexion deviation of 27.5 μm (allowed profile form deviation, see fig. 5) and to an allowed passive cutting force of 200 N. Inserting the theoretical quantities in the cutting force law, the starting values of the process parameters are calculated to a=0.35 mm and f=0.36 mm (fig.8).

Figure 5:
Workpiece for the subsequently experiments.

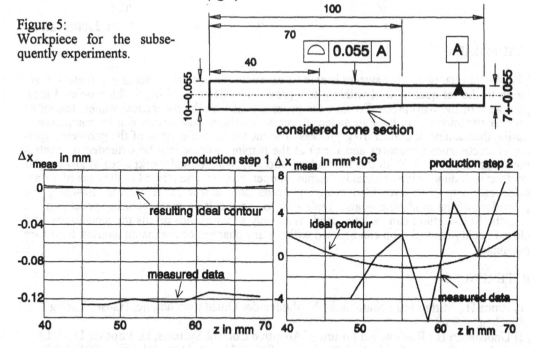

Figure 6: Form deviation of the cone section in the first and second production step.

In the first production step, the cone section is overlapped by an extreme offset deviation due to contouring errors of the NC-control (fig. 6). In the example, the flexion deviation is too large as well. For the next production step the controller compensates the offset deviation, and the flexion deviation meets the predicted amount (fig.6) due to changed values of a and f (fig.7). The remaining deviations from the ideal geometry mainly results from the

measuring incertainty of the CMM (fig. 6). The good control characteristic is expressed by
the difference between the calculated passive-force from the allowed passive-force (fig. 7).
Fig. 7 also shows a similar behaviour of the process parameters a and f. In the third production step, all changes are nearly 0. The process parameters tend to stable values.
After the second or third workpiece, the process runs for the next 50-60 workpieces with
this set of process parameters until the tool wear affects the production result. The next
workpieces should be measured and the results should transferred to the controller to read-
just the process parameters. Especially this procedure allows a easy adaption of SPC-methods.

Figure 8:
Estimated
passive-force-
deviation,
process para-
meters depth
of cut a and
forward feed f.

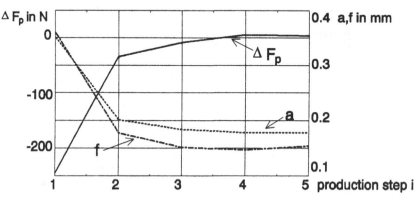

## 6. SUMMARY

The shown quality control system is able to calculate the optimal process parameters in relation to the set quality level within a few production steps, resulting in the predicted ideal
geometry of the workpiece. To start the process based on the theoretical values, the initial
process parameters can be determined. Process deviations and -errors can be compensated
during the running manufacturing process. The model for the control of the geometric quality characteristics (measures and form) at the turning process can be extended to control
the roughness as well. With this method, rotationally symmetrical workpieces of any regular
contour including inside or outside machining operations can be treated in the case of a one-
sided clamping. The two-sided clamping needs only a few modifications. Basically the
procedure of the modelling is also applicable to other cutting manufacturing processes. At
last, the SPC-method can be integrated. With only a few modifications the advantages of a
closed quality control loop can be used without any changes of the manufacturing hardware
equipment.

## REFERENCES

[1] Rinne H., Mittag H.J.: Statistische Methoden der Qualitätssicherung, Hanser Verlag,
München, 1995
[2] Unbehauen H.: Review and Future of Adaptive Control Systems; in: Popovic D. (Ed.):
Analysis and Control of Industrial Processes, Vieweg Verlag, Braunschweig, 1991, 3-22
[3] Tönshoff H.-K.: Spanen, Springer Verlag, Berlin, 1995
[4] Meyer K.F.: Der Einfluß der Werkzeuggeometrie und des Werkstoffes auf die Vor-
schub- und Rückkräfte des Drehens, Industrie-Anzeiger, Essen, 86 (1964), 835-844
[5] Sachs E.W.: Control Applications of Reduced SQP Methods; in: Bulirsch, R., Kraft, D.
(Ed.): Computational Control, Birkhäuser Verlag, Basel, 1994, 89-104
[6] Großmann Ch., Terno J.: Numerik der Optimierung, Teubner Verlag, Stuttgart, 1993

# THE EFFECT OF PROCESS EVOLUTION
# ON CAPABILITY STUDY

**A. Passannanti and P. Valenti**
**University of Palermo, Palermo, Italy**

KEY WORDS: Quality, Process Capability, Process Evolution

ABSTRACT: In today manufacturing scenario the trend to produce with lower and lower tolerance and continuously decrease the defectives rate seems to be irreversible. Then the evaluation of system performances is essential to determine if the process is able to respect our tolerances. With the classical approach this evaluation is realised by considering the process in the in control state and evaluating the capability indices and, consequently, the defectives rate. In this paper the evolution of the process is considered in terms of the mean out-of-control-time, and the real defective rate is evaluated by proposing a new capability index.

## 1. INTRODUCTION

The modern quality methodology considers prevention as a fundamental aspect because it reduces quality costs greatly. In manufacturing this means to establish a priori if the production process is able to manufacture the product, i.e. if the nonconformity rate is acceptably low.

In other words two important aspects must be considered:
- the process variability;
- the product tolerance range.

Published in: E. Kuljanic (Ed.) *Advanced Manufacturing Systems and Technology*,
CISM Courses and Lectures No. 372, Springer Verlag, Wien New York, 1996.

The output of any process is not deterministic but it is affected by various causes of variability. At the same time any product has a nominal dimension and a tolerance range. $C_p$ index, the ratio between tolerance range and natural variability, expresses the process capability to respect the tolerance range. This approach is equivalent to compute the minimum number of nonconformities the process can produce; the index can be calculated only if the process is in an in-control state.

The process variability has two different aspects [1]:
- the natural or inherent variability at a specified time, that is "instantaneous variability";
- the variability over time.

In fact any process is dynamic in nature because the parameters of the process tend to change for different reasons such as raw materials, human errors, environmental conditions, etc.

The $C_p$ index takes into account only the first process variability cause. The aim of this paper is to propose a capability index that takes into account both of the causes.

When the out-of-control state is reached, a greater rate of nonconformities is produced. In other words the $C_p$ index is meaningful only when hypothetically the mean time to out-of-control is infinite but, in the real case, a more complete analysis has to be realised.

In the present paper starting with the calculation of the real nonconformities rate during a production cycle, a new index $C_{pd}$ is calculated. This new index is always less than $C_p$ and depends on the mean time to out-of-control ($T_m$) and on the rapidity to detect the out of control state.

The results are presented for different values of $T_m$, of the control chart parameters and of the shift value.

## 2. THE CAPABILITY INDICES

In literature it is possible to find various indices to quantify the capability of a process. The more used index is $C_p$ that is equal to $\dfrac{T}{6\sigma}$ where T is the tolerance range and $\sigma$ is the standard deviation of the process. A process with an higher $C_p$ has better characteristics because it produces a lower defective rate. But the knowledge of $C_p$ is not enough to consider acceptable the process. In fact with $C_p$ we have only verified if the instantaneous variability of the process is not too large to retain a process not adequate to respect the tolerance range. A further analysis must be conducted to verify if the process mean is different from the product mean and, then, if it is useful to recenter the process. This is realised using the $C_{pk}$ index defined as

$$C_{pk} = \min\left\{\frac{(\overline{X} - LS)}{3\sigma}; \frac{(US - \overline{X})}{3\sigma}\right\}$$

where LS and US are the lower and the upper specification values.

The correspondence between $C_p$ values and defective rate is explained in the following table. The process is retained adequate if $C_p > 1.33$.

| Process → | Excellent | | Adequate | | Adequate with reserve | | Not adequate | |
|---|---|---|---|---|---|---|---|---|
| Characteristic ↓ | Cp-Cpk | Defective rate | Cp - Cpk | Defective rate | Cp - Cpk | Defective rate | Cp - Cpk | Defective rate |
| Critique | ≥ 1.67 | 0.00031 % | ≥ 1.33 | 0.0063 % | ≥ 1.00 | 0.27 % | < 1 | > 0.27 % |
| Important | ≥ 1.55 | 0.00033 % | ≥ 1.14 | 0.07 % | ≥ 0.94 | 0.5 % | < 0.94 | > 0.5 % |
| Secondary | ≥ 1.41 | 0.0025 % | ≥ 1.00 | 0.27 % | ≥ 0.71 | 0.71 % | < 0.71 | > 3 % |

Before calculating capability an analysis of the values obtained from the process must be realised. The various steps are described below.

First of all the in-control state is verified by using control charts. The first chart verifies that the mean of the process has not changed during the time. Instead the second chart verifies that the variability of the process has not changed.

A second analysis concerns the normality of data that can be verified, for example, using a normal probability plot.

## 2. QUALITY CYCLE AND DEFECTIVE RATE

To calculate the real defective rate it is useful to define the quality cycle, i.e. the time between two successive in control period.

When the process goes out of control, it cannot return to an in control state without intervention.

The control chart methodology consists of sampling from a process over the time and charting some process measurement. If the sample measurement is beyond a specified value, the process is supposed to be out-of-control.

The cycle time is the sum of the following: (a) the time until the assignable cause occurs, (b) the time until the next sample is taken, (c) the time to analyse the sample and chart the result, (d) the time until the chart gives an out-of-control signal and (e) the time to discover the assignable cause and repair the process.

For simplicity b, c and e are supposed equal to zero.

The readiness to single out the out-of-control state depends on the control chart parameters that are: the sample size n; the sample period h; the control limit L. In particular, ARL is defined as the average number of samples to point out the out of control state for given shift value and chart configuration. The product of ARL for the sample period is ASN (average time to signal) that is the mean time to find the out of control state.

The first step is to calculate the defectives rate during a cycle.

A product is considered defective if the measurement is beyond the tolerance range. During the in control state the defectives rate is:

$$2*P(x > UTL) = 2*P\left(\frac{x-\mu}{\sigma} > \frac{UTL - \mu}{\sigma}\right) = 2*P\left(z > \frac{\left(\mu + \frac{T}{2}\right) - \mu}{\sigma}\right) = 2*P(z > 3C_p)$$

where P indicates probability and $\mu$ is the target of the process (considering the process centered).

If $\varepsilon$ is supposed positive and $P\left(z < \frac{LTL - \mu}{\sigma}\right)$ equal to zero the defectives rate during the out-of-control state is:

$$P\left\{z > \left[\frac{\left(\mu + \frac{T}{2}\right) - (\mu + \varepsilon)}{\sigma}\right]\right\} = P\left\{z > \left(3C_p - \frac{\varepsilon}{\sigma}\right)\right\} = P\{z > 3C_{pk}\}$$

The total number of defectives in a cycle is:

$$\left[2 \cdot P(z > 3C_p) + ARL(\varepsilon) \cdot h \cdot P(z > 3C_{pk})\right]$$

where U is the productivity (piece/minute).
The cycle length is:

$$\left[T_m + ARL(\varepsilon) \cdot h\right]$$

and the total defective rate is

$$d_t = \frac{2 \cdot P(z > 3C_p) + ARL(\varepsilon) \cdot h/T_m \cdot P(z > 3C_{pk})}{1 + ARL(\varepsilon) \cdot h/T_m}$$

This last value depends on: the capability of the process in the in control state ($C_p$); the shift value ($C_{pk}$); the control chart parameters (h and ARL) and the mean time to out-of-control ($T_m$).

## 3. DYNAMIC CAPABILITY INDEX

The total defective rate can be used to propose a new capability index. Starting from the total defective rate the $C_{pd}$ is defined as:

$$Cpd = \frac{invnorm\left(1 - \frac{d_t}{2}\right)}{3}$$

where invnorm is the inverse normal distribution and $d_t$ is the total defective rate. This index is always less or equal than $C_p$. In fact, in the hypothetical case, when the mean time to out of control tends to infinity, $C_{pd}$ tends to $C_p$ and if $T_m$ tends to zero, $C_{pd}$ tends to $C_{pk}$. The values obtained for different combination of $T_m$, $C_p$ and $\Delta/\sigma$ are reported in the following figures from which it can be pointed out, as is obvious, that $C_{pd}$ increases with $C_p$ and $T_m$ and decreases with $\Delta/\sigma$.

Figure 1 shows that, for fixed values of $T_m$ and $\Delta/\sigma$, $C_{pd}$ has a linear trend with the change of $C_p$. The linearity is maintained for various combinations of n and h.

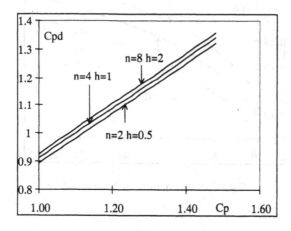

Fig 1. Cpd=f(Cp) for $\Delta/\sigma$=1 and Tm=30

Instead figure 2 confirms that it is not possible to select the control chart parameters n and h without an a priori knowledge of $\Delta/\sigma$ values. In fact, the figure shows that the optimal values of n and h are different when $\Delta/\sigma$ changes.

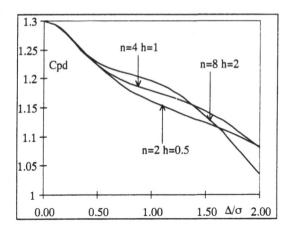

Fig.2 Cpd=f(Δ/σ) for Cp=1.3 and Tm=30

Finally, figure 3 shows that the $C_{pd}$ is an increasing function of $T_m$ that, starting from $C_{pk}$ for $T_m$ equal to zero, tends to $C_p$ while the $T_m$ value tends to infinity.

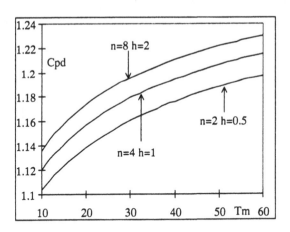

Fig.3 Cpd=f(Tm) for Cp=1.3 and Δ/σ=1

Even if $C_{pd}$ values can be easily calculated knowing control chart parameters and system evolution, an expression, able to give a good approximation, is proposed.
In fact considering the trend of the curves in the figures and with a statistical software the approximations of $C_{pd}$ are listed below:

for n=2 and h=0.5 $\qquad C_{pd} = -0.0232 + 0.842 \cdot C_p - 0.0925 \cdot \Delta/\sigma + 0.0561 \cdot \ln(T_m)$

for n=4 and h=1 $$C_{pd} = -0.0131 + 0.843 \cdot C_p - 0.094 \cdot \Delta/\sigma + 0.0559 \cdot \ln(T_m)$$

for n=8 and h=2 $$C_{pd} = 0.0137 + 0.845 \cdot C_p - 0.127 \cdot \Delta/\sigma + 0.0572 \cdot \ln(T_m).$$

In figures 4-5-6 the error curves as a function of $C_p$, $T_m$ and $\Delta/\sigma$ and for various combination of control chart parameters are reported.

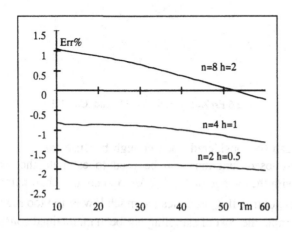

Fig.4 Err%=f(Tm) for Cp=1.1 and Δ/σ=1

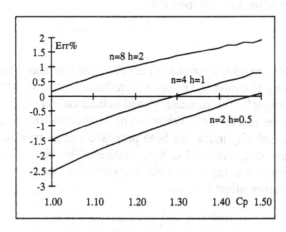

Fig.5 Err%=f(Cp) for Tm=30 and Δ/σ=1

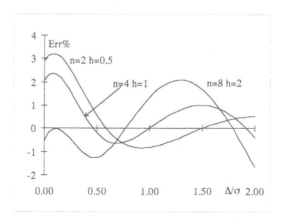

Fig.6 Err%=f(Δ/σ) for Tm=30 and Cp =1.3

The approximation can be considered good enough because the error is always less than 3% and then the proposed functions can be used to estimate the real capability of a production process with the change of $C_p$, $T_m$, Δ/σ and control chart parameters.

The approximation is not true if $T_m$ assumes too much low or too too much high values.

In the developed example the cost of sampling has been maintained constant.

The equation shape is also valid if this parameter changes; in fact simulation analysis has shown that the errors are always less than 3%.

Considering the aim of the paper no more analysis has been considered necessary to improve the precision of the interpolation curves.

## 4. CONCLUSIONS

The modern quality methodology considers prevention as a fundamental aspect. In such a condition it is necessary to evaluate a priori if a production process is capable that is if the defective rate is low enough. The commonly used indices can only give a partial answer to this question because the real dynamic behaviour of the process isn't taken into account.

In this paper a new capability index has been proposed and an investigation of its values for various parameters combination has been realised. The aim is to propose a simple approximation to estimate real capability able to assume decisions with greater safe instead of finding a precise mathematical formula.

## REFERENCES

1. Douglas C. Montgomery "Introduction to Statistical Quality Control"
2. T.J Lorenzen L.C. Vance, "The Economic Design of Control Charts: A Unified Approach", Technometrics, 1986, Vol.28, No.1

# MEASURING QUALITY RELATED COSTS

**J. Mrsa and B. Smoljan**
**University of Rijeka, Rijeka, Croatia**

KEY WORDS: Quality Costs, Heat Treatment, Quality Management

ABSTRACT: The cost aspect of heat treatment quality management has been presented. Quality management cannot be successful without quality cost management. The activity of heat treatment quality system as cost centers have been defined. The heat treatment quality costs have been classified in categories.

## 1. INTRODUCTION

Quality management includes all activities of overall management function that determine the quality policy, objectives and responsibilities, and implement them by means such as quality planning, quality control, quality assurance, and quality improvement within the quality system. Total quality management is management approach of an organization centered on quality, based on the participation of all its members and aiming at long-term success trough customer satisfaction, and benefits to all members of the organization and to society [1].

Quality management cannot be successful without quality cost management. The costs of quality are substantial and a source of significant savings, although most firms spend about

Published in: E. Kuljanic (Ed.) *Advanced Manufacturing Systems and Technology*, CISM Courses and Lectures No. 372, Springer Verlag, Wien New York, 1996.

20 percent of annual sales on costs of quality. The 2.5 percent benchmark is the amount quality experts estimate should be spent at the optimal quality level [2].

First step in any quality improvement program is realistic cost/benefit analysis in terms of the effort needed or justified obtaining the desired levels or assurance of quality. This analysis can be successfully done on the base of: true factory man hour rate in terms of overhead and salary, quality/quality control activity costs both in terms of man hours and money and quality related costs of any care study item. The quality cost figures are quite straightforward in respect of: prevention costs, appraisal costs, internal failure costs, external failure costs in any cost center [3].

Companies on different ways use quality cost variables in their performance measures of managers. For these purpose managers have to explicit goals in their performance measures for reducing the number of products rejected for quality reasons and savings in quality costs.

## 2. QUALITY MANAGEMENT OF HEAT TREATMENT

By heat treatment the stabilizing or adding new characteristic to the metal performance and function can be done. The effectiveness of heat treatment is affected by process selection, process parameter selection, system control, in one side, and metal selection, material selection, heat treatment equipment, handling equipment, measuring instruments, tools, etc., in other side. Quality management is important in heat treatment. Quality management of heat treatment consists on [4]:

(1) Quality policy
(2) Action of quality policy
     - Quality planing
     - Quality control
     - Quality assurance
     -Quality improvement

In Fig. 1 the schema of all activities which are dealing with heat treatment is presented. The quality of heat treatment is influenced by the combination of any involved activity.

The process design should use appropriate methods for the effects of heat treatment. Quality assurance for heat treatment must follow the ISO 9000 series standards. Quality assurance is all the planed and systematic activities implemented within the quality system, and showed as needing, to provide adequate confidence that an entity will fulfil requirements for quality (ISO 8402 :1994).

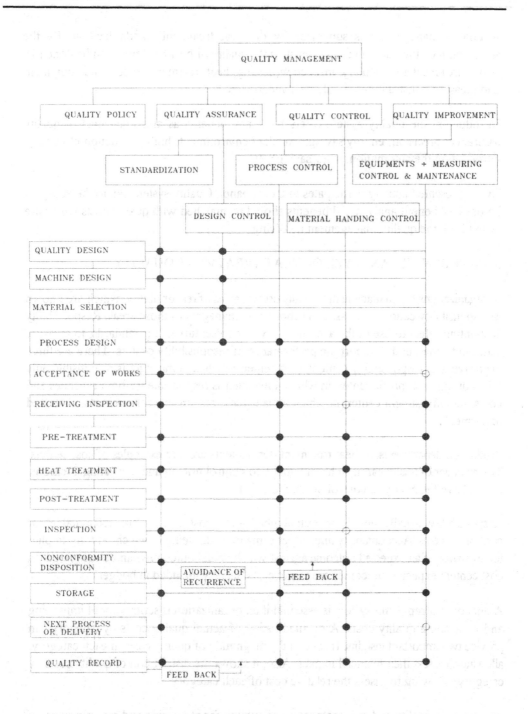

Fig. 1 Quality management of heat treatment

A quality management is supported by the heat treatment standardization. By the standardization the stabilizing and evaluating the quality of heat treatment can be done [4]. In heat treatment a vocabulary, inspection, testing, heat treatment process and equipment and measuring instruments should be standardized.

Introduction the quality system offers the many benefits as are: high level of quality assurance; precision; energy saving; excellent environment; high production efficiency; introduction the automation; labor savings, etc.

Every presented activity participates in a cost panel. Quality system cannot be accepted before cost benefit analyze. All benefits should be matched with quality costs which are raised from the quality management involving.

## 3. COST BENEFIT ANALYZE OF HEAT TREATMENT QUALITY

In organizations top management typically creates areas of responsibility, which are known as responsibility centers, and assigns subordinate managers to those areas. Responsibility accounting refers to use of the accounting system to set targets or standards, to measure actual outcomes, and to report the performance of responsibility centers. There are three major types of responsibility centers: cost center, in which a manager is responsible only for incurring cost, profit center, in which a manager is responsible for both revenues and costs, and investment center, in which a manager is responsible for revenues, costs and investments.

Production departments, as heat treatment departments are, are examples of cost centers. The manager of this department has no ability to control pricing and marketing decisions, but he have the ability to control production costs.

Responsibility usually entails accountability. Some cost centers are accountable for controlling costs. Accountability implies performance measurement, which, in turn, implies the existence of an expected outcome against which actual outcomes can be compared. In cost centers expected outcomes are target costs, usually included in budget.

A quality cost reporting system is essential if an organization is serious about improving and controlling quality costs. A detailed listing of actual quality costs by category can provide two important insights. It reveals the magnitude of quality costs in each category, allowing to assess their financial impact. Also, it shows the distribution of quality costs by category, allowing to assess the relative cost of each category.

Investment in quality and management responsibility for measuring and reporting quality costs must be monitored in responsibility costs centers. In Fig. 1. two types of quality are recognized:

-quality of design;
-quality of conformance.

Quality of design is a function of products specifications. Quality of conformance is a measure of how a product meets its requirements or specifications. If the product meets all of the designed specifications, it is fit for use. Of the two types of quality, quality of conformance should receive the most emphasis.

## 4. COSTS OF HEAT TREATMENT QUALITY

The costs of heat treatment quality are the costs that exist because poor heat treatment quality may or does exist. Quality costs are the costs associated with the creation, identification, repair, and prevention of defects. These costs can be classified into four categories: prevention costs, appraisal costs, internal failure costs, and external failure costs. Sum of all these costs have to be minimal.

Prevention costs are incurred to prevent defects in the products or services being produced. Prevention costs are incurred in order to decrease the number of nonconforming units. These costs can be recognized in any part of quality system in heat treatment: quality design, machine design, material selection and process design.

Appraisal costs are incurred to determine whether products are conforming to their requirements. These costs include acceptance of works, receiving inspection and testing materials, supervising appraisal activities, process acceptance, supplier verification and field testing. Process acceptance, very important in heat treatment, involves sampling post treatment material while in process to see if the process is in control and producing nondefective material.

Internal failure costs are incurred because nonconforming products are detected before being shipped to outside parties. These are the failures detected by appraisal activities. Internal failure costs are scrap, rework, downtime, reinspection and retesting.

External failure costs incurred after products are delivered to customers. These costs include lost sales, returns, allowances, warranties, repair, and complaint adjustment. External failure costs concern to whole company and they are not recorded at heat treatment cost centers.

Nonfinancial quality costs in heat treatment are: percentage of products passing quality tests first time, outgoing quality level for each product line, percentage of shipments returned from next production phase because of poor quality, and percentage of shipment made on the scheduled delivery date or percentage of delay. At some circumstances average delay costs have to be calculated, but in heat treatment it is complicated to calculate and some assumptions have to be made. It comprises the loss of use of productive manufacturing resources because a product is taking longer to manufacture than planned, i.e., loss of potential revenue.

The report of quality cost has to be used to examine interdependencies across the four categories of quality costs. On the base of a quality report can be examined both, how investment in heat treatment prevention is associated with reductions in appraisal, internal failure, or external failure costs realated with heat treatment and how increase expenditures in product design, one of prevention costs, is associated with decrease expenditures on customer service warranty costs, an external failure cost category.

## 4. CONCLUSION

Quality management cannot be successful without quality cost management. The quality cost information is quite straightforward in respect of: prevention costs, appraisal costs, internal failure costs, external failure costs in any cost center.

Quality management of heat treatment consists on quality policy and quality planing, quality control, quality assurance and quality improvement. Every activity participates in a cost sum. Heat treatment quality costs can be classified into four categories: prevention costs, appraisal costs, internal failure costs, and external failure costs. Sum of all these costs have to be minimal.

Nonfinancial quality costs in heat treatment are important, and they have to be monitored to.

## REFERENCES

[1] ..., ISO 8402: 1994.

[2] Hansen, D. and Mowen, M.: Management Accounting, South Western Publishing Co, Cincinnati, 1992.

[3] Horngren, C.: Cost Accounting, Prentice Hall, London, 1991.

[4] Kanetake, N.: Total Quality Management is the Key Word in Heat Treatment, 5th World Seminar on Heat Treatment and Surface Engineering IFHT`95, Isfahan, 1995.

[5] Rooney, E., Measuring Quality Related Costs, CIMA, London, 1992.

[6] Mrša, J.: Reporting and Using Quality Cost Information, 30th Symposium HZRFR`95, Zagreb, 1995. (In Croatian)

# CERTIFICATION OF QUALITY ASSURANCE SYSTEMS

G. Meden
University of Rijeka, Rijeka, Croatia

KEY WORDS: Certification, Quality Assurance, Standard, Systems

ABSTRACT: The paper discuss about objectives and reasons to implement the ISO 9000 Quality System Standard series and an adequate certification process. Quality Management Systems are being enlightened in accordance with international directives, standards, and policy.
There are many issues which must be addressed by a company in making the decision whether or not to become certified. Defining customer requirements allows the most important decision to be made, and requirements met.

> *"Some mighty forces now drive*
> *this scaling up of effort."*
> J. M. Juran, 1993.

## 1. INTRODUCTION

Quality is no accident, especially consistently high quality.
When paying a price for a commodity or service the customer expects certain quality assured from the producer. To satisfy the customer demand for an adequate design, the product must be designed either with the customer directly involved or with one in mind. All

Published in: E. Kuljanic (Ed.) *Advanced Manufacturing Systems and Technology*,
CISM Courses and Lectures No. 372, Springer Verlag, Wien New York, 1996.

products, even with vastly different complexity, must be designed to create ultimate satisfaction when used for the intended purpose. The only practical information available to a producer is usually the customer's requirement.

Production processes and operations that are predetermined by the design influence, to a large extent, the actual quality of products and services. Quality of design permeates through the entire production process and further into external supplies of material, production equipment, labor, technological knowledge, and so on. Therefore designing and assuring quality must extend as a managerial activity into the manufacturing and other spheres too. Modern quality assurance management implies that the product or service is designed for attainment of the required quality.

A regimented program of control, inspection, and certification, further enhanced by timed and random audits, helps meet the specification, and assures the customer that a superior quality product or services has been purchased. Besides, the regulatory environment may require to have an internationally standardized and certified quality system implemented.

In the meanwhile, the most powerful is an intense new global competition in quality. This competition has produced a major shift in world economic priorities. So, while the twentieth century has been the Century of Productivity, the 21st century will be the Century of Quality [1].


## 2. QUALITY APPRAISAL

Quality is a very popular subject of discussion in today's world. Indeed, many of us think of quality only when we purchase something, and we expect top value for our money.

The written quality specification are as old as recorded history. In the absence of measurement standards, the quality characteristics were described in words. The rule was *"caveat emptor"* i.e. let the buyer beware [2].

What and how quality is measured depends upon how quality is defined. Hopefully, these definitions are suitable to quantification and are broad enough to encompass the concepts of quality being *"a degree of customer satisfaction"* and *"a degree of fit'ness for use"* [3].

The competition is going to force everybody to give the customer a positive experience compared with the products and services quality he expects. Price has no meaning without a measure of quality being purchased and, consequently, without adequate measures of quality, business drifts to the lowest bidder, low quality, and high cost being the inevitable result.

One way to control and guide operations in a way that guarantees the right quality is to create and introduce a quality assurance system. Introducing a quality assurance system is one of the most important, and the most profitable investment, because the right-first-time quality avoids waste, saves time and money.

Industry is striving to improve quality in all of its products, and the dynamic nature of certification and accreditation schemes throughout the world is a reason for concern as well.

In many industries customers and regulatory bodies demand formal quality assurance, which can mean a kind of quality evaluation and promotion of standardized quality assurance system certification [4].

### 3. WHY THE ISO 9000 CERTIFICATION ?

In the 1987 the International Organization for Standardization published a series of quality standards known as the ISO 9000 Series. These quality standards, in the mean time, gained an international acknowledgment and acceptance.

The ISO 9000 series of standards include the following:

- ISO 9001 - A quality systems model for design, development, production, installation, and servicing.

- ISO 9002 - A quality systems model for production and installation.

- ISO 9003 - A quality systems model for final inspection and testing.

- ISO 9004 - A quality systems guideline for improving company quality without contractual obligation to its customers.

The objective of the ISO series of standards is to allow companies to create proper quality systems that will enable them to consistently deliver products or services at a desired level of quality. Using ISO 9000 to implement a basic quality system for the purpose of internal processes improvement can benefit a company in many ways.

Improving the quality of operations and thereby increasing their efficiency as well as increasing control over operation can serve to decrease costs. The pay-off will come in terms of things like less scrap, reworks, delays, complaints, etc. An improvement in quality should result as improvement in relative market share which should eventually result in increased return on investment and, subsequently, increased profits.

ISO 9000 may also be used as a measure for supplier control. The benefits of having suppliers with certified quality system are that, usually, the level of receiving inspection and vendor qualification efforts can be reduced, and resulting in increased savings.

The basis of an ISO 9000 quality system implementation is the company management commitment to the quality process emphasis completely throughout the company. If the company management elects to pursue the ISO 9000 quality system implementation and subsequently certification, and truly commits to all processes and operations as it is by Figure 1, they should see the benefits mentioned under both internal improvements and markets positioning.

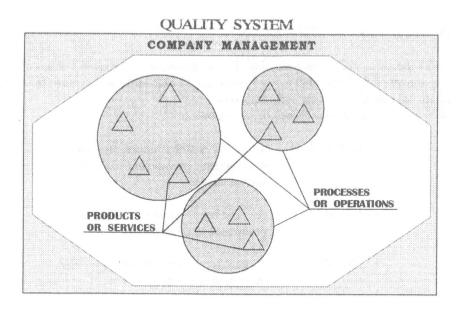

Figure 1 - An overwhelming quality system

The ISO 9000 certification provides a company with a number of major benefits including elimination of expensive and time consuming multiple audits by prospective customers, enhanced efficiency in production and distribution, and demonstrated compliance with the internationally accepted quality standards on global market, Figure 2 [5]. Therefore, the quality system certification in conjunction with a quality mark signifying the ISO 9000 certification is a key sales and marketing advantage too.

Figure 2 - Quality Passport logotype

In particular, it is the European Community approach to use certified quality system as a method of assuring the quality of goods freely moving within the EC. Relative directives contain essential requirements as well as conformity assessment requirements.

The essential requirements address product related requirements, e.g. in gas appliances, pressure vessels, constructions materials, tools, machinery, ships, telecommunication equipment, cable ways, lifting appliances, measuring and testing equipment, medical devices, toys, etc.

Conformity assessment requirements address how a company proves that it complies with the essential requirements of the directive. Conformity assessment procedure includes type testing of the design and, possibly, periodic surveillance inspections or the quality system certification.
It should be emphasized that the quality system certification is only one of possibilities. Quality Management Systems are being developed to comply with the international directives, standards, and laws.

## 4. CERTIFICATION PROCESS

The certification process involves an application, documentation review, possibly a pre-assessment, and a final assessment, followed by certification and ongoing surveillance. Because of the ongoing surveillance, it is important for the company to select a certifier with whom they can maintain a long-term relationship. A schematized outline of the certification hierarchy is shown in Figure 3.

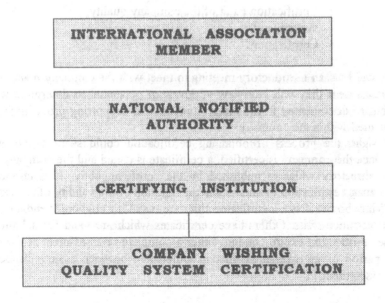

Figure 3 - Hierarchy of certification

Once the application required is completed and main information on the company, its size, its scope of operations, and desired time frame for certification have been determined, the certifying organization will request the company to submit documentation of its quality system. This is usually in the form of what is known as Quality Manual. What they are looking for is the overall documentation which describes the quality system in order to compare it to ISO 9000 standard to determine compliance.

In some cases the pre-assessment is an optional step which a company wishing certification may elect to undergo or to by-pass. The actual certification process can vary from company to company, depending on practical circumstances and specific particulars, but usually the main steps are [6]:

- Management commitment
- Steering team installation
- Gap analysis i.e. a study of the existing quality system
- Training in the ISO 9000 quality system, documentation preparing, and auditing techniques
- Documentation writing
- New procedures implementation
- Certifying body selection
- Examination of the quality system documentation by the certifying body
- Certification audit of the company quality system, and
- Certification.

The auditors will hold an introductory meeting to meet with the company management, and during the assessment they will interview all levels of personnel to determine whether the quality system as documented in the quality manual and supporting procedures have been fully implemented within the company.

At the first sight, the process of obtaining certification could seem to be quite over-exhaustive. Once the company is certified, a certificate is issued and the company is listed in a register or directory which is published by the certifying body. It is important for a company pursuing certification to understand the duration or the validity of its certification.

Some certifying bodies have certificates that are valid throughout pending continuing successful surveillance visits. Others have certificates which are valid for a limited period. Those whose certification expire conduct either a complete re-assessment at the end of the certification period or an assessment that is somewhere between a surveillance visit and complete re-assessment.

## 5. OBTAINING CERTIFICATION

The company must have a positive attitude about the quality system requirements because the cooperation of every department and everyone within each department is predominant. It is found that the fundamental for successful certification is a team effort.

Moreover, this is necessary for maintaining the quality system because ISO standard requires an audit every six months. Each of these audits is completely random and takes just the time to sample the company's procedures and processes.

For example, the auditor may ask a operator to show the operating process and procedures, and may look for logs to establish traceability, signatures, responsibilities, and closing of the operating loop.

In welding operation, for instance, welding rods must be stored at a certain temperature and in a dry space for quality welds. The auditor may ask a welder about the place and temperature the welding rods are stored, or what is the welding procedure, and usually will ask to see his log to check when he was last certified.

The auditor may ask for a log of training, lists of employees send for training, list of courses taken, and copies of the training policy.

Thus, obtaining certification should not be taken for easy. In a practical case, a list of activities for obtaining the ISO certification could be as follows [7]:

- Serialization, calibration, and traceability of all measuring equipment
- Revision level control and signature logs for all documentation
- Operators certification and procedure qualification
- Route sheets and work instructions
- Retrieval of quality records
- Handling and disposition of nonconforming material and products
- Preparation of quality assurance manual and formal procedures
- Implementation of quality system
- Performing internal audits and coaching each department about possible questions and answers
- Scheduling pre-assessment and/or final assessment audits

By obtaining and maintaining the ISO certification, a company can improve its quality control system, cut waste, and motivate its employees. In so doing, it improves its competitive position on both international and local markets.

Perhaps the most important feature of the ISO 9000 quality system standards is that, unlike many previous quality control standards, it goes far beyond the attempt to ensure product quality through an inspection of finished end products rolling off a production line. Instead, the ISO standards attempt to build in quality through an examination of the entire design, development, and manufacturing process together with shipping and after-sales service.

In practical terms, the implementation of the quality system in a process includes all operations from sales and marketing, design and engineering, customer order entry, receipt of raw materials, all areas of manufacturing, assembly, and calibration, to final inspection and testing, shipping, and an extended product service support.

## 6. BOTTOM LINE

The bottom line in regards to whether a company really does need a certified ISO 9000 quality system implemented revolves around four basic points [5].

First, it must take the time to know its products or services market requirements.

Second, it must also know its competition. If competitors elect to pursue certification, they may be perceived in the market place as having a higher level of quality, and may adversely impact the company's operation.

Third, know its certifier. It is a long term relationship, and the company must be sure to selects someone who has resources to support its needs.

Finally, above all, management must talk to their customers and meet their requirements.

Companies that have attained world class quality have begun requiring their suppliers to move toward world class quality as well. In this way, quality criteria spread gradually within the entire supplier chain. A company desiring to upgrade the quality of its products or services must consider the contribution of its suppliers and all personnel to the quality of each product or service.

The manufacturing industry, firstly in Japan, then in the USA, and now widely in Europe, has notably benefited from the deployment of quality function and development of quality management. Although, the quality management practically is nothing more than systematically applied common sense for a quality.

Various programs are being studied and implemented, but the most successful implementation appear to be at those companies that require each individual to be fully responsible for his own work and quality approach.

## REFERENCES

1. Juran, J. M.:        Made in U.S.A. : A Renaissance in Quality, Harvard Business Review,    July - August 1993, 47

2. Juran, J. M., ed.:  Quality Control Handbook, McGraw - Hill, New York, 1962.

3. Meden, G.:          Quality characteristic measurement and evaluation, 3rd International Conference on Production Engineering CIM'95 , Proceedings, Zagreb, 1995, E1 - 7

4. Meden, G.:          Quality Performance Evaluation, 3rd International Conference AMST'93, Proceedings, Vol. II, Udine, 1993, 231 - 238

5. Potts, L.:          What is ISO 9000 and why should I care ? , ABS Quality Evaluations, Inc., Houston, 1992.

6. Purcell, D.:        ISO 9000 : Putting QA to work in the marine market place, Marine Log, 100 (1995) 5, New York, 1995, 38 - 40

7. Vermeer, F. J.:     ISO certification pays off in quality improvement, Oil & Gas Journal, 90 (1992) 15, Tulsa, Oklahoma, 1992, 47 - 52

# QUALITY ASSURANCE USING INFORMATION TECHNOLOGICAL INTERLINKING OF SUBPROCESSES IN FORMING TECHNOLOGY

D. Iwanczyk
University of Bochum, Bochum, Germany

**KEYWORDS:** Quality Assurance, Quality Control Loops, Interlinked Forming Processes

**ABSTRACT:** Combined hot forming production processes are characterized by the fact that the quality features of workpieces are changed in successive production stages. In each stage of production the newly changed features have a direct influence on the production quality of the ensuing stages as well as on the quality of the final product. Quality assurance for combined hot forming production processes must take this fact into consideration. Therefore, extended horizontal quality control loops have been designed which enhance the information technological interlinking of the production stages for combined hot forming production processes. On the one hand, using these quality control loops allows preventative measures to be taken to ensure that quality is achieved, thus permitting the results of the previous stages to be individually tailored to the ensuing stages of production. On the other hand, controlling the quality of the production stages is made possible and its effects are dependent on the quality of the ensuing stages. The entire production line can be covered by using varying ranges of horizontal control loops and optimal quality control, and, from an economic point of view, optimized quality control of the whole production can take place.

## 1. INTRODUCTION

The success of a company is mainly dependent on its powers of innovation, its productivity, and the quality of its goods. In particular, quality has become extremely important [1]. When ensuring quality, it is no longer enough to carry out measurements and tests after the completion of the product. The quality of the product should be planned,

Published in: E. Kuljanic (Ed.) *Advanced Manufacturing Systems and Technology*, CISM Courses and Lectures No. 372, Springer Verlag, Wien New York, 1996.

guided, controlled and manufactured in all the production stages, keeping in mind the demand that

"quality should not be controlled but produced"[2].

Apart from the legal aspects resulting from increased safety, regulations (product liability laws) [3], interests in economic and competitive viability primarily force today's forming technology companies to pay particular attention to ensuring product quality. The growing tendency to manufacture products as near-net-shaped as possible has an influence on the shaping of future product quality and the corresponding quality assurance measures that must be undertaken [4]. Near-net-shape production is the manufacturing of workpieces using forming technology whose functional surfaces require either no or very little precision work at a later date. This places far greater demands on the dimensional, shape and locational tolerances which have to be maintained, as well as requiring that all the quality features are consistently maintained. These greater demands on the quality of the forming workpieces can only be attained if the correspondingly high requirements are fulfilled by all the production stages in the process. Quality orientated management is necessary throughout the entire process.

## 2. AIMS

As a rule production processes in hot forming technology consist of several successive stages. Combining parts of the processes over the material flow in line production is a typical feature of hot forming processes. The workpiece is worked on in various stages where certain changes to the features are dealt with in turn. In this kind of combined production the quality at each step of the process is crucial for the quality of the end product. Therefore, if quality orientated production is to be achieved, not only the influence of each stage on the result of the production process as a whole must be recognised and taken into consideration but also the reciprocal influence of the individual stages [5].

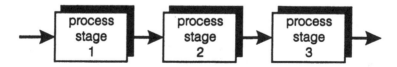

*Fig. 1: Typical production sequence of a of combined production process*

As a preview to the above mentioned background a methodical approach was devised to ensure quality assurance in combined forming production processes [6].

Besides the necessity to optimize the individual stages in the combined hot forming production processes once more, horizontal control loops were devised and employed in interlinking processes which covered more than one stages. On the one hand, by using a horizontal control loop, preventative measures to ensure quality can be brought into play, which allow ensuing production stages to be individually tailored to the results of previous stages. On the other hand, it has become possible to control the quality of the production stages whose effects are dependent on the quality of the ensuing stages. The whole production line can be covered by employing horizontal control loops with varying ranges and partial overlapping.

The quality control measures which are restricted to the production level are complemented by vertical control loops, which bring higher authorities into play in the production process. Integrating the horizontal and vertical control loops guarantees that the comprehensive quality control measures are optimally suited. Apart from sound knowledge of technology, the basis of all quality controlling activities is suitable communication structures for exchanging information on quality. One should be at production level and the other should transcend all the levels. A joint central quality databank used by all the participants is essential for integrating all the quality control measures.

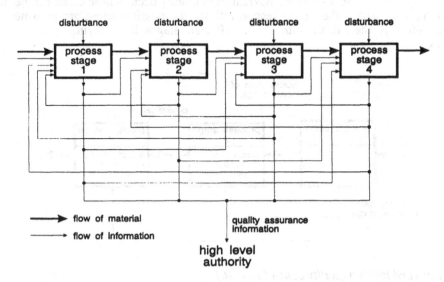

*Fig. 2: Informational concatenation of the process stages of combined production processes [6]*

In the following the most important components of the production orientated combined hot forming concept for quality assurance will be presented.

## 3. EXTENDED HORIZONTAL CONTROL LOOPS ON PRODUCTION LEVELS

The whole process can only be optimized, from an economic and quality-orientated point of view, if there is a mutual harmony of the complementing production subprocesses. Therefore there are established horizontally integrated control loops on the process level which cover more than one stage in the combined production process.

The first kind of extended horizontal control loop is the forward-looking in character. This kind of quality control loop makes it possible for the specific results concerning the quality of previous stages to be acted upon in later stages of the process. When producing each workpiece in the individual stages, quality features are created and corresponding quality data relating to the workpieces is established. These features, which are to a certain degree unique to each workpiece, are important for later work on the workpieces in the ensuing stages and are therefore recorded. If the established quality data allows further work to take place, the course of the process of the ensuing stages is adapted. Thus it is possible to react to individual differences of initial quality, to avoid defects continuing through the whole line of production and to minimize the influence of differences in quality which might occur. The result is a production structure, tolerant of defects.

The second type of quality control components introduced onto the process level was the quality control loop with feedback capabilities. Often in the course of combined production the degrees of deviation in the features of intermediate products only becomes obvious after the intermediate products have reached the next stage of production. This means that quality defects which occurred at an earlier stage and have not been discovered later become evident when the process does not run as planned. Further quality control loops which take these facts into consideration, are realised by feeding back information from previous stages of the process. When deviations in quality occur whose cause can be traced back to previous stages, the parties responsible for the possible cause are informed. The existing defect is then dealt with in the affected stages by applying quality control measures.

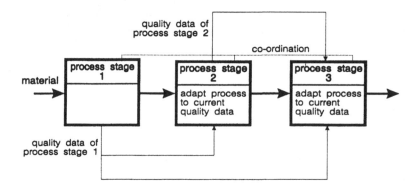

*Fig. 3: Forward looking quality control loop [6]*

## 4. VERTICAL QUALITY CONTROL LOOPS

It is well-known that the earlier quality assurance measures are built into the process of developing a new product, the more effective they are. Therefore the quality control activities presented here are not just limited to the production stage but higher operational spheres are brought into play by employing larger vertical control loops.

At production level it only makes sense for higher authorities to intervene if disturbances occur which could affect the quality and whose causes can be found in complex influences determined by the system. In these cases, long term quality assurance closely linked to the process cannot produce satisfactory results. Therefore the aims of long term vertical control loops in the quality assurance concept using combined hot forming processes are:

- analyse and eliminate weak spots
- optimize production of improved products
- analyse the extent of influence
- ensure the quality of planning activities
- co-ordinate and optimize horizontal control loops

The higher levels in the production stage were previously only brought into play with the "large control loops" in the quality assurance concepts which have been covered so far. Their aims are to eliminate primary and secondary causes of quality defects as well as to promote process optimisation in the long term. This naturally leads to relatively long reaction times. Besides that other cases are also feasible in which fast reactions in the higher levels can contribute to stabilizing existing production processes whose quality is disrupted. This is particularly relevant if the cause of the disrupted process shows a discrepancy between the existing quality of the initial data and the participants' expectations. For this reason fast vertical control loops were designed allowing influence which meets the demands of stages in combined production processes. The basic pre-requisite for this is that there are direct communication channels between the higher levels and the production level.

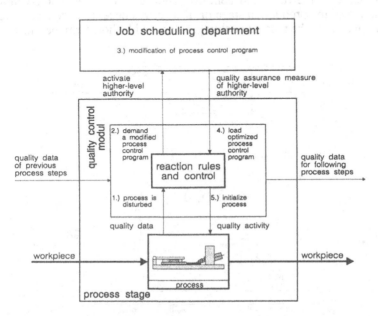

*Fig. 4: Fast vertical quality control loop [6]*

Example: a higher level, for example job scheduling, is informed, while production is in progress, of existing quality defects whose recognised causes are to be found in the insufficiently consulted process control data in the wrong or present production situation. At the same time all the necessary process data of the work preparation is made available. The higher level's first reaction is to analyse the existing situation and immediately redefine the process control data. The new process control data is then fed back to the stage of the production where the defect occurred.

## 5. QUALITY CONTROLLERS

Similar to the classic control loops, quality control loops can also have controlled systems and controlling means. Quality control functions are also required on the production level to effect the extended horizontal quality control loops. The tasks to be accomplished by these controllers are determined by the intended behaviour of the control loops. As an example of this, the quality controller is seen as a horizontal quality control loop, with a preventative effect. The preventative pattern of behaviour found in extended quality control loops assumes that there is a quality controller module in each stage of the process, which has suitable strategies in store to vary the production programme within the stage of the process so that it will react to changed initial qualities.

A first set of tasks is characterized by its ability to ascertain and react to changed initial qualities. For this purpose the detailed quality data and information of the stored process stages are analysed and categorized. The size and effects of the deviations between the quality actually attained and what had been predicted for the intermediary products must be evaluated. In the case of a large degree of deviation, the control programme of the production stage will be modified on-line so that the desired end product quality, or at least a minimisation of the negative influences, will be reached, despite a change in the basic conditions. The decisions as to the necessary quality assurance reactions will be made based on knowledge of the rules and controls. This results in direct intervention in the control programme of the production stage or, if it proves necessary, demands are made to higher levels to overcome the present quality problems with short term measures.

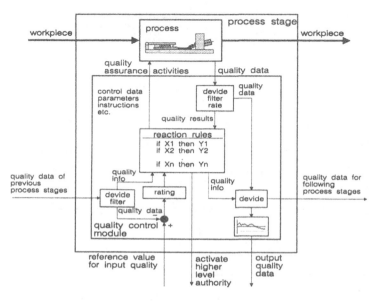

*Fig. 5: Preventative quality control loop of a production stage [6]*

On the process control level, approaches based on the knowledge acquired are particularly suitable for effecting the reaction controls in the quality control function. The tasks to be dealt with by the reaction controls are generally:

- to comprehend and evaluate the present production situation

- to introduce suitable measures during a production stage

- to activate other components of the quality control

These reaction controls have to be individually tailored to each application (subprocess, type of control). By doing this, progressive methods based on the knowledge acquired from analyses and decisions can be used, such as expert systems and fuzzy logic.

## 6. COMMUNICATION STRUCTURES AND ELEMENTS

The most decisive precondition enabling quality control loops to function is the existence of a well-functioning communication system to exchange data and information relevant to quality. Taking the different demands into consideration which arise in the flow of quality data to be realized on the process level, two different approaches for exchanging information were followed up.

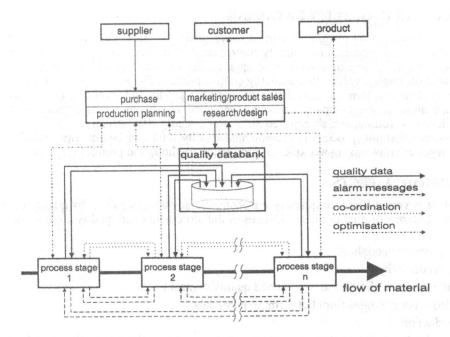

*Fig. 6: Quality-important information and data flow in combined production processes [6]*

One of them is direct and fast communication between individual process stages by directly transmitting information from stage to stage. The data to be transferred is mostly high priority, temporary or does not feature in the normal plans. With this kind of data transmission it is necessary to ensure that important quality data from a process station is punctually transmitted to the stored process stations and meets all the demands. The demands which are made in a direct process station- to- process station- communication are the following:

- priority controlled data transmission

- transmission not dependent on direction

- can be directly addressed

- short transmission time

- variable message contents

A suitable possibility to communicate directly between the several production stages is to send messages using the message-passing-principle [7]. Every participant of this communication has to be equipped with a combined "sending and receiving module".

A great deal of the quality data is also important for long-term analysis, optimisation tasks, and documentary purposes. It is therefore essential to introduce a quality databank suited to these purposes. This databank can automatically be used for simultaneously exchanging information between the stages in all the cases where direct transmission is not absolutely necessary. The user only requires a transmitting and receiving function as a communication interface for the quality databank. They contain all the necessary access functions. The functions (writing and reading operations) are carried out via a simple transparent databank interface.

## 7. THE CENTRAL QUALITY DATABANK

A quality database is an integrative part of every quality control. Therefore, an important component of the aforementioned quality control concept is the central quality databank. Its main tasks are to collect data relevant to quality, as it is needed, and to make this data available to all those involved in the quality control process. Because of its central position the quality databank forms a natural link between the production level and the previously connected areas of the company and the other CIM components. An object orientated approach was favoured, which resulted in a user-friendly data structure. In the field of combined hot forming processes, accumulated quality data can be roughly divided into three categories emphasising workpieces, production methods and production processes.

## 8. ECONOMIC ASPECTS

As a rule it is very difficult to quantify the economic advantages gained by quality control measures. In terms of quality the advantages of the aforementioned quality control concept are:
- higher production quality
- higher productivity
- controlled process chains due to extended quality control loops
- extended product ranges from hot forming processes
- error reduction

The special advantages of this concept can also be seen in the error tolerant behaviour engendered by using control loops which are forward-looking in character. This fact can also be taken into consideration when setting up the individual stages. Individual stages no longer have to submit to strict tolerances if it is possible to individually compensate for certain fluctuations in the features of the intermediary products in the course of production. This leads to simple and cost-effective production processes.

## REFERENCES

1. *Rinne, H.; Mittag, H.-J.:* Statistische Methoden der Qualitätssicherung,
Karl Hanser Verlag, München 1989

2. *Warnecke, H.-J.:* „CIM - Die Unternehmen vernetzen sich",
Märkte im Wandel, Bd. 14: CIM, Spiegel-Verlag, Hamburg 1990

3. *Bauer, O.:* „Vorbeugen ist besser - Qualitätssicherung im Schmiedebetrieb erfordert viele Maßnahmen im gesamten Fertigungsablauf",
MM Maschinenmarkt, Vogel Verlag und Druck KG, Würzburg 96 (1990) 43

4. *König, W.:* „Fertigungsverfahren Band 4 - Massivumformen", 3. Aufl.,
VDI-Verlag GmbH, Düsseldorf 1993

5. *Maßberg, W.; Iwanczyk, D.:* „Ausschuß vermeiden - Rechnergestütztes Qualitätssicherungssystem für das Warmumformen",
MM Maschinenmarkt, Vogel Verlag und Druck KG, Würzburg 99 (1993) 46

6. *Iwanczyk, D.:* Präventive Qualitätssicherung mittels informationstechnischer Verkettung von Teilprozessen in der Umformtechnik
Dissertation, Ruhr-Universität Bochum, Bochum 1994

7. *Kühn, R.:* Architektur von hierarchischen Betriebssystemen für die Prozeßautomatisierung in der Umformtechnik,
Dissertation, Ruhr-Universität Bochum, Bochum 1992

# ANALYSING QUALITY ASSURANCE AND MANUFACTURING STRATEGIES IN ORDER TO RAISE PRODUCT QUALITY

C. Basnet and L. Foulds

University of Waikato, Hamilton, New Zealand

I. Mezgar

Hungarian Academy of Sciences, Budapest, Hungary

KEY WORDS: Manufacturing management strategies, product quality, simulation, Taguchi method.

ABSTRACT: The different manufacturing strategies can influence production to a significant extent and, through this, manufacturing process and product quality. A stochastic simulation-based approach has been developed and carried out for estimating product quality on strategic level. The methodology is based on the Taguchi method, and consists of a simulation of the manufacturing system with a view to identifying manufacturing strategies that are relatively insensitive to process fluctuations and are cost effective. The experimental analysis is carried out with the strategies as controllable factors, and process fluctuations as noise factors. The response variable of the experiment is the unit cost of the finished product. While the Taguchi method is usually applied to the tactical design of production systems, we have shown how it can be applied to the strategic design and analysis of manufacturing systems.

## 1. INTRODUCTION

Traditional manufacturing has focused on a low cost strategy to capture the market, and consequently, to improve the bottom line. The techniques of flow design, work design, and automation have been the traditional mainstay in accomplishing this strategy. Recently, competitive pressures have pushed quality to the centre stage of manufacturing management. Adoption of a quality strategy has entailed worker involvement, statistical process control, just in time, automation etc. In designing manufacturing systems with the goal of high quality, we need to adopt quality strategies which can cope with the vagaries of manufacturing such as process variabilities. This does not imply that cost is no longer a

Published in: E. Kuljanic (Ed.) *Advanced Manufacturing Systems and Technology*, CISM Courses and Lectures No. 372, Springer Verlag, Wien New York, 1996.

consideration in manufacturing. These types of manufacturing systems has to be designed in a way that yield consistently high quality and, which are, at the same time, cost effective.

There are many factors that influence product quality during the product life-cycle. The technical related variables during the design and manufacturing phase can be called as primary factors (e.g. tolerances), but the management-related factors have significant influence on product quality as well. The realisation of both the quality- and manufacturing management strategies need a longer introductory time, and the real results of these technologies will appear only after a longer period.

So, the thorough analysis and selection of the different quality- and manufacturing management strategies have a very important role. The stochastic simulation is a proper tool to do this analysis and evaluation before introducing any quality, or manufacturing management strategies in a manufacturing system. An additional advantage of the preliminary analysis can be that a robust strategy can be selected by combining the Taguchi methodology with the design of experiment approach. By doing such analysis industrial firms can make more reliable decisions concerning quality assurance and manufacturing strategies. The Taguchi methodology can be applied also in the parallel technical design of products and manufacturing systems as described in [1].

Taguchi has proposed a method of parameter design which provides a means of selecting process parameters which are least sensitive to process variations, i.e., are robust. This process selection is at the tactical level - detailed day to day operation of the manufacturing plant. In this paper we extend the Taguchi method to the strategic design and analysis of a manufacturing system and present a methodology for selecting quality strategies which are robust and cost effective. In evaluating quality strategies, we consider not only the cost of such strategies, but also the benefits (improvement in process capability). This provides a quantitative basis for the selection of quality strategies. The manufacturing strategies of SPC, JIT, and automation are tested in a simulated manufacturing scenario, and their effects on quality costs are investigated. From these, the strategies are evaluated on their robustness from the viewpoint of quality.

## 2. STRATEGIES FOR QUALITY- AND MANUFACTURING MANAGEMENT

The quality assurance and manufacturing management strategies have a great importance in the market competition especially in economies that are in transient period from any aspect (like in Central Europe [2]), or that intends to raise their competitiveness radically [3]. New forms of manufacturing architectures, structures have been developed as well that can give more appropriate answers for the market demands.

In order to keep their market positions manufacturing enterprises have a strong motivation to move from large, hierarchical organisations to small, decentralised, partly autonomous and co-operative manufacturing units, which can respond quickly to the demands of the customer-driven market. Parallel with these organisational changes the type of production is also changing from mass production techniques to small lots manufacturing. The autonomous production units geographically could be found both inside the enterprise, but outside as well, physically in a long distance. The latter case can be called as the "World Class Manufacturing" (WCM) concept. In all of the new forms of the manufacturing organisations different optimal quality- and manufacturing management strategies can be applied.

## 2.1 Quality assurance strategies

The quality of the product is an aggregate of the quality of individual features (geometrical characteristics) and properties (e.g. material). The main difference between traditional and new quality philosophies is that traditional quality approaches focus on correcting mistakes after they have been made while the new philosophies concentrate on preventing failures. For measuring the quality (based on tolerances) the process capability index and the quality loss function can be used. Quality should be designed into the product and into the processes involved when designing, fabricating and maintaining it through its life cycle. Total quality management (TQM) means the continuous satisfaction of user requirements at lowest cost by minimum effort of the company.

Taguchi is noted chiefly for his work on designing products and processes with a view to making the quality of the product robust (insensitive) to process or product variations [4]. His approach have found widespread dissemination and application [5].

The main reasons why product quality characteristics can deviate from the predetermined values during manufacturing are the inconsistency of material, tool wear, unstable manufacturing process, operator errors, low level system management. In on-line quality control those methods are applied that help to raise the efficiency and stability of the production process (e.g. SPC, diagnosis). In discrete part manufacturing one of the most important tools of process control for quality is statistical process control (SPC). SPC is based on the hypothesis that if a process is stable, than one can measure a product characteristic that reflects the behaviour of the process. That is, different sample groups of the product will have the identical statistical distribution of the characteristic. Drucker in [6] highlights the role of statistical quality control (SQC) in manufacturing. He sees SQC as improving not only the quality and productivity of an organisation, but also the dignity of labour. In this article he also discusses the modular factory concept.

## 2.2 Manufacturing management strategies

The automation which is based on computer technology can give a manufacturing company a big competitive advantage. The different forms and levels (e.g. high-level integrated automation in Computer Integrated Manufacturing Systems) of automation needs different manufacturing strategies.

The human resource management strategy influences all important competitive factors. The education level of the workers, employees can increase/decrease the quality, the time-to-market as well.

The inventory management strategies has a big influence on cost and on throughput time. The introduction of Just-In-Time (JIT), or other strategies need a careful analysis because there is a big financial risk which one to choose.

## 3. THE TAGUCHI METHOD

The methodology for quality control developed by G.Taguchi [4] covers the quality assurance involving both product design and process design stages. Taguchi follows an experimental procedure to set parameters of processes. In Taguchi's method of parameter design, the factors affecting the outcome of a process are separated into controllable factors and noise factors. Controllable factors are the parameters whose value we seek to determine. Noise factors are factors which cause variation in the response variable, but are normally not amenable to control (but they are controlled for the sake of the experiment). Treatment combinations for controllable factors are set by an experimental design called the inner array, treatment combinations for noise factors are called the outer array. Usually, to save cost and time, a full factorial design is not used for these arrays. Instead, a fractional design is used, as illustrated in Figure 1, where there are three controllable factors and three noise factors. The controllable factors are set in a full factorial design, but the noise factors are set in an L4 array. Usually, an analysis of variance is not the object of Taguchi-style experiment. The object is to identify the set of parameter values that is most robust to the variations in the noise factors. Taguchi proposes a measure called Signal to Noise (S/N) Ratio. When the response variable is such that a lower value is desirable, one measure of S/N ratio is:

$$S/N \text{ Ratio } = -10\log_{10}\left(\frac{1}{n}\right)$$

where $y_i$ is the response variable, and there are $n$ observations in a run. A high value of S/N ratio indicates that the corresponding combination of controllable factors is robust.

## 4. STRATEGIC DESIGN AND ANALYSIS OF MANUFACTURING SYSTEM

With the development of modern simulation languages, simulation has established its place in the design of manufacturing systems [7]. Simulation permits modelling of manufacturing systems at any desired level of detail. Controlled experiments can be carried out on a simulated environment at a low cost.

We suggest the use of this environment for the selection of manufacturing strategies following the Taguchi method. The effect of these strategies on the quality levels can be modelled, and experimented upon. For example, the continuous improvement aspect of JIT can be modelled by using a learning curve model. The Taguchi method permits us to select robust strategies. An application of this simulation methodology is presented next.

### 4.1 Selecting performance measure and define the target

The goal of the experiment was to identify levels of the manufacturing strategies which were most robust to quality problems. We used cost per unit of good finished product as the surrogate for the cost effective quality level achieved by the manufacturing system. Consequently, this cost was the response variable. The analysis gave an answer that on what levels of the noise factors were the production cost on minimum while the selected management strategies (factors) are the most insensitive on the noise in the given manufacturing system.

### Inner array of noise factors (L4 layout)

| | | | | |
|---|---|---|---|---|
| Noise 1 | 2 | 2 | 1 | 1 |
| Noise 2 | 2 | 1 | 2 | 1 |
| Noise 3 | 1 | 2 | 2 | 1 |

1 = Low
2 = High

| Run No. | Param 1 | Param 2 | Param 3 | | | | | Avg | S/N Ratio |
|---|---|---|---|---|---|---|---|---|---|
| 1 | 1 | 1 | 1 | | | | | | |
| 2 | 1 | 1 | 2 | | | | | | |
| 3 | 1 | 2 | 1 | | | | | | |
| 4 | 1 | 2 | 2 | | | | | | |
| 5 | 2 | 1 | 1 | | | | | | |
| 6 | 2 | 1 | 2 | | | | | | |
| 7 | 2 | 2 | 1 | | | | | | |
| 8 | 2 | 2 | 2 | | | | | | |

Outer array of controllable factors
(full factorial design)

Figure 1. Example of a Taguchi Experimental Design

## 4.2 Identify factors

The controllable factors are the manufacturing strategies that we wish to investigate. The three strategies and their levels are:
1. Statistical process control  (Yes /No)
2. Automation (Yes /No)
3. JIT learning (Yes/No)
A full factorial design is used for these factors.
The noise factors  that cause variation in the quality of the product and their levels are:
1. Nominal process capability of the processes (Higher/Lower)
2. Complexity - number of serial processes (Low / High)
3. Process shift (Low/High)
As is usual with Taguchi designs, all the factors are  at two levels.  An L4 design is used for the noise factors.

## 4.3 Conduct experiment

A simulated environment of a manufacturing system was created using the simulation language VAX/VMS SIMSCRIPT II.5. [8]. The following assumptions apply to the system. It comprises a number of serial processes. All the processes work to their nominal process capability. The processes are liable to a process shift which causes the process capability to decrease. The distribution of the interval (number of parts) between process shifts is Poisson. However, if nothing is done to the shifted process, it will reset to the nominal capability after another interval, which has the same distribution. There is a nominal cost of processing for each part. After all the operations (serial) are carried out, the quality of the part is checked against the upper (USL) and the lower (LSL) specification limits for all the processes, and if any measurement is out of specification, the product is scrapped.

Three quality strategies are considered: statistical process control (SPC), just-in-time (JIT), and automation. When SPC strategy is adopted, samples of size $n$ are taken at an interval of $f$ parts. Upper and lower control limits (UCL and LCL) are calculated on the basis of 3 sigma, and if the process is found out of control, the process is reset to the nominal process capability. There is an additional cost (Cspc) of manufacturing, attributable to the adoption of SPC.

As pointed out earlier, many benefits of automation strategy have been identified in the literature. As far as the quality of the product is concerned, the adoption of automation strategy results in an increase in the process capability. There is an additional cost (Cauto) per part associated with this strategy.

Similarly, the advocates of JIT have claimed many advantages for this strategy. It is agreed that JIT is both a philosophy and a discipline which results in a continuous improvement of the processes of a manufacturing organisation. In our simulated environment, the adoption of JIT causes the process capability to increase over time, following a learning curve . Again, there is an additional cost (Cjit) incurred for each part. The investigation of the manufacturing strategies was carried out for a manufacturing system with the following parameters.

Mean quality level of all processes = 100
Upper Specification Limit = 160, Lower Specification Limit = 140
Nominal cost (Cnominal) = $1
Nominal process capability:  0.9 (high) and  0.8 (low)
Mean of the interval between process shifts: 4000
Amount of process shift: 0.4 (high) and 0.3 (low). When the process shifts, the process capability decreases by this amount.
SPC sample size, $n$: 4, SPC Sampling interval, $f$: 300
SPC cost Cspc (additional to the nominal cost of $1): $0.15
Increase in the process capability due to automation: 0.2
Automation cost Cauto (additional to other costs): $0.20
Learning curve under JIT (percentage): 99.5%. Doubling of the number of products causes 99.5% increase in process capability. JIT cost Cjit(additional to other costs):  0.1
Complexity of the manufacturing: 6 (high) or 5 (low) serial processes

## 4.4 Analyse Results and Select Manufacturing Strategy

The Taguchi method and inside this the S/N ratio formula for the smaller-is-better has been used since the goal of the experiment was to minimise cost. The result of the simulation experiment is shown in Table 1.

| | | | Nominal $C_p$ | 2 | 2 | 1 | 1 | | |
| | | | Process-shift | 2 | 1 | 2 | 1 | | |
| | | | Complexity | 1 | 2 | 2 | 1 | | |
| Run | SPC | JIT | Auto | Average cost/unit for 3 replications | | | | Average | S/N |
|---|---|---|---|---|---|---|---|---|---|
| 1 | 1 | 1 | 1 | 1.43 | 1.27 | 2.08 | 1.47 | 1.56 | -4.04 |
| 2 | 1 | 1 | 2 | 1.45 | 1.35 | 1.84 | 1.47 | 1.52 | -3.73 |
| 3 | 1 | 2 | 1 | 1.37 | 1.27 | 1.85 | 1.42 | 1.48 | -3.50 |
| 4 | 1 | 2 | 2 | 1.45 | 1.39 | 1.72 | 1.47 | 1.51 | -3.59 |
| 5 | 2 | 1 | 1 | 1.59 | 1.45 | 2.24 | 1.66 | 1.73 | -4.91 |
| 6 | 2 | 1 | 2 | 1.60 | 1.59 | 1.97 | 1.63 | 1.68 | -4.55 |
| 7 | 2 | 2 | 1 | 1.54 | 1.43 | 2.01 | 1.59 | 1.64 | -4.38 |
| 8 | 2 | 2 | 2 | 1.61 | 1.55 | 1.89 | 1.64 | 1.67 | -4.49 |

Table 1. Results of the Experiment

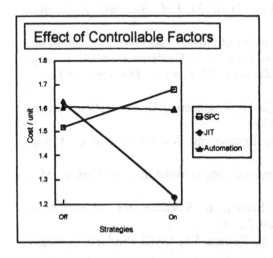

Figure 2. Main Effects of Controllable Factors

From this table, it can be seen that the most robust strategy for this scenario is the use of JIT, without the adoption of SPC and automation strategies. This gives the lowest average cost per unit of $1.48, and the highest S/N Ratio of -3.50. The main effects of the controllable factors are shown graphically in Figure 2. Obviously, JIT has a profoundly desirable effect on quality measured as cost per good units. The effect of automation is to decrease the cost only slightly, while the use of SPC strategy increases costs.

## 5. CONCLUSIONS

We have presented a methodology for the selection of manufacturing strategies. This methodology is based on the Taguchi's method of parameter design and takes into account the current process capability, possible process shifts, the complexity of the existing processes, and the costs and the benefits of the proposed strategies. The methodology was demonstrated on an example hypothetical manufacturing system.

The Taguchi method is normally applied to the tactical design of the production system. Our methodology presents an application of the method to the strategic design of manufacturing systems. This methodology is extendable to include other manufacturing strategies such as total quality control and supplier certification.

## ACKNOWLEDGEMENTS

The third author is grateful for the Academic Research Visitors Grant that he received from the School Research Committee of the University of Waikato, which made it possible for him to carry out the work.

## REFERENCES

1. Mezgar, I.: Parallel quality control in manufacturing system design, in the Proc. of Int. Conf. of Industrial Engineering and Production Management, April 4-7., 1995. Marrakech, Morocco, Eds. IEPM-FUCAM, pp. 116-125.
2. Foulds, L.R., Berka, P.: The achievement of World Class manufacturing in Central Europe, Proc. of the IFORS SPC-2 Conference on "Transition to Advanced Market Economies", June 22-25, 1992, Warsaw, Eds.: Owsinski, J.W., Stefanski, J., Strasyak, A., pp139-145.
3. Hyde, A., Basnet, C., Foulds, L.R.: Achievement of world Class Manufacturing in new Zealand: Current Status and Future Prospects, Proc. of the Conference on "NZ Strategic Management Educators", Hamilton, New Zealand, NZ Strategic Management Society Inc., pp.168-175.
4. Taguchi, G., Elsayed, E., Hsiang, T.C.: Quality Engineering In Production Systems, McGraw-Hill Book Company, New York, 1989.
5. Ross, P.J.: Taguchi Techniques for Quality Engineering, McGraw-Hill Book Comp., New York-Auckland,1988.
6. Drucker, P.F.: The Emerging Theory of Manufacturing, Harvard Business Review, May-June, 1990, 94-102.
7. Law, A.M., Haider, S.W.: Selecting Simulation Software for Manufacturing Applications, Industrial Engineering, 31, 1989, 33-46.
8. VAX/VMS SIMSCRIPT II.5 User's Manual, Release 5.1, CACI Products Company, February 1989.

# DEVELOPMENT OF EXPERT SYSTEM
# FOR FAULT DETECTION IN METAL FORMING

**P. Cosic**

**University of Zagreb, Zagreb, Italy**

KEY WORDS : Development, Expert System, Fault Detection, Metal Forming

ABSTRACT : The development of an off-line expert system prototype dedicated to quality defect diagnosis in deep drawing technology is described. Database creating, database search strategy, defining of logical relationships and reasoning system reliability are outlined. The diagnosis of an observed defect or failure, applying a set of rules, gives the expert system user an insight into the potential classified and graded faults. After the multilevel validation of potential causes of the observed faults, the expert system chooses the solution which is based on the greatest probability of the evaluation function.

## 1. INTRODUCTION

*"As a rule, the greater a system's generality the lower its efficiency."* [1]

The majority of complex and interesting problems do not have clear algorithmic solutions. Thus, many important tasks are realised in a complex environment which interferes with making the precise description and rigorous analysis of a particular problem. Traditional algorithmic methods are not suitable enough for successful connecting of design and manufacturing processes due to their complex nature (creativity, intuition, heuristics) [2].

Published in: E. Kuljanic (Ed.) *Advanced Manufacturing Systems and Technology*,
CISM Courses and Lectures No. 372, Springer Verlag, Wien New York, 1996.

Therefore, in this paper the use of expert systems for the purpose of detection and elimination of faults in the technology of deep drawing is considered. This technology is selected because of its wide use in production, long-time application in practice, and relatively suitable validation of selected variables by simulation or design of experiments.

## 2. SETTING PROBLEM SCOPE

Generally speaking, deep drawing is a process of forming sheet metal between an edge-opposing punch and a die (draw ring) to produce a cup, cone, box or shell-like part. The flat blank is prepared from a strip or sheet. In the first stage of the expert system development, the problem field is restrained by the following features:

- steel sheet
- axi-symmetrical product shape
- no change of blank thickness
- workpieces with or without flanges
- multiphase production of workpieces.

The production of high-quality workpieces depends on [3, 4, 5]:

- chemical structure (percentage of carbon and alloy elements),
- crystallographical structure (percentage of ferrite, grain size of steel, ageing of material),
- state of sheet or strip surface (clean and deoxydational surface),
- mechanical properties (ratios of ultimate strength and conventional yield limit, extension properties, deformability),
- draw ratios ,
- product shape and sizing,
- lubrication (to prevent galling, wrinkling or tearing, ease of application and removal, corrosivity),
- effect of rolling direction (elongation of metal during rolling introduces directionality, anisotropy or fibering),
- die and punch radius,
- punch-to-die clearance and tolerance,
- elements for relieving and aggravating of material flow, etc.

## 3. EXPERT SYSTEM DEVELOPMENT

The developed expert system includes: *rules, frames* (classes), *inheritance, graphics, truth maintenance* and *alternative views*. Rules are used for capturing the knowledge of an expert. The grouping of rules shows the intent to control which rules are used (fired) when a new fact is introduced into the knowledge base. *Priorities* of the rules define the ordering of sequence in which rules are invoked. *Certainty factors* allow users to implement their own interpretation of how certainty factors should be used. *Frames* (often called classes) are the basic building blocks of an object system, representing the generic classifications of things that make up the observed domain. A frame consists of a set of *attributes*, often called "slots". Good *inheritance* mechanism (often referred to as class hierachies) allows multiple layers of parent/child relationships. *Truth maintenance* has the facilities that store the links between asserted values and the rules that made that assertion. *Forward chaining* as a control rule is defined using IF-THEN syntax that logically connects one or more antecedent clauses ( or premises) with one or more consequents (or conclusions). *Backward chaining* as a control rule is used if we begin with a conclusion (or hypothesis) and want to know all possible pieces of information that led to that conclusion. The *inference engine* (Figure 1) fires one rule and asserts a new fact (the rule's consequent), and the new fact matches the antecedent of another rule.

As a consequence of the inference engines linking together chains of rules, the expert system has the ability to explain the chain of reasoning that led to the final conclusion. This explanatory capability is the most useful feature to trust the advice the system is providing, because it removes the black-box mystery from the process of converting raw facts into expert advice. This feature also makes expert systems excellent training tools, because the novice can examine step by step an expert's thought process. Expert systems have to allow the developer to stop the inference engine temporarily at *predefined breakpoints*. Breakpoints are useful for seeing if rule is fired at all, and for seeing whether the rules are fired in the anticipated order.

To build the required robust domain model, we use the structures that cluster and organise our facts concerning the observed process-structures called *frames*. The replacement of our simple  assertions with a representation that uses frames provides :

- *children* - the direct descendents, inheriting by default all the attributes of the *parents*
- a child may have more than one parent
- the related ability to constrain the allowable values that an attribute can take on.
- modularity of information
- a mechanism that will allow us to restrict the scope of facts considered during forward or backwaed chaining
- access to a mechanism that supports the inheritance of information down a class hierarchy.

In the considered model each object type will be implemented as a frame. These frames will be organised into a hierarchy that containts all components. In fact, the parent /child relationship of frames in a hierarchy are often refered to as class/subclass.

Validation techniques [6, 7 ] are used to assure the correctness of functionality and internal logic, to check if the correct data or information is passed between the internal components of an expert system. The validation set contains the following criteria:

- completeness
- consistency
- robustness
- system aspects
- user aspects
- expert aspects.

Each of these criteria should be observed and implemented during the building of the expert system.

The developed expert system [8] can be described as a diagnosis system (Figure 1). Diagnosis systems relate observed behavioural irregularities with underlying causes. The area of fault elimination causes includes the elements of design systems and debugging systems. Design systems construct descriptions of objects in various relationships with each other, and verify that these configurations conform to stated constraints. Debugging systems are based upon planning, design, and prediction capabilities to create specifications or recommendations for correcting a diagnosed problem. By using simulation and design of experiments the developed expert system includes the elements of prediction systems.

## 4. IMPLEMENTATION OF EXPERT SYSTEM IN DEEP DRAWING

The developed expert system works in the off-line mode as a help for the elimination of the observed faults during the technological process. The expert system is used in the way that the selected sketch or photograph of the observed faults (42 faults up to now), suggests us the possible cause of the faults (66 causes so far) and the possible ways of the fault elimination (up to now 59 ways of fault elimination).

The evaluation function estimates the possible causes of the faults and suggests actions taking into account the following criteria : price of the product, time of use, level of disturbing technological process, and standardisation of tool parts and technological parameters.

The expert system can eliminate the observed faults in three groups of actions. The actions can be related to workpiece material, tool, technological parameters, and to their interactions (Figure 2).

The first selected example can explain the way for observing the kinds of faults on workpieces and classificitation of actions for possible corrections. Analysing the observed faults (Figure 3), fractures at the vessel's bottom can be noticed. The possible causes of the faults can be multiple. They might include : sheet thickness tolerance, ultimate strength, grain orientation and size, surface faults, roughness level of die and blankholder, punch-to-die clearance, too high blankholder pressure, too large draw ratios, too small punch radius,

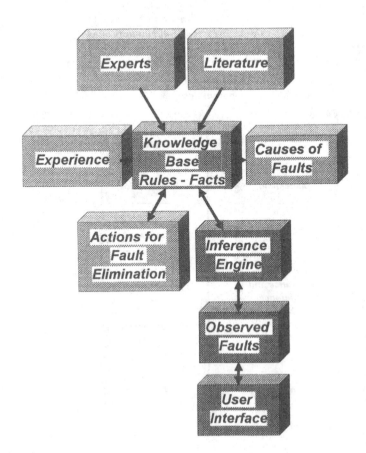

inadequate lubrication, too thin sheet , too large ratios of ultimate strength and conventional yield limit , the deep drawing tool behaving as a blanking tool, etc. The possible actions for faults eliminations are classified with respect to the complexity of criteria and the cost of actions. For the implementation of measures it is necessary to group the steps related to workpiece material, technological parameters, process of anneling, and tool constructional properties.Analysing the selected photograph or sketch (Figure 3), three types of faults and possibly causes of the faults are offered to us when using forward chaining rules as selected

Figure 1. The components of an expert system for fault dignosis

items (Figure 4). When we use the input data (blank radius, cup radius, thickness of cup's wall, ultimate strength of sheet material, yield stress, coefficient of friction, quality of sheet material surface, direction of sheet rolling) to determine the possible causes of faults by backward chaining rules, the system suggests us multiple potential solutions using forward chaining. In the prototype phase of the expert system the end user must contact expert system and work intensively and interactivly to cut time of the inference process of finding out the solution with the highest probability. However, in the mentioned process some difficults can be encountered. Sometimes some interactions or a few causes can have equal importance in the solving process. Therefore, special selected cases should be used in special areas of reseach : design of experiments (factorial experiments, analysis of variance, regression analysis) and simulation. The developer and, in the final instance, the end user can obtain a special corrective method for the validation of the selected actions in order to choose the optimal solution.

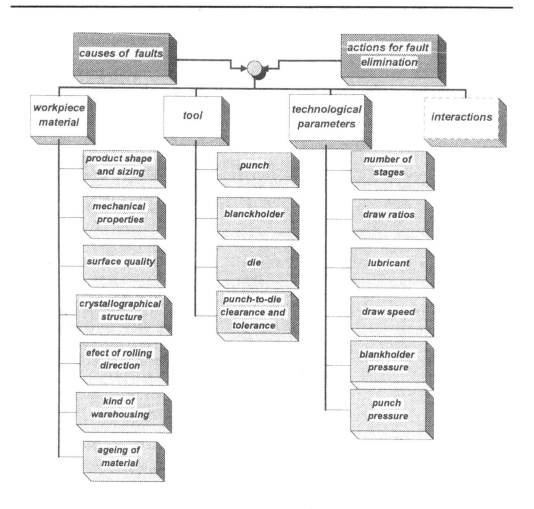

Figure 2. Possible ways of solving problems

Using the evaluation function with the penality values based on heuristics, analytical evaluation, design of experiments and simulation, and applying the criteria of complexity and action costs, the most probable measure would be selected out of the suggested ones. It often proves to be the decreasing of the blankholder pressure.

Figure 3. Flaw and fracture at vessel's bottom

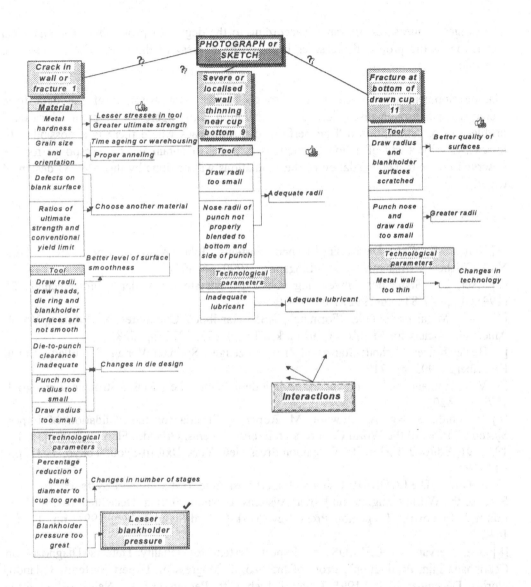

Figure 4. Types of faults, possibly causes of faults and potential actions for fault elimination

## CONCLUSION

The developed expert system can be a help in the process of elimination of observed faults at the workpiece during the technological process. The observed difficulties in the visual recognition of faults can influence the adequacy of specificied actions for the elimination of fault causes. So, the expert system will be supplemented with more input data related to material properties, chosen manufacturing technology and constructional tool properties.

The suggested measures are ranked according to the degree of probability. The analytical evaluation of the proposed measures helps in the process of the most efficient measure selection.

The development will be continued through further sistematisation of the data from literature, and through discussions and cooperation with experts of different educational profiles. Also, simulations will be performed in the cases where there is a possibility to lessen the subjectivity of action estimations in order to obtain the best response for the observed problem. The validation of the simulations will be done by the selected design of experiments.

REFERENCES

[1] Payne, E. C. : Developing Expert Systems, John Wiley & Sons, Inc., New York/Chichester/Brisbane/Toronto/Singapore, 1990, pp. 401.

[2] Cser, L. : Stand der Anwendung von Expertensystemen in der Umformtechnik, 25 (1991) 4 pp. 77-83 ; 26(1992) 1 pp. 51-60.

[3] . - , Metal Handbook, "Forming", ASM Handbook Committee, 8th Edition, Vol 4, American Society for Metals, Metals Park, Ohio, 44073, 1970, pp. 528.

[4] Oehler Kaiser, "Schnitt-Stanz-und Ziehwerkzeuge", Springer-Verlag, 7. Auflage, Berlin Heidelberg, 1993, pp. 719.

[5] V. P. Romanovskij, "Spravočnik po holodnoj štampovke", Mašinostroenie, Leningrad, 1979, pp. 520.

[6] P. Smith, S. Ng, A. Steward, M. Roper, "Criteria for the Validation of Expert Systems", Proc. of the World Congress on Expert Systems, Orlando, Florida, December 16-19, 1991, Editor J. Liebowitz, Pergamon Press New-York-Oxford-Seoul-Tokyo, 1991, pp. 980 - 988.

[7] A. Greb, "The LOOKER: Using an Expert System to Test an Expert System", Proc. of the World Congress on Expert Systems, Orlando, Florida, December 16-19, 1991, Editor J. Liebowitz, Pergamon Press New-York-Oxford-Seoul-Tokyo, 1991, pp. 1005 - 1012.

[8] J.A. Cervantes, "CELLOS, an Expert System for Quality Defects Diagnosis on Cellophane Film Production", Proc. of the World Congress on Expert Systems, Orlando, Florida, December 16-19, 1991, Editor J. Liebowitz, Pergamon Press New-York-Oxford-Seoul-Tokyo, 1991, pp. 476 - 483.

[9] Nilsson, N. J. : Principles of Artificial Intelligence, Tioga Publishing Company Palo Alto, California, 1980, pp. 476.

[10] Hayes-Roth, F., Waterman, D.A., Lenat, D.B. : Building Expert Systems, Addison-Wesley Publishing Company, Inc., 1983, pp. 444.

# MANAGERIAL ISSUES FOR DESIGN OF EXPERIMENTS

**F. Galetto**
**Polytechnic of Turin, Turin, Italy**

KEY WORDS: (DOE), Development, Quality Management, ANOVA, Taguchi Method

ABSTRACT: Design is the most important phase for Quality. Many times managers know little about Statistics and Probability Theory; nevertheless they have to make decisions based on few data analysed with statistical methods. Recently western nations have bee using blindly methods imported from Japan, just because they are Japanese methods. Taguchi Methods are called upon by managers and scientists to support them in making decisions; these methods are very appealing for them as they are claimed to be less expensive (i.e. less testing is needed) and easier (i.e. no knowledge is needed). The paper shows that Logic and the Scientific Approach are able to provide the right route toward the good methods for Quality. The absence of the statistical approach before, during and after the experiments typically results in relatively uninformative output of questionable general validity. Based on actual industrial applications, theory and simulations we will show most of the important differences between the proposed methods for DOE and the pitfalls managers are making. Managers have to learn Design of Experiments to draw good decisions. Management must understand that they have a new job: learning to use Quality methods for Quality.

## 1. INTRODUCTION

Quality is a fashionable subject at the present time: many top managers are giving this matter greater importance; journalists as well are dealing with it; many Gurus are providing their precepts for making Quality; many managers either attend several seminars on Quality looking for "cooking books" they can apply immediately with no effort from themselves or they buy packaged programs and off-the-shelf solutions.

Published in: E. Kuljanic (Ed.) *Advanced Manufacturing Systems and Technology*, CISM Courses and Lectures No. 372, Springer Verlag, Wien New York, 1996.

Too many of them think that Quality is growing and linked with Certification (based on ISO 90001-2-3 standards). Only short-sighted managers have been relating Quality to high cost; moreover products coming from Certified suppliers are often not better than from other suppliers; Companies can get Quality from qualified suppliers meeting their needs, not from paper-certified ones.

If managers meditate upon these facts, they must acknowledge that "low Quality can be a very costly luxury for a company" and that "Quality has always been a competitive advantage". Unfortunately top managers often think, wrongly, that "all the problems of disquality are originated by the workers, either in the manufacturing or in other areas". They do not understand the important idea (shared by Deming, Juran, ..., and myself ) that "more than 90% of times poor quality depends on the managers".

"Anybody can make a commitment to Quality at the boardroom table" (L. Iacocca)

Unfortunately _management commitment to Quality is not enough_; managers must understand and learn Quality ideas.

Too many companies are well behind the desired level of Quality management practices.

Quality is a serious and difficult business; it has to become an integral part of management.

The paper is addressed to managers because they are decisions markers: "managers have the responsibility of major decisions in a company and the soundness of their decisions affects the Quality of the products and the customer satisfaction". In order to make sound decisions managers have to be aware of the consequences of their decisions; in relation to Quality matters, managers have to commit themselves to assure that the concepts and disciplines associated with Quality will be introduced into the developments programs of the company.

Looking at the decisions of many companies, managers (if they are intellectually honest) have to admit that, in many western countries, there is a general "lack of credible executive action giving people permission to do things right and the help that such permission requires".

Many times **managers** know little about Statistics and Probability Theory; nevertheless they **have to make decisions based on few data analysed with statistical methods** (devised by Statistics experts).

**Managers do not like to ask themselves whether a method is good or bad especially when a method provides them with results that are appealing**; so-called experts do the same several times.

Quality has always been a competitive advantage. Japanese recognised that and made the right decision: to learn. They called American Gurus to teach them Quality ideas and methods. So they broke the "Disquality Vicious Circle".

Recently western nations have recognised that education and training are essential, but in some way they are not making Quality decisions: they use blindly methods imported from Japan, just because they are Japanese methods (e.g. Taguchi Methods).

The paper shows that Logic and the Scientific Approach are able to provide the right route toward the good methods for Quality.

We show some methods, in order to invite managers to break the "Vicious Circle" (IGNORANCE-PRESUMPTUOUSNESS-PRESUMPTUOUSNESS-IGNORANCE) that prevents Companies from getting the Quality their customers need.

Managers and scientists who will understand the core of the following ideas will help their nations to reduce the Quality gap, and therefore the disquality costs.

Quality achievement is not a matter of statistics, but of sound engineering practice.

If a manufacturer is able to produce Quality items at minimum costs, he can sell them at lower prices than competitors and then he certainly is bound to win the competitiveness fight and increase its market share.

This certain route led the Japanese to their present dominance.

## 2. THREE ACTUAL CASES

We show three cases published by Taguchi Methods experts; the scientific analysis of data provides different conclusions. Since the scientific analysis is correct, it follows that a huge amount of money was wasted.

Managers, at every level, have to meditate upon these facts, decide to learn, and to climb the ladder of knowledge: from

ignorance → awareness → simple knowledge → know-how → full understanding.

Quality is a competitiveness factor that must be integrated timely in all the company activities in order **to prevent failures**; the only way is to give due importance to Quality in each phase of the product **development cycle**.

That does not happen overnight, and needs a management metamorphosis, from "weather-managers" to Rational Managers .

Starting from the seed of any knowledge "I know that I don't know", intellectually honest people breaks the "Disquality Vicious Circle" and climb the ladder of know ledge.

Managers are decision makers and therefore the need the tools for thinking in decision-making in order to be rational managers: recognise problems, collect information, set priorities accurately, find causes, consider all factors and other people's views, consider alternative courses of actions, consider consequences and sequel to troubleshoot the future, consider risks to any choice, .....

To make full use of the thinking ability of people there is a basic approach, the ITE approach to decisions: every time a managers has to make a step in the decision process, he must ask himself

If *I do this*    Then *I'll have this consequence*    Else *I'll have this other consequence.*

ITE is an integrated, holistic approach that releases intellectual resources that have been hidden, unused or underused, opening channels of communication among people.

The first step toward this serious learning is: Intellectual Honesty.

There are many Quality techniques useful during the development phase; only two are mentioned here, FMECA and DOE.

FMECA is to be used in order to identify potential failures and take preventive actions; unfortunately managers either do not know the technique or they use the silly rule of making decisions based on a priority index which is the product of 3 or 4 indexes; so doing they do not base their behaviour on a rational approach and they do not make full use of the thinking ability of people. There is no space here to pursue further the matter.

DOE helps a lot in *preventing problems*. The only way an engineer can "communicate" with a phenomenon (failures, defects, yield, ROI, ...) is through "data" (measurements on a relevant characteristic manifestation of the phenomenon).

The manager/engineer has to draw "objective" conclusions about the phenomenon, and, since generally the information available in the data is incomplete, he has to measure the degree of uncertainty associated with such conclusions.

A golden rule has to be understood by managers: tests have to be designed in order to get the information needed to take good decisions.

Statistics provides suitable tools for "Design of Experiments" (DOE) in order to get maximum of information at stated costs and risks.

The G-method is a method of DOE based on the Scientific Approach, originated by Galileo, and on the Gauss-Markov theorem (that states optimum properties of linear estimators).

Mathematics, Logic and Physics can prove that Taguchi approach is wrong especially when he writes ".... when there is interaction, it is because insufficient research has been done on the characteristic values".

Managers have to learn Design of Experiments ideas to draw good decisions.

Quality is number one management objective, not only for product and services, but for Quality methods as well.

### The Case of the "electrolytic capacitors"

Data and analysis are related to an experiment carried out in 1985 in UK (mentioned by G. Pistone at annual Conference of the SIS, Italian Society of Statistics, 1990, Padova).

| T | R | S | V | n. data | mean | s | S/N |
|---|---|---|---|---|---|---|---|
| -1 | -1 | -1 | -1 | 4 | -8,99786 | 0,288428 | 29,882039 |
| 1 | -1 | -1 | -1 | 4 | -9,79303 | 0,389932 | 27,998568 |
| -1 | 1 | -1 | -1 | 4 | -8,80422 | 0,250446 | 30,919533 |
| 1 | 1 | -1 | -1 | 4 | -7,94556 | 0,162311 | 33,795531 |
| -1 | -1 | 1 | -1 | 4 | -9,12618 | 0,055851 | 44,265161 |
| 1 | -1 | 1 | -1 | 4 | -9,21131 | 0,029337 | 49,938118 |
| -1 | 1 | 1 | -1 | 4 | -9,07666 | 0,027821 | 50,271063 |
| 1 | 1 | 1 | -1 | 4 | -7,77125 | 0,230403 | 30,560055 |
| -1 | -1 | -1 | 1 | 4 | -10,3254 | 0,165593 | 35,897299 |
| 1 | -1 | -1 | 1 | 4 | -10,1674 | 0,06934 | 43,324507 |
| -1 | 1 | -1 | 1 | 4 | -8,79386 | 0,377056 | 27,355476 |
| 1 | 1 | -1 | 1 | 4 | -8,05024 | 0,467217 | 24,725800 |
| -1 | -1 | 1 | 1 | 4 | -9,31471 | 0,162271 | 35,178572 |
| 1 | -1 | 1 | 1 | 4 | -8,97796 | 0,222795 | 32,105445 |
| -1 | 1 | 1 | 1 | 4 | -9,10306 | 0,143904 | 36,022291 |
| 1 | 1 | 1 | 1 | 4 | -7,57433 | 0,088606 | 38,637624 |

4 factors at 2 levels:

      R : radiation flow     T : temperature      S : size of capacitors  V : voltage

The response is "current intensity" ($\mu A$).

For each treatment are shown: the mean of 4 replications, the standard deviation (the square root of the variance), and the S/N, Signal to Noise ratio.

The analysis of S/N is:

| Source | df | SS | MS | F | liv.prob. F |
|--------|-----|-----------|-----------|--------|-------------|
| T | 1 | 4,73692 | 4,73692 | 0,067 | 0.8063 |
| R | 1 | 43,2383 | 43,2383 | 0,610 | 0.4700 |
| S | 1 | 248,6896 | 248,6896 | 3,510 | 0.1199 |
| V | 1 | 37,15833 | 37,15833 | 0,525 | 0.5014 |
| T*R | 1 | 39,04037 | 39,04037 | 0,551 | 0.4913 |
| T*S | 1 | 25,71987 | 25,71987 | 0,363 | 0.5731 |
| T*V | 1 | 18,89046 | 18,89046 | 0,267 | 0.6276 |
| R*S | 1 | 12,79817 | 12,79817 | 0,181 | 0.6885 |
| R*V | 1 | 10,93448 | 10,93448 | 0,154 | 0.7106 |
| S*V | 1 | 109,1914 | 109,1914 | 1,541 | 0.2695 |
| Error | 5 | 354,2195 | 70,8439 | | |
| total | 16 | 21273,41 | | | |

**Significance level was not stated.**
*If significance level were stated at 10%,*

  1.   R should be not significant         2.  T should be not significant
  3.   V should be not significant         4.  S should be not significant
  5.   all interactions of 1st order should be not significant,

i.e. nothing have influence on the response variable (with 10% risk of wrong decision).
**Analysing scientifically the data with the G–Method,** whitout being blinded by S/N, it is
found, **_with a significance level of 1% (i.e. 10 times better !!!)_**,
  • **_R is significant_**  • **_T is significant_**  • **_V is significant_**  • **_S is significant_**
  • **_all interactions of 1st order are significant !!!_**

**The Case of "tensile strength"**
Data and analysis are related to an experiment mentioned by G. Taguchi, in his book
*System of Experimental Design (vol. 1, pagg. 36-40).* 2 factors:
    A : supplier (3 levels, supplier not-Japanese, Japanese1, Japanese2)
    T : test temperature (4 levels, -30°C, 0°C, 30°C, 60°C)
The response is the "tensile strength" ($kg/mm^2$). Here are shown the values - *80.*

|          |     |    | values |     |     |
|----------|-----|-----|-----|-----|
| Supplier | B1 | B2 | B3 | B4 |
| A1 | 20 | 8 | 0 | -9 |
| A2 | 22 | 12 | -2 | -12 |
| A2 | 25 | 8 | 0 | -14 |
| A2 | 28 | 10 | 3 | -13 |
| A2 | 25 | 9 | 0 | |
| A2 | 26 | 12 | | |
| A3 | 17 | 6 | -8 | -20 |
| A3 | 23 | 8 | -6 | -18 |
| A3 | 20 | 4 | -3 | -22 |

Taguchi carried out the analysis of the means and considered the
* linear effect $B_l$ of the temperature
* and two contrasts
    $L_1(A)$ between the suppliers not-Japanese and the 2 Japanese
    $L_2(A)$ between the suppliers Japanese1 and Japanese2
* and the interaction of suppliers-contrasts with linear effect, as well.
The book marks with ** the effects *significant at 1%*.

| Source | df | SS | MS | |
|--------|----|----|----|----|
| $L_1(A)$ | 1 | 9.50 | 9.50 | |
| $L_2(A)$ | 1 | 62.16 | 62.16 | ** |
| $B_l$ | 1 | 2056.86 | 2056.86 | ** |
| $B_l L_1(A)$ | 1 | 36.63 | 36.63 | ** |
| $B_l L_2(A)$ | 1 | 1.26 | 1.26 | |
| Error | 21 | 48.97 | 2.332 | |
| total | 32 | 2234.28 | | |

**Analysing scientifically the data with the G-Method**, whitout being blinded by the means, it is found, *with a significance level of 1%*
   * $L_1(A)$ *is significant*
   * $L_2(A)$ *is significant*
   * linear effect $B_l$ *is significant*
   * *interaction of suppliers and linear effect of temperature is not significant*

**The Case of "Polysilicon Deposition Process"**
Data and analysis are related to an experiment mentioned by M. Phadke (a disciple of G. Taguchi, and manager at AT&T, the company that used a lot Taguchi Method and fired 40000 people in march 1996), in his book *Quality Engineering using Robust Design (cap. 4)*. 6 factors at 3 levels:
   A : temperature    B : pressure       C : nitrogen flow
   D : silane flow    E : setting time   F : cleaning method
A 18 test plan (*fractional experiment*) was carried out: *no alias structure was provided !!!!*. The response is the "surface deposition thickness" (Å) measured on 3 wafers ( on 3 points). The table shows the data.
In the analysis of S/N [only factors considered] **significance level was not stated**.
*If significance level were stated at 10%,*
   1.    A [+ ...] should be *significant*        2. B [+ ...] should be not significant
   3     C [+ ...] should be *significant*        4. D [+ ...] should be *significant*
   5.    E [+ ...] should be not significant      6. F [+ ...] should be *significant*
**Analysing scientifically the data with the G-Method**, whitout being blinded by S/N, it is found, *with a significance level of 1%*
   * *linear effect of all factors <u>aliased</u> are all significant*
   * *quadratic effect of all factors <u>aliased</u> are all significant*
   * *several interactions of linear effects of one factor (<u>aliased</u>) with linear*

*effect and quadratic effect another factor (<u>aliased</u>) are significant*

| | A | B | C | D | E | F | test Wafer 1 | | | test Wafer 2 | | | test Wafer 3 | | |
|---|---|---|---|---|---|---|---|---|---|---|---|---|---|---|---|
| 1 | -1 | -1 | -1 | -1 | -1 | -1 | 2029 | 1975 | 1961 | 1975 | 1934 | 1907 | 1952 | 1941 | 1949 |
| 2 | -1 | 0 | 0 | 0 | 0 | 0 | 5375 | 5191 | 5242 | 5201 | 5254 | 5309 | 5323 | 5307 | 5091 |
| 3 | -1 | 1 | 1 | 1 | 1 | 1 | 5989 | 5894 | 5874 | 6152 | 5910 | 5886 | 6077 | 5943 | 5962 |
| 4 | 0 | -1 | -1 | 0 | 0 | 1 | 2118 | 2109 | 2099 | 2140 | 2125 | 2108 | 2149 | 2130 | 2111 |
| 5 | 0 | 0 | 0 | 1 | 1 | -1 | 4102 | 4152 | 4174 | 4556 | 4504 | 4560 | 5031 | 5040 | 5032 |
| 6 | 0 | 1 | 1 | -1 | -1 | 0 | 3022 | 2932 | 2913 | 2833 | 2837 | 2828 | 2934 | 2875 | 2841 |
| 7 | 1 | -1 | 0 | -1 | 1 | 1 | 3030 | 3042 | 3028 | 3486 | 3333 | 3389 | 3709 | 3671 | 3687 |
| 8 | 1 | 0 | 1 | 0 | -1 | -1 | 4707 | 4472 | 4336 | 4407 | 4156 | 4094 | 5073 | 4898 | 4599 |
| 9 | 1 | 1 | -1 | 1 | 0 | 0 | 3859 | 3822 | 3850 | 3871 | 3922 | 3904 | 4110 | 4067 | 4110 |
| 10 | -1 | -1 | 1 | 1 | 0 | -1 | 3227 | 3205 | 3242 | 3468 | 3450 | 3420 | 3599 | 3591 | 3535 |
| 11 | -1 | 0 | -1 | -1 | 1 | 0 | 2521 | 2499 | 2499 | 2576 | 2537 | 2512 | 2551 | 2552 | 2570 |
| 12 | -1 | 1 | 0 | 0 | -1 | 1 | 5921 | 5766 | 5844 | 5780 | 5695 | 5814 | 5691 | 5777 | 5743 |
| 13 | 0 | -1 | 0 | 1 | -1 | 0 | 2792 | 2752 | 2716 | 2684 | 2635 | 2606 | 2765 | 2786 | 2773 |
| 14 | 0 | 0 | 1 | -1 | 0 | 1 | 2863 | 2835 | 2859 | 2829 | 2864 | 2839 | 2891 | 2844 | 2841 |
| 15 | 0 | 1 | -1 | 0 | 1 | -1 | 3218 | 3149 | 3124 | 3261 | 3205 | 3223 | 3241 | 3189 | 3197 |
| 16 | 1 | -1 | 1 | 0 | 1 | 0 | 3020 | 3008 | 3016 | 3072 | 3151 | 3139 | 3235 | 3162 | 3140 |
| 17 | 1 | 0 | -1 | 1 | -1 | 1 | 4277 | 4150 | 3992 | 3888 | 3681 | 3572 | 4593 | 4298 | 4219 |
| 18 | 1 | 1 | 0 | -1 | 0 | -1 | 3125 | 3119 | 3127 | 3567 | 3573 | 3520 | 4120 | 4088 | 4138 |

| Source | df | SS | MS | F |
|---|---|---|---|---|
| A | 2 | 440 | 220 | 16 |
| B | 2 | 7 | 3.5 | |
| C | 2 | 134 | 67 | 5.0 |
| D | 2 | 128 | 64 | 4.8 |
| E | 2 | 18 | 9 | |
| F | 2 | 181 | 90.5 | 6.8 |
| Error | 9 | 121 | 13.4 | |
| total | 17 | 1004 | | |

## 3. SIMULATION STUDIES

Form the previous cases it is evident that a lot of disquality was generated and a huge amount of money was wasted. Were they unfortunate? Absolutely not, they were a-scientific.

In order to analyse the stupid statement "Taguchi methods works well" a simulation study was carried out: a known model with 4 factors, A, B, C, D, was chosen and we analysed how many times the Taguchi Method and G-Method "found the known truth".

In the model the "true values" for the *significant* effect of factors A, B, C, D, and for the interactions AB, AD, BD, CD, were stated. 1000 simulation runs were carried out for each of the following situation, by dividing the values of significant effect and by increasing the variability (four values of standard deviation s); a complete plan was generated and a fractional plan (8 runs) was used ( confounding pattern I+ABCD):

| | effect divided by | | | | | | |
|---|---|---|---|---|---|---|---|
| | 1 | 2 | 5 | 10 | 50 | 100 | 1000 |
| s=1 | * | * | * | * | * | * | * |
| s=2 | * | * | * | * | * | * | * |
| s=5 | * | * | * | * | | | |
| s=10 | * | * | * | * | | | |

*When the importance of factor and interaction is reduced,* a  s o u n d  m e t h o d
h a s  t o  f i n d  t h a t, *according the statistic theory.*

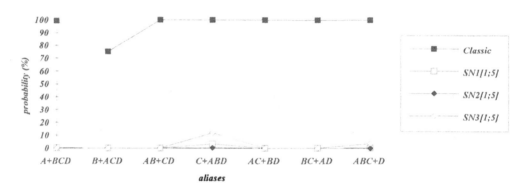

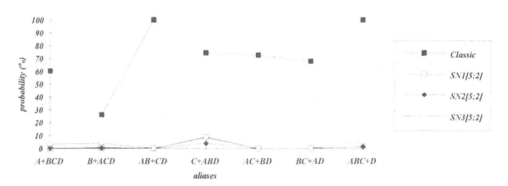

*When the variability is increased,* a  s o u n d  m e t h o d  h a s  t o  f i n d  t h a t,
*according the statistic theory.*
<u>G-Method finds that.</u>

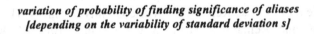

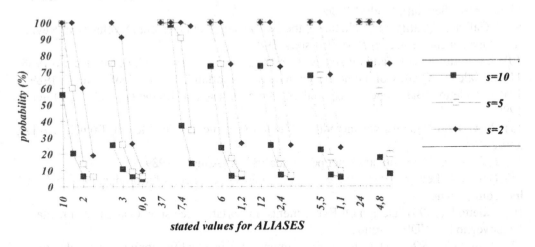

variation of probability of finding significance of aliases
[depending on the variability of standard deviation s]

stated values for ALIASES

*Taguchi Method did not find* that, in spite of the fact that it was used as "error" the value of the pooling of "truly not significant effects". The three most popular formulae for the Signal/Noise ratio, devised by Taguchi, were used for the analysis.
Some graph are presented to show the probability of "finding the truth".
**Does Taguchi Method works ???     *It is really robust in FAILURE!!!***
"Signal/Noise ratios" used in connection with the so called Robust Design are nonsense from a scientific point of view: these ratios are multifunctional transformations of the data, and at the end the transformed data must be normally distributed if, logically, the F ratio resulting from ANOVA and shown in  the "Quality Engineering using Robust Design" books should have any statistical sense).

## 4. CONCLUSION

It is left to the Intellectually Honest reader.
"Quality of methods for Quality is important" and that some methods are misleading (e.g. Taguchi Methods, Bayes Methods, ...)

REFERENCES
1)  D.H. Besterfield, "Quality Control", Prentice-Hall
2)  P.B. Crosby, "Quality is free", McGraw-Hill
3)  A.V. Feigenbaum "Total Quality Control", McGraw-Hill
4)  Juran, "Quality Control Handbook", McGraw-Hill
5)  W.E. Deming, "Out of the crisis", Cambridge
6)  F. Galetto, "An application of  experimental  design in the automotive field", SIA

Congress, Paris 1978

7) F. Galetto, "Assessment of Product Reliability" World Quality Congress '84, Brighton

8) F. Galetto, "Quality and Reliability: How to get results", 10th EOQC Seminar (Automotive Section), Madrid 1986

9) F. Galetto, "Quality and Reliability, the Iveco way", Management Development Review (by Management Centre Europe, Brussells, 1987)

10) F. Galetto. "Quality and reliability. A must for the industry", ISATA, Montecarlo 1988

11) F. Galetto, "Quality of methods for Quality is important", EOQC Conf., Vienna 1989

12) F. Galetto, "Basic and managerial concerns on Taguchi Methods", ISATA, Florence 1990

13) F. Galetto "Qualità dei metodi: il Metodo G è meglio dei Metodi Taguchi", ATA, aprile 1989

14) F. Galetto, "Elementi strategici per la qualità", ATA luglio 1989.

15) Galetto,F. Levi,R (1993) "Planned Experiments -Key factors for product Quality ", 3rd Int. Conf., Udine

16) Galetto,F (1993) "Design Of Experiments. Importanti idee sulla Qualità per i manager ", Convegno DEINDE, Torino

17) Galetto,F (1994) "Design Of Experiments per la Qualità: applicazioni industriali", Roma, 6 maggio 1994

18) Levi, R., (1988), "Pianificazione ed analisi degli esperimenti nella ricerca applicata", Convegno int. su metodologie e mezzi innovativi per la sperimentazione nel settore automotoristico, 163-171, Università di Firenze - ATA, 163-171

19) Levi, R. (1991), "Piani sperimentali e metodi Taguchi: luci e ombre", ATA, vol 44, n. 11, 777- 781

20) Galetto,F (1995) "Qualità, alcuni strumenti da Manager", CUSL

21) Taguchi G. (1987) - System of Experimental Design - UNIPUB/Kraus Int. and American Supplier Institute Inc.

22) Taguchi G. (1986) - Introduction to Quality Engineering

# SPLIT-PLOT DESIGN: A ROBUST ANALYSIS

**R. Guseo**
**University of Udine, Udine, Italy**

KEY WORDS: Split–plot design, error misspecification robustness.

ABSTRACT: This paper is devoted to the examination of the robustness of standard split–plot analysis for a two factors design with complete blocks under normality assumptions.
For instance, in ceramics firing, the main factors are the oven temperature $A$ and the clay mixture $B$. With a block factor it is possible to control different geometrical shapes of each piece.
In a factorial design each piece would be treated separately and, in this case, at fixed temperature $A$ for a particular clay mixture $B$. If the design is a *split–plot*, a batch *(whole–plot)* is defined with different clay mixture pieces *(sub–plots)* of the same shape (block) and is simultaneously treated at a fixed temperature in the oven.
The simultaneous application of treatements to a group of experimental units violates the independence of error terms. The usual equicorrelation assumption, characterizing *split–plot* designs, may no longer be assumed when an unknown spatial dependence is acting on the error term.
Under this general assumption it is proved that the usual tests for block and main effects of $A$ are exact and equivalent to the standard case. The standard tests for the main effect of $B$ and interaction $A \times B$ are not $F$–distributed. Nevertheless, with an approximation it is possible to recover robustly the well–known procedure.

Published in: E. Kuljanic (Ed.) *Advanced Manufacturing Systems and Technology*,
CISM Courses and Lectures No. 372, Springer Verlag, Wien New York, 1996.

## 1. INTRODUCTION

The replicated *split–plot* design within experimental blocks is well–known and it was
introduced as a special version of the factorial design when the level of a main factor (whole) is simultaneously applied to a set of experimental units. This particular
framework requires a special dependence claim, i.e. the equicorrelation among assumed
multinormal experimental errors.

This paper is devoted to the general case where the dependence among errors is arbitrary. In paragraph 2 the standard two factors split–plot design with random blocks
is examined. Paragraphs 3 and 4 generalize the error structure under which an exact
analysis of a block effect and a main effect of the whole factor $A$ is proved. In paragraph
5, under an arbitrary multinormal assumption, the ratios between $MS$ related to the
effects $B$ and $A \times B$ are proved not to be $F$–distributed as in the uncorrelated case.
Via simple approximations, it is shown, in paragraph 6, that the classical standard
tests (split–plot) for main effect $B$ and for interaction $A \times B$ may be controlled, under
the null hypothesis, with a semi–parametric argument that allows to recover robustly
the well–known decisional procedure.

## 2. SPLIT–PLOT DESIGN WITH TWO FACTORS WITHIN COMPLETE BLOCKS

Let us consider the simplest situation with two main factors and a block factor. Unlike
the usual assumptions of a factorial design, a *single* and *randomized* application of a
design treatment is not easily performed. Economical and physical reasons suggest the
application of a factor $A$ to a class of experimental units (*whole–plots*). For a fixed
level $a_j$ of the main factor $A$, the levels of a second factor $B$ may be randomly assigned
to the elements (sub–plots) of the whole–plots.

For instance, in ceramics firing, the main factors are the oven temperature $A$ and the
clay mixture $B$. With a block factor it is possible to control different geometrical
shapes of each piece. In a factorial design each piece would be treated separately and,
in this case, at fixed temperature $A$ for a particular clay mixture $B$. If the design is a
*split–plot*, a batch *(whole–plot)* is defined with different clay mixture pieces *(sub–plots)*
of the same shape (block) and is simultaneously treated at a fixed temperature in the
oven. Following the terminology of the example, the independence assumption of the
errors related to the same batch for fixed shape and temperature of the oven has low
credibility. It seems more appropriate to suppose them correlated. The standard split–
plot design assume an *equicorrelation*.

The reference model is only apparently analogous to the corresponding one used under
the factorial assumptions. To be more precise, the model is

$$y_{ijk} = \mu + \tau_i + \alpha_j + \beta_k + (\alpha\beta)_{jk} + \varepsilon_{ijk}, \tag{1}$$

where $i = 1, 2, \ldots, r$ is the block index; $j = 1, 2, \ldots, a$ is the level index of factor $A$; $k =$

$1, 2, \ldots, b$ is the level index of factor $B$; $y_{ijk}$ is the observed response. The explanatory components of (1) represent: $\mu$, the grand mean response; $\tau_i$, the random effect of the $i$-th block, $\tau_i \sim \mathcal{N}(0, \sigma_\tau^2)$; $\alpha_j$, the fixed effect of the $j$-th level of $A$, $\sum_{j=1}^{a} \alpha_j = 0$; $\beta_k$, the fixed effect of the $k$-th level of $B$, $\sum_{k=1}^{b} \beta_k = 0$; $(\alpha\beta)_{jk}$, the fixed level of the interaction $A \times B$ at levels $j$ for $A$ and $k$ for $B$, $\sum_{j=1}^{a}(\alpha\beta)_{jk} = \sum_{k=1}^{b}(\alpha\beta)_{jk} = 0$.

The stochastic term, $\varepsilon_{ijk}$, represents the experimental error and is assumed to have null mean and normal distribution, $\varepsilon_{ijk} \sim \mathcal{N}(0, \sigma^2)$. Independence between $\tau_i$ and $\varepsilon_{ijk}$ is allowed, $\tau_i \perp \varepsilon_{ijk}$, nevertheless, unlike canonical factorial model, the correlation among errors is

$$E(\varepsilon_{ijk} \cdot \varepsilon_{i'j'k'}) = \begin{cases} \rho\sigma^2 & \text{if } i = i', \ j = j', \ k \neq k'; \\ \sigma^2 & \text{if } i = i', \ j = j', \ k = k'; \\ 0 & \text{elsewhere.} \end{cases}$$

## 3. A GENERALIZATION

The equicorrelation condition characterizing split–plot designs is not acceptable in many applied contexts. In order to overcome this restriction by modelling the dependence locally, new designing techniques, such as the *nearest neighbour* (NN) technique have been proposed.

In [1], for instance, the experimental error is assumed to be such that $E(\varepsilon_{ijk}\varepsilon_{ijk'}) = \rho^{|k-k'|}\sigma^2$ if $|k - k'| = 1$, $E(\varepsilon_{ijk}^2) = \sigma^2$ and zero elsewhere.

In [2] a first order NN balanced design is introduced in order to model a particular spatial autocorrelation: $E(\varepsilon_{ijk}\varepsilon_{ijk'}) = \rho^{|k-k'|}\sigma^2$, $E(\varepsilon_{ijk}^2) = \sigma^2$ and zero elsewhere. More recently, in [3] optimal tests are proposed for nested factorial designs under a special circular dependence structure.

In this paper the dependence assumption of paragraph 2 is extended to a more general form, $E(\varepsilon_{ijk}\varepsilon_{ijk'}) = \sigma^2 q_{kk'}$, where $Q = (q_{kk'})$ is an *unknown* full rank correlation matrix. This very weak assumption is of interest because a specifically patterned matrix $Q$ may be *a priori* unjustified. The other assumptions remain unchanged.

Let us consider the following vectorization of the model (1), $y^* = (y_{111}, y_{112}, \ldots, y_{11b}, y_{121}, y_{122}, \ldots, y_{12b}, \ldots y_{ra1}, y_{ra2}, \ldots, y_{rab})'$, based upon the Kronecker product $\otimes$

$$y^* = (1_r \otimes 1_a \otimes 1_b)\mu + (1_r \otimes I_a \otimes 1_b)\alpha + (1_r \otimes 1_a \otimes I_b)\beta + (1_r \otimes I_a \otimes I_b)v + t^* + \epsilon^*, \quad (2)$$

where, $t^* = \tau \otimes 1_a \otimes 1_b \sim \mathcal{N}(0, \sigma_\tau^2 abW)$, $W = (I_r \otimes U_a \otimes U_b)$ with $U_n = \frac{1}{n}J_n = \frac{1}{n}1_n 1_n'$, because $\text{Var}(t^*) = E(t^* t^{*'}) = E(\tau\tau' \otimes aU_a \otimes bU_b)$, and $\epsilon^* \sim \mathcal{N}(0, \sigma^2 V)$, $V = (I_r \otimes I_a \otimes Q)$. By the previous assumptions, $t^*$ and $\epsilon^*$ are stochastically independent, $t^* \perp \epsilon^*$. Note that, via this representation, the standard split–plot design is characterized by $Q = (1 - \rho)I_b + \rho J_b$. Vectors $\alpha, \beta$ and $v$ summarize the effects of the factors $A$, $B$ and the interaction effects $A \times B$ in (1).

## 4. BLOCK AND FACTOR $A$ EFFECTS ANALYSIS

The random effect of the block factor $\tau$ is evaluated by $SSR$,

$$SSR \;=\; \sum_i y_{i..}^2/ab - y_{...}^2/rab = \sum_{ijk} (\tau_i - \tau_./r + \varepsilon_{i..}/ab - \varepsilon_{...}/rab)^2 . \qquad (3)$$

The vector, whose squared norm is equal to $SSR$, is $[(\boldsymbol{I}_r - \boldsymbol{U}_r) \otimes \boldsymbol{U}_a \otimes \boldsymbol{U}_b](\boldsymbol{\epsilon}^* + \boldsymbol{t}^*)$. Let us define $\boldsymbol{e}^* = \boldsymbol{\epsilon}^* + \boldsymbol{t}^*$ and $\boldsymbol{R} = [(\boldsymbol{I}_r - \boldsymbol{U}_r) \otimes \boldsymbol{U}_a \otimes \boldsymbol{U}_b]$, then $SSR = \boldsymbol{e}^{*\prime}\boldsymbol{R}\boldsymbol{e}^*$ and, by the independence between $\boldsymbol{\epsilon}^*$ and $\boldsymbol{t}^*$, $\boldsymbol{e}^* \sim \mathcal{N}(\boldsymbol{0}, \sigma^2 \boldsymbol{V} + \sigma_\tau^2 ab \boldsymbol{W})$. Let us consider, now, $\tilde{\boldsymbol{R}} = \boldsymbol{R}/\phi$, $\phi = \frac{s\sigma^2}{b} + \sigma_\tau^2 ab$, $s = \mathbf{1}_b'\boldsymbol{Q}\mathbf{1}_b$ and, in particular,

$$\frac{SSR}{\phi} = \boldsymbol{e}^{*\prime}\tilde{\boldsymbol{R}}\boldsymbol{e}^* . \qquad (4)$$

In order to determine the $SSR/\phi$ distribution, let us examine $\tilde{\boldsymbol{R}}\mathrm{Var}(\boldsymbol{e}^*)$,

$$\tilde{\boldsymbol{R}}\mathrm{Var}(\boldsymbol{e}^*) = \frac{1}{\phi}\left\{(\boldsymbol{I}_r - \boldsymbol{U}_r) \otimes \boldsymbol{U}_a \otimes \boldsymbol{U}_b(\sigma^2 \boldsymbol{Q} + \sigma_\tau^2 ab \boldsymbol{I}_b)\right\} . \qquad (5)$$

Let us define $\boldsymbol{Z}_0 = \boldsymbol{U}_b(\sigma^2\boldsymbol{Q}+\sigma_\tau^2 ab\boldsymbol{I}_b)$, then $\boldsymbol{Z}_0^2 = \phi\boldsymbol{Z}_0$, because $\boldsymbol{U}_b(\sigma^2\boldsymbol{Q}+\sigma_\tau^2 ab\boldsymbol{I}_b)\boldsymbol{U}_b = \frac{\sigma^2}{b^2}\mathbf{1}_b\mathbf{1}_b'\boldsymbol{Q}\mathbf{1}_b\mathbf{1}_b' + \sigma_\tau^2 ab\boldsymbol{U}_b = \left(\frac{s\sigma^2}{b} + \sigma_\tau^2 ab\right)\boldsymbol{U}_b = \phi\boldsymbol{U}_b$. The matrix $(\tilde{\boldsymbol{R}}\mathrm{Var}(\boldsymbol{e}^*))^2$ is idempotent, $(\tilde{\boldsymbol{R}}\mathrm{Var}(\boldsymbol{e}^*))^2 = \frac{1}{\phi^2}[(\boldsymbol{I}_r - \boldsymbol{U}_r) \otimes \boldsymbol{U}_a \otimes \boldsymbol{Z}_0^2] = \tilde{\boldsymbol{R}}\mathrm{Var}(\boldsymbol{e}^*)$.

By a theorem about normal quadratic forms, see e.g. [4] p. 57, and because the rank of the correspondig matrix in the (4) is $r(\tilde{\boldsymbol{R}}) = r - 1$, the ratio $SSR/\phi$ is centrally chi squared distributed

$$\frac{SSR}{\phi} = \boldsymbol{e}^{*\prime}\tilde{\boldsymbol{R}}\boldsymbol{e}^* \sim \mathcal{X}^2_{[r-1,0]} . \qquad (6)$$

Obviously, the mean value of $MSR = SSR/(r-1)$ is $E(MSR) = \frac{s\sigma^2}{b} + \sigma_\tau^2 ab$.

The fixed effect of the factor $A$ is evaluated by the quadratic form $SSA$,

$$SSA \;=\; \sum_j y_{.j.}^2/rb - y_{...}^2/rab = \sum_{ijk}(\alpha_j + \varepsilon_{.j.}/rb - \varepsilon_{...}/rab)^2 . \qquad (7)$$

Let us define $\psi = \frac{s\sigma^2}{b}$, where $s = \mathbf{1}_b'\boldsymbol{Q}\mathbf{1}_b$, then, following an analogous proof, we attain

$$\frac{SSA}{\psi} \sim \mathcal{X}^{2\prime}_{[a-1, \frac{b^2 r}{2\sigma^2 s}\sum_{j=1}^a \alpha_j^2]} , \qquad (8)$$

and $E(MSA) = \frac{s\sigma^2}{b} + rb\sum_{j=1}^a \frac{\alpha_j^2}{a-1}$.

Let us consider now the error linked to the form $SSEA$,

$$SSEA \;=\; \sum_{ij} y_{ij.}^2/b - \sum_j y_{.j.}^2/rb - \sum_i y_{i..}^2/ab + y_{...}^2/rab$$

$$=\; \sum_{ijk}(\varepsilon_{ij.}/b - \varepsilon_{.j.}/rb - \varepsilon_{i..}/ab + \varepsilon_{...}/rab)^2 . \qquad (9)$$

By a similar proof we have

$$\frac{SSEA}{\psi} \sim \chi^2_{[(r-1)(a-1);0]},\tag{10}$$

and, therefore, $E(MSEA) = \frac{s\sigma^2}{b}$.

For a fixed $t^*$ and by a theorem in [5] p. 178, $SSEA$, $SSR$ and $SSA$ are independent. These arguments justify exactly (in probability) the traditional use of the first part the analysis of variance table, Tab. 1, typically applied in a standard split–plot under the more general dependence conditions here assumed for the errors within the *whole–plots*.

## 5. FACTOR $B$ AND INTERACTION $A \times B$ EFFECTS ANALYSIS

The factor $B$ fixed effect is recognized by $SSB$,

$$SSB \;=\; \sum_k y^2_{..k}/ra - y^2_{...}/rab = \sum_{ijk}(\beta_k + \varepsilon_{..k}/ra - \varepsilon_{...}/rab)^2.\tag{11}$$

Let us define $b^* = \mathbf{1}_r \otimes \mathbf{1}_a \otimes \beta$. The vector, whose squared norm is equal to $SSB$, is

$$[U_r \otimes U_a \otimes (I_b - U_b)](b^* + \epsilon^*).\tag{12}$$

Let us define, now, $B = [U_r \otimes U_a \otimes (I_b - U_b)]$ and, for simplicity, redefine $e^* = b^* + \epsilon^*$, so that $SSB = e^{*\prime}Be^*$, with $e^* \sim \mathcal{N}(b^*, \sigma^2 V)$.

Let $SSB/\gamma = e^{*\prime}\tilde{B}e^*$ be a suitable quadratic form, where $\tilde{B} = B/\gamma$ and $\gamma$ is a constant to be defined later. It is easy to prove the following identity

$$\tilde{B}\mathrm{Var}(e^*) = \frac{\sigma^2}{\gamma}[U_r \otimes U_a \otimes (I_b - U_b)Q],\tag{13}$$

Therefore, the matrix $\tilde{B}\mathrm{Var}(e^*)$ is idempotent if and only if

$$(I_b - U_b) = \frac{\sigma^2}{\gamma}(I_b - U_b)Q(I_b - U_b).\tag{14}$$

Such an equation has usually no solution for all matrices $Q$. A case which satisfies the (14) is the following one. If $Q = (1 - \rho)I_b + \rho J_b$ (standard split–plot) then $(I_b - U_b)[(1 - \rho)I_b + \rho b U_b](I_b - U_b) = (I_b - U_b)(1 - \rho)$ and, therefore, (14) is true with $\gamma = \sigma^2(1 - \rho)$. Excluding rare situations, $SSB$ is not a chi squared random variable.

It is convenient, in the sequel, to determine the mean value and the variance of $SSB$ by exploiting the known theorem on moments and cumulants of normal quadratic forms. See, e.g., theorem 1 in [4] p. 55. Its adapted version gives rise to

$$E(SSB) \;=\; \sigma^2\left(\mathrm{tr}Q - \frac{s}{b}\right) + ra\sum_{k=1}^{b}\beta_k^2,\tag{15}$$

and, therefore, $E(MSB) = \frac{\sigma^2}{b-1}\left(\mathrm{tr}Q - \frac{s}{b}\right) + \frac{ra}{b-1}\sum_{k=1}^{b}\beta_k^2.$
The variance of $SSB$ is

$$\mathrm{Var}(SSB) = 2\left\{\sigma^4\mathrm{tr}([(I_b - U_b)Q]^2) + 2ra\sigma^2\beta'Q\beta\right\}. \tag{16}$$

Let us consider now the error linked to the form $SSEAB$,

$$\begin{aligned}
SSEAB &= \sum_{ijk} y_{ijk}^2 - \sum_{ij} y_{ij.}^2/b - \sum_{jk} y_{.jk}^2/r + \sum_{j} y_{.j.}^2/rb \\
&= \sum_{ijk}\left(\varepsilon_{ijk} - \varepsilon_{ij.}/b - \varepsilon_{.jk}/r + \varepsilon_{.j.}/rb\right)^2. \tag{17}
\end{aligned}$$

The vector, whose squared norm is equivalent to $SSEAB$, is $[(I_r - U_r)\otimes I_a \otimes (I_b - U_b)]\epsilon^*$, so that, if $E_{AB} = [(I_r - U_r)\otimes I_a \otimes (I_b - U_b)]$, then $SSEAB = \epsilon^{*'}E_{AB}\epsilon^*$. As previously stated, define similarly $\tilde{E}_{AB} = E_{AB}/\delta$, with $\delta$ a real constant to be determined. It is easily proved that $\frac{SSEAB}{\delta} = \epsilon^{*'}\tilde{E}_{AB}\epsilon^*$, is not chi squared distributed. The mean value of $SSEAB$ is

$$E(SSEAB) = \sigma^2\mathrm{tr}[(I_r - U_r)\otimes I_a \otimes (I_b - U_b)Q] = \sigma^2(r-1)a\left(\mathrm{tr}Q - \frac{s}{b}\right) \tag{18}$$

and, obviously, $E(MSEAB) = \frac{\sigma^2}{b-1}\left(\mathrm{tr}Q - \frac{s}{b}\right).$
The variance of SSEAB has the following form

$$\mathrm{Var}(SSEAB) = 2\sigma^4 a(r-1)\mathrm{tr}([(I_b - U_b)Q]^2). \tag{19}$$

The fixed effect due to interaction between factors $A$ and $B$ may be detected via the quadratic form $SSAB$,

$$\begin{aligned}
SSAB &= \sum_{jk} y_{.jk}^2/r - \sum_{j} y_{.j.}^2/rb - \sum_{k} y_{..k}^2/ra + y_{...}^2/rab \\
&= \sum_{ijk}\left((\alpha\beta)_{jk} + \varepsilon_{.jk}/r - \varepsilon_{.j.}/rb - \varepsilon_{..k}/ra + \varepsilon_{...}/rab\right)^2. \tag{20}
\end{aligned}$$

Let us define $m^* = 1_r \otimes v$, where $v_{ab\times 1}$ is a vectorized version of the columns of interaction effects matrix $((\alpha\beta)_{jk})$: $v_i = ((\alpha\beta)_{1i}, (\alpha\beta)_{2i}, \ldots, (\alpha\beta)_{ai})'$, $i = 1, 2, \ldots, b$, so that $v = (v_1', v_2', \ldots, v_b')'$.
The vector, whose squared norm is equivalent to $SSAB$, is $[U_r \otimes (I_a - U_a) \otimes (I_b - U_b)](m^* + \epsilon^*)$. Let us define $D = [U_r \otimes (I_a - U_a) \otimes (I_b - U_b)]$, $\tilde{D} = D/\vartheta$ with $\vartheta$ a real constant and, redefine for reasons of simplicity, $e^* = (m^* + \epsilon^*)$, then $\frac{SSAB}{\vartheta} = e^{*'}\tilde{D}e^*$, with $e^* \sim \mathcal{N}(m^*, \sigma^2 V)$ is not chi squared distributed.
The mean value of $SSAB$ is $E(SSAB) = \sigma^2(a-1)\left(\mathrm{tr}Q - \frac{s}{b}\right) + r\sum_{jk}(\alpha\beta)_{jk}^2$ and, therefore, $E(MSAB) = \frac{\sigma^2}{(b-1)}\left(\mathrm{tr}Q - \frac{s}{b}\right) + \frac{r}{(a-1)(b-1)}\sum_{jk}(\alpha\beta)_{jk}^2.$
The variance of $SSAB$ is $\mathrm{Var}(SSAB) = 2\sigma^4(a-1)\mathrm{tr}([(I_b - U_b)Q]^2) + f(v)$, where, $f(v) = 0$, if the null hypothesis holds true $(\alpha\beta)_{jk} = 0$, $j = 1, 2, \ldots, a$; $k = 1, 2, \ldots, b$.

## 6. APPROXIMATIONS AND ROBUSTNESS ASPECTS

Independence among $SSB$, $SSAB$ and $SSEAB$ is based upon a theorem in [5] p. 178. Now, let $X$ and $Y$ be two stochastically independent random variables and denote by $Z$ the ratio between $Y$ and $X$, $Z = Y/X$.

Due to independence, $E(Z) = E(Y)E(1/X)$, while, due to Jensen's theorem,

$$E(Z) \geq E(Y)/E(X). \tag{21}$$

The variance of $Z$, $V(Z)$, may be approximated by

$$\text{Var}(Z) = E\left(\frac{Y - \mu_Z X}{X}\right)^2 \simeq \frac{\text{Var}(Y) + \mu_Z^2 \text{Var}(X) + (\mu_Y - \mu_X \mu_Z)^2}{E(X^2)}. \tag{22}$$

If $E(X) = \mu_X = E(Y) = \mu_Y$, then $\mu_Z = E(Z) \simeq 1$ by (21) and

$$\text{Var}(Z) \simeq \frac{\text{Var}(Y) + \text{Var}(X)}{\text{Var}(X) + (E(X))^2}. \tag{23}$$

This argument may be useful in determining, through Tchébychev's inequality, the quantile controlling the significance of the usual ratios $\tilde{F}_B = MSB/MSEAB$ and $\tilde{F}_{AB} = MSAB/MSEAB$.

Under the null hypothesis, $H_0$, the mean values pertaining to numerator and denominator of $\tilde{F}_B$ are perfectly equal. Therefore, let us define

$$T_1 = (\text{tr}((I_b - U_b)Q))^2 = \left(\text{tr}Q - \frac{s}{b}\right)^2, \quad \text{and} \quad T_2 = \text{tr}\left([(I_b - U_b)Q]^2\right),$$

so that the approximate variance of $\tilde{F}_B$, given (23), is

$$\text{Var}(\tilde{F}_B) \simeq \frac{T_2\left(1 + \frac{1}{a(r-1)}\right)}{\frac{T_2}{a(r-1)} + \frac{1}{2}T_1}. \tag{24}$$

Let us examine, now, the ratio between the first order approximations of $T_1$ and $T_2$ in a neighborhood of $Q = (1 - \rho)I_b + \rho J_b$. We attain preliminarly

$$T_2 \simeq (b-1)(1-\rho)^2 + 2(1-\rho)\text{tr}((I_b - U_b)(dQ)), \tag{25}$$

$$T_1 \simeq (b-1)^2(1-\rho)^2 + 2(1-\rho)(b-1)\text{tr}((I_b - U_b)(dQ)), \tag{26}$$

and, therefore, within the assumed approximations $\xi = \frac{1}{2}T_1/T_2 = \frac{b-1}{2}$. Variance (24) may be approximated further,

$$\text{Var}(\tilde{F}_B) \simeq \frac{\left(1 + \frac{1}{a(r-1)}\right)}{\frac{1}{a(r-1)} + \xi} \simeq \frac{1}{\xi}. \tag{27}$$

Let us refer, now, to the ratio $\tilde{F}_{AB}$. By an analogous argument the variance is

$$\mathrm{Var}(\tilde{F}_{AB}) \simeq \frac{\left(\frac{1}{a-1} + \frac{1}{a(r-1)}\right)}{\frac{1}{a(r-1)} + \xi} \simeq \frac{1}{(a-1)\xi}. \tag{28}$$

Let us select a three-$\sigma$ protection. The corresponding quantile, $h$, may be estimated by $\hat{h} = \xi^{-1/2}3 + 1$, so that,

| $\hat{h}$ | $\xi$ | | | | | |
|---|---|---|---|---|---|---|
| $a$ | 0.5 | 1 | 1.5 | 2 | 2.5 | 3 |
| 2 | 5.2 | 4 | 3.45 | 3.12 | 2.90 | 2.73 |
| 3 | 4 | 3.12 | 2.73 | 2.50 | 2.34 | 2.22 |
| 4 | 3.45 | 2.73 | 2.41 | 2.22 | 2.10 | 2.0 |

As a concluding remark, we have attained again the well–known classic criterion according to which a type $\tilde{F}$ ratio is near to the significance threshold if its values belong to the interval (2 – 4). Nevertheless, the proposed tables allow a much more flexible choice. A motivated access to the second part of the split–plot analysis of variance table, Tab. 1, may be allowed by exploiting an approximation even if the matrix $Q$ has an unknown covariance pattern, different from the standard one, $(1 - \rho)I_b + \rho J_b$.

| Tab. 1 | ANOVA: Generalized Split–Plot with two Factors and Complete Blocks | | |
|---|---|---|---|
| Source | $MS$ | $E(MS)$ | $F$ |
| BLOCK | $MSR = \frac{SSR}{(r-1)}$ | $\frac{s\sigma^2}{b} + ab\sigma_\tau^2$ | $F_R$ |
| FACT. $A$ | $MSA = \frac{SSA}{(a-1)}$ | $\frac{s\sigma^2}{b} + br \sum_j \frac{\alpha_j^2}{(a-1)}$ | $F_A$ |
| ERROR $(A)$ | $MSEA = \frac{SSEA}{((a-1)(r-1))}$ | $\frac{s\sigma^2}{b}$ | |
| FACT. $B$ | $MSB = \frac{SSB}{(b-1)}$ | $\frac{\sigma^2}{b-1}\left(\mathrm{tr}Q - \frac{s}{b}\right) + ra \sum_k \frac{\beta_k^2}{(b-1)}$ | $\tilde{F}_B$ |
| INT. $A \times B$ | $MSAB = \frac{SSAB}{((a-1)(b-1))}$ | $\frac{\sigma^2}{b-1}\left(\mathrm{tr}Q - \frac{s}{b}\right) + r \sum_{jk} \frac{(\alpha\beta)_{jk}^2}{((a-1)(b-1))}$ | $\tilde{F}_{AB}$ |
| ERROR $(A \times B)$ | $MSEAB = \frac{SSEAB}{(a(b-1)(r-1))}$ | $\frac{\sigma^2}{b-1}\left(\mathrm{tr}Q - \frac{s}{b}\right)$ | |

## REFERENCES

1. Kiefer, J. and Wynn, H.P.: Optimum balanced block and Latin square design for correlated observations, The Annals of Statistics, 9 (1981), 737–757

2. Cressie, N.A.C.: Statistics for Spatial Data, Wiley, New York, 1991

3. Khattree, R. and Naik, D.N.: Optimal tests for nested designs with circular stationary dependence, J. of Statistical Planning and Inference, 41 (1994), 231–240

4. Searle, S.R.: Linear Models, Wiley, New York, 1971

5. Rao, C.R. and Mitra, S.K.: Generalized Inverse of Matrices and its Applications, Wiley, New York, 1971

# ROBUSTNESS OF CANONICAL ANALYSIS
# FOR SOME MULTIRESPONSE DEPENDENT EXPERIMENTS

## C. Mortarino
## University of Padua, Padua, Italy

KEY WORDS AND PHRASES: Fractional factorial designs, Extended $V$–robustness.

ABSTRACT: Multiresponse (fractional) factorial experiments are essential instruments for quality improvement of industrial processes. The usual assumption requires however uncorrelated response components and, more important, independent runs. Canonical analysis may be very inefficient if that assumption is not verified. Results have however been recently proved showing optimality of this analysis also in presence of a quite general dependence pattern here described. These results are here illustrated for application purposes also with the help of a simple example, which can be nethertheless extended to cover more complex situations.

## 1. INTRODUCTION

Off-line process optimization and quality assurance usually refer to multiresponse situations and efficient statistical methods should preserve this multiresponse feature. Between-responses correlations could be exploited to improve the quality of the analysis, although it is often difficult to explicit them or to give for them sensible approximated values.

In the first steps of a study, the actual behaviour of the system to be analyzed is usually approximated by a simple model linear in its parameters. In order to provide estimates for those parameters, many experimental designs are available. Among

Published in: E. Kuljanic (Ed.) *Advanced Manufacturing Systems and Technology*, CISM Courses and Lectures No. 372. Springer Verlag, Wien New York, 1996.

them, because of their simple use and good properties, two-level factorial designs are extensively used. With a general factorial design, many explanatory variables, *factors,* each having two or more levels, can be simultaneously handled. These experiments provide the opportunity to estimate not only individual effects of each factor, *main effects,* but also eventual interactions.

Full factorial designs provide for an experimental trial, *run,* corresponding to each possible combination of factors' levels. This, however, leads quickly to a large experiment, often in contrast with expenditure links. Some reduced plans from those designs, called *fractions,* may thus represent a useful compromise.

For two-level designs with $n$ factors, fractions are formed by $t \cdot 2^{n-m}$ runs, where $m$ and $t$ are indexes of reduction of the plan, which results in a reduction of effects (typically higher order interactions) which can be estimated. Notice that $t$ can be any integer in $\{1, 2, \ldots, 2^m\}$ : if $t$ equals a power of 2, the resulting plan is called a *regular* fraction, otherwise this plan is an *irregular* one.

Section 2 will introduce the model here used with a description of symbols and notations. A simple multiresponse example will be also presented in order to give an immediate idea of a situation fitting the context here described and to explain in a more intuitive way some of the techniques used. In section 3, the main question will be discussed: usual assumption of independence among runs and between responses measured at the same run is often unacceptable; a new one much weaker than the previous one will be described and a new sufficient condition in order to verify it will be presented. Results proved with this assumption in previous papers [1] and [2] will be also recalled, proving that a standard estimation method, i.e. ordinary least squares method, gives minimum variance linear unbiased parameters' estimators. Through the example introduced in previous section a possible source of dependence will be finally examined. Section 4 is devoted to the examination of the behaviour of above-mentioned estimation method when the true dependence pattern only lies in a neighbourhood of the assumption presented in section 3.

## 2. CANONICAL MODEL

Let

$$y_{iu} = \boldsymbol{x}'_u \boldsymbol{\beta}_i + \varepsilon_{iu}, \qquad i = 1, 2, \ldots, h, \quad u = 1, 2, \ldots, N, \tag{1}$$

be the model we assume as an approximation of the unknown behaviour of the target system; here $N$ is the number of experimental units, $h$ is the number of responses observed on each unit, $y_{iu}$ represents the $i$-th response corresponding to the $u$-th unit, $\boldsymbol{x}'_u$ is the $u$-th row of $\boldsymbol{X}_{N \times k}$, which is the matrix associated to a (fractional) factorial design with $N$ runs; $\boldsymbol{x}'_u$ contains factors' levels used for the $u$-th run, levels expressed in coded form, i.e. as elements of the subset $\{-1, +1\}$; $\boldsymbol{\beta}_i$ is here a $k$-dimensional parameters' vector whose estimation is required; finally $\varepsilon_{iu}$ are stochastic terms which are assumed to have zero expectation and, at least in this canonical model, common variance and independent distribution.

Model (1) can be alternatively represented in matrix form

$$y^*_{Nh \times 1} = X^*_{Nh \times kh} \beta^*_{kh \times 1} + \varepsilon^*_{Nh \times 1}, \tag{2}$$

obtained by stacking the $N$ $h$-dimensional vectors corresponding to the responses measured on each unit. Note that $X^* = X \otimes I_h$ where $\otimes$ denotes Kronecker product and $I_h$ is the identity matrix of order $h$.

Remark that this model structure entails the assumption of a common model matrix for each component of the response: parameters are however free to change from component to component and, as will be explained with the help of the following example, factors assumed to influence each response component need not to be the same.

**Example** (adapted from [3]). In the manifacture of a certain dyestuff for textile use, the purpose is to obtain a product of desired hue $(Y_1)$ and brightness $(Y_2)$, such that it is maximally resistant to standard washing cycles $(Y_3)$. According to the knowledge of the manifacturing process, six factors were suspected to influence the response variables; these are listed below with an indication of the two levels to be tested for each one:

$$
\begin{array}{llll}
\xi_1 : & \text{polysulfide index} & \{ & 6, & 7\} \\
\xi_2 : & \text{reflux rate} & \{ & 150, & 170\} \\
\xi_3 : & \text{moles polysulfide} & \{ & 1.8, & 2.4\} \\
\xi_4 : & \text{time (in minutes)} & \{ & 24, & 36\} \\
\xi_5 : & \text{solvent (cm}^3) & \{ & 30, & 42\} \\
\xi_6 : & \text{temperature (}^\circ\text{C)} & \{ & 120, & 130\}
\end{array}
$$

Assignment of coded level $(-1)$ to the lowest level of each factor and $(+1)$ to the higher level allows description of a 6-factor 2-level full factorial design $(2^6)$ through the following table:

| runs | coded levels | $\xi_1$ | $\xi_2$ | $\xi_3$ | $\xi_4$ | $\xi_5$ | $\xi_6$ | factors |
|---|---|---|---|---|---|---|---|---|
| | 1 | +1 | +1 | +1 | +1 | +1 | +1 | |
| | 2 | +1 | +1 | +1 | +1 | +1 | −1 | |
| | 3 | +1 | +1 | +1 | +1 | −1 | +1 | |
| | ⋮ | ⋮ | ⋮ | ⋮ | ⋮ | ⋮ | ⋮ | |
| | 64 | −1 | −1 | −1 | −1 | −1 | −1 | |

Matrix $X$ is obtained from this table as follows: columns corresponding to main effects are *equal* to the columns of this table; columns corresponding to required interactions can be obtained as *products* of columns of this table. Model matrix $X^*$ is finally calculated from $X^* = X \otimes I_3$.

Conversely, if we knew that not *all* factors are expected to influence *all* response components, we could use a smaller plan: supposing that each response component can be influenced only by a factor subset, we could construct a plan with as many factors as the cardinality of the greatest among those subsets and that could be done without violating the assumption of a common matrix model. If, for example, there are, reliable indications that

a) $\xi_1$ and $\xi_6$ are not expected to influence $Y_1$,
b) $\xi_1$ and $\xi_4$ are not expected to influence $Y_2$,
c) $\xi_4$ and $\xi_5$ are not expected to influence $Y_3$,

in this simple case each response component is influenced only by four factors; then we could use a $2^4$ design with only 16 runs, described by

| | coded levels | $\xi_1$ | $\xi_2$ | $\xi_3$ | $\xi_4$ | $\xi_5 \simeq \xi_1$ [a] | $\xi_6 \simeq \xi_4$ | factors |
|---|---|---|---|---|---|---|---|---|
| runs | 1 | +1 | +1 | +1 | +1 | +1 | +1 | |
| | 2 | +1 | +1 | +1 | −1 | +1 | −1 | |
| | 3 | +1 | +1 | −1 | +1 | +1 | +1 | |
| | ⋮ | ⋮ | ⋮ | ⋮ | ⋮ | ⋮ | ⋮ | |
| | 16 | −1 | −1 | −1 | −1 | −1 | −1 | |

[a] $\simeq$ denotes that equality holds if levels are expressed in coded form.

From previous table it's easy to see that for each response component columns corresponding to factors assumed to influence it form the basis of a $2^4$ design: columns corresponding to $(\xi_2, \xi_3, \xi_4, \xi_5)$ are main effects columns of a $2^4$ design – used for the description of $Y_1$– as well as columns corresponding to $(\xi_2, \xi_3, \xi_5, \xi_6)$ and to $(\xi_1, \xi_2, \xi_3, \xi_6)$ do for the description of $Y_2$ and $Y_3$, respectively.                                □

Observe that the subject of previous example could have been chosen from many other fields, as metallurgy or chemistry, for example, since, as already anticipated, multiresponse problems arise from very different situations. This example was not intended to be exhaustive of the range of possible applications, but it was only proposed to illustrate in a quick way the techniques here referred to.

Once model (2) is assumed, it's easy to calculate the minimum variance best linear unbiased estimates for $\boldsymbol{\beta}^*$ from

$$\tilde{\boldsymbol{\beta}}^*_{\text{OLS}} = (\boldsymbol{X}^{*\prime}\boldsymbol{X}^*)^{-1}\boldsymbol{X}^{*\prime}\boldsymbol{y}^* = [(\boldsymbol{X}'\boldsymbol{X})^{-1}\boldsymbol{X}' \otimes \boldsymbol{I}_h]\boldsymbol{y}^*. \tag{3}$$

## 3. DEPENDENT EXPERIMENTS

Previous model (2) relied on the strong assumption of uncorrelation between responses measured on different experimental units and between different response components measured on the same experimental unit. This second kind of uncorrelation, in particular, is almost always denied by real situations. Uncorrelated runs prove also to be quite difficult to realize in experimentation: precise rules given to an operator in order to perform correctly an experiment may not be fully understood and, consequently, partially violated. In this case the analyst should not proceed along the standard way, but he/she may not be able to know how the experiment was performed and analyze data accordingly. That's why it is important to explore which situations can be treated through the canonical methods described in section 2.

Consider model (2)

$$y^* = X^* \beta^* + \varepsilon^* ,$$

where, instead of assumption about $\varepsilon^*$ previously stated we assume

$$E(\varepsilon^*) = 0 \quad \text{and} \quad Cov(\varepsilon^*) = \Sigma^* . \tag{4}$$

In other words *nothing is assumed:* we leave the dependence pattern completely free. We now partially restrict this broad range of patterns introducing for $\Sigma^*$ the assumption of extended $V$–robustness property with respect to the model matrix $X^*$, property proposed for the uniresponse case in [4] and extended to the multiresponse case by [1].

**Definition.**    A covariance matrix

$$\Sigma^*_{Nh \times Nh} = \begin{bmatrix} \Sigma^*_{11} & \Sigma^*_{12} & \cdots & \Sigma^*_{1N} \\ \Sigma^*_{21} & \Sigma^*_{22} & \cdots & \Sigma^*_{2N} \\ \vdots & \vdots & & \vdots \\ \Sigma^*_{N1} & \Sigma^*_{N2} & \cdots & \Sigma^*_{NN} \end{bmatrix} ,$$

is extended $V$–robust with respect to $X^*_{Nh \times kh} = X \otimes I_h$ if a function $g : \mathbb{R}^k \to C_h$, where $C_h$ denotes the space of all $h \times h$ matrices, exists such that

$$\forall u, w = 1, 2, \ldots, N \qquad \Sigma^*_{uw} = g(x'_u \odot x'_w) , \tag{5}$$

where $\odot$ denotes Hadamard product.

This condition simply requires that the $h \times h$ covariance matrix between responses measured on two experimental units depends on Hadamard product between vectors describing the experimental setting used for those units. Note that, since we use factors' levels coded form, $x'_u \odot x'_w$ simply tells us which factors changed their level from run $u$ to run $w$ and which didn't. We are just linking the covariance among responses measured on two experiments to the difference of set-up between those two experiments. The most important thing is that *nothing* is assumed about the *link function g*: this leaves a sufficient range for dependence among different runs and a totally unconstrained range for dependence among response components.

A useful result could be obtained turning expression (5), which is in terms of $\boldsymbol{X}$, the coded matrix, into an equivalent expression in terms of $\boldsymbol{T}^{(i)}$ matrices: $\boldsymbol{T}^{(i)}$ is, for $i = 1, 2, \ldots, h$, the matrix of a $2^{n-m}$ factorial design whose $N = 2^{n-m}$ rows represent experimental runs through the original co-ordinates of levels of factors supposed to influence the $i$-th response component.

For every $\boldsymbol{\gamma}' = [\gamma_1, \gamma_2, .., \gamma_k]$ and $\boldsymbol{\delta}' = [\delta_1, \delta_2, .., \delta_k]$, define $\mathcal{T}(\boldsymbol{\gamma}, \boldsymbol{\delta}) \subset \mathbb{R}^k$ as the subset

$$\mathcal{T}(\boldsymbol{\gamma}, \boldsymbol{\delta}) = \{ (\gamma_1 \pm \delta_1, \gamma_2 \pm \delta_2, \ldots, \gamma_k \pm \delta_k) \}, \tag{6}$$

of cardinality $2^k$. Within the class

$$\mathcal{F} = \{ f(\cdot, \cdot) : \mathbb{R}^k \times \mathbb{R}^k \to C_h \}$$

let us consider a particular subset:

$$\tilde{\mathcal{F}}^{|\tau} = \{ f \in \mathcal{F} \mid \exists g_f(\cdot) : \mathbb{R}^k \to C_h : \tag{7}$$
$$\forall (\boldsymbol{\gamma}, \boldsymbol{\delta}) \ \forall \boldsymbol{x}, \boldsymbol{y} \in \mathcal{T}(\boldsymbol{\gamma}, \boldsymbol{\delta}), \quad f(\boldsymbol{x}, \boldsymbol{y}) \equiv g_f(|\boldsymbol{x} - \boldsymbol{y}|) \},$$

where $|\boldsymbol{x} - \boldsymbol{y}|$ denotes the vector whose $j$-th element is $|x_j - y_j|, \quad j = 1, 2, \ldots, k$.

We are now able to prove the sufficient condition previously mentioned, which is an extension of a previous one proved in [1] for the case $\boldsymbol{T}^{(i)} = \boldsymbol{T} \ \forall i = 1, 2, \ldots, h$.

**Theorem 1.** If $\forall u, w = 1, 2, \ldots, 2^{n-m}$, any function $l$ exists such that

$$\boldsymbol{\Sigma}^*_{uw} = l[m_1(\boldsymbol{t}^{(1)}_u, \boldsymbol{t}^{(1)}_w), m_2(\boldsymbol{t}^{(2)}_u, \boldsymbol{t}^{(2)}_w), \ldots, m_h(\boldsymbol{t}^{(h)}_u, \boldsymbol{t}^{(h)}_w)] \tag{8}$$
$$m_i(\cdot, \cdot) \in \tilde{\mathcal{F}}^{|\tau} \ \forall i = 1, 2, \ldots, h,$$

where $(\boldsymbol{t}^{(i)}_u)'$ is $\boldsymbol{T}^{(i)}$'s $u$-th row, then matrix $\boldsymbol{\Sigma}^* = \text{cov}(\boldsymbol{\varepsilon}^*)$ is extended $V$–robust with respect to $\boldsymbol{X}^* = \boldsymbol{X} \otimes \boldsymbol{I}_h$.

Proof.
It is easy to see that for every $i$ $\{ \boldsymbol{t}^{(i)}_u, u = 1, 2, \ldots, 2^{n-m} \}$ is a subset of the family (6). Because of (8) and of (7), it follows that $\forall u, w = 1, 2, \ldots, 2^{n-m}$

$$\boldsymbol{\Sigma}^*_{uw} = l[m_1(\boldsymbol{t}^{(1)}_u, \boldsymbol{t}^{(1)}_w), m_2(\boldsymbol{t}^{(2)}_u, \boldsymbol{t}^{(2)}_w), \ldots, m_h(\boldsymbol{t}^{(h)}_u, \boldsymbol{t}^{(h)}_w)]$$
$$= l[g_1(|\boldsymbol{t}^{(1)}_u - \boldsymbol{t}^{(1)}_w|), g_2(|\boldsymbol{t}^{(2)}_u - \boldsymbol{t}^{(2)}_w|), \ldots, g_h(|\boldsymbol{t}^{(h)}_u - \boldsymbol{t}^{(h)}_w|)].$$

Since each of the arguments of $g_i(\cdot)$ can assume only two values, denoting by $I_A$ the indicator function of the set $A$, there exist a function $\tilde{g}_i$ such that

$$
\begin{aligned}
g_i(|t_u^{(i)} - t_w^{(i)}|) &= g_i\left(\left[\,|t_{u1}^{(i)} - t_{w1}^{(i)}|,\ |t_{u2}^{(i)} - t_{w2}^{(i)}|,\ \ldots,\ |t_{uk}^{(i)} - t_{wk}^{(i)}|\,\right]\right) \\
&= \tilde{g}_i\left(\left[\,I_{\{t_{u1}^{(i)}=t_{w1}^{(i)}\}},\ I_{\{t_{u2}^{(i)}=t_{w2}^{(i)}\}},\ \ldots,\ I_{\{t_{uk}^{(i)}=t_{wk}^{(i)}\}},\,\right]\right) \\
&= \tilde{g}_i\left(x_{u1}x_{w1},\ x_{iu}x_{jw},\ \ldots,\ x_{u(k+1)}x_{w(k+1)}\right) \\
&= \tilde{g}_i(x_u' \odot x_w').
\end{aligned}
$$

It follows that

$$
\Sigma_{uw}^* = l[\tilde{g}_1(x_u' \odot x_w'),\tilde{g}_2(x_u' \odot x_w'),\ldots,\tilde{g}_h(x_u' \odot x_w')] = \tilde{l}(x_u' \odot x_w')
$$

as claimed.                                                                                              □

We want to emphasize that to $\tilde{\mathcal{F}}^{|\tau}$ belong very different functions: regardless how their components are "added" (giving the maximum freedom to the pattern of covariance within responses), each component separately can be some general distance function (or "similarity" index) among the design points expressed through the factors' original co-ordinates.

The emphasis given to $V$–robust covariance patterns is motivated by the strong results proved through it. In [1] it has been proved that, if in (2) model matrix $X^*$ is associated to a *regular* fraction of a factorial design and $\Sigma^*$ is extended $V$–robust w.r.t. $X^*$, optimal estimates for $\beta^*$ can be obtained with the same method used in section 2, i.e. with the method used when responses were completely uncorrelated because

$$
\hat{\beta}_{\text{WLS}}^* = \tilde{\beta}_{\text{OLS}}^*;
$$

here $\hat{\beta}_{\text{WLS}}^* = (X^{*\prime}\Sigma^* X^*)^{-1}X^{*\prime}\Sigma^* y^*$ is weighted least squares estimator, which however could not be calculated since $\Sigma^*$ is, except for trivial cases, an unknown matrix. For irregular fractions a further step is necessary: these designs are, by construction, formed by several smaller (regular) subfractions: in this case it has been proved in [2] that the same result above described can be obtained only if we also assume independece among runs belonging to different subfractions; this makes that result particularly suitable for sequential experiments.

**Example (continued).** In this situation a $2^6$ or a smaller plan was performed. Runs order should have been randomized, but, in practice, can we really be sure? Maybe we should consider the possibility of correlated runs: for example, a greater similarity between experimental set-up in two runs could entail a greater similarity between factors that cannot be controlled and, hence, we may have a greater covariance between corresponding responses than we have for responses measured at runs with a more different experimental set-up. Besides, we cannot imagine why $Y_1$ and $Y_2$ should be uncorrelated and nothing might be said about the link of those two components with $Y_3$. But even if we can think that those guesses are right, how can we explicit $\Sigma^*$ to calculate $\beta^*_{WLS}$? That's a real task, but we do not need to face it, because in this case, we now that that estimator equals (3), which does not depend at all upon $\Sigma^*$.

## 5. APPROXIMATED V–ROBUSTNESS

When using results proved via $V$–robustness property we may wonder what would be our loss if this property doesn't hold *exactly* but only in an approximated way. In particular an additive perturbation, of small amount but generical direction was studied in [5] and the following result was there proved.

**Thorem 2.** Let $C\tilde{\beta}$ and $C\hat{\beta}$ estimators obtained through ordinary least squares and weighted least squares methods, respectively, for $C\beta$ for model (2). Let $\Sigma^* = \mathrm{Cov}(\varepsilon^*) = W + \rho\Gamma$, where $W$ is an extended $V$–robust matrix with respect to $X^*$. Here $\Gamma$ is a generical symmetrical matrix, such that

$$|\rho| < \frac{1}{\|\Gamma W^{-1}\|},$$

for any matrix norm (so that $\Sigma^*$ is positive definite).
Then as $\rho \to 0$

$$\mathrm{Cov}(C\tilde{\beta}) - \mathrm{Cov}(C\hat{\beta}) = D(O(\rho^2)), \qquad \{D(O(\rho^2))\}_{ij} = O(\rho^2) \qquad \forall(i,j);$$

in other words the difference between those covariance matrices, or the variance loss due to the use of the ordinary least squares estimator in place of the weighted least squares one, is of order $\rho^2$.

## REFERENCES

1. Guseo, R. and Mortarino, C.: Multiresponse dependent experiments: robustness of $2^{k-p}$ fractional factorial designs, 1995 (submitted to J. Statist. Plann. Infer.)
2. Mortarino, C.: Multiresponse irregular fractions of two-level factorial designs with dependent experimental runs, Proceedings XI International Workshop on Statistical Modelling - Poster Session, Orvieto, July 15-19 1996
3. Box, G.E.P. and Draper, N.R.: Empirical model building and response surfaces, Wiley, New York, 1987
4. Krouse, D.P.: Patterned matrices and the missspecification robustness of $2^{k-p}$ designs, Comm. Statist. Theory Meth., 23 (1994), 3285–3301
5. Guseo, R. and Mortarino, C.: $V$–robustezza approssimata, Atti della XXXVIII Riunione Scientifica della Società Italiana di Statistica, Rimini, 9–13 aprile 1996, vol.2, 219–226.

# AUTHORS INDEX

Printed in the United States
By Bookmasters

ted in the United States
ɔokmasters